Introduction to
Materials
Science

McGraw-Hill Series in Materials Science and Engineering

AVITZUR *Metal Forming: Processes and Analysis*
AZÁROFF *Introduction to Solids*
BARRETT AND MASSALSKI *Structure of Metals*
BLATT *Physics of Electronic Conduction in Solids*
BRICK, GORDON, AND PHILLIPS *Structure and Properties of Alloys*
BUERGER *Contemporary Crystallography*
BUERGER *Introduction to Crystal Geometry*
DE HOFF AND RHINES *Quantitative Microscopy*
DRAUGLIS, GRETZ, AND JAFFEE *Molecular Processes on Solid Surfaces*
ELLIOTT *Constitution of Binary Alloys, First Supplement*
GILMAN *Micromechanics of Flow in Solids*
GORDON *Principles of Phase Diagrams in Materials Systems*
GUY *Introduction to Materials Science*
HIRTH AND LOTHE *Theory of Dislocations*
KANNINEN, ADLER, ROSENFIELD, AND JAFFEE *Inelastic Behavior of Solids*
MURR *Electron Optical Applications in Materials Science*
PAUL AND WARSCHAUER *Solids under Pressure*
ROSENFIELD, HAHN, BEMENT, AND JAFFEE *Dislocation Dynamics*
RUDMAN, STRINGER, AND JAFFEE *Phase Stability in Metals and Alloys*
SHEWMON *Diffusion in Solids*
SHEWMON *Transformations in Metals*
SHUNK *Constitution of Binary Alloys, Second Supplement*
WERT AND THOMSON *Physics of Solids*

A. G. Guy

Professor of Metallurgy
University of Florida

Introduction to Materials Science

With the collaboration of

M. E. Bever
Professor of Metallurgy
Massachusetts Institute of Technology

L. L. Hench
Head Professor of Ceramics
University of Florida

A. Peterlin
Director, Camille Dreyfus Laboratory
Research Triangle Institute

on topics involving metallurgy, ceramics, and polymers, respectively

McGraw-Hill Book Company

New York
St. Louis
San Francisco
Düsseldorf
Johannesburg
Kuala Lumpur
London
Mexico
Montreal
New Delhi
Panama
Rio de Janeiro
Singapore
Sydney
Toronto

Introduction to Materials Science

Library of Congress Catalog Card Number 75–139553

007–025310–2

1 2 3 4 5 6 7 8 9 0 HDMM 7 9 8 7 6 5 4 3 2 1

This book was set in Baskerville, was printed by Halliday Lithograph Corporation, and bound by The Maple Press Company. The designer was Marsha Cohen; the drawings were done by Felix Cooper. The editors were B. J. Clark and Peter Karsten. Robert R. Laffler supervised production.

To my wife
Baroness Ingeborg von der Brüggen

Künstlerin — Raconteuse — Любимая — Conselheira — Sjuksköterska

Contents

Preface

This is a textbook for students in engineering, the physical sciences, and the biological sciences. It is also intended for the practicing engineer or scientist who wishes to obtain a background in the broad field of materials. Its aims are to give an understanding of the nature and behavior of metals, ceramics, and polymers, to compare the characteristics of these materials, and to furnish a sound foundation for further study and use of specific materials in technology and science.

The effective assistance and collaboration of specialists on all three types of materials has resulted in a usefully integrated treatment. Consequently, the student should find a helpful unity in the concepts, rather than a confusing diversity. The integrated approach is also useful pedagogically in reducing the amount of material to be covered in a subject as large as materials. A unified treatment of the important topic of phase diagrams, for example, is given for both ceramics and metals. The corresponding aspects of polymers are discussed in the same context. Treatments of such subjects as atomic structure and crystal structure are built on the student's prior knowledge from college chemistry and physics; needless repetition is avoided, and a fuller coverage is achieved. To treat the numerous quantitative aspects of materials, use is made of calculus and some elementary ideas from differential equations. A few special mathematical concepts, such as the *simplex* from topology, are explained and applied. The problems at the end of each chapter are closely tied to the development of the subject matter. One Fortran IV problem in each chapter gives the student an opportunity to practice some simple computer techniques.

The integration of subject matter implied by the term *materials science* has been built on a suitable generalization and broadening of the framework successfully used for physical metallurgy. The important chapter, Imperfections in Crystals, typifies the integration of subject matter. A unified treatment of vacancies in metals and of doping in semiconductors has been based on the methods of *chemistry of defects in solids.* Dislocations are also treated in a general manner using examples of both ceramics and metals. The major imperfections in polymers, *kinks* in an otherwise regular chain molecule, are contrasted with dislocations. The effect of imperfections on some reactions and properties of materials is discussed in the latter half of the book. Useful integration of subject matter has also been facilitated by the introduction of a few relatively advanced concepts frequently met in the literature of materials science. An example is the concept of reciprocal lattice, which is first employed to treat crystallography in a general manner. This same concept is then used to explain the conditions for diffraction (by x-rays, electrons, or neutrons) that apply to metals, ceramics, and polymers in many possible states of aggregation. Diffraction theory is the basis for interpreting the patterns produced by dislocations in transmission-electron microscopy. The reciprocal-lattice concept is also applied to stereographic projections, which have useful applications for all types of materials.

In view of the present strong trend toward a universal system of units the recommended system, known as SI (Système International), has been adopted. Only for the treatment of topics in chemistry are the traditional units employed, in keeping with the current practice in this field. As an aid to the reader who is accustomed to such units as lbs/in^2 or kg/mm^2, for example, the appropriate conversion (from newtons per square meter, in this case) is shown in many of the graphs and tables. The form employed is:

$$(M\ \mathrm{N/m^2})\ 1.450 \times 10^{-4} = N\ \mathrm{lb/in^2}$$

$$(M\ \mathrm{N/m^2})\ 1.020 \times 10^{-7} = N\ \mathrm{kg/mm^2}.$$

Furthermore, the quantity in SI units is expressed using a suitable power of 10 to facilitate comparison with the customary units. For example, a strength value of 20×10^7 N/m^2 is essentially 20 kg/mm^2 or $20 \times 1.45 \approx 30$ kpsi (1,000 lbs/in^2).

It is a pleasure to thank those engineers and scientists who furnished recent data, supplied illustrative material, critically read portions of the text, or suggested appropriate topics or methods of integration. Professor A. J. McEvily created a treatment of fracture, Chapter 10, well integrated with the balance of the text.

Special acknowledgement is made to the following persons who have read entire chapters or have made a comparable contribution to the final form of the book: Dr. H. I. Aaronson, Professor Leo Brewer, Dr. J. E. Burke, Professor J. G. Byrne, Dr. R. W. Cahn, Professor R. L. Coble, Professor R. T. DeHoff, Professor M. A. Eisenberg, Dr. N. A. Gjostein, Professor R. W. Gould, Professor J. R. C. Guimarães, Dr. Hsun Hu, Professor R. A. Huggins, Professor R. E. Hummel, Dr. K. A. Jackson, Professor J. J. Jonas, Professor W. A. Mannheimer, Professor G. P. Mohanty, Professor J. P. Nielsen, Professor W. D. Nix, Dr. Toru Okabe, Professor G. M. Pound, Professor Marcel Pourbaix, Dr. V. Ramachandran, Professor Gerhard Schmid, Professor J. C. Shyne, Dr. P. G. Shewmon, Professor C. S. Smith,

Dr. Thor L. Smith, Professor Gareth Thomas, Professor W A. Tiller, Professor W. van Gool, Professor E. D. Verink, Professor C. M. Wayman, Professor J. J. Wert, Professor Charles A. Wert, and Dr. J. H. Westbrook.

The integration of subject matter would have been impossible without the expert guidance and the active participation of my three collaborators, Professor M. E. Bever, Professor L. L. Hench, and Dr. A. Peterlin. Professor Morris Cohen aided greatly in the initial planning of the book, and Professor M. E. Fine contributed substantially to many sections. Mr. Robert S. Cohen formulated and tested the Fortran IV problems. Mr. Felix Cooper skillfully converted preliminary sketches into illustrations that convey much of the message of the book.

A. G. Guy

Chapter One

Atomic Bonding and Crystal Structure

INTRODUCTION

Modern machines are not simply *made of* a metal, a plastic, or a ceramic; the behavior of the material *during fabrication and in service* is part of the total design. A striking example is an integrated circuit in an electronic device (cheaper than and only one-thousandth the size of the previous vacuum-tube version). The circuit exists in the material itself, usually silicon, by virtue of certain changes in chemical composition, which we consider in Chap. 6. Modern ceramic materials are also specifically designed to have the desired combination of physical and mechanical properties for a given application. An example (Fig. 1-1) is the ceramic substrate for electronic devices, on which electrical circuits are deposited. In the manufacture of polymer fibers such as nylon, the drawing operation that produces the final fiber also transforms the internal structure of the polymer, greatly increasing the strength. Similarly, a sheet of a copper-base alloy, already strengthened by suitable cold rolling during manufacture, improves in structure and strength as it is stamped and formed into a useful machine part. Clearly, effective design requires an understanding of the behavior of various materials; it may even demand the creation of a new material with the required response during manufacture and the desired properties in operation.

1-1. Structure of Materials

The variety of materials is enormous. Merely to list the commercially available metals, ceramics, and polymers would require most of this book. The technical literature on a single important material, such as polyethylene, can easily fill a bookshelf. Fortunately, the broad concept of *structure* gives a basis for studying materials in general and provides a background for reading the literature on a specific material. The structures of interest vary widely in size and character (see Fig. 1-1). *Constitution,* the aspect of structure of primary

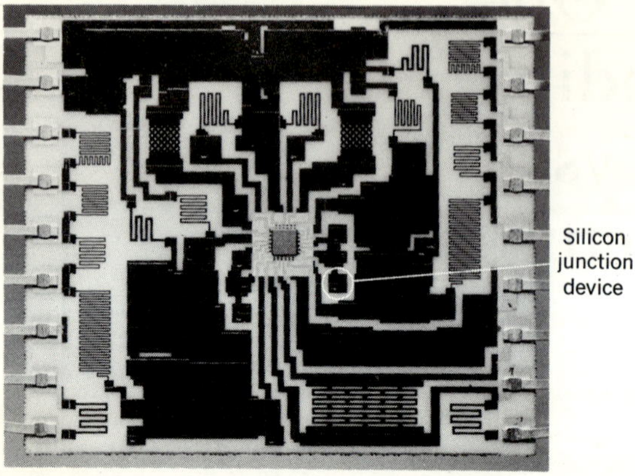

(*a*) Photograph of an integrated circuit for a telephone. The base is an alumina (Al_2O_3) ceramic.

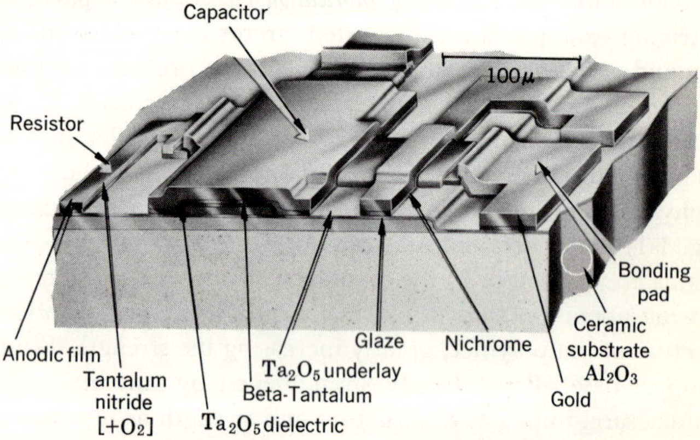

(*b*) Artist's sketch of a typical cross-section, showing various materials including Al_2O_3 ceramic substrate.

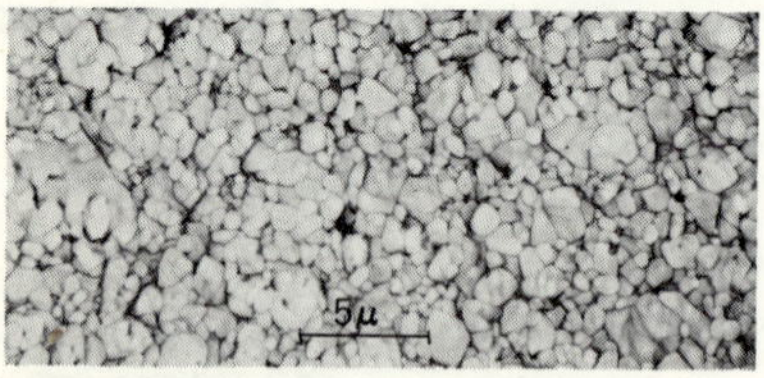

(*c*) The constitution of the Al_2O_3 ceramic is revealed in this electron micrograph.

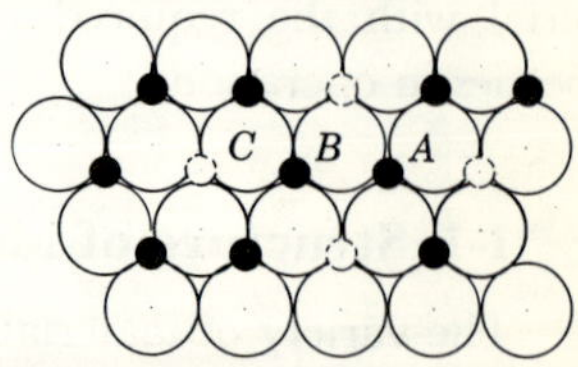

(*d*) The aluminum and oxygen atoms in Al_2O_3 form this type of crystal structure.

FIGURE 1-1 An example of the use of an alumina ceramic in an overall design. Three aspects of the structure of this ceramic material are its ordinary macrostructure (*a*) and (*b*), its constitution (*c*), and the crystal structure formed by its atoms (*d*). (*Courtesy Bell Telephone Laboratories.*)

interest to materials science (Fig. 1-1*c*), is intermediate in size between macrostructure (Fig. 1-1*a*) and crystal structure (Fig. 1-1*d*). Two factors determine the constitution of a given material: (1) the phases comprising it, and (2) their distribution (as separate pieces, as a continuous matrix, etc.). A photomicrograph may reveal what phases are present and how they are distributed. Other techniques, such as x-ray diffraction, permit a study of *phase structure*, that is, the state of aggregation of the atoms or molecules within a given phase. The vapor phase differs sharply in structure from the condensed phases (liquid, amorphous, glassy, and crystalline). Chapters 1 through 5 consider various aspects of the constitution of materials, beginning with the nature of crystalline phases. This chapter gives a quantitative description of crystal structures which is adequate for the many applications (especially to crystalline defects) studied in the remainder of the book. We consider first the bonding in solids, augmenting the treatment usually given in a course on general chemistry.

BONDING IN SOLIDS

The manner in which the properties of individual atoms determine the bonding in materials can be shown in several ways. Quantitative calculations are often useful and are illustrated in the final chapter by a quantum-mechanical approach for treating electrical properties. More often, qualitative concepts derived from quantum mechanics are combined with empirical data to explain the effect of bonding on the properties of complex materials. This approach often considers only one type of bonding (ionic, for example), although theory shows that several types are present in a typical material.

Modern theory pictures the electrons of a free atom as occupying atomic orbitals having characteristic energies and spatial distributions. The eleven electrons of sodium, for example, occupy orbitals of increasing energy, $1s^2$, $2s^2$, $2p^6$, and $3s$; the superscripts indicate the number of electrons in a given type of orbital. The spatial representation of an orbital (Fig. 1-2) indicates the probability of finding the given type of electron at various locations in the atom. An important feature of *s*-type orbitals is their spherical symmetry (Fig. 1-2*a*). The entire array of six $2p$ electrons also has spherical symmetry, but each of the individual $2p$ orbitals, shown in Fig. 1-2*b*, has a dumbbell shape. Each *s*- or *p*-type orbital may contain either one electron or two electrons of opposite spin. *s*- and *p*-type electrons both contribute to bonding, provided that the entire octet of electrons is not already present, as in filled inner shells or in the outer shell in the rare gases. In atoms with unfilled *d*-shells (the transition elements) *d*-orbitals also contribute to bonding. The directional character of the *p*- and *d*-orbitals is evident in covalent bonding, but spherically symmetrical electron distributions characterize the "filled shells" produced in ionic bonding.

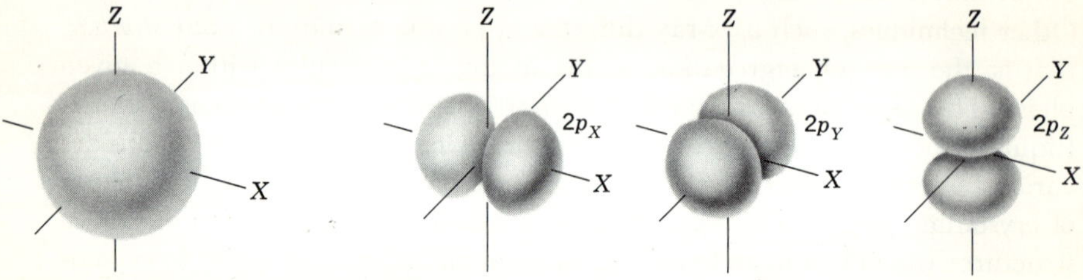

(*a*) 3*s* orbital.
(*b*) The three types of 2*p* orbitals.

FIGURE 1-2 Three-dimensional representations of *s*- and *p*-type atomic orbitals.

1-2. Ionic Bonding

The criterion for ionic bonding is difference in electronegativity (the tendency to acquire electrons). Therefore, the purest type of ionic bonding is found in substances composed of two types of atoms. One, such as sodium, has a low ionization energy and easily loses an electron; the other, such as chlorine, has a high electron affinity and strongly tends to acquire an electron. The sodium gives up its valence electron to the chlorine, each of the resulting ions then has a stable filled shell of outer electrons, and a strong electrostatic attraction is set up that bonds the Na^+ cation and the Cl^- anion into a very stable crystal structure. A useful measure of the strength of bonding is the *cohesive energy*, defined as the energy required to separate 1 mole of crystal into ions at infinite separation. If bonding is purely ionic, then the cohesive energy can be calculated as the energy of the ions at the equilibrium spacing in the crystal structure, using the state of infinite separation of the ions as the zero of potential energy. Some of the features of the calculation[1] are illustrated in Prob. 1-1. At separation distances less than that actually observed in NaCl crystals, the repulsive energy generated by deformation of the ionic cores causes the total energy of the crystal to rise.

Two factors determine the actual crystal structure adopted by the sodium and chloride ions. First, the structure must contain the proper ratio of ions to balance the electric charges. In NaCl this ratio is 1:1, whereas in CaF_2 it is two anions, F^-, to one cation, Ca^{2+}. Second, each anion should have as many nearest neighboring cations (as high a coordination number) as possible,

[1] The calculation is presented in detail in the book: Charles Kittel, "Introduction to Solid State Physics," 3d ed., p. 91ff, John Wiley & Sons, Inc., New York, 1966.

thus providing many sites for bonding. The schematic illustration of the NaCl structure in Fig. 1-3*a* shows that its coordination number is only 6, whereas the CsCl structure in Fig. 1-3*b* has a coordination number of 8. To see why sodium chloride cannot assume a higher coordination number, consider the size of the opening (interstice) that would exist if the anions were touching in the NaCl structure (Fig. 1-3*c*) and in the CsCl structure (Fig. 1-3*d*). At this

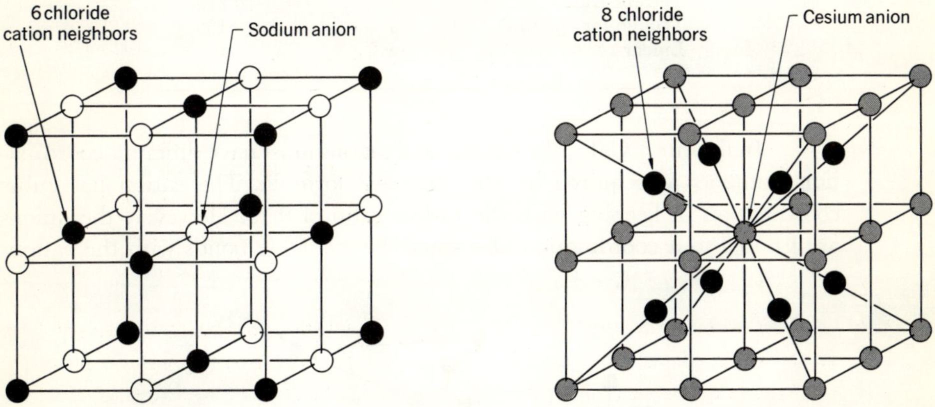

(*a*) Sodium chloride.

(*b*) Cesium chloride.

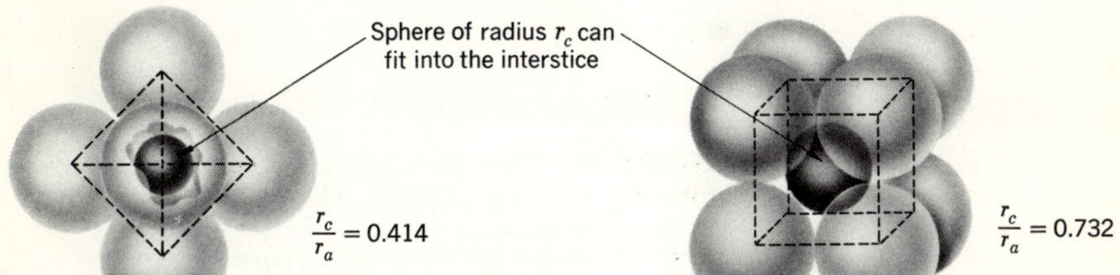

(*c*) Limiting radius ratio for an octahedral interstice.

(*d*) Limiting radius ratio for a cubic interstice.

FIGURE 1-3 Schematic illustration of the crystal structures of two ionic solids.

limiting radius ratio, electrical repulsion between adjacent anions interferes with bonding; consequently, the cation must actually have a larger radius to prevent contact between neighboring anions. Since the radius ratio of the sodium ion to the chloride ion is 0.525, the chloride ions are separated in the NaCl structure but would impinge in a CsCl-type structure. For ionic substances with even smaller ratios of cation to anion radii, the following arrangements of the anions must occur in the crystal structure:

Arrangement	**Coordination number**	**Limiting radius ratio**
Tetrahedral	4	0.225
Triangular (planar)	3	0.155
Linear	1 or 2	0

In fluorite,† CaF_2, the cations and anions must have different coordination numbers as required by the chemical formula. The cation has cubic coordination in keeping with the radius ratio of 0.73. However, the anions are limited to a coordination of 4 since the available bonds with the cations

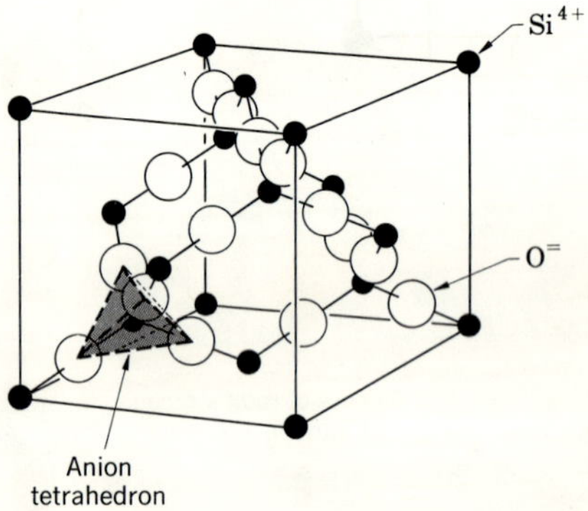

FIGURE 1-4 Illustration of the low coordination numbers in SiO_2. The crystal structure shown is cristobalite, the polymorphic form that is stable at high temperatures.

must be divided among twice as many anions. What happens if a small radius ratio limits the *cation* coordination to 4? This is the case for SiO_2, where the radius ratio is only 0.29. The small silicon ion is then at the center of a tetrahedral grouping of oxygen ions (Fig. 1-4), but each oxygen is coordinated with only two silicon ions.

† This important crystal structure is illustrated in the summary of crystal structures which is on the inside back cover.

1-3. Covalent Bonding

When electrons are shared between two atoms (rather than being transferred to make two different ions), the bonding is known as *covalent*. Unlike ionic bonding, where a simple electrostatic calculation can give an estimate of the strength of bonding, there is no elementary method for determining the characteristics of covalent bonds. Quantum mechanics furnishes quantitative solutions for a few of the simpler molecules, and it also provides a logical framework for the construction of empirical methods for predicting covalent structures. The molecular orbital (MO) approach has been especially succcessful.

MOLECULAR ORBITALS. The familiar atomic orbitals (AO) for free atoms ($1s$, $2s$, $2p$, etc.) are the elements from which an MO is constructed. When two (or more) atoms join together to form a molecule, the AOs combine to form the corresponding MOs. The valence electrons of the atoms in the molecule actually "fill up" only the MOs of lowest energy. Figure 1-5 gives a simple

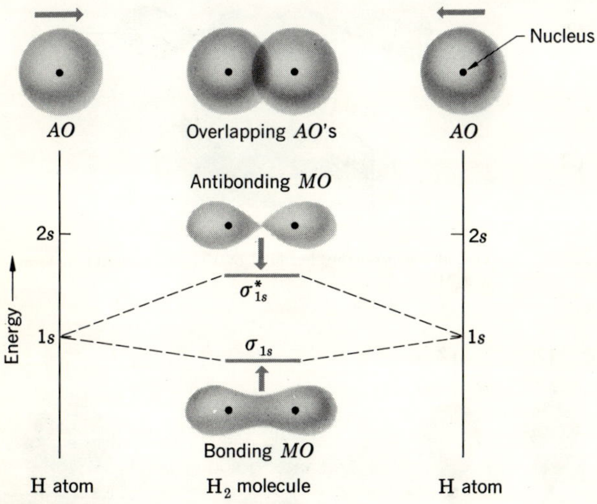

FIGURE 1-5 Schematic illustration of the formation of the MOs σ^*_{1s} and $\sigma_{\epsilon\epsilon}$ (antibonding) in a hydrogen molecule from the $1s$ AOs in free atoms of hydrogen. The two free atoms are pictured as approaching each other until their AOs overlap.

illustration of this process for the formation of a molecule of hydrogen from two hydrogen atoms. While the two hydrogen atoms are separated, each has the usual AOs. As the atoms come together, MOs form from the AOs. Recall that various AOs exist in a hydrogen atom even though only the $1s$ orbital is usually occupied. Similarly, a whole series of MOs can be constructed for the hydrogen molecule, although the two shown in the figure, σ_{1s} and σ^*_{1s}, are more than sufficient for our present purpose. The σ_{1s} MO can be pictured as the overlapping of a $1s^{(+)}$ AO (the + sign refers to electron spin) and a $1s^{(-)}$ AO; the quantum-mechanical calculation is roughly equivalent to adding the electron densities of these two AOs to give the (bonding) MO shown. The high electron density in the center serves to bond the two positively charged nuclei and makes the molecule more stable than two free atoms. The overlapping of two $1s^{(+)}$ AOs (or two $1s^{(-)}$ AOs) produces the σ^*_{1s} (antibonding) MO, corresponding to the subtraction of the electron density of one AO from the other. Since there are only two electrons in a hydrogen molecule, they occupy σ_{1s}, the MO of low energy, the σ^*_{1s} MO being vacant. In a helium molecule it would be necessary to use both types of MOs (see Prob. 1-3).

When we consider $2p$ (Fig. 1-6), and $3d$ types of AO we find that overlap

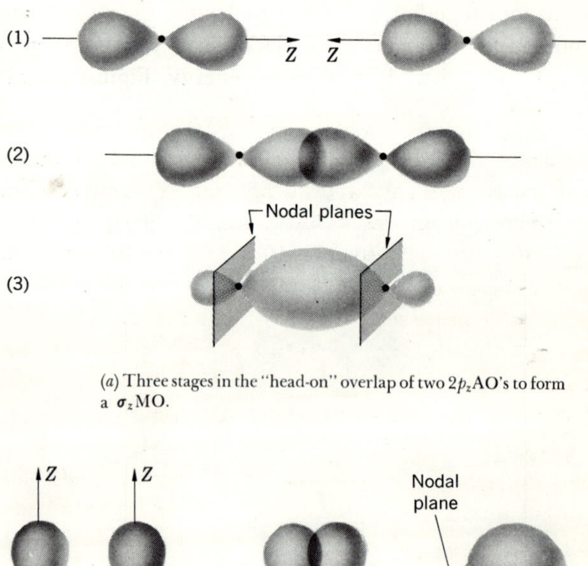

(*a*) Three stages in the "head-on" overlap of two $2p_z$AO's to form a σ_zMO.

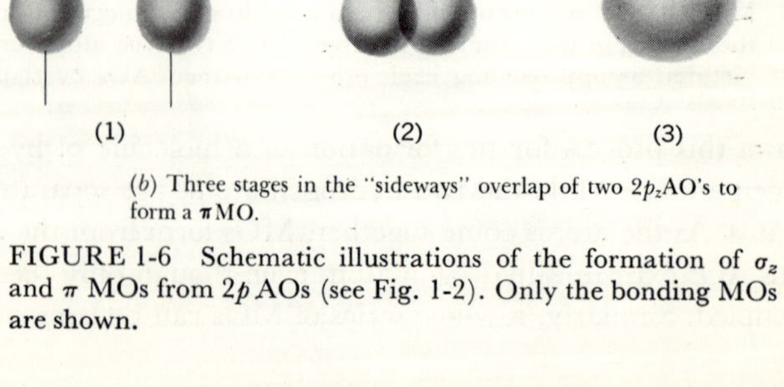

(*b*) Three stages in the "sideways" overlap of two $2p_z$AO's to form a πMO.

FIGURE 1-6 Schematic illustrations of the formation of σ_z and π MOs from $2p$ AOs (see Fig. 1-2). Only the bonding MOs are shown.

can now occur in various ways. If two $2p$ AOs of the same type (both $2p_z$, for example) overlap "head-on," as in Fig. 1-6*a*, the resulting MO is known as σ_z since the nodal planes are perpendicular to the z axis. On the other hand, if the overlap of $2p_z$ AOs occurs "sideways," as in Fig. 1-6*b*, then a π MO is formed, having one nodal plane containing the (internuclear) z axis. Overlapping of $3d$-type AOs can produce σ or π MOs and also a third type, the δ MO. In atoms that have certain energy levels close together, these levels sometimes interact (as a preliminary to the formation of a molecule with other atoms) and form hybrid AOs. For example, a $2s$ and a $2p$ AO can interact to form two sp hybrid AOs (Fig. 1-7*a*) which lie along a straight line. A triangular pattern of sp^2 hybrid AOs in a single plane is produced when a second type of $2p$ AO also takes part, as in Fig. 1-7*b*. Finally, a tetrahedral, spatial pattern of sp^3 hybrid AOs (Fig. 1-7*c*) can also occur. These strongly directional AOs account for the oriented character of many covalent bonds.

The MOs of typical substances (Fig. 1-8) illustrate hybridization and directional bonding. If the carbon atom in methane, CH_4, used its unhybridized AOs ($2s$, $2p_x$, $2p_y$, and $2p_z$) to form MOs with the hydrogen atoms, two different kinds of bond would be formed. Preliminary hybridization to the tetrahedral sp^3 leads to σ MOs in the molecule (Fig. 1-8*a*) and to a lower energy. Carbon, silicon, and germanium can each form a crystalline modification in which each atom is bonded to four other atoms. Again the four equivalent tetrahedral AOs are preferred and lead to the diamond-cubic crystal structure shown in Fig. 1-8*b*. Actually, the stable condition of carbon under ordinary conditions is graphite (Fig. 1-8*c*). In this case each atom bonds primarily with only three other carbon atoms (the sp^2 configuration forms), and thus graphite has a sheet-type structure. The separate sheets are held together by weak van der Waals bonding. The final example, C_2H_4 (Fig. 1-8*d*), involves partial hybridization. Three of the AOs of each carbon atom form the planar sp^2 hybrids, leaving one $2p$ AO perpendicular to this plane. The four hydrogen atoms form σ MOs, and the carbon atoms form a σ-type C–C bond, all employing the sp^2 hybrids. Simultaneously the two $2p$ AOs undergo sideways overlap (Fig. 1-6*b*) and form a π MO, constituting a second C–C bond. The unsymmetrical nature of the π MO hinders rotation about the carbon double bond and stabilizes the planar arrangement of ethylene.

The strength of individual bonds in covalent molecules depends on such factors as degree of ionic character, bond order (single bond, double bond, etc.), and resonance of bonds between alternative positions. Table 1-1 gives some examples of experimentally determined bond energies for neighboring atom pairs in typical molecules. Using these values, one can estimate the total bonding energy of materials such as polymers and judge their relative stabilities.

POLYMER STRUCTURES. An important example of covalent bonding is found in polymeric materials, which include rubbery *elastomers*, high-strength *fibers*, and the variety of structural materials denoted by the term *plastics*. For

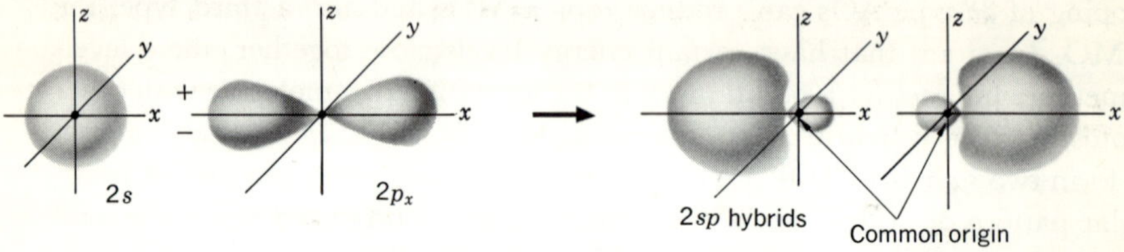

(*a*) Formation of two, linear *sp* hybrid AO's.

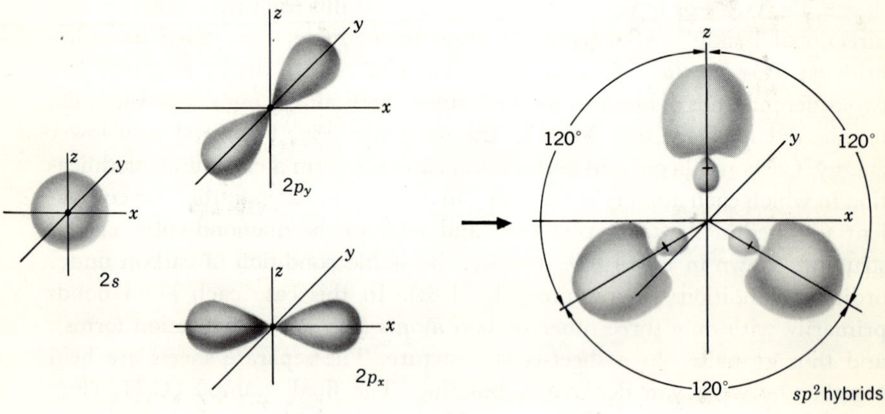

(*b*) Formation of three, planar sp^2 hybrid AO's.

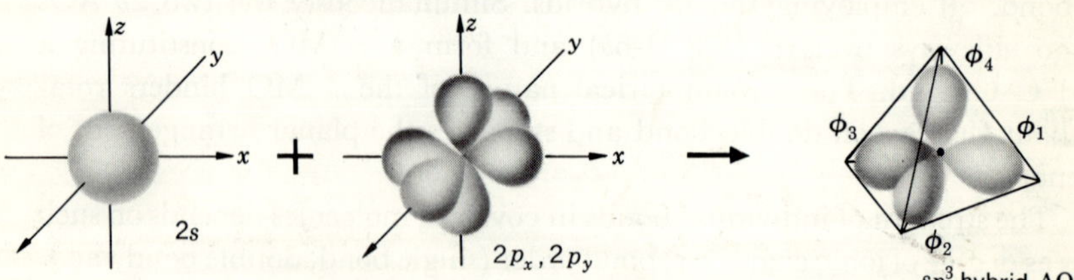

(*c*) Formation of four, tetrahedral sp^3 hybrid AO's.

FIGURE 1-7 Schematic illustrations of the formation of hybrid AOs from the 2*s* and 2*p* AOs of a given atom. For clarity, most of the orbitals are displaced from their common origin.

simplicity the following discussion of polymeric structures focuses on polymers made from only one type of monomer; many commercial polymers are actually mixtures of several different monomers.

An important plastic, polyethylene, is made by polymerizing ethylene molecules, a single one of which is called a *monomer*. Since the monomer has doubly bonded carbon atoms (one σ and one π MO, as in Fig. 1-8*d*), a more

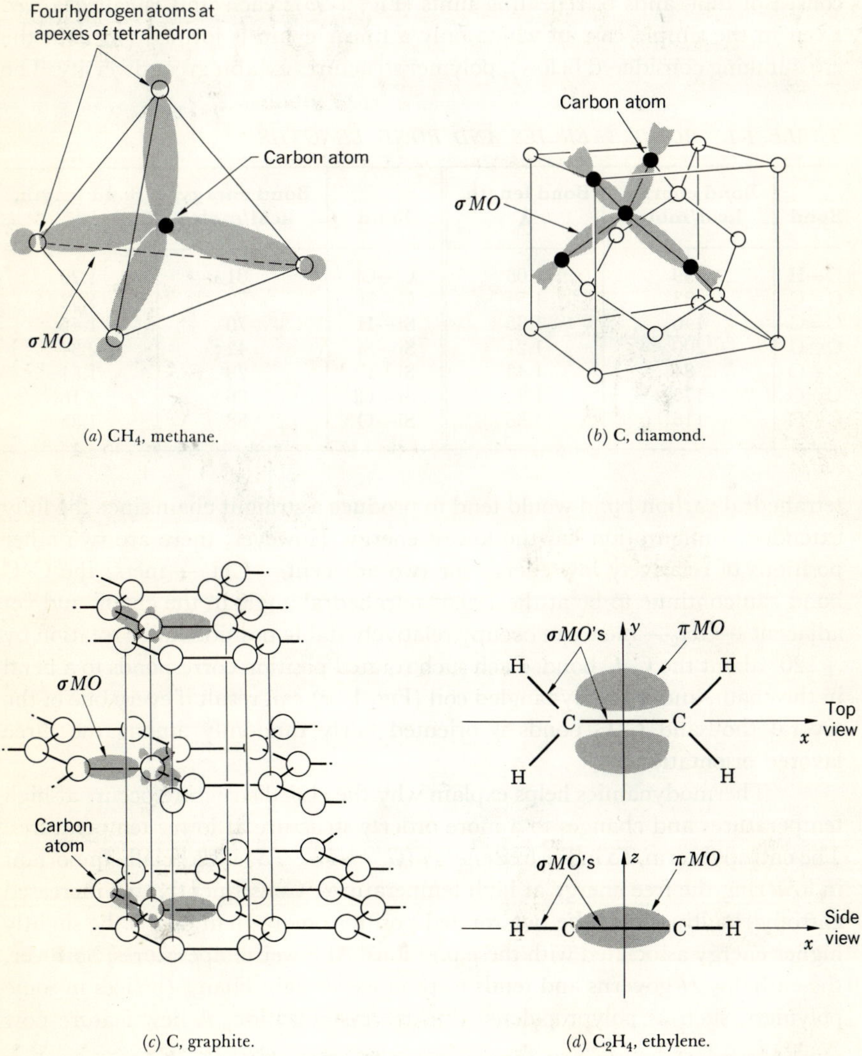

FIGURE 1-8 Schematic illustrations of the MOs in several structures having a strongly oriented character.

stable structure results when the weak π bond is replaced by another σ bond (see Prob. 1-4). In practice, however, the reaction (polymerization) of two ethylene molecules occurs only under special circumstances, for example, in the presence of a free-radical initiator such as a peroxide (Fig. 1-9*a*). Polymerization continues by the addition of single ethylene molecules to the chain and terminates by the linking of two shorter chains; an impurity incorporated into the chain may also determine growth. The final polymer molecule may consist of thousands of repeating units (Fig. 1-9*b*); each unit is called a *mer*. Even in the simple case in which only a linear chain is formed (without the cross-linking considered below), polymer structures exhibit great diversity. The

TABLE 1-1. BOND ENERGIES AND BOND LENGTHS

Bond	Bond energy, kcal/mole	Bond length, Å	Bond	Bond energy, kcal/mole	Bond length, Å
C—H	99	1.08	C—Cl	81	1.76
C—C	83	1.54	C—Si	72	1.93
C=C	146	1.35	Si—H	70	1.45
C≡C	200	1.21	Si—Si	42	2.34
C—O	84	1.43	Si—F	129	1.81
C=O	175	1.22	Si—Cl	86	2.16
C—F	116	1.36	Si—O	88	1.83

tetrahedral carbon bond would tend to produce a straight chain since the fully extended configuration has the lowest energy. However, there are two other positions of relatively low energy for two adjacent —CH_2— mers; the C–C bond can continue to be at the 109.5° tetrahedral angle in the chain, and yet adjacent —CH_2— mers can occupy relatively stable positions after rotation by $\pm 120°$ about the C–C bond. Each such rotated position corresponds to a bend in the chain, and a grossly tangled coil (Fig. 1-9*c*) can result if every one of the several thousand C–C bonds is oriented fairly randomly among the three favored orientations.

Thermodynamics helps explain why the coiled structure occurs at high temperatures and changes to a more orderly structure at lower temperatures. The entropy[1] term TS in the free energy ($G = H - TS$) is especially important in lowering the free energy at high temperatures. Consequently, the increased entropy resulting from the two rotated positions more than offsets the slightly higher energy associated with these positions. At lower temperatures, however, the enthalpy H governs and tends to produce straight chains (helices in some polymers, such as polypropylene) during crystallization. A new feature now

[1] Definitions and examples of the quantities G, H, and S are given in Sec. 2-6 for those unfamiliar with thermodynamics.

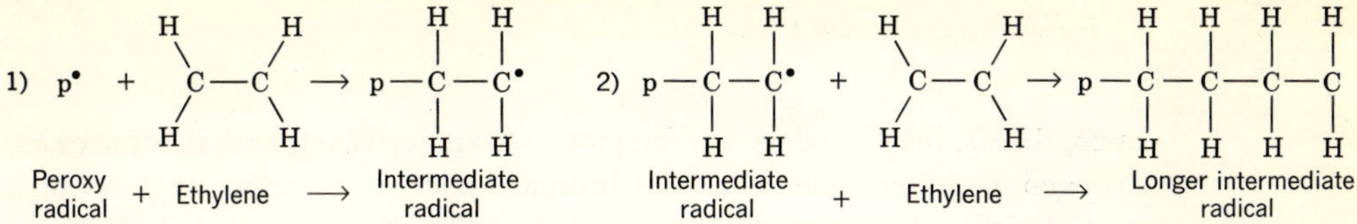

(*a*) The role of free radicals (indicated by "•") in the initial steps of polymerization of ethylene.

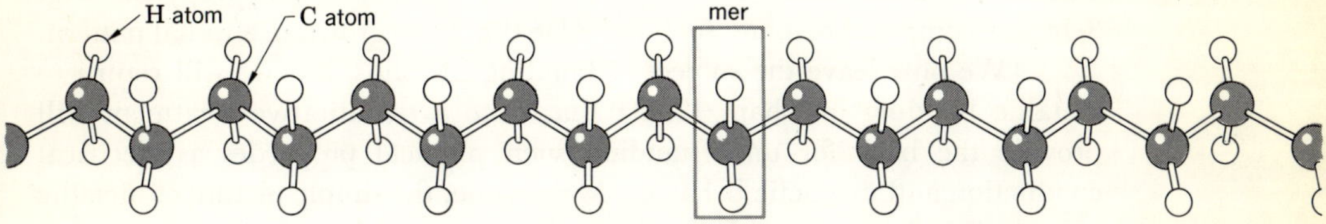

(*b*) A (straight) portion of a typical chain of polyethylene. The entire chain may contain 50,000 monomer units.

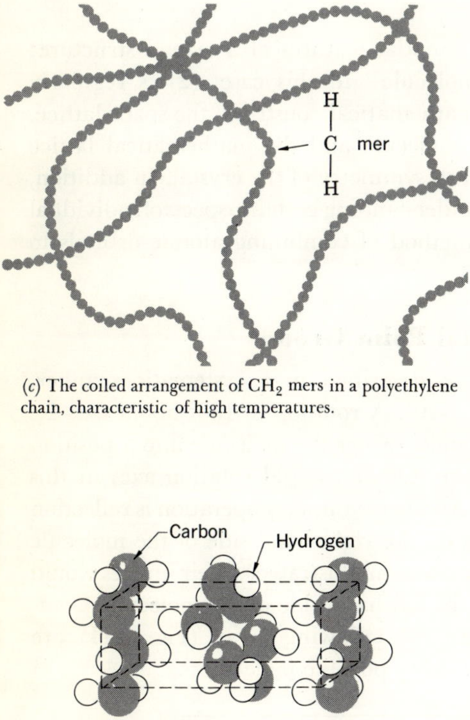

(*c*) The coiled arrangement of CH_2 mers in a polyethylene chain, characteristic of high temperatures.

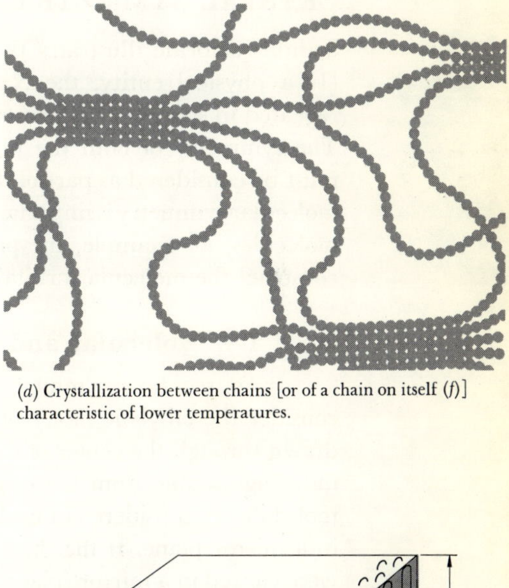

(*d*) Crystallization between chains [or of a chain on itself (*f*)] characteristic of lower temperatures.

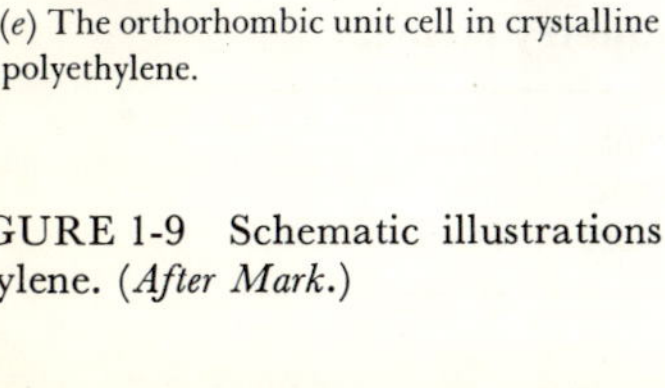

(*e*) The orthorhombic unit cell in crystalline polyethylene.

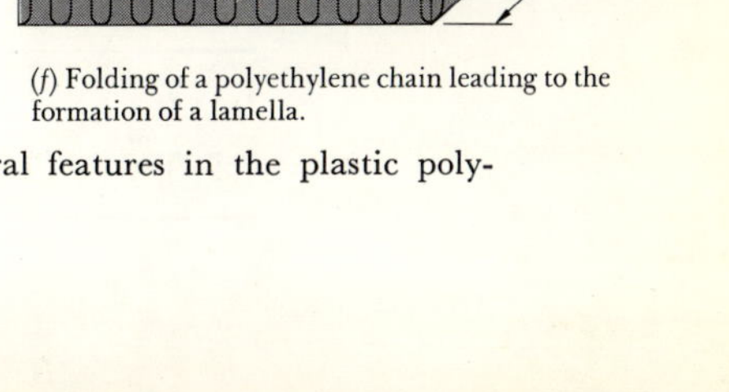

(*f*) Folding of a polyethylene chain leading to the formation of a lamella.

FIGURE 1-9 Schematic illustrations of some structural features in the plastic polyethylene. (*After Mark.*)

enters, that is, the secondary bonding forces (van der Waals and, in some cases, hydrogen bonding) that favor the intimate packing of adjacent parts of a given chain or of separate chains (Fig. 1-9*d*). The closely packed chains form a crystal structure, Fig. 1-9*d*, of the kind described in the next section. When repeated folding occurs in a given chain, as in Fig. 1-9*f*, the result is the formation of a crystalline lamella. The spherulites in commercial polypropylene (Fig. 3-7*b*) are composed of many such lamellae that have grown in a radial pattern.

We now leave the subject of bonding in solids, but we will return to metallic bonding in Chap. 12. At that time a quantitative treatment will provide the basis for understanding such physical properties as electrical conduction and magnetic behavior. Meanwhile, the simple picture of metallic bonding involving a cloud of valence electrons will be adequate for understanding many aspects of the behavior of metals and alloys.

CRYSTAL SYMMETRY

Sodium chloride illustrates the two essential features of a crystal structure: (1) a physical entity, the NaCl "molecule" in this case, (2) is regularly repeated in space on the points of a mathematical construct, the space lattice. The symmetry of both the physical molecule and the mathematical lattice must be considered as part of the overall symmetry of the crystal. In addition, molecular symmetry is important for understanding certain aspects of individual molecules, for example, the proper method of combining atomic orbitals to construct the molecular orbitals.

1-4. Molecular and Crystal Point Groups

To see how symmetry can be described in a systematic manner, consider the ethylene molecule (Fig. 1-10). A rotation of 180° about axis 2_z, drawn through the center of the molecule, brings the molecule into a position indistinguishable from the original one. Other twofold rotation axes in this molecule are considered in Prob. 1-5. Another symmetry operation is reflection in a mirror plane. If the three atoms on the right-hand side of the molecule were viewed in a mirror (placed at the position illustrated), their images would coincide with the identical atoms on the left-hand side of the molecule. Combinations of symmetry elements from the following array (Fig. 1-11) are adequate to describe the symmetry of every possible molecule:

Symbol	Symmetry element
$1, 2, 3, \ldots, n$	Axes of n-fold rotation
m	Mirror plane
$\bar{1}, \bar{2}, \bar{3}, \ldots, \bar{n}$	Axes of n-fold rotation-inversion

The operation of an axis of rotation-inversion, say $\bar{2}$ in Fig. 1-11*c*, consists of a 180° rotation plus inversion through the center of the molecule (see Prob. 1-5).

A group of mutually compatible symmetry operations constitute a *molecular point group*. The molecular point group for ethylene, for example, consists of the operations 1, 2_x, 2_y, 2_z, m_{xy}, m_{yz}, m_{zx}, and $\bar{1}$. Two different notations are commonly used for the 44 possible molecular point groups. In

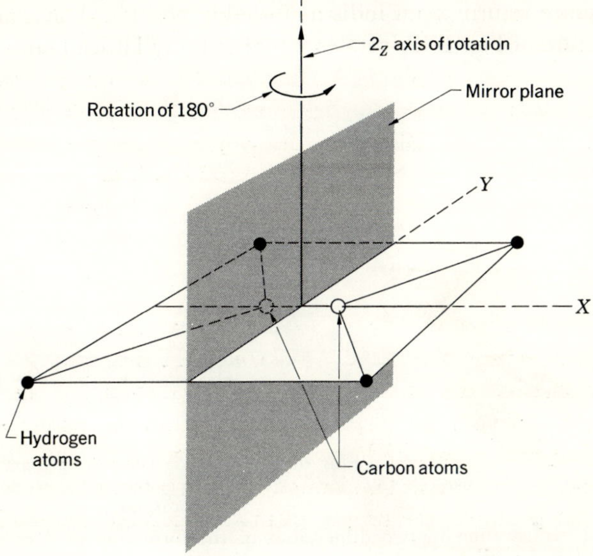

FIGURE 1-10 Schematic illustration of one of the twofold rotation axes and the associated mirror plane in the ethylene molecule.

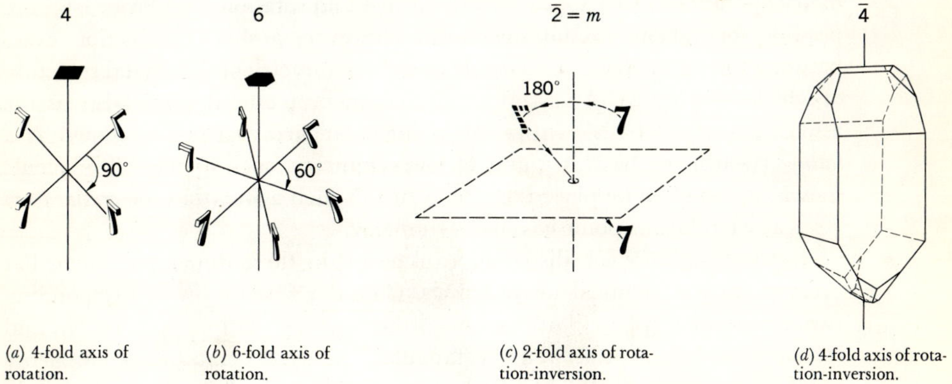

(*a*) 4-fold axis of rotation.

(*b*) 6-fold axis of rotation.

(*c*) 2-fold axis of rotation-inversion.

(*d*) 4-fold axis of rotation-inversion.

FIGURE 1-11 Schematic illustrations of typical elements of symmetry. (*After L. V. Azároff.*) A crystal of the mineral calcopyrite, $CuFeS_2$, exhibits a fourfold rotation-inversion axis. (*After W. L. Braggs.*)

the Schoenflies system the above symmetry is denoted $\mathscr{D}_{2h}$, whereas in the Herman-Maugin system (Table 1-2) the notation is *mmm*. The number of molecular point groups is limited to 44 because sixfold symmetry is the highest observed in actual molecules.

The same symmetry operations can operate at a point in a three-dimensional crystal structure, but there is now the requirement that the entire crystal structure return to an indistinguishable position. For example, if the crystal structure of Fig. 1-12*c* rotates 90° about any lattice point, such as the

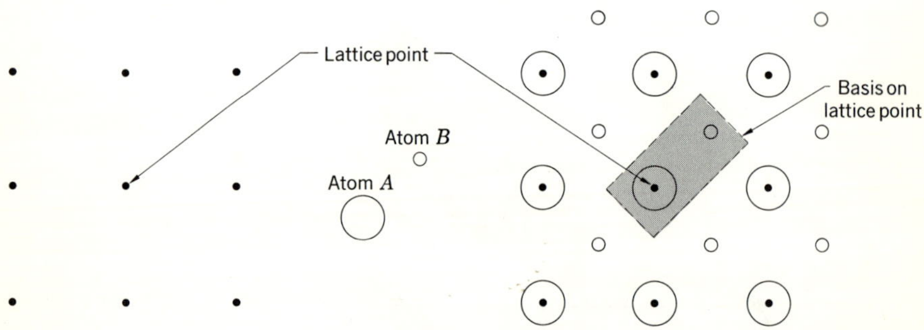

(*a*) Portion of an infinite (plane) lattice.

(*b*) A typical basis, the molecule *AB*.

(*c*) Portion of a corresponding infinite crystal structure.

FIGURE 1-12 Illustration in two dimensions of the formation of a crystal structure having fourfold symmetry using a basis of lower symmetry.

one indicated (simply rotate the figure to see the result), all of the atoms move to positions previously occupied by an atom of the same type. All of the symmetry operations listed above, except for five-fold rotations (see Prob. 1-6), are possible for various crystal structures. However, a sharp distinction exists between the symmetry of a molecule and the symmetry of the crystal structure of which it is a part. A crystal structure can be pictured (Fig. 1-12) as the association of a "basis" with each point of an array of points known as a *lattice*. Even if the basis is a unit of low symmetry, such as the AB molecule shown, it can often be placed on the lattice in such a way that the *crystal point group* at each lattice point has high symmetry.

A *space lattice* is a distribution of points in three dimensions such that every point has identical surroundings. Consider first the five corresponding two-dimensional (plane) lattices. One is the "square" lattice (Fig. 1-12*a*), and it is obviously different from a "rectangular" array of points or from an array involving an inclined axis (see Prob. 1-7). Similarly, in three dimensions only 14 qualitatively different arrangements of points are possible. Seven of these 14 space lattices are shown in Table 1-2, the seven primitive lattices corresponding exactly with the geometric figures of the seven crystal systems. The seven remaining lattices have additional points in the center or on the faces of the geometric figures; the face-centered cubic and body-centered cubic space lattices (Fig. 1-13) are important examples. Alternatively, these two

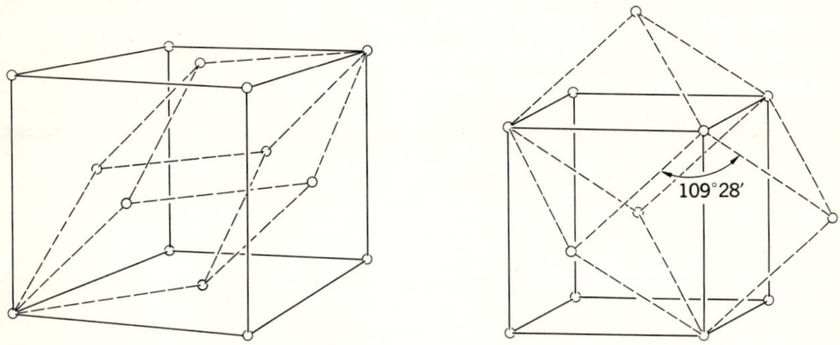

(*a*) Face-centered cubic space lattice. The rhombohedral axes of the primitive cell are shown by dashed lines.

(*b*) Body-centered cubic space lattice. The rhombohedral axes of the primitive cell are shown by dashed lines.

FIGURE 1-13 Alternate descriptions of two space lattices that have cubic symmetry.

lattices are described by primitive cells with rhombohedral crystal axes. Cubic axes are convenient for computations and for showing the symmetry of the space lattice. Their disadvantage is the existence of extra lattice points (three in the face-centered cubic, one in the body-centered cubic) in addition to the minimum of one[1] lattice point per cell in a primitive cell.

An additional concept, the *space group*, is important for the detailed study of crystal structures. A space group is a repetition of symmetry elements at the points of an infinite space lattice. The repetition is accomplished by the set of translation vectors

$$\mathbf{T}_u = \sum_{i=1}^{3} u_i \mathbf{a}_i = u_1\mathbf{a}_1 + u_2\mathbf{a}_2 + u_3\mathbf{a}_3 \tag{1-1}$$

Here the u_i are integers, and $\mathbf{a}_1$, $\mathbf{a}_2$, and $\mathbf{a}_3$ are the base vectors of the crystal axes (Table 1-2), specified by their lengths and interaxial angles. A typical operation of the vectors $\mathbf{T}_u$ (Fig. 1-14) takes a unit of structure, such as a molecule, located at the origin and repeats it at the points of the (primitive) space lattice. The symmetry of the unit of structure helps determine the crystal point group of the complete crystal structure (see Prob. 1-8). In addition to the

[1] A primitive cell has eight corner points, but each of these points is shared with seven adjoining cells.

TABLE 1-2. DESCRIPTIONS OF THE CRYSTAL SYSTEMS, SPACE LATTICES, AND POINT GROUPS

Sketches of the primitive lattice	Crystal systems	Space lattices	Point groups†
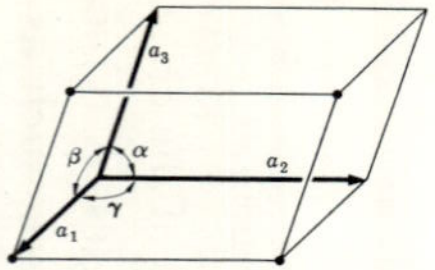	Triclinic $a_1 \neq a_2 \neq a_3$ $\alpha_1 \neq \alpha_2 \neq \alpha_3$	Primitive	1 $\bar{1}$
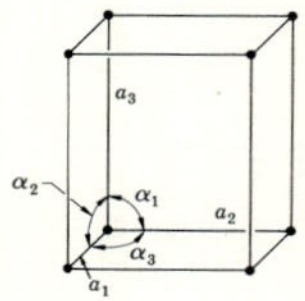	Monoclinic $a_1 \neq a_2 \neq a_3$ $\alpha_1 = \alpha_3 = 90^\circ \neq \alpha_2$	Primitive Base-centered	m 2 $2/m$
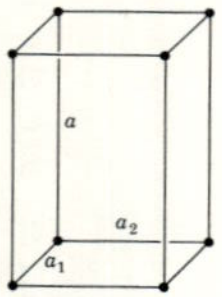	Orthorhombic $a_1 \neq a_2 \neq a_3$ $\alpha_1 = \alpha_2 = \alpha_3 = 90^\circ$	Primitive Face-centered Body-centered Base-centered	mm 222 mmm
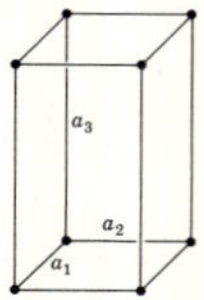	Tetragonal $a_1 = a_2 = a \neq a_3$ $\alpha_1 = \alpha_2 = \alpha_3 = 90^\circ$	Primitive Body-centered	$\bar{4}$ 4 $4/m$ $\bar{4}2m$ $4mm$ 42 $4/mmm$
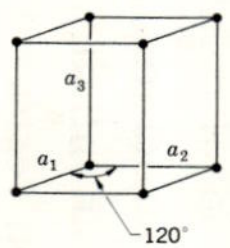	Hexagonal $a_1 = a_2 = a \neq a_3$ $\alpha_1 = \alpha_2 = 90^\circ$ $\alpha_3 = 120^\circ$	Primitive	$\bar{6}$ 6 $6/m$ $\bar{6}2m$ $6mm$ 62 $6/mmm$
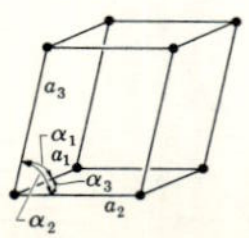	Rhombohedral $a_1 = a_2 = a_3 = a$ $\alpha_1 = \alpha_2 = \alpha_3 = \alpha \neq 90^\circ$	Primitive	3 $\bar{3}$ $3m$ 32 $\bar{3}m$
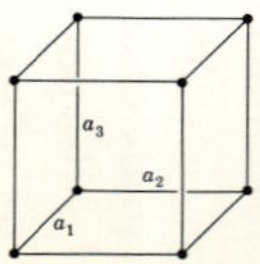	Cubic $a_1 = a_2 = a_3 = a$ $\alpha_1 = \alpha_2 = \alpha_3 = 90^\circ$	Primitive Body-centered Face-centered	23 $m3$ $\bar{4}3m$ 43 $m3m$

point groups, two other symmetry operations are repeated at the points of a space lattice. They are called *microscopic* symmetry operations because their effect cannot be seen in the *macroscopic* symmetry of the crystal. These opera-

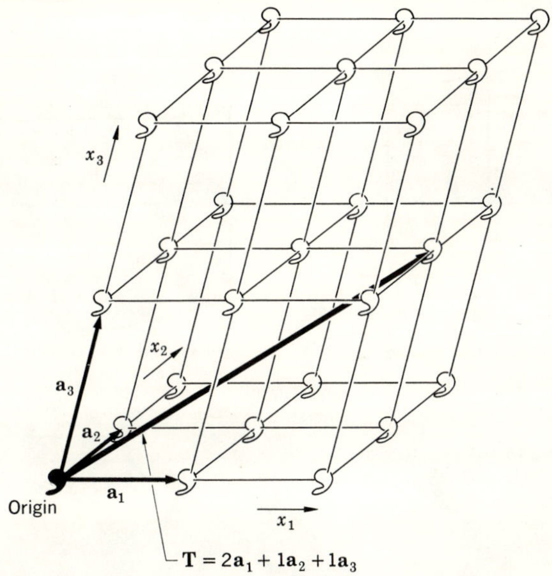

FIGURE 1-14 Schematic illustration of the operation of translation vectors to repeat a structural unit with no symmetry (the 1 molecular point group) at the points of a triclinic space lattice.

tions are *screw rotation*, a rotation accompanied by a translation along the axis, and *glide-plane reflection*, a mirror reflection accompanied by a translation along the mirror plane.

1-5. Metallic Structures

Some of the simplest crystal structures consist of metal atoms at the points of a space lattice. A crystal of copper, for example, is a *face-centered cubic* (FCC) array of copper atoms resembling a close packing of hard spheres (Fig. 1-15*a*). Only one type of atomic plane in the FCC structure is close-packed, the octahedral planes shown in Fig. 1-15*b*. These planes form the surfaces of the primitive FCC unit cell (Fig. 1-13*a*). Adjacent octahedral planes are stacked as compactly as possible (Fig. 1-15*c*), occupying successive positions commonly denoted as *A*, *B*, and *C*. Tetrahedral and octahedral interstices (see Sec. 1-2) occur in close-packed structures (Fig. 1-15*d*). Each interstice exists within a coordination polyhedron, formed by lines connecting the centers of the (four or six) atoms surrounding the given interstice.

Metals such as magnesium, zinc, and titanium have the other close-

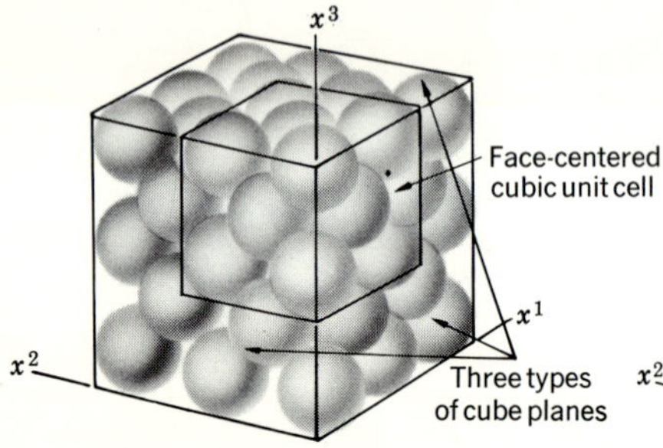

(*a*) A portion of the structure showing the three cube planes.

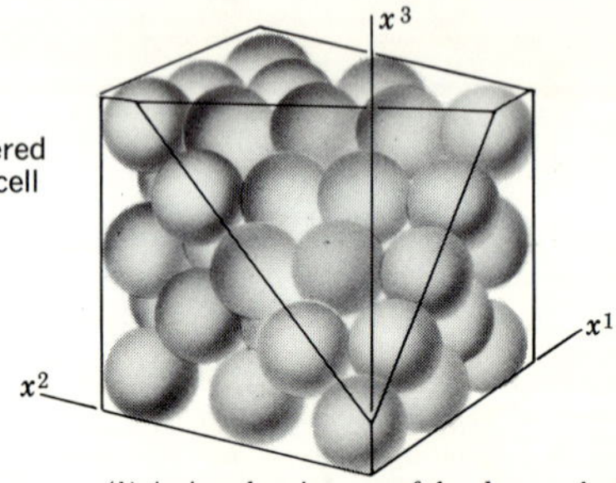

(*b*) A view showing one of the close-packed octahedral planes.

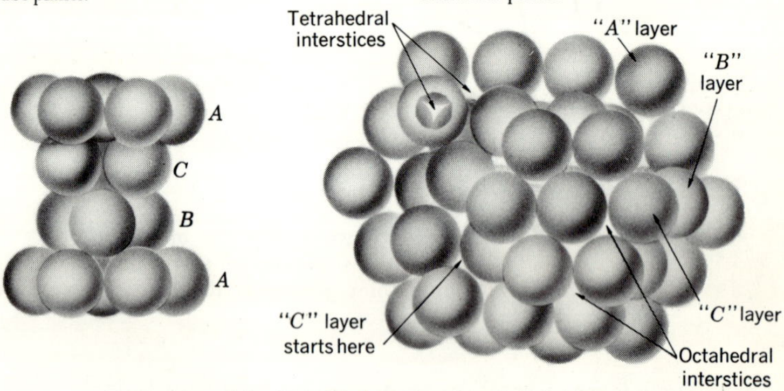

(*c*) Two views of the *ABCABC*...stacking of octahedral planes.

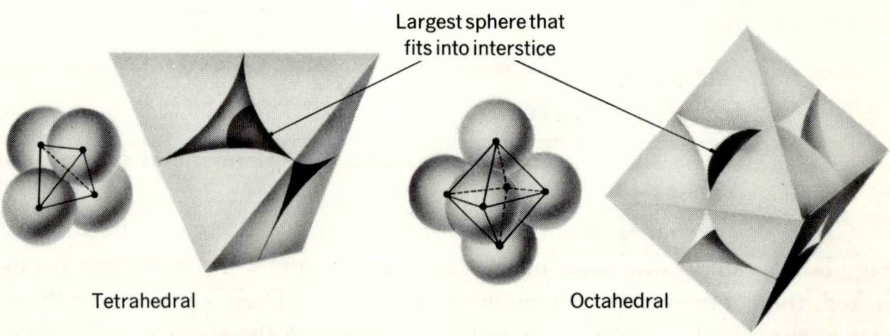

(*d*) Two types of interstitial spaces within coordination polyhedra.

FIGURE 1-15 Several aspects of the hard-sphere model of a face-centered cubic metal, such as copper. (*After W. D. Kingery and L. V. Azároff.*)

packed structure, *hexagonal close-packed* (HCP), consisting of a group of two (identical) atoms at each point of the hexagonal lattice, as shown in Fig. 1-16*a*. The enlarged unit cell (indicated by thin lines) involving the extra crystal axis

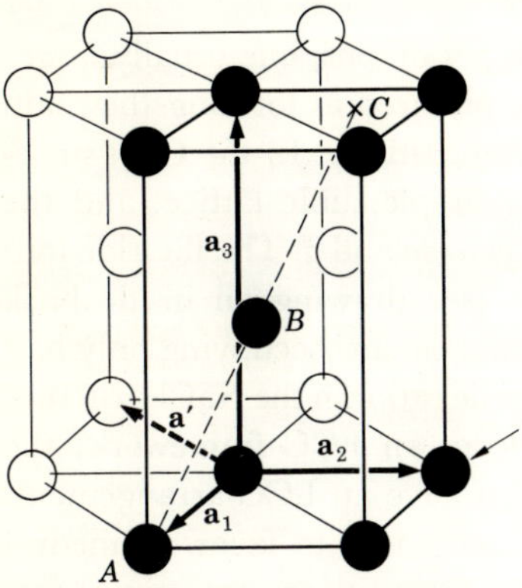

(*a*) The crystal structure consists of *AB* atom pairs on the points of the hexagonal space lattice.

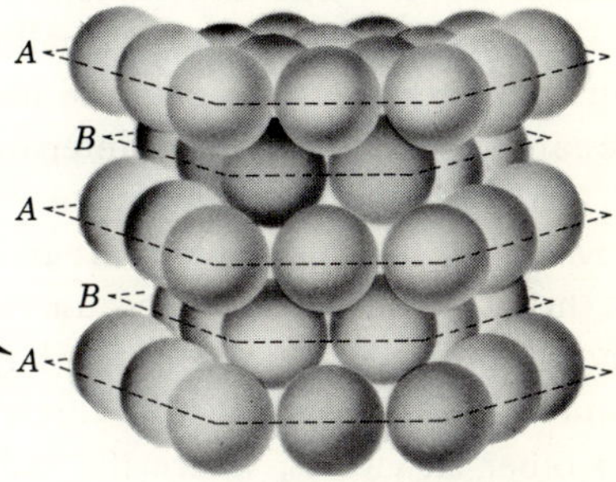

(*b*) Successive basal planes are stacked in *ABABAB* arrangement.

FIGURE 1-16 Important features of the hexagonal close-packed crystal structure.

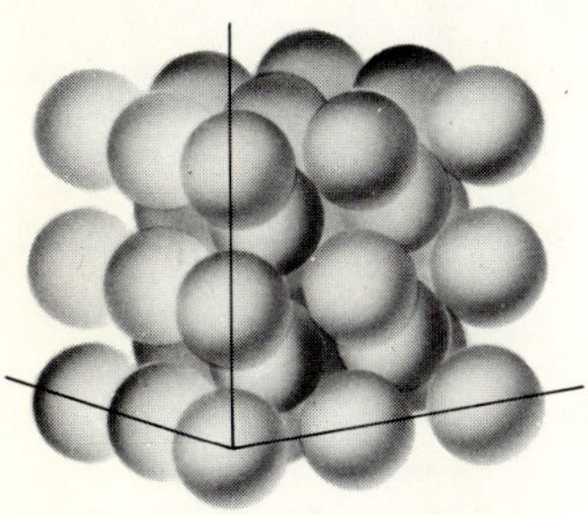

(*a*) A portion of the structure showing the three cube planes.

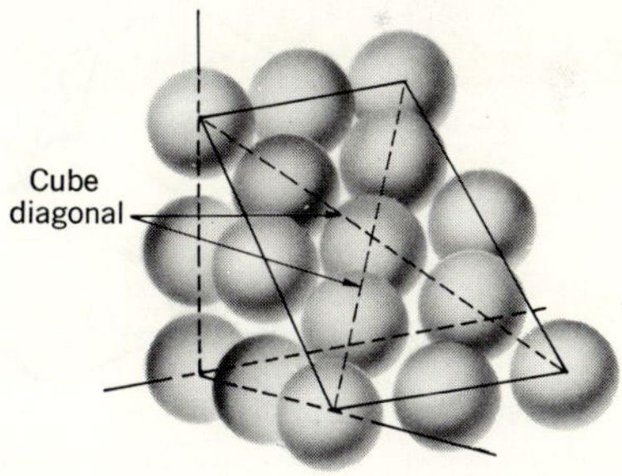

(*b*) A view showing one of the dodecahedral planes containing lines of atoms in contact.

FIGURE 1-17 Hard-sphere models of a body-centered cubic metal, such as α-iron.

a′ shows the hexagonal symmetry more clearly. The same close-packed layer found in the FCC structure also occurs here. It forms the basal plane (Fig. 1-16*b*) and is stacked in the *ABABAB* arrangement. Because no atoms exist at *C*-type positions (Fig. 1-16*a*), the atomic positions in the HCP structure do not themselves form a space lattice (Prob. 1-10). The third common metallic structure, *body-centered cubic* (BCC) (Fig. 1-17) is not a close-packing of atoms.

The atoms are in contact only along the directions of the three cube diagonals (Fig. 1-17*b*) which are also the crystal axes of the primitive cell (Fig. 1-13*b*).

1-6. Ionic and Covalent Structures

Ionic bonding involves two (or more) types of ions, one usually larger than the other. The large anions can often be pictured as forming the main framework within which the small cations fit interstitially. In the CsCl structure, for example, the chloride ions are on a simple-cubic lattice, and the cesium ions are in the cubic interstices, as in Fig. 1-3*b* and *d*. The fluoride ions in CaF_2 also form a simple-cubic framework (see drawing on inside back cover), but the calcium ions are in an FCC arrangement, occupying only half of the available interstices because of the 1:2 atomic ratio. In the NaCl structure shown in Fig. 1-3*a* and *c*, the large anions form an FCC framework; the cations are in the octahedral interstices and also have an FCC arrangement. In other structures, where the radius ratio is too small to favor octahedral coordination, the anions and cations can still form interpenetrating face-centered cubic arrays but with the two arrays displaced only one-quarter of the cell diagonal. The cubic zinc sulfide (ZnS, zinc blende) structure (Fig. 1-18) is of this type; it has fourfold coordination.

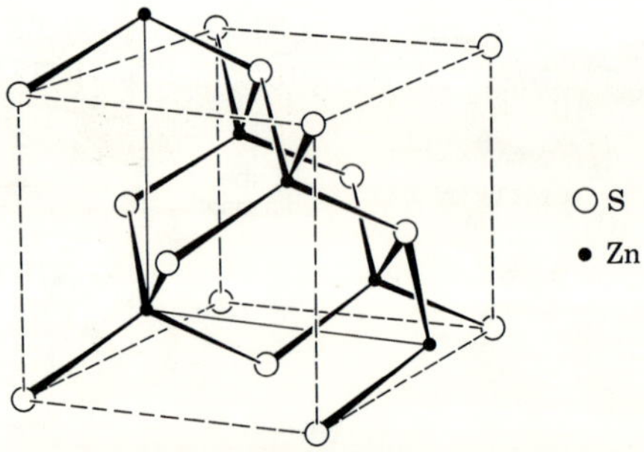

FIGURE 1-18 View of the cubic zinc sulfide (zinc blende) structure emphasizing the FCC arrangement of the sulfur ions. The zinc ions form a parallel, interpenetrating FCC arrangement, two axes of which are shown.

In covalently bonded crystals (germanium or antimony, for example) an "8 minus N" rule determines the number of bonds that a given atom makes with its neighbors; that is, to complete the octet of electrons needed for atomic stability, an element having N valence electrons obtains $8-N$ additional electrons by sharing with this number of neighboring atoms. The crystal structures of diamond and graphite, shown in Fig. 1-8, illustrate this rule on the one hand and show an exception, caused by hybridization, on the other hand. The diamond-cubic structure (Fig. 1-8*b*) is identical with that of zinc blende (Fig. 1-18), except that both types of sites are occupied by one kind of covalently bonded atom, carbon (or silicon or germanium, etc.). In the arsenic group of semimetals (As, Sb, and Bi), an atom has five valence electrons and therefore forms covalent bonds with three neighbors. The resulting crystal structure (see drawing on inside back cover) is almost face-centered cubic, but the bonding produces puckered layers of covalently bonded atoms parallel to the octahedral plane. Consequently, a crystal will easily break along this type of plane, a phenomenon known as *cleavage*. An example of two different atoms forming a covalently bonded crystal is SiC, useful as an abrasive and as an electronic material. One of its crystalline forms, β-SiC, has the cubic ZnS structure (Fig. 1-18) and is therefore analogous to diamond. Other modifications, termed α-SiC, have hexagonal structures consisting of various packing sequences of close-packed layers.

Polymerization, which is important for organic materials like polyethylene, also occurs with inorganic structural units such as the SiO_4^{4-} ion, shown in Fig. 1-19*a*. In a compound such as Mg_2SiO_4, all four of the negative

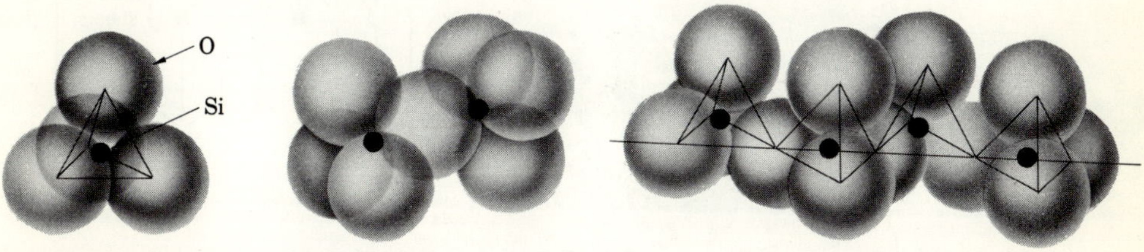

(*a*) The tetrahedral SiO_4^{4-} ion. (*b*) Sharing of one oxygen atom in the $Si_2O_7^{6-}$ ion. (*c*) Portion of a polymerized chain.

FIGURE 1-19 Illustration of various types of polymerization involving the SiO_4^{4-} ion as the structural unit. (*After L. H. Van Vlack.*)

charges on the SiO_4^{4-} ion are used for ionic bonding with Mg^{2+} ions. If fewer positive charges are available, as in the compound $Ca_2MgSi_2O_7$, then two SiO_4^{4-} groups polymerize by sharing a common oxygen atom (Fig. 1-19*b*) to form the $Si_2O_7^{6-}$ structural unit. Polymerization can continue with the formation of an entire chain when two oxygen ions are shared, a structural unit then being SiO_3^{2-}, as shown in Fig. 1-19*c*. $MgSiO_3$ is an example of a crystal structure involving a silicate chain. Both $Ca_2MgSi_2O_7$ and $MgSiO_3$ have rather complex orthorhombic crystal structures.[1]

[1] L. Bragg and G. F. Claringbull, "Crystal Structures of Minerals," G. Bell & Sons, Ltd., London, 1965.

When three oxygen ions in each SiO_4^{4-} tetrahedron are shared with a neighboring tetrahedron, an $Si_2O_5^{2-}$ unit acts as a building block for a sheet-like structure (Fig. 1-20*a*). The tetrahedra in this silicate sheet appear as a row of triangles in the side view (Fig. 1-20*b*). Bonding of two adjacent sheets occurs by ionic bonding of a cation (Mg^{2+}) with O^{2-} ions in octahedral coordination and produces a "sandwich." Succeeding layers of the sandwich

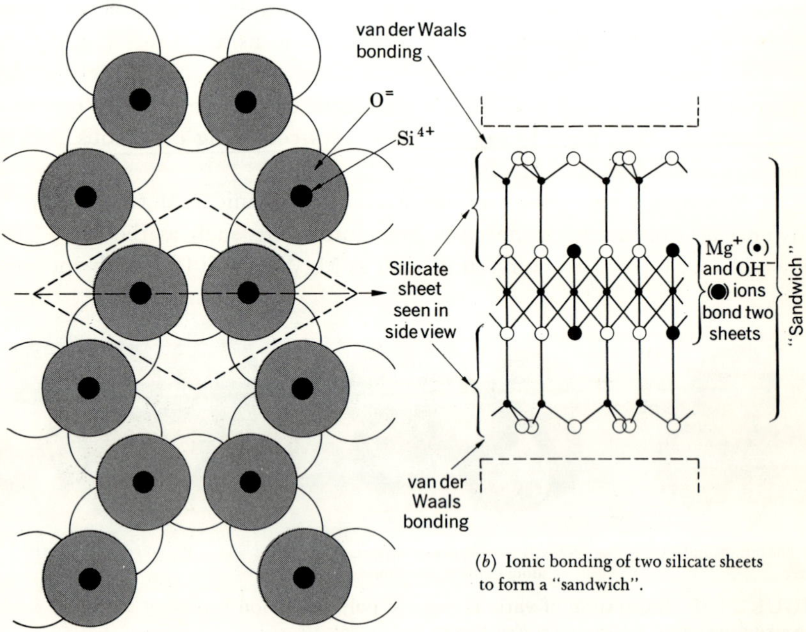

(*b*) Ionic bonding of two silicate sheets to form a "sandwich".

(*a*) An Si_2O_5 unit in part of a silica-sheet lattice.

FIGURE 1-20 Talc, $Mg_3(Si_2O_5)_2(OH)_2$, is a planar structure constructed of polymerized silicate sheets, ionically bonded as "sandwiches," the latter being held together only by van der Waals bonding.

are held together only by van der Waals bonding, and thus materials of this type (talc, clay, etc.) are usually soft and easy to cleave. The cristobalite modification of crystalline SiO_2 (Fig. 1-4) is an example of the structures produced when polymerization occurs in a three-dimensional pattern. Glass is a related structure but without the regularity that characterizes the crystalline state.

QUANTITATIVE OPERATIONS WITH CRYSTAL STRUCTURES

To treat various changes that occur in crystal structures—deformation, transformation, precipitation, etc.—we often need to describe precisely the relative locations of atoms, of planes of atoms, or of two crystal structures. In this section, we present two of the most useful approaches: the vector method for specifying directions and planes in a crystal, and the stereographic projection for treating orientations in three dimensions.

1-7. Directions and Planes in Crystal Structures

The vector **r** from the origin to a point P in the crystal (Fig. 1-21) specifies a particular direction in the crystal. For many purposes the length

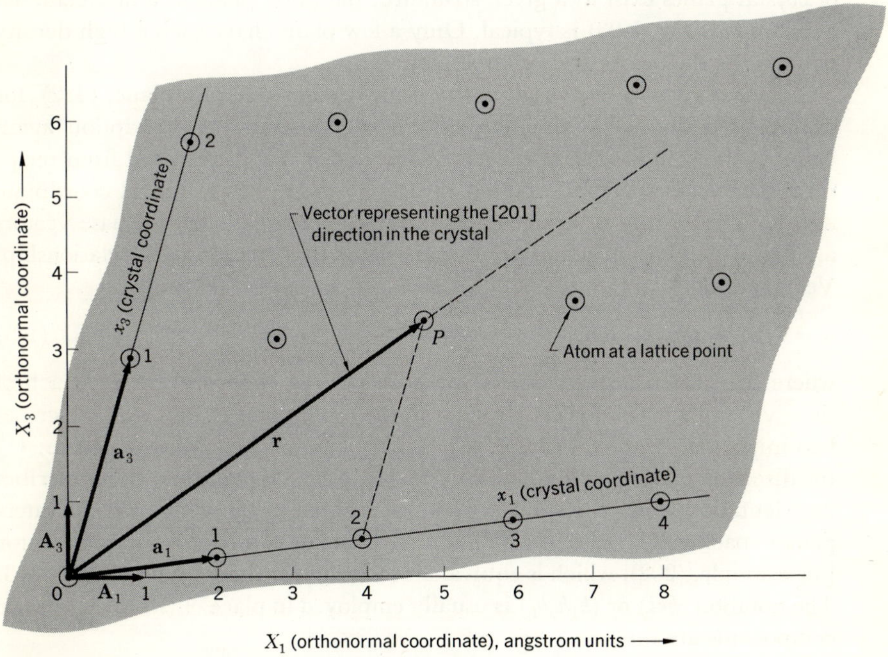

FIGURE 1-21 Method for specifying a direction in a crystal. This sketch shows a two-dimensional section through a monoclinic crystal in a particular relation to cartesian axes X_1. The X_2 and x_2 axes coincide and are directed perpendicularly into the page.

of the vector is unimportant. The crystal base vectors ($\mathbf{a}_1$, $\mathbf{a}_2$, and $\mathbf{a}_3$) permit a simple quantitative description of **r** using three components (r_1, r_2, and r_3); thus

$$\mathbf{r} = r_1\mathbf{a}_1 + r_2\mathbf{a}_2 + r_3\mathbf{a}_3 \tag{1-2}$$

The three r_i are the crystal-lattice components of $\mathbf{r}$ and are read on the x_1, x_2, and x_3 crystal coordinate axes. Usually point P is at a lattice point, as in Fig. 1-21, and then the r_i are integers (2, 0, 1 in this case). Thus

$$\mathbf{r} = 2\mathbf{a}_1 + 0\mathbf{a}_2 + 1\mathbf{a}_3 \tag{1-3}$$

The three components, by convention enclosed in square brackets, as [201], denote the crystal direction in question. Crystal directions could be described by components (R_i) referred to ordinary cartesian coordinates (X_i), but Prob. 1-11 shows that such components are nonintegral and generally inconvenient.

For many purposes the atoms in a crystal structure are pictured as lying in parallel planes that pass through the centers of the atoms. A stacking of any one type of plane, such as one of the cube planes (or an octahedral plane) shown in Fig. 1-15, builds up the entire crystal structure. Many types of crystal planes exist in a given structure; the (432) plane in a face-centered cubic metal (Fig. 1-23) is typical. Only a few planes have such a high density of atoms as the octahedral or cube planes.

As suggested by the similarity of the designation of a plane, (432), for example, to that of a direction, [201], the quantitative description again employs the components of a vector. This vector $\mathbf{s}$ is constructed from reciprocal base vectors ($\mathbf{b}_1$, $\mathbf{b}_2$, and $\mathbf{b}_3$) measured along the reciprocal coordinate axes y_i. The vectors $\mathbf{b}_i$ are geometrically related[1] to the crystal base vectors $\mathbf{a}_i$; Fig. 1-22 gives a quantitative example of this "reciprocal" relationship. Vectors of the type

$$\mathbf{s} = s_1\mathbf{b}_1 + s_2\mathbf{b}_2 + s_3\mathbf{b}_3 \tag{1-5}$$

where the s_i are integers, specify the points of the reciprocal lattice. Each of these vectors, such as the vector $\mathbf{s}$ out to the (101) point in Fig. 1-22, describes two important characteristics of a possible plane in the crystal structure: (1) the direction of $\mathbf{s}$ is perpendicular to the (101) crystal plane and thus describes its orientation, and (2) the magnitude $|\mathbf{s}|$ is equal to $1/d$, where d is the interplanar spacing. The case in which the components of $\mathbf{s}$ have a common factor [for example, (202), which is equivalent to 2(101)] is the subject of Prob. 1-13. The notation (hkl) or $(h_1h_2h_3)$ is usually employed in place of $(s_1s_2s_3)$ when the components are integers.

[1] A given crystal coordinate system, described by the base vectors $\mathbf{a}_i$, is associated with a given reciprocal coordinate system $\mathbf{b}_i$, defined by the equations

$$\mathbf{b}_1 = \frac{\mathbf{a}_2 \times \mathbf{a}_3}{\mathbf{a}_1 \cdot \mathbf{a}_2 \times \mathbf{a}_3} \qquad \mathbf{b}_2 = \frac{\mathbf{a}_3 \times \mathbf{a}_1}{\mathbf{a}_1 \cdot \mathbf{a}_2 \times \mathbf{a}_3} \qquad \mathbf{b}_3 = \frac{\mathbf{a}_1 \times \mathbf{a}_2}{\mathbf{a}_1 \cdot \mathbf{a}_2 \times \mathbf{a}_3} \tag{1-4}$$

The scalar (dot) product is $\mathbf{a} \cdot \mathbf{b} = ab \cos\theta$, where θ is the angle between the positive directions of the two vectors. The vector (cross) product is a vector perpendicular to the plane of the two vectors and has the magnitude $|\mathbf{a} \times \mathbf{b}| = ab \sin\theta$. The direction of the product $\mathbf{a} \times \mathbf{b}$ is such that $\mathbf{a}$, $\mathbf{b}$, and $\mathbf{a} \times \mathbf{b}$ form a right-hand system.

The intercept form of the equation of a plane,

$$\sum_{i=1}^{3} h_i x_i = h_1 x_1 + h_2 x_2 + h_3 x_3 = k \tag{1-6}$$

where k is an integer, gives a simple representation of the parallel set of $(h_1h_2h_3)$ planes; x_i are the crystal axes. For example, Fig. 1-24 shows the

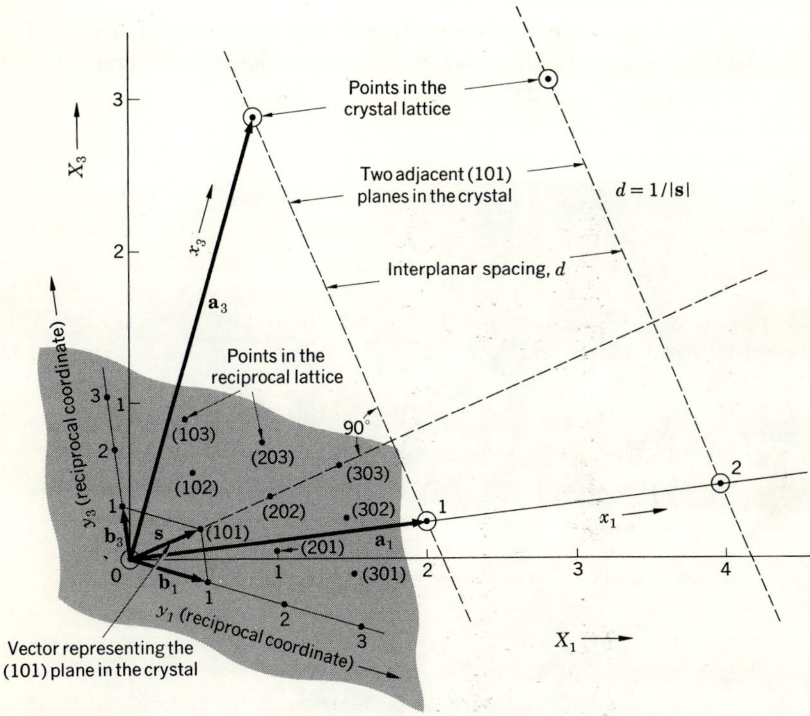

FIGURE 1-22 Use of a reciprocal-lattice vector **s** to describe a plane in the crystal lattice of Fig. 1-21. The crystal-lattice vectors $\mathbf{a}_1$ and $\mathbf{a}_3$ have been enlarged to show the small reciprocal-lattice vectors $\mathbf{b}_1$ and $\mathbf{b}_3$ more clearly. The $\mathbf{b}_i$ are related to the $\mathbf{a}_i$ by Eq. (1-4).

particular (221) plane given by $k = 1$. The intercept on any axis (say, x_1) is easily determined since the other two coordinates are zero ($x_2 = x_3 = 0$). Equation (1-6) becomes $h_1x_1 = k$ or $2x_1 = 1$, and thus the x intercept is $\frac{1}{2}$. The three intercepts give the (221) plane shown. Other (parallel) planes of the set are obtained when $k = 2, 3, 4, \ldots$. Sometimes we have a clear picture of integral intercepts, as in Fig. 1-25*a*, and wish to determine the corresponding $(h_1h_2h_3)$ values, also known as *Miller indices* after their originator. For this purpose we can use the following four-step procedure:

	Crystal axis x_1	x_2	x_3
1. Determine intercepts of plane on crystal axes	2	3	1
2. Take reciprocals	$\frac{1}{2}$	$\frac{1}{3}$	1
3. Reduce to smallest integers in the same ratio	3	2	6
4. Enclose in parentheses		(326)	

The plane sketched in Fig. 1-25*a* corresponds to $k = 6$ in Eq. (1-6); thus five other planes of this array are closer to the origin. Some important types of planes are shown in Fig. 1-25*b* (see also Prob. 1-15).

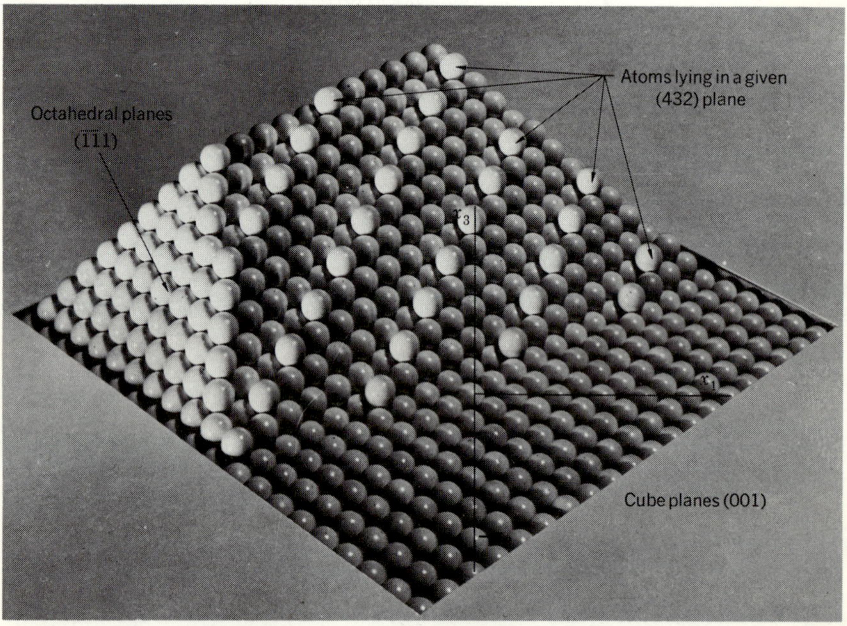

FIGURE 1-23 Ball model showing the atomic distribution on the (432) plane of a face-centered cubic metal. The octahedral and cube planes are shown for comparison. (*By permission from J. F. Nicholas, "An Atlas of Models of Crystal Structures," Gordon and Breach, Science Publishers, Inc., 1965.*)

A *zone axis* **z** is the line of intersection of two (or more) planes and is often of special interest. In view of its geometry (see Prob. 1-16), the direction of **z** is given by the cross product $\mathbf{z} = \mathbf{h} \times \mathbf{h}'$ of the reciprocal lattice vectors of the two intersecting planes. The form of the vector cross product employing indices[1]

$$z_i = e_{ijk}h_j h'_k \tag{1-7a}$$

is especially convenient in this case because the z_i are crystal-lattice components even though the h_i are in the reciprocal lattice. The e_{ijk} are defined to have the value zero if any two of the indices are equal, the value $+1$ if ijk is an even

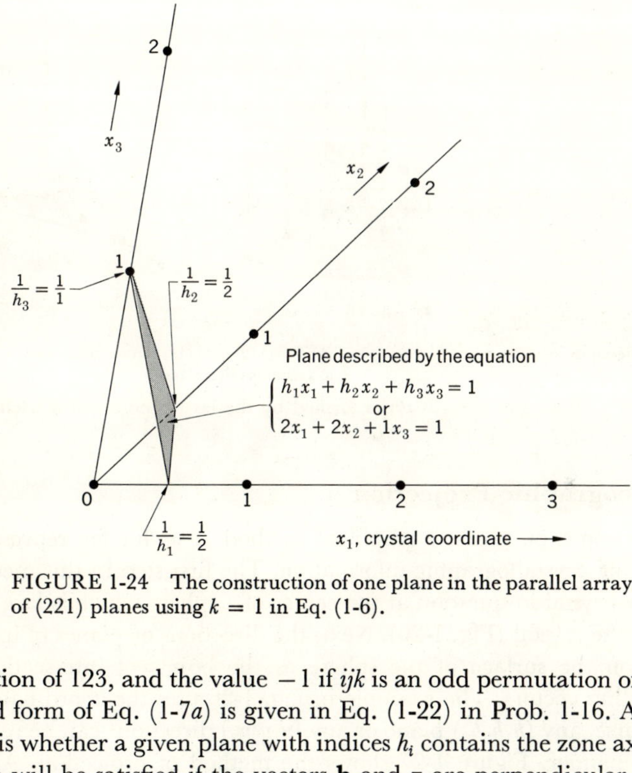

FIGURE 1-24 The construction of one plane in the parallel array of (221) planes using $k = 1$ in Eq. (1-6).

permutation of 123, and the value -1 if ijk is an odd permutation of 123. The expanded form of Eq. (1-7*a*) is given in Eq. (1-22) in Prob. 1-16. A common question is whether a given plane with indices h_i contains the zone axis z_i. This condition will be satisfied if the vectors **h** and **z** are perpendicular, that is, if their scalar product is equal to zero. Since a given reciprocal base vector is perpendicular to two crystal base vectors, see Eq. (1-4), six of the terms of the

[1] The index notation can also replace summation signs. When a letter, such as j and k in Eq. (1-7*a*), occurs twice in a given term, the *summation convention* provides that summation will be performed with respect to this index. Thus, Eq. (1-7*a*) is equivalent to

$$z_i = \sum_{j=1}^{3} \sum_{k=1}^{3} e_{ijk}h_j h'_k \tag{1-7b}$$

The index i provides three equations, one each for z_1, z_2, and z_3. A computer program for use with the e_{ijk} function is part of Prob. 4-19.

scalar product are zero. The remaining three terms determine the condition in question, namely,

$$h_i z_i = h_1 z_1 + h_2 z_2 + h_3 z_3 = 0 \tag{1-8}$$

Problem 1-17 applies this equation to determine planes belonging to a given zone.

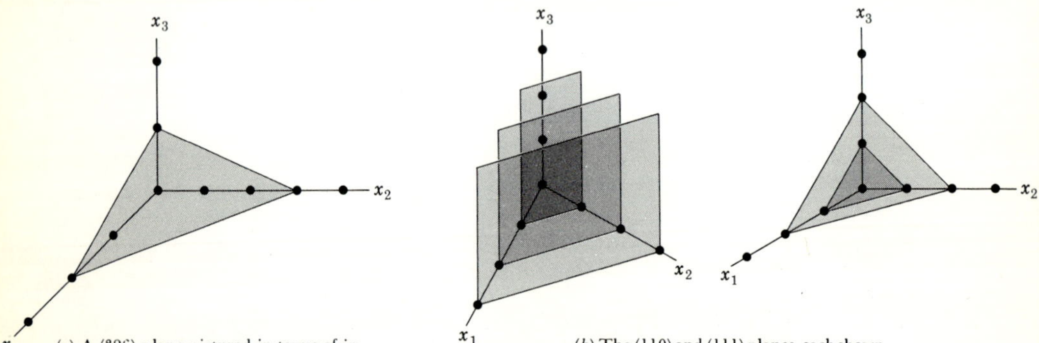

(*a*) A (326) plane pictured in terms of integral intercepts.

(*b*) The (110) and (111) planes, each shown as a set of parallel planes.

FIGURE 1-25 Planes of various types formed by the lattice points of a triclinic lattice.

1-8. Stereographic Projection

For some purposes a graphical method is useful for representing or manipulating crystallographic information. The first step in this method is to picture the crystal in question at the center of a sphere which is large in comparison to the crystal (Fig. 1-26). Next, the directions or planes of interest are recorded on the surface of the sphere as the points of intersection of the corresponding vectors. The example in Fig. 1-26 uses the coordinate vectors, but, of course, any $(h_1h_2h_3)$ plane or any $[u_1u_2u_3]$ direction[1] can be recorded in the same manner. Figure 1-27 shows the method of projection. Any point, such as p, on the "front" half of the sphere projects as point P in the projection plane. The cartesian coordinates (X,Y) of P are conveniently expressed in terms of the spherical coordinates (θ,φ) of p (see Prob. 1-18) as follows:

$$X = \frac{2 \sin \theta \cos \varphi}{1 + \cos \theta \cos \varphi} \qquad Y = \frac{2 \sin \varphi}{1 + \cos \theta \cos \varphi} \tag{1-9}$$

The useful properties of this stereographic mode of projection include the fact

[1] The usual notation for a direction in a crystal is $[uvw]$ or $[u_1u_2u_3]$, provided that the components are integers. The notation $[r_1r_2r_3]$ in Eq. (1-2) applies for both integral and nonintegral values, as in Eq. (1-11).

that the angle between any two vectors remains unchanged during projection and also during the "rotations" of the projection that will now be discussed.

To change the orientation of a crystal (in a laboratory device, for example), we usually rotate the crystal about one of the three cartesian axes. Consider first a rotation about the y axis in Fig. 1-27. The point p follows a latitude circle (Fig. 1-28) during such a rotation, while a second point q, having the same θ value remains on the same longitude circle as p. Clearly, the latitude and longitude circles are convenient references for such a rotation,

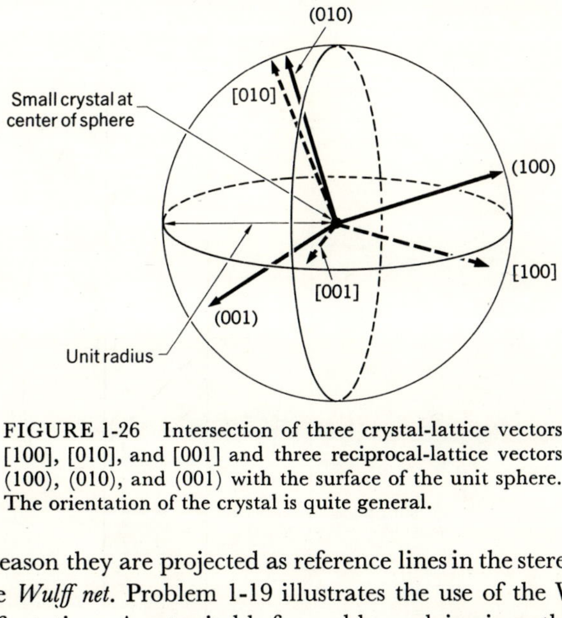

FIGURE 1-26 Intersection of three crystal-lattice vectors [100], [010], and [001] and three reciprocal-lattice vectors (100), (010), and (001) with the surface of the unit sphere. The orientation of the crystal is quite general.

and for this reason they are projected as reference lines in the stereographic net known as the *Wulff net*. Problem 1-19 illustrates the use of the Wulff net for other types of rotations. A net suitable for problem solving is on the inside back cover.

The most common operation with a stereographic projection is the measurement of the angle between two points A and B (Fig. 1-29). These points might represent experimental data (from an x-ray study, for example) on two crystal planes and lie in apparently random positions. To measure the angle between the two points, we must first put them on the same great circle[1] by rotating the stereographic projection about the central pin (Fig. 1-29*b*). Using the underlying Wulff net, we can then measure the angle, namely, 71°. The fiducial arrow permits us to return the stereographic projection to its original position.

[1] The great circles on a sphere are those of maximum diameter, including all longitude lines and the equatorial latitude line.

A typical problem is to determine the orientation of a crystal from metallographic information. Crystal planes of a given type may be visible microscopically because a second phase has precipitated preferentially on this plane or because plastic deformation has occurred on the plane. For example,

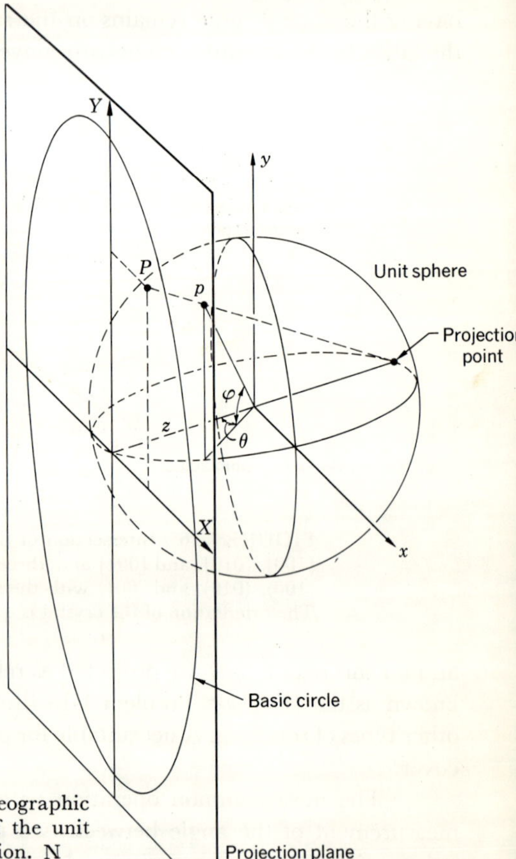

FIGURE 1-27 Principle of the stereographic projection of point p on the surface of the unit sphere as point P in the plane of projection. N that a point, such as p, can be plotted direcuy on the Wulff net using its (θ,φ) coordinates.

α-uranium (orthorhombic; $a_1 = 2.85$ Å, $a_2 = 5.87$ Å, $a_3 = 4.96$ Å) undergoes the "slip" process of plastic deformation on (010) planes. After slight deformation of a previously polished specimen, the intersections (traces) of the slip planes on the surface appear as parallel straight lines (Fig. 1-30a). An analysis of the slip traces on one surface (Fig. 1-30b) gives only limited information; namely, the slip plane is in the zone of planes whose zone axis is the slip trace. This zone of planes lies on the great circle shown in Fig. 1-30*b*. Analysis of the second surface also gives a great circle (Fig. 1-30*c*) which becomes a curved line after a 90° rotation into the plane of surface 1. The point

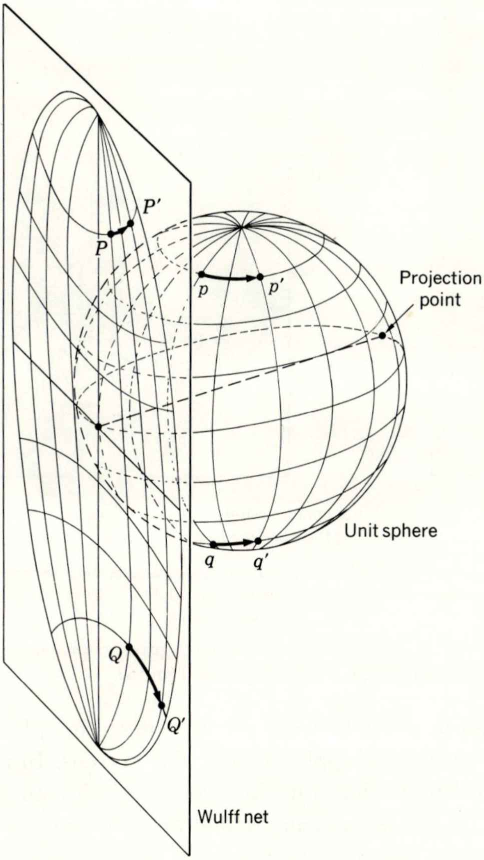

FIGURE 1-28 The relation of the Wulff net to a ruled globe rotating about the y axis.

of intersection of the great circles obtained from surfaces 1 and 2 represents the location of the (010) slip plane in the grain being studied.

A useful reference projection for a given crystal structure is the *standard projection* of important planes and directions when the crystal axes are conveniently oriented with respect to the axes of the unit sphere. A standard (100) projection for cubic crystals is shown on the inside back cover of this book. Projections of cubic crystals are especially simple for two reasons: (1) the vector representing a plane, such as (213), coincides in direction with the vector representing the direction [213], having the same indices, and consequently

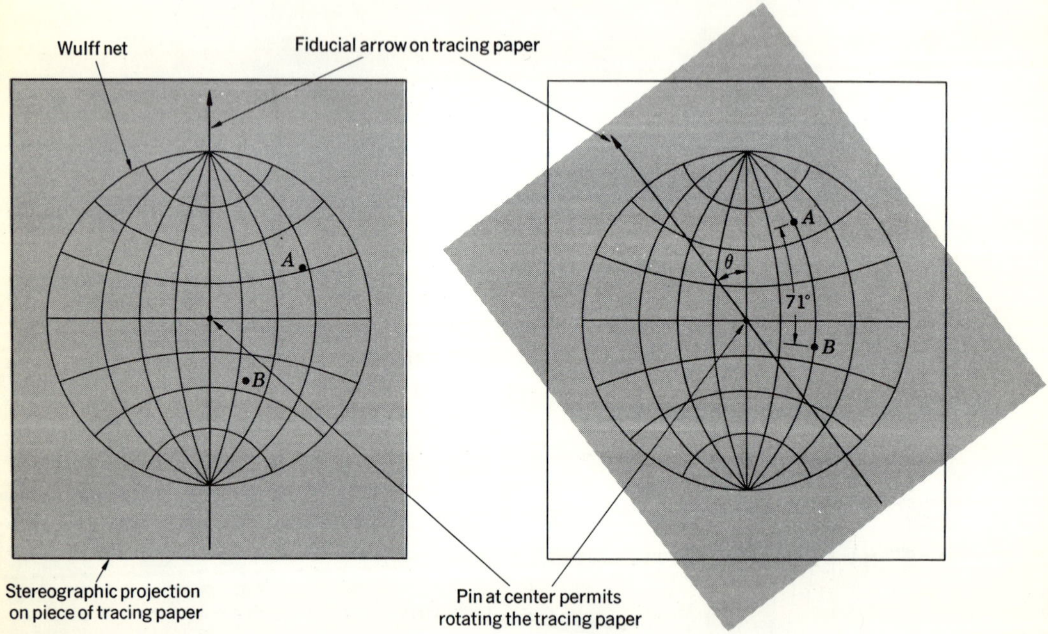

(*a*) Initial positions of points *A* and *B* as recorded on the stereographic projection from experimental data.

(*b*) Rotation of θ degrees about the *z*-axis puts *A* and *B* on the same great circle for measurement of the angle between them.

FIGURE 1-29 A typical manipulation of a stereographic projection using the Wulff net: the measurement of the angle between points *A* and *B*.

no distinction need be made between points representing planes or directions; and (2) the same projection applies to all cubic crystals. In noncubic crystals a separate standard projection must be made for each combination of lattice parameters (including interaxial angles that are different than 90°). See Prob. 1-20.

DIFFRACTION STUDIES OF CRYSTAL STRUCTURES

Although modern microscopy is beginning to show us actual crystal structures directly (see Fig. 5–14), most quantitative crystallographic studies involve the indirect interpretation of diffraction patterns. A standard topic in elementary physics is the diffraction of light waves having wavelengths of about 5000 Å by

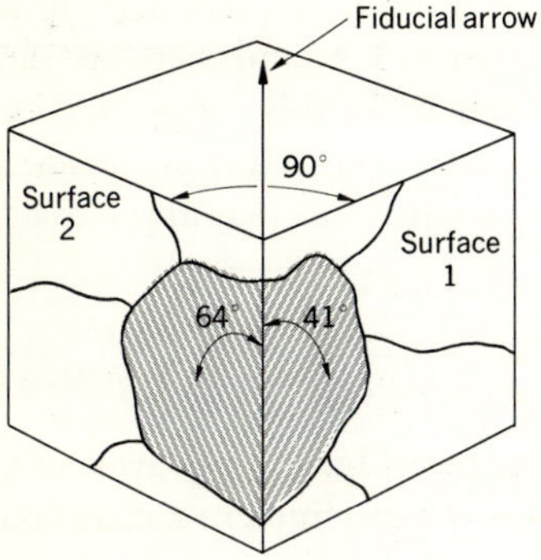

(*a*) Specimen of α-uranium polished on two surfaces and then deformed to produce slip traces within the grain of interest.

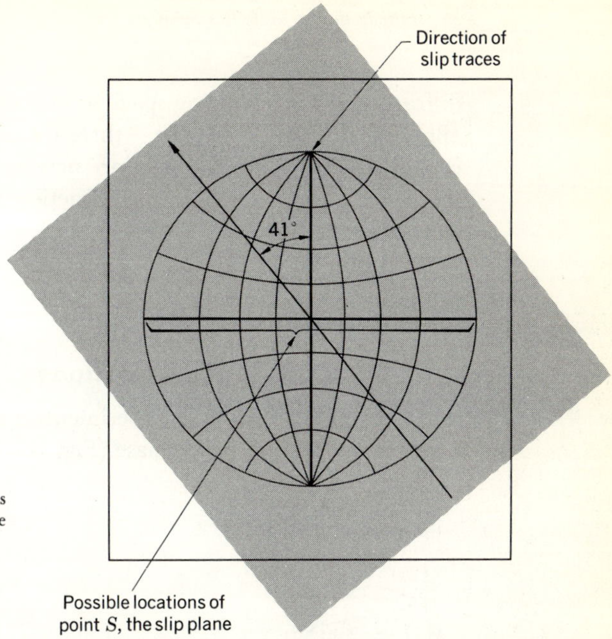

(*b*) The stereographic projection of surface 1 showing that the slip plane, *S*, must lie somewhere along the equator when the slip traces are vertical.

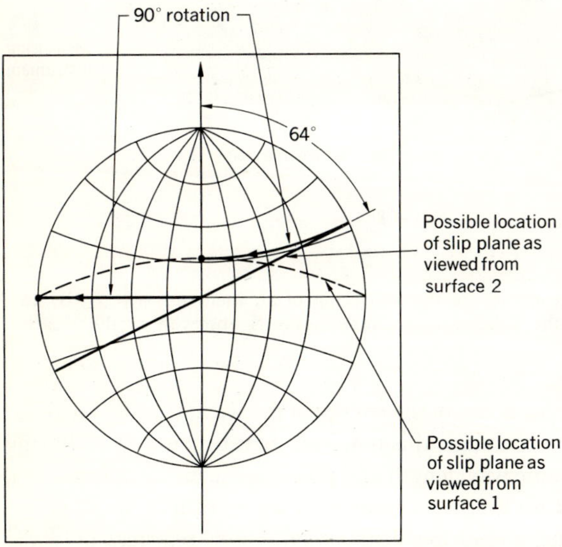

(*c*) After the method of (*b*) gives the straight line for surface 2, the view from surface 1 can be obtained by rotation about the (vertical) *y*-axis.

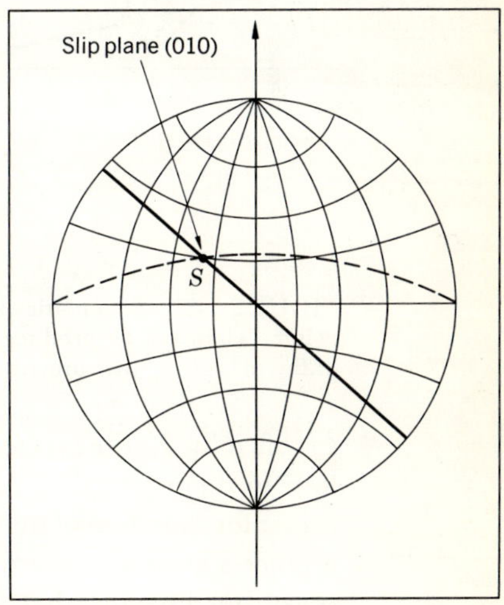

(*d*) Superposition of the projection of (*c*) (after rotation) on the projection of (*b*).

FIGURE 1-30 Method of determining the orientation of a plane when its traces can be measured on two surfaces of a specimen.

diffraction gratings having spacings of a comparable size. The same principles apply to diffraction by the "three-dimensional grating" represented by a crystal structure. The spacings are now on the order of 1 Å, and therefore the waves that are susceptible to diffraction must also be in this range; x-rays, electrons, or neutrons can serve as the waves for diffraction studies on crystals. Let us begin by considering the general diffraction equations that apply to all diffraction studies.

1-9. Diffraction Equations

In most diffraction experiments a parallel beam having a wavelength λ and with all wavelets in phase (Fig. 1-31) strikes the specimen in a direction

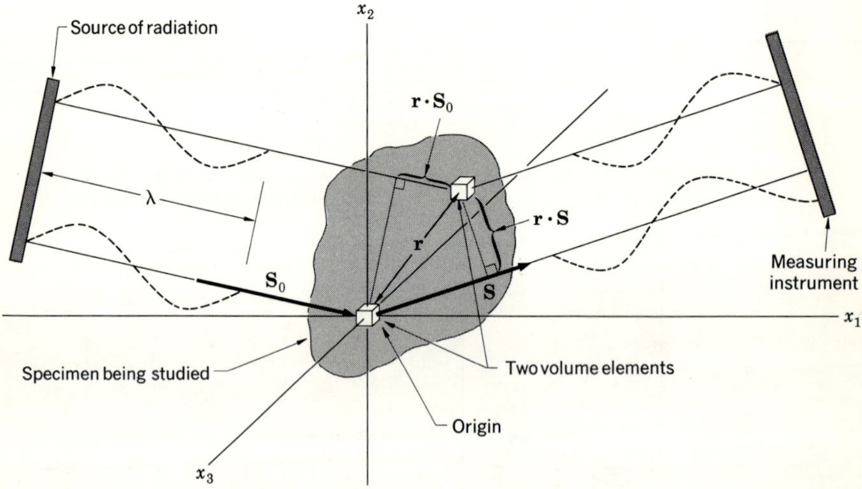

FIGURE 1-31 Schematic illustration of the scattering of an incident wave $\mathbf{S}_o$ by two volume elements to produce the coherently scattered wave $\mathbf{S}$ observed at the location of the measuring instrument.

specified by the unit vector $\mathbf{S}_0$. Some of the energy of the wave transfers without loss[1] to the material of the specimen; this material then becomes a scattering center for radiation of the same wavelength and phase as the incident radiation, a process known as *coherent scattering*. The scattered waves radiate in all directions, but imagine that the measuring instrument shown registers only the intensity of scattering in the direction specified by the unit vector $\mathbf{S}$. Since the path length for the volume element at $\mathbf{r}$ is shorter than that for the volume element at the origin by $(\mathbf{S} - \mathbf{S}_o) \cdot \mathbf{r}$, the phase difference is $(2\pi/\lambda)(\mathbf{S} - \mathbf{S}_o) \cdot \mathbf{r}$.

[1] This is termed an *elastic* collision. When *inelastic* collisions occur, there is loss of energy, and the *incoherent* scattered radiation has a longer wavelength than the incident radiation.

This expression is more conveniently written $2\pi\mathbf{s} \cdot \mathbf{r}$, where $\mathbf{s} = (\mathbf{S} - \mathbf{S}_o)/\lambda$. With reference to the amplitude of the wave scattered at the origin, the amplitude $dA(\mathbf{s})$ scattered by the volume element dV at $\mathbf{r}$ is[1]

$$dA(\mathbf{s}) = \rho(\mathbf{r})e^{-2\pi i\mathbf{r}\cdot\mathbf{s}}\,dV \tag{1-10}$$

The minus sign occurs because of the definition used for path length. The quantity $\rho(\mathbf{r})$ is some suitable measure of the effectiveness with which the material in the volume dV at the end of vector $\mathbf{r}$ scatters the incident wave.

Now it might appear that Eq. (1-10), the rather complicated expression for a single element of volume, would become unmanageable when integrated over the entire volume of the crystal. Fortunately, the resulting equation,

$$A(\mathbf{s}) = \int_V \rho(\mathbf{r})e^{-2\pi i\mathbf{r}\cdot\mathbf{s}}\,dV \tag{1-11}$$

is especially convenient for two reasons. First, the integral is a well-known quantity, the *Fourier transform* of the function $\rho(\mathbf{r})$; numerous tables[2] give the Fourier transform $A(\mathbf{s})$ for various values of $\rho(\mathbf{r})$. Second, the vector $\mathbf{s}$ that determines the amplitude is not in the crystal but is referred to the reciprocal lattice. Consequently the dot product $\mathbf{s} \cdot \mathbf{r}$ can be written in the simple form $s_i r^i = s_1 r^1 + s_2 r^2 + s_3 r^3$ for all crystal systems.

The amplitude function $A(\mathbf{s})$ depends on the distribution in three dimensions, $\rho(\mathbf{r})$, of the individual unit cells $[\rho_u(\mathbf{r})]$ which cause the diffraction in question. In a crystal of sulfur, for example, $\rho_u(\mathbf{r})$ is the scattering effectiveness of a single atom of sulfur in the primitive cell. For an infinite crystal (Fig. 1-32*a*) the repetition of $\rho_u(\mathbf{r})$ extends to infinity in all directions (in practice, infinity means more than about 10^3 unit cells in a given direction). In this case $A(\mathbf{s})$ consists simply of a peak function (here, the delta function) repeated at the points of the reciprocal lattice. These peak functions explain diffraction by ordinary sulfur crystals. Sometimes a crystal structure is infinite in two directions but only 10 unit cells or so in extent in the third direction (Fig. 1-32*b*). Precipitates of copper in an aluminum-5 w/o copper alloy may have such a structure; thousands of precipitate platelets will then be parallel to each other and separated by some average distance. In this case the peak function is a repetition in the reciprocal lattice of line segments, the direction of each segment corresponding to the direction in which the crystal has a finite size.

Once we know the amplitude function $A(\mathbf{s})$ in the reciprocal lattice, we can predict the type of diffraction effect that will be observed for any

[1] A sinusoidal variation can be represented by $ei\theta = \cos\theta + i\sin\theta$. The advantage of this representation is mathematical convenience for such operations as integration.

[2] "Tables of Integral Transforms," vol. 1, Bateman Manuscript Project, McGraw-Hill Book Company, New York, 1954.

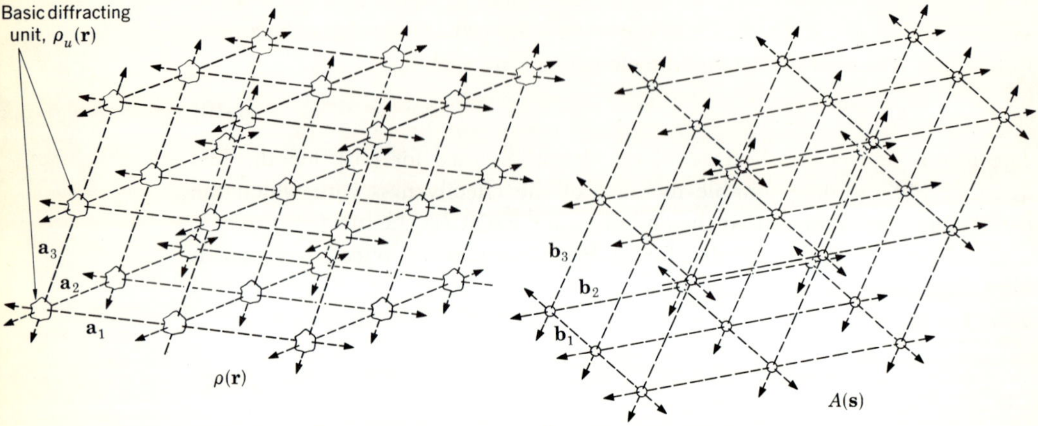

(*a*) For a perfect infinite crystal (approximated by a single crystal of sulfur, for example). The basic unit, $\rho_u(\mathbf{r})$ (a sulfur atom) repeats at each point of an infinite space lattice to form the scattering function $\rho(\mathbf{r})$. $A(\mathbf{s})$ consists of infinitesimal points of finite magnitude at the points of the reciprocal lattice.

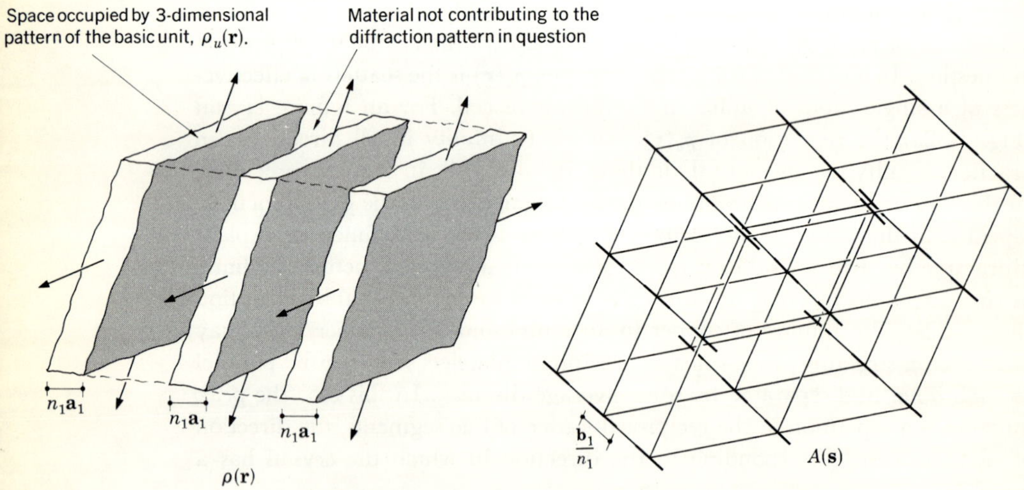

(*b*) For a parallel array of crystallites, each finite in the direction of $\mathbf{a}_1$. The basic unit, $\rho_u(\mathbf{r})$, repeats as in (*a*) except that in the direction of $\mathbf{a}_1$ the number of repetitions, n_1, is finite in each of an infinite number of parallel regions. $A(\mathbf{s})$ consists of lines of length b_1/n_1 in the $\mathbf{b}_1$ direction through the points of the reciprocal lattice.

FIGURE 1-32 Schematic illustrations of the amplitude function $A(\mathbf{s})$ corresponding to two different distributions $\rho(\mathbf{r})$ of the basic diffracting unit $\rho_u(\mathbf{r})$ ($\mathbf{a}_i$ and $\mathbf{b}_i$ are base vectors in the crystal lattice and in the reciprocal lattice, respectively; a_i and b_i are their magnitudes).

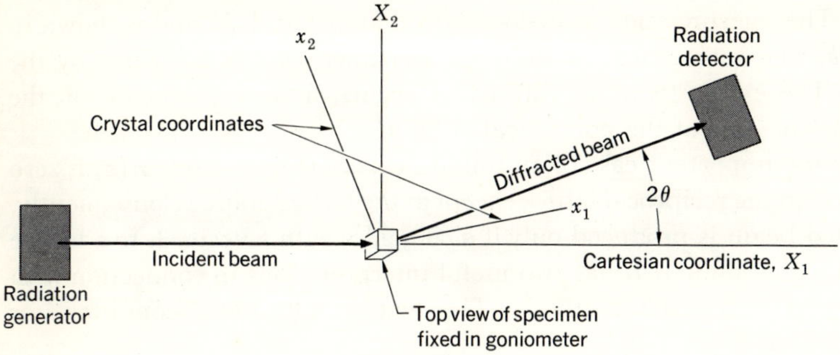

(*a*) A typical experimental arrangement that fixes the values of $\mathbf{S}_0$, $\mathbf{S}$, and λ (and therefore determines the vector $\mathbf{s}$).

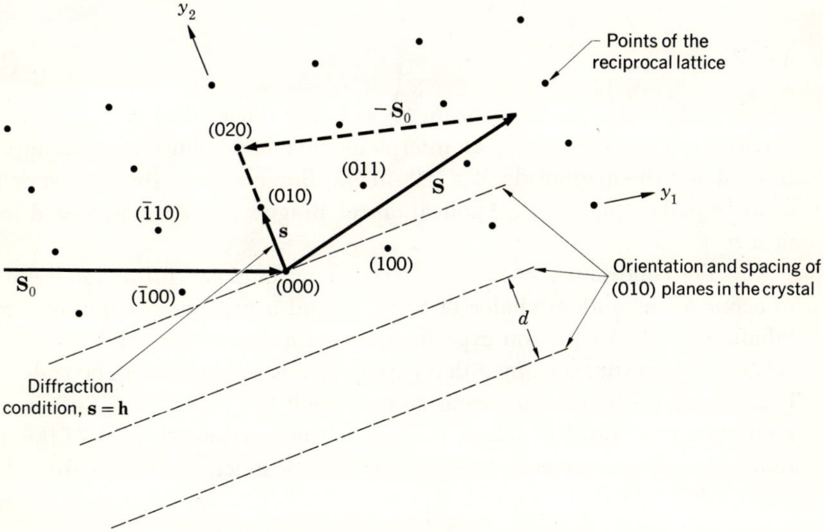

(*b*) A diffracted beam is produced if $\mathbf{s}$ touches a point, such as $(\mathbf{h}_1\mathbf{h}_2\mathbf{h}_3)$, in the reciprocal lattice where $A(\mathbf{s})$ has a finite value.

FIGURE 1-33 The relation between the experimental arrangement for diffraction and the interpretation of the amplitude function in the reciprocal lattice. The x_3 crystal axis and the y_3 reciprocal axis are directed out of the plane of the page, making the present case quasi-two-dimensional.

particular type of diffraction experiment. Consider an experiment in which a wave having wavelength λ (Fig. 1-33*a*) strikes a crystal of arbitrary orientation and produces a diffracted beam whose intensity is measured by a suitable detector. The interpretation is analogous to that of Fig. 1-31 and is shown in Fig. 1-33*b*. The geometrical condition of the experiment is described by the vector **s**. The extent to which diffraction occurs, however, depends on the magnitude[1] of $A(\mathbf{s})$ at the point located by **s**.

In the important case of an infinite crystal (Fig. 1–32*a*), $A(\mathbf{s})$ is zero everywhere in the reciprocal lattice except at the lattice points. Consequently, a diffracted beam is produced only if **s** coincides with a vector **h** to a lattice point. The relation $\mathbf{s} = \mathbf{h}$ has two useful interpretations in connection with experimental diffraction methods. The first is the Laue condition for diffraction (see Prob. 1–21),

$$\begin{aligned} a_1(\cos\alpha - \cos\alpha_0) &= h_1\lambda \\ a_2(\cos\beta - \cos\beta_0) &= h_2\lambda \\ a_3(\cos\gamma - \cos\gamma_0) &= h_3\lambda \end{aligned} \qquad (1\text{–}12)$$

Here the a_i are the crystal lattice constants, and $\alpha_0, \beta_0, \gamma_0$ and α, β, γ are the angles that the incident and reflected beam, respectively, make with the crystal axes. The second is Bragg's law,

$$\lambda = 2d \sin\theta \qquad (1\text{-}13)$$

which pictures the incident beam $\mathbf{S}_0$ as being reflected at the angle θ from "reflecting planes"[2] having an interplanar spacing d. Since the magnitude of **h** is $1/d$ and the magnitude of **s** is $2 \sin\theta/\lambda$, Bragg's law is simply a statement of the equality $|\mathbf{s}| = |\mathbf{h}|$. Applications of Bragg's law are considered in the next section.

The condition $\mathbf{s} = \mathbf{h}$ is such a restrictive one that diffraction is unlikely to occur for a random choice of $\mathbf{S}$, $\mathbf{S}_0$, λ, and a fixed orientation of a given infinite crystal. A common experimental technique to relax this restriction is the use of a powder specimen that exposes about 10^5 individual crystals to the beam, each of the crystals being large enough to be effectively infinite. The amplitude function $A(\mathbf{s})$ is now the sum of the functions for each of the many randomly oriented crystals and has the form of spheres of radius $|\mathbf{h}|$, labeled (100), (010), (001), (110), etc., in Fig. 1-34*a*. When the detector of a diffrac-

[1] The *intensity* of diffraction, I, is proportional to $A(\mathbf{s}) \cdot A^*(\mathbf{s})$, where $A^*(\mathbf{s})$ is the complex conjugate of $A(\mathbf{s})$. For real quantities this expression reduces to $I = A^2(\mathbf{s})$.

[2] A *reflecting* plane with indices (222), for example, can be reduced to a *geometrical* Miller plane by dividing the indices by 2. The notation $(222) = 2(111)$ is sometimes employed to emphasize that the physical phenomenon is a second-order reflection from the (111) Miller plane. Bragg's law then becomes $n\lambda = 2d \sin\theta$, where n is the order of reflection.

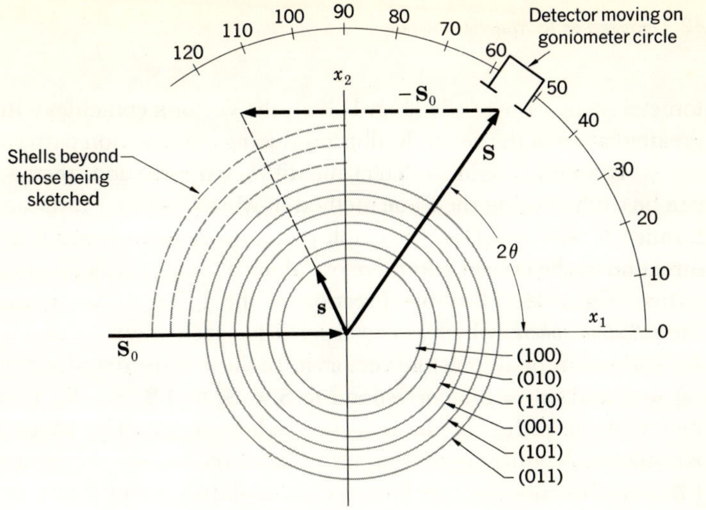

(*a*) A section on the x_1x_2 plane through $A(\mathbf{s})$ for a powder specimen (or a polycrystalline metal).

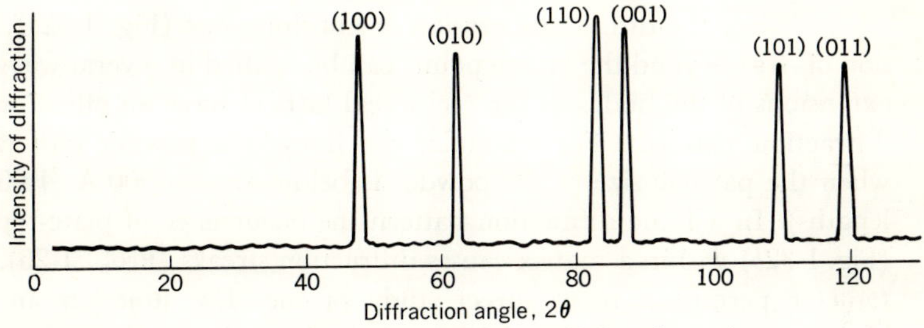

(*b*) Diffractometer pattern from a typical powder specimen.

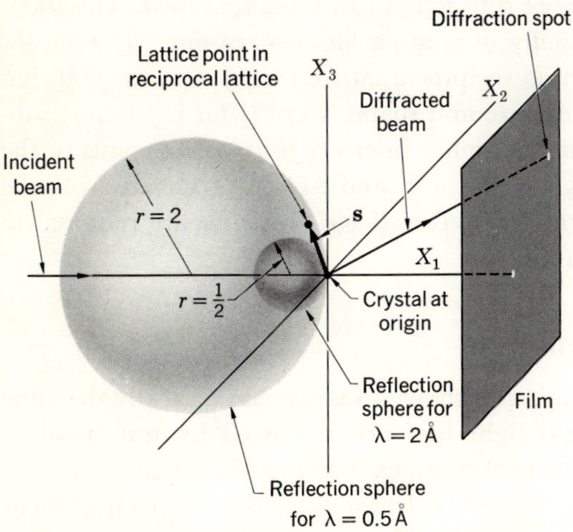

(*c*) Position of the reflection spheres in the reciprocal lattice of a single crystal for wavelengths ranging from = 0.5 to = 2.0

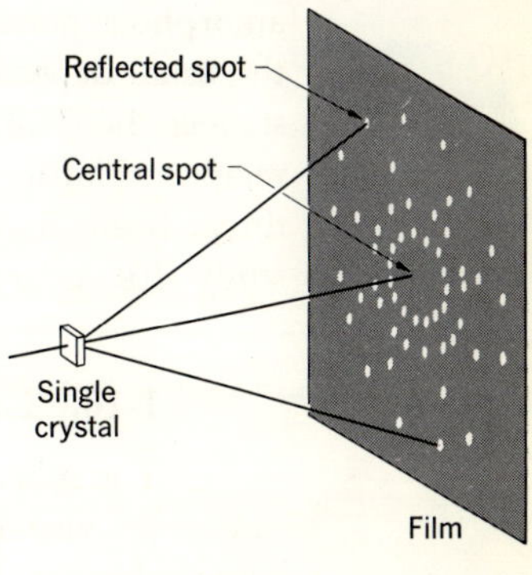

(*d*) Typical transmission Laue pattern from a single crystal.

FIGURE 1-34 Two techniques for obtaining diffraction patterns from infinite crystals, the powder method and the Laue method.

tometer scans a range of θ (Fig. 1-35*a*), the vector **s** coincides with successively greater values of the vector **h**, thus producing a diffraction pattern (Fig. 1-34*b*).

If a film is used to detect the diffracted radiation, then a single crystal can be studied using the Laue method in which "white" radiation (containing a range of wavelengths λ) is employed. The incident beam is along a certain direction in the crystal and therefore along a certain direction in the reciprocal lattice (Fig. 1-34*c*). Possible directions of the diffracted beam are predicted by the *reflection sphere*, which is constructed for each wavelength λ as the locus of the end points of all possible vectors **s**. In view of the definition of **s**, the values $|\mathbf{s}| = 0$ and $|\mathbf{s}| = 2/\lambda$ correspond to $\mathbf{S} = \mathbf{S}_0$ and $\mathbf{S} = -\mathbf{S}_0$, respectively, and therefore the radius of the reflection sphere is $1/\lambda$. The range of λ in white *x*-radiation produces a continuous series of reflection spheres, and a reflection (in some direction) occurs for a reciprocal-lattice point that is touched by one of these spheres. Certain of these reflections have the proper direction to be recorded on a flat film (Fig. 1-34*d*). The reflection sphere is also useful in the rotating-crystal method for diffraction studies (see Prob. 1-22).

When the diffracting crystals are of finite size (Fig. 1-32*b*), the extension of $A(\mathbf{s})$ beyond the lattice points can be studied in several ways. First, the extensions of the nodes in the reciprocal lattices have an effect on the usual diffraction patterns. For example, the lines in a powder pattern broaden when the particle size of the powder is below about 3000 Å (1,000 unit-cell lengths). In a Laue diffraction pattern the occurrence of plate-type particles (Fig. 1-32*b*) within a matrix causes diffraction streaks (Prob. 1-23). A diffractometer permits a more direct study of the $A(\mathbf{s})$ function since the end of the vector **s** can be placed at any desired location in the reciprocal lattice, and the corresponding intensity of diffraction can be determined. The (000) reciprocal-lattice point is especially interesting for two reasons: (1) even for amorphous particles (for which no reciprocal lattice exists) $A(\mathbf{s})$ at (000) has the usual dependence on particle size and shape, and (2) for crystalline substances the (000) point has a much higher intensity than other points in the reciprocal lattice. Since the angle θ is near 0° and **S** almost coincides with the direct beam $\mathbf{S}_0$, the special techniques of *small-angle scattering* are required to study $A(\mathbf{s})$ near the (000) point.

1-10. X-ray Diffraction

The short-wavelength radiation known as x-rays has basically the same nature as visible light. Whereas light rays are produced by transitions of *valence* electrons between various energy states, x-rays are generated only by transitions involving *inner* (core) electrons. In an x-ray tube a beam of high-speed electrons ejects inner electrons from the atoms of the target metal. The electrons from outer levels then fall into vacant positions in the inner levels and liberate energy in the form of x-rays. Of principal importance are the $K\alpha$ characteristic x-rays, examples of which are given in Table 1-3 for some common metals. There are also $K\beta$, $L\gamma$, and $L\beta$ characteristic radiations and a weaker *white* radiation, so called because it covers a range of wavelengths like those for ordinary white light instead of just a few characteristic values. Each of the characteristic radiations is actually a multiplet, and the separate effects of the $K\alpha_1$, $K\alpha_2$ doublet can often be seen in diffraction patterns.

When x-rays are used for diffraction studies, various factors must be considered (Prob. 1-24), but the essential one is the scattering effectiveness $\rho(\mathbf{r})$ of the material being studied. Scattering of x-rays is caused by the electrons in the atoms, and therefore $\rho(\mathbf{r})$ increases directly with atomic number Z. When the scattering is referred to that produced by a single electron, the *atomic scattering factor* $f(\mathbf{s})$ is equal to Z at $\theta = 0$, although it drops off as $(\sin\theta)/\lambda$ increases. $A(\mathbf{s})$ is identical to $f(\mathbf{s})$ for a structure consisting of one atom of the same kind per unit cell. If the unit cell contains n atoms, each located by the vector $\mathbf{r}_k$, then the structure factor, $F(\mathbf{s}) = A(\mathbf{s})$, is given by

$$F(\mathbf{s}) = \sum_{k=1}^{n} f_k(\mathbf{s}) e^{-2\pi i \mathbf{s} \cdot \mathbf{r}_k} \tag{1-14}$$

The exponential term takes account of the difference in phase of the radiation scattered at various locations in the unit cell. For example, in the CsCl crystal structure (Fig. 1-3*b*) the chloride ion has coordinates (000), and the cesium is at the $(\frac{1}{2}\,\frac{1}{2}\,\frac{1}{2})$ position in the unit cell. Since diffracted beams are produced only for $\mathbf{s} = \mathbf{h}$, the real parts of the two terms in Eq. (1-14) are

$$f_{\text{Cl}} \cos 2\pi(0) = f_{\text{Cl}} \tag{1-15}$$

and

$$f_{\text{Cs}} \cos 2\pi\tfrac{1}{2}(h_1 + h_2 + h_3) = \begin{cases} f_{\text{Cs}} & \Sigma h_i \text{ even} \\ -f_{\text{Cs}} & \Sigma h_i \text{ odd} \end{cases} \tag{1-16}$$

Therefore, the structure factor for CsCl is the sum $f_{\text{Cl}} + f_{\text{Cs}}$ for planes such as (110) and (200) but is the difference $f_{\text{Cl}} - f_{\text{Cs}}$ for planes such as (100) and (111). When nonprimitive unit cells are used for the crystal structure of a single element, say body-centered cubic iron, certain diffraction lines are completely absent. Problem 1-25 considers the face-centered cubic structure.

A widely used x-ray technique, the powder method, identifies an unknown substance by determining its interplanar spacings d and the intensity

TABLE 1-3. STRUCTURE AND PROPERTIES OF REPRESENTATIVE METALS, SEMICONDUCTORS, CERAMICS, AND POLYMERS

Material	Crystal structure	Lattice constants	$K\alpha_1$ x-ray wavelength, Å	X-ray K absorption edge, Å	Melting temperature, °C	Enthalpy (heat) of fusion, kcal/mole
Aluminum	FCC	4.050	8.339	7.948	660	2.570
α-Al_2O_3	$D5_1$ (Rhombohedral)	$a_1 = 5.118$ $\alpha_1 = 55.3°$			2,040	26.00
Copper	FCC	3.615	1.541	1.381	1,084	3.120
Germanium	A4(DC)	5.658	1.254	1.117	937	
Graphite (carbon)	HCP	$a_1 = 2.461$ $a_3 = 6.708$	44.7	43.68		
Iron	BCC	2.866	1.936	1.743	1,530	3.700
Kaolinite clay, $Al_2(Si_2O_5)(OH)_4$	Triclinic	$a_1 = 5.14$ $\alpha_1 = 91.8°$ $a_2 = 8.93$ $\alpha_2 = 105°$ $a_3 = 7.37$ $\alpha_3 = 90°$			d. 985	
Magnesium	HCP	3.209	9.890	9.512	650	2.160
Nickel	FCC	3.524	1.658	1.488	1,452	4.210
Silicon	A4(DC)	5.431	7.125	6.738	1,410	11.10
SiC (β)	B3(ZnS)	5.00			d. > 2,720	
SiO_2 (quartz)	Hexagonal	$a_1 = 4.913$ $a_3 = 5.405$			1,610	2.040
Silver	FCC	4.086	0.5594	0.4859	961	2.855
Talc, $Mg_3(Si_2O_5)(OH)_2$	Monoclinic	$a_1 = 5.26$ $a_2 = 9.10$ $\alpha_2 = 100°$ $a_3 = 18.81$			d.910	

Titanium	HCP	$a_1 = 2.951$ $a_3 = 4.679$		2.749	2.497	1,725	4.50
Tungsten	BCC	3.165		0.2090	0.1784	3,380	8.42
UO_2	Cl(CaF_2)	5.469				~2,800	
Zinc	HCP	$a_1 = 2.665$ $a_3 = 4.947$		1.435	1.283	419.6	1.765
				Repeating unit (mer) and conformation			
Polyhexamethylene-adipamide (e.g., Nylon 6,6)	Triclinic	$a_1 = 4.9$ $a_2 = 5.4$ $a_3 = 17.2$	$\alpha_1 = 48°$ $\alpha_2 = 77°$ $\alpha_3 = 63°$	$—NH(CH_2)_6NH\overset{\overset{O}{\parallel}}{C}(CH_2)_4\overset{\overset{O}{\parallel}}{C}—$ Planar zigzag chain		265	0.135
Polyethylene	Orthorhombic	$a_1 = 7.36$ $a_2 = 4.92$ $a_3 = 2.534$		$—CH_2—$ Planar zigzag chain (Fig. 1-9*e*)		135	0.067
Polyethylene-terephthalate (e.g., Mylar)	Triclinic	$a_1 = 4.56$ $a_2 = 5.94$ $a_3 = 10.75$	$\alpha_1 = 98°$ $\alpha_2 = 118°$ $\alpha_3 = 112°$	$—O(CH_2)_2O\overset{\overset{O}{\parallel}}{C}C_6H_4\overset{\overset{O}{\parallel}}{C}O—$ Nonplanar chain		265	0.066
Polypropylene	Monoclinic	$a_1 = 6.65$ $a_2 = 20.96$ $a_3 = 6.50$	$\alpha_2 = 99.3°$	$—\underset{\underset{CH_3}{\vert}}{C}HCH_2—$ Helical chain		176	0.029

† Important types of crystal structure are summarized and illustrated in the table on the inside back cover.

I/I_1 of each diffraction line relative to that of the strongest line I_1. The experimental equipment may be either an x-ray powder camera employing photographic film (see Prob. 1-26) or a diffractometer (Fig. 1-35*a*). In a diffractometer a beam of (mostly) $K\alpha$ x-rays strikes a flat polycrystalline sample at an angle θ. If the diffraction conditions of Fig. 1-34*a* are satisfied, the primary beam is diffracted through the angle 2θ and enters the counter tube. A recorder automatically plots the intensity of the diffracted beam (Fig. 1-35*b*) as the counter moves on the goniometer circle (in synchronization with the specimen) over a range of 2θ values. The interplanar spacing d corresponding to a given value of θ can be calculated from Bragg's law [Eq. (1-13)] since the wavelength

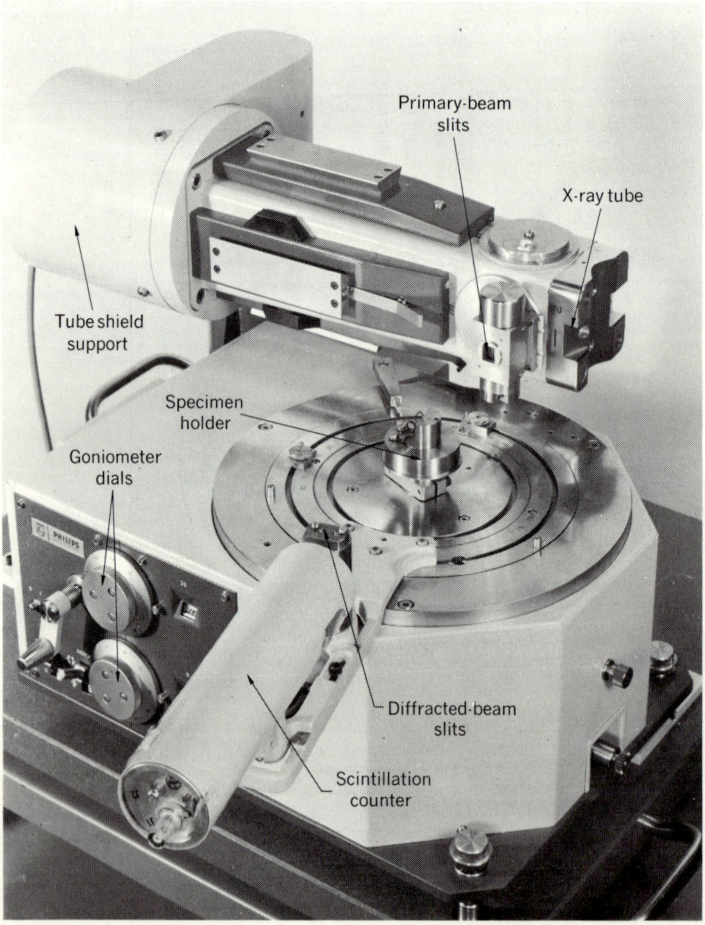

(*a*) An x-ray diffractometer (with x-radiation shield removed).

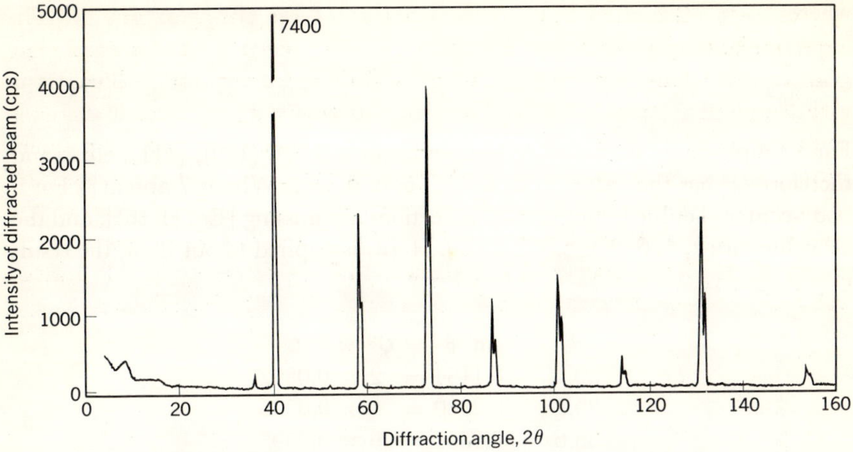

(*b*) Record of diffraction lines obtained from an "unknown metal" (see text) using copper radiation.

FIGURE 1-35 An x-ray diffractometer and a powder-type diffraction pattern obtained by its use. (*Courtesy Philips Electronic Instruments.*)

is known. In this case $\lambda = 1.541$, Cu $K\alpha$ radiation. The first six lines of the pattern of Fig. 1-35*b* give the following data:

θ	$d = \dfrac{\lambda}{2 \sin \theta}$	I/I_1 (relative intensity)
20.1	2.238	1.00
29.2	1.582	0.31
36.6	1.292	0.54
43.6	1.119	0.16
50.2	1.001	0.20
57.4	0.914	0.07

If the pattern were a complex one, the best procedure for identification would be to compare it with the 10,000 reference patterns in the ASTM (American Society for Testing Materials) Powder Diffraction Data File. These patterns are filed according to the *d* value of the most intense line.

To illustrate the method for solving simple diffraction patterns, we begin by assuming that our pattern is from a cubic crystal. The *d* values are then related to the reflection indices by the following equation:

$$d = \frac{a}{[(h_1)^2 + (h_2)^2 + (h_3)^2]^{1/2}} = \frac{a}{Q} \tag{1-17}$$

where a is the lattice constant. Therefore, Eq. (1-13) can be put in the form

$$\sin^2 \theta = Q^2 C \tag{1-18}$$

where

$$C = \frac{\lambda^2}{4a^2} \tag{1-19}$$

For a simple-cubic lattice all reflections occur—(100), (110), (111), etc.—and therefore Q^2 has the values 1, 2, 3, 4, 5, 6, 8, 9, etc. (Why is 7 absent?) For a body-centered cubic lattice certain reflections are missing [Eq. (1-16)], and the Q^2 values are 2, 4, 6, 8, etc. When Eq. (1-18) is applied to our data, the result is

θ	$\sin^2\theta$	$= Q^2 \times$	C
20.1	0.1184	= 2 ×	0.0592
29.2	0.2380	= 4 ×	0.0595
36.6	0.3555	= 6 ×	0.0593
43.5	0.4738	= 8 ×	0.0592
50.2	0.5903	= 10 ×	0.0593
57.4	0.7097	= 12 ×	0.0593
		Average	0.0593

The Q^2 values for the body-centered cubic lattice were used in this trial since simple-cubic crystal structures are much less common. The value of a is calculated from Eq. (1-19) using $\lambda = 1.541$; thus

$$a = \frac{\lambda}{2\sqrt{C}} = \frac{1.538}{2 \times 0.2436} = 3.16 \text{ Å} \tag{1-20}$$

Table 1-3 shows that tungsten is a body-centered cubic metal that has the lattice constant 3.16 Å.

Examples of x-ray diffraction from finite crystals and from imperfectly crystalline materials are found in the study of polymers. Electron microscopy shows that individual lamellae (Fig. 1-9*f*), each having a common thickness t and width w, are intimately packed in various ways (Figs. 3-7 and 3-8), depending on the previous history of the polymer. A single lamella produces a "two-point" region of diffuse scattering around the direct x-ray beam, corresponding to the condition near the (000) point in the reciprocal lattice. If the lamellae are oriented at random (Fig. 3-7*b*), then the x-ray pattern takes the form of a diffuse ring. By a suitable analysis, even the size and shape of colloidal particles can be determined from the shape of the curve of intensity versus θ for small-angle scattering. Other important aspects of materials that are studied using x-rays include preferred orientation, crystal size, crystal perfection, residual stress, thermal expansion, nature of solid solutions, and compositional analysis. We consider some of these topics in later chapters.

1-11. Electron Diffraction

The scattering function $\rho(\mathbf{r})$ for electron diffraction is analogous to that for x-ray diffraction. The quantity corresponding to the x-ray atomic scattering factor f in electron units is

$$f_e = (3.6 \times 10^6)(Z - f)\left(\frac{\lambda}{\sin\theta}\right)^2 \tag{1-21}$$

where Z is the atomic number and λ is the wavelength[1] of the electron beam in angstrom units. The structure factor F_e for electron diffraction is calculated by Eq. (1-14), except that f_e is substituted for f. Although Eq. (1-21) is not applicable at exactly $\theta = 0$ (where $f = Z$), it predicts that f_e is about 10^4 times larger than f at the small angles employed in electron diffraction. Consequently, diffraction patterns can be obtained from extremely thin specimens and also from small areas (in the *selected-area diffraction* technique). On the other hand, magnetic and electric fields in the material alter f_e, and thus quantitative structural studies are more difficult than with x-rays.

In a typical apparatus (Fig. 1-36*a*), electrons from a heated filament are accelerated by a potential of 50,000 volts. In view of the short wavelength of the resulting beam, the useful diffraction angles can be small, covering a range of just a few degrees. Patterns can be obtained in transmission, but the specimen must then be less than 10^3 Å in thickness since the electrons are strongly absorbed. If patterns are obtained from electrons diffracted from surface layers (the reflection method), the difficulties of preparing thin sections are avoided. In this case the electron beam grazes the surface of the specimen, and the diffraction pattern is characteristic of a surface layer less than 100 atoms deep. The reflection method is useful for investigating surface films such as those formed by corrosion.

The diffraction pattern from a single crystal (Fig. 1-36*b*) is simply the pattern of (h_1h_20) points in the reciprocal lattice. The reflection sphere has such a large radius ($r = 1/\lambda \approx 20$ Å) that it approximates a plane through the origin perpendicular to the direction of the incident beam. If the beam is in the [001] direction (in the crystal), then the sphere passes through many (h_1h_20) points that lie close to the origin. These "points" are actually rods extended in the $\mathbf{b}_3$ direction in the reciprocal lattice since the crystal is finite in this direction. The interpretation of a powder pattern (Fig. 1-36*c*) is essentially the same as for x-rays, except that the lines are always broadened.

The strong absorption of an electron beam is the basis for the LEED (low-energy electron diffraction) technique for studying single atomic layers

[1] From the relation $\lambda = h/mv$ (in Chap. 12), it follows that $\lambda = (150/V)^{1/2}$, where V is the accelerating voltage; relativistic effects are neglected. Note that λ is 1 Å at 150 volts and is only 0.054 Å at the 50,000 volts commonly used for electron diffraction.

on the surface of materials. With an accelerating voltage of about 500 volts (instead of 50,000 volts), the electrons can penetrate only the surface layer of atoms, and therefore the diffraction pattern gives information primarily about the surface. Some results of this type of study are discussed in Sec. 5-3.

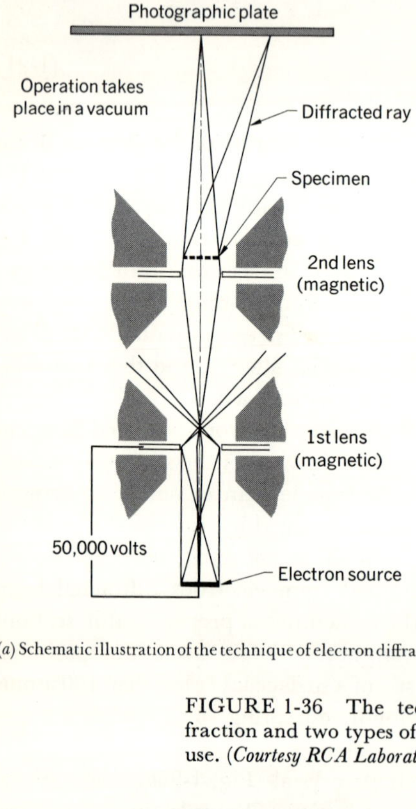

(*a*) Schematic illustration of the technique of electron diffraction.

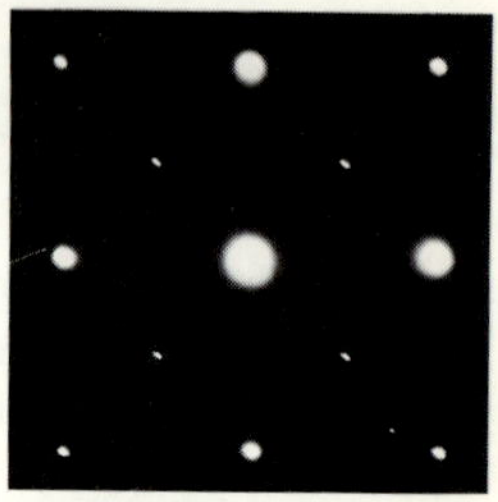

(*b*) Diffraction pattern from a single crystal of GaAs, electron beam along the (001) crystal direction. (Courtesy of M. S. Abrahams)

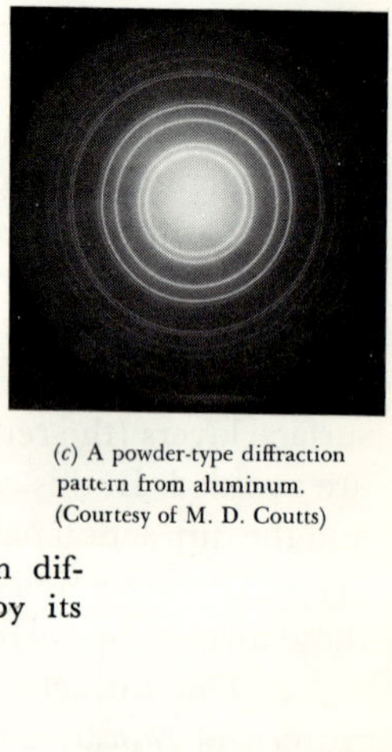

(*c*) A powder-type diffraction pattern from aluminum. (Courtesy of M. D. Coutts)

FIGURE 1-36 The technique of electron diffraction and two types of patterns obtained by its use. (*Courtesy RCA Laboratories.*)

1-12. Neutron Diffraction

In contrast to x-ray and electron diffraction, which are relatively common laboratory techniques, neutron diffraction is a specialized research tool. The massive equipment operates on the same principles as other diffraction techniques but requires a strong neutron flux from a reactor, heavy shielding, and also a monochromator to select a certain wavelength λ from the white radiation that characterizes thermal neutrons.[1]

[1] Neutrons produced in a reactor are slowed down through collisions with a moderator (heavy water or graphite) and assume the energies characteristic of the ambient temperature, say 300°K. Ideal gas theory gives an average value of $\frac{1}{2}mv^2 = \frac{3}{2}kT$. From the relation $\lambda = h/mv$, it follows that $\lambda = (h^2/3mkT)^{1/2}$, and thus λ is about 1.5 Å at room temperature.

Neutrons have a magnetic moment, and therefore they interact with those parts of the atom that have a magnetic moment, primarily the nucleus. For those materials whose electrons also exhibit strong magnetic properties, a further interaction arises from this cause. Additional experimental effects include scattering related to the size of the nucleus, strong incoherent scattering from special "isotope" and "spin" effects, and also high absorption in some cases. The net result of these many effects is an atomic scattering factor b for neutrons that follows no particular pattern with increase in atomic number (Table 1-4), in contrast to the regular increase of the scattering factor f for

TABLE 1-4. ATOMIC SCATTERING FACTORS FOR SEVERAL ELEMENTS IN ORDER OF INCREASING ATOMIC NUMBER

Element	**Neutron atomic scattering factor,† $b \times 10^{14}$ m**	**X-ray atomic scattering factor, f, (electron units)**
Hydrogen	−0.38	1
Deuterium	0.65	1
Beryllium	0.77	4
Carbon	0.66	6
Magnesium	0.52	12
Aluminum	0.35	13
Titanium	−0.38	22
Iron	0.96	26
Cobalt	0.25	27
Molybdenum	0.66	42
Tungsten	0.47	74

Conversion of units: (b in meters) × 10^2 = b in centimeters.
† Values from G. E. Bacon, "Applications of Neutron Diffraction in Chemistry," The Macmillan Company, New York, 1963.

x-rays. When the structure factor F_n for neutron diffraction is calculated by Eq. (1-14), a new feature occurs because some of the b values can be negative. As a result, a solid-solution alloy may have zero intensity for all of the usual diffraction lines.

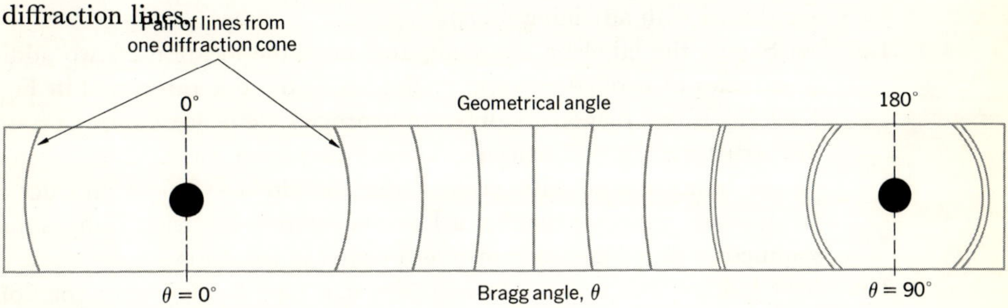

FIGURE 1-37 Diffraction pattern produced in an x-ray powder camera using CuKα x-rays (λ = 1.541). (*Courtesy Philips Electronic Instruments.*)

For some problems neutron diffraction is much superior to x-ray diffraction. Materials such as hydrides and beryllides consist of a light element in a crystal structure with a heavy element. In x-ray studies the intensity scattered by the light element may be undetectable since the ratio of f values is so small (6/74 for tungsten carbide, for example). In contrast the b values are certain to be comparable (0.66/0.47 for tungsten carbide). The opposite problem occurs in the study of a structure containing atoms of almost the same atomic number, say iron and cobalt. It is then difficult to distinguish the locations of these two atoms in x-ray studies since their f values are almost the same; however, their b values differ by a factor of 4. Chapter 12 gives an example of the use of neutron diffraction in magnetic ceramics.

PROBLEMS

1-1. (*a*) Calculate the potential energy of one Na^+ ion and one Cl^- ion when their centers are 2.81 Å apart, the separation in a crystal of NaCl. (*b*) Perform a similar circulation but for two pairs of ions (four ions) in a square arrangement such that two Na^+ ions are diagonally opposite each other. Show that the potential energy per pair of ions is lower than for case *a*. (*c*) For what size crystal would the potential energy per pair of ions be a minimum?

1-2. Make trigonometric calculations to verify the limiting radius ratios given in Fig. 1-3.

1-3. Apply Fig. 1-5 to investigate the possibility of forming a helium molecule, He_2, from two free helium atoms. Recall that the increase in energy on forming a σ_{1s}^* MO from two 1*s* AOs is slightly greater than the decrease on forming a σ_{1s} MO.

1-4. Using the bond energies of Table 1-1, show that the ethylene molecule has less bonding energy than the ($-CH_2-CH_2-$) unit that it forms in in polyethylene (your calculations must include the two *end* bonds which are shared with adjoining units).

1-5. (*a*) Sketch the ethylene molecule, and show the location of two additional axes of twofold rotation analogous to the one illustrated in Fig. 1-10. (*b*) If two diagonally opposite hydrogen atoms were missing from this structure but the molecule retained its planar form, would mirror planes be associated with each of the twofold axes? (*c*) Construct a (nonplanar) molecule that has a $\bar{1}$ axis of rotation-inversion. Why is this symmetry operation also known as a center of inversion?

1-6. Sketch a few unit cells of plane lattices that have the following rotation axes: (*a*) twofold, (*b*) threefold, (*c*) fourfold, (*d*) sixfold. (*e*) Using a suitable sketch, prove that a fivefold axis is impossible for a crystal (even though it can occur in molecular point groups). [HINT: Consider four adjacent lattice points (1, 2, 3, and 4) spaced a distance a apart along an axis of the crystal. Clockwise operation of an n-fold rotation axis through point 2 takes point 1 to a new position 1′. Counterclockwise operation of the same rotation through point 3 takes point 4 to 4′. Points 1′ and 4′, being lattice points, must be a distance ma apart, where m is an integer. Therefore $ma = a + 2a \cos(360/n)$.]

1-7. (*a*) Sketch the five two-dimensional (plane) lattices [analogous to the cubic, tetragonal, rhombohedral, monoclinic, and (base of) hexagonal, respectively]. (*b*) Explain why each is a *distinct* distribution of points in two dimensions. (*c*) Why is the "centered square" not a separate plane lattice?

1-8. (*a*) What is the crystal point group in Fig. 1-14? (*b*) If the structural unit were of the $\bar{1}$ molecular point group (rather than the 1 group), what would the crystal point group be?

1-9. During heat treatment, a steel (iron containing 5 A/o carbon) may have the *austenite* structure. This crystal structure is face-centered cubic with a lattice constant of about 3.65 Å; the carbon atoms are in interstitial positions. (*a*) What are the sizes of the octahedral and tetrahedral interstices? (*b*) The radius of the carbon atom (or ion) in austenite is about 0.77 Å. Which type of interstice will probably be occupied by a carbon atom?

1-10. Apply the requirement of "identical surroundings" to the atomic positions in the HCP structure, and show that these positions do not satisfy the definition of a space lattice.

1-11. In Fig. 1-21, (*a*) estimate the numerical (nonintegral) values of the orthonormal coordinates R_1 and R_3 of the vector **r**. (*b*) If the crystal rotates about the origin until the X_3 and x_3 axes coincide, what are the new values of R_1 and R_3?

1-12. Using $\mathbf{a}_2 = 4$ Å, sketch in three dimensions the monoclinic crystal structure of Fig. 1-21, and show in it the direction [312].

1-13. Show that the (202) reciprocal-lattice point in Fig. 1-22 corresponds to a set of planes in the crystal parallel to the (101) planes but with an interplanar spacing only half as large. Even if the extra (interleaved) planes contain no atoms, these planes are useful for describing second-order diffraction.

1-14. (*a*) Construct the three-dimensional reciprocal lattice corresponding to the crystal lattice of Prob. 1-12, and show in it the vector **s** representing the (312) crystal planes. (*b*) In a *cubic* crystal the vector [312] is parallel to the vector (312). Does this condition exist in the present case?

1-15. (*a*) Sketch the (100) plane and the ($\bar{1}00$) (can be read, "minus one, zero, zero") plane. These planes differ only in their location relative to the origin. (*b*) Sketch all six members of the {100} family of planes [(100), (010), etc.]. (*c*) Show that the [101] direction is perpendicular to the (101) plane in a cubic crystal but not in other crystal systems, such as the tetragonal.

1-16. (*a*) On a drawing analogous to Fig. 1-25*a*, sketch the (112) and (110) planes. Identify the zone axis, and explain why it must be perpendicular to the **h** vector of both planes. (*b*) Show that Eq. (1-7*a*) gives the following values of z_i, expressed as determinants:

$$z_1 = \begin{vmatrix} h_2 & h_2' \\ h_3 & h_3' \end{vmatrix} \qquad z_2 = \begin{vmatrix} h_3 & h_3' \\ h_1 & h_1' \end{vmatrix} \qquad z_3 = \begin{vmatrix} h_1 & h_1' \\ h_2 & h_2' \end{vmatrix} \tag{1-22}$$

(*c*) Use Eq. (1-22) to determine the zone axis in part *a*.

1-17. Use Eq. (1-8) to determine which of the following planes belong to the [111] zone, that is, which contain [111] as a zone axis: $(1\bar{1}0)$, (100), $(1\bar{2}1)$, $(\bar{2}11)$, $(1\bar{1}2)$, $(\bar{3}21)$.

1-18. (*a*) By the use of similar triangles, in Fig. 1-27 prove the following relation between the *cartesian* coordinates (x,y,z) of p and (X,Y) of P:

$$\frac{x}{X} = \frac{y}{Y} = \frac{1+z}{2} \tag{1-23}$$

(*b*) Verify the following standard relations between cartesian and spherical coordinates for a unit sphere

$$x = \sin\theta\cos\varphi \qquad y = \sin\varphi \qquad z = \cos\theta\cos\varphi \tag{1-24}$$

(*c*) Using the equations of part *b* in the results of part *a*, obtain Eq. (1-9).

1-19. (*a*) Show that a rotation about the x axis can easily be performed, provided that the Wulff net is rotated 90°. (*b*) A rotation about the z axis is so simple geometrically that the corresponding *polar net* is seldom employed. Sketch a polar net, using Eq. (1-9) to calculate the location of the latitude lines for $\varphi = 20$, 45, and 70°.

1-20. Make a (001) standard projection for α-uranium that shows planes of the {100} family, that is, (100), (010), (001), $(\bar{1}00)$, $(0\bar{1}0)$. [The $(00\bar{1})$ plane would plot on the "back" half of the unit sphere and therefore is usually not shown.] Plot also the {110} and {111} families of planes. Plot the ⟨110⟩ and ⟨111⟩ families of directions (use square brackets). α-uranium is orthorhombic; $\mathbf{a}_1 = 2.85$ Å, $\mathbf{a}_2 = 5.87$ Å, and $\mathbf{a}_3 = 4.96$ Å.

1-21. Derive the Laue condition [Eq. (1-12)] by resolving **h** and $\mathbf{s} = (\mathbf{S} - \mathbf{S}_0)/\lambda$ into components along the crystal axes and equating the magnitudes of the corresponding components. [HINT: The magnitude of the component of **h** along the a_1 axis, for example, is given by the scalar product $\mathbf{h} \cdot (\mathbf{a}_1/a_1) = h_1\mathbf{b}_1 \cdot (\mathbf{a}_1/a_1) = h_1/a_1$, the other two terms of the scalar product being zero. Similarly, the component of **S** along the $\mathbf{a}_1$ crystal axis is $\mathbf{S} \cdot (\mathbf{a}_1/a_1) = \cos\alpha$.]

1-22. Adapt Fig. 1-34*c* to the analysis of the rotating-crystal method in which a beam of constant λ is incident on a single crystal continuously rotated about a vertical axis perpendicular to the incident beam. One of the *crystal* axes is chosen as the axis of rotation, and the diffracted beams are recorded on a vertical, cylindrical film with the specimen midway along the length of the cylinder axis. (*a*) What orientation of the reciprocal lattice results from choosing a crystal axis as the rotation axis? (*b*) Show that the reflection sphere sweeps out a "doughnut without hole"-shaped region in the reciprocal lattice. (*c*) Sketch the type of pattern that would appear on the film (laid out on a flat surface).

1-23. In the early stage of precipitation hardening of an Al-5 $^{w}/_{0}$ Cu solid solution, platelets of almost pure copper about 2 unit cells in thickness form on the {100} crystal planes of the face-centered cubic solid solution. Both the precipitate and the matrix have the same orientation, and both lattice constants are about 4.0 Å. A Laue pattern is made of a crystal oriented with the [001] direction toward the x-ray beam and with the [$\bar{1}$00] direction vertical. (*a*) Sketch the location on the film of the four symmetric diffraction spots (602), (062), ($\bar{6}$02), and (0$\bar{6}$2) if the specimen-to-film distance is 3 cm. [NOTE: The Miller indices of the planes in question are (301), (031), etc.] (*b*) Show the direction of the streaks through these spots and their approximate length (determined by the extension of the points in the reciprocal lattice of Fig. 1-32*b*).

1-24. If a specimen of a metal (take iron as an example) is being studied by x-ray diffraction, the generation of *fluorescent x-rays* in the specimen by the x-ray beam should be avoided. (*a*) Explain with the aid of a sketch why iron would produce Fe $K\alpha$ fluorescent x-rays if Ni $K\alpha$ x-rays were employed for a diffraction study. (*b*) Would you recommend either copper or cobalt x-rays instead? Why? For Co $K\alpha$, $\lambda = 1.789$ Å.

1-25. Use Eq. (1-14) to show that diffraction will occur in a face-centered cubic metal only if the h_i values are either all odd integers or all even integers (including zero).

1-26. Determine the lattice parameter of the cubic metal that produced the diffraction pattern recorded on the film of Fig. 1-37. An x-ray powder camera is a short cylinder; the powdered specimen is on a glass fiber at the cylinder axis, and the film wraps inside the cylinder. Since $A(\mathbf{s})$ consists of spherical shells (Fig. 1-34*a*), the incident beam produces a cone of radiation for each shell. Two segments of each cone are recorded on the film, separated by a total angle of 4θ. [HINT: The "x-ray centers" of the holes in the film are 180° apart. This reference distance permits precise calculation of the *d* values of the several diffraction lines.]

A portion of this problem can be done with a digital computer if desired. First, determine *N* different values of THETA, θ, in radians by the appropriate measurements on the film of Fig. 1-37. The following Fortran IV program then determines the C values [Eq. (1-19)] by statement 6 if the unknown metal is body-centered cubic. The program next calculates the average value of C (CAV) and finally calculates the lattice constant *A* by Eq. (1-20). The wavelength λ is designated XLAMB. Start by recording on the data card the value of λ (1.541) followed by the *N* values of θ in radians.

```
C  CALCULATION OF LATTICE PARAMETER, A
   1 FORMAT (F6.4)
C  READ DATA
   2 READ (5,1) XLAMB
C  CALCULATION
     N = 0
     SUM = 0
   3 READ (5,1,END = 9) THETA
   4 N = N + 1
   5 B = (SIN(THETA)) ** 2
   6 C = B/(2 * N)
   7 WRITE (6,1) C
   8 SUM = SUM + C
     GO TO 3
   9 CAV = SUM/N
  10 A = XLAMB/(2 * SQRT(CAV))
C  OUTPUT SECTION
  11 WRITE (6,1) A
  12 STOP
     END
```

Statement 7 causes all of the C values to be printed. If they are not almost identical, then the metal is not body-centered cubic; statement 6 must then be rewritten in a form suitable for the face-centered cubic lattice.

REFERENCES

Azároff, L. V.: "Elements of X-ray Crystallography," McGraw-Hill Book Company, New York, 1968.

Barrett, C. S., and T. B. Massalski: "Structure of Metals," 3d ed., McGraw-Hill Book Company, New York, 1966.

Cohen, J. B.: "Diffraction Methods in Materials Science," The Macmillan Company, New York, 1966.

Companion, A. L.: "Chemical Bonding," McGraw-Hill Book Company, New York, 1964.

Cullity, B. D.: "Elements of X-ray Diffraction," Addison-Wesley Publishing Company, Inc., Reading, Mass., 1956.

DiBenedetto, A. T.: "Structure and Properties of Materials," McGraw-Hill Book Company, New York, 1967.

Guinier, A., and D. L. Dexter: "X-ray Studies of Materials," Interscience Publishers, New York, 1963.

"International Tables for X-ray Crystallography," vol. 1 (1952), vol. 2 (1959), vol. 3 (1962), Kynoch, Birmingham, England.

Lipson, H., and C. A. Taylor: "Fourier Transforms and X-ray Diffraction," G. Bell & Sons, Ltd., London, 1958.

Mahan, B. H.: "University Chemistry," Addison-Wesley Publishing Company, Inc., Reading, Mass., 1965.

Chapter Two

Phase Diagrams

A phase diagram shows the phases that exist in a material of given chemical composition under specified, useful conditions. For example, the iron-carbon diagram (Fig. 2-22) shows that a 0.8 ᵂ/₀† carbon steel contains 89.4 ᵂ/₀ ferrite

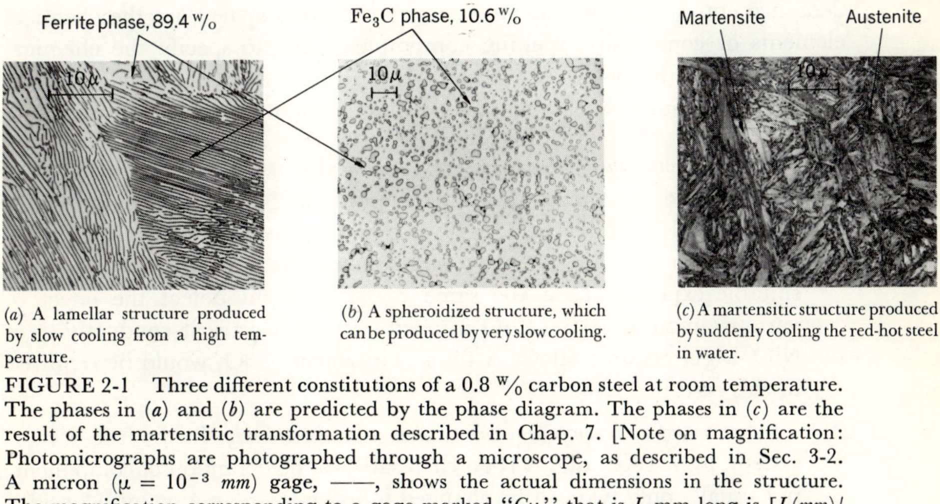

(*a*) A lamellar structure produced by slow cooling from a high temperature.

(*b*) A spheroidized structure, which can be produced by very slow cooling.

(*c*) A martensitic structure produced by suddenly cooling the red-hot steel in water.

FIGURE 2-1 Three different constitutions of a 0.8 ᵂ/₀ carbon steel at room temperature. The phases in (*a*) and (*b*) are predicted by the phase diagram. The phases in (*c*) are the result of the martensitic transformation described in Chap. 7. [Note on magnification: Photomicrographs are photographed through a microscope, as described in Sec. 3-2. A micron ($\mu = 10^{-3}$ *mm*) gage, ——, shows the actual dimensions in the structure. The magnification corresponding to a gage marked "$G\mu$" that is L mm long is $[L(mm)/G(\mu)] \times 10^3$ = magnification. The magnification of Fig. 2-1*b* is 1000, usually written ×1000.) (*Courtesy Edgar C. Bain Laboratory for Fundamental Research, U.S. Steel Corporation.*)

phase and 10.6 ᵂ/₀ Fe_3C phase after slow cooling from a high temperature (Sec. 2-3 describes the method of calculation). The pair of photomicrographs in Fig. 2-1*a* and *b* demonstrate that the phase diagram determines only two factors in constitution [(1) the nature and (2) the amount of each phase] but not

† A composition expressed in percent (%) usually refers to percent by weight, but percentage of atoms, of molecular weights (moles), and of volume are also useful. The symbols ᵂ/₀, ᴬ/₀, ᴹ/₀, and ᵛ/₀ serve to distinguish among these possible types of percentage.

the distribution of the phases. This third factor, which depends on the thermal and mechanical history of the specimen, is the subject of several later chapters. Furthermore, unusual mechanical or thermal treatments may produce constitutions whose microconstituents are not even the stable phases found in the phase diagram. The martensite phase in a quenched steel (Fig. 2-1*c*) is a typical example. Nevertheless, phase diagrams serve as an indispensable starting point for studying the constitutions of metallic and ceramic materials.

INTERPRETATION OF PHASE DIAGRAMS

2-1. Systems, Components, and Phases

Some of the terms used in the study of phase diagrams have special meanings. The *system* under consideration consists of a definite amount (100 g or 1 mole, for example) of a given material, perhaps 50 M/o NiO plus 50 M/o MgO. The surroundings, such as the piston-and-cylinder container in Fig. 2-2*a*, are not part of the system. The *components* of a system are the chemical elements or congruently melting compounds[1] used to specify the chemical composition of the system (NiO and MgO in the present case). Usually the choice of components is obvious and satisfies the following three requirements:

1. *The components must be adequate to describe the experiment in question.* For example, CoO is almost always an impurity in NiO, but in most experiments involving the NiO–MgO system its effect is undetectable. The system then contains only two components, NiO and MgO. However, suitable experiment on the same material might detect the effect of various amounts of the CoO impurity on the electrical conductivity of NiO–MgO ceramic alloys. A third component, CoO, would be required in this case.
2. *The amount of each component can be changed independently.* In the iron-carbon system, iron, Fe_3C, and carbon each satisfies the definition of component, but usually only two can be changed independently in amount. Consider the initial system to be 100 g of the alloy of Fig. 2-1*a* at a high temperature; 99.2 g of iron can be added, and independently, 0.8 g of carbon, thus increasing the system to 200 g total. The amount of Fe_3C necessarily increases by about 10.6 g (dependently), even though no Fe_3C compound itself was added to the system. At low temperatures, where the reaction $3Fe + C \rightleftarrows Fe_3C$ cannot occur, a mixture of iron, carbon, and Fe_3C might behave like a three-component system (Prob. 2-1).

[1] A compound that melts in the usual way without changing its composition is said to melt *congruently*. *Incongruently* melting compounds decompose in the process of melting.

3. *A component must be a measurable (not a hypothetical) quantity.* For example, perhaps the concept of a short-lived molecule, $NiMgO_2$, explains perfectly an unusual behavior in a NiO–MgO alloy. Nevertheless, if the amount of the $NiMgO_2$ molecule cannot actually be measured, this species cannot be employed as a component. On the other hand, such an unorthodox quantity as a vacant position in a crystal lattice (a vacancy) may serve as a component (see Sec. 4-3).

Two terms, *phase* and *microconstituent*, are useful for describing the substances in typical constitutions, such as in Fig. 2-1. A *phase* is a homogeneous portion of a system, separated from any different phase by a bounding surface. Thermodynamic descriptions of a system require this definition of a phase. When microscopic examination of a material reveals constituents that do not satisfy this definition, the more general term *microconstituent* applies. For example, a microconstituent may lack a definite boundary (Guinier-Preston zones, Sec. 7-9) or may be a fine mixture of two phases (pearlite in Fig. 2-22). A phase may be physically a single entity, but usually it consists of many particles, such as the particles of the Fe_3C phase in Fig. 2-1*b*.

2-2. One-component Systems

An especially simple example of the above ideas is a one-component system such as MgO or carbon or copper. Imagine that the hypothetical experimental arrangement of Fig. 2-2*a* subjects the system to a range of the experimental variables, temperature and pressure. Then at equilibrium the system exists as the solid phase, the liquid phase, or the vapor phase (Fig. 2-2*b*). For example, at temperatures below the melting point T_f and at ordinary pressures (near 1 atm), copper exists as the solid phase, usually in the form of many small grains (polycrystalline copper). To read this or any other phase diagram, one first locates the point, such as A, determined by the experimental values of interest ($T = 1200°C$, $P = 10^2$ N/m^2, in this case). Since A is in the region of the diagram known from previous experimental data to be liquid, we know that the system exists as the liquid phase.

Merely from a knowledge of the triple point (T_t and P_t), one can sketch a qualitative one-component diagram for substances, like copper, that have only one type of solid phase. A line drawn vertically upward from the triple point forms the boundary between the solid and liquid phases, and the two sloping lines complete the diagram. This rough diagram has many uses (Prob. 2-2). Precise experimental data on the sloping lines of one-component diagrams are available,[1] but a good estimate can be made using the

[1] O. Kubaschewski and E. LL. Evans, "Metallurgical Thermochemistry," Pergamon Press, New York, 1958; and A. N. Nesmeyanov, "Vapor Pressure of the Chemical Elements," Elsevier Publishing Company, Amsterdam, 1963.

Clausius-Clapeyron equation,

$$\frac{dT_e}{dP} = T_e \frac{\Delta V}{\Delta H} \tag{2-1}$$

T_e (in degrees Kelvin) is the temperature of the equilibrium in question

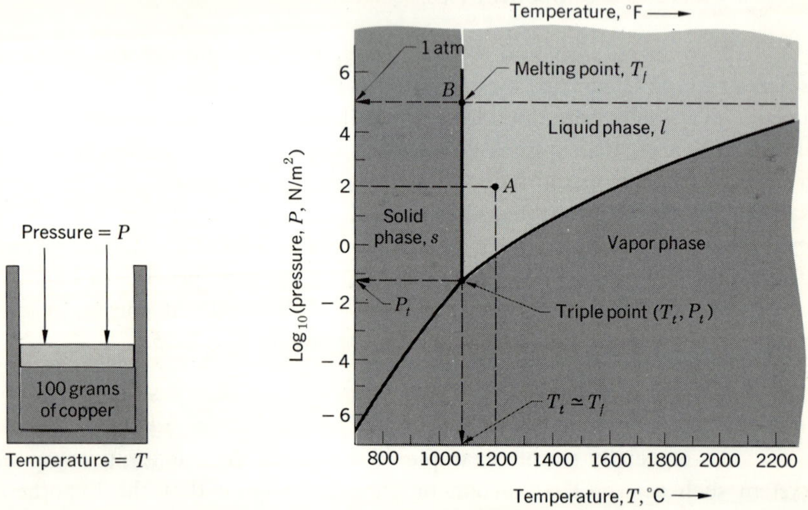

(*a*) Hypothetical heated piston and cylinder for visualizing a system at pressure P and temperature T.

(*b*) The one-component diagram for copper concisely records the experimental behavior of this element at various pressures and temperatures. (Conversion of units: (M N/m^2)0.987×10^{-5} $= N$ atm.)

FIGURE 2-2 Several aspects of the one-component diagram for copper.

(solid-vapor, for example), P is the pressure in atmospheres, ΔV is the difference in molar volume of the two phases [roughly 22.4 $(T_e/273)(1/P_e)$, in liters], and ΔH is the heat of reaction (enthalpy), which is about 81 kcal/mole for the sublimation of copper at 1000°C (1273°K).

THE PHASE RULE. In using the apparatus of Fig. 2-2*a* to investigate a single-phase region (liquid, for example), we exercise two degrees of freedom by fixing both T and P. Study of a *line* in the phase diagram requires a different technique since two phases coexist in equilibrium. The phase rule, first derived[1] by J. Willard Gibbs, requires that at equilibrium

$$p + f = c + 2 \tag{2-2}$$

[1] Because of the thermodynamic requirements for equilibrium between phases, an increase in the number of phases (p) restricts the system and reduces the freedom (f) of the experimenter to adjust the temperature, pressure, and phase compositions.

where p, f, and c are, respectively, the number of phases, degrees of freedom, and components in the system. In a one-component system ($c = 1$) containing two phases ($p = 2$) at equilibrium, $f = c + 2 - p = 1 + 2 - 2 = 1$, and there is only one degree of freedom. If, in a system consisting of a mixture of the solid and liquid phases, we employ this degree of freedom to fix the pressure at 1 atm (Fig. 2-2*b*), then the temperature is determined automatically to be that at point B on the solid-liquid line. If three phases coexist, the system automatically determines the temperature and pressure (T_t and P_t) (Prob. 2-3). If the experimenter chooses to exercise one degree of freedom (by changing the temperature, for example), one phase disappears.

POLYMORPHISM AND METASTABILITY. Many elements and compounds have a more complex one-component diagram than copper because the solid can exist in more than one crystalline form. This property is called *polymorphism* (or *allotropy*). Iron, titanium, and uranium are examples of metals in which polymorphism is crucial in technical applications. An especially striking instance of polymorphism occurs in SiO_2 (quartz), which can exist in four solid modifications under equilibrium conditions. The actual behavior of SiO_2 on being cooled (or heated) is further complicated by the phenomenon of *metastability*. For example, cristobalite, a high-temperature modification of SiO_2, frequently continues to exist below 1470°C, the temperature at which the tridymite modification becomes the equilibrium phase.

When very high pressures are applied to metals or other materials, changes of two types usually occur. First, the temperatures of phase equilibrium shift even for condensed phases (liquid and solid) (see Prob. 2-4). For example, the melting temperature (solid-liquid equilibrium) of copper appears to be independent of pressure in Fig. 2-2*b* (over a pressure range of about 10^6 N/m^2 or 10 bar),† but it is actually increased 200°C by a pressure of 30 kbar (30,000 bar). The Clausius-Clapeyron equation [Eq. (2-1)] permits an estimate to be made of the magnitude and direction of this pressure effect (Prob. 2-5). A second effect of pressure is to cause polymorphism in many substances. Iron, which exhibits the α and γ phases at ordinary pressures, changes into a new crystalline modification, the hexagonal ϵ phase, at pressures above 100 kbar.

2-3. Two-component Systems

The change from a one-component system to a two-component system involves the addition of another variable, the chemical composition (X) of one

† 1 bar (1.000×10^5 N/m^2) is practically equal to the pressure of 1 atm (1.013×10^5 N/m^2).

of the components. The two-dimensional *PT* diagram now becomes the three-dimensional *PTX* diagram. Generally *PTX* diagrams are complex, but the diagrams discussed in this section, solid-solution and eutectic, are relatively simple.

SOLID-SOLUTION SYSTEMS. A simple *PTX* diagram describes those few systems in which the two components are mutually soluble in the solid phase (and also in the liquid and vapor phases). The two one-component (*PT*) diagrams form the front and back faces of the *PTX* figure, and in this case the phase fields of the one-component diagrams simply extend the length of the *X* coordinate, although separated by two-phase regions. Since most technological applications of phase diagrams are made at atmospheric pressure,[1] the corresponding *TX* section is usually adequate and has the advantage of two-dimensional convenience and simplicity (Fig. 2-3).

The interpretation of a two- (or higher) component diagram usually begins with the choice of a specific system, such as (1 mole of) the 20 M/o MgO alloy, X_2, shown in Fig. 2-3. To completely specify the state S of the system, we must also choose a value of the second variable, T. If the point $S(X, T)$ lies in a one-phase region of the diagram, $S(X_2, T_3)$ for example, then the system consists entirely of this phase and the phase condition is simply

Point:	20 M/o MgO alloy at temperature T_3	
Phase:	*Liquid*	
Composition:	20 M/o MgO	(2-3)
Amount:	1 mole	

A convenient schematic illustration of this phase condition is simply the blank circle shown since no structural features would be seen in a (hypothetical) microscopic examination.

If composition X_2 cools slowly from temperature T_3, the solid phase, α, first begins to form when the temperature reaches the liquidus curve. The solidification of a solid-solution alloy differs from that of a pure component in two important respects: (1) solidification occurs over a range of temperatures rather than at a single "melting temperature" (see Prob. 2-6), and (2) the composition of the solid phase in equilibrium with the liquid changes during the course of solidification. Consideration of a typical point, $S(X_2, T_2)$, explains this behavior. Since this point is in the two-phase region of the diagram, the system consists of two phases, the liquid phase l and the solid-solution phase† α. The two points $l(X_1, T_2)$ and $\alpha(X_3, T_2)$ representing these two phases are at the ends of the *tie line* drawn through the system point $S(X_2, T_2)$. The phase

[1] Variations in pressure (of oxygen, for example) are sometimes important, especially in ceramic materials, as discussed in Sec. 4-5.

† Since both MgO and NiO have the NaCl structure (Fig. 1-3*a*), the solid solution also has essentially this structure (Fig. 2-3).

analysis in this case is

Point:	20 $^{\mathrm{M}}\!/_{0}$ MgO (X_2) alloy at temperature T_2	
Phase:	*Solid solution*	*Liquid*
Composition:	36 $^{\mathrm{M}}\!/_{0}$ MgO (X_3)	15 $^{\mathrm{M}}\!/_{0}$ MgO (X_1) (2-4)
Amount:	$\dfrac{X_2 - X_1}{X_3 - X_1} = \dfrac{20 - 15}{36 - 15} = 0.24$ mole	$1.00 - 0.24 = 0.76$ mole

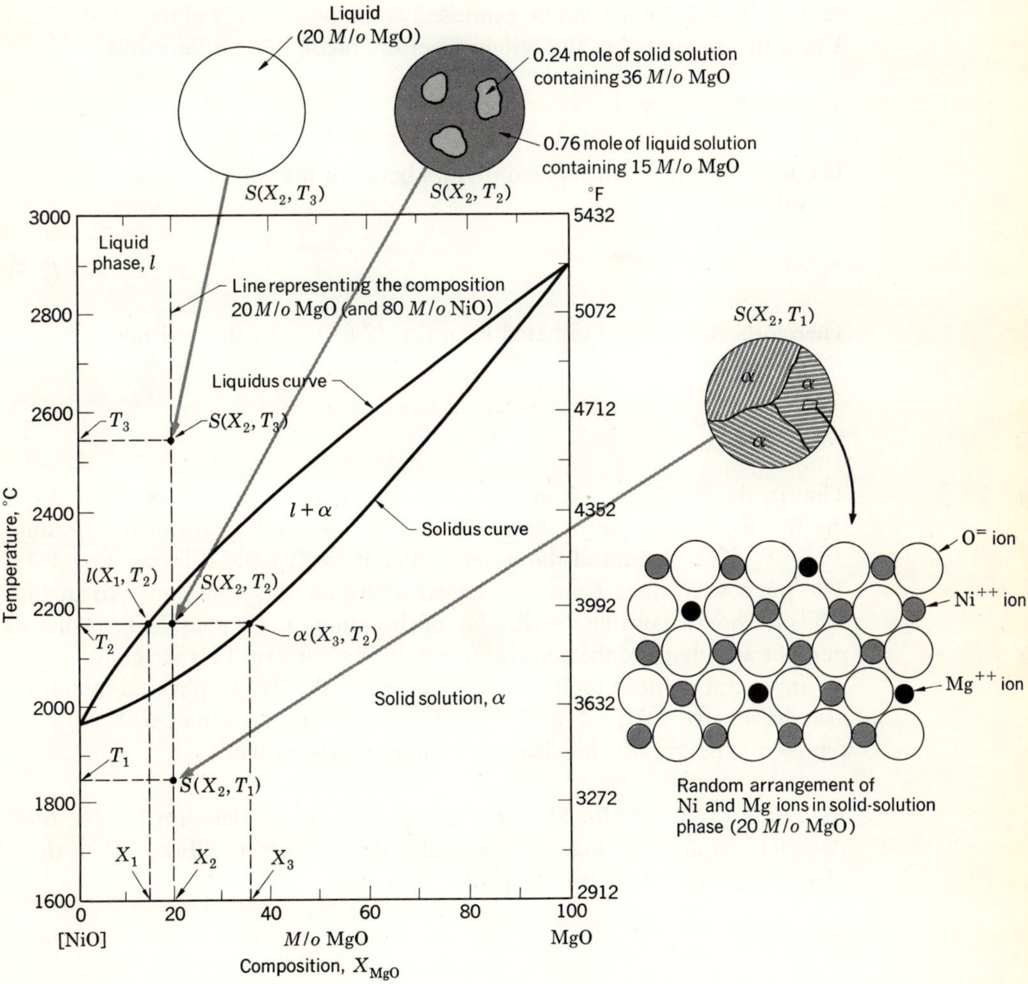

FIGURE 2-3 T-X diagram at $P = 1$ atm for the NiO–MgO system, which forms a complete series of solid solutions. This phase diagram shows the nature of the two-phase (liquid and solid) region. Three schematic illustrations of possible microstructures are shown.

The phase compositions X_1 and X_3 are simply read off the composition axis as indicated by the vertical dashed lines in Fig. 2-3. These compositions then determine the amounts of the two phases by the following argument. The total amount of a given component (MgO, for example) in the system becomes divided between the two phases; thus

$$(\mathrm{MgO})_S = (\mathrm{MgO})_\alpha + (\mathrm{MgO})_l \tag{2-5}$$

Each of these amounts can be expressed as the product XA of the composition X and the amount A of the system or of the phase in question; thus

$$X_2A_S = X_3A_\alpha + X_1A_l \tag{2-6}$$

The total amount of the system divides between the two phases, and we have the relation

$$A_S = A_\alpha + A_l \tag{2-7}$$

Therefore, A_l can be eliminated from Eq. (2-6) to give the *lever rule*,

$$\frac{A_\alpha}{A_S} = \frac{X_2 - X_1}{X_3 - X_1} \tag{2-8}$$

That is, the fraction A_α/A_S of α phase in the system is the portion $X_2 - X_1$ of the tie line *opposite* the α phase, divided by the total length of the tie line, $X_3 - X_1$. The amount of the second phase is easily obtained from Eq. (2-7). The purpose of the schematic microstructure for this phase analysis in Fig. 2-3 is to show a possible distribution of this amount of α phase. The choice of number and shape of the α grains (crystals) is arbitrary. This sketch illustrates an important point in multiphase equilibrium; that is, the physical quantities are the phases $l(X_1,T_2)$ and $\alpha(X_3,T_2)$, but not the system $S(X_2,T_2)$. X_2 represents the overall chemical composition of the system.

EUTECTIC SYSTEMS. Most two-component phase diagrams involve both two-phase equilibria (Fig. 2-4*a*) and also three-phase equilibria. When three phases exist, the phase rule [Eq. (2-2)] becomes

$$f = c + 2 - p = 2 + 2 - 3 = 1 \tag{2-9}$$

The ordinary phase diagram is isobaric (meaning at constant pressure), usually at 1 atm. Therefore, this one degree of freedom has been used, and a three-phase system is invariant. Since neither the temperature nor any of the chemical compositions X of the three phases can vary (without loss of one of the phases), the phases are represented by three points at constant temperature, that is, along a horizontal line (Fig. 2-4*b*). If the central phase meets the

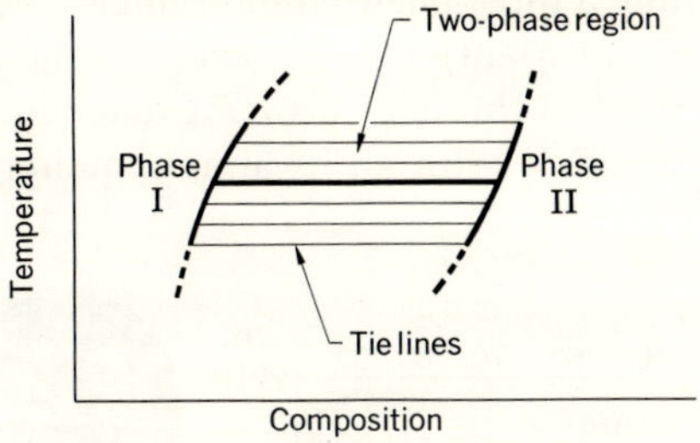

(*a*) A two-phase equilibrium is univariant. Either the temperature or the composition of one of the phases can be varied.

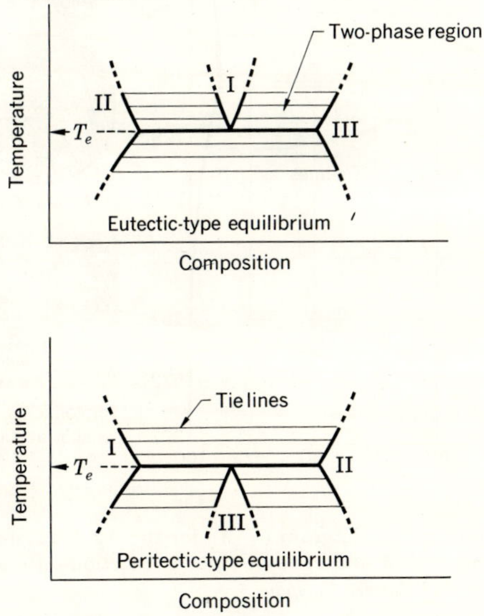

(*b*) A three-phase equilibrium is invariant. The three phases must be at the equilibrium temperature, T_e, and have fixed compositions.

FIGURE 2-4 Schematic illustrations of the types of phase equilibria that can occur in ordinary (isobaric) two-component phase diagrams. I, II, or III each represents a single-phase region, which may be solid, liquid, or vapor.

line from above, the equilibrium is said to be of the *eutectic* type.[1] The alternative arrangement is called a *peritectic*-type equilibrium (Sec. 2-11).

[1] Strictly, a eutectic has a liquid as phase I and two solids as phases II and III.

The silver-copper system can serve as an example of the principles involved in an isobaric, three-phase equilibrium. The complete experimental behavior of these alloys is shown by the *PTX* diagram. Since pressure is also a variable in this diagram, a three-phase equilibrium is now univariant (Prob. 2-8), and a four-phase (invariant) reaction exists in the diagram. When a section at $P = 1$ atm is taken through this *PTX* diagram, a typical eutectic diagram is obtained (Fig. 2-5) with an invariant equilibrium involving the

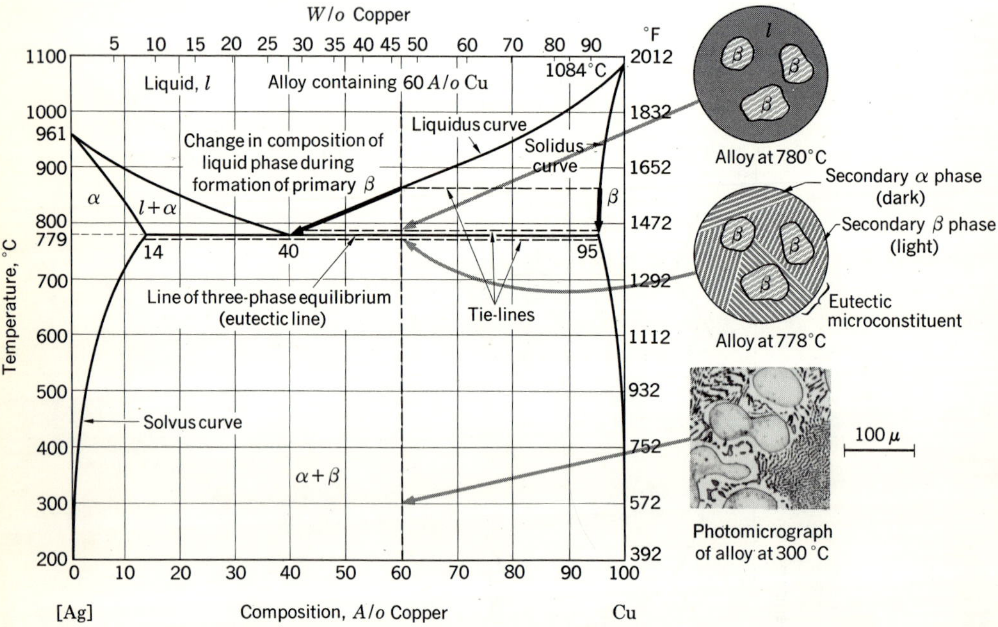

FIGURE 2-5 The usual phase diagram (*T-X*) for the Ag–Cu system at 1 atm pressure. The liquid phase in this system undergoes a eutectic reaction on reaching 779°C. (*Photomicrograph courtesy Handy and Harman Company.*)

phases l, α, and β. For alloys having compositions either below 14 A/o Cu or above 95 A/o Cu, the phase analyses are similar to those described in the previous section. That is, at temperatures between the liquidus and solidus curves the liquid solution changes to the α (or β) solid solution; when the solvus temperature is reached, the β phase begins to form in the grains of α phase (see Prob. 2-9). The term *limited solid solution* describes phases like α and β that extend over only a portion of the compositional range of the phase diagram.

Any alloy whose composition passes through the line of three-phase equilibrium, such as the 60 A/o Cu alloy shown, undergoes the eutectic reaction

$$l(40\ ^{A}\!/_{0}\ \text{Cu}) \rightarrow \alpha(14\ ^{A}\!/_{0}\ \text{Cu}) + \beta(95\ ^{A}\!/_{0}\ \text{Cu}) \tag{2-10}$$

as part of its solidification process. First, however, this liquid alloy forms some (primary) crystals of the β phase by solid-solution–type freezing in the temperature range from the liquidus curve to the eutectic line. The phase analysis at 780°C, just above the eutectic temperature, is

Point:	60 A/o Cu alloy at 780°C		
Phase:	*Beta* (β)	*Liquid* (l)	
Composition:	~95 A/o Cu	~40 A/o Cu	(2-11)
Amount:	$\frac{60 - 40}{95 - 40} = 0.36$ mole	$1 - 0.36 = 0.64$ mole	

The liquid phase has now the eutectic composition and upon reaching the eutectic temperature can undergo the reaction of Eq. (2-10). At the completion of the reaction, say, at 778°C, all of the eutectic liquid has changed into a fine mixture of the α and β phases, the eutectic microconstituent. The amount of this microconstituent is 0.64 mole, the same as the amount of the initial liquid. If the solidified alloy is very slowly cooled to lower temperatures, both the α and β phases change in composition along the solvus lines. At 300°C, for example, the phase analysis is

Point:	60 A/o Cu alloy at 300°C		
Phase:	*Alpha* (α)	*Beta* (β)	
Composition:	1 A/o Cu	~100 A/o Cu	(2-12)
Amount:	$\frac{40}{99} = 0.40$ mole	0.60 mole	

The corresponding photomicrograph in Fig. 2-5 shows two microconstituents, primary β phase and the eutectic. The small particles of α phase and β phase that form the eutectic microconstituent are often called *secondary* α and β, but they are identical in composition and structure to the primary phases.

2-4. Phases in Organic Polymers

Most of the phases considered in the previous sections are agglomerations of simple units, molecules such as SiO_2 or even individual atoms such as copper or silver. In contrast, the individual molecules that form a liquid or solid phase in an organic polymer may consist of about 10^4 mers. An example is the long-chain molecule in polyethylene (Fig. 1-9*d*). Furthermore, the individual molecules are not identical but vary in molecular weight (typically over a tenfold range), in configuration, and often in such structural features as branching of the main chain. Nevertheless, a homopolymer such as polyethylene or Nylon 6,6 (on the right in Fig. 2-6) has a well-defined melting point for a certain average molecular weight. If two types of monomers, such as two nylon monomers (Fig. 2-6), are *copolymerized* to make long-chain molecules containing the two monomers in random positions along the chain, the melting point of the random copolymer phase passes through a minimum

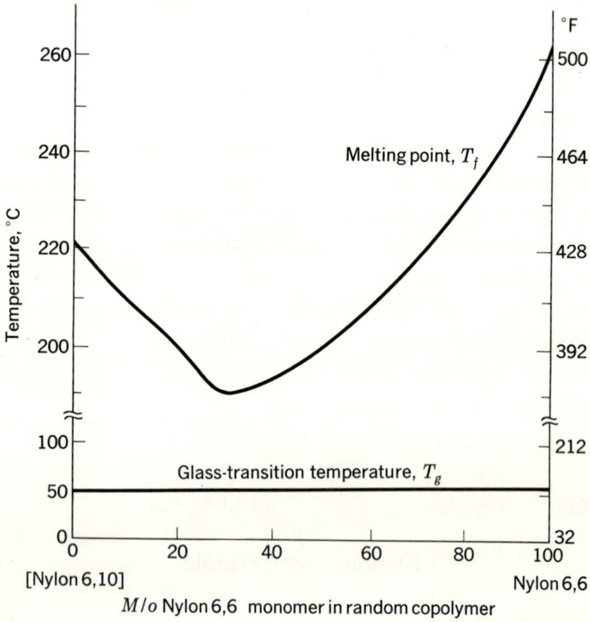

FIGURE 2-6 A "component" diagram useful for polymer systems. Variation of T_f and T_g with composition in a continuous series of components, each composed of long-chain molecules of the polyamide (nylon) type.

at some ratio of the monomers. Although this diagram has a superficial resemblance to a *binary* phase diagram for simple components, its interpretation is quite different. Each of the compositions actually represents a distinctive component, each with a characteristic melting point. The construction and interpretation of a usual phase diagram involving two polymer components are complicated by the tendency for chemical reaction between the two components, especially in the molten condition. Immiscibility in the solid state is characteristic of virtually all homopolymers. Consequently, the "alloying" of homopolymers is accomplished by mechanical mixing, a process known as *blending* of polymers.

Immiscibility explains the phenomenon of ***phase separation*** observed in some polymer systems. For example, if two types of homopolymers are blended, they will tend to separate at the elevated temperature of the molding or casting process into the two pure phases. The minor phase sometimes occurs as relatively large spheres, rods, or plates in the major phase, but usually the morphology is relatively formless. Phase separation can be caused on a much finer scale and with useful "tie molecules" connecting the two phases by the use of block copolymers. The ***block*** technique of polymerization permits alternate sections of a single chain to be constructed of two different monomers, with many mers in each section. The molecule used in making Kraton 101 consists of three blocks (Fig. 2-7*a*), a central region of butadiene with a

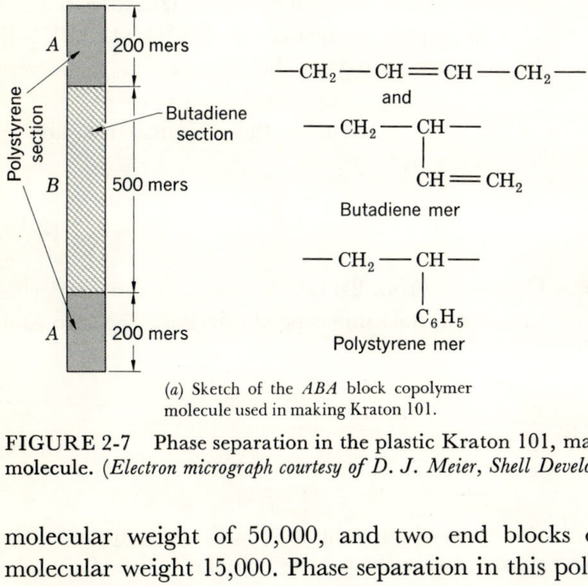

(*a*) Sketch of the *ABA* block copolymer molecule used in making Kraton 101.

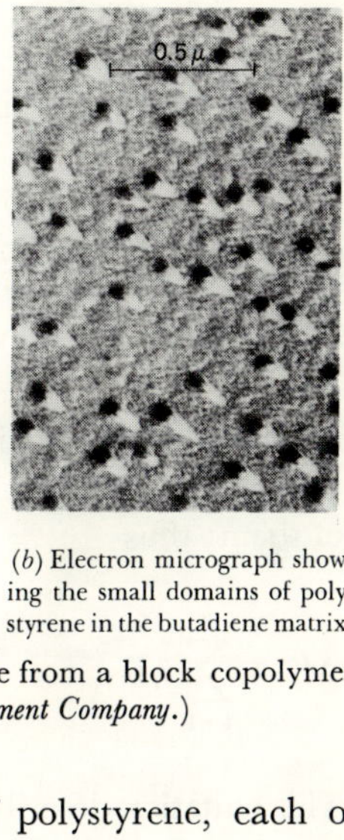

(*b*) Electron micrograph showing the small domains of polystyrene in the butadiene matrix.

FIGURE 2-7 Phase separation in the plastic Kraton 101, made from a block copolymer molecule. (*Electron micrograph courtesy of D. J. Meier, Shell Development Company.*)

molecular weight of 50,000, and two end blocks of polystyrene, each of molecular weight 15,000. Phase separation in this polymer occurs only to the extent of producing small spherical domains of polystyrene embedded in the polybutadiene matrix (Fig. 2-7*b*). Chain molecules of both phases are intermixed in the interface. The rigid polystyrene is a glass below 100°C (T_g) and acts as crosslink for the polybutadiene rubbers, thus contributing to the strength of the final structure.

The solid phases and the glass-transition temperature indicated in Fig. 2-6 are discussed in Chap. 5. The dominant characteristic of polymer structures, often even in the molten state, is the absence of thermodynamic equilibrium. For this reason, phase diagrams are seldom useful. Kinetic factors are decisive in determining the structures assumed by polymeric systems.

THERMODYNAMIC BASIS OF PHASE DIAGRAMS

Thermodynamics underlies phase diagrams and is useful for many aspects of their study. The following treatment is not a substitute for a course in thermodynamics but is intended (for those who have not yet had such a course) to illustrate the application of thermodynamic principles to phase diagrams. The "compositional" approach is used, in which thermodynamic behavior is described simply in terms of the chemical composition of the system. The complementary "structural" approach involves crystal structure, valence electrons, etc.

2-5. Introductory Example

To introduce the thermodynamic approach, let us consider the following problem: A specimen of copper is heated at 927°C (1200°K) in vacuum. How low must the pressure of oxygen be maintained to prevent oxidation?

The first step is to write the equation for the chemical reaction in question. If Cu_2O forms, the reaction is

$$4Cu + O_2 \rightarrow 2Cu_2O \tag{2-13}$$

[Another possible oxide is CuO (see Prob. 2-11)]. Although many properties of a system change during a reaction, the same type of relation applies to each of them; thus

$$\Delta V = \sum_{i=1}^{3} \nu_i V_i \tag{2-14}$$

This equation describes the total change in volume, ΔV, that accompanies the reaction. V_i is the molar volume of the ith species; ν_i is the number of moles taking part in the reaction and is positive for products of the reaction but negative for species that disappear. For the reaction of Eq. (2-13), Eq. (2-14) becomes

$$\Delta V = 2V_{Cu_2O} - 4V_{Cu} - V_{O_2} \tag{2-15}$$

Since the molar volume of gaseous oxygen is 22.4 liters at STP., whereas that of the solid species is only 10^{-2} liter, the disappearance of oxygen is the main factor in the change in volume.

The Gibbs free energy G, although more difficult to visualize than volume, can be treated in a similar manner. Roughly, G can be compared to

TABLE 2-1. *STANDARD VALUES AT 25°C OF ENTHALPY $H°$, FREE ENERGY $G°$, AND ENTROPY $S°$ OF SEVERAL SOLID, LIQUID, AND GASEOUS SUBSTANCES.*†

Substance	$H°$, kcal/mole	$G°$, kcal/mole	$S°$, cal/deg-mole	Substance	$H°$ kcal/mole	$G°$, kcal/mole	$S°$, cal/deg-m
Al	0	0	6.75	H_2O (liquid)	−68.32	−56.69	16.75
Al_2O_3	−399	−377	12.2	Hg(liquid)	0	0	18.5
C(graphite)	0	0	1.36	HgO	−21.7	−14.0	17.2
CO(gas)	−26.4	−32.8	47.30	Mg	0	0	7.8
CO_2(gas)	−94.05	−94.26	51.06	MgO	−144	−136	6.4
Ca	0	0	9.95	Mn	0	0	7.6
CaO	−151.9	−144.4	9.5	MnO	−92	−87	14.4
Cu	0	0	7.97	Ni	0	0	7.2
Cu_2O	−40	−35.0	24	NiO	−54.4	−51.7	9.2
CuO	−37	−30.5	10.4	O_2(gas)	0	0	49.0
Fe	0	0	6.5	Pb	0	0	15.5
Fe^{2+} (solution)	−21	−20.3	−27.1	PbO	−52.4	−45.2	16.2
FeO	−64	−58	13.4	Si	0	0	4.5
Fe^{3+} (solution)	−11.4	−2.53	−70.1	SiO_2	−205.4	−192.4	10.0
Fe_2O_3	−197	−177	21.5	Ti	0	0	7.2
H^+ (solution)	0	0	0	TiO_2	−225.8	−212.5	12.0
H_2(gas)	0	0	31.21	Zn	0	0	9.95
H_2O(gas)	−57.80	−54.63	45.11	ZnO	−83.2	−76.0	10.5

† The designations $\Delta H°$, $\Delta G°$, and $\Delta S°$ are frequently used in the literature, emphasizing the fact that reference values have been subtracted.

potential energy in a mechanical system, the system tending to move to decrease its G value. The analog of Eq. (2-14) for free energy is

$$\Delta G^{\circ} = 2G^{\circ}_{Cu_2O} - 4G^{\circ}_{Cu} - G^{\circ}_{O_2} \tag{2-16}$$

where G_i° is the molar free energy of substance i (under specified standard conditions indicated by the superscript °). The change ΔG° in free energy for these conditions is given by the values from Table 2-1. Thus

$$\Delta G^{\circ} = 2(-35.0) - 4(0) - 0 = -70.0 \text{ kcal} \tag{2-17}$$

That is, if the reaction of Eq. (2-13) occurs under the specified conditions (at 25°C with the oxygen at 1 atm pressure), the change in free energy is −70 kcal. Since the system can be pictured as dropping to a lower level of energy, clearly oxidation of the copper is possible under these conditions.

To obtain the answer to the original question, one must consider the reaction when it occurs at a high temperature and under the low pressure of oxygen represented by a good vacuum. The temperature factor is easily taken into account using the Ellingham diagram (Fig. 2-8), in which ΔG° for various reactions is plotted as a function of temperature. For example, the line corresponding to the reaction of Eq. (2-13) passes through the point ($\Delta G^{\circ} = -70$, $T = 25°\text{C}$) in accordance with the data of Table 2-1. ΔG° values corresponding to other temperatures (even above the melting point of copper) can also be read from the diagram; at 927°C the value is $\Delta G^{\circ} = -44$ kcal. The second factor, oxygen pressure, can be evaluated when the system is at equilibrium. Since a chemical reaction can occur only if the corresponding change in free energy, ΔG, is negative, a reaction for which $\Delta G = 0$ has no tendency to occur, either in the forward or in the reverse direction; the system is at equilibrium. We have already seen that the reaction of Eq. (2-13) tends to occur under standard conditions, particularly with the oxygen at 1 atm pressure. Imagine now that the three substances of Eq. (2-13) are in a state of equilibrium (obviously *not* all in their standard states). The change in free energy for the reaction may be written (without the superscript °)

$$\Delta G = \Sigma v_i G_i = 0 \tag{2–18}$$

Here the molar free energy G_i for each substance is the value for its actual state rather than for its standard state.

Further development of Eq. (2-18) is facilitated by the concept of *activity*, a_i, where

$$a_i = \gamma_i N_i \tag{2-19}$$

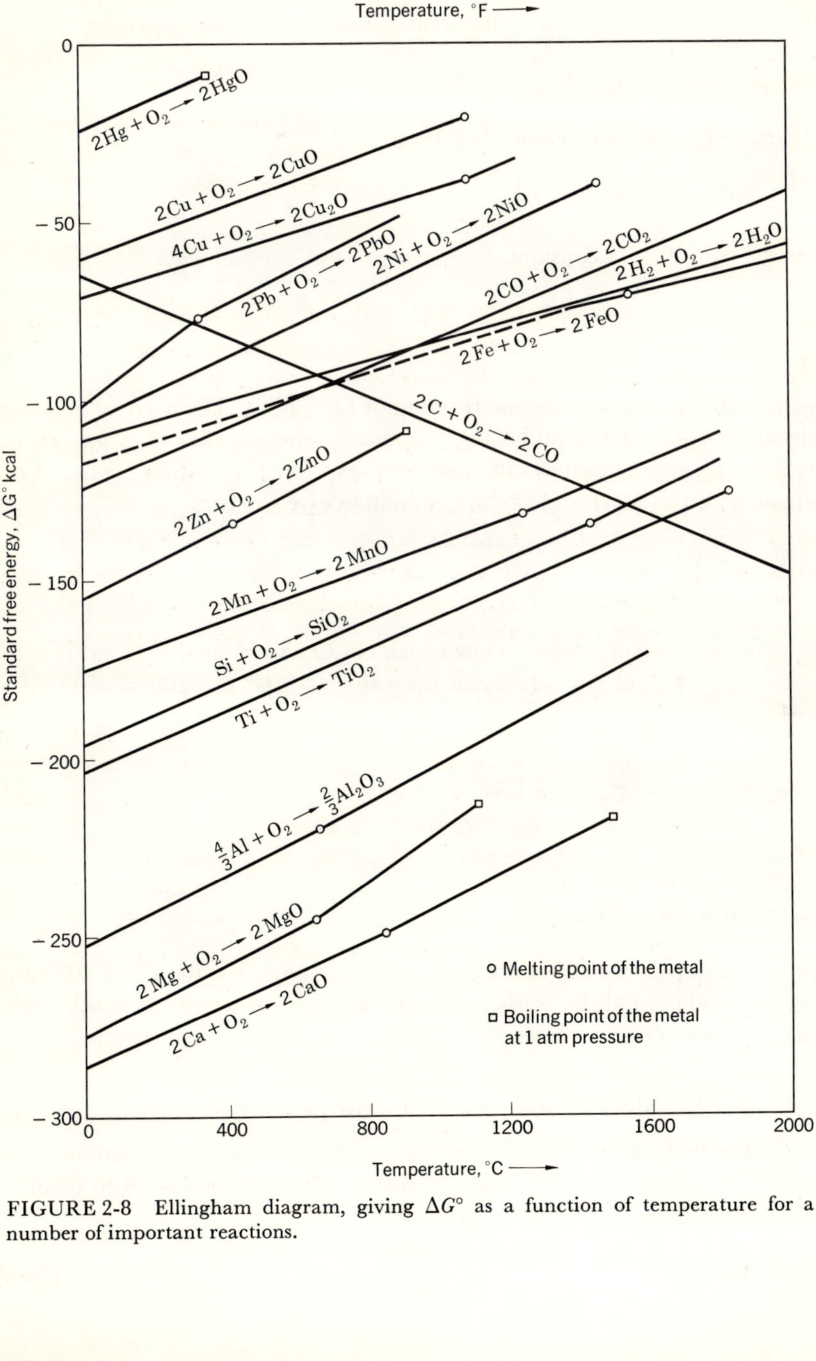

FIGURE 2-8 Ellingham diagram, giving ΔG° as a function of temperature for a number of important reactions.

a_i is essentially a corrected measure of the (mole fraction) concentration N_i of component i; γ_i is the activity coefficient. The relation between G_i and a reference value G_i^o is given by the standard thermodynamic equation

$$G_i = G_i^o + RT \ln a_i \tag{2-20}$$

Therefore Eq. (2-18) becomes (see Prob. 2-12)

$$\Delta G = \Delta G^o + RT \ln K = 0 \tag{2-21}$$

K is the equilibrium constant,

$$K = \frac{a_{Cu_2O}^2}{a_{Cu}^4 a_{O_2}} \tag{2-22}$$

in the form appropriate for the reaction of Eq. (2-13). The activity of a pure substance (such as Cu and Cu_2O) is unity, whereas that of a gas at low pressure is approximately its pressure, expressed in atmospheres. Consequently, Eqs. (2-21) and (2-22) combine to give

$$\ln \frac{1}{P_{O_2}} = -\frac{\Delta G^o}{RT} \tag{2-23}$$

P_{O_2} is the pressure of oxygen below which Cu_2O will be unable to form on the heated specimen of copper. Since the value of ΔG^o at 1200°K (927°C) is -44 kcal,

$$\ln P_{O_2} = -\frac{44{,}000}{2 \times 1{,}200} = -18.4 \tag{2-24}$$

and

$$P_{O_2} = 1.0 \times 10^{-8} \text{ atm}$$

Since ordinary air is one-fifth oxygen, a total pressure less than 5×10^{-8} atm prevents oxidation of the copper. Similar calculations determine possible conditions for reduction of ores (Prob. 2-13).

2-6. Stable Phases in a One-component System

The free-energy G can be used in two different ways to determine which phase(s) is stable under specified conditions. The first method is to consider a reaction between two possible phases; for example,

$$Cu(s) \rightarrow Cu(l) \tag{2-25}$$

in the case of the copper one-component system (Fig. 2-2*b*). The value of the change in free energy,[1]

$$\Delta G = G^l - G^s \tag{2-26}$$

then determines which phase is stable. If $\Delta G > 0$, the reaction cannot occur, and thus the solid phase is stable. If $\Delta G < 0$, the liquid phase is stable. If $\Delta G = 0$, both phases coexist at equilibrium; that is, copper is at its melting point. This first procedure has the advantage that only a *difference* in free energy [Eq. (2-26)] need be determined rather than the separate values.

The second method involves the estimation of an entire curve of G values as a function of an experimental variable, usually temperature or composition. The free-energy curves for the solid and liquid phases of copper are shown in Fig. 2-9*a* as a function of temperature. The important aspect of these curves is their intersection at the melting point of copper, T_f. Below T_f the solid phase has the lower value of free energy and is the stable phase. Above T_f, G^l is the lower value, and thus the liquid phase is stable.

The shape of a given free-energy curve is obviously important and can be determined from the two terms used in the definition of the free energy,

$$G = H - TS \tag{2-27}$$

H, the enthalpy, represents the heat Q added to a system.[2] The entropy S is defined by the equation

$$dS = \frac{dQ}{T} \tag{2-28}$$

For some purposes the entropy is divided into configurational entropy S_c plus entropy from other causes, S_v; thus

$$S = S_c + S_v \tag{2-29}$$

The configurational entropy is defined in terms of Boltzmann's constant k and the number w of configurations in which the atoms of the system can exist; thus

$$S_c = k \ln w \tag{2-30}$$

Therefore, entropy often measures the *randomness* of the system. For example,

[1] The symbols l and s are superscripts identifying the phase in question. Subscripts are employed, as in Eq. (2-19), to identify the components within a given phase.

[2] Textbooks on thermodynamics (references at end of chapter) specify the conditions under which these relations are valid.

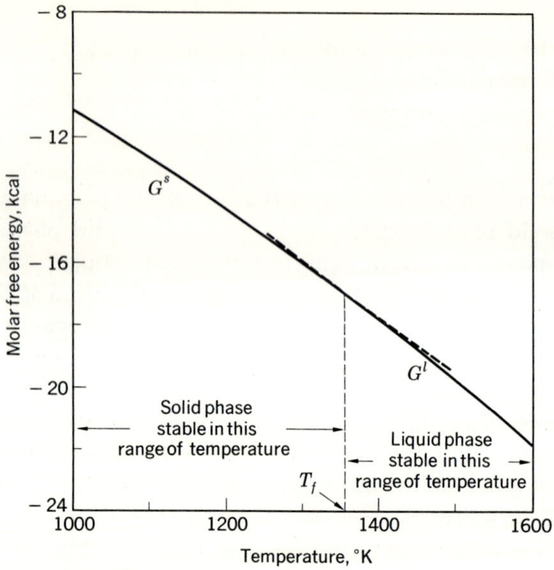

(*a*) The two free-energy curves intersect at the melting point. At a given temperature the phase with lower free energy is stable.

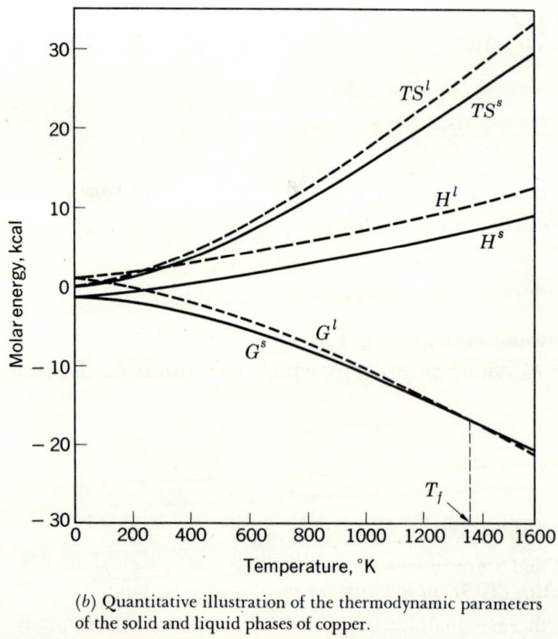

(*b*) Quantitative illustration of the thermodynamic parameters of the solid and liquid phases of copper.

FIGURE 2-9 Illustration of the free-energy curves for the solid and liquid phases of copper.

at $T = 0°K$, a perfect crystal of copper has zero entropy, whereas hypothetical liquid copper would have a contribution from the S_c term in view of the randomness of atomic positions in a liquid phase.

The above relations permit sketching the free-energy curves for liquid and solid copper, even in the region of temperature where a given phase is unstable. The starting place in Fig. 2-9*b* is the reference value $H^s = 0$ at 298°K (25°C), from Table 2-1. The curve of H^l lies above H^s; in particular, at the melting point the separation is the heat of fusion, 3.120 kcal/mole. The change in enthalpy with temperature is determined by the heat added to the system,

$$dH = dQ = C_P\, dT \tag{2-31}$$

where C_P is the heat capacity at constant pressure. Generally, the heat capacity is greater for the phase that is less strongly bound (because of larger thermal vibrations at a given temperature), and therefore H^l rises more rapidly with increasing temperature than does H^s. Both TS curves must be zero at $T = 0$, even though S is greater than zero. The increase of S with temperature [Eqs. (2-28) and (2-31)] depends on C_P, and therefore S^l increases more rapidly than does S^s. The desired curves of G^s and G^l are obtained by subtracting the TS values from the H values. However, Prob. 2-14 shows that the trend of G is determined only by the C_P value through the equation

$$G = H_{(0°K)} - \int_0^T \left(\int_0^T \frac{C_P}{T}\, dT \right) dT \tag{2-32}$$

2-7. Calculation of a Simple Eutectic Diagram

In some cases the entire equilibrium diagram for a two-component system can be calculated fairly accurately by the first method of Sec. 2-6, using only the latent heats of fusion, ΔH_f, and the melting temperatures of the components. The present example requires that a eutectic reaction occur (Fig. 2-5) with α and β being almost the pure components. The liquid solution is assumed to be ideal, so that the activity coefficients are unity, and the activities can be replaced by the mole fractions N_i of Eq. (2-19). The initial ΔH_f and T_f data for a typical system, bismuth-cadmium, are given in Fig. 2-10. The arrow in this figure illustrates the type of "thought experiment" for which the following thermodynamic analysis applies. Imagine that the amount of liquid alloy is very large and thus at temperature T, where bismuth first begins to solidify from the alloy, 1 mole of bismuth can solidify without appreciably changing the composition N_{Bi} of the liquid alloy. This solidification reaction can be written

$$[\mathrm{Bi}]_{N_{Bi}}(l) \rightarrow \mathrm{Bi}(s) \tag{2-33}$$

where the bracketed symbol [Bi] indicates a component in solution and the subscript $[\ ]_{N_{Bi}}$ designates the concentration of the solution. When the reaction occurs under equilibrium conditions,

$$\Delta G = 0 = G_{Bi}(s) - G_{Bi}(l)_{N_{Bi}} \tag{2-34}$$

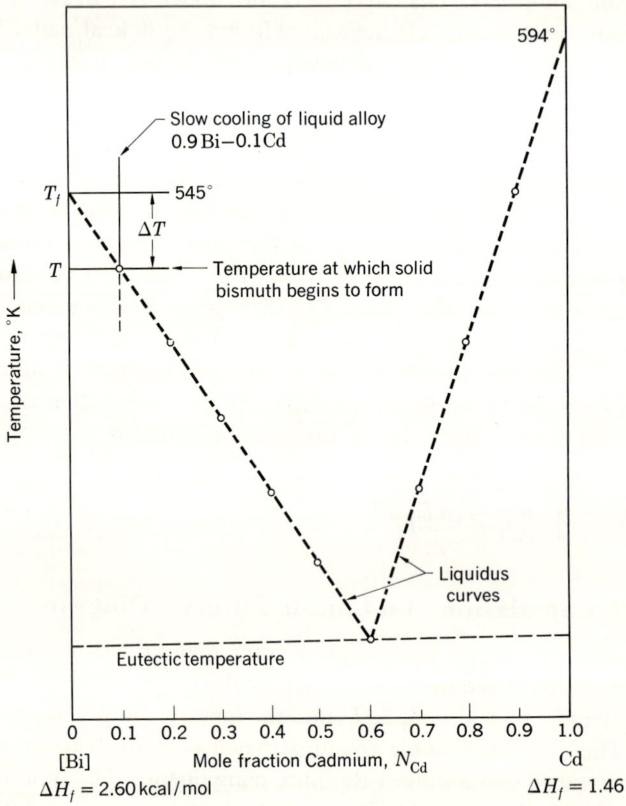

FIGURE 2-10 Procedure for calculating points on the liquidus curves of the eutectic diagram of the bismuth-cadmium system.

The reference free-energy value is usually chosen to be that of the pure substance in its stable state, solid bismuth in this case. Therefore, we set $G_{Bi}(s) \equiv 0$ at temperature T. Equation (2-34) requires that the free energy $G_{Bi}(l)_{N_{Bi}}$ of bismuth in the liquid solution of composition N_{Bi} be zero also. The free energy of pure liquid bismuth, $G_{Bi}(l)$, is given by Eq. (2-20) as

$$G_{Bi}(l) = G_{Bi}(l)_{N_{Bi}} - RT \ln N_{Bi} = -RT \ln N_{Bi} \tag{2-35}$$

A second expression for $G_{Bi}(l)$ is obtained from Eq. (2-27) using the approximations

$$H_{Bi}(l) \approx \Delta H_{f(Bi)} \tag{2-36}$$

$$S_{Bi}(l) \approx \frac{\Delta H_{f(Bi)}}{T_f} \tag{2-37}$$

The approximation occurs in using these relations at temperature T rather than at T_f. When the two expressions for $G_{Bi}(l)$ are equated, the result is

$$-RT \ln N_{Bi} = \Delta H_{f(Bi)} - \frac{T\,\Delta H_{f(Bi)}}{T_f} \tag{2-38}$$

Therefore,

$$T = \frac{\Delta H_{f(Bi)}}{\left(\frac{\Delta H_{f(Bi)}}{T_f} - R \ln N_{Bi}\right)} \tag{2-39}$$

The quantity $\Delta T = T_f - T$ is more precise if it is calculated directly from the equation

$$\Delta T = T_f - T = T_f - \frac{\Delta H_{f(Bi)}}{\left(\frac{\Delta H_{f(Bi)}}{T_f} - R \ln N_{Bi}\right)} = \frac{-RT_f^2 \ln N_{Bi}}{\Delta H_{f(Bi)} - RT_f \ln N_{Bi}} \tag{2-40}$$

For the liquid alloy of Fig. 2-10 with $N_{Bi} = 0.9$, this equation becomes

$$\Delta T = \frac{(-2)(545)^2(-0.105)}{2.60 \times 10^3 - 2 \times 545(-0.105)} = 23° \tag{2-41}$$

Data of this type plot to form two liquidus curves (Fig. 2-10). The calculated eutectic temperature and composition are at the point of intersection of the two curves (Prob. 2-15).

2-8. Thermodynamics of Solutions

To work with phase diagrams containing solid solutions, we must be able to describe the properties of each of the components in a given solid-solution phase. For example, how big is an atom of magnesium when it is in a solid solution with cadmium? If it were always the size that it is in pure magnesium (and if cadmium behaved similarly), the molar volume would plot as the dashed line in Fig. 2-11 (Prob. 2-16). Obviously the experimental data in this case require that one or both of the metals in solution be smaller than in the pure state, but the determination of these values requires an ingenious procedure. Let the molar volume V of the solution at composition N_{Mg} be composed of two parts; thus

$$V = N_{Mg}\bar{V}_{Mg} + N_{Cd}\bar{V}_{Cd} \tag{2-42}$$

$\bar{V}_i$ is known as the *partial* molal volume of component i and is determined by the intercept of the tangent line shown in Fig. 2-11. The two parts of Eq.

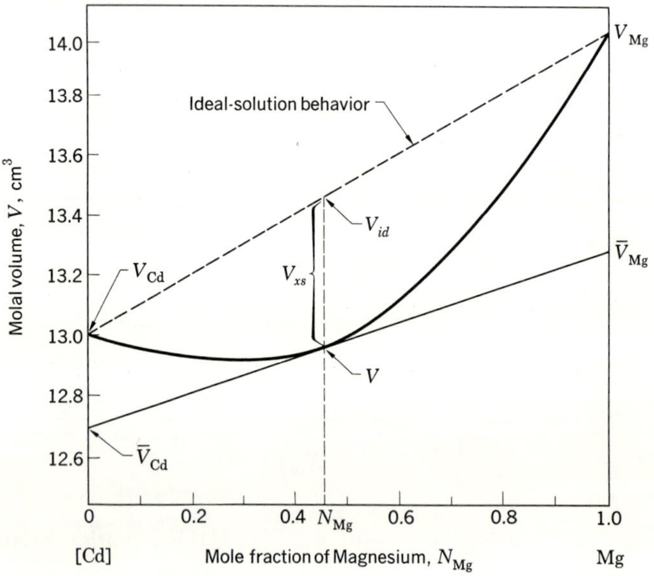

FIGURE 2-11 The molal volumes in the magnesium-cadmium solid solutions at room temperature.

(2-42) sum to give the molar volume V at the composition N_{Mg} (in the manner illustrated in Prob. 2-16). $\bar{V}_{Mg}$ is the increase in volume of a solution of concentration N_{Mg} when 1 mole of magnesium is added. In this "thought experiment" the amount of solution must be large enough to prevent the addition of magnesium from appreciably changing the concentration. For some purposes the *deviation* from ideal behavior is of primary importance and is expressed in terms of a (positive or negative) excess quantity V^{xs}, such that

$$V = V^{id} + V^{xs} \tag{2-43}$$

The ideal volume V^{id} involves the *molar* quantities V_i; thus

$$V^{id} = N_{Mg}V_{Mg} + N_{Cd}V_{Cd} \tag{2-44}$$

Volume was chosen as an example of properties of solutions because it can be visualized easily, but other properties are described by the same method. The molar enthalpy H is especially interesting because in an ideal-solution system this value remains unchanged when the two components are mixed to form a solution; that is, the *change*[1] in enthalpy on mixing, ΔH_m, is zero (see Prob. 2-17). Only chemically similar components approximate ideal-solution behavior. A more common type of solution behavior is the *regular solution* for which

$$\Delta H_m = RT\alpha N_A N_B \tag{2-45}$$

where α is a constant, but most metallic systems cannot be described by simple equations of this kind.

The entropy change on mixing is usually dominated by the configurational part [see Eq. (2-29)], arising from the greatly increased value of w in Eq. (2-30) when components A and B mix together and form a random solid solution. The quantitative expression is (Prob. 2-18)

$$\Delta S_m^{id} = -R(N_A \ln N_A + N_B \ln N_B) \tag{2-46}$$

This value of the entropy of mixing applies to both ideal and regular solutions. In general,

$$\Delta S_m = \Delta S_m^{id} + \Delta S^{xs} \tag{2-47}$$

The definition of free energy [Eq. (2-27)] leads to the relation

$$\Delta G_m = \Delta H_m - T\,\Delta S_m \tag{2-48}$$

This equation can be developed for solutions of various types by substitution of appropriate expressions for ΔH_m and ΔS_m.

2-9. Calculation of an Ideal Solid-solution Diagram

With this background in the thermodynamics of solutions, we can now proceed to the interpretation of equilibrium between solution-type phases. If in a binary system a solid solution α (having composition N_A^α) is in equilibrium with a liquid solution l, then an infinitesimal amount of the reaction

$$\alpha\ (N_A^\alpha) \rightarrow l\ (N_A^l) \tag{2-49}$$

[1] ΔH_m is used in the literature rather than the equivalent notation H^{xs}.

must be accompanied by zero change in free energy. Therefore,

$$dG = dG^l - dG^\alpha = dN_A^l\bar{G}_A^l + dN_B^l\bar{G}_B^l + dN_A^\alpha\bar{G}_A^\alpha + dN_B^\alpha\bar{G}_B^\alpha = 0 \qquad (2\text{-}50)$$

where dN_A^l, etc., are infinitesimal changes in the compositions of the liquid and alpha phases. Following established custom, the symbol $\bar{G}_i$ is replaced by μ_i, for which the name *chemical potential* is employed. Since $dN_A^l = -dN_A^\alpha$ and $dN_B^l = -dN_B^\alpha = 1$, Eq. (2-50) will be equal to zero if (Prob. 2-19)

$$\mu_A^l = \mu_A^\alpha \qquad \text{and} \qquad \mu_B^l = \mu_B^\alpha \qquad (2\text{-}51)$$

That is, two phases (at constant temperature and pressure) are in equilibrium when each component has the same chemical potential in both phases. The following calculation of a solid-solution diagram illustrates the use of this important relation.

The NiO–MgO diagram (Fig. 2-3) can be calculated fairly accurately using only the T_f and ΔH_f values for the two components and assuming ideal behavior of the liquid and solid solutions. The free-energy curves in this case (Fig. 2-12) are analogous to those for a one-component system (Fig. 2-9) but are plotted as a function of composition (at constant temperature). The curve for the solid-solution phase α can begin at pure *solid* MgO with the reference value $G^\alpha = 0$, whereas that for the liquid phase can begin at pure *liquid* NiO with $G^l = 0$. Since $\Delta H_m = 0$, ΔG^α for the solid solution containing N_B mole fraction of NiO is

$$\Delta G^\alpha = RT(N_A \ln N_A + N_B \ln N_B) + \left[T\frac{\Delta H_{f(\text{NiO})}}{T_{f(\text{NiO})}} - \Delta H_{f(\text{NiO})}\right]N_B \qquad (2\text{-}52)$$

The first term is the ideal entropy of mixing of solid NiO and MgO, and the second term is an evaluation of $T\,\Delta S_f$ and ΔH_f for the conversion of N_B mole fraction of NiO from liquid to solid. Calculation of Eq. (2-52) at 0.1 increments of N_B permits plotting the curve of G^α shown in Fig. 2-12. An analogous calculation determines G^l (Prob. 2-20).

To see how the criterion of equilibrium [Eq. (2-51)] can be read from these two curves, recall the tangent method (Fig. 2-11) for determining partial molal quantities. A common tangent to both free-energy curves (Fig. 2-12) determines the two chemical potentials μ_A (or μ_{MgO}) and μ_B (or μ_{NiO}), which apply equally to the α solid solution (composition $N_B = 0.48$) and to the liquid solution ($N_B = 0.72$); that is, $\mu_A^\alpha = \mu_A^l$, and $\mu_B^\alpha = \mu_B^l$. The experimental phase diagram has its solidus and liquidus points (at 2600°K) within a few percent of these values.

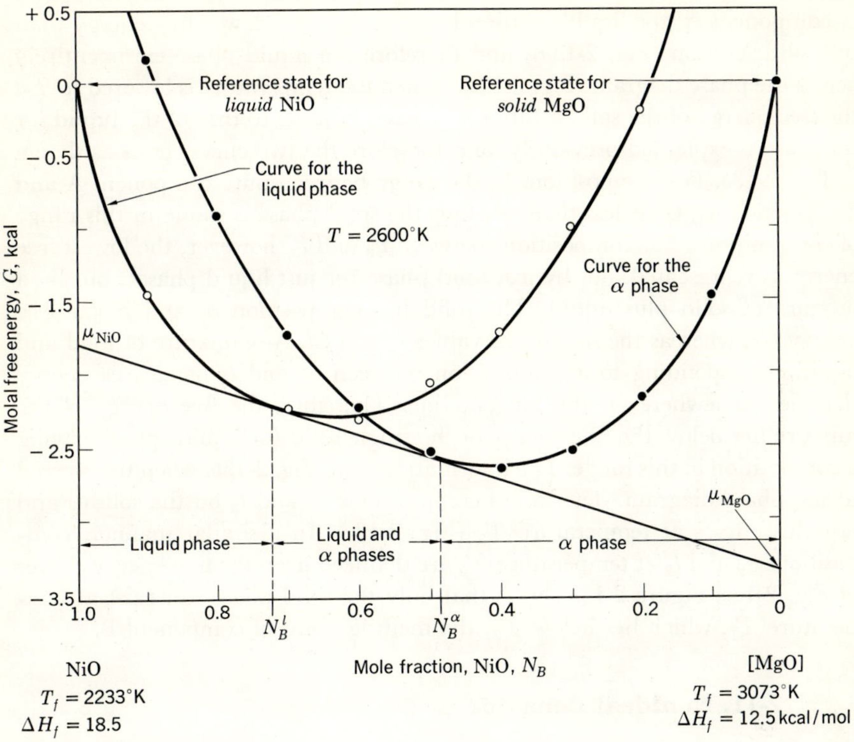

FIGURE 2-12 Calculation of the free-energy curves for the alpha solid solution and the liquid solution in the NiO–MgO system at 2600°K. (*Adapted from J. Mackowiak; data of O. Kubaschewski.*)

DERIVATION OF PHASE DIAGRAMS FROM FREE-ENERGY CURVES

The previous section showed some simple applications of free-energy to phase equilibria. This topic will now be developed further by considering how phase diagrams are determined by the free-energy curves of the individual phases. We begin with ideal-solution phases, but also consider various types of nonideal behavior.

2-10. Ideal-solution Behavior

In a system (such as NiO–MgO in Sec. 2-9) that has ideal liquid and solid solutions, the free-energy curves for the liquid and solid phases must change their relative positions with change in temperature in the manner shown in Fig. 2-13. At temperature T_4, which lies above T_A, the melting point of component A, the liquid solution has everywhere a lower free energy than the solid solution (Fig. 2-13*a*), and therefore the liquid phase exists entirely across the phase diagram (Fig. 2-13*e*). When the temperature is lowered to T_3, the free energy of the solid solution decreases relative to that of the liquid for the reasons explained previously, and therefore the two curves cross as shown in Fig. 2-13*b*. For compositions in the range between pure component A and composition α_3, G^α is less than G^l; thus the solid phase is stable in this range of compositions. For compositions between α_3 and l_3, however, the lowest free energy is represented, not by just solid phase (or just liquid phase), but by a mixture of solid plus liquid. The solid has composition α_3 and molar free energy G^α, whereas the liquid has values l_3 and G^l. Any mixture of solid and liquid corresponding to a composition between α_3 and l_3 has a free energy that lies somewhere on the tangent line. Therefore, the free energy of the mixture lies below the free energy of the single solid (or liquid) phase having a composition in this range. The information from Fig. 2-13*b*, when transferred to the phase diagram, determines compositions α_3 and l_3 on the solidus and liquidus curves at temperature T_3 (Fig. 2-13*e*). In a similar manner, compositions α_2 and l_2 at temperature T_2 are deduced from the free-energy curves of Fig. 2-13*c*. Figure 2-13*d* shows that only the solid phase can exist at temperature T_1, which lies below T_B, the melting point of component B.

2-11. Nonideal Behavior

SOURCES OF FREE-ENERGY DATA. Since the phases in most systems behave nonideally, their free-energy curves must be obtained from experimental data in one of two ways. First, if the system in question has received extensive thermodynamic study so that all of the thermodynamic properties of each of the phases (liquid, α solid solution, β solid solution, etc.) are known as functions of composition and temperature, then quantitative free-energy curves can, of course, be plotted. Limited data of this kind are available[1] for the simpler two-component systems, but almost none are available for three or more components. The second possibility is to estimate free-energy curves from theories employing only limited data. Although the ideal-solution theory is applicable to only a few systems, the regular-solution theory[2] characterized by the expression

$$G_m^{xs} = a_{AB} V \varphi_A \varphi_B \tag{2-53}$$

[1] R. Hultgren, et al., "Selected Values of Thermodynamic Properties of Metals and Alloys," John Wiley & Sons, Inc., New York, 1963.

[2] G. N. Lewis, et al., "Thermodynamics" 2d ed., McGraw-Hill Book Company, New York, 1961.

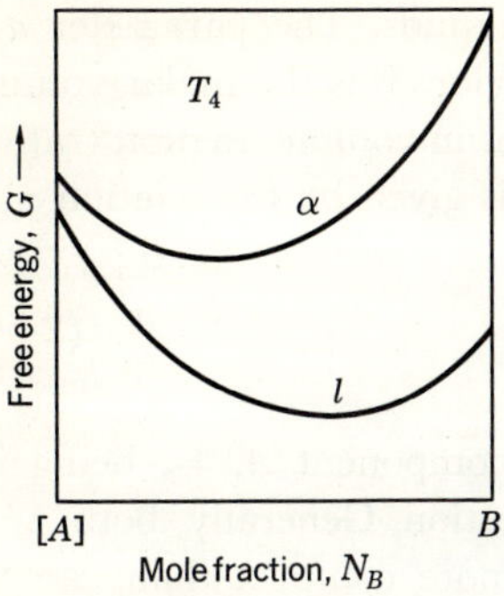

(*a*) Free-energy curves at temperature T_4 where only the liquid is stable.

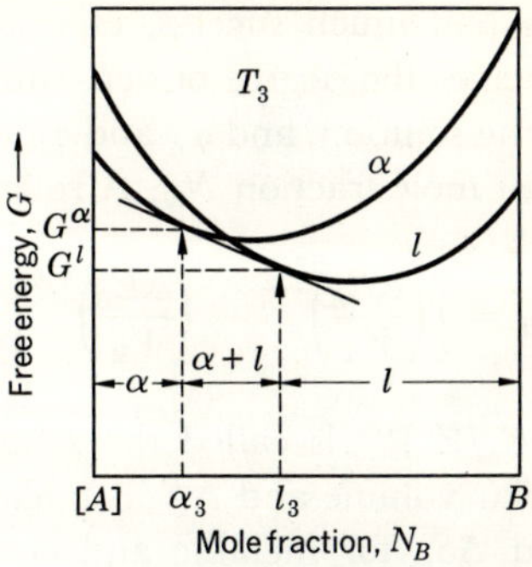

(*b*) Free-energy curves at temperature T_3 where the solid solution is stable over a small range of composition.

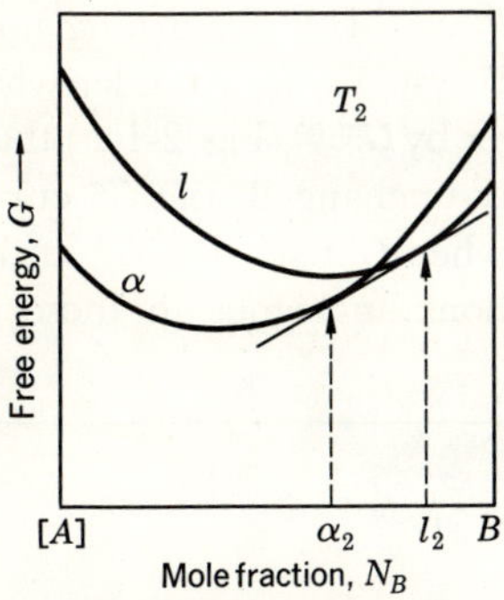

(*c*) Free-energy curves at temperatures T_2 where the solid solution is stable over a wide range of composition.

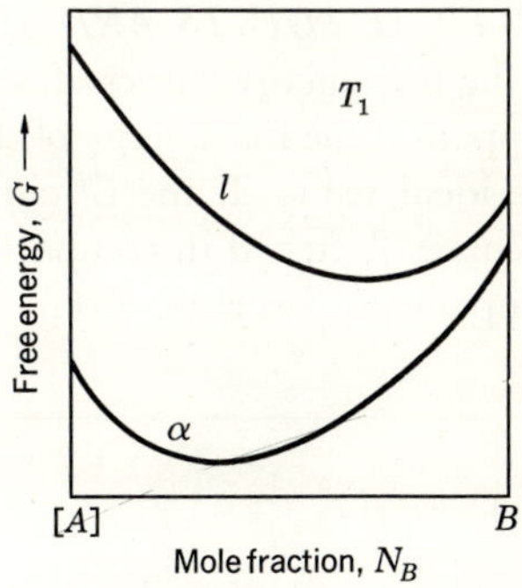

(*d*) Free-energy curves at temperature T_1 where only the solid solution is stable.

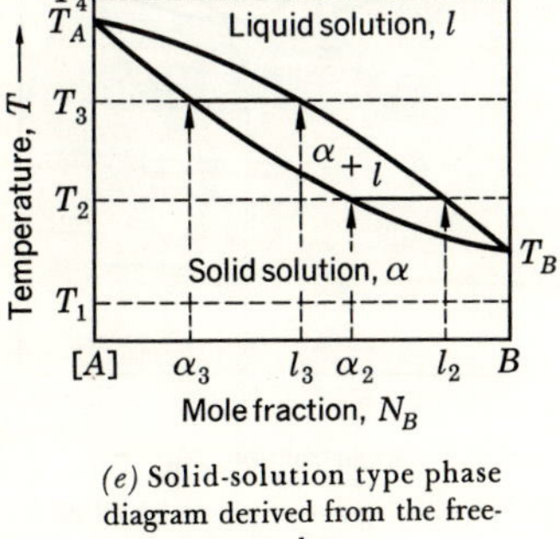

(*e*) Solid-solution type phase diagram derived from the free-energy curves above.

FIGURE 2-13 Schematic free-energy curves at four temperatures, (*a*) to (*d*), for ideal solutions, leading to a solid-solution-type phase diagram (*e*) (*Adapted from G. V. Raynor.*)

has had much success, especially for nonpolar liquids. The parameter a_{AB} specifies the degree of departure from ideal behavior, V is the molar volume of the solution, and φ_A and φ_B are the compositions in volume fraction (rather than mole fraction N_i). A rough estimate of a_{AB} is given by the relation

$$a_{AB} = \left[\left(\frac{\Delta E_A}{V_A}\right)^{1/2} - \left(\frac{\Delta E_B}{V_B}\right)^{1/2}\right]^2 \quad \text{cal/cc} \tag{2-54}$$

$(\Delta E_A/V_A)^{1/2}$ is called the *solubility parameter* for component A, V_A being its molar volume and ΔE its molar energy of sublimation. Generally, both ΔH_m and ΔS_m for metallic and ceramic phases are more complex than can be accounted for by the regular-solution theory. Furthermore, when intermediate phases appear in a phase diagram, these phases are clearly beyond the scope of theories of this type.

CRITICAL POINTS AND SPINODAL DECOMPOSITION. Returning now to the free-energy curves of real (rather than ideal) systems, we consider what happens if the free energy of the solid phase is larger by $G^{\alpha(xs)}$ (Fig. 2-14*a*) than the ideal value. If the G^l curve remains relatively unchanged, the G^α curve becomes flattened in comparison. Consequently, when G^l rises past G^α as the temperature decrease from T_4 to T_1, constructions analogous to those of

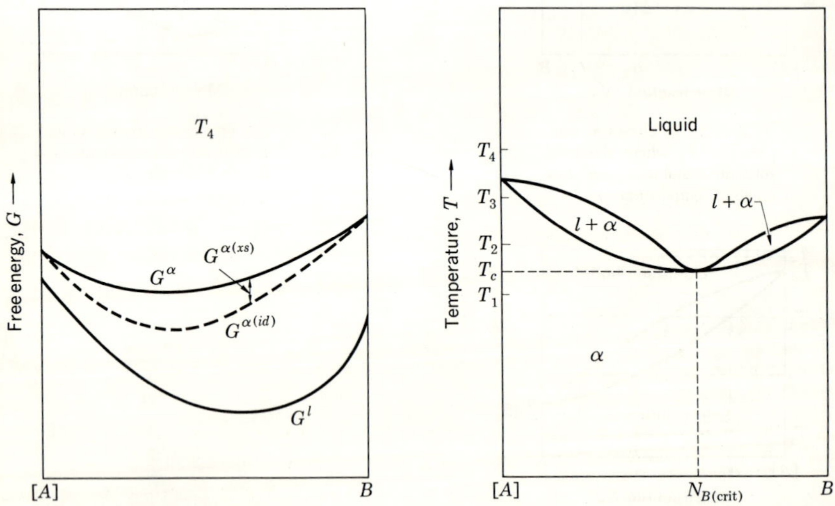

(*a*) Free energy curves when $G^{\alpha(xs)}$ is large and positive.

(*b*) Phase diagram of the solid-solution type with a minimum point.

FIGURE 2-14 Example at temperature T_4 of free-energy curves that can be used to deduce a phase diagram containing a critical point.

Fig. 2-13 produce the diagram of Fig. 2-14*b*. At the critical temperature T_c, the two free-energy curves are tangent at the composition $N_{B(crit)}$. Problem 2-21 considers the conditions under which the critical point will be a maximum rather than a minimum.

The solid phase in Fig. 2-14*b* may undergo a further reaction, called *phase separation*, on further cooling. Usually nonideality of the enthalpy H, rather than of the entropy term TS [see Eq. (2-27)], is decisive in determining the character of the free-energy curve. Furthermore, as T becomes smaller, H becomes even more dominant. For example, at T_4 (Fig. 2-14*a*) the effect of $G^{\alpha(xs)}$ is only to flatten somewhat the parabolic free-energy curve. At the lower temperature T_1, a portion of the curve is almost horizontal (Fig. 2-15*a*); at

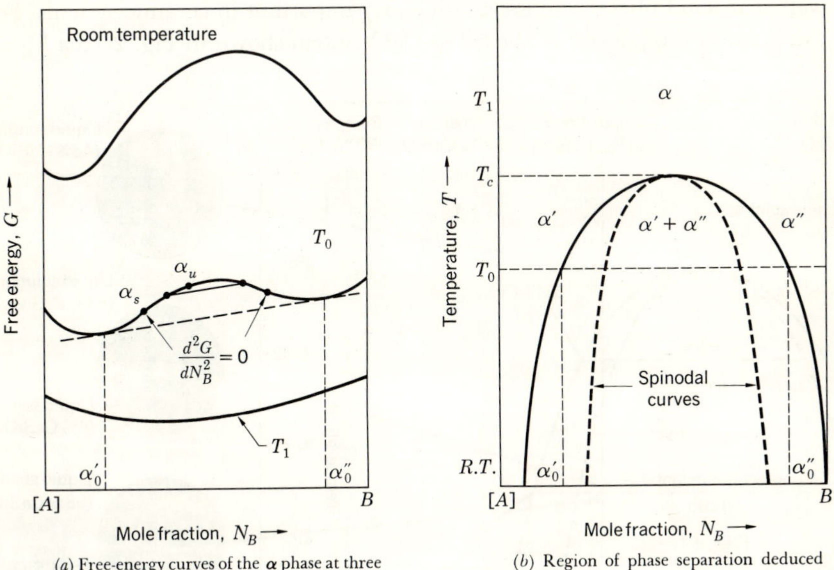

(*a*) Free-energy curves of the α phase at three different temperatures.

(*b*) Region of phase separation deduced from the free-energy curves.

FIGURE 2-15 Explanation in terms of the free-energy curve of the occurrence of a miscibility gap in a solid (or liquid) solution.

T_0 a hump has appeared in the center, and at room temperature the hump is a prominent feature of the curve. Since a tangent can be drawn once the hump has appeared, a mixture of phases α' and α'' is stable in the central region of the diagram at temperatures below a critical temperature T_c.

The inflection points, where $d^2G/dN_B^2 = 0$, are termed *spinodes* and determine an important characteristic of compositions lying between them. Imagine that a solution α_u, whose composition lies near the middle of this range, undergoes a local fluctuation of concentration in any small volume to produce two subvolumes, one slightly enriched in component B and the other slightly impoverished in B. You can deduce, from the position of the line joining these two compositions, that the free energy of the volume has decreased; therefore, such a partial decomposition can occur and progress. In contrast to the *unstable* composition α_u in the spinodal region, a composition α_s lying between a spinode and α_0' (or α_0'') is *metastable.* The reasoning employed above shows that in this case a slight fluctuation produces two subvolumes having a higher total free energy than the solution α_s. Consequently, the solution may remain indefinitely in the metastable condition if the atoms are unable to make the major rearrangement necessary to produce the equilibrium phases α_0' and α_0'', a mixture of which has a lower free energy and is therefore stable.

THE MONOTECTIC REACTION. Miscibility gaps occur in liquid phases as well as in solid phases and are particularly important in ceramic systems. For example, in the portion of the SiO_2–CaO system shown in Fig. 2-16 a liquid

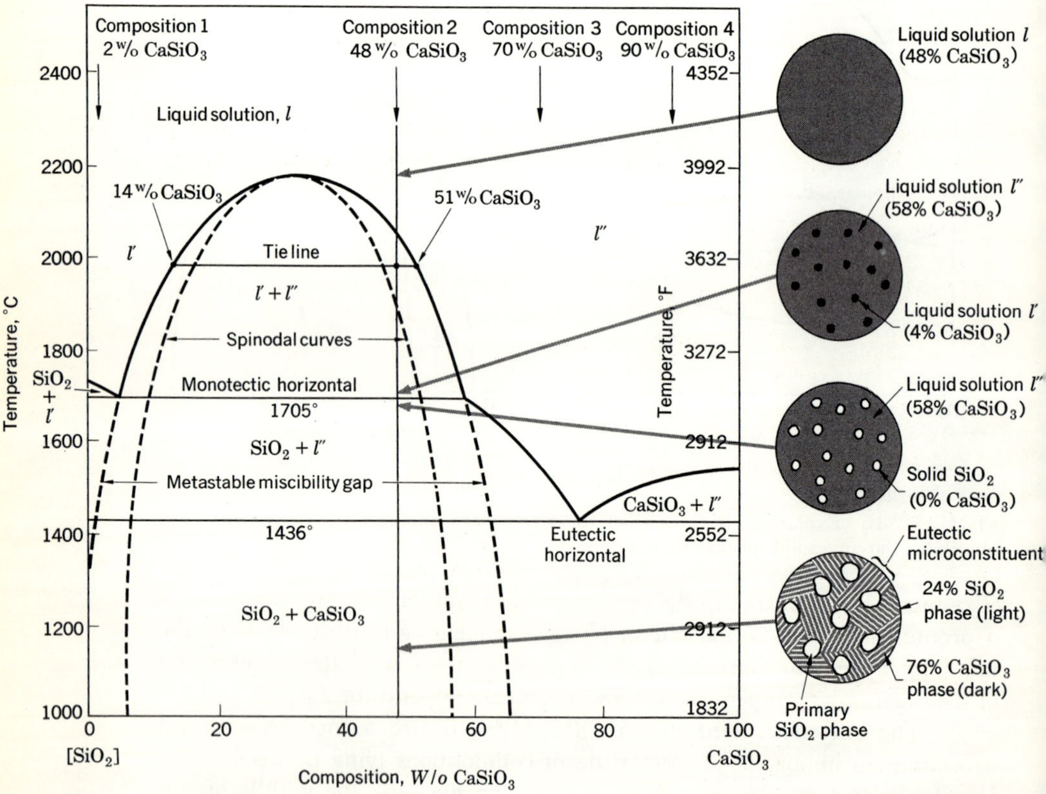

FIGURE 2-16 The phase diagram of the partial system SiO_2–$CaSiO_3$, illustrating a miscibility gap and spinodal curves. The diagram has been simplified by omitting the polymorphic changes of the solid phases.

miscibility gap lies above 1705°C, the temperature of the monotectic horizontal [a type of three-phase equilibrium (Fig. 2-4*b*)]. Ignoring for the moment the dashed lines in the diagram, let us consider the changes that occur during cooling of various compositions in this system under equilibrium conditions. The four compositions indicated on the diagram are examples of the four qualitatively different types of behavior. Compositions 3 and 4 do not pass through the monotectic horizontal, but instead experience only eutectic-type solidification, discussed previously in connection with the Ag–Cu system.

Composition 2 is especially significant since its first equilibrium reaction on cooling is the separation of the initial liquid l into two different liquid phases l' and l''. For example, at 2000°C the phase analysis is

Point:	48 w/o $CaSiO_3$ composition at 2000°C	
Phase:	*Liquid* l'	*Liquid* l''
Composition:	14 w/o $CaSiO_3$	51 w/o $CaSiO_3$ (2-55)
Amount:	$\frac{51 - 48}{51 - 14} = 0.08$ weight fraction	0.92

l' normally exists as small drops distributed throughout the matrix liquid l''. On further cooling the compositions of l' and l'' continue to follow the two boundaries of the miscibility gap. At 1706°C, just above the monotectic reaction, the phase analysis is

Point:	48 w/o $CaSiO_3$ composition at 1706°C	
Phase:	*Liquid* l'	*Liquid* l''
Composition:	4 w/o $CaSiO_3$	58 w/o $CaSiO_3$ (2-56)
Amount:	$\frac{58 - 48}{58 - 4} = 0.185$ weight fraction	0.815

The monotectic reaction is of the eutectic type and can be represented by the equation

$$l' \rightarrow \alpha + l'' \tag{2-57}$$

That is, the drops of l' decompose into a solid phase (almost pure SiO_2) and l''. At a temperature just below the monotectic reaction the phase analysis is

Point:	48 w/o $CaSiO_3$ composition at 1704°C	
Phase:	SiO_2	*Liquid* l''
Composition:	~0 w/o $CaSiO_3$	58 w/o $CaSiO_3$ (2-58)
Amount:	$\frac{58 - 48}{58 - 0} = 0.17$ weight fraction	0.83

As the temperature is lowered almost to the eutectic horizontal, the amount of SiO_2 dispersed in the liquid phase increases in amount because of the change in l'', determined by the sloping liquidus line. The phase analysis at 1437°C is

Point:	48 w/o $CaSiO_3$ composition at 1437°C	
Phase:	*SiO_2*	*Liquid l''*
Composition:	~0 w/o $CaSiO_3$	76 w/o $CaSiO_3$ (2-59)
Amount:	$\frac{76 - 48}{76 - 0} = 0.37$ weight fraction	0.63

At the eutectic temperature the liquid phase decomposes into a eutectic mixture of SiO_2 and $CaSiO_3$ in a manner similar to that described previously for Ag–Cu alloys. Composition 1 (which is the subject of Prob. 2-22), although not involving the miscibility gap, does undergo the monotectic and eutectic reactions.

Ceramic materials are often in a nonequilibrium state. To explain metastable structures analogous to those occurring in glass,[1] additional (dashed) lines have been added to the diagram of Fig. 2-16. First, the miscibility gap has been extended by lines of metastable equilibrium to temperatures below the monotectic horizontal. This construction reflects the fact that the process of forming a solid phase may proceed so slowly that the liquid (or glass) phases may continue to exist at low temperatures on ordinary cooling. Second, spinodal curves have been added, including their extensions to low temperatures. Although the exact position of the region of spinodal decomposition has not been determined, the reasoning of Fig. 2-15 assures that this region will exist. Since the stable phases often form slowly in ceramic systems, spinodal decomposition offers a means of establishing small regions having two different compositions, a condition having much practical importance (see Chap. 7).

THE PERITECTIC REACTION. The second type of three-phase equilibrium illustrated in Fig. 2-4*b* is the peritectic,

$$l + \beta \rightarrow \alpha \tag{2-60}$$

The rather strange solidification behavior produced by this reaction is shown schematically for two silver-platinum alloys in Fig. 2-17. Alloy 1, on cooling from the liquid region, begins to form the β solid solution. At 1186°C this alloy consists of liquid phase plus 0.35 atomic fraction of β phase. During cooling through the peritectic horizontal at 1185°C, the alloy undergoes the reaction

[1] Commercial glasses (Table 5-2) are related to compositions on the left side of Fig. 2-16, but contain in addition such oxides as Na_2O, B_2O_3, and Al_2O_3.

of Eq. (2-60) and becomes entirely α phase. The sequence of sketches in Fig. 2-17 illustrating this solidification behavior includes one showing an intermediate stage; the initially formed α, necessarily at the original interface

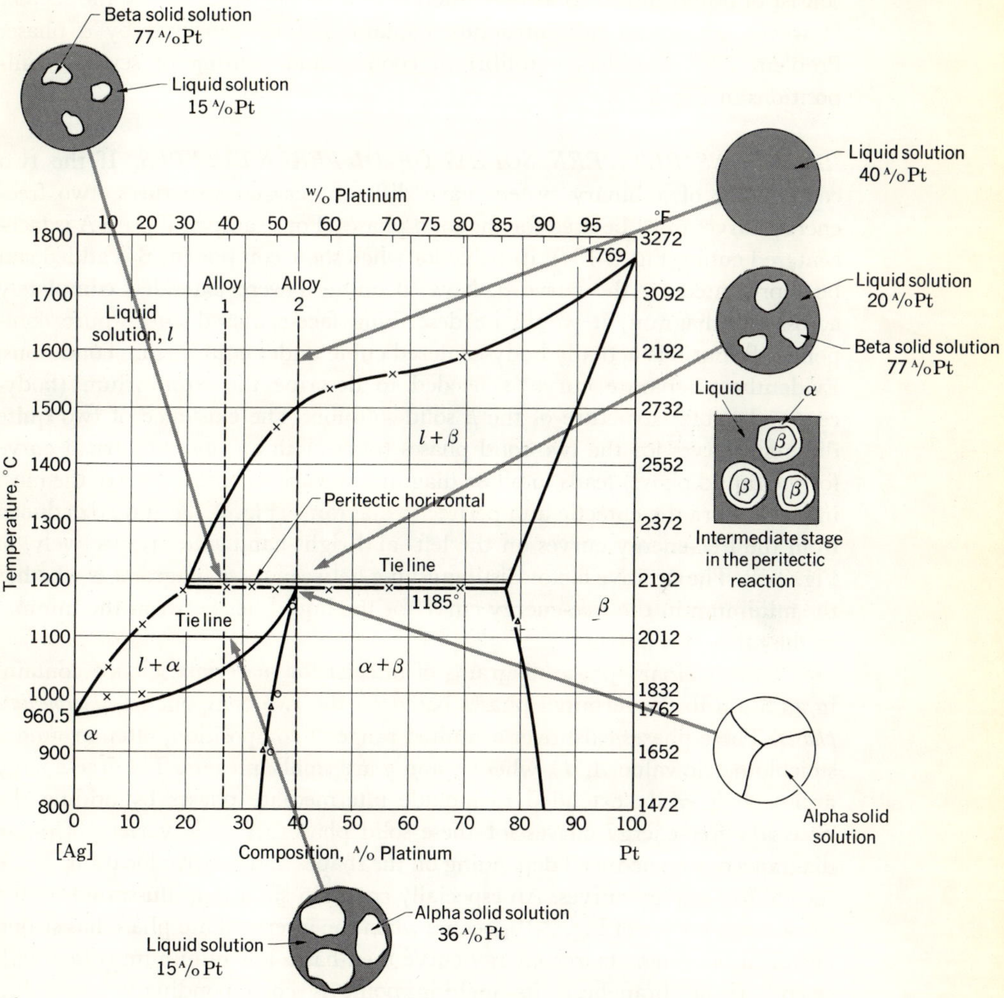

FIGURE 2-17 The silver-platinum phase diagram, used to illustrate the character of peritectic reactions. The lower part of the diagram has been simplified by omitting several solid-state reactions.

between the liquid and β phases, encases the β grains. Further reaction between the liquid and β phases is now difficult; thus equilibrium is seldom achieved in peritectic reactions (Fig. 2-21*c*).

Usually the alloy being considered is not of peritectic composition (alloy 1) but is some general composition passing through the peritectic horizontal. Alloy 2, for example, forms 0.12 atomic fraction of β phase on cooling to 1186°C. At 1184°C, after the peritectic reaction, this alloy should consist of liquid plus 0.35 atomic fraction of α phase, as shown by the sketch. However, an actual microstructure contains β phase encased by α phase. Problem 2-23 considers equilibrium cooling and heating of several compositions in Fig. 2-17.

REACTIONS BETWEEN SOLIDS OF DIFFERENT TYPES. If the two components of a binary system have different crystal structures, two free-energy curves must be used for the solid phases. For example, if pure A is face-centered cubic (Fig. 2-18*a*), its behavior when some component B is added can be represented by the curve α. Now, if curve α were extended completely across the diagram, it would be describing face-centered cubic pure component B, but B is actually body-centered cubic under equilibrium conditions. Evidently a separate curve is needed to describe the equilibrium (body-centered cubic) structure of the β solid solution. The existence of two quite different curves for the two solid phases (often with an unsymmetrical curve for the liquid phase) leads to phase diagrams of various kinds. Two of the most important are the eutectic and peritectic diagrams (Fig. 2-18*c* and *f*) deduced from the free-energy curves on the left- and right-hand sides, respectively, of Fig. 2-18. The decisive factor distinguishing between these diagrams is whether the minimum in the free-energy curve for the liquid lies between the minima in the other two curves.

Most binary phase diagrams of interest for materials science contain, in addition to the terminal phases based on the two components, *intermediate phases.* These phases exist over a limited range of composition, often around a stoichiometric value A_xB_y, where x and y are small integers. The free-energy approach is easily extended to include intermediate phases by adding the necessary free-energy curves for these solid phases. A wide variety of phase diagrams can be deduced depending on the shapes and relative locations of the several free-energy curves. An especially common situation, illustrated by the free-energy curves of Fig. 2-19, occurs when the intermediate phase has strong chemical bonding. Its free-energy curve then has a low minimum (often with steeply rising branches), its melting point is correspondingly high, and frequently two eutectic reactions are produced in the diagram (Fig. 2-19*d*). The phase diagram of Fig. 2-19*e* is important for technical applications such as high-temperature alloys; its deduction is the subject of Prob. 2-24. The NiAl phase has the CsCl structure, whereas Ni_3Al has the ordered structure of Cu_3Au (Fig. 2-23).

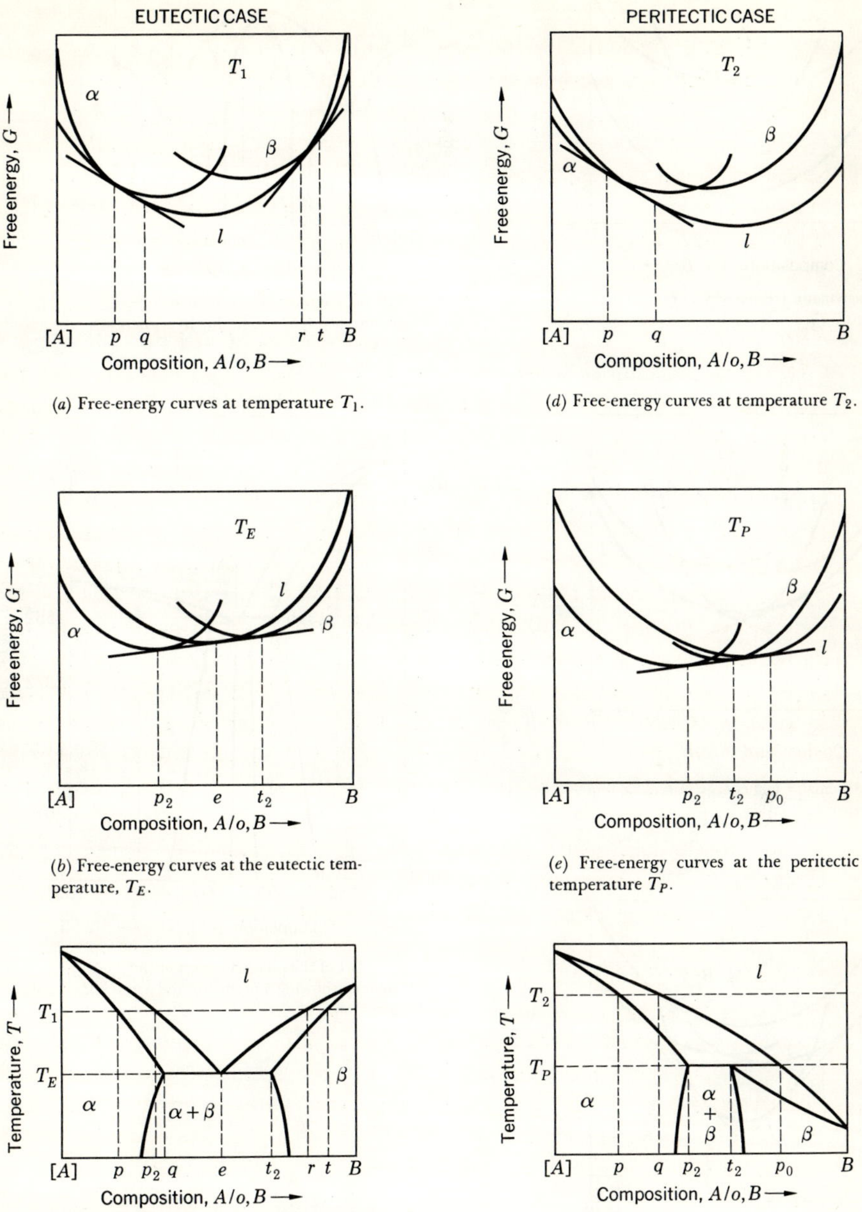

(*a*) Free-energy curves at temperature T_1.

(*b*) Free-energy curves at the eutectic temperature, T_E.

(*c*) Eutectic diagram deduced from the free-energy curves.

(*d*) Free-energy curves at temperature T_2.

(*e*) Free-energy curves at the peritectic temperature T_P.

(*f*) Peritectic diagram deduced from the free-energy curves.

FIGURE 2-18 Two important configurations of free-energy curves when the component metals have different crystal structures; for example, *A* being face-centered cubic and *B* body-centered cubic. (*After G. V. Raynor.*)

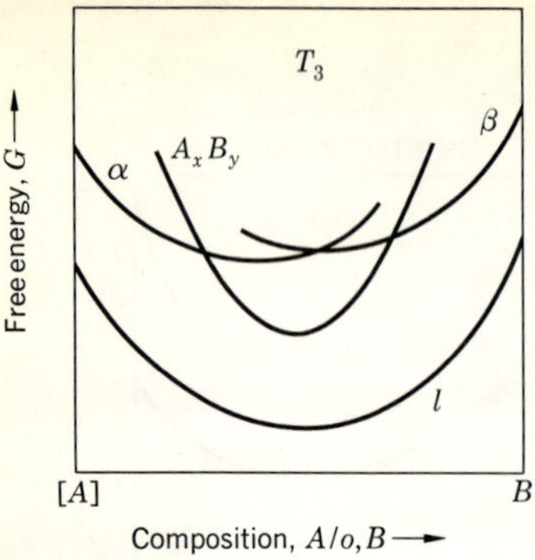

(*a*) Schematic free-energy curves at temperature T_3.

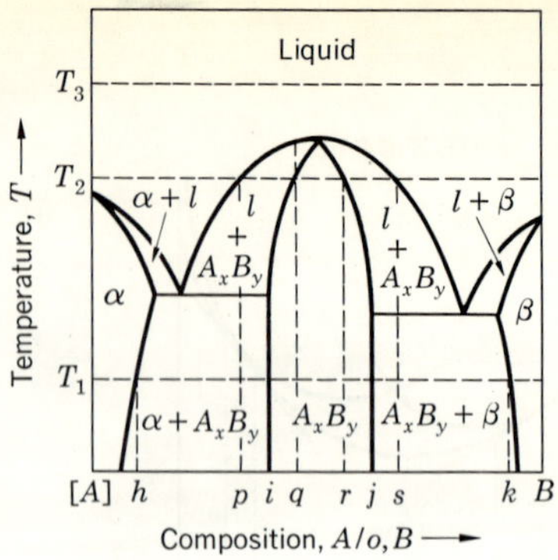

(*d*) Phase diagram containing two eutectic reactions deduced from the free energy curves.

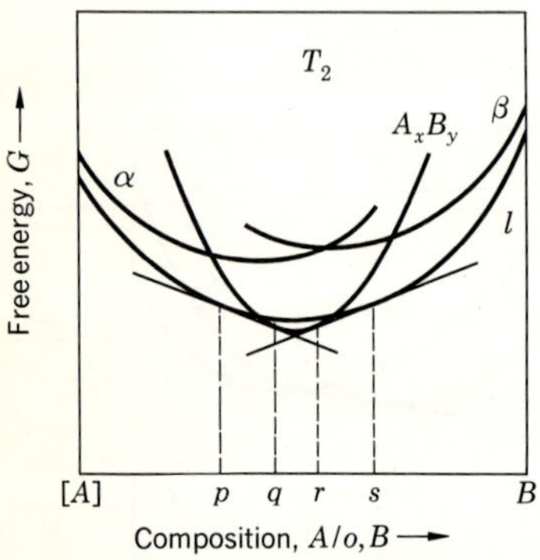

(*b*) Schematic free-energy curves at temperature T_2.

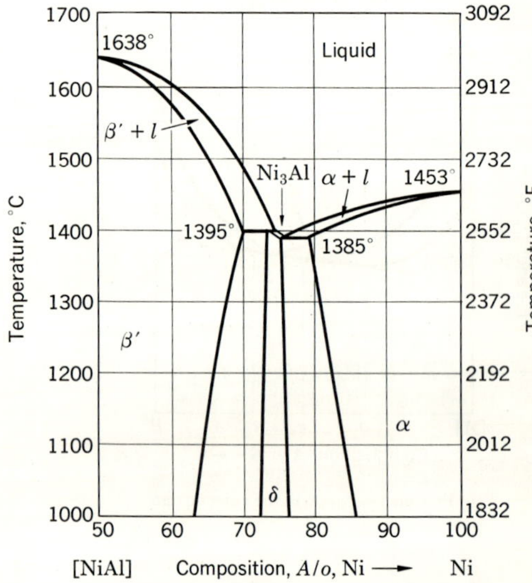

(*e*) Part of the phase diagram of the Al-Ni system, showing a peritectic and a eutectic reaction.

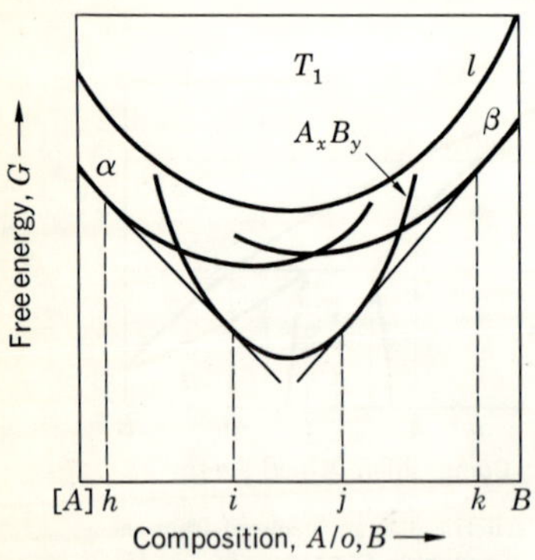

(*c*) Schematic free-energy curves at temperature T_1.

FIGURE 2-19 Examples of phase diagrams that can be deduced from free-energy curves when an intermediate phase occurs in the binary system. (*After G. V. Raynor.*)

NONEQUILIBRIUM ASPECTS OF PHASE DIAGRAMS; SOLID-STATE REACTIONS

As far as thermodynamic principles (such as the phase rule) are concerned, a solid phase and its reactions are no different than the corresponding liquid case. For example, the γ solid solution can undergo the *eutectoid* (meaning *like a eutectic*) reaction

$$\gamma \rightarrow \alpha + \beta \tag{2-61}$$

and decompose into two other solid phases. However, atoms in solid phases are less mobile by factors of about 10^4 than those in liquid phases. Nonequilibrium states are therefore common in solid phases. A few examples are given here, and others are considered in later chapters.

2-12. Nonequilibrium Solidification

An important type of nonequilibrium behavior is the segregation of solute that occurs during the solidification of a liquid to form a solid solution. As an amount A_0 of liquid gradually solidifies, as shown in the lower part of Fig. 2-20*c*, the compositions C^s and C^l of the solid and liquid phases at the solidifying interface are essentially those given by the phase diagram. Consider the system of Fig. 2-20*a* in which the fraction g of the total amount A_0 of the system has solidified during cooling to temperature T. An additional fraction dg solidifies on cooling to $T - \Delta T$. The amount of solute, $C^lA_0(1 - g)$, that was in the liquid at T becomes divided at $T - \Delta T$ between the remaining liquid (of changed composition) and the fractional amount $A_0\,dg$ of solid; thus

$$C^lA_0(1 - g) = (C^l + dC^l)A_0(1 - g - dg) + C^sA_0\,dg \tag{2-62}$$

This equation can be put in the form (the term $dC^l\,dg$ is negligible)

$$\frac{dC^l}{C^s - C^l} = -\frac{dg}{1 - g} \tag{2-63}$$

This equation permits a calculation of C^l (and therefore of C^s) as a function of the fraction g of solidification.

The manipulation of Eq. (2-63) is especially simple when the liquidus and solidus curves can be approximated by the straight lines shown in Fig. 2-20*b*. A constant ratio $C^s/C^l = k$ then exists, and thus C^l can be eliminated to give the result

$$\frac{dC^s}{C^s} = -\frac{(k - 1)\,dg}{1 - g} \tag{2-64}$$

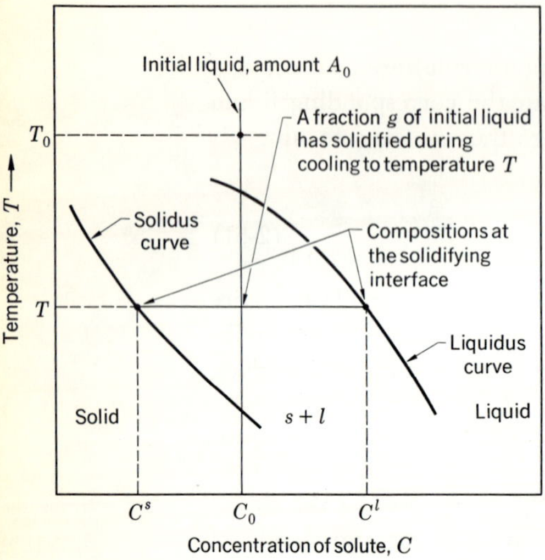

(*a*) General analysis for arbitrary liquidus and solidus curves.

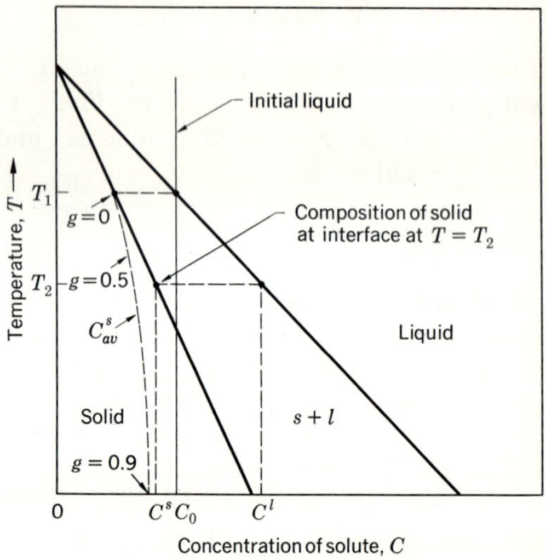

(*b*) Analysis for special case when $C^s/C^l = k$. The curve for C^s_{av} corresponds to the process of nonequilibrium solidification with $k = 0.5$, part (*c*).

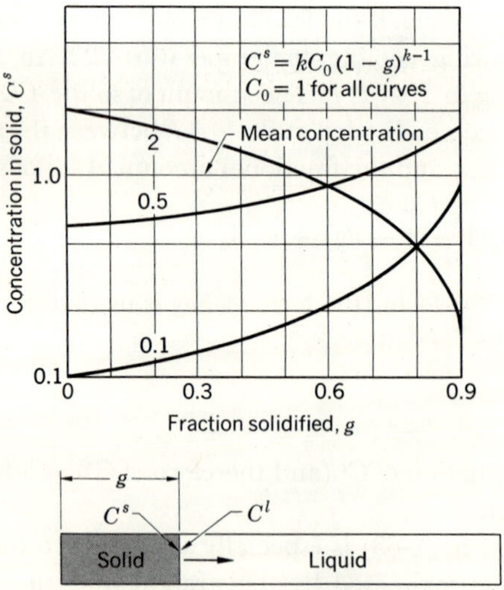

(*c*) Calculated segregation curves along a solidified bar for k values of 0.1, 0.5, and 2. (After Pfann).

FIGURE 2-20 **Theoretical analyses of nonequilibrium solidification.**

Integration of this equation between $g = 0$ (where $C^s = kC_o$) and a general value g gives

$$C^s = kC_o(1 - g)^{k-1} \tag{2-65}$$

Representative curves of C^s along a solidified bar are given in Fig. 2-20*c* for three k values. During *equilibrium* solidification the solid phase is assumed to be uniform and to follow the solidus line. The average solid composition $C^s_{(av)}$ during *nonequilibrium* solidification lies below the solidus line, as shown by the curve for $k = 0.5$ in Fig. 2-20*b*. The final stage of solidification, beyond about $g = 0.9$, cannot be described by Eq. (2-65), which loses physical significance.

Few solid materials are in a state of perfect equilibrium, and many have constitutions significantly affected by nonequilibrium solidification. The segregation that occurs during solid-solution freezing produces "coring" in the final solid phase. The Cu–Ni system, like the NiO–MgO system (Fig. 2-3), is a complete series of solid solutions. The solid solutions that form in the early stages of solidification are rich in nickel (Fig. 2-21*a*) and are easily distinguished from the remainder of the microstructure. Because nonequilibrium

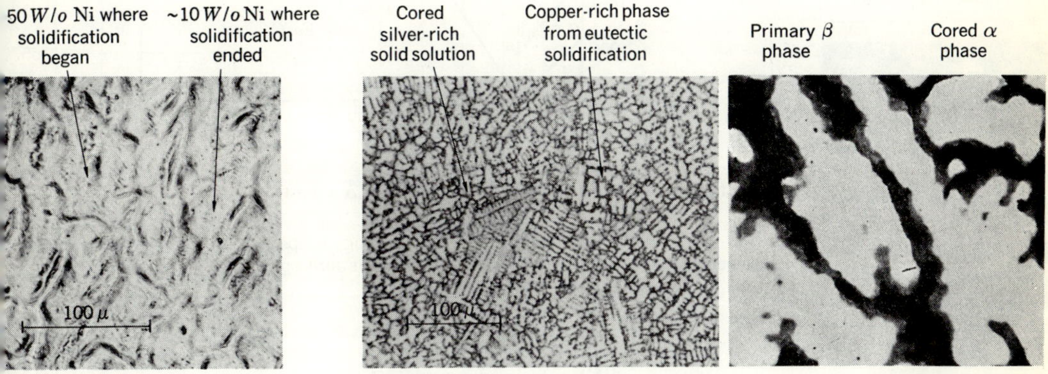

(*a*) Coring in the solid-solution alloy, 30 *W/o* Ni in the Cu-Ni system. Photographed using phase contrast; Sec. 3-2. (Courtesy Anaconda American Brass Company)

(*b*) Nonequilibrium eutectic microconstituent in sterling silver, 7.5 *W/o* Cu, in the Ag-Cu system, Fig. 2-5 (*b*). (Courtesy Handy and Harman Company)

(*c*) Encasement of primary β phase during the peritectic reaction in the 27 *A/o* Pt alloy in the Ag-Pt system, Fig. 2-19. (Courtesy F. N. Rhines, University of Florida)

FIGURE 2-21 Photomicrographs illustrating nonequilibrium solidification in several types of phase diagram.

solidification lowers the temperature at which the liquid phase continues to exist, anomalous three-phase reactions may occur. An example is the occurrence of a nonequilibrium eutectic reaction (Fig. 2-21*b*) in a composition that does not pass through the eutectic horizontal. In peritectic systems an additional cause of nonequilibrium behavior is the encasement of the primary crystals by the second phase (α), which forms during the peritectic reaction [Eq. (2-60)]. Even extremely slow cooling seldom prevents nonequilibrium retention of the primary crystals.

2-13. Metastability in Solid-state Reactions

For reactions such as the eutectoid reaction, Eq. (2-61), involving only solid phases, metastability is a primary phenomenon. The technically important iron-carbon diagram (Fig. 2-22) contains a peritectic and a eutectic

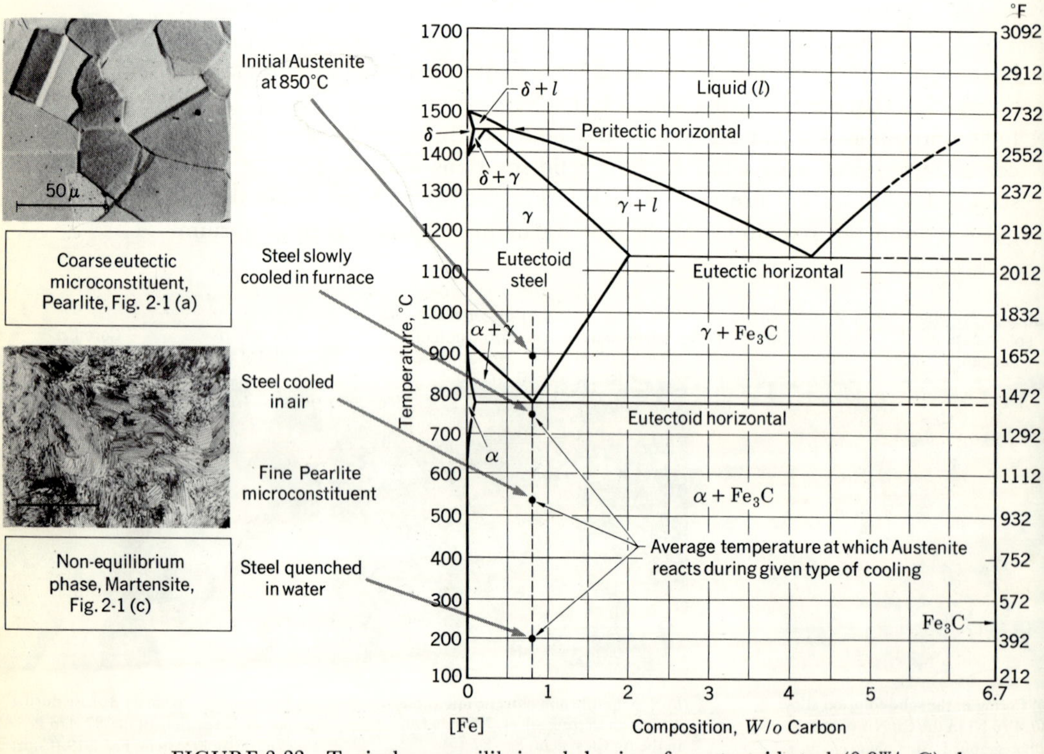

FIGURE 2-22 Typical nonequilibrium behavior of a eutectoid steel ($0.8^{W}/_{0}$ C) shown in relation to the iron-carbon phase diagram. (*Photomicrographs courtesy Edgar C. Bain Laboratory for Fundamental Research, U.S. Steel Corporation.*)

reaction at higher temperatures, but the eutectoid reaction is of major importance for the heat treatment of steels. For slowly cooled steels, phase analyses involving the eutectoid reaction can be made in the same manner as for a eutectic system (Sec. 2-3). The eutectoid microconstituent in steels, a lamellar mixture of the ferrite (α) and cementite (Fe_3C) phases, is called *pearlite*. The crystal structure of Fe_3C is shown in Fig. 5-22. When steels are cooled more rapidly, metastable austenite (γ) continues to exist hundreds of degrees below the eutectoid horizontal but finally changes ("decomposes") into a finer aggregate of α plus Fe_3C or into the nonequilibrium phase, martensite. Other materials react differently in the solid state, but pronounced metastability is a common characteristic.

For most purposes a solid solution, such as the α phase in the Cu–Au system (Fig. 2-23), can be considered to be a random mixture of the two kinds

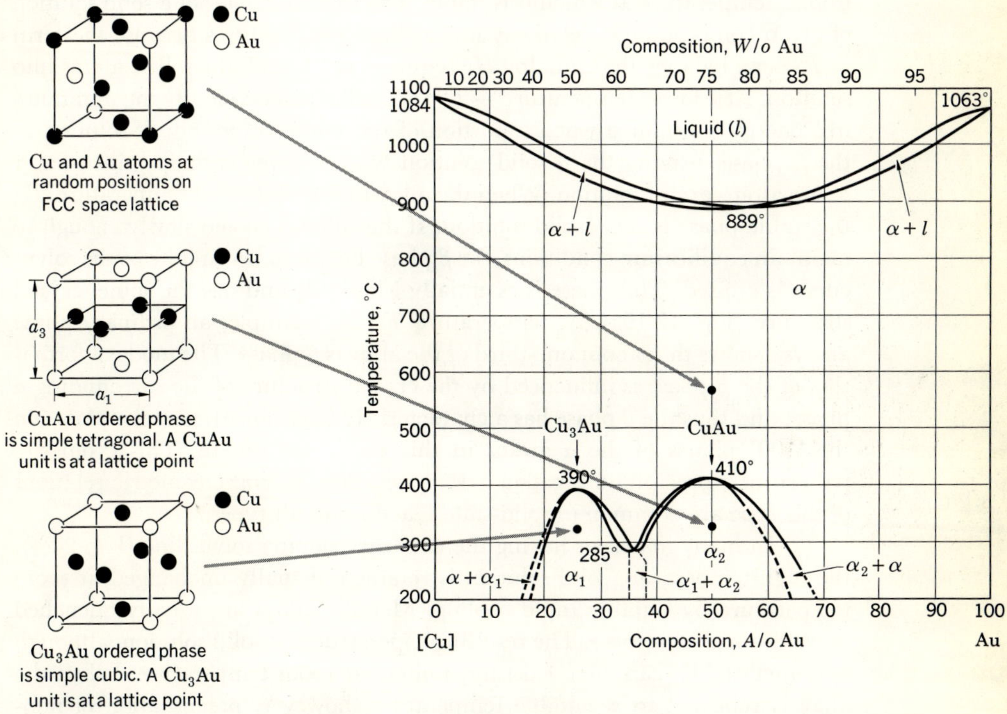

FIGURE 2-23 The copper-gold phase diagram (simplified) exhibiting the formation of ordered phases. In the range of composition between about 20 and 60 A/o Au, the α solid solution transforms into ordered phases at lower temperatures.

of atoms. Actually, neighboring atoms usually arrange themselves preferentially to some degree, and several aspects of such "ordering" are treated in Chaps. 5 and 7. In this chapter we consider only the one type of ordered phase that behaves normally in phase diagrams, for example, CuAu or Cu_3Au in the Cu–Au system. The 50 A/o Au alloy is an ordinary solid solution α in the range of temperatures below the solidus curve but above 410°C. Below 410°C the ordered phase CuAu is more stable, and thus the reaction

$$\alpha \text{ (solid-solution phase)} \rightarrow \text{CuAu (ordered phase)} \tag{2-66}$$

takes place on cooling. For a composition slightly different than the stoichiometric one a solid-solution–type reaction occurs involving a two-phase region (Prob. 2-27).

A solid-state reaction of great technical importance is the formation of a second substance within a solid solution. Thermodynamics explains why such *precipitation* of the second phase tends to occur when a solid solution α is cooled from a temperature at which it is stable. The free energy G^α of a solid solution of given composition is relatively low at high temperatures because the term $-TS^\alpha$ can balance the enthalpy H^α required to "force" the solute atoms into solution. At a lower temperature $-TS^\alpha$ is smaller; consequently the minimum free energy is attained when a portion of the solute comes out of solution as the β phase, leaving the α solid solution with a lower enthalpy since fewer solute atoms are in solution. When the alloy in Fig. 2-24 is at temperature T_2, the stable phase is the α solid solution. If this alloy is cooled slowly enough to maintain equilibrium conditions, the β phase begins to precipitate as the solvus curve is crossed. This phase is essentially $MgFe_2O_4$ and has the spinel crystal structure (Fig. 12-19*a*). At temperature T_1, for example, an ordinary phase analysis shows that about one-third of the alloy is β phase. The mode of formation of the β phase is influenced by the crystal structure of the surrounding α phase, and thus the β phase has a characteristic *habit* (form), namely, plates on the {100} planes of the α grains in this particular system. This geometric pattern of a precipitate is called a *Widmannstätten pattern*. Geometric relations of this type are common in solid-state reactions of all types.

In many materials having the necessary sloping solvus line (Fig. 2-24), the high-temperature solid solution is retained virtually unchanged at room temperature by suitably rapid cooling. Metallic alloys are usually quenched in water for this purpose. The resulting supersaturated solid solution, although only metastable, can often exist indefinitely at room temperature. When the alloy is reheated to a suitable temperature, however, precipitation of nonequilibrium phases occurs in the manner described in Sec. 7-9 and gives the alloy useful properties.

MULTICOMPONENT SYSTEMS

Many useful materials, such as steels and ceramic porcelains, contain not just two important components, but three or more. The principles of binary equilibria apply also for these multicomponent systems, but difficulties of two types are experienced in the practical use of these principles. First, the difficulty of determining data for a complete *n*-component system increases greatly as *n* increases, and therefore relatively few three-component diagrams and almost no higher-component diagrams exist. Second, visualization of multi-component equilibria involves multidimensional space and becomes increasingly difficult as *n* increases beyond 3.

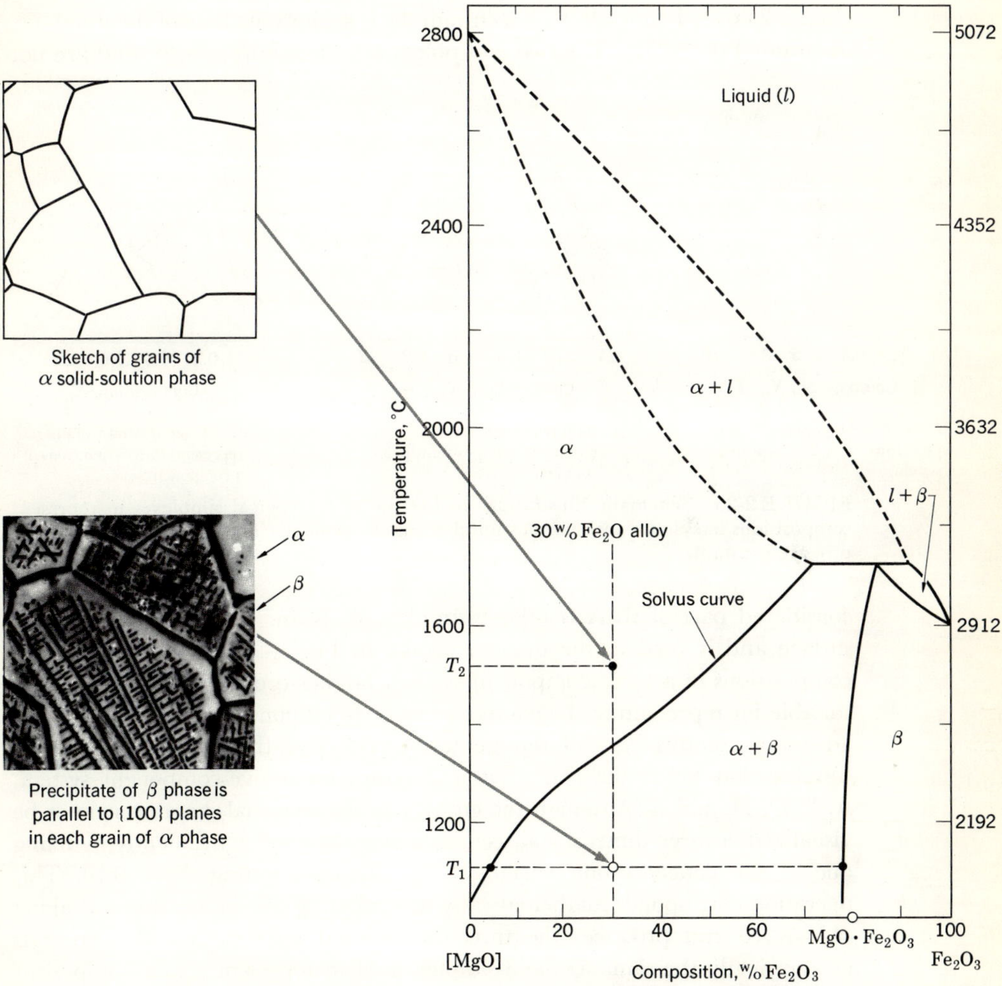

FIGURE 2-24 Illustration of precipitation of a second phase during slow cooling of a solid-solution phase. The schematic microstructures show the geometric (Widmannstätten) pattern of the precipitated phase within the initial solid solution. (*Diagram adapted and simplified from Levin, E. M., et al., "Phase Diagrams for Ceramists," American Ceramic Society, Columbus, Ohio,* 1964. *Micrograph courtesy H. L. McCollister, Owens-Illinois Technical Center; by permission from American Ceramic Society Bulletin* **44** (1965), 915.)

2-14. Methods of Representation and Analysis

To simplify the representation of multicomponent systems, both pressure and temperature are usually considered fixed, and thus only the concentration variables need be considered. For this purpose the mathematical concept of *simplex* is especially convenient, as shown by the sketches of Fig. 2-25. A zero-simplex, s^0, is a single point and can represent a given component (Fig. 2-25*a*). A 1-simplex, s^1, is a straight line, such as the base of the usual *TX* diagram (Fig. 2-25*b*). The two end points are zero-simplexes, s^0, and are not

(*a*) A point, s^0, represents a 1-component system.

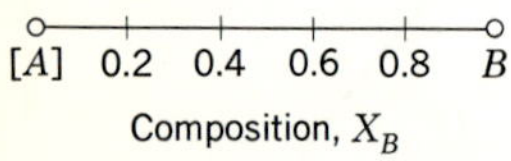

(*b*) A line, s^1, represents all compositions in a 2-component system.

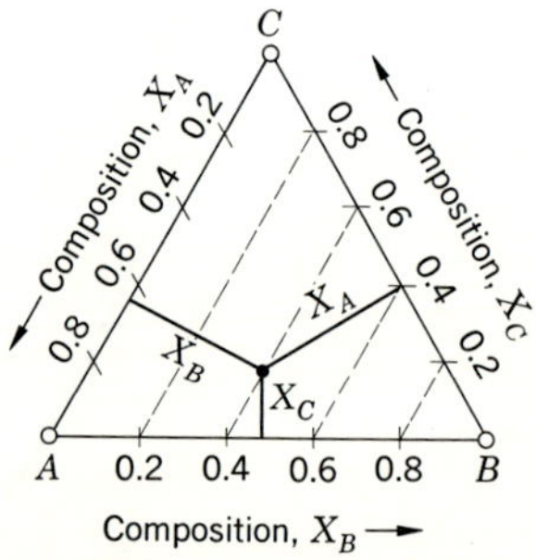

(*c*) A triangular surface, s^2, represents the compositions in a 3-component system.

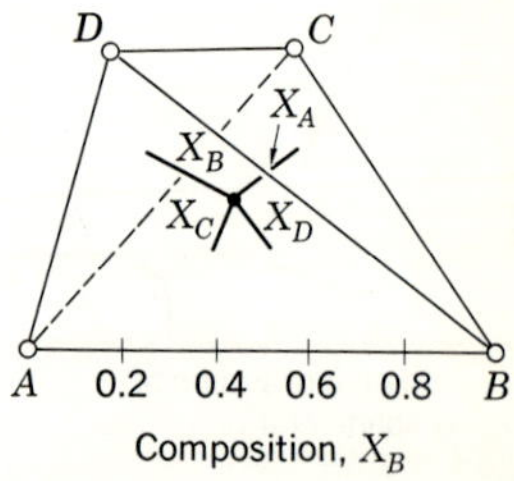

(*d*) A tetrahedral volume, s^3, represents the compositions in a 4-component system.

FIGURE 2-25 Schematic illustrations of the use of topological simplexes to represent compositions in systems with various numbers of components. The pressure and temperature are constant.

considered part of s^1. A regular 2-simplex, s^2, is an (equilateral) triangular surface and is used in the manner shown in Fig. 2-25*c* to plot all possible compositions in a three-component system. Higher-order simplexes, s^{n-1}, are suitable for representing the compositions in *n*-component systems. To see the advantage of this type of representation, consider the four-dimensional s^4 simplex that forms the composition diagram for a five-component system, A, B, C, D, and E. Although the entire four-dimensional diagram cannot be visualized, a three-dimensional section can be obtained conveniently by fixing one of the concentration variables, for example, setting $X_E = 0.10$. This operation corresponds mathematically to sectioning the hypervolume s^4 along $X_E = 0.10$ and produces the three-dimensional hypersurface s^3. This s^3 is geometrically the same as the 3-simplex used to represent a four-component system (Fig. 2-25*d*). Similarly, sectioning the three-dimensional s^3 diagram of a four-component system produces a convenient two-dimensional s^2 diagram.

A representative TX_AX_B (three-component) diagram is shown in Fig. 2-26*a*. Actual phase analyses of a given alloy (such as the 0.35A–0.5B composition shown) are made at constant temperature using the s^2 isothermal

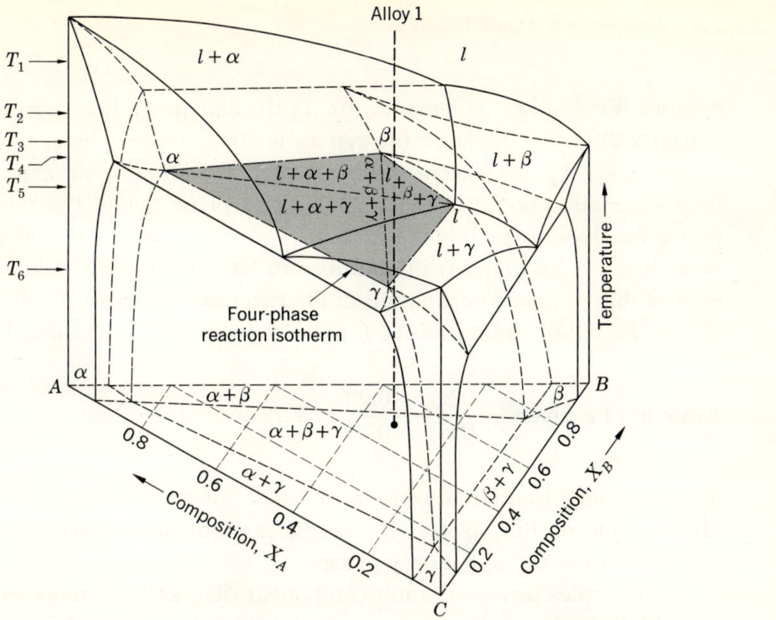

(*a*) The "space model" of the diagram. The location of alloy 1 (0.35 *A*-0.5 *B*) is shown.

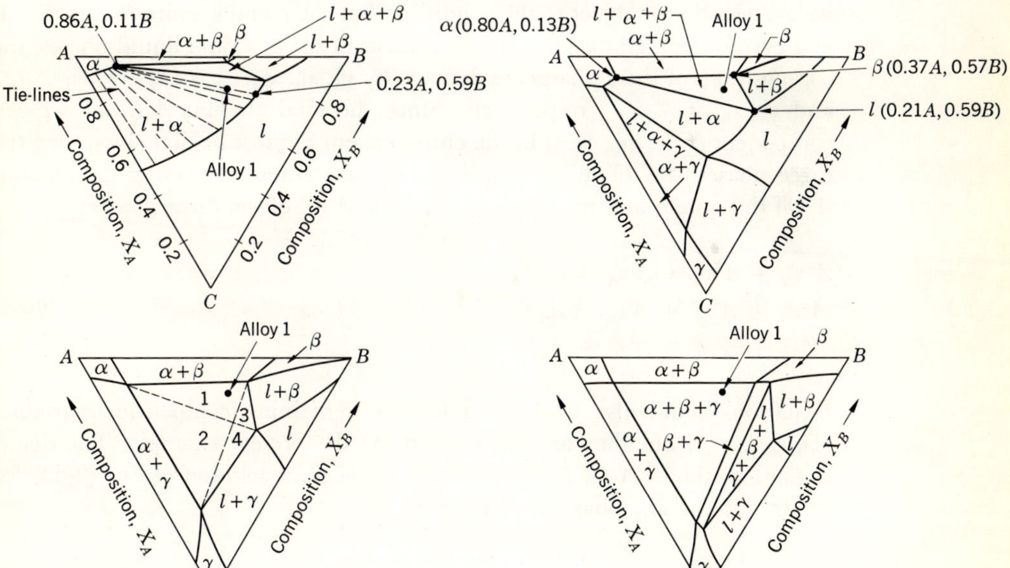

(*b*) Isothermal sections (s^2 diagrams) through the space figure at four temperatures of interest.

FIGURE 2-26 A three-dimensional T-X_A-X_B phase diagram and four s^2 sections through it. (*Adapted from F. N. Rhines, "Phase Diagrams in Metallurgy," McGraw-Hill Book Company, New York, 1956.*)

sections (Fig. 2-26*b*). At temperature T_1 the alloy point lies in the liquid phase (Prob. 2-29), and therefore the system is 100% liquid phase. On cooling to temperature T_2, the system passes through the liquidus surface and crystallizes a certain amount of the α phase. The phase analysis in this two-phase region involves the tie line that passes through the composition of alloy 1. The two phases (α and liquid) in equilibrium have their compositions at the two ends of this tie line. The amounts of the two phases are calculated by the lever rule of Eq. (2-8), using the X_A (or the X_B) composition values. In this case,

$$\text{Amount of } \alpha \text{ phase} = \frac{0.35 - 0.23}{0.86 - 0.23} = 0.19 \text{ weight fraction} \tag{2-67}$$

The remaining isothermal sections of Fig. 2-26*b* show that on further cooling, alloy 1 (1) forms the β phase, (2) undergoes a four-phase reaction, and (3) ends as a mixture of the α, β, and γ phases.

Complex aspects of multicomponent diagrams are more easily treated by an algebraic (rather than a geometric) method. In this procedure an *n*-component alloy is described by its *n*-composition variables X_A, X_B, X_C, . . . , which are subject to the condition that their sum is unity (or 100%). If at a certain temperature the system consists of the α, β, and liquid phases, the compositions of these phases are denoted by x_A^α, x_B^α, x_C^α, . . . , x_A^β, x_B^β, x_C^β, . . . , and x_A^l, x_B^l, x_C^l, . . . , respectively. Since the total amount $A_S X_C$ of a given component (C in this case) in the entire system S, must be divided among the three phases, the following equations can be written connecting the amount A_S of the system and the amounts A^l, A^α and X^β of the three phases:

$$\begin{aligned} A^\alpha x_A^\alpha + A^\beta x_A^\beta + A^l x_A^l &= A_S X_A \\ A^\alpha x_B^\alpha + A^\beta x_B^\beta + A^l x_B^l &= A_S X_B \\ A^\alpha x_C^\alpha + A^\beta x_C^\beta + A^l x_C^l &= A_S X_C \end{aligned} \tag{2-68}$$

Equations could also be written for the remaining composition variables ($D, E,$ . . .), but they are not needed for the present example. The determinant method (Prob. 2-30) permits a convenient solution of Eq. (2-68) for the fraction of any phase; for example,

$$\frac{A^\alpha}{A_S} = \frac{D_\alpha}{D} \tag{2-69}$$

$$D = \begin{vmatrix} x_A^\alpha & x_A^\beta & x_A^l \\ x_B^\alpha & x_B^\beta & x_B^l \\ x_C^\alpha & x_C^\beta & x_C^l \end{vmatrix} \qquad D_\alpha = \begin{vmatrix} X_A & x_A^\beta & x_A^l \\ X_B & x_B^\beta & x_B^l \\ X_C & x_C^\beta & x_C^l \end{vmatrix}$$

This procedure, the generalized lever rule, is applicable for a system consisting of n phases or less. Use of a digital computer permits easy solution of Eq. (2-69). The three-phase state of alloy 1 at T_3 (Fig. 2-26*b*) is the subject of Prob. 2-30.

When an isobaric ($P = 1$ atm) n-component system consists of $n + 1$ phases, the phase rule [Eq. (2-2)] shows that there are no degrees of freedom. A constant-temperature reaction must occur among the n phases until one of them disappears. If the liquid phase is present and if the last phase is designated ν, the n possible reactions can be written as follows:

$$l \rightarrow \alpha + \beta + \cdots + \nu \qquad \text{(eutectic reaction)}$$
$$l + \alpha \rightarrow \beta + \cdots + \nu \qquad (2\text{-}70)$$
$$\cdots\cdots\cdots\cdots$$
$$l + \alpha + \beta + \cdots \rightarrow \nu \qquad \text{(peritectic reaction)}$$

By analogy with the behavior of a two-component system, the first of these reactions is the eutectic, and the last is the peritectic. The intermediate reactions are called *quasi-peritectic*. Alloy 1 of Fig. 2-26 at T_4 undergoes the quasi-peritectic invariant reaction

$$l + \alpha \rightarrow \beta + \gamma \qquad (2\text{-}71)$$

During this reaction all of the liquid phase and part of the α phase are used up, leaving $\alpha + \beta + \gamma$ as the phase state at temperature T_5. A phase analysis (Prob. 2-30) verifies that the amount of α phase decreased during the reaction.

2-15. Liquidus Projections

A valuable procedure for obtaining information on the solidification behavior of complex ternary systems employs a projection of the liquidus surface onto the composition triangle. Although such a projection, in the case of the simple diagram of Fig. 2-26*a* is merely three lines meeting at point L, quite complex projections characterize real systems, especially for ceramic materials (Fig. 2-27). The interpretation of such diagrams[1] can be illustrated by the following simple example. Consider composition 1, initially in the molten condition. (Point B is the binary eutectic involving the pure solid phases MgO and $MgO \cdot Al_2O_3$. A series of pseudobinary-eutectic compositions form the line BT that leads down to the ternary eutectic involving $2MgO \cdot SiO_2$ as the third phase.) When melt 1 reaches the liquidus surface on cooling, MgO

[1] E. M. Levin, et al., "Phase Diagrams for Ceramists," pp. 14–25, American Ceramic Society, Columbus, Ohio, 1964.

will be the first (primary) phase to crystallize. During primary crystallization the liquid composition will change from composition 1 to point P on the pseudobinary eutectic. The amount of primary MgO can be calculated by the usual

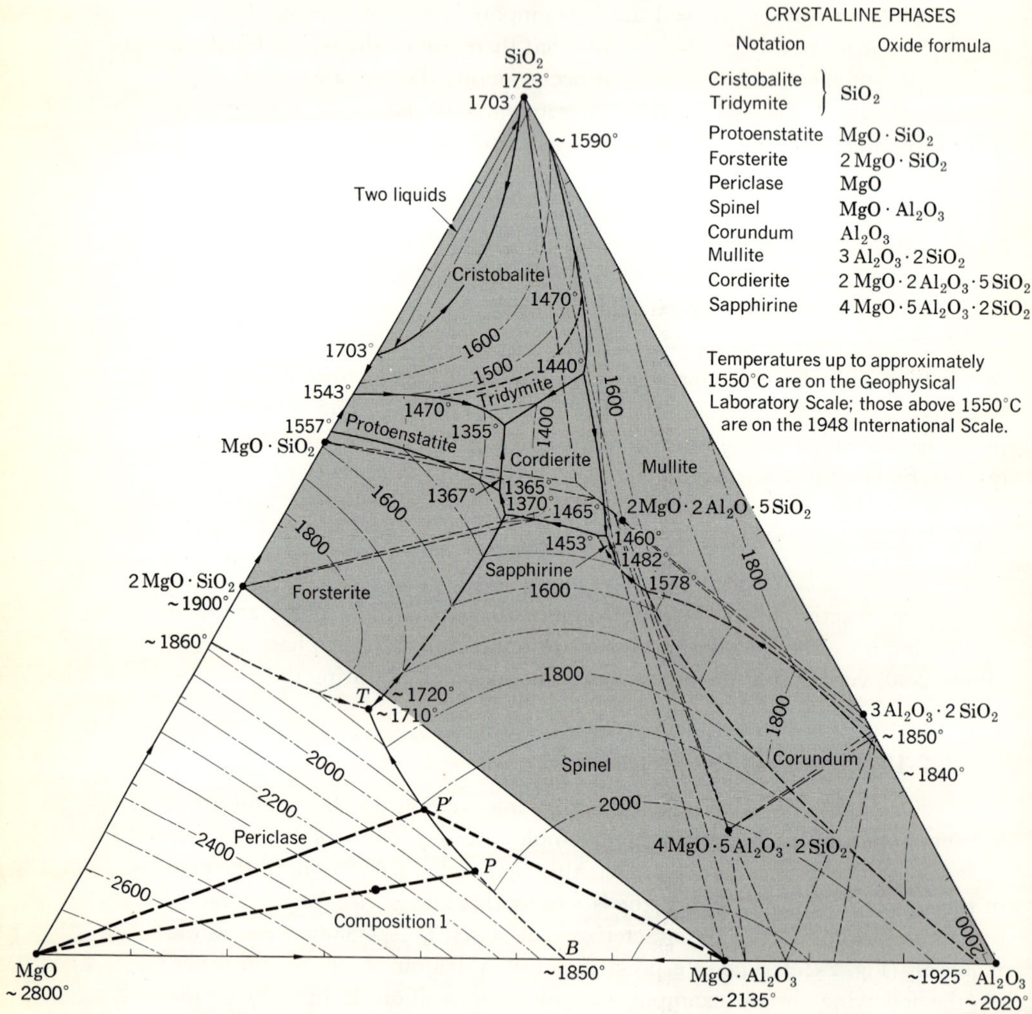

FIGURE 2-27 Projection onto the composition triangle of the liquidus surface for the system $MgO–Al_2O_3–SiO_2$. (*Diagram reproduced from "Phase Diagrams for Ceramists", Fig. 712, by E. M. Levin, et al., American Ceramic Society, Columbus, Ohio, 1964.*)

lever rule at any stage in this process of crystallization. The next stage of freezing is the (secondary) solidification of the pseudoeutectic constituent as the remaining liquid cools from P to T. A phase analysis can be made at any intermediate point such as P' (see Prob. 2-31). Final (tertiary) solidification occurs at constant temperature (1710°C), with the formation of the ternary eutectic microconstituent.

PROBLEMS

2-1. Three powders, iron, carbon (graphite), and Fe_3C, are mixed together in various proportions at room temperature and then compressed cold to produce specimens suitable for magnetic testing. Explain why such specimens represent a three-component system.

2-2. The triple point of chromium is $T_t = 1875°C$ and $P_t = 10^3$ N/m². (*a*) Sketch a qualitative one-component diagram. (*b*) What is the vapor pressure of molten chromium just above the melting point? (*c*) Copper is sometimes purified by melting in a vacuum furnace having a pressure of 10 N/m². Explain whether this treatment is suitable for chromium.

2-3. Consider a system consisting of 100 g of copper occupying half of the total volume of an evacuated, sealed quartz tube. (*a*) If the tube is slowly heated, how many phases exist shortly after the melting point is reached? (*b*) Explain how the system has automatically fixed (1) T_t and (2)P_t. (*c*) If the experimenter lowers the temperature to 10° below T_t, explain why the system changes to two phases (*not* to just the solid phase).

2-4. The two lines in Fig. 2-2*b* representing an equilibrium involving the vapor phase have slopes that differ by about 10^3 from the slope of the line involving only condensed phases. Show that Eq. (2-1) explains this significant difference in behavior.

2-5. Use Eq. (2-1) to estimate the change in the melting point of copper caused by increasing the pressure from 0 to 13 bar. The molar volume of liquid copper is 0.0080 liter, and that of solid copper is 0.0071 liter. The change in enthalpy (heat of fusion) is 3.120 kcal/mole. [NOTE: A useful conversion factor is 0.02422 kcal = 1 liter-atm.]

2-6. When a liquid system loses heat to its surroundings, the system may decrease in temperature and/or transform to the solid state with liberation of its heat of fusion, ΔH_f. (*a*) Consider 1 mole of pure liquid NiO, 50°C above its freezing point. Sketch the curve of temperature versus time if this system loses 1 kcal/sec to its surroundings. Use $C_P \approx 15$ cal/(mole)(°C) for the molar specific heat and $\Delta H_f = 18.5$ kcal/mole for the molar heat of fusion. (*b*) The corresponding cooling curve for a solid-solution alloy is significantly different since solidification occurs over a range of temperatures. Make a similar sketch for the 20% MgO alloy of Fig. 2-3, initially 50°C above the liquidus curve, using the same data as in part *a*.

2-7. Consider a ceramic material consisting of 30 $^{M}\!/_{0}$ MgO and 70 $^{M}\!/_{0}$ NiO (Fig. 2-3). (*a*) At 2500°C: (1) What phase(s) are present? (2) What is the chemical composition of each phase? (3) What amount of each phase is present? (*b*) Make a similar phase analysis at 1825°C. (*c*) Make a similar phase analysis at 1815°C. (*d*) Make a similar phase analysis at 1800°C. (*e*) Make a similar phase analysis at room temperature. (*f*) Sketch a possible microstructure for each of these five temperatures.

2-8. (*a*) Use the phase rule to show that four phases can be in equilibrium in a two-component *PTX* diagram. (*b*) Why must the phases exist along a straight line in the diagram? (*c*) Imagine that a one-dimensional "*X* diagram" is constructed by taking a constant-temperature section $T = T_1$ through a *TX* diagram (which is already at $P = 1$ atm). What is the maximum number of phases that can be in equilibrium in such a diagram?

2-9. Consider a system consisting of 1 mole of a 90 Ag–10 $^{A}\!/_{0}$ Cu alloy. Make phase analyses at the following temperatures: (*a*) 1000°C, (*b*) 875°C, (*c*) 800°C, and (*d*) 300°C. (*e*) Sketch a possible microstructure for each temperature. For part *d*, assume that three grains of β phase form in each grain of α phase. (*f*) Why does no eutectic microconstituent appear in any of the microstructures?

2-10. When compositions are in weight percent ($^{W}\!/_{0}$), the amount of the system is usually a weight (grams, pounds, etc.). Consider a system consisting of 100 g of a 53 Ag–47 $^{W}\!/_{0}$ Cu alloy. Make phase analyses at the following temperatures: (*a*) 900°C, (*b*) 780°C, and (*c*) 778°C. (*d*) For the phase analysis of Eq. (2-11) the statement is sometimes (loosely!) made that the system contains "36 percent" β phase. By comparing the corresponding value in part *b*, suggest an adequate terminology for expressing the *amount* of a phase in percentage.

2-11. Carry out a thermodynamic analysis for CuO analogous to that for Cu_2O illustrated in the text. (*a*) What is the equilibrium pressure of O_2 for this reaction? (*b*) Under what conditions would Cu_2O be the only oxide forming on copper?

2-12. (*a*) Derive Eq. (2-21) from Eqs. (2-18) and (2-20). (*b*) Show the general expression for K in terms of ν_i and ΔG_i.

2-13. Some metals are produced from their oxide ores by being heated in contact with carbon (in the form of coke). Not only does the coke burn and thus help keep the reacting substances at a high temperature, but the carbon (or carbon monoxide) reacts with the oxygen in the metal oxide and liberates the metal. Combine two suitable reactions in Fig. 2-8 and the corresponding ΔG° values to show quantitatively that (*a*) copper ore can be reduced using carbon monoxide with liquid copper as product, (*b*) iron ore is more difficult to reduce and carbon is required to produce liquid iron, (*c*) aluminum ore cannot be reduced by carbon at reasonable temperatures (slightly above the melting point of aluminum).

2-14. Derive Eq. (2-32) by taking the derivatives of both sides of Eq. (2-27) and showing by substitution that $dG = -S\,dT$. Integrate the latter equation and replace $G_{(0°K)}$ by an equivalent quantity.

2-15. Calculate ΔT for N_{Bi} values of 0.8, 0.7, and 0.6; also make the corresponding calculations for N_{Cd} values of 0.9, 0.8, 0.7, and 0.6. Plot these values as indicated in Fig. 2-10, and determine the calculated eutectic temperature and composition. The actual values are 417°K and $N_{Cd} =$ 0.55.

2-16. Prove that the volume of a solution of N_B mole fraction of B in A lies on the dashed line in Fig. 2-11 if each component retains its molar volume V_i unchanged. [HINT: Show that the expression for the total volume V has the form of the equation of a straight line, $y = mx + b$.]

2-17. Explain with the aid of a suitable sketch why V^{xs} is sometimes referred to as the "change of volume on mixing," ΔV_m.

2-18. Derive Eq. (2-46) by writing the change in free energy, ΔG_m, for the process of mixing pure solid A and solid B, which is described by the reaction

$$N_A A(s) + N_B B(s) \rightarrow \alpha \qquad (2\text{-}72)$$

[1 mole of solid solution of A in B]. [HINT: Use Eq. 2-20 to express the free energy $N_A \bar{G}_A$ of component A in the solid solution; do the same for B. The free energies $N_A G_A^\circ$ and $N_A G_A^\circ$ appear twice and cancel. Since the solution is assumed to be ideal, $a_i = N_i$, and $\Delta H_m = 0$.]

2-19. By making the indicated substitutions in Eq. (2-50), prove the relations of Eq. (2-51).

2-20. (*a*) Set up an equation for ΔG^l similar to Eq. (2-52). The principal difference is that the MgO phase must be melted in preparation for mixing with liquid NiO. (*b*) Evaluate your equation at $N_B = 0.72$ and Eq. (2-52) at $N_B = 0.48$, the points of tangency. (*c*) Plot these values and determine μ_A and μ_B.

2-21. Show that a solid-solution diagram with a maximum point can be produced if $G^{\alpha(xs)}$ has a sufficiently larger negative value (compared to the values for the liquid phase, which is ordinarily also nonideal). Use the method of Fig. 2-13 to deduce the phase diagram.

2-22. Make phase analyses at the following temperatures for composition 1 (2 $^w/_o$ $CaSiO_3$) in Fig. 2-16: (*a*) 1706°C, (*b*) 1704°C, (*c*) 1437°C, and (*d*) 1435°C. Sketch a possible microstructure for each of these four temperatures.

2-23. (*a*) Use Fig. 2-17 to make phase analyses of a 10 A/o Pt alloy at (1) 1200°C, (2) 1050°C, and (3) 900°C. (*b*) Consider the heating of a 60 A/o Pt alloy, initially in an equilibrium condition. Make phase analyses at (1) 800°C, (2) 1184°C, (3) 1186°C, and (4) 1600°C. (5) In view of the nature of the peritectic reaction on heating, explain whether equilibrium would be easily established in this case.

2-24. Use sketches of free-energy curves at about seven different temperatures to deduce the phase diagram of Fig. 2-19*e*. The relative positions of the curves for the three solid phases can be estimated from the melting temperatures of NiAl, Ni_3Al, and nickel. The minimum in the curve for the liquid must satisfy the rule illustrated in Fig. 2-18.

2-25. (*a*) Sketch a diagram similar to Fig. 2-20*b*, except with the liquidus and solidus lines sloping upward, and show that k is then greater than 1. (*b*) Explain why the curve for $k = 2$ decreases with increase in g in Fig. 2-20*c*.

2-26. (*a*) Using $k = 0.5$, determine the nonequilibrium solidus line (Fig. 2-20*b*) in the Ag–Cu diagram (Fig. 2-5) for the 7.5 W/o Cu alloy. [HINT: Use Fig. 2-20*c* to determine C^s_{av} for a given fraction of solid, say, $g = 0.3$. The lever law then gives C^l, and therefore the temperature of this C^s_{av} value.] (*b*) What percentage of the alloy undergoes the eutectic reaction on being rapidly cooled from the liquid state?

2-27. Make phase analyses for the 57 A/o Au alloy in Fig. 2-23 at the following temperatures: (*a*) 400°C, (*b*) 350°C, and (*c*) 300°C. Sketch possible microstructures. (*d*) Explain why the ordered phase differs from the ideal CuAu structure.

2-28. On the s^2 diagram (Gibbs triangle) for the Pb–Sn–Sb system (*a*) show the location of the alloy 20 W/o Pb–60 W/o Sn. (*b*) At 189°C the equilibrium alloy consists of β phase (5 W/o Pb–50 W/o Sn), γ phase (3 W/o Pb–90 W/o Sn), and δ phase (85 W/o Pb–13 W/o Sn). Plot the locations of each of these phases.

2-29. (*a*) Sketch isothermal sections at T_1 and T_6, analogous to those shown in Fig. 2-26*b* for the other four temperatures. (*b*) Make phase analyses for alloy 1 at T_1 and T_6.

2-30. (*a*) Recall the following definitions concerning determinants, illustrated here for a determinant of order 3 (a 3 × 3 determinant):

$$\text{Determinant of } C_{ij} \equiv |C_{ij}| \equiv \begin{vmatrix} C_{11} & C_{12} & C_{13} \\ C_{21} & C_{22} & C_{23} \\ C_{31} & C_{32} & C_{33} \end{vmatrix} \qquad (2\text{–}73)$$

Sign convention for position in a determinant is as follows:

$$\begin{vmatrix} + & - & + \\ - & + & - \\ + & - & + \end{vmatrix}$$

To determine the cofactor of a typical component, say C_{21}, in a determinant (1) use the sign associated with the position of the component, i.e. $(-)$, (2) eliminate the row and column containing the component in question, giving

$$\begin{vmatrix} & C_{12} & C_{13} \\ C_{21} & & \\ & C_{32} & C_{33} \end{vmatrix}$$

(3) evaluate the subdeterminant that remains, and (4) obtain the numerical value, which in this case would be

$$-(C_{12} \times C_{33} - C_{13} \times C_{32})$$

The value, DET, of an $n \times n$ determinant is

$$\text{DET} = \sum_{i=1}^{n} C_{ij}(\text{cofactor of } C_{ij}) \qquad j \text{ has any fixed value} \qquad (2\text{–}74)$$

(*b*) The following computer program can be used to evaluate a determinant:

```
C  CALCULATION OF VALUE, DET, OF 3-BY-3 DETERMINANT
C  DIMENSION
      DIMENSION C(3,3)
    1 FORMAT (9F7.3)
C  READ DATA
    2 READ (5,1) ((C(K,I), I = 1,3), K = 1,3)
C  CALCULATION†
```

† Since statement 3 contains more than 66 characters, a continuation card must be used.

```
    3 DET = C(1,1) * C(2,2) * C(3,3) - C(1,1) * C(3,2) * C(2,3)
    *         - C(2,1) * C(1,2) * C(3,3) + C(2,1) * C(1,3) * C(3,2)
    *         + C(3,1) * C(1,2) * C(2,3) - C(3,1) * C(1,3) * C(2,2)
C  OUTPUT SECTION
    4 WRITE (6,1) DET
    5 STOP
      END
```

(*c*) Use the method of Eq. (2-69) to determine the amounts of the α, β, and liquid phases in alloy 1 at T_3 (Fig. 2-26*b*). (*d*) Verify that at T_5 alloy 1 contains less α phase than at T_3.

2-31. Make phase analyses for composition 1 in Fig. 2-27 at the following temperatures: (*a*) 2500°C, (*b*) when the liquid phase is at *P*, (*c*) when the liquid phase is at *P*′, and (*d*) 1600°C. Sketch a possible microstructure for each of these four temperatures.

REFERENCES

Elliott, R. P.: "Constitution of Alloys," first suppl., McGraw-Hill Book Company, New York, 1965.

Gordon, Paul: "Principles of Phase Diagrams in Materials Science," McGraw-Hill Book Company, New York, 1968.

Guertler, W.: "Konstitution der Ternären Metallischen Systeme," Dahlem, Berlin, 1956.

Hansen, M., and K. Anderko: "Constitution of Binary Alloys," McGraw-Hill Book Company, New York, 1958.

Levin, E. M., et al.: "Phase Diagrams for Ceramists," vol. 1 (1964), vol. 2 (1969), American Ceramic Society, Columbus, Ohio.

Mackowiak, J.: "Physical Chemistry for Metallurgists," American Elsevier Publishing Company, New York, 1966.

Prince, A.: "Alloy Phase Equilibria," American Elsevier Publishing Company, New York, 1966.

Raynor, G. V.: Phase Diagrams and Their Determination, in R. W. Cahn (ed.), "Physical Metallurgy," p. 309, North Holland Publishing Company, Amsterdam, 1970.

Rhines, F. N.: "Phase Diagrams in Metallurgy," McGraw-Hill Book Company, New York, 1956.

Shunk, F. A.: "Constitution of Alloys," second suppl., McGraw-Hill Book Company, New York, 1969.

Wall, F. T.: "Chemical Thermodynamics," W. H. Freeman and Company, San Francisco, 1965.

Chapter Three

Microstructures and Stereometric Microscopy

The most common method for determining the constitutions of materials is by examining a suitable specimen with an optical microscope or an electron microscope. The structures revealed in this way are called *microstructures*. Microstructures are significant because of the information they give about the properties of a given sample of material. Much of the interpretation of microstructures is qualitative, but techniques are available that permit quantitative evaluations as well.

Ordinary microscopy (metallography) reveals the constitution at the surface of a specimen, but other techniques examine the entire volume directly. X-ray diffraction, considered in Chap. 1, is an example. Other x-ray techniques are radiography (analogous to medical applications) and various methods for investigating crystalline imperfections. Sonic, electrical, and magnetic methods can also give information about internal structure, such as the presence of cracks and other defects. An especially important technique, transmission electron microscopy, is described in Chap. 4. Stereometric microscopy provides methods for obtaining quantitative data on three-dimensional phase structures from analyses of ordinary two-dimensional microstructures.

MICROSCOPY

3-1. Preparation of the Specimen

Figure 3-1 shows a method for preparing a specimen for observation under the microscope. Often the specimen is a sample about 1 cm on a side cut from a large piece of the material and mounted for ease in handling (Fig. 3-1*a*). First, the cut surface is ground flat and relatively smooth using abrasives. Next, polishing is done on rotating discs covered with a suitable cloth to which a paste [containing fine diamond, alumina (Al_2O_3), or magnesia (MgO)] is applied. Rough polishing produces a surface that appears smooth to the eye, but microscopic observation reveals scratches characteristic of the last abrasive employed (Fig. 3-1*b*). Final polishing leaves the specimen with a mirror finish (Fig. 3-1*c*). Although one can see certain structural features in such a specimen, a chemical etching procedure is required to bring out the structural detail (Fig. 3-1*d*), including grain boundaries and grain contrast in this case.

(*a*) Specimen mounted in plastic, ½ normal size.

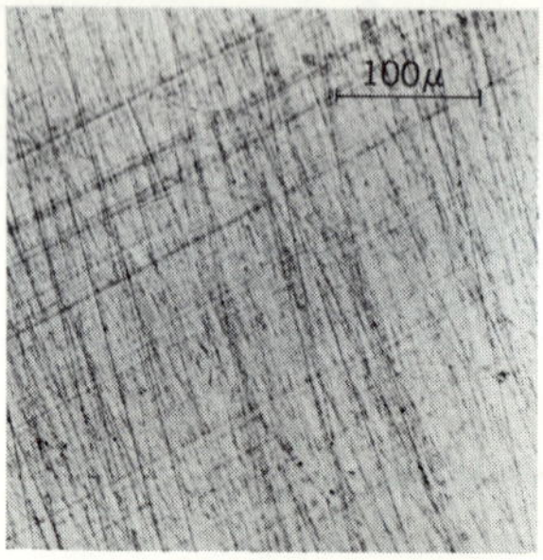

(*b*) Appearance after rough polishing with 6μ diamond paste.

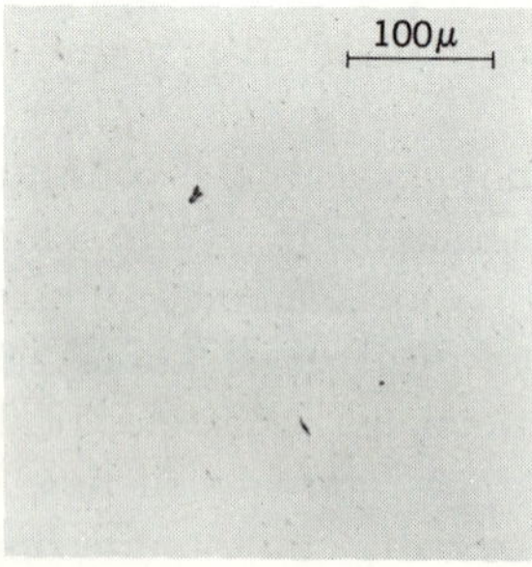

(*c*) Appearance after fine polishing with γ-alumina paste.

(*d*) Appearance after chemical etching with a dilute solution of NH_4OH and H_2O_2.

FIGURE 3-1 Illustration of a typical procedure for the preparation of a specimen for examination under the microscope. The material is cartridge brass (70 w/o Cu–30 w/o Zn). (*Courtesy Buehler, Ltd.*)

The method described here is the most common one for polishing both metallic and nonmetallic materials, but other methods deserve mention. Grinding and polishing have been mechanized in various ways, especially for preparation in quantity of a given type of specimen. Other methods for surface preparation are basically different from the usual polishing procedures. An ultramicrotome is able to cut off thin layers from some relatively soft materials. These layers require no polishing and can be observed either with transmitted light or with reflected light after etching, as described below. Another method that eliminates most of the grinding and hand-polishing operations is electropolishing, in which the specimen is made the anode in an electrolytic cell. As the electrolytic process removes successive atomic layers from the surface of the specimen, an excellently polished surface is produced on suitable materials. Electropolishing has an important advantage over customary polishing in that it effectively removes the thin ($\sim 10^2$ Å) layer of disturbed material that is unavoidably produced by grinding and polishing with abrasives. The usual chemical etching also removes this disturbed layer.

3-2. The Light Microscope

Many different functions can be performed by a microscope that employs visible light. A basic function is to magnify an image. The maximum *useful* magnification usually attainable is about 600 diameters ($\times 600$) (see Prob. 3-1), although photomicrographs are frequently made at $\times 1000$ or higher for ease of viewing. Light is pictured as a "ray" for determining the magnifying properties of optical lenses. Other important uses of the microscope depend on the wave character of light. Monochromatic (usually, green) light used in microscopy can be pictured as electromagnetic waves vibrating in all directions. A polarizing device, such as a Nicol prism, filters out all waves except those vibrating in one plane, say the vertical. Such polarized light has the useful property that the plane of polarization is rotated during transmission through or reflection from many materials. Another useful characteristic of light waves (polarized or unpolarized) is their phase. Light reflected from a surface has a different phase depending on whether it is reflected from the main surface or from a depression.

Various types of microscopes are used, depending on the material being observed. The multipurpose instrument of Fig. 3-2*a* permits the study of microstructures using phase contrast or polarized light, and will accommodate either transparent or opaque materials. For observation of opaque materials such as metals, light from the source is reflected onto the surface of the specimen by a glass reflector. The image returning from the specimen passes through the glass reflector and reaches the eyepiece. The features of the microstructure are visible because of differences in the intensity of the light reflected from various parts of the surface of the specimen. These differences arise from two different causes, absorption and scattering. As explained in Chap. 12, substances differ in the degree to which they absorb light. A polished specimen consisting of particles of graphite embedded in a metal will show dark graphite particles under the microscope because light incident on the graphite is largely absorbed. On the other hand, a homogeneous material absorbs light uniformly and shows only a featureless appearance (for example, the unetched specimen in Fig. 3-1*c*). Etching produces the shading of the various grains (Fig. 3-1*d*) and reveals the grain boundaries by creating submicroscopic facets within the grains or slight differences in level at grain boundaries. These irregularities scatter a portion of the light rays (see Prob. 3-2) and thus cause darker areas to appear in the observed structure.

Chemical etching is only one of several methods used to reveal microstructures. Electrolytic etching, mentioned above in connection with electrolytic polishing, is also useful for mechanically polished specimens. Thermal etching involves heating the polished specimen to a high temperature so that changes in surface contours or surface composition occur by one or more of the following processes: evaporation of atoms or molecules from the surface, diffusion of atoms along the surface (see Chap. 6), or chemical reaction at the surface, such as the formation of an oxide or sulfide. Many materials also become etched when a polished surface is bombarded by ions, conveniently performed in an apparatus used for sputtering.

Figure 3-2*b* illustrates the application of polarized light to transparent specimens such as minerals and some ceramics. A "thin section" for this purpose is prepared by cementing a thin, flat specimen to a glass slide and then grinding it to a thickness of about 30 μ. In a specimen with large grains, some individual grains occupy the entire thickness of such a specimen. As the light passes through such a grain, the direction of polarization of the light changes if the material is optically anisotropic. Since the polarizer gives incident light having a known direction of polarization, the analyzer can determine the change in direction caused by the grain in question. Both the orientation and the optical properties of the grain can be determined using a polarizing microscope.

Phase-contrast microscopy (Fig. 3-2*c*) creates contrast by detecting differences in phase, rather than differences in reflected intensity. Nonuniformity in level of the surface of a specimen produces variations in phase. A specimen with areas having different refractive indexes shows the same effect. A special annular diaphragm permits the optical system to convert differences in phase into differences in intensity. The matched diaphragm for each of several objective lenses is positioned by the phase drum. Phase-contrast microscopy using reflected light is suitable only for surfaces with very slight unevenness.

The photographing of microstructures can be done in two ways. Camera attachments are available for most desk microscopes (Fig. 3-2*a*), using either conventional or Polaroid film. For photomicrographs of the highest quality, however, special metallographs are employed. Typically a metallograph consists of a versatile, high-quality microscope combined with a camera that minimizes vibration during the exposure time, usually several seconds. Glass plates are frequently employed rather than film. Using special attachments (or an entire special metallograph), one can examine materials under unusual conditions. For example, a hot-stage attachment permits direct

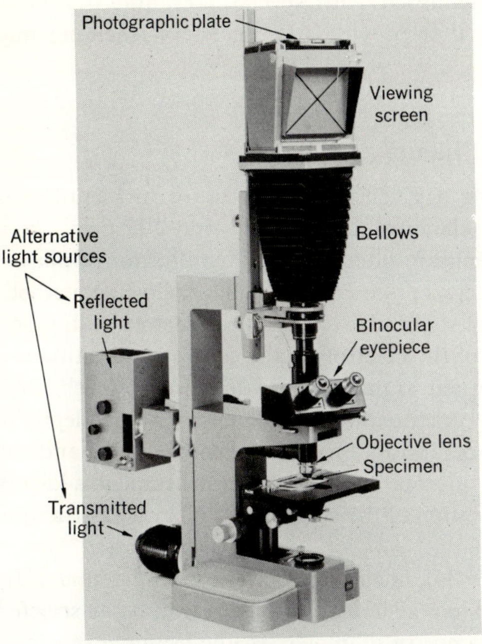

(a) Microscope equipped with camera for photomicrography.

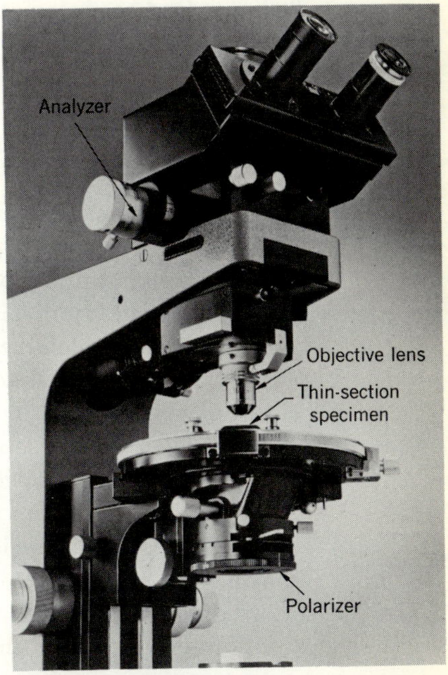

(*b*) View of microscope attachments for use of polarized light in transmission.

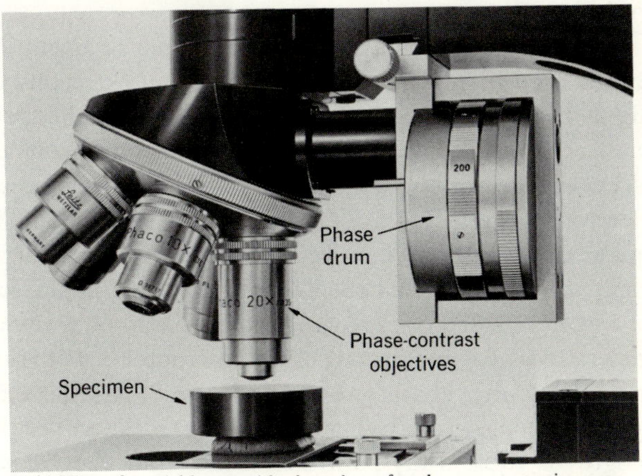

(*c*) View of special lenses with phase drum for phase-contrast microscopy with reflected light.

FIGURE 3-2 Versatile microscope for examining specimens using either reflected or transmitted light. (*Courtesy E. Leitz, Inc., Wetzlar, Germany.*)

observation of structural changes occurring in a material heated as high as 2000°C. Radioactive materials, polished with special precautions, can be examined using a metallograph that provides the necessary shielding and facilities for remote control.

3-3. Identification of Phases

The first step in analyzing a given microstructure is to identify the phases that are present. If the chemical composition and "history" of a material are known, the phase diagram often permits identification of phases, even in nonequilibrium microstructures (see Prob. 3-3). The following methods permit the direct identification of some phases: x-ray or electron diffraction (see Chap. 1), chemical analysis of the electrolytically separated constituents, electrical and magnetic tests adapted to microscopic dimensions, mechanical differentiation of hard and soft constituents by differences in the depth of standard scratches or by differences in microhardness (see Chap. 9), and the x-ray emission microanalyzer (or microprobe). The latter instrument is ideally suited to the quantitative identification of the constituents in a microstructure, as will now be described.

Using a microprobe (Fig. 3-3*a*), one can examine a specimen under the built-in microscope and then obtain a chemical analysis (Fig. 3-3*c*) of selected microscopic areas only a few microns in diameter. In the example shown, a series of chemical analyses for silicon and calcium were made at 1-μ intervals across a particular grain boundary in a specimen of relatively pure magnesium oxide. The results indicated that SiO_2 and CaO had segregated to this grain boundary.

Figure 3-3*b* illustrates the principle of operation of the microprobe. Suitable magnetic lenses act to produce an extremely fine beam of electrons, about 1 μ in diameter. When this beam strikes the specimen (at the point of intersection of the two cross hairs), two useful phenomena occur: some of the electrons are simply scattered back from the specimen, and some of the electrons generate x-rays by bombarding the atoms in the specimen. The quantitative results of each of these phenomena depend on the chemical composition at the particular point struck by the microbeam.

Let us first consider an analysis based on x-rays, since this is the more generally useful application of the microprobe. If the incident electron beam generates x-rays characteristic of silicon, Si $K\alpha$, these x-rays are emitted from the specimen. An analyzing crystal (Fig. 3-3*b*), set for the characteristic wavelength of these x-rays, diffracts them into the Geiger counter. Since the intensity of the Si $K\alpha$ x-rays is in proportion to the percentage of silicon in the specimen, quantitative x-ray counting gives the chemical analysis of the microscopic volume in question.

Another use of the microprobe is to give a *micrograph* of the distribution of certain chemical elements in the alloy. The principle is the same as that just described, but the analyzing microbeam, rather than remaining at one point,

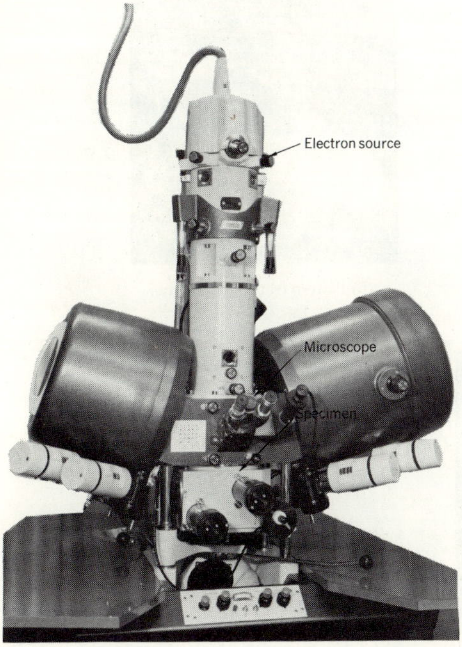

(*a*) A commercial instrument.

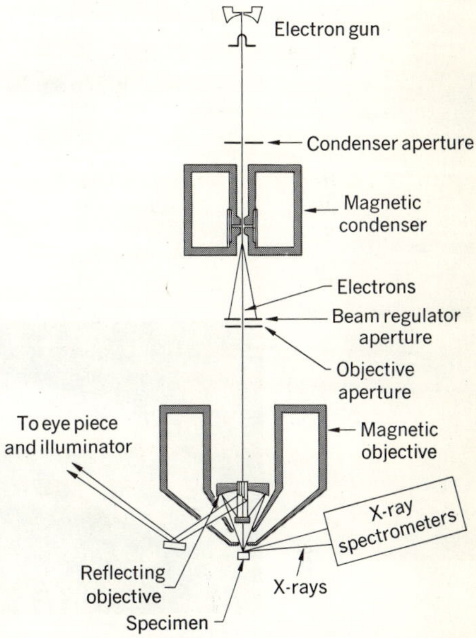

(*b*) Principle of operation.

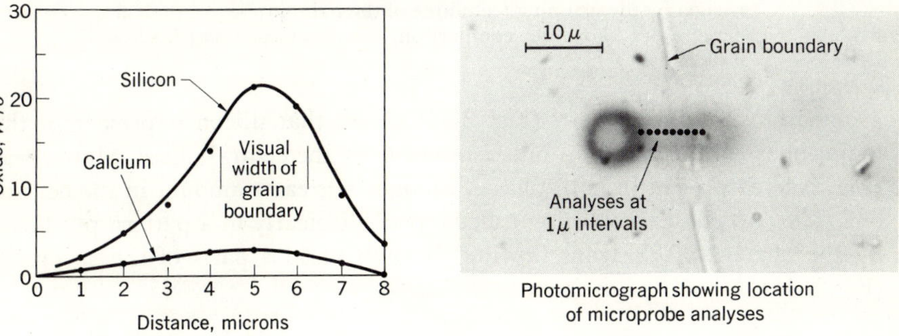

(*c*) Determination of impurities of SiO_2 and CaO near a grain boundary in magnesium oxide. (Courtesy of M. H. Leipold, University of Kentucky)

FIGURE 3-3 Use of the x-ray emission microanalyzer (microprobe) as a tool for quantitative microscopic analyses. (*Courtesy Consolidated Electrodynamics, Inc.*)

is scanned over the surface of the specimen. The measured x-ray intensity is correspondingly scanned over the surface of a television tube and produces an image of the specimen (in terms of its chemical composition). For example, when an oxide inclusion in a piece of steel (Fig. 3-4*a*) is examined for silicon

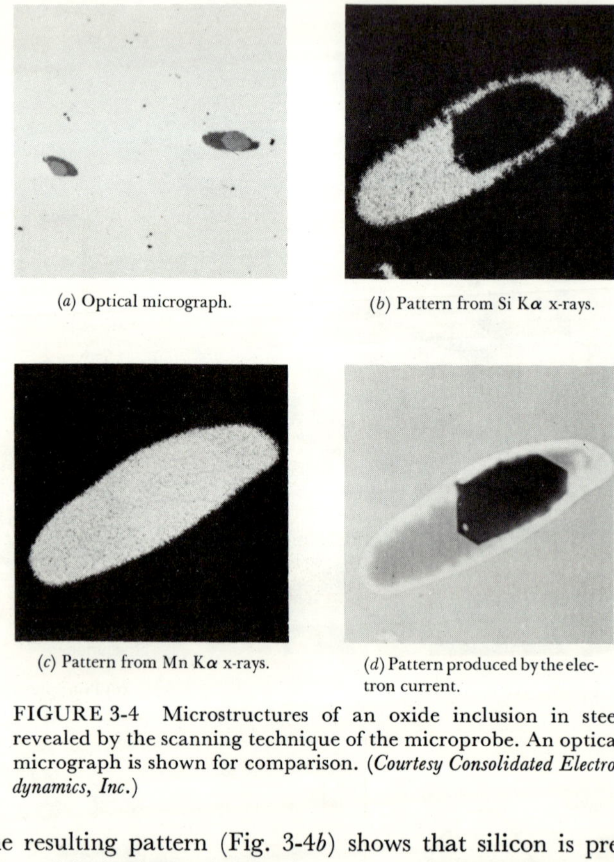

(*a*) Optical micrograph. (*b*) Pattern from Si Kα x-rays.

(*c*) Pattern from Mn Kα x-rays. (*d*) Pattern produced by the electron current.

FIGURE 3-4 Microstructures of an oxide inclusion in steel revealed by the scanning technique of the microprobe. An optical micrograph is shown for comparison. (*Courtesy Consolidated Electrodynamics, Inc.*)

x-rays, the resulting pattern (Fig. 3-4*b*) shows that silicon is present in the major phase of the inclusion but is absent from the central phase. Manganese (Fig. 3-4*c*) is present in both phases, though in greater amount in the central one. The two phases are distinguished especially clearly in a pattern produced by the current of electrons flowing through various parts of the specimen (Fig. 3-4*d*).

3-4. Electron Microscopes

Electrons can often be used to form a "photographic" image of the surface of a specimen. Three of the more useful techniques are described here.

REPLICA ELECTRON MICROSCOPY. The resolving power of a microscope is roughly equal to the wavelength of the light that it uses. Since typical electron beams have wavelengths less than 1 Å, electron micrographs can be obtained with very high magnifications. Figure 3-5*a* shows the analogy between the focusing of light rays by lenses and the focusing of the electron beam by

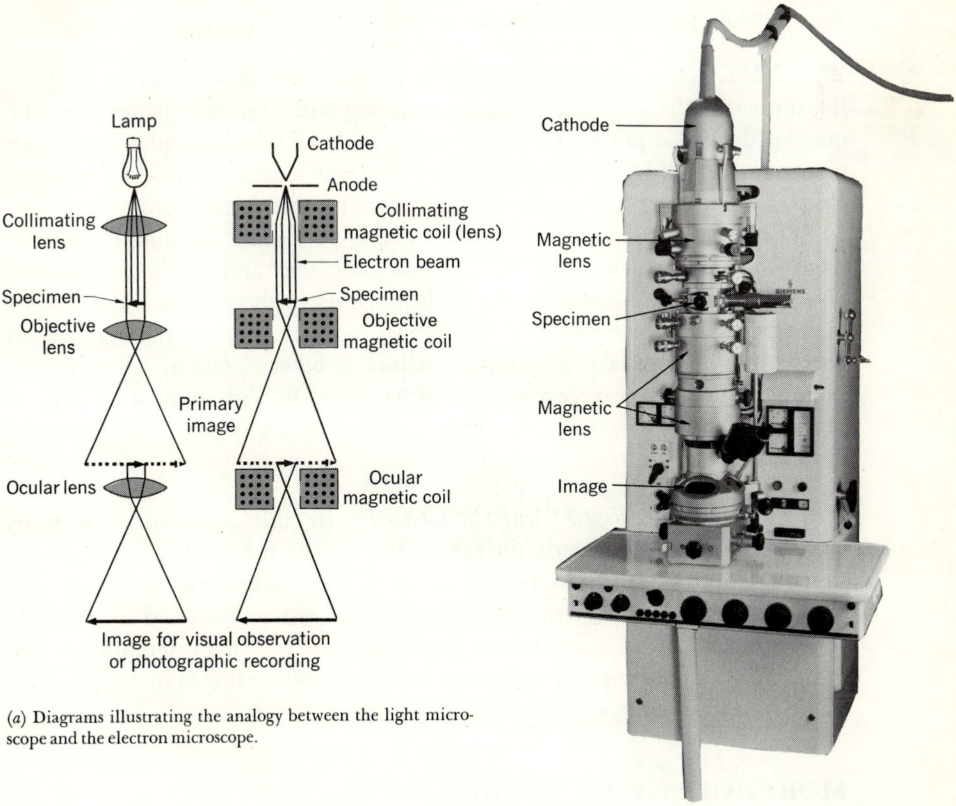

(*a*) Diagrams illustrating the analogy between the light microscope and the electron microscope.

(*b*) An electron microscope suitable for both replica and transmission techniques, Elmiskop Ia.

FIGURE 3-5 Illustrations of (*a*) the principle of operation of an electron microscope and (*b*) a commercial instrument. (*Courtesy Siemens Corporation.*)

magnetic coils. Electrons produced by the hot-wire cathode are accelerated through the anode by a potential of about 50,000 to 100,000 volts. After the electron beam has been collimated, it passes through the specimen and is selectively absorbed or diffracted by different parts of the specimen. The beam is then magnified by the objective and ocular magnetic coils before it is viewed on a fluorescent screen or is photographed to produce the analog of the ordinary photomicrograph, but with greatly improved resolution and magnification. The electron micrographs in Fig. 3-7 are examples. In this type of electron microscopy the specimen is not the material itself but instead a thin film, the *replica*, on which fine details of the material's surface have been recorded. Although the usual electron microscope has a resolving power of roughly 5 Å, experimental difficulties limit the resolving power to about 10 times this value when replicas are employed.

SCANNING ELECTRON MICROSCOPE. This type of microscope observes the surface of the specimen directly, scanning with the electron beam in the manner described previously for the microprobe. An image with great depth of focus is formed by the back-scattered electrons plus the secondary-emission electrons. The irregular surface produced during fracture of materials (Fig. 10-3) is revealed especially clearly by this technique. Scanning electron microscopes have great flexibility in magnification and in their ability to observe specimens directly during such processes as thermal etching or other high-temperature treatments. If differences in electrical potential are created on the specimen surface (of semiconductors or other electrical materials, for example), the image records these differences. In this way, such usually invisible features as *p-n* junctions in semiconductors (see Sec. 4-5) can be revealed.

THERMIONIC-EMISSION MICROSCOPE. In this apparatus, electrons emitted from the activated[1] surface of the heated material are focused by suitable lenses to form a magnified image of the specimen surface (Fig. 3-6) on a fluorescent screen. Grain contrast (similar to that obtained by etching) occurs because the ease of electron emission varies with the orientation of the grains. This microscope is especially useful for observing high-temperature phase transformations as they occur.

MORPHOLOGY OF MICROSTRUCTURES

Three aspects of its microstructure greatly affect the properties of a material: (1) the nature of each phase, (2) the relative amounts of the phases, and (3) the morphology (shape and arrangement) of the phases. Morphology includes such features as geometric shape, crystallographic orientation, and mode of distribution (for example, lamellar mixture or grain-boundary precipitate) of the various phases. In this section we consider both the qualitative aspects and the quantitative characterization of morphology. An important factor determining the final morphology is the manner of formation of the solid—by

[1] An activator such as barium beryllate is applied to the surface of the specimen to facilitate the emission of electrons.

(*a*) Grain structure after heating for 3 minutes.

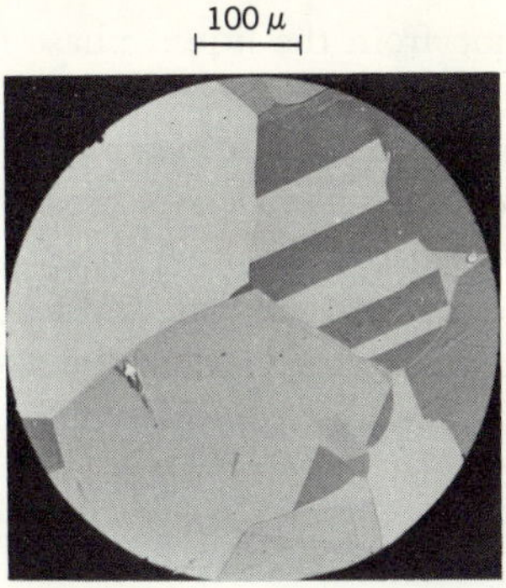

(*b*) Grain structure after heating for 30 minutes.

FIGURE 3-6 Frames from a motion picture of the growth of austenite grains at 1066°C in a commercial steel, taken with a thermionic-emission microscope. (*Courtesy W. L. Grube and S. R. Rouze, General Motors Research Laboratories.* By permission from "*High Temperature, High Resolution Metallography,*" *Gordon and Breach,* (*Science Publishers, Inc., New York, 1967.*)

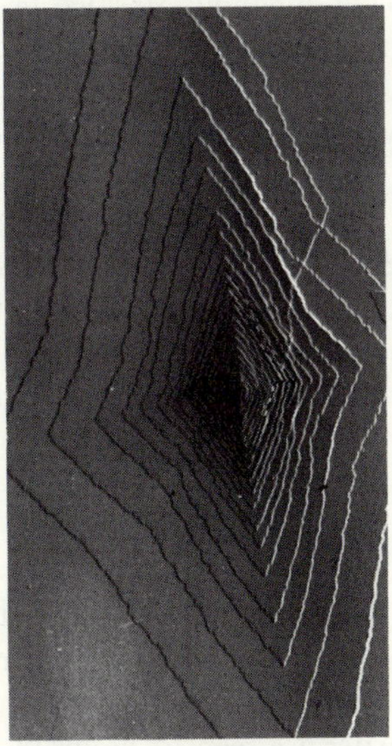

(*a*) Crystals of polyethylene grown from dilute solution. Dashed line indicates spiral form of growth. (Replica electron micrograph).

(*b*) A radial pattern of lamella forms macrospherulites.

FIGURE 3-7 Two morphologies of crystalline lamallae in polyethylene. (*Courtesy A. Peterlin, Camille Dreyfus Laboratory.*)

solidification from the liquid phase (Chap. 2), by deposition from the vapor phase, or by other processes.

Morphology is a significant factor in every material, but organic polymers exhibit an especially broad spectrum and variability. Many polymers are able to exist in an amorphous (glassy) state at room temperature; their microstructures are uniform and featureless. The basic morphological feature of crystalline polymers is a lamella about 10^2 Å in thickness (Fig. 3-7*a*). During slow crystallization from polymer melts, a group of lamellae grows as a spherulite (Fig. 3-7*b*). Macrospherulites produce an objectionable milky appearance in an otherwise clear plastic and also cause poor mechanical properties.

The properties of a crystalline polymer are greatly improved if macrospherulites are prevented from forming. If a melt is rapidly cooled, a common occurrence during industrial processing, the crystalline lamellae form only small sheaves (Fig. 3-8*a*). This structure is called a *microspherulite* because it is similar to the initial stage in the growth of a macrospherulite. If a polymer melt is kept in motion during cooling, crystallization of the longer molecules occurs at a relatively high temperature in the form of an aligned "thread." Oriented crystallization of irregular platelets then occurs along such a central thread to form a "shish-kebab" pattern (Fig. 3-8*b*), also termed *row nucleation*. The formation of a polymer fiber is believed to occur by the unfolding of the chains in crystalline lamellae (Fig. 3-8*c*).

Significant changes in morphology, and therefore in properties, are also produced by the thermal, mechanical, or physical treatments applied to finished or semifinished materials of all types. For example, the changes in steel produced by heat treatment (Fig. 2-1) are well known. Many examples of morphological changes in various materials are discussed in later chapters. We now consider the types of microstructures commonly encountered in metals and ceramics and the quantitative methods for describing their characteristics.

3-5. Classification of Microstructures

The simplest morphologies are those of single-phase materials and typically consist of an aggregate of grains. The structure of cartridge brass (Fig. 3–1*d*) and of the austenite phase in steel (Fig. 3–6) are representative examples. The morphological characteristics are seen to be grain size, grain shape, and crystallographic orientation of the grains relative both to neighboring grains and to the geometry of the specimen as a whole. Boundaries, of great importance for morphology, include grain boundaries in polycrystalline phases and also twin boundaries (Sec. 5-3), which appear as straight lines in brass and in austenite. Other boundaries are discussed in Chap. 5.

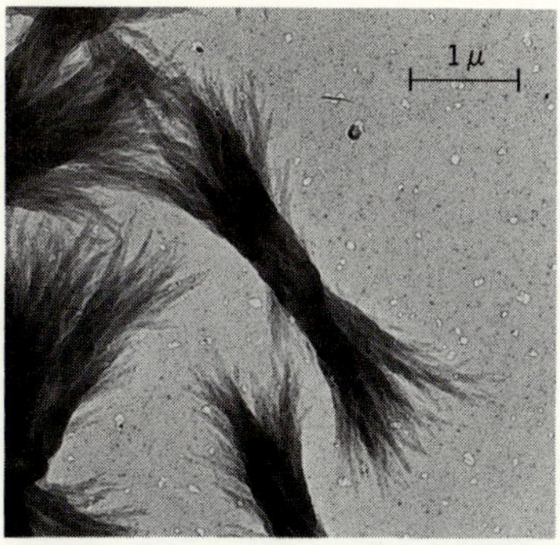

(*a*) Nylon 4 forms microspherulites during rapid cooling from the melt. (Courtesy A. Peterlin, Camille Dreyfus Laboratory)

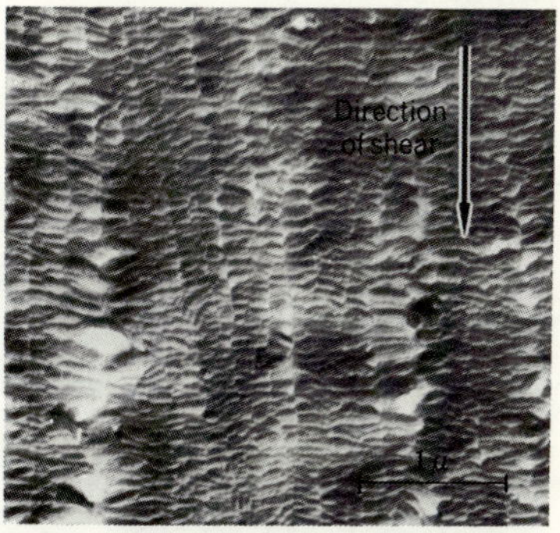

(*b*) A polyethylene melt forms this oriented structure as a result of shearing motion during crystallization. (Courtesy R. B. Williamson, University of California, Berkeley)

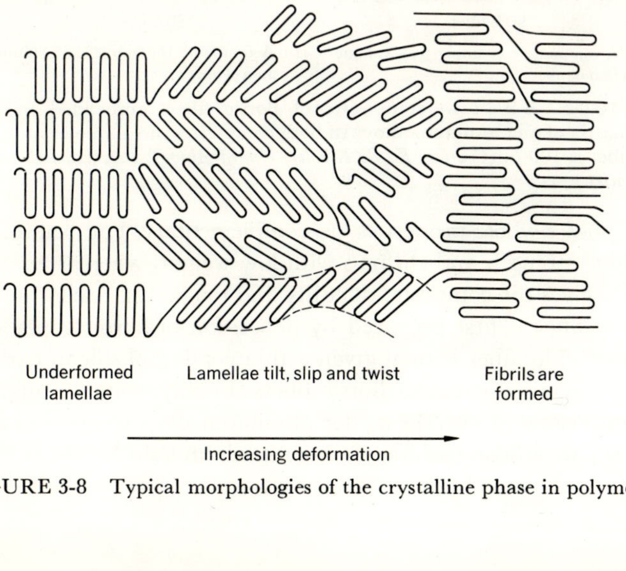

FIGURE 3-8 Typical morphologies of the crystalline phase in polymers.

When a microstructure contains two (or more) phases, several new features become significant. An obvious one is the relative amounts of the phases, such as the amounts of the carbide (Fe_3C) and ferrite phases in steel (Fig. 2-1*b*). The type of distribution of the phases is often of even greater importance. For example, if the carbide phase occurs as a continuous layer in the grain boundaries (known as a *grain-boundary network*), a sharp blow may cause the steel to break in a brittle manner by cracking in the carbide phase. The *principle of the continuous phase* states that the properties of a material will be essentially those of the phase(s) that forms a continuous path through the microstructure. The types of distributions of two phases [one example is pearlite (Fig. 2-1*a*)] vary greatly in the size and shape of the microregions of each phase.

A *composite* material (Fig. 3-9) results from the "artificial" joining of

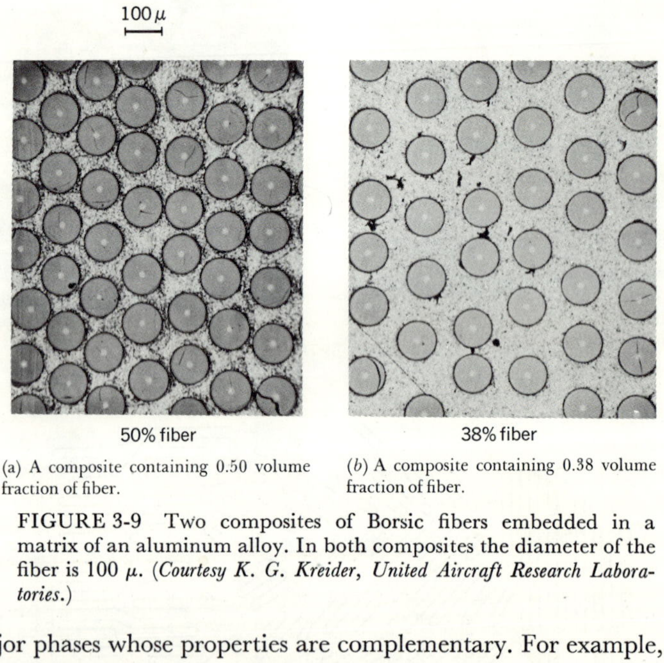

(a) A composite containing 0.50 volume fraction of fiber.

(*b*) A composite containing 0.38 volume fraction of fiber.

FIGURE 3-9 Two composites of Borsic fibers embedded in a matrix of an aluminum alloy. In both composites the diameter of the fiber is 100 μ. (*Courtesy K. G. Kreider, United Aircraft Research Laboratories.*)

two major phases whose properties are complementary. For example, a boron fiber has high strength and stiffness with low weight, whereas an aluminum-alloy matrix gives the composite ductility and toughness. A continuous length of the boron fiber is first prepared by deposition of boron vapor on a 10-μ tungsten wire. The fiber is then given a thin coating of silicon carbide (SiC) to complete the production of Borsic fibers. Finally, suitably aligned Borsic fibers are impregnated with the molten aluminum alloy. During use at elevated temperatures, the silicon-carbide coating retards undesirable reactions between the boron and the aluminum.

3-6. Concepts from Topology

To begin our quantitative study of microstructures, we will use some elementary concepts from topology to describe the shapes assumed by a given portion of a phase. Consider first the three-dimensional structure of a one-phase material consisting of equiaxed grains (Fig. 3-10). The topological features of interest are

Feature	Dimensions	Related simplex	Complex
Grain volume	3	s^3	c^3
Grain face	2	s^2	c^2
Grain edge	1	s^1	c^1
Grain vertex	0	s^0	(s^0)

A vertex (the junction of four grains in space) is a point and therefore is identical to the zero-simplex s^0, previously described in Sec. 2-14. A grain edge (where three grains meet) can be represented by joining several 1-simplexes to form a curved line, the 1-complex c^1. Similarly, the triangular s^2 serves as the unit for building the curved, many-sided surface c^2 that represents a grain face (Prob. 3-4). The 3-complex c^3 can represent a grain having any number n_2 of faces, each face being of arbitrary shape and curvature.

Consider a specimen consisting of a certain number n_3 of grains (3-complexes c^3). The number n_2 of grain faces (c^2) and of other topological features must then satisfy the Euler-Poincaré relation,[1]

$$n_0 - n_1 + n_2 - n_3 + \cdots (-1)^k n_k = 1 \tag{3-1}$$

The reader may verify this equation by applying it to the two-dimensional pattern of the microstructure in Fig. 3-10 or to a three-dimensional grain structure. C. S. Smith[2] has shown how this general equation can be interpreted for the conditions met in many real structures. In an ordinary (two-dimensional) microstructure, for example, exactly *three* grain boundaries (c^1) meet at a grain junction (s^0). Consequently, the number $n_2(r)$ of two-dimensional polygons having r edges must satisfy the relation

$$\sum_{r=3}^{9} (6 - r)n_2(r) = 0 \tag{3-2}$$

[1] This form of the equation applies to configurations of any number of dimensions; for example, it applies to the composition diagram for a $(k + 1)$-component system discussed in Sec. 2-14.

[2] C. S. Smith, Some Elementary Principles of Polycrystalline Microstructures, *Metallurgical Reviews*, vol. 9, no. 33, 1964.

in a large array of variously shaped grain outlines (ranging from triangular to nine-sided polygons) as seen in a two-dimensional microstructure. Problem 3-5 applies this equation to a typical structure and illustrates the fact that the average grain outline has six edges.

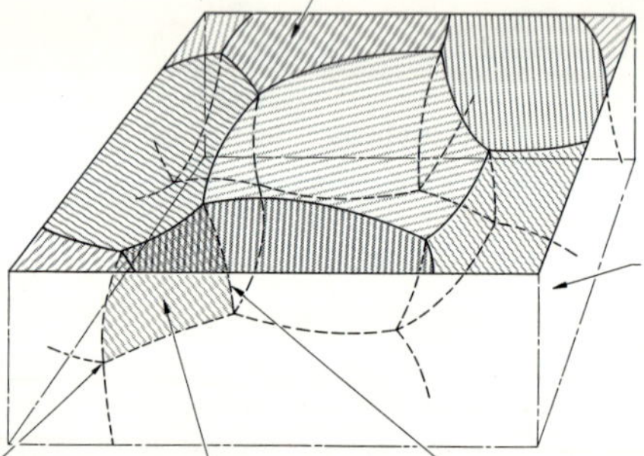

FIGURE 3-10 Sketch showing the relation of the two-dimensional microstructure of a material to the underlying spatial structure. Note that only portions of the total volume and total face of any one grain are shown.

Examination of three-dimensional grain structures (Fig. 3-10) shows that four grain edges c^1 meet at a grain junction c^0. Also three faces c^2 meet at an edge c^1. This latter configuration corresponds in a two-dimensional microstructure to three grain boundaries c^1 meeting at a junction s^0. The analog of Eq. (3-2) for three dimensions also involves n_3, the number of grains (i.e., polyhedral volumes c^3), and takes the form

$$\sum_{r=3}^{9} (6 - r)n_2(r) = n_3 + 1 \tag{3-3}$$

where r still represents the number of edges of a face of a polyhedron. Since a positive quantity now appears on the right-hand side of the equation, the average value of r will be less than 6; a value slightly greater than 5 characterizes the grains in typical alloys (Fig. 3-11). The number of faces on the average grain is about 13, and the number of vertexes is about 23. Theoretically, if each grain were a regular tetrakaidecahedron, an assembly of grains could pack together to fill space and also would nearly satisfy the surface-energy requirements discussed in Sec. 5-3. Actual grains, however, have usually the irregular shapes shown in Fig. 3-11.

These topological concepts for a single phase can be generalized for any number of phases. In a two-phase ($\alpha + \beta$) structure, for example, two types of polyhedral volume exist, c^3_α and c^3_β. Since a face is common to two polyhedra, the three possible types of faces are c^2_α, $c^2_{[(1/2)\alpha(1/2\beta)]}$, and c^2_β. Three polyhedra

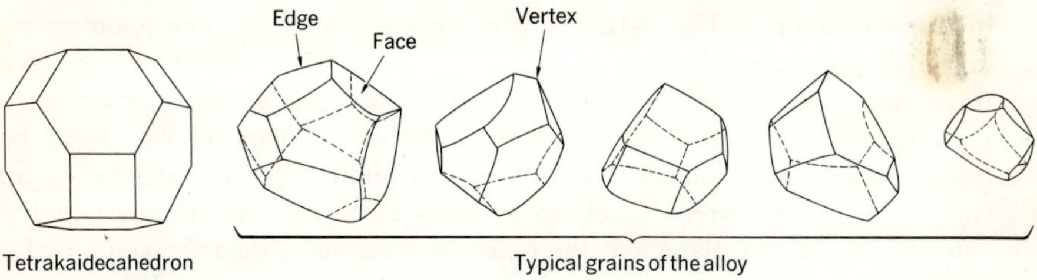

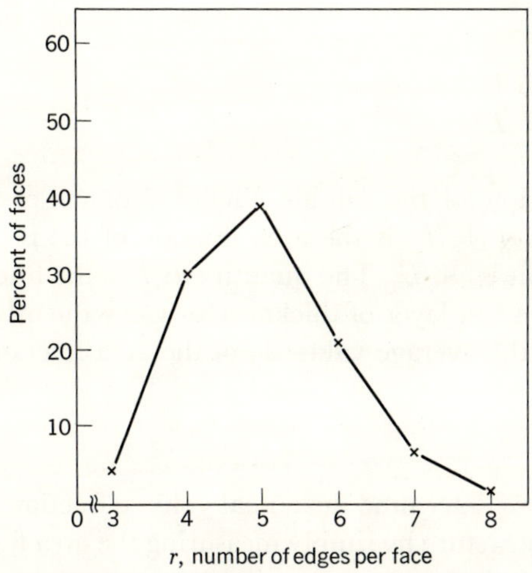

FIGURE 3-11 Percentage of grain faces having r edges in a sample of 100 grains of an aluminum-tin alloy. (*After W. M. Williams and C. S. Smith.*)

meet along an edge and produce the following types of edges: c^1_α, $c^1_{[(2/3)\alpha(1/3)\beta]}$, $c^1_{[(1/3)\alpha(2/3)\beta]}$, and c^1_β. Finally, the junction of four polyhedra may be any of the following: s^0_α, $s^0_{[(3/4)\alpha(1/4)\beta]}$, $s^0_{[(2/4)\alpha(2/4\beta)]}$, $s^0_{[(1/4)\alpha(3/4)\beta]}$, and s^0_β. These topological features are the subject of Prob. 3-6.

3-7. Stereometric Microscopy

If a material were transparent, its three-dimensional morphology would be immediately evident. In actual opaque materials, stereometric microscopy (also known as *quantitative metallography*) obtains information on the spatial structure by analyzing ordinary two-dimensional microstructures. These microstructures may be obtained by metallography, thin-section techniques, electron microscopy, or other suitable means. A typical problem is the determination from ordinary photomicrographs of the amount of β phase in a mixture of α and β (Fig. 3-12*a*). Only one test plane (i.e., one photomicrograph) is shown in this illustration, but obviously an adequate analysis of a three-dimensional structure requires the number of test planes sufficient for a representative sample of the structure. Let the test plane of Fig. 3-12*b* be typical of many test planes, all parallel and uniformly "testing" the entire volume of the structure. If $(\Delta x)_i$ is the thickness of the layer of material lying between the *i*th test plane and the plane next above it, then the volume of β phase in this layer is approximately $V_i = A_i(\Delta x)_i$, where A_i is the area of β phase measured on the photomicrograph of the *i*th plane. The approximation becomes better the more closely spaced the test planes are taken, until in the limit,

$$V_V = \frac{\int V_i}{L^3} = \int_{x=0}^{x=L} \frac{A_x}{L^2}\frac{dx}{L} \tag{3-4}$$

is exactly the expression for the volume fraction (volume of β phase per unit volume). The quantity A_x/L^2 is the area fraction of the β phase in the test plane at height x, that is, A_i/L^2. The quantity dx/L is the fraction of the total volume occupied by a test layer of thickness dx. Consequently, the integral of Eq. (3-4) represents the average value A_A of the area fraction, and thus

$$V_V = A_A \tag{3-5}$$

In other words, we can determine the actual volume fraction V_V of β phase in a three-dimensional structure by simply measuring the area fraction of β phase in a sufficient number of properly selected photomicrographs to determine the true area fraction A_A.

Actual measurements in quantitative metallography generally involve lineal analysis or point counting, rather than the direct measurement of areas. In lineal analysis many test lines sample each surface. Figure 3-12*c* shows a typical line and the fraction l/L of the line that lies in the β phase. A derivation analogous to that given above shows that (see Prob. 3-8)

$$A_A = L_L \tag{3-6}$$

Here L_L is the average lineal fraction of a test line corresponding to the β phase. Point counting is usually done with a grid of lines, the crossing points of which form a set of points (Fig. 3-12*d*) convenient for "testing" a plane.

Reasoning similar to that used above gives the useful relations

$$P_P = A_A = V_V \tag{3-7}$$

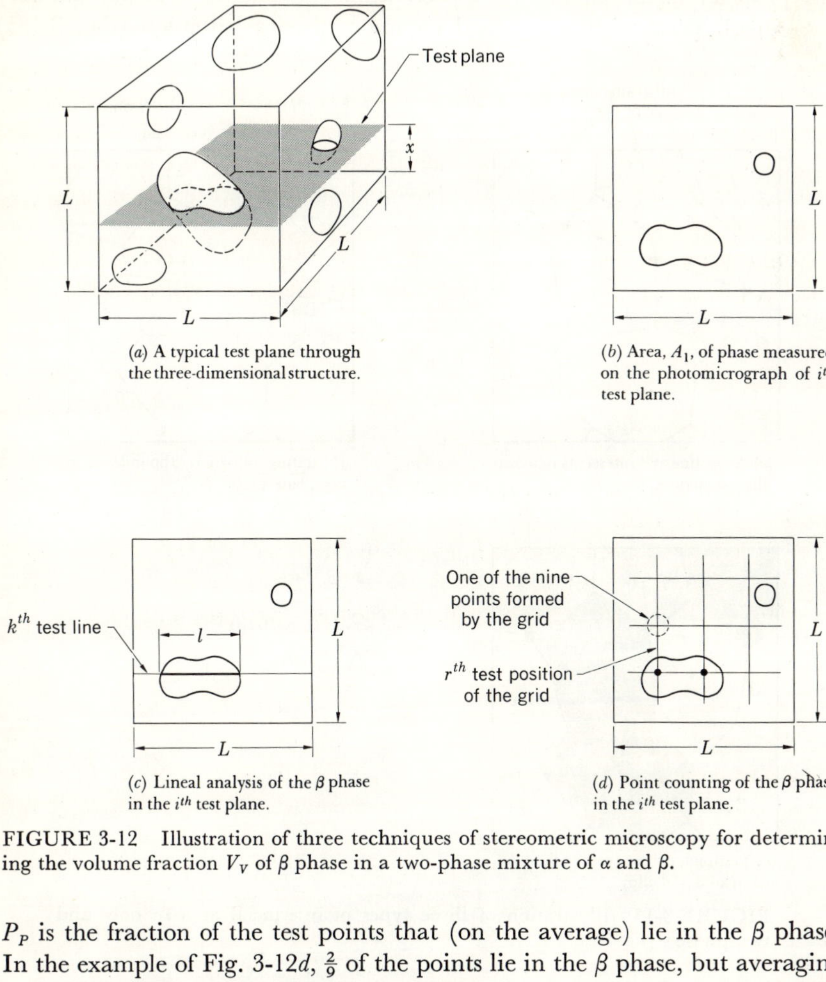

(*a*) A typical test plane through the three-dimensional structure.

(*b*) Area, A_1, of phase measured on the photomicrograph of ith test plane.

(*c*) Lineal analysis of the β phase in the ith test plane.

(*d*) Point counting of the β phase in the ith test plane.

FIGURE 3-12 Illustration of three techniques of stereometric microscopy for determining the volume fraction V_V of β phase in a two-phase mixture of α and β.

P_P is the fraction of the test points that (on the average) lie in the β phase. In the example of Fig. 3-12*d*, $\frac{2}{9}$ of the points lie in the β phase, but averaging with other test positions of the grid would give a value near $P_P = 0.1$, that is, $(V_V)_\beta = 0.1$. In a structure containing k phases, this same procedure determines the volume fraction $(V_V)_i$ corresponding to each type of volume c_i^3.

Surfaces (c^2) such as grain boundaries and interphase boundaries are usefully measured as the quantity S_V of surface area per unit volume of the material. Consider first a one-phase material where all of the grain faces are of the c_α^2 type. A test line (Fig. 3-13*a*) then determines an average N_L (number of intercepts per unit length) for use in the relation[1]

$$S_V = 2N_L \tag{3-8}$$

A typical use of this equation is to determine the amount of grain boundary in a metal (Prob. 3-9). A second common physical situation (Fig. 3-13*b*)

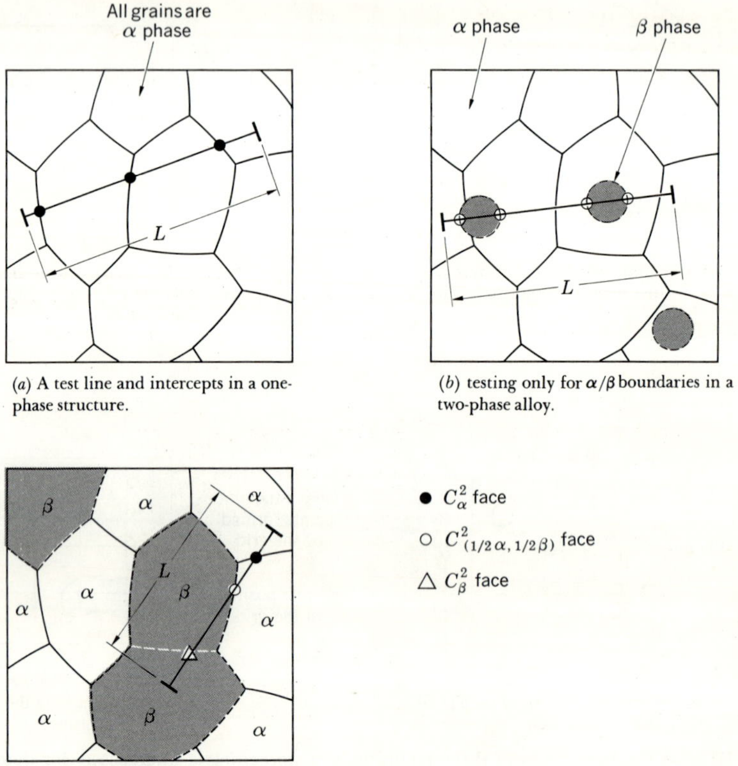

(*a*) A test line and intercepts in a one-phase structure.

(*b*) testing only for α/β boundaries in a two-phase alloy.

(*c*) Testing for three types of boundaries in an $\alpha + \beta$ alloy.

FIGURE 3-13 Illustration of three types of interfacial area in one- and two-phase structures.

involves only the surface separating the α and β phases (faces of the $c^2_{[(1/2)\alpha(1/2)\beta]}$ type). Equation (3-8) is valid for this situation also, S_V now being the amount of $c^2_{[(1/2)\alpha(1/2)\beta]}$ boundary. The most general microstructure involving the α and

[1] Derivations of this and other useful relations are given in R. T. DeHoff and F. N. Rhines (eds.), "Quantitative Microscopy," McGraw-Hill Book Company, New York, 1968; also in E. E. Underwood, "Quantitative Stereology," Addison-Wesley Publishing Company, Inc. Reading, Mass., 1970.

β phases (Fig. 3-13*c*) contains c_α^2, c_β^2, and $c^2_{[(1/2)\alpha(1/2)\beta]}$ boundaries. The three corresponding types of intercepts can be tabulated separately, and Eq. (3-8) can be applied three times to determine the amount of each type of boundary.

If structural features, such as grains, are elongated (along the axis of the specimen, for example) or are otherwise nonuniform in various directions, the test lines used for Eq. (3-8) must test uniformly in all directions through the structure to obtain an average value of S_V. On the other hand, purposely testing a structure with separate sets of lines (parallel to the x, y, or z direction, for example) gives quantitative data on the nonuniformity of the microstructure. Problem 3-10 illustrates a typical application of this procedure.

Linear features such as grain edges are specified by an L_V value (length of c^1-type structure in unit volume). The relation

$$L_V = 2P_A \tag{3-9}$$

permits a convenient determination of L_V by a count of P_A (points per unit area); a *point* is an intersection of a c^1 feature with the two-dimensional test plane. Only one type of linear feature exists in a one-phase material (Fig. 3-14*a*). As many as four different c^1 features occur in a two-phase structure

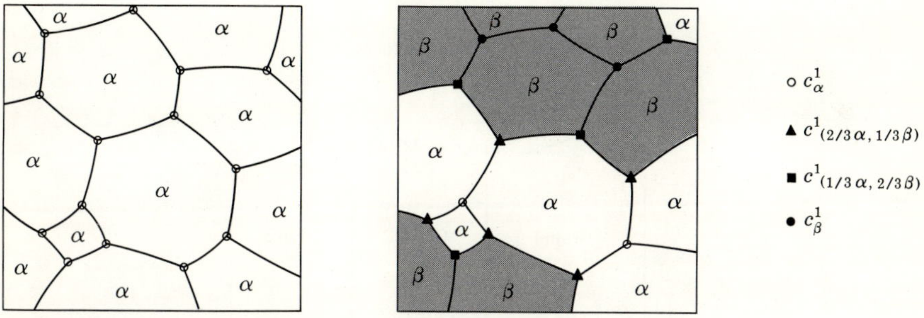

(*a*) Counting to determine the length of grain edge in an α-phase structure.

(*b*) Counting to determine the lengths of four types of edges in an $\alpha+\beta$ alloy.

FIGURE 3-14 Illustration of the point-counting technique for determining L_V in two types of microstructure.

(Fig. 3-14*b*) and require four separate counts to apply Eq. (3-9) for each type of edge.

Since a very large number of counts on many sample microstructures is required to obtain good precision in stereometric microscopy, mechanized procedures of various kinds are employed, often in connection with automatic digital computing. Commercial instruments examine a specimen directly through a microscope and determine the quantities A_A, N_L, N_A (number of features per unit area), distribution of particle sizes, and other useful data.

PROBLEMS

3-1. Elementary physics texts give the equation $R = \lambda/(2n \sin \theta)$ for the resolving power R of a lens in terms of the wavelength λ of the light, the refractive index n of the medium between the lens and object being observed, and θ, the semiangle of the lens. R is the minimum distance between two points for which distinct images can be formed. (*a*) Use a rough sketch of a lens operating in air with $\theta = 30°$ to show that R is then about 0.5×10^{-6} m. (*b*) Show how the use of cedar oil ($n = 1.5$) and of a closer lens-to-object distance can increase $n \sin \theta$ (termed the *numerical aperture*) to about 1.5. (*c*) What is the corresponding value of R? (*d*) The human eye can resolve fine details separated by 10^{-4} m (at a viewing distance of 0.25 m). Use this fact in calculating the useful magnification that can be used with the R value in part *c*. (*e*) The employment of special lenses and lighting system permits the use of ultraviolet light, for example, $\lambda = 3650$ Å (rather than the usual filtered green light of 5400 Å). What is the useful magnification in this case?

3-2. Chemical etching of a polished surface produces low-index crystal facets (Fig. 3-15) whose orientation varies from one grain to the next. Although

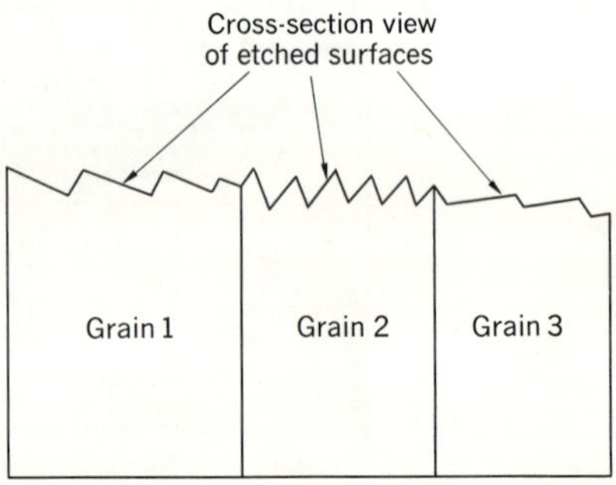

FIGURE 3-15 Schematic illustration of the crystal facets produced by etching typical grains.

these facets are usually not easily resolved using a light microscope, they alter the reflectivity of the surface and produce the contrast between grains seen in photomicrographs. (*a*) Use a schematic drawing of light rays in a microscope to show that one of the two grains will appear bright under the microscope and the other somewhat dark. (*b*) Why will the grain boundary be visible?

3-3. Using the copper-zinc phase diagram (Fig. 3-16) make phase analyses of Muntz metal at the following temperatures: (*a*) 800°C, (*b*) 500°C, (*c*) 260°C, and (*d*) 250°C, and sketch possible microstructures for each of these four temperatures. (*e*) If a specimen of Muntz metal is rapidly cooled from 800°C, its microstructure consists of a small amount of a dark-etching phase that has formed at the grain boundaries of the major phase. Identify the two phases and explain your reasoning.

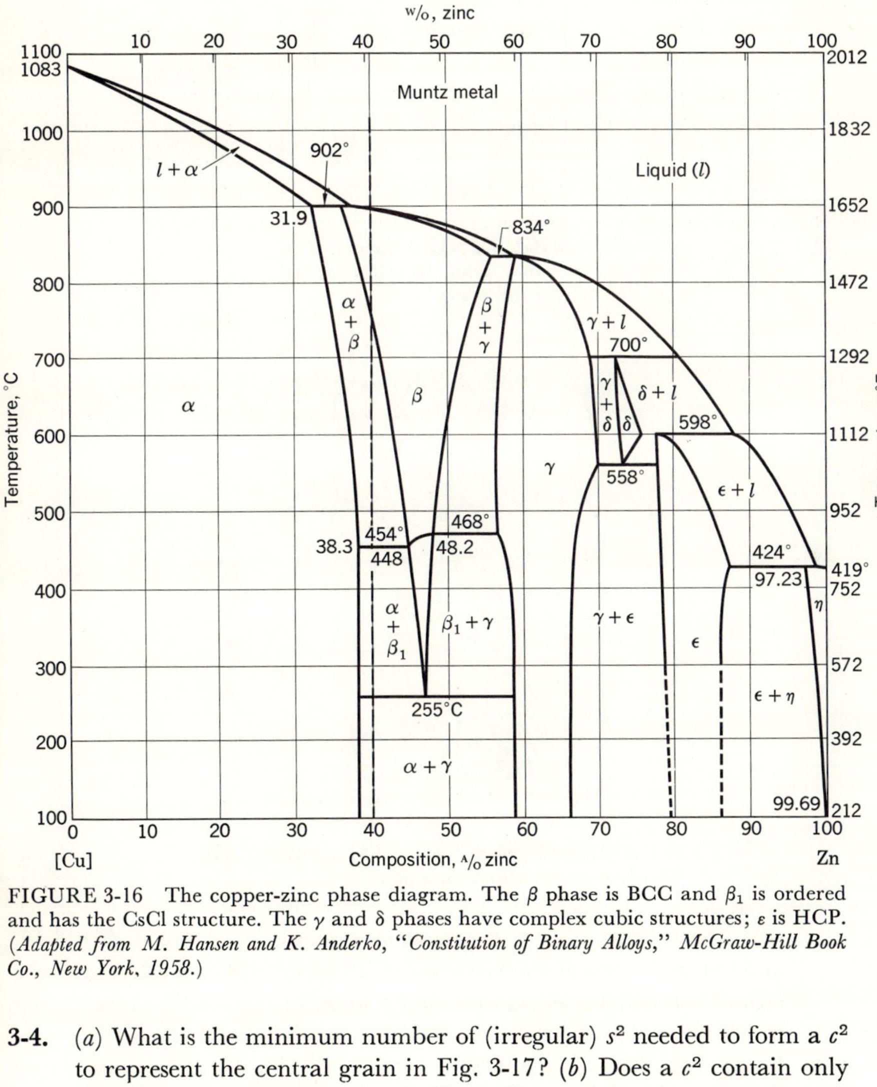

FIGURE 3-16 The copper-zinc phase diagram. The β phase is BCC and β_1 is ordered and has the CsCl structure. The γ and δ phases have complex cubic structures; ε is HCP. (*Adapted from M. Hansen and K. Anderko, "Constitution of Binary Alloys," McGraw-Hill Book Co., New York, 1958.*)

3-4. (*a*) What is the minimum number of (irregular) s^2 needed to form a c^2 to represent the central grain in Fig. 3-17? (*b*) Does a c^2 contain only 2-simplexes or also some s^1 and s^0? (*c*) If one of the edges were curved, as shown by the dashed line, sketch a modified c^2 that would approximate the grain face. (*d*) Using a simple sketch, explain how a grain volume c^3 can be represented by a number of s^3.

3-5. (*a*) Write out Eq. (3-2) in full. (*b*) According to this equation, could an array consist entirely of five-sided grains? (*c*) Of six-sided grains?

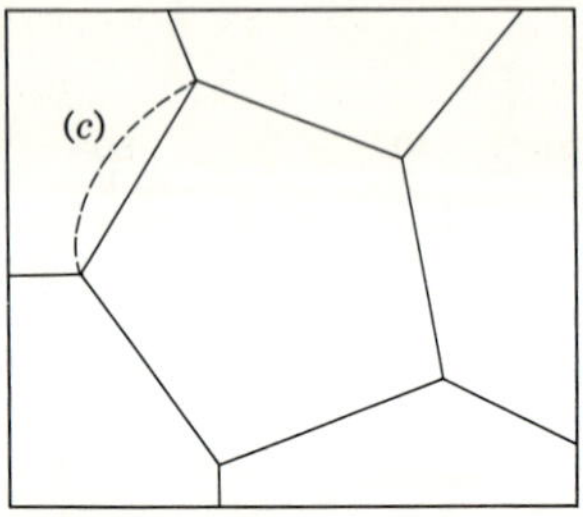

FIGURE 3-17 Sketch of a planar grain face formed by straight edges.

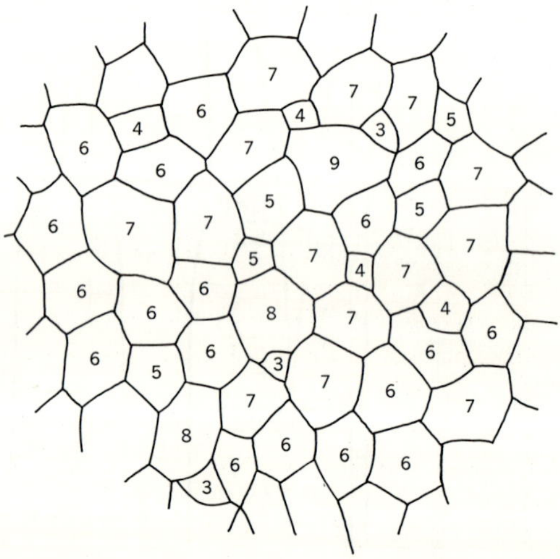

FIGURE 3-18 Sketch of ASTM grain-size standard for steel. (*After C. S. Smith.*)

(*d*) Count the number p_n of polygonal grains of each type in Fig. 3-18, and test the equation for this rather small array of grains. (*e*) What is the *average* number of sides per grain in the array in Fig. 3-18? (*f*) According to Eq. (3-2)?

3-6. Using Fig. 3-13*c* as a model, make a few simple, three-dimensional sketches on which you can identify many of the topological features occurring in a two-phase structure.

3-7. Equations (3-5) and (3-6) show that lineal analysis can determine the volume fraction of a second phase. (*a*) Describe a feasible experimental program for obtaining 10 equally spaced test planes (photomicrographs) and 10 equally spaced test lines on each plane, thus providing a reasonable sample of the structure in question. (*b*) Make lineal analyses of the two photomicrographs of Fig. 3-9 to verify the volume fraction of reinforcing fiber in each. (*c*) Let N be the number of separate readings (about 10) that you made in analyzing one of the photomicrographs in part *b*. A convenient statistical measure of the precision of your readings is the standard deviation (SD). If X_i is the value of the ith reading, then the mean value (XBAR) is

$$\bar{X} = \frac{\sum_{i=1}^{N} X_i}{N} \tag{3-10}$$

and the mean square value (X2BAR) is

$$\overline{X^2} = \frac{\sum_{i=1}^{N} X_i^2}{N} \tag{3-11}$$

The variance (VAR) is the difference

$$\text{VAR} = \overline{X^2} - (\bar{X})^2 \tag{3-12}$$

and the standard deviation is the square root of the variance. The statistical interpretation of the standard deviation is that 95 percent of the total number of readings differ from the mean value $\bar{X}$ by no more than twice the standard deviation (± 2 SD). Use the following computer program to determine the standard deviation of your N readings:

```
C  CALCULATION OF STANDARD DEVIATION, SD
   1 FORMAT (I2)
   2 FORMAT (F6.4)
     SQUARE = 0
     SUM = 0
C  READ N VALUE
   3 READ (5,1) N
C  CALCULATION
   4 DO 7 I = 1, N
   5 READ (5,2) X
   6 SUM = SUM + X
   7 SQUARE = SQUARE + X ** 2
   8 XBAR = SUM/N
   9 X2BAR = SQUARE/N
```

```
   10 VAR = X2BAR - XBAR ** 2
   11 SD = SQRT (VAR)
C  OUTPUT SECTION
   12 WRITE (6,2) SD
   13 STOP
      END
```

3-8. Devise a proof of Eq. (3-6) involving random increments $(\Delta x)_i$ between test lines.

3-9. Use Eq. (3-8) to determine the "density" of all boundaries, S_V, in the steel specimen of Fig. 3-6*a*. A test line L, 0.03 m long, is conveniently drawn on a piece of transparent paper. The values of N_L to be averaged are determined by placing L in about 10 different random positions on the photomicrograph.

3-10. A cylindrical specimen of pure aluminum was found to have the value $S_V = 10^5$ m^2/m^3 for test lines parallel to the cylindrical axis and $S_V = 3 \times 10^5$ for test lines perpendicular to the axis. Sketch a three-dimensional grain structure consistent with these data.

3-11. The specimen of Prob. 3-10 is sectioned 10 times perpendicular to the cylindrical axis at equally spaced positions along its length. Discuss whether the L_V value given by the average P_A count on these 10 sections is a reliable estimate of the average amount of grain edge in the specimen.

3-12. A useful quantitative measure of grain size for equiaxed (i.e., roughly spherical) grains is the ASTM grain-size number n, defined by the relation

$$N = 2^{n-1} \tag{3-13}$$

Here N is the number of grains per square inch in a photomicrograph of the specimen at $\times 100$. What is the grain size of the specimen in Fig 3-18?

REFERENCES

Biggs, W. D.: Elements of Metallography, in R. W. Cahn (ed.), "Physical Metallurgy," 2d ed., p. 655, North Holland Publishing Company, Amsterdam, 1970.

Brandon, D. G.: "Modern Techniques in Metallography," D. Van Nostrand Company, Inc., Princeton, N.J., 1966.

Brick, R. M., et al.: "Structure and Properties of Alloys," 3d ed., McGraw-Hill Book Company, New York, 1965.

Clark, G. L.: "Encyclopedia of Microscopy," Reinhold Publishing Corporation, New York, 1961.

DeHoff, R. T., and F. N. Rhines (eds.): "Quantitative Microscopy," McGraw-Hill Book Company, New York, 1968.

Hartshorne, N. H., and A. Stuart: "Crystals and the Polarizing Microscope," Edward Arnold (Publishers) Ltd., London, 1960.

Kehl, G. L.: "Principles of Metallographic Laboratory Practice," 3d ed., McGraw-Hill Book Company, New York, 1949.

"Metallography 1963," Iron and Steel Institute, Special Report 80, London, 1964.

Nutting, J., and R. G. Baker: "Microstructure of Metals," Institute of Metals, London, 1965.

Rack, H. J., and R. W. Newman: Microstructures, in R. W. Cahn (ed.), "Physical Metallurgy," 2d ed., p. 705, North Holland Publishing Company, Amsterdam, 1970.

Rostoker, W., and J. R. Dvorak: "Interpretation of Metallographic Structures," Academic Press Inc., New York, 1965.

Smallman, R. E., and K. H. G. Ashbee: "Modern Metallography," Pergamon Press, New York, 1966.

Underwood, E. E.: "Quantitative Stereology," Addison-Wesley Publishing Company, Inc., Reading, Mass., 1970.

Wood, E. A.: "Crystals and Light," D. Van Nostrand Company, Inc., Princeton, N.J., 1964.

Chapter Four

Inperfections in Crystals

The properties of materials can be divided into two classes, depending on their sensitivity to the presence of defects in crystalline structures. Properties such as density and the elastic constants are *structure-insensitive*. Many of the properties of greatest technical importance (mechanical strength and ductility, for example) are termed *structure-sensitive* because they are greatly affected by the relatively minor changes in crystal structure caused by defects. Perfect crystals may occur in special cases (in some filamentary growths known as *whiskers*, for example), and they are known to possess remarkably high strength. Ordinary materials, however, contain imperfections of various kinds that produce both useful properties and also such undesirable effects as causing the strength to decrease below that of a perfect lattice. *Dislocations* are the most striking defects and are responsible for the useful property of ductility in metals, ceramics, and crystalline polymers. *Point defects* such as vacancies are the vehicle for most solid-state reactions; impurity atoms considered as defects in a perfect lattice are responsible for the functioning of most semiconductor devices. In fact, the science of materials is in large part a study of relatively small amounts of imperfections imbedded in a crystalline matrix. We first consider how thermal vibrations disturb the perfections of this matrix.

THERMAL VIBRATIONS

A solid material behaves quite differently from a monatomic gas on being heated. The kinetic theory of gases shows that the thermal energy simply increases the velocity of the average atom. However, the atoms in a solid are bound to their neighbors in the crystal lattice and can only oscillate slightly as an expression of their thermal energy. The coordinated oscillations of many neighboring atoms can be pictured as a "wave" moving in the crystal. Because the energy of a particular type of wave can be measured in units known as *phonons*, the thermal energy of the entire solid specimen is represented quite simply by a certain collection of phonons, each one of which is pictured as a special kind of "particle." The present section explains this useful phonon approach to thermal vibrations of solid materials. The relation of phonons to heat capacity is considered in Sec. 12-4.

4-1. Lattice Vibrations

When an atom attempts to vibrate as an expression of its thermal energy, it tends to pull its neighbors along also. Consequently, at a given instant, neighboring atoms have related but different displacements in their lattice plane (Fig. 4-1*a*) and/or transverse to this plane (Fig. 4-1*b*). The relative

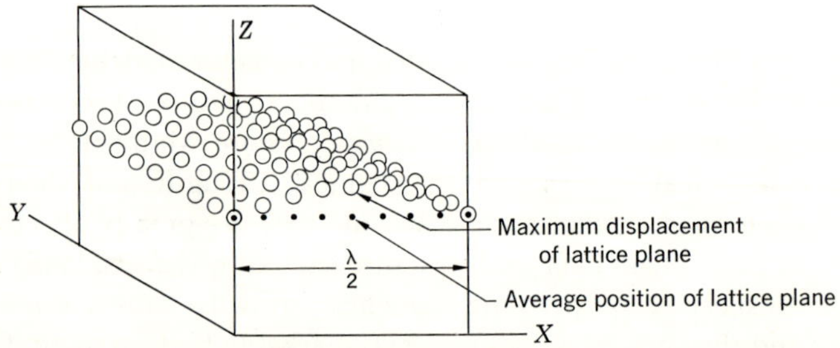

(*a*) A transverse wave in the *Z* plane propagating in the *X* direction.

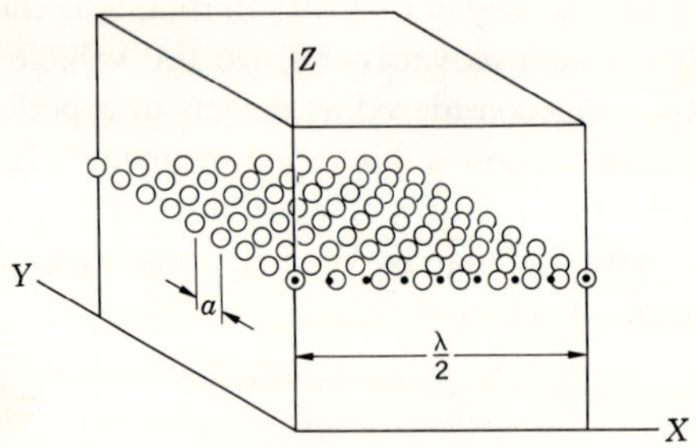

(*b*) A longitudinal wave propagating in the *X* direction. Its frequency is $= \frac{v_s}{\lambda} = \frac{v_s}{16a}$.

FIGURE 4-1 Schematic illustration of a longitudinal and a transverse vibrational mode in a simple-cubic lattice. Only one plane of the array of parallel planes is shown, but the other planes are vibrating similarly.

displacements of neighboring atoms can be described by a sine wave. Since longitudinal waves can move in each of the three coordinate directions, the vibrational spectrum of a pure metal contains three types of longitudinal modes; similarly, it contains six transverse modes (Prob. 4-1). If the displacements are constrained to be zero at the surfaces of the crystal, the wavelengths for waves propagating in the X direction lie within fixed limits. The maximum value occurs when half a wavelength spans the entire length L of the crystal. The minimum is $\lambda = 2a$, where a is the atomic spacing. Below this value the description of a wave in terms of the motion of individual atoms loses physical significance. This entire spectrum of wavelengths can be described by the set of equations

$$\lambda_q = \frac{2L}{q} \qquad q = 1, 2, 3, \ldots, \frac{L}{a} \tag{4-1}$$

We now use a wave vector $\mathbf{K}_q$ to indicate the direction of propagation of a wave having wavelength λ_q. The magnitude is $|\mathbf{K}_q| = K_q = 2\pi/\lambda_q$. Therefore, an alternate description of the spectrum of vibrations of Eq. (4-1) is

$$K_q = \frac{\pi}{L} q \tag{4-2}$$

where K_q is the X component of the wave vector. In three dimensions (see Prob. 4-2), the wave vector $\mathbf{K}$ has components in all three directions. A crystal containing N atoms, with one atom in the primitive cell, has N longitudinal modes and $2N$ transverse modes.

The frequency ν of a given mode of vibration is an important property that can be determined by various experimental techniques. The data for aluminum are shown in Fig. 4-2. $K_{\max}$ is $2\pi/a$ in a face-centered cubic lattice (see Chap. 12) compared to π/a in a simple-cubic lattice. A simplified calculation of the (dispersion) curve of ν versus $\mathbf{K}$ for atoms of mass M yields the one-quarter sine curve shown, but the maximum point must be adjusted to take account of the long-range interactions that actually exist in solid materials. At low values of $\mathbf{K}$ (long wavelengths) a linear relation exists between $\mathbf{K}$ and ν analogous to the propagation of sound waves in an isotropic medium,

$$\nu = \frac{v_s}{\lambda} \tag{4-3}$$

The velocity of sound, v_s, is $(E/\rho)^{1/2}$ for longitudinal waves and $(G/\rho)^{1/2}$ for transverse waves, E being Young's modulus, G the shear modulus, and ρ the density. A typical value of v_s (in aluminum) is 5×10^3 m/sec for longitudinal waves. The phase velocity of short-wavelength vibrations is also v_s, but the physically meaningful group velocity approaches zero as $\mathbf{K}$ approaches $\mathbf{K}_{\max}$.

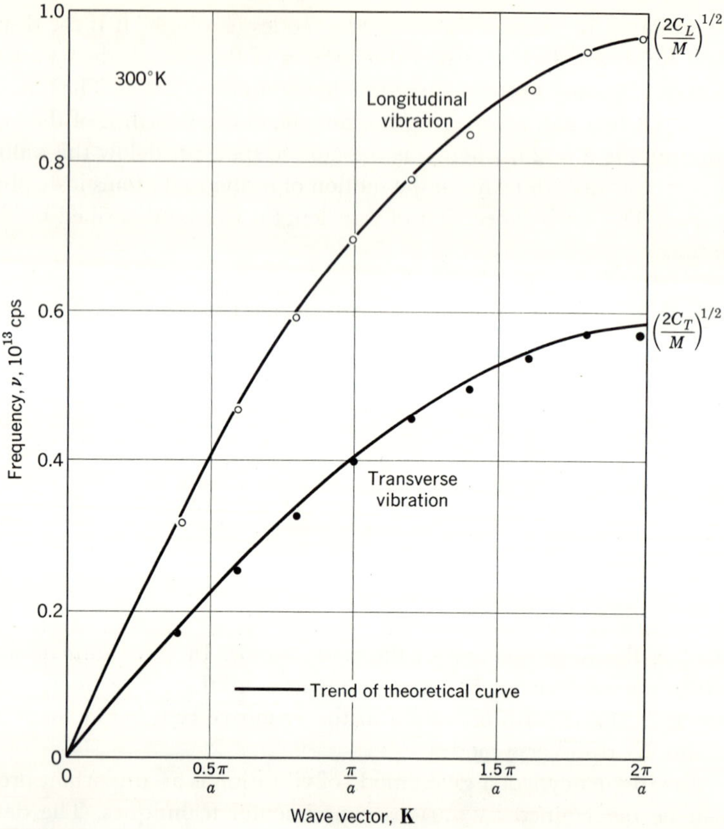

FIGURE 4-2 Variation of frequency ν with increase in wave vector $\mathbf{K}_2$ for longitudinal and transverse waves propagating in the [100] direction in aluminum. When the constants C_L and C_T are suitably chosen, the theoretical curve based on nearest-neighbor interactions fits the experimental data quite well. (*Data of R. Stedman and G. Nilsson.*)

4-2. The Nature of Phonons

Thermal vibrations in solids are conveniently described by quasi-particles called *phonons*, each having an energy $h\nu$, where ν is the frequency of the vibrational mode and h is Planck's constant. A given mode may contain $0, 1, 2, \ldots, n$ phonons, and thus the total energy of a mode having frequency ν_q is

$$E_q = (n + \tfrac{1}{2})h\nu_q \tag{4-4}$$

The integer n is the number of phonons, and $\frac{1}{2}h\nu_q$ is the zero-point energy that exists even at 0°K. At a given temperature T (in degrees Kelvin), the number of phonons in a given vibrational mode varies with time, but the average is given by the Bose-Einstein distribution as

$$n_{av} = \frac{1}{e^{(h\nu_q/kT)} - 1} \tag{4-5}$$

Since $e^x = 1 + x$ when x is small compared to unity,

$$n_{av} \approx \frac{kT}{h\nu_q} \tag{4-6}$$

when kT is large compared to $h\nu_q$. At lower temperatures n_{av} rapidly approaches zero.

For quantitative calculations involving phonons, the spectrum of lattice vibrations is conveniently replaced by the corresponding collection of phonons, pictured as particles existing within the crystal. Although the energies of phonons vary by 10^8 within the usual crystal, the average energy corresponds to a frequency ν_D, close to ν_{max}. The energy of a typical phonon in a material is conveniently characterized using the relation

$$h\nu_D = k\theta \tag{4-7}$$

where the Debye temperature θ has a characteristic value for each material (Table 4-1). The number of phonons in a material at a given temperature is

TABLE 4-1. THE DEBYE TEMPERATURE OF SEVERAL IMPORTANT SUBSTANCES

Substance	**θ, °K**	**Substance**	**θ, °K**
Ag	220	Hg	80
Al	428	In	110
Au	165	K	90
Be	1000	Li	340
C (diamond)	1840	Mg	410
Ca	230	Na	155
Cd	210	Ni	460
Cs	40	Pb	90
Cu	330	Sn	200
Fe	420	Ti	430
Ga	325	W	310
		Zn	308

easily calculated from Eq. (4-5). For aluminum ($\theta = 428°K$) at heat-treating temperatures (800°K), for example,

$$n_{av} = \frac{1}{e^{(k428/k800)} - 1} = 1.4 \tag{4-8}$$

since there are $3N$ vibrational modes, the total number of phonons is $4.2N$ in an aluminum crystal containing N atoms. Equation (4-6) shows that the number of phonons increases approximately linearly with temperature above the Debye temperature θ.

POINT DEFECTS

Figure 4-3 illustrates the common types of point defects in solid materials. These defects can be divided into three classes: (1) lattice-site defects, such as vacancies and interstitialcies; (2) compositional defects, that is, impurity atoms; and (3) electronic defects, described later using Fig. 4-6. Nonequilibrium lattice-site defects can be produced by deformation or irradiation (Sec. 8-2), but the present treatment considers only defects existing under equilibrium conditions at a given temperature and begins with the relatively simple behavior of metals. In nonmetallic materials the three types of defects are often interrelated, as described in Sec. 4-5. Typically, a material contains several kinds of point defects, but usually one kind dominates a certain phenomenon. For example, vacancies are responsible for diffusion in most metallic systems. Because of their importance, vacancies have been chosen as the example in the following treatment, but similar analyses can be made for the other point defects.

4-3. Equilibrium Concentration of Vacancies in Metals

The fraction N_V° of vacant lattice sites (i.e., vacancies) increases with temperature.[1] The quantitative relation is determined by measuring simultaneously the changes in lattice constant a and in length L as the specimen is slowly heated from low temperatures (where $N_V^\circ \approx 0$) almost to the melting point (see Prob. 4-4). The lattice constant measures the uniform thermal expansion, which is not affected by the presence of vacancies; the overall length of the specimen includes the volume occupied by the vacancies, which have the same partial molar volume as atoms of the metal. Curves of vacancy concentration (Fig. 4-4) typically reach a value of about 10^{-4} at the melting point of the metal.

For thermodynamic calculations, vacancies are conveniently considered to be a component, V; empty space is the corresponding "pure vacancy" phase. At equilibrium at temperature T (in degrees Kelvin) a solid component A is pictured as a solid solution α containing a certain fraction N_V° of vacant lattice sites. Imagine that an additional mole of vacancies, V, is

[1] The superscript (°) indicates the equilibrium concentration of the vacancies.

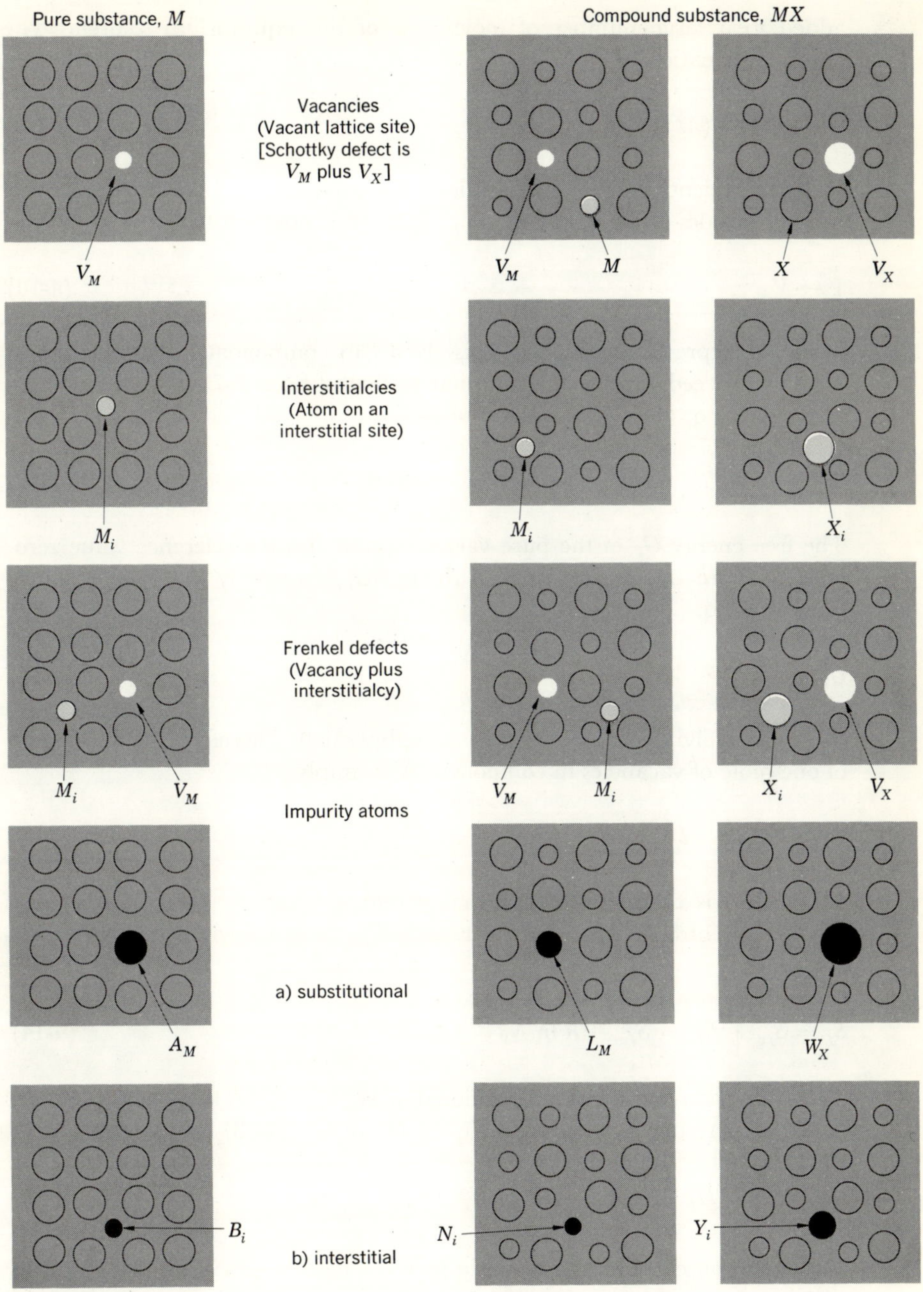

FIGURE 4-3 Schematic illustrations of simple point defects that occur in the crystal lattices of pure substances M and of compounds MX (M represents a metal and X a nonmetal). The notation V_M, etc., is explained in Sec. 4-5. Electronic defects are illustrated separately in Fig. 4-6.

added to a large number of moles, M, of this equilibrium solution. The chemical equation is

$$V + M\alpha \rightarrow (M + 1)\alpha' \tag{4-9}$$

If M is sufficiently large, α' is essentially the same as α, and $M\alpha$ can be omitted from both sides of the equation. The equation is now written

$$V \rightarrow V_A \tag{4-10}$$

where V_A represents vacancies "dissolved" in component A. Calculation of N_V° now proceeds by a method similar to that of Secs. 2-7 and 2-8. Since the reaction of Eq. (4-10) occurs at equilibrium,

$$\Delta G = 0 = G_{V_A} - G_V \tag{4-11}$$

The free energy G_V of the pure vacancy phase has the reference value zero, and therefore G_{V_A} is also zero. Although G_{V_A} is strictly the total quantity [compare Eq. (2-42)],

$$G_{V_A} = 1\bar{G}_V + M(\bar{G}'_A - \bar{G}_A) \tag{4-12}$$

the term involving component A is negligibly small. Therefore, the free energy of one mole of vacancies in component A is simply

$$G_{V_A} = \bar{G}_V = H_f - TS_f \tag{4-13}$$

H_f (or $\bar{H}_V$) is the increase in enthalpy (energy) and S_f (or $\bar{S}_V$) is the total increase in entropy. According to Eq. (2-29), the entropy has two parts, and thus

$$S_f = S_{fv} + S_{fc} = S_{fv} - R \ln N_V^\circ \tag{4-14}$$

The expression for the configurational part, $S_{fc} = -R \ln N_V^\circ$, is from Eq. (2-46). The result of substituting Eq. (4-14) in Eq. (4-13) is

$$G_{V_A} = 0 = H_f - TS_{fv} + RT \ln N_V^\circ \tag{4-15}$$

This equation can be rearranged to give the desired expression for N_V°; thus

$$N_V^\circ = e^{-(H_f - TS_{fv})/RT} = e^{-G_f/RT} \tag{4-16}$$

Although G_f is usually referred to as the *free energy of formation* of vacancies, it lacks the term for configurational entropy. Equation (4-16) correctly describes the exponential increase of vacancy concentration with temperature observed experimentally (Fig. 4-4). Since H_f and S_{fv} can be determined from an experimental curve, Eq. (4-16) is often written

$$N_V^{\circ} = e^{S_{fv}/R} e^{-H_f/RT} \tag{4-17}$$

The first exponential is virtually independent of temperature and typically has a value on the order of e^1, that is, about 3.

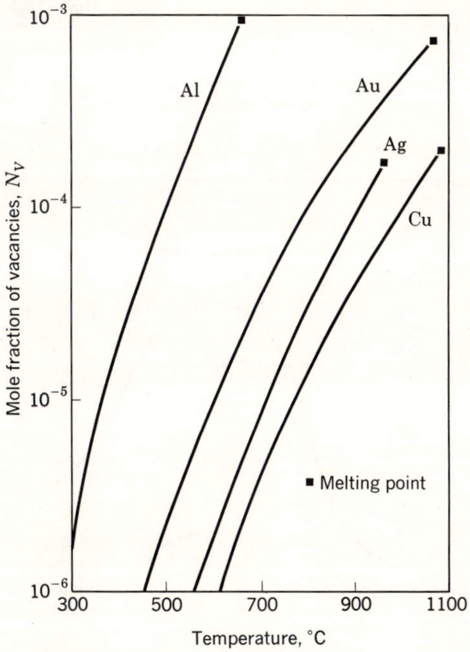

FIGURE 4-4 Representative vacancy concentrations in several metals as a function of temperature. (*Data of R. O. Simmons and R. W. Balluffi.*)

4-4. Sources and Sinks for Vacancies in Metals

Vacancies have both an equilibrium aspect, represented by N_V°, and a kinetic aspect, the net rate R of formation. A vacancy can form most easily at the surface of a metal, and in particular at a kink (step) (Fig. 4-5) in one of the ledges that form the surface (Sec. 5-3). Atom I can move to the new position on the surface (and thus create a vacancy) only by overcoming the energy "hill" of magnitude† ΔG_f^A, a quantity known as the *free energy of*

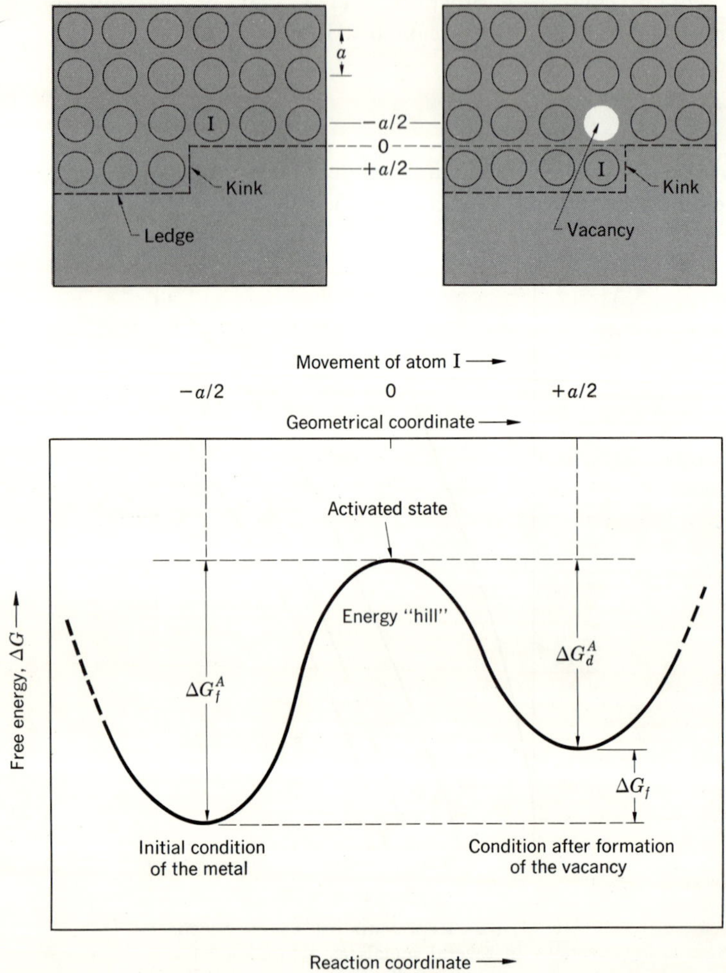

FIGURE 4-5 Use of the activated-complex approach to treat the formation of a vacancy at a kink on a surface ledge. ΔG_f^A and ΔG_f refer to one vacancy (rather than to a mole of vacancies) in this illustration.

activation. When atom I is at the midpoint of its movement (at the top of the energy hill), this local region in the metal is an *activated complex* and has the

† $\Delta G = G - G_{\text{(reference)}}$. If the reference value is zero, ΔG and G are identical. The ΔG notation is the one generally used in the literature.

additional free energy ΔG_f^A that characterizes the activated state. The activated-complex method of chemical kinetics gives the number of vacancies formed per square centimeter of surface per second as

$$R_f = N_s K_f e^{-\Delta G_f^A/RT} \tag{4-18}$$

N_s is the number of kinks per square centimeter of surface, and K_f is a factor dependent on the details of the reaction. The rate R_d of destruction of vacancies is the subject of Prob. 4-5. The net rate R of formation of vacancies is

$$R = R_f - R_d = N_s K_f (1 - N_V/N_V^0) e^{-\Delta G_f^A/RT} \tag{4-19}$$

where N_V^0 is the equilibrium mole fraction of vacancies. Both the creation and destruction of vacancies are occurring constantly, but at equilibrium the two rates are equal, and $R = 0$.

Although vacancies form most easily at the surface (or grain boundary) of a metal, they can form relatively easily at dislocation lines (especially at jogs, discussed below), and with a great expenditure of energy they can form (as a Frenkel defect, shown in Fig. 4-3) at a normal lattice site. The total rate of vacancy formation should therefore be written for a typical metal as

$$R = (C_s e^{-30T_m/RT} + C_l e^{-40T_m/RT} + C_n e^{-120T_m/RT})\left(1 - \frac{N_V}{N_V^0}\right) \tag{4-20}$$

where T_m (in degrees Kelvin) is the melting point of the metal. The ΔG_f^A values in this equation (expressed as multiples of T_m) have been estimated quantitatively to facilitate the following discussion. C_s, C_l, and C_n are rate constants (for example, $C_s = N_s K_{fs}$) for surfaces, dislocation lines, and normal atoms, respectively. The C values depend strongly on the number of sites at which the formation reaction can occur; in a typical specimen $C_s \ll C_l \ll C_n$. At sufficiently low temperatures, below about $0.5T_m$, the exponential term for surface sources is dominant, and thus vacancies tend to originate or be destroyed at surfaces (including grain boundaries). At higher temperatures the large value of C_l makes dislocations the major source of vacancies, except for regions immediately adjacent to surfaces. The term involving C_n, which has a very high energy, can contribute only at very high temperatures; an example is radiation damage, in which neutron irradiation produces a thermal spike and causes momentary local melting of a small portion of the metal.

Although Eq. (4-19) is satisfactory for treating thermal effects involving relatively small departures from N_V^0, the problem of large supersaturation (or undersaturation) of vacancies requires a different approach. The chemical potential μ, which was introduced in Eq. (2-51) to describe phase equilibria, can be expressed quantitatively for vacancies as follows:

$$\mu_V = \mu_V^\circ + RT \ln \frac{N_V}{N_V^\circ} \tag{4-21}$$

an expression analogous to that for G in Eq. (2-20). Usually the reference value μ_V° is set equal to zero so that $\mu_V = 0$ when $N_V = N_V^\circ$, the equilibrium vacancy concentration. If aluminum, for which $H_f = 18$ kcal/mole, is heated at 600°C and then quenched in water near 0°C to retain the vacancy concentration characteristic of 600°C, Eq. (4-21) gives $N_V/N_V^\circ = e^{23}$ as the supersaturation. The chemical potential is then $\mu_V = 23RT$, an energy value much larger than that available thermally. Consequently, vacancies are able to precipitate from such supersaturated solutions by mechanisms not dependent on either the surface or dislocations. The formation of vacancy disks under these conditions is discussed later in connection with Fig. 4-12.

4-5. Point Defects in Nonmetallic Crystals

In contrast to metals, in which the electrons are free to move through the entire specimen, nonmetals are characterized by electrons bound in local configurations, either in covalent bonding or in ionic bonding. In the state of lowest energy, all of the electrons are in the valence band (Fig. 4-6) according to the standard theory described in Chap. 12. None of the electrons can contribute to electrical conductivity as long as the valence band is full. When an electron gains energy E_g (Fig. 4-6*a*) by thermal activation, it enters the conduction band and simultaneously leaves a positive hole in the valence band. Both the electron and the hole contribute to electrical conductivity. These particles are conveniently pictured as being point defects for the purpose of calculating their concentrations in the crystal and their effects on the properties of materials. Impurity (dopant) atoms, in amounts less than 0.1 percent, are also considered as defects in nonmetallic crystals because of their role in creating special energy levels, E_d and E_a (Fig. 4-6*b* and *c*), and in causing the formation of vacancies or interstitialcies. "Doping" of pure compounds with impurities permits the development of (electronic) semiconductivity in crystals, such as KBr, that ordinarily conduct only by the much slower motion of ions. To furnish an adequate framework for the treatment of these important topics, the methods of defect chemistry are introduced in this section.

PURE SILICON AND DOPING. The process of thermal activation of an electron e into the conduction band could be expressed as the reaction

$$e \text{ (valence band)} \rightarrow e \text{ (conduction band)} \tag{4-22}$$

This notation, however, fails to emphasize the importance of the empty electronic state (the hole, h) in the valence band created by this reaction. This accompanying process could be written

$$h \text{ (conduction band)} \rightarrow h \text{ (valence band)} \tag{4-23}$$

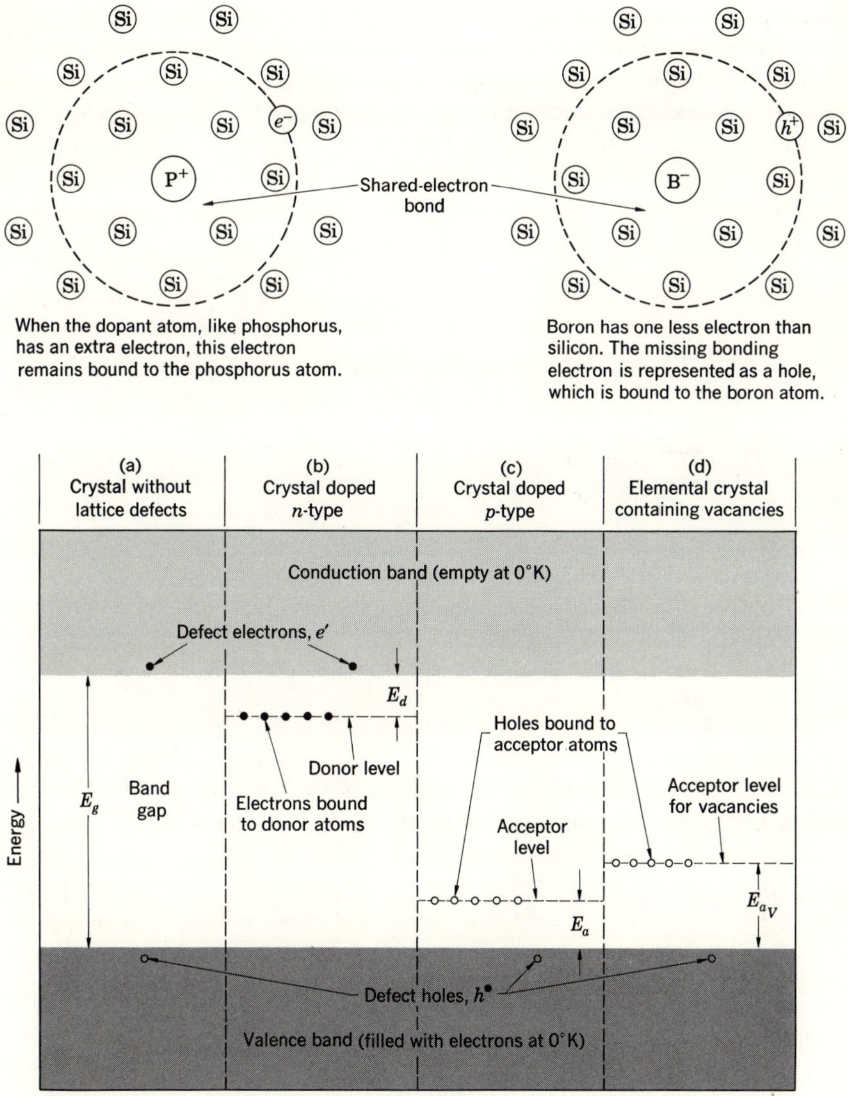

FIGURE 4-6 Schematic illustration of the energy levels of electrons (and of holes) in a nonmetallic crystal. (*Crystal models after J. S. Blakemore, "Solid State Physics," W. B. Saunders Company, Philadelphia, 1969.*)

In the notation of defect chemistry a species in its normal electronic state is denoted by a cross ($^\times$); for example, $e^\times$ represents an electron in the (nearly) full valence band, and $h^\times$ represents a hole in the conduction band (which is nearly empty of electrons). A prime (′) indicates that a species has one electronic unit of *negative* charge relative to the species it replaced; for example, an electron in the conduction band (defect electron) is denoted e' since it has unit negative charge relative to the hole, $h^\times$, that is replaced in the conduction band. Similarly, a filled circle ($^\bullet$) indicates a unit of *positive* charge, so that $h^\bullet$ denotes a hole in the valence band (defect hole) since the hole is positively charged relative to the electron it replaced in the valence band. This notation of defect chemistry permits writing the sum of Eqs. (4-22) and (4-23) in the following convenient form

$$e^\times + h^\times \rightarrow e' + h^\bullet \tag{4-24a}$$

Since only the defect species, e' and $h^\bullet$, need be considered in treating the electronic properties of nonmetals, Eq. (4-24*a*) is usually written

$$0 \rightarrow e' + h^\bullet \tag{4-24b}$$

The fraction of defect electrons, N_e, and of defect holes, N_h, are related by an equilibrium constants, as follows:

$$K_g = N_e N_h \tag{4-25}$$

By analogy with the corresponding equilibrium constant for a chemical reaction [Eq. (2-21)], K_g can be expressed as

$$K_g = \exp \frac{-\Delta G'}{kT} \tag{4-26}$$

The value of $\Delta G'$ depends on the energies of both electrons and holes. For simple band structures, such as that of silicon, K_g has approximately the value

$$K_g = (0.9 \times 10^{-12}) T^3 \exp \frac{-E_g}{kT} \tag{4-27}$$

where E_g is the energy-band gap (Fig. 4-6*a*), which is equal to 1.14 ev for silicon. The numerical factor is discussed in the references at the end of the chapter. The extremely low defect concentration at room temperature is shown in area II (intrinsic range) of Fig. 4-7. A temperature of about 1000°K would be required to produce a technically useful concentration of defects, $N_e = 10^{-6}$, corresponding to appreciable intrinsic semiconductivity.

The temperature range for semiconductivity can be lowered to room temperature by using extrinsic semiconductors (containing impurity or "dopant" atoms), rather than intrinsic semiconductors like pure silicon or germanium. When an atom such as phosphorus having five valence electrons substitutes for a silicon atom in the crystal lattice, it provides one more electron than required for the usual covalent binding. The excess positive charge on the phosphorus tends to keep the extra electron in its vicinity at low temperatures, but the energy E_d of the corresponding donor state lies only 0.044 ev below the conduction band, as shown schematically in Fig. 4-6*b*. In the notation of defect chemistry, $P_{Si}^{\times}$ represents the phosphorus atom (including its extra electron) on a lattice site normally occupied by a silicon atom. Thermal energy can cause the ionization of the phosphorus atom according to the reaction

$$P_{Si}^{\times} \rightarrow P_{Si}^{\bullet} + e' \tag{4-28}$$

in which the defect electron e' is "donated" to the conduction band, leaving the phosphorus atom with unit positive charge ($^{\bullet}$). Since the number of $P_{Si}^{\bullet}$ and of e' is almost equal, the equilibrium for this reaction is conveniently expressed as

$$(N_e)^2 = N_d(0.48 \times 10^{-7})T^{3/2} \exp \frac{-E_d}{kT} \tag{4-29}$$

N_d is the mole fraction of dopant added, and E_d is the energy separating the donor and the conduction levels (Fig. 4-6*b*). Complete ionization corresponds to $N_e = N_d$ and occurs below room temperature for the usual concentrations of phosphorus in silicon, $N_d = 10^{-6}$ (see Prob. 4-6). The electrical conduction provided by the e' electrons is called *n-type* semiconductivity because conduction is by negative carriers.

The addition of an impurity such as boron (having a valence of $+3$) to silicon or germanium causes *p-type* semiconductivity in which positive carriers, the defect holes $h^{\bullet}$, are responsible for electrical conductivity. The appropriate analysis is similar to that for *n*-type semiconductors (see Prob. 4-7) and leads to an equation analogous to Eq. (4-29), except that the energy E_a of the acceptor level (Fig. 4-6*c*) replaces E_d, and the concentration N_a of acceptor impurity atoms is used in place of N_d. When appreciable concentrations of both donor and acceptor impurities occur simultaneously, the behavior is more complex (see references at end of chapter).

Vacancies form in silicon in the same manner as in metals, but in the notation of defect chemistry the reaction is written in the slightly different form

$$V_V \rightarrow V_{Si}^{\times} \tag{4-30}$$

The subscript ($_V$) indicates that the vacancy V_V was originally in the "vacancy phase" (empty space). After the reaction the vacancy V_{Si} is on a silicon site, indicated by the subscript ($_{Si}$). The cross ($^\times$) emphasizes that a *neutral* silicon atom has been replaced by the vacancy. The concentration of vacancies, N_V, is given by Eq. (4-17) with $H_f = 2.32$ ev. As shown in Fig. 4-6*d*, an acceptor level E_{aV} is associated with vacancies in silicon, and therefore the ionization reaction

$$V_{Si}^\times \rightarrow V_{Si}' + h^\bullet \tag{4-31}$$

can occur. The corresponding equilibrium constant is

$$K = \frac{[V_{Si}'][h^\bullet]}{[V_{Si}^\times]} \tag{4-32}$$

where the square brackets indicate concentration in mole fraction of the bracketed species. The symbol N_h is usually employed for $[h^\bullet]$.

Although the total spectrum of defect reactions in a crystal like silicon appears complex, useful calculations can be made by treating only the reaction that is dominant under given conditions. Consider the change in electronic-carrier concentrations N_e and N_h produced at room temperature in pure silicon (the central composition in Fig. 4-7) by additions of either phosphorus or boron. In pure silicon K_g can be calculated using Eq. (4-27) and gives values $N_e = N_g \approx 10^{-12}$. The electron-hole reaction [Eq. (4-24*b*)] maintains this value constant within the entire intrinsic range of the diagram of Fig. 4-7. Small amounts of phosphorus, such as 10^{-14} mole fraction, even if fully ionized, could change N_e by only 1%. At concentrations above 10^{-12}, however, fully ionized phosphorus atoms are a decisive factor since $N_e = N_d$, and the concentration of carriers increases linearly with increase in phosphorus. The product $N_h N_e$ is a constant [Eq. (4-25)], and therefore the value of N_h decreases in this region. If boron is the dopant, rather than phosphorus, a similar behavior occurs except that holes become the dominant carrier. At 300°K the vacancy concentration is only 10^{-39} and obviously can be neglected.

In addition to the pronounced variations in carrier concentrations caused by doping, changes in temperature can radically change the behavior of a given material, *n*-type silicon for example. Problem 4-8 considers the four regions that occur in this case. Other aspects of the electrical properties and doping of semiconductors are considered in Chaps. 6 and 12.

PURE MX-TYPE COMPOUNDS. Nonmetallic compounds, whether an ionic crystal like NaCl or a covalent one like SiC, are conveniently classified as belonging to the MX group, where M represents the metallic atom and X represents the nonmetallic atom. Other groupings such as MX_2 (for example $BaCl_2$) are discussed in the next chapter. Defect reactions in compounds are sufficiently complex that full use must be made of the defect notation introduced previously; for example,

Chemical species ↘   Relative charge ↙

$$V^{\bullet}_{\text{Na}}$$

↖ Lattice site

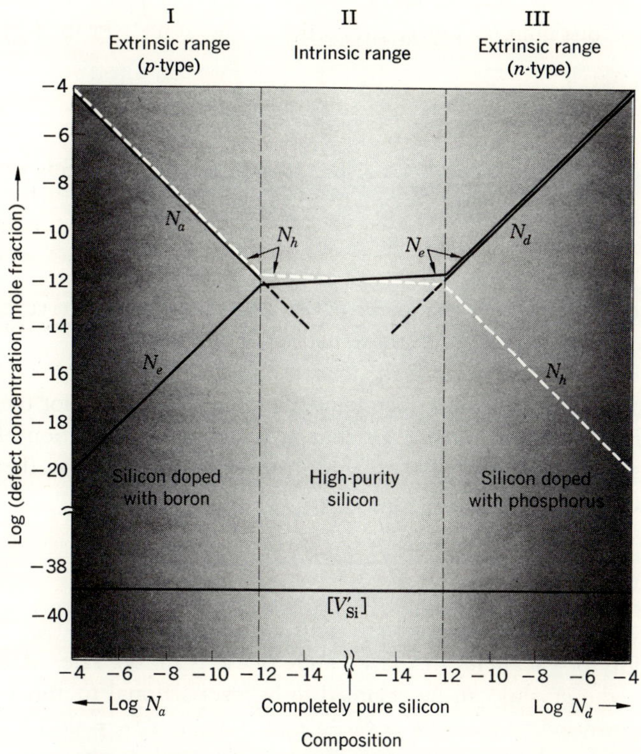

FIGURE 4-7 Logarithmic plots of the concentrations of several defects versus the concentration for a typical donor (phosphorus) and acceptor (boron) dopant in silicon at room temperature, 300°K. Note that coincident lines have been displaced for clarity.

The three features recorded by this notation can be used to verify that mass, electrical charge, and lattice sites are conserved in a given set of defect reactions.

Consider the defect reaction for the formation of vacancies in a crystal of PbS, which has the NaCl crystal structure

$$2V_V \rightarrow V_{\text{Pb}}^{\times} + V_{\text{S}}^{\times} \qquad (4\text{-}33)$$

The vacancies are pictured as coming from the vacancy phase ($_V$) and forming a *pair* of vacancies, thus conserving the 1:1 ratio of sites in the PbS lattice. Since the symbol $V_{Pb}^{\times}$ denotes the vacancy created by removing an entire lead atom (Pb^{2+} and two electrons), this region of the lattice is deficient in electrons. The vacancy pulls an electron from the valence band,

$$V_{Pb}^{\times} \rightarrow V_{Pb}' + h^{\bullet} \tag{4-34}$$

becomes negatively ionized, and creates a hole $h^{\bullet}$. The $V_S^{\times}$ vacancy behaves in an analogous manner but creates a defect electron e' (Prob. 4-9); thus,

$$V_S^{\times} \rightarrow V_S^{\bullet} + e' \tag{4-35}$$

Therefore, the usual vacancy-pair (Schottky) reaction in PbS is actually

$$2V_V \rightarrow V_{Pb}' + V_S^{\bullet} \tag{4-36}$$

The electrons and holes are pictured as recombining by the reverse of Eq. (4-24*b*) and are omitted from this equation.

The defect reaction for the formation of vacancies [Eq. (4-36)] can be treated thermodynamically in the same manner as Eq. (2-13) for the oxidation of copper. The molar free-energy change ΔG_p for the formation of a pair of vacancies can be related to the concentration of vacancies by the analog of Eq. (2-21); thus

$$\Delta G_p = \Delta G_p^\circ + RT \ln \frac{a_{V_{Pb}'} a_{V_S^{\bullet}}}{1^2} \tag{4-37}$$

The activity of vacancies in the vacancy phase is unity. The activities vacancies in the crystal can be assumed to be proportional to the corresponding concentrations (Henry's law). At equilibrium, $\Delta G_p = 0$, and by analogy with Eq. (4-17) the equilibrium vacancy concentrations are

$$[V_{Pb}'][V_S^{\bullet}] = K_p = K_p^\circ \exp \frac{-E_p}{RT} \tag{4-38}$$

Here the concentration in mole fraction (indicated by square brackets) was substituted for the activity since the activity the constants of proportionality being included in the constant K_p or K_p°. The value of E_p for PbS, 1.78 ev, is typical of nonmetallic crystals; Prob. 4-10 shows that the vacancy concentration is very low at normal temperatures.

Interstitialcies (see Fig. 4-3) occur less commonly than vacancies in crystals. Since the atoms are usually large compared to the interstices in the lattice, the energy to form an interstitialcy is large. Notable exceptions are materials with an open structure (ZnS, CaF_2) and the silver halides, such as AgBr. The energy required to form a silver interstitialcy Ag_i is lower than that required to form a bromine vacancy V_{Br}. Interstitial sites are pictured as being either vacant, V_i, or occupied by silver, and thus the reaction of forming a Frenkel defect [vacancy plus interstitialcy, both ionized (Fig. 4-3)] is written

$$V_i + Ag_{Ag} \rightarrow Ag_i^{\bullet} + V'_{Ag} \tag{4-39}$$

Here the usual practice has been followed by omitting the ($^{\times}$) since the meaning of Ag_{Ag}, for example, is unambiguous. By analogy with Eq. (4-38), the concentration of defects is (see Prob. 4-12)

$$[Ag_i^{\bullet}][V'_{Ag}] = K_F^{\circ} \exp \frac{-E_F}{RT} \tag{4-40}$$

where $E_F \equiv \Delta H_F$ is the enthalpy of formation of Frenkel defect pairs. E_F is about 1.0 ev, and K_F° is about 10^3.

Compounds can be classed as *stoichiometric* and *nonstoichiometric*. A pure, stoichiometric compound has a negligible concentration of lattice defects at room temperature; its composition must usually be adjusted to produce a useful electronic material. Such adjustment can be done either by adding impurity atoms (doping in a manner similar to that already described for silicon) or by displacing the composition of the crystal from the stoichiometric composition MX to a nonstoichiometric one, $M_{1-\delta}X$, where δ is usually a small number on the order of 10^{-5}. Nonstoichiometric compounds have a significant value of δ in their normal conditions; δ ranges from very small numbers to as high as 10^{-1} (for $Fe_{1-\delta}O$). Adjustments of composition are also made in nonstoichiometric compounds; illustrative calculations for PbS are given here.

The composition of a macroscopic specimen can be adjusted only if there is a suitable source of the species to be added and if diffusion of this species is rapid. For diffusion, a temperature of about $\frac{2}{3}T_m$ is required, as described in Chap. 6. The addition of atomic species can be done in various ways, but a common industrial procedure involves the vapor phase; the theory of this method will now be described. If pure PbS is placed in a sealed, evacuated container and heated to 1100°K (827°C), sublimation will occur by the reaction

$$PbS[= Pb_{Pb} + S_S] \rightarrow Pb_V + \tfrac{1}{2}(S_2)_V \tag{4-41}$$

The equilibrium constant† K_s for this reaction is

$$K_s = p_{Pb}(p_{S_2})^{1/2} = (1.7 \times 10^9) \exp \frac{-3.48}{kT} \tag{4-42}$$

At 1100°K the value is

$$K_s = 2.7 \times 10^{-7} \text{ atm}^{3/2} \tag{4-43}$$

Since lead occurs as atoms and sulfur as S_2 molecules in the gas phase, the partial pressure of cadmium is twice that of sulfur. Consequently,

$$p_{Pb} = 2p_{S_2} = 2^{1/3}K_s^{2/3} \tag{4-44}$$

These partial pressures have the significant property that they represent the minimum value of the total pressure

$$p_{\text{total}} = p_{Pb} + p_{S_2} \tag{4-45}$$

of all possible combinations of values of p_{Pb} and p_{S_2} satisfying the equilibrium relation [Eq. (4-42)].

If an additional source of lead is available to the vapor surrounding the PbS crystal, an excess of lead atoms will be incorporated into the lattice, accompanied by an equal number of vacancies V_S to maintain the 1:1 ratio of lattice sites. The corresponding defect reaction, including ionization, is

$$Pb_V + V_V \rightarrow Pb_{Pb} + V_S^{\bullet} + e' \tag{4-46}$$

with equilibrium constant

$$K_{Pb} = \frac{[V_S^{\bullet}]N_e}{p_{Pb}} = (1.05 \times 10^9) \exp \frac{1.34}{kT} \tag{4-47}$$

When an excess of sulfur is present in the gas phase, the defect reaction is

$$\tfrac{1}{2}(S_2)_V + V_V \rightarrow S_S + V'_{Pb} + h^{\bullet} \tag{4-48}$$

for which

$$K_S = \frac{[V'_{Pb}]N_h}{(p_{S_2})^{1/2}} = (1.85 \times 10^{-5}) \exp \frac{-0.64}{kT} \tag{4-49}$$

† The experimental data for PbS and much additional information are given in F. A. Kröger, "The Chemistry of Imperfect Crystals," p. 500, North Holland Publishing Company, Amsterdam, 1964.

The value of K_g [Eq. (4-25)] for PbS is

$$K_g = (1.37 \times 10^{-3}) \exp \frac{-1.0}{kT} \tag{4-50}$$

These values of the equilibrium constants permit one to plot the concentrations of defects in PbS in a manner analogous to Fig. 4-7 to give the *Brouwer diagram* for this crystal (Fig. 4-8). Again, a useful starting place is the central section of the diagram where the stoichiometric composition exists. Since the reaction of forming an electron-hole pair is the dominant one in this region, $N_e = N_h = K_g^{1/2} = 2.0 \times 10^{-4}$ at 1100°K. This value determines the horizontal line in the diagram. In region III, $N_e = [V_S^{\bullet}]$ according to Eq. (4-46), and therefore the boundary of regions II and III is at

$$[V_S^{\bullet}]N_e = 4 \times 10^{-8} \tag{4-51}$$

The value of K_{Pb} [Eq. (4-47)] is 1.4×10^{-3} at 1100°K, and thus p_{Pb} at the boundary of regions II and III is

$$p_{Pb} = \frac{[V_S^{\bullet}]N_e}{K_{Pb}} = \frac{4.0 \times 10^{-8}}{1.4 \times 10^{-3}} = 2.9 \times 10^{-5} \text{ atm} \tag{4-52}$$

This value determines the location of the vertical dashed line on the right. Similarly, the boundary of regions I and II (see Prob. 4-13) is $p_{S_2} = 3.2$ atm, which corresponds to a partial pressure of lead of 1.5×10^{-7} atm. The values of $[V_S^{\bullet}]$ in region I are calculated using Eq. (4-49); those of $[V'_{Pb}]$ in region III are calculated using Eq. (4-52).

The concentration of V'_{Pb} and $V_S^{\bullet}$ in region II can be estimated starting with the reaction for the creation of a pair of ionized vacancies,

$$2V_V = V'_{Pb} + V_S^{\bullet} \tag{4-53}$$

The equilibrium constant is

$$K_p = [V'_{Pb}][V_S^{\bullet}] = 2.4 \times 10^{-2} \exp \frac{-1.78}{kT} \tag{4-54}$$

and has the value 1.9×10^{-10} at 1100°K. At the stoichiometric composition $[V'_{Pb}] = [V_S^{\bullet}] = 1.4 \times 10^{-5}$ mole fraction. This is the final piece of data that permits sketching the full diagram.

The diagram shows that a crystal of stoichiometric composition would be in equilibrium with a pressure of 2.0×10^{-6} atm of lead. Under normal conditions of preparation a crystal is under the minimum pressure,

5.3×10^{-5} atm of lead and 2.6×10^{-5} atm of sulfur. The corresponding excess of $V_S^{\bullet}$ (and therefore of lead atoms) is $\delta = (K_{Pb}p_{Pb})^{1/2} = (1.4 \times 10^{-3} \times 5.3 \times 10^{-5})^{1/2} = 2.7 \times 10^{-4}$. This value is for the crystal in equilibrium with

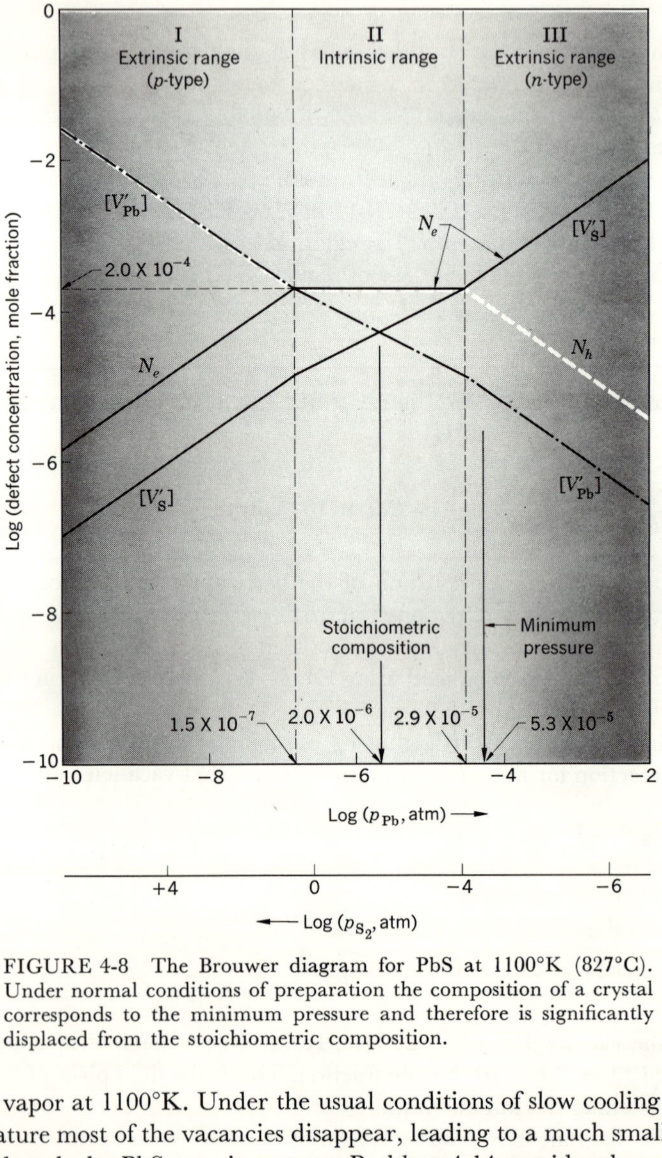

FIGURE 4-8 The Brouwer diagram for PbS at 1100°K (827°C). Under normal conditions of preparation the composition of a crystal corresponds to the minimum pressure and therefore is significantly displaced from the stoichiometric composition.

its own vapor at 1100°K. Under the usual conditions of slow cooling to room temperature most of the vacancies disappear, leading to a much smaller value of δ, although the PbS remains *n*-type. Problem 4-14 considers how PbS can be made *p*-type.

THE NATURE OF DISLOCATIONS

Point defects, such as a vacancy, and two-dimensional defects such as ordinary grain boundaries are not difficult to visualize, but the intermediate one-dimensional defects (dislocations) require special attention. Dislocations were first introduced as theoretical "constructs," but today they are studied directly in microscopes of various kinds. An effective introduction to dislocations is accomplished by way of the electron micrographs obtained from the electron microscope.

4-6. Diffraction Studies of Lattice Defects

If an electron microscope is operated with an accelerating potential of at least 100 kv, the electron beam can penetrate a reasonable "sample" of a

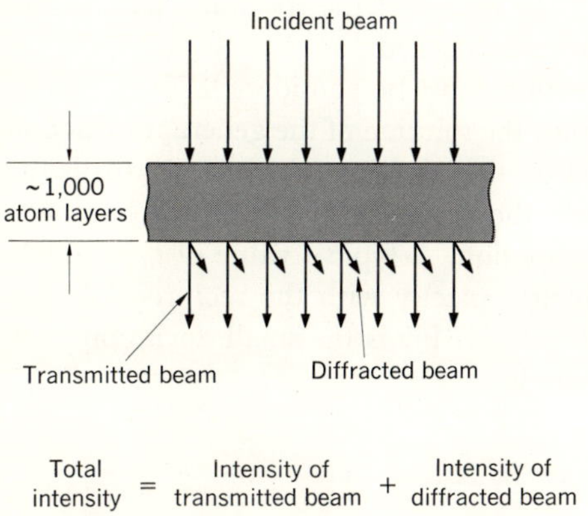

(*a*) General character of the specimen and the electron beams.

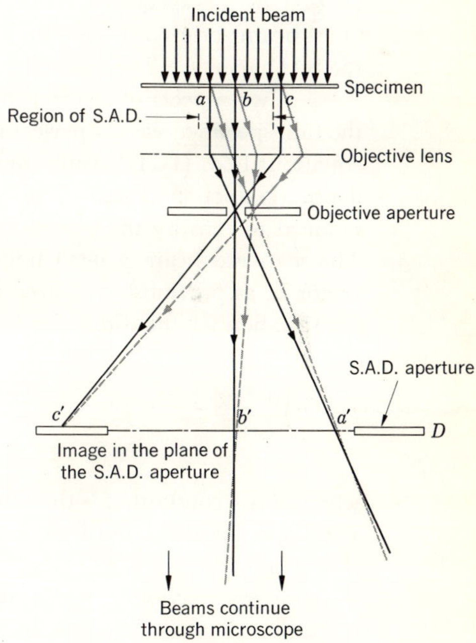

(*b*) Sketch of beam paths showing the use of the objective aperture and the S.A.D. aperture.

FIGURE 4-9 Schematic illustration of electron diffraction by transmission through a thin specimen. The technique of selected-area diffraction (SAD) is of crucial importance. (*Adapted from R. E. Smallman and K. H. G. Ashbee, "Modern Metallography," Pergamon Press, New York, 1966.*)

crystal lattice (Fig. 4-9*a*). The usual diffraction pattern, (Fig. 1-36*b*) could be obtained if desired, but a significantly different technique is employed to reveal defects of various kinds. The final image on the screen of the electron microscope is produced both by the transmitted portion and by the diffracted portion of each incident ray; the paths of three incident rays are shown in Fig. 4-9*b*. An objective aperture of the type illustrated can block the diffracted portion of the beam and permit only the transmitted beam to reach the screen. The result is a bright-field image. A dark-field image, produced by the diffracted rays alone, is obtained by substituting for the aperture shown one that blocks only the transmitted beam. We will see below that lattice defects can produce local changes in lattice geometry so that point *a* in the specimen of Fig. 4-9*b* may produce strong diffraction while point *b* causes little diffraction. Since the same total intensity emanates from both points, the intensity at point a' in a bright-field picture will be low since much of the intensity is in the diffracted beam, which was blocked from reaching the screen. A second aperture (the SAD aperture in Fig. 4-9*b*) permits selecting any small area ($\sim 2\mu^2$) of the specimen for study—the technique of *selected-area diffraction.* Only beams originating within a limited region of the specimen pass within the SAD aperture. For example, beam *c* is blocked.

Two aspects of electron diffraction must be considered. First, even if the thin specimen were a perfect crystal, the solution of the general diffraction equation[1] [Eq. (1-11)] would depend on the effective thickness t_0; this is the depth (extinction distance) at which the intensity of the beam has been reduced to zero by the process of diffraction. A typical value of t_0 is 500 Å. The scattered beam **s** need not coincide exactly with the reciprocal lattice vector **h**, as previously required in Fig. 1-33. If s is the small deviation in **s**, the equation for the diffracted intensity is

$$I = B\frac{\sin^2 \pi z s}{(\pi s)^2} \tag{4-55}$$

where B is a constant, z is the distance below the surface of the specimen, and $s \approx 1/t_0$ is the deviation from **s** in the z direction. The sine function is zero at $z \approx nt_0$, where n is an integer, and therefore the diffracted intensity shows the strange behavior of oscillating between zero and a maximum value with increasing depth in the specimen. The second aspect concerns the effect of local imperfections in the specimen. If the atom at **r** has been displaced by the amount **R**, then Eq. (1-11) becomes

$$A(\mathbf{s}) = \int_V \rho(\mathbf{r})e^{-2\pi i\mathbf{s}\cdot(\mathbf{r}+\mathbf{R}])}\,dV \tag{4-56}$$

[1] The kinematical theory is the basis of the approximate treatment presented here. Applications of the rigorous dynamical theory are discussed in the references listed at the end of the chapter.

Since **s** is near a reciprocal lattice point **h** when appreciable diffraction occurs, the scalar (dot) product can be expanded to give

$$A(\mathbf{s}) = \int_V \rho(\mathbf{r}) e^{-2\pi i \mathbf{s}\cdot\mathbf{r}} e^{2\pi i \mathbf{h}\cdot\mathbf{R}}\, dV \tag{4-57}$$

This equation differs from Eq. (1-11) only by the phase factor $e^{-2\pi i \mathbf{h}\cdot\mathbf{R}}$, and therefore this factor can be pictured as modifying the diffracted intensity that would be produced by a perfect crystal [Eq. (4-55)].

4-7. Stacking Faults

To see how the diffraction equations can be used for qualitative interpretations of electron micrographs, consider the pattern of fringes shown in Fig. 4-10*c*. One possibility for producing such a pattern is a wedge-shaped specimen (Fig. 4-10*a*) since the maximum value of z in Eq. (4-55) varies along

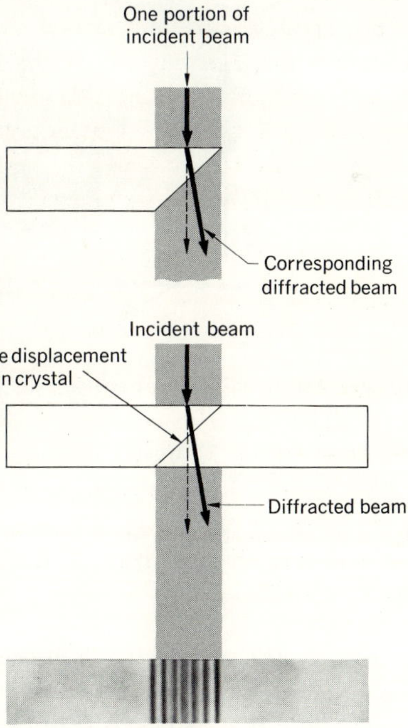

(*a*) Wedge-shaped crystal as specimen.

(*b*) Specimen containing a lattice displacement along an inclined plane.

(*c*) Electron micrograph showing a pattern of parallel diffraction fringes , x 30,000.

FIGURE 4-10 Two types of specimens that can produce a pattern of diffraction fringes in an electron micrograph. The micrograph of (*c*) shows stacking faults in a silicon crystal produced during crystal growth. (*Courtesy G. Thomas, University of California, Berkeley.*)

the wedge, and therefore the diffracted intensity exhibits a periodic variation. A more likely possibility (Fig. 4-10*b*) is the occurrence of a lattice displacement along an inclined plane in the crystal. A common type of displacement is a *stacking fault*, where two adjacent planes of atoms are not in the sequence characteristic of a perfect lattice. For example, the (111) planes of a face-centered cubic metal are normally in the sequence *ABCABC* illustrated in Fig. 1-15*c*. If, as shown in Fig. 4-12*b*, the sequence in a given location is

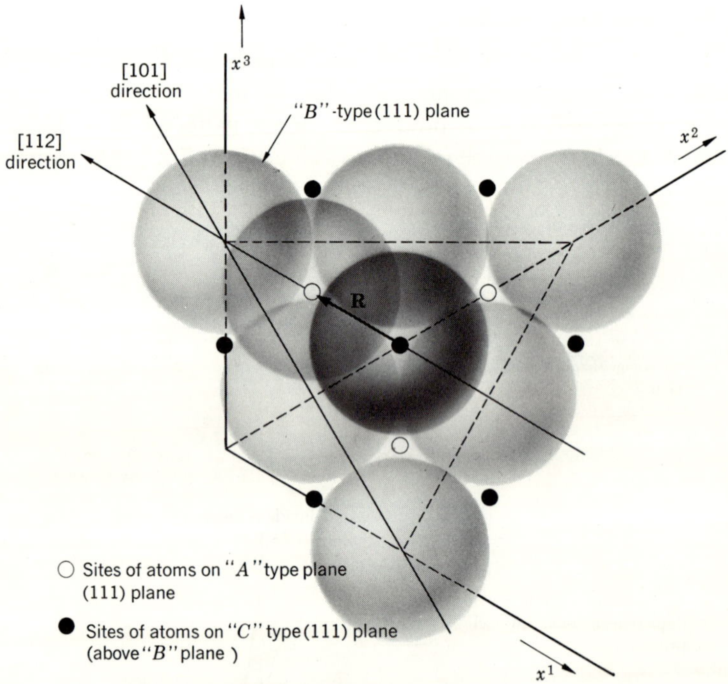

FIGURE 4-11 Geometry of a stacking fault in a face-centered cubic metal. **R** is the displacement vector that converts (normally) "*C*" atoms to "*A*" positions.

changed to *ABCAB*⋮*ABC*, a stacking fault occurs at the indicated location. To visualize the lattice displacement **R** across this stacking fault, we begin with a *B*-type (111) plane (Fig. 4-11). An *A*-type plane lies below this plane, the atoms occupying the sites indicated by open circles. A *C*-type plane lies above the *B* plane, its atom sites being the closed circles. The displacement that changes the atoms in a (normally) *C*-type plane to *A* sites is in the $[\bar{1}\bar{1}2]$

direction (or $[\bar{1}2\bar{1}]$ or $[2\bar{1}\bar{1}]$) and is one-sixth of the length of the vector described by these components; consequently, $\mathbf{R} = \frac{1}{6}[\bar{1}\bar{1}2]$. The phase factor in Eq. (4-57) is determined by $\mathbf{h} \cdot \mathbf{R}$ and will be zero (and therefore without effect) only if this product is an integer, 0, 1, 2, etc. Problem 4-15 treats the (111) reflection in this regard. Consider what happens when the specimen is in position for a (200) reflection. Then $\mathbf{h} \cdot \mathbf{R} = 2(-\frac{1}{6}) + 0 + 0 = -\frac{1}{3}$, and consequently there is a difference in phase between that portion of the beam scattered above the fault (Fig. 4-10*b*) and that portion scattered below the fault. A periodic variation in diffracted intensity then develops, depending on the height at which a particular portion of the beam encounters the stacking fault.

An important connection exists between excess vacancies and dislocations. When aluminum is quenched from high temperatures, small faulted

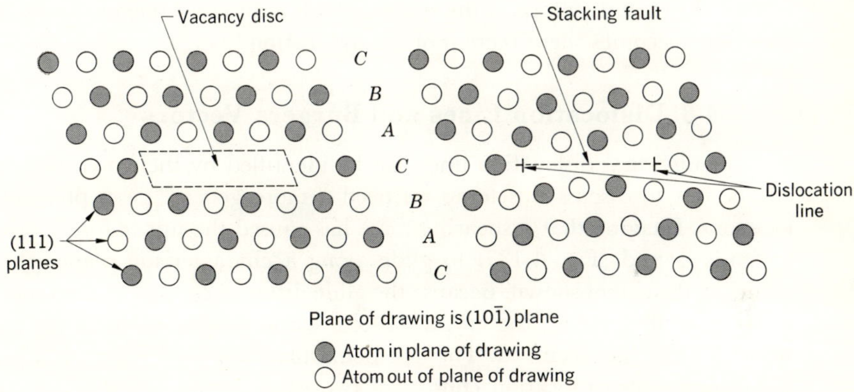

(*a*) Cross-section view of disc formed by excess vacancies on a (111) plane.

(*b*) Adjacent (111) planes collapse and form a loop of dislocation-line surrounding an area of stacking fault.

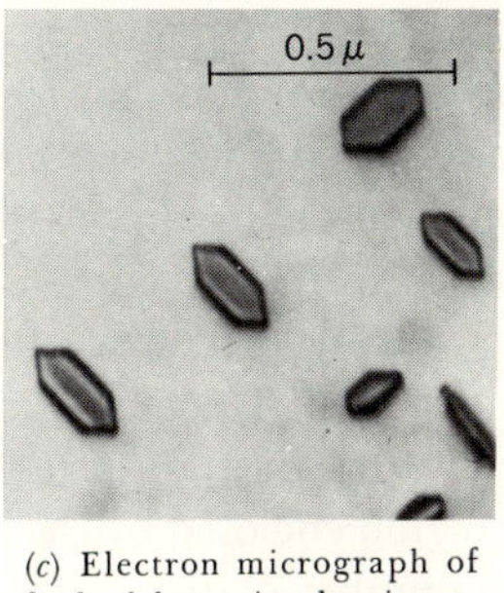

(*c*) Electron micrograph of faulted loops in aluminum. (Courtesy of G. Thomas, University of California, Berkeley)

FIGURE 4-12 The formation of faulted loops in a face-centered cubic metal.

loops (Fig. 4-12*c*) are observed in electron micrographs. The diffraction fringes indicate that this region of the lattice consists of a stacking fault. The mechanism by which these loops form involves the precipitation of vacancies as thin disks (Fig. 4-12*a*). The lattice then collapses (Fig. 4-12*b*) eliminating a platelet of vacancies on a *C*-type (111) plane and producing the stacking fault illustrated in Fig. 4-11. An important phenomenon occurs at the boundary of the stacking fault, where the stacking fault makes a transition into the surrounding perfect lattice; an extra plane of atoms is on one side of the boundary compared to the other side. The line marking the end of the extra plane of atoms is a *dislocation line* (Fig. 4-12*b*), which in Fig. 4-12*c* forms a prismatic loop around the stacking fault. Consideration of the atomic arrangement on the (111) planes explains why the dislocation loop consists of six straight segments. When the *C*-type plane ends, the dividing line is along a $\langle 1\bar{1}0 \rangle$ direction since these are the directions of close packing. At various positions around the circumference of the stacking fault a different one of the six $\langle 1\bar{1}0 \rangle$ directions becomes the direction of the dislocation line.

4-8. Dislocation Lines and Burgers Vectors

Individual dislocation lines can be identified by their own diffraction effects, but to discuss this topic we need a more general description of dislocations. Imagine that a shearing force has caused the upper front corner of an ionic crystal (Fig. 4-13*a*) to glide along a characteristic plane (the slip plane) to the extent shown. Because the glide distance is equal to the repetition distance in the crystal, the crystal lattice retains its perfection in the region where gliding has occurred. The line bounding the glided region is a perfect dislocation; this dislocation consists of two extra planes whereas the partial dislocation of Fig. 4-12*b* has only one extra plane. The *direction* of a dislocation line at a given point is specified by a unit vector $\boldsymbol{\xi}$, as shown in Fig. 4-13*b*.

A second characteristic of a dislocation is the displacement associated with it, denoted by the Burgers vector **b**. Although in the present case we see that **b** is $\frac{1}{2}[\bar{1}10]$, the formal definition that leads to this value is illustrated in Fig. 4-13*a*. First, an atom-to-atom closed circuit is made in the region of interest (around the dislocation in this case) in the direction that would advance a right-hand screw in the direction $\boldsymbol{\xi}$. This circuit is drawn in such a manner that a closed circuit would be produced if the crystal were perfect. The circuit around a dislocation does not close, however, and the vector needed to close the circuit [drawn from start (*S*) to finish (*F*) as shown] is the Burgers vector **b** of the region enclosed by the original circuit.[1] In general, the region may contain any number of dislocations, and **b** is then the resultant of

[1] The opposite sense [(−) instead of (+)] would be assigned to this Burgers vector by the other convention employed in the current literature. The convention employed here has the advantage of being consistent with the usual notation in continuum mechanics.

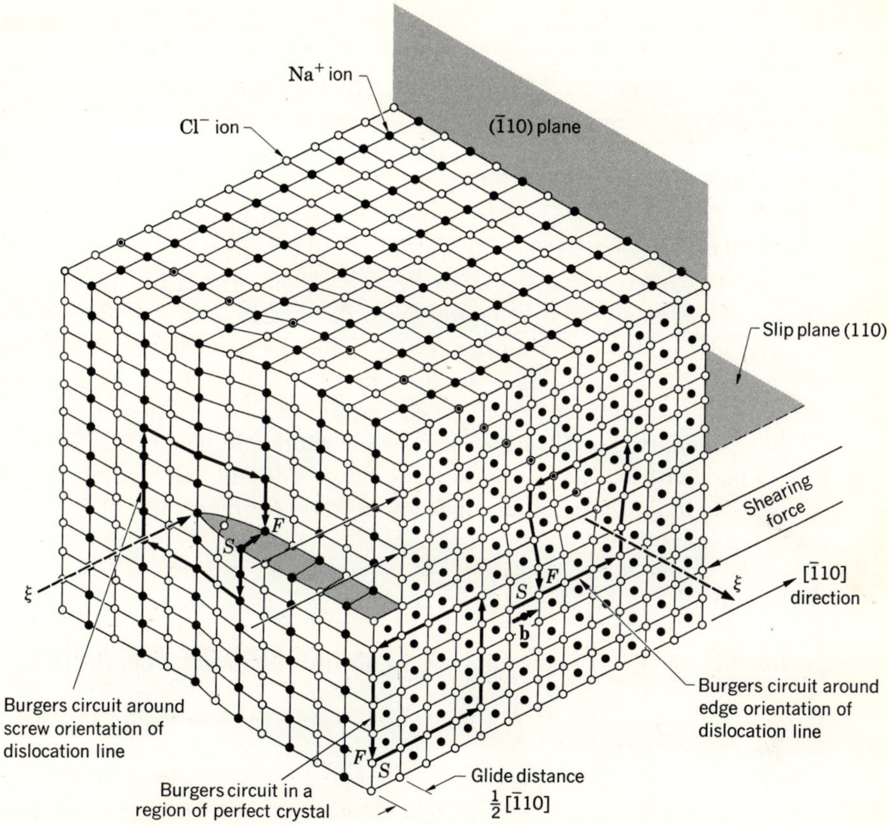

(*a*) Schematic view of the exterior of a block of crystal lattice containing a curved length of dislocation line.

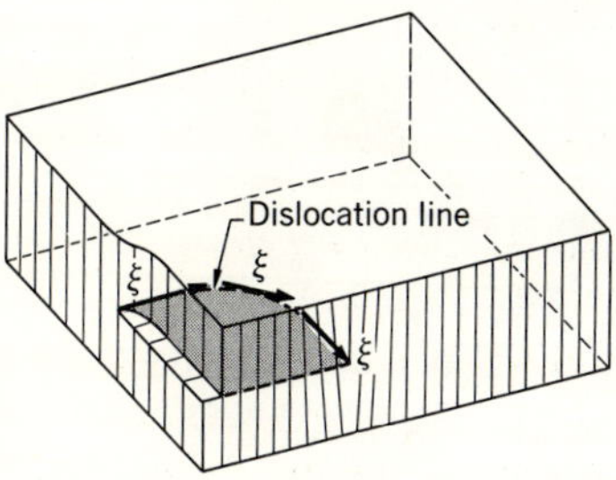

(*b*) The direction of the dislocation line at any point is described by the unit vector, $\boldsymbol{\xi}$, tangent to the line.

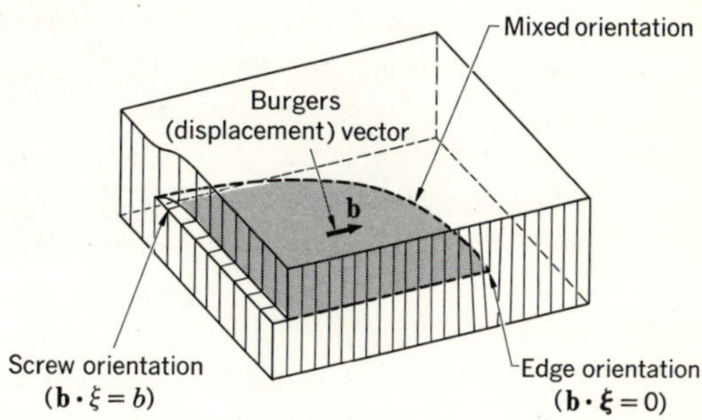

(*c*) Description of the dislocation line in terms of its orientation relative to the Burgers vector.

FIGURE 4-13 A perfect dislocation forming the boundary of a region of plastic glide in a crystal having the NaCl crystal structure. (*After J. J. Gilman.*)

the individual Burgers vectors. Problem 4-17 considers the Burgers circuit for the screw dislocation.

Although only a single dislocation is illustrated in Fig. 4-13, the appearance of this dislocation is quite different on the two sides of the crystal. In the place where $\mathbf{b} \cdot \boldsymbol{\xi} = 0$, the dislocation line is said to be in the *edge orientation* (Fig. 4-13*c*) [previously met in the prismatic loop (Fig. 4-12*b*)]. Where $\mathbf{b} \cdot \boldsymbol{\xi} = \pm b$, the dislocation has a *screw orientation* (+, right-hand, −, left-hand). In intermediate orientation the dislocation is called *mixed* (edge and screw). Problem 4-18 offers a chance to visualize the various orientations of a dislocation in a simplified crystal structure. From the definition of a dislocation line in terms of a glided region in the crystal, it follows that the line cannot end inside the crystal. Instead there are three possibilities: (1) it can form a closed loop, like a prismatic loop; (2) it can emerge at the surface of the crystal (or at a grain boundary); or (3) it can join two or more additional dislocations to form a node or junction. Since the same Burgers circuit can be drawn about the dislocation line entering the node and the two (or more) lines leaving the node, it follows that

$$\sum_{i=1}^{n} \mathbf{b}_i = 0 \qquad (4\text{-}58)$$

provided all the $\boldsymbol{\xi}_i$ are chosen positive in the direction away from the node.

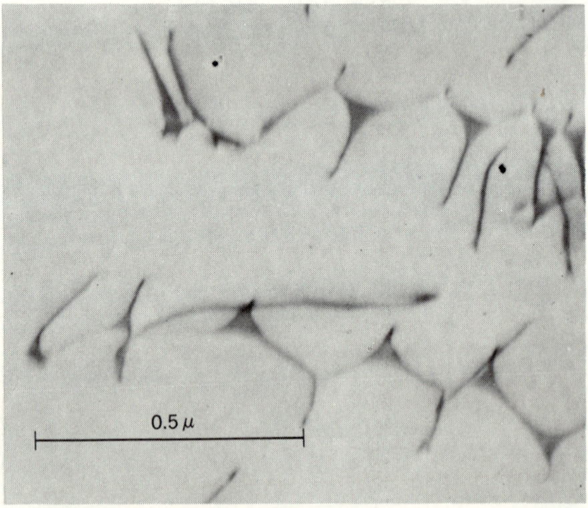

FIGURE 4-14 A network of dislocations in a face-centered cubic crystal of α-brass (33 w/o Zn in Cu) that has been plastically deformed. Each dislocation consists of two partials joined by a ribbon of stacking fault. Where three dislocation lines meet at a node, the region of stacking fault is often greatly enlarged. (*Courtesy G. Thomas. By permission from J. Australian Institute of Metals, vol. 8 p. 80, 1963.*)

The bending of lattice planes in the vicinity of a dislocation line is responsible for the diffraction effects from dislocations. For example, if the $(\bar{1}10)$ planes in Fig. 4-13*a* are almost in position for diffraction, then the planes either to the right or to the left of the edge dislocation will be tilted into the exact position for strong diffraction. As a result, the intensity in the usual bright-field micrograph is lower in this region, and the dislocation is visible as a dark line (slightly offset from the actual position of the dislocation line). This effect occurs for any orientation of the given dislocation line and depends only on the direction of the Burgers vector. Since the displacement **R** of any atom in the vicinity of the dislocation line is some fraction of $\pm\mathbf{b}$, the phase factor in Eq. (4-57) contains the quantity $\mathbf{h} \cdot \mathbf{b}$. The relation $\mathbf{h} \cdot \mathbf{b} = 0$ is especially

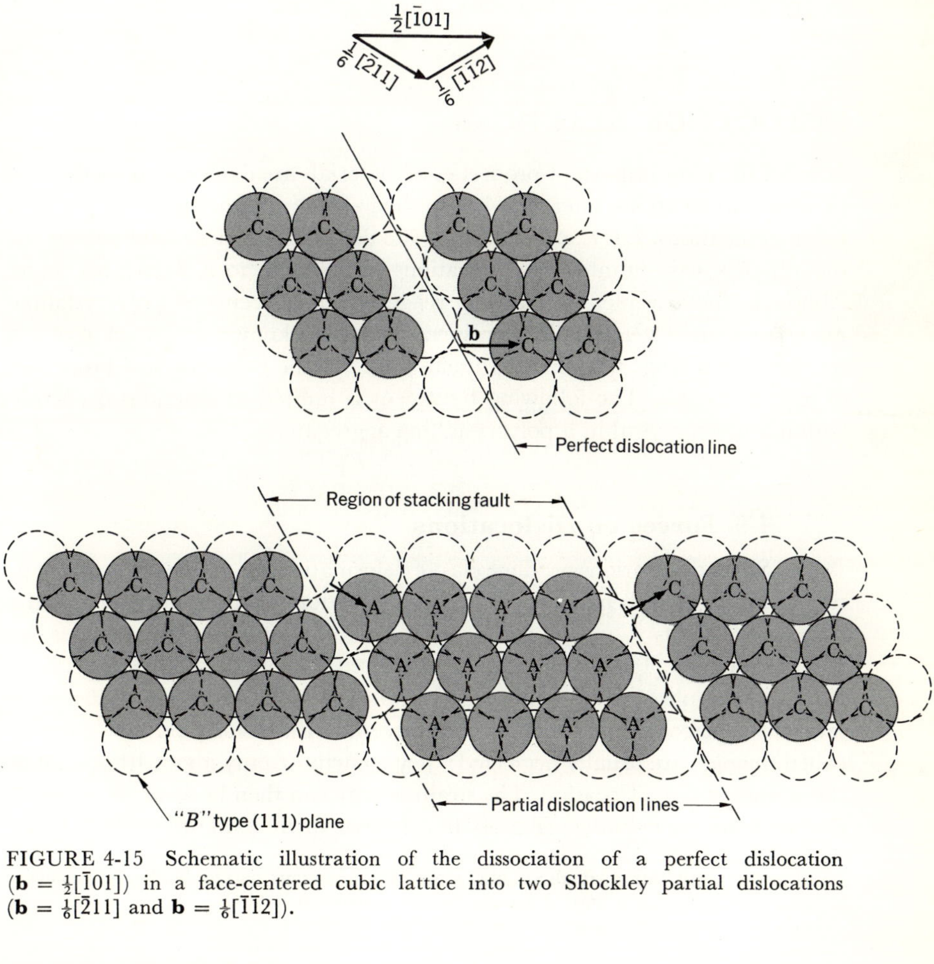

FIGURE 4-15 Schematic illustration of the dissociation of a perfect dislocation ($\mathbf{b} = \frac{1}{2}[\bar{1}01]$) in a face-centered cubic lattice into two Shockley partial dislocations ($\mathbf{b} = \frac{1}{6}[\bar{2}11]$ and $\mathbf{b} = \frac{1}{6}[\bar{1}\bar{1}2]$).

useful in determining the Burgers vector of a dislocation; a dislocation line is invisible[1] if its Burgers vector is parallel to the diffracting plane.

Although it was convenient to begin our discussion using perfect dislocations as examples, in most substances the energy decreases when a perfect dislocation dissociates into partial dislocations (Fig. 4-14). The elastic energy of a dislocation is proportional to b^2 [see Eq. (4-61)], and therefore a unit length of a perfect dislocation, such as $\frac{1}{2}[\bar{1}01]$ (Fig. 4-15) has higher energy than a pair of partial dislocations, each of unit length. The distance separating the two partial dislocations is limited by the energy to create the intervening region of stacking fault. Consequently, the typical *extended dislocation* consists of two partial dislocations joined by a ribbon of stacking fault, the width of the ribbon being larger the lower the stacking-fault energy of the material. The extended dislocations in Fig. 4-14 have a width of about four atom distances, except at nodes where the extension is much greater.

DISLOCATION REACTIONS

Some of the most important properties of materials are determined by the way in which dislocations interact with applied stresses, with vacancies, or with other dislocations. For example, the plastic deformation of a metal reflects the motion of a vast number of dislocations in response to a deforming force. Although the complex dislocation reactions that occur in polycrystalline materials cannot yet be fully analyzed, enough is known about some of the simpler interactions to provide a useful guide to the behavior and properties of most materials. The following treatment is limited to reactions occurring within a given crystal in a polycrystalline aggregate.

4-9. Forces on Dislocations

Since dislocations produce elastic deformations in the crystal, a description of the forces and energies involved in dislocation reactions must include the stress components σ_{ij}, strain components ε_{ij}, and elastic constants such as the shear modulus G. These quantities are discussed in more detail in Sec. 9-1ff. As a simple example consider a unit length of a screw dislocation (Fig. 4-16*a*). The severely displaced atoms immediately adjacent to the dislocation line (a region called the *core*) are usually neglected; they contribute only about 10 percent of the energy of the dislocation. The strain energy can then be approximated by that in a hollow cylinder, (Fig. 4-16*b*) sheared in the same manner as is the

[1] In the edge orientation, **R** also contains minor components perpendicular to **b**. Therefore, an edge dislocation may be barely visible when $\mathbf{h} \cdot \mathbf{b} = 0$.

crystal outside of the core. As can be verified for the infinitesimal shell of radius r, the dislocation introduces only the strain component $\varepsilon_{\phi z} = \varepsilon_{z\phi}$ where

$$\varepsilon_{\phi z} = \frac{b}{2\pi r} \tag{4-59}$$

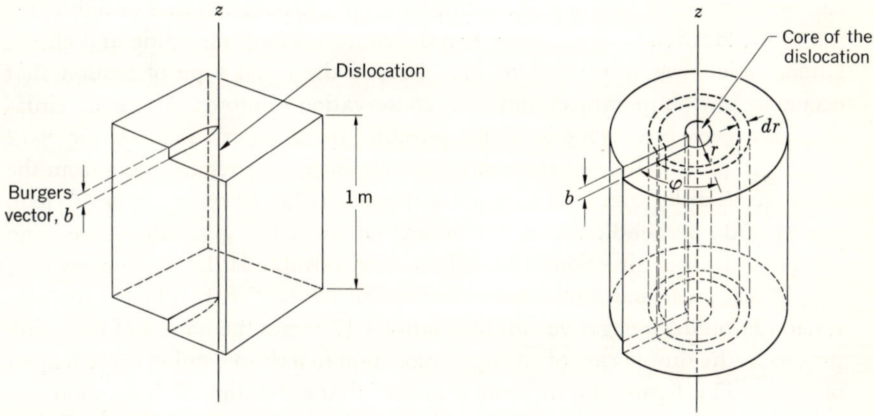

(*a*) An isolated screw dislocation in a crystal.

(*b*) Representation of the dislocation by a sheared hollow cylinder.

FIGURE 4-16 Illustration of the shear strain $\varepsilon_{\varphi z}$ associated with the portion of a screw dislocation that lies outside of the core.

and b is the magnitude of the Burgers vector. The corresponding stress component $G\varepsilon_{\phi z}$ is

$$\sigma_{\phi z} = \frac{Gb}{2\pi r} \tag{4-60}$$

Its magnitude decreases linearly with increase in the distance r from the dislocation line. The elastic strain energy, involving the product $\varepsilon_{\phi z}\sigma_{\phi z}$, has the form,

$$E = \alpha G b^2 \tag{4-61}$$

where α is a constant on the order of unity.

The stress distribution around an edge dislocation is more involved, but important features are a region of compression on the side containing the interleaved plane and a region of tension on the other side. The strain energy per unit length of edge dislocation is only slightly larger than that for a screw dislocation [Eq. (4-61)]. Consequently, this expression, with $\alpha = 1$, can be used to approximate the energy of any type of dislocation line. Since this amount of energy must be supplied per unit increase in length of a dislocation line, a line tension

$$T = Gb^2 \tag{4-62}$$

can be pictured as acting and tending to keep a dislocation line straight.

Dislocations can move by two different mechanisms, glide and climb. Glide, which was illustrated in Fig. 4-13, is the usual type of motion that occurs at low temperatures and with conservation of atoms. There are situations, however, in which glide is not possible; the dislocation loop of Fig. 4-12 is an example. Such dislocations are called *sessile* to distinguish them from the usual *glissile* dislocations. In view of the direction of its Burgers vector this loop should glide perpendicular to the area it encloses, but generally there is no slip plane in this direction. The only motion possible is the nonconservative one, climb, in which atoms are either added to the "extra plane" (or subtracted to produce negative climb). Figure 4-17 shows the nature of the climb process in the simple case of an edge dislocation in a simple-cubic arrangement of atoms. This figure also illustrates the fact that a "straight" dislocation line contains a certain number of jogs, and that these jogs are the preferred sites at which vacancies interact with the dislocation line.

The forces that cause dislocations to move are conveniently divided into three categories: (1) external stresses acting on straight dislocations, (2) line-tension effects on curved dislocations, and (3) the effect of vacancy concentration. The effect of an external stress, generally specified by the stress components σ_{ij} (Sec. 9-1) is compactly described in index notation [see Eq. (1-7*a*)] by the Peach–Koehler equation

$$F_k = -\, e_{ijk}\xi_i\sigma_{jl}b_l \tag{4-63}$$

F is the force exerted on a unit length of dislocation line; the dislocation has direction $\boldsymbol{\xi}$ and Burgers vector **b**. This equation can be used to show, for example (see Prob. 4-19), that an edge dislocation experiences a force tending to make it climb when a tensile stress acts parallel to the Burgers vector of the dislocation. The effect of line tension is conveniently described in terms of the radius of curvature, R, of the dislocation line. The force

$$F_T = \frac{T}{R} = \frac{Gb^2}{R} \tag{4-64}$$

acts perpendicular to the dislocation line in a direction tending to eliminate the curvature.

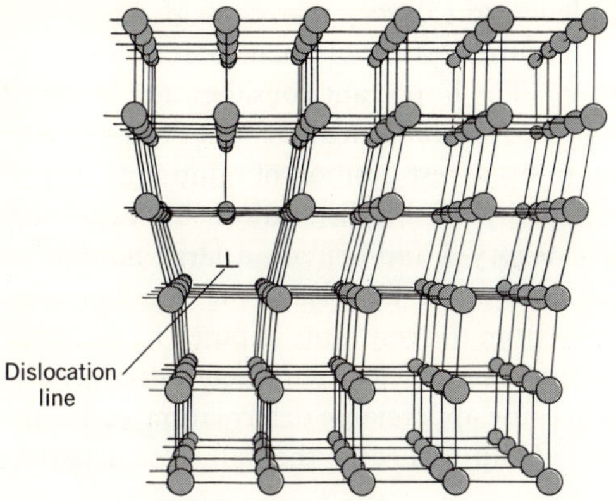

(*a*) Three-dimensional sketch of an edge dislocation containing a jog.

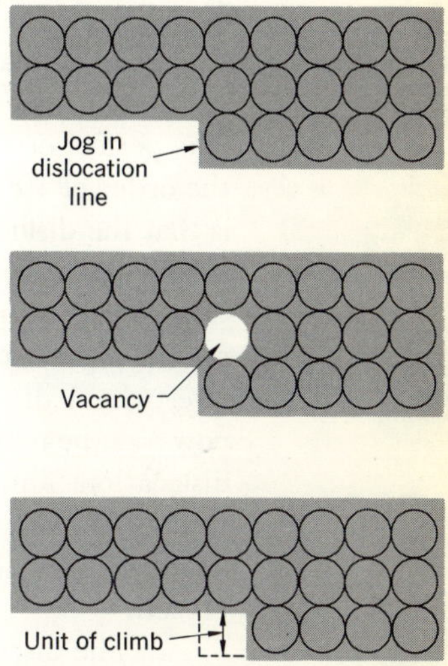

(*b*) Side view of "extra plane". A vacancy diffuses to a jog in the dislocation line and causes one-atom length of line to climb one-atom distance.

FIGURE 4-17 Schematic illustration of the (negative) climb of a dislocation in a simple-cubic lattice. The number of atoms is not conserved locally in this process.

The effect of a nonequilibrium vacancy concentration N_V on a dislocation line arises from the free energy [see Eq. (4-21)]

$$\Delta G = kT \ln \frac{N_V}{N_V^\circ} \tag{4-65}$$

required to form a vacancy; N_V° is the equilibrium concentration of vacancies. By the jog mechanism illustrated in Fig. 4-17, the excess vacancies will cause the line to climb. To move a unit length of line forward approximately one atom distance λ requires the destruction of n/λ vacancies, where n is the number of "interleaved planes" constituting the dislocation. The corresponding change in energy is $-(nkT/\lambda) \ln (N_V/N_V^\circ)$. In view of the relation (energy = force × distance) a force

$$F_c = \frac{-nkT}{\lambda^2} \ln \frac{N_V}{N_V^\circ} \tag{4-66}$$

can be pictured as acting on a unit length of dislocation line. For the case in which $N_V < N_V^\circ$ (undersaturation of vacancies) the sign of F_c changes, and the direction of climb becomes positive.

4-10. Motion of Dislocations

The stress necessary to cause plastic deformation in a material by slip changes drastically as the number of dislocations in the material changes from (1) zero, to (2) a few on parallel planes, to (3) very many, tangled together. The perfect crystal of case 1 would require a stress of about $G/15$, a value on the order of 10^{10} N/m^2 (or 10^6 lb/in^2). The important consideration in case 2 is that the dislocations are free to glide. A minimum stress, the Peierls stress,[1] is required to remove a dislocation from its rest position of minimum energy. Once this stress is exceeded, the velocity of the dislocations increases very rapidly. Most materials belong to category 3 and contain a large number of intertwined dislocations. The minimum stress for plastic deformation must now free the more mobile dislocations from the restraints imposed by the other dislocations and obstacles (like precipitate particles) purposely introduced to strengthen the material. The stress to cause appreciable deformation, called the *yield stress*, usually ranges from $10^{-2}\,G$ up to nearly the value for a perfect crystal.

The discussion above dealt with glide of dislocations, but at high temperatures a dislocation can move by the (usually much slower) process of climb, even though it is obstructed from gliding. For example, in a situation where a material deforms by high-temperature creep (discussed in Chap. 9), the following analysis applies in the vicinity of a unit length of dislocation line. The externally applied loads produce a local stress state σ_{ij} at the dislocation, which in turn exerts a force per unit length $\mathbf{F}$ [Eq. (4-63)] on the dislocation line. The component F_c of this force in a possible direction of climb gives a velocity u to the dislocation line. Since n/λ vacancies must be produced to advance the unit line by one atom distance λ, the rate of vacancy production is

$$R = \frac{nu}{\lambda^2} \qquad \frac{\text{vacancies}}{\text{m-sec}} \tag{4-67}$$

These excess vacancies must diffuse away in the manner described in Chap. 6. For diffusion radially from the line and if the concentration is reduced to the equilibrium value N_V° in about 10^{-4} cm, the rate of diffusion away from a unit length of dislocation line is approximately $D_V(N_V - N_V^\circ)/V_V$. Here D_V is the diffusion coefficient of the vacancies, and V_V is the atomic volume of a single vacancy. Under steady-state conditions,

$$R = \frac{nu}{\lambda^2} = \frac{D_V(N_V - N_V^\circ)}{V_V} \tag{4-68}$$

The reasoning that led to Eq. (4-66) shows that the same equation (with

[1] The Peierls stress is only about $10^{-4}\,G$ for simple dislocations, but it can be much larger in other cases.

change in sign) describes the steady-state vacancy concentration N_V at a dislocation line on which the force F_c is acting. From Eq. (4-66),

$$N_V = N_V^o e^{F_c \lambda^2 / nkT} \tag{4-69}$$

Substitution of this expression in Eq. (4-68) gives the climb velocity as

$$u = \frac{D_V N_V^o \lambda^2}{n V_V} (e^{F_c \lambda^2 / nkT} - 1) \tag{4-70}$$

or, since $F_c \lambda^2 / nkT \ll 1$,

$$u = \frac{D_V N_V^o F_c \lambda^4}{n^2 V_V kT} \tag{4-71}$$

For reasonable values of the quantities in this equation, the climb velocity has a value on the order of 10^{-6} m/sec. In Chap. 9 we will see that the overall creep process is an alternation on the atomic level of slow climb and rapid glide.

4-11. Interaction of Dislocations

The occurrence of intersecting slip planes in most materials permits a variety of behavior for dislocations. In face-centered cubic metals, for example, the four {111} slip planes have the tetrahedral geometric relationship shown schematically by the four surfaces of the Thompson tetrahedron (Fig. 4-18).

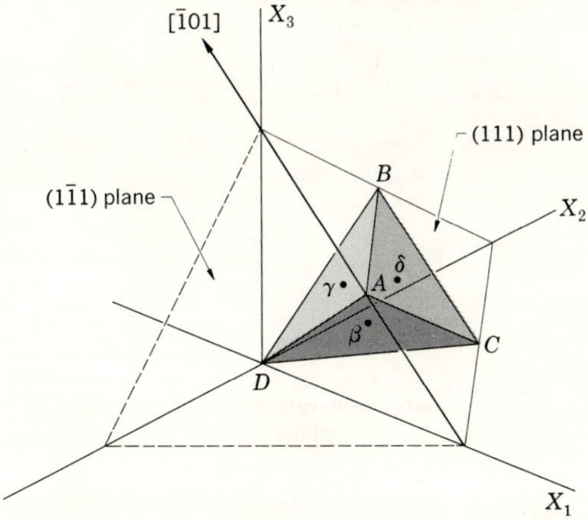

FIGURE 4-18 The Thompson tetrahedron shown in relation to the (111) plane. Burgers vectors are conveniently described employing the letters shown; for example, $CB \equiv \frac{1}{2}[\bar{1}01]$. A Greek letter ($\beta$) marks the center of the triangular face opposite a given apex (B).

Consider first the motion of a single dislocation line. If the dislocation is moving in a given plane, say (111), both its Burgers vector, say $\mathbf{b} = CB \equiv \frac{1}{2}[\bar{1}01]$ for a perfect dislocation, and its direction vector $\boldsymbol{\xi}$ lie in this plane. If the dislocation is extended, then the Burgers vectors of both partials, $C\delta$ and δB, lie in the slip plane. A dislocation line ordinarily cannot transfer to an intersecting slip plane, say $(1\bar{1}1)$ (plane BCD in Thompson tetrahedron), but this *cross-slip* process can occur if both $\mathbf{b}$ and $\boldsymbol{\xi}$ lie in a direction that is common to both slip planes. In the present example this is the $[\bar{1}01]$ direction; Fig. 4-19

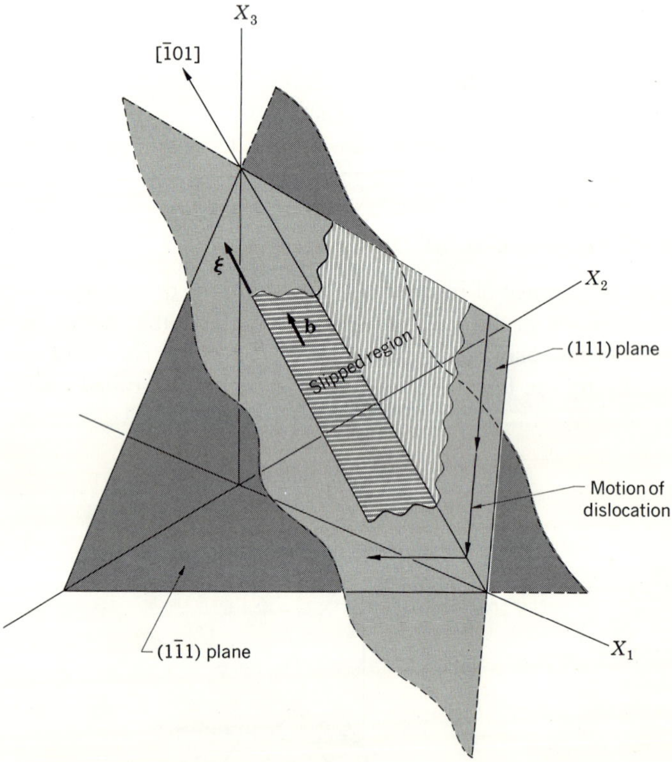

FIGURE 4-19 Schematic illustration of the cross-slip of a perfect screw dislocation from the (111) plane to the $(1\bar{1}1)$ plane.

illustrates the process of cross slip in this case. Recall that $\mathbf{b}$ and $\boldsymbol{\xi}$ are parallel only in the screw orientation of a dislocation line. An extended screw dislocation must constrict (bring the two partial dislocations together) before cross slip is possible (see Prob. 4-20).

If two dislocations meet while moving on separate slip planes, the resulting interaction depends on the orientation of the two Burgers vectors.

An extreme example can be illustrated using Fig. 4-19. If two dislocations, one on the (111) plane and the other on the $(1\bar{1}1)$ plane, approach the line $[\bar{1}01]$, they will annihilate each other on meeting if their Burgers vectors are equal but of opposite sign. The behavior when the Burgers vectors have the same sign is considered later. Interesting results occur when an energetically favorable interaction converts two glissile dislocations into a sessile dislocation. Since the energy of a dislocation is proportional to $|\mathbf{b}|^2$ [Eq. (4-61)], the dislocation reaction

$$\mathbf{b}_1 + \mathbf{b}_2 \rightarrow \mathbf{b}_3 \tag{4-72}$$

is energetically favorable if

$$|\mathbf{b}_1|^2 + |\mathbf{b}_2|^2 > |\mathbf{b}_3|^2 \tag{4-73}$$

The Thompson tetrahedron shows that a dislocation with $\mathbf{b} = BA$ can move along the (111) plane and that a dislocation DB can move along the $(1\bar{1}1)$ plane. Vector addition (see Prob. 4-21) shows that the interaction of these two dislocations along the $[\bar{1}01]$ direction produces the dislocation DA, and obviously the reaction is energetically favorable, according to Eq. (4-73). Since the Burgers vector $\mathbf{b} = DA$ lies in neither the (111) nor $(1\bar{1}1)$ planes, this interaction (called the *Lomer lock*) produces a sessile dislocation.

In most materials, partial dislocations play an important role in the interactions that change the character of dislocations. Consider the Frank sessile dislocation of Fig. 4-12*b*. Its Burgers vector can be represented by a vector such as $B\beta$ since $\mathbf{b}$ is perpendicular to a (111) plane. Reaction with one of three possible Shockley partials, βC for example, leads to a perfect dislocation by a reaction of the type

$$B\beta + \beta C \rightarrow BC \tag{4-74}$$

Of course, this perfect dislocation is mobile and can glide either extended or unextended.

The Lomer lock discussed above has a direct analogy in the case of extended dislocations. The Burgers vectors of the two interacting dislocations have the relation shown in Fig. 4-20*a*. Considering the intersecting planes shown in Fig. 4-20*b*, we see that the dissociation of BA is $B\delta + \delta A$, the order shown on the (111) plane. Correspondingly, the dissociation of DB is $\alpha B + D\alpha$ on the $(1\bar{1}1)$ plane (Why not the reverse order?). When the two dislocations meet at the $[\bar{1}01]$ line shown, the two leading partials undergo the reaction

$$\alpha B + B\delta \rightarrow \alpha\delta \tag{4-75}$$

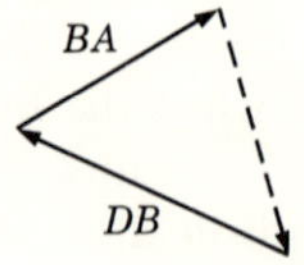

(*a*) Relation of the Burgers vectors of the two perfect dislocations.

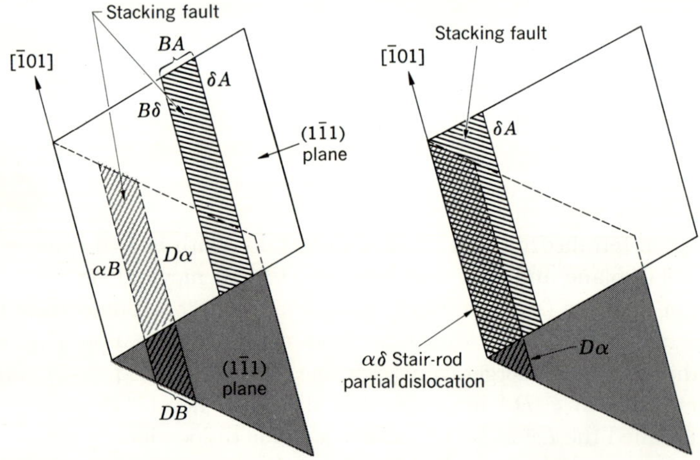

(*b*) The two dislocations dissociated in their slip planes.

(*c*) Interaction of two of the partial dislocations to form a stair-rod dislocation.

FIGURE 4-20 Illustration of the formation of a Lomer-Cottrell block by the intersection of two extended dislocations.

to form a sessile stair-rod dislocation $\alpha\delta$, connecting two regions of stacking fault. This type of dislocation reaction, called the *Lomer-Cottrell block*, plays an important role in decreasing the mobility of dislocations in many materials.

Since dislocations are on all four of the {111} slip planes, a common occurrence is the cutting of a given dislocation by one of the *forest* dislocations, the term applied to the dislocations on intersecting slip planes. The effect of such cutting of one dislocation by another is easily predicted since each dislocation produces in its own slip plane a lattice displacement equal to its Burgers vectors $\mathbf{b}_1$. Consequently, the dislocation line of the second dislocation experiences a displacement equal to $\mathbf{b}_1$ at the point of intersection. Since the intersection is mutual, the dislocation line of dislocation 1 experiences a displacement $\mathbf{b}_2$. In the example shown in Fig. 4-21 of the intersection of two edge dislocations, only line 2 is displaced out of its initial slip plane. This type of displacement is the jog we have considered previously. If the line is merely displaced in its own slip plane, the term *kink* is employed (see Prob. 4-22).

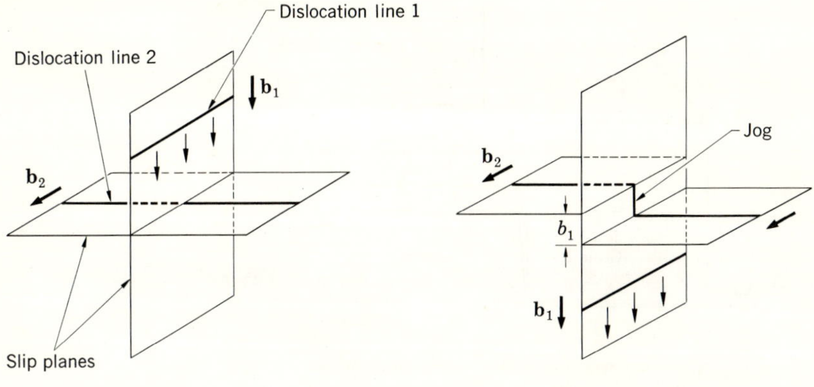

(*a*) Edge dislocations on different planes before intersection.

(*b*) After intersection a jog is present on dislocation line 2.

FIGURE 4-21 Schematic illustration of two edge dislocations on slip planes before and after intersection. The intersection process itself (not illustrated) is a complex one, involving the local repulsion and bending of the two dislocation lines.

Displacement parallel to the dislocation line leaves the line straight; the accompanying strain is dissipated by glide along the dislocation line.

4-12. Dislocation Arrays

During various treatments of materials, especially as a result of plastic deformation, the quantity of dislocations can increase greatly. Measured in units of meters of dislocation line per cubic meter, the dislocation density in a metal may increase from an initial value of 10^{12} to a final value of 10^{16}. The complex tangle of dislocations resulting from plastic deformation of aluminum is shown in Fig. 8-7*a*. Much of the energy in such a material is represented by the interaction of the long-range stress fields of the individual dislocations. To the extent that thermal activation permits climb to occur, the high-energy arrangements of dislocations are replaced by lower-energy arrangements. A typical arrangement produced by a high-temperature anneal is shown in Fig. 8-7*c*. The dislocations are now arranged in characteristic grid patterns in boundary regions; the cells enclosed by the boundaries are relatively free of dislocations. For many purposes, then, the dislocation structure of a material consists of large cells of an essentially perfect lattice surrounded by a boundary layer in which the dislocations form a regular, repetitive pattern.

As preparation for considering a few of the simpler boundary arrangements, let us consider the nature of the most general boundary. The character of the boundary must depend first, on the relative orientation of one grain with respect to the other (across the boundary), described by the deformation

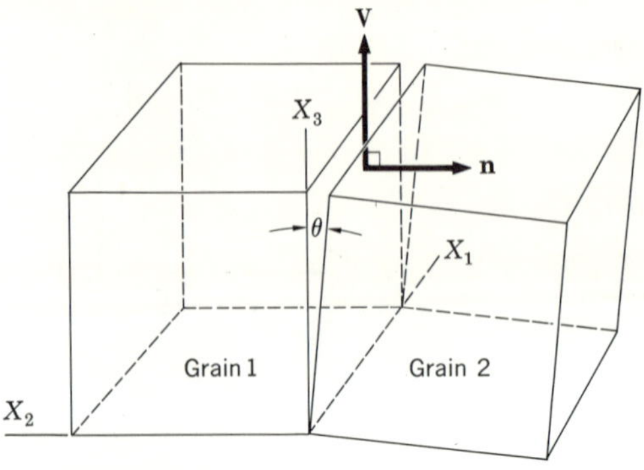

(*a*) The geometrical parameters describing a symmetrical tilt boundary. Use of these parameters in Eq. (4-77) gives $B_2 = \boldsymbol{\theta}$.

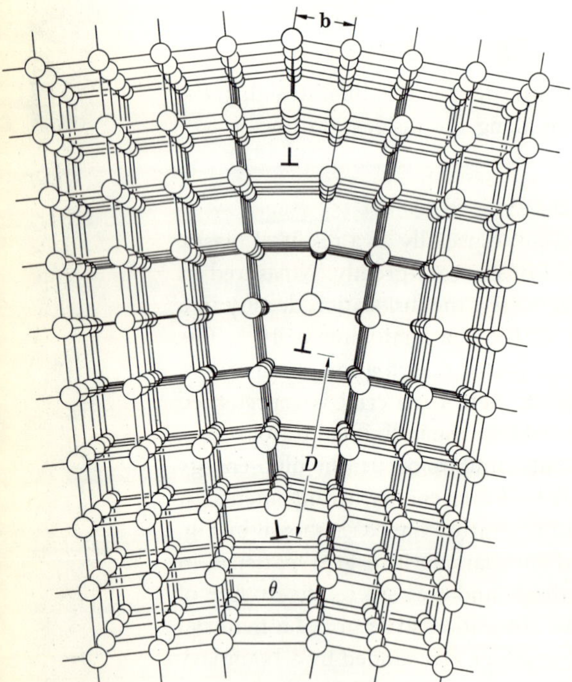

(*b*) The array of edge dislocations that constitutes this tilt boundary.

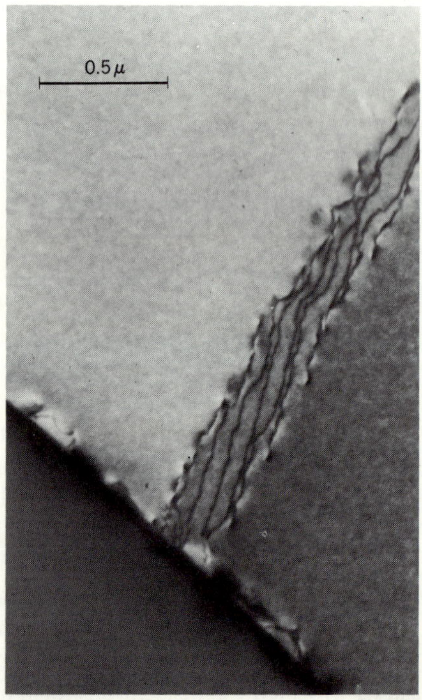

(*c*) An electron micrograph of a nonsymmetrical tilt boundary in aluminum. The boundary ends on a grain boundary at the bottom of the micrograph. (Courtesy of G. Thomas, University of California, Berkeley)

FIGURE 4-22 **Example of the deduction of the dislocation structure of a boundary in the simple case of a tilt boundary.**

components δ_{ij}, and second, on the orientation of the boundary, specified by its unit normal **n**. The resultant **B** of all the Burgers vectors cut by a vector **V** in a given direction in the boundary has the components

$$B_i = \beta_{ij} e_{jkl} V_k n_l \tag{4-76}$$

where

$$\beta_{ij} = e_{jkl} n_k \, \delta_{il} \tag{4-77}$$

As an example of the application of these equations, consider the simple tilt boundary shown in Fig. 4-22. In this case the significant values of δ_{ij} are $\delta_{23} = \delta_{32} = \theta$, and Eq. (4-77) then gives $\beta_{21} = \theta$ as the only nonzero value. This value in turn gives $B_2 = \theta$ when **V** is the unit vector shown; therefore, all the dislocations must be in edge orientation. If B_2 is the resultant of an array of dislocations, each with Burgers vector b and spaced a distance D apart, then

$$D = \frac{b}{\theta} \tag{4-78}$$

Figure 4-22*b* shows the lowest-energy dislocation array that satisfies this condition in a simple-cubic crystal. If the plane of the boundary is not symmetrical with respect to the slip planes in the two crystals, then dislocations in a second set of slip planes are present in the boundary array.

Another simple dislocation array exists in the *twist boundary* that separates two crystals whose orientations differ as shown in Fig. 4-23*a*. To understand why at least two sets of screw dislocations must be present in the array at this boundary, first consider the effect that would be produced by a single set of screw dislocations, each with Burgers vector **b** and separated by the spacing D. Since each dislocation produces a displacement **b** (Fig. 4-13*a*), an array uniformly spaced at distance D (Fig. 4-23*b*) would produce an angle of shear θ, given by Eq. (4-78). Such a shear could account for the displacement of point P in the y direction, but evidently a second shear is required in the x direction. Consequently, the simplest pattern is a cross-grid of two sets of screw dislocations. In face-centered cubic metals twisted on a (111) plane, all three sets of screw dislocations cooperate in forming a hexagonal net (Fig. 4-23*c*).

The preceding treatment of dislocations is the background necessary for understanding discussions of the effect of dislocations on the behavior and properties of materials. This background will be used in later chapters in studies of such topics as deformation, fracture, and phase transformations. Although metals were used as the principal examples in the present treatment,

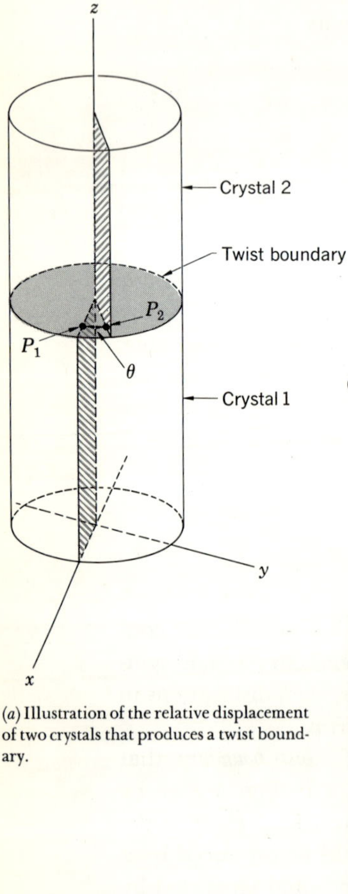

(*a*) Illustration of the relative displacement of two crystals that produces a twist boundary.

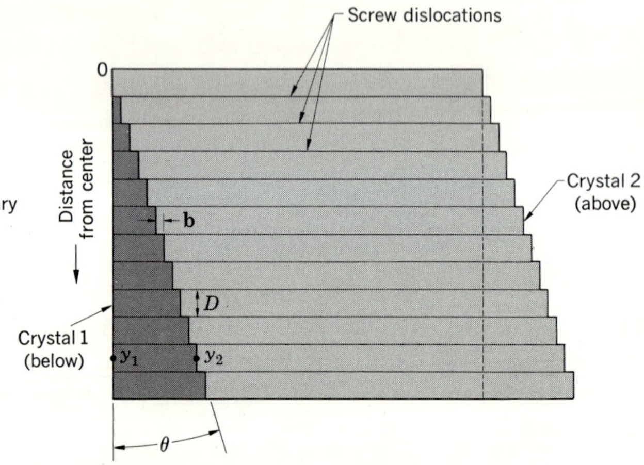

(*b*) Schematic illustration of the action of one set of screw dislocations in producing a displacement.

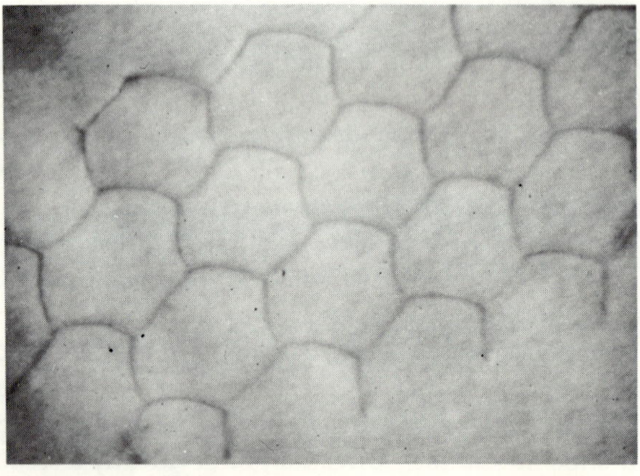

(*c*) Electron micrograph of a hexagonal network of screw dislocations on a twist boundary in aluminum. (Courtesy of G. Thomas. By permission from Transmission Electron Microscopy of Metals, John Wiley & Sons, New York, 1962)

FIGURE 4-23 Illustrations of the formation of twist boundaries on a slip plane common to the two adjacent crystals.

the same concepts apply with suitable modifications to other inorganic crystals. For example, the existence of electrical charges on the various ions in ionic materials is an additional factor influencing the behavior of dislocations. On the other hand, imperfections in organic polymers are radically different from those in metals and ceramics and are considered briefly in the following section.

IMPERFECTIONS IN ORGANIC POLYMERS

Although the same terms—point defect and dislocation—are commonly employed, the imperfections in crystalline polymers are basically different from those considered above. The primary bonds in the polymer chain are not broken during structural changes in solid polymers or as part of the principal mechanisms of plastic deformation. The main function of a dislocation (Fig. 4-13*a*) is to permit the easy breaking and reforming of primary bonds. Since this action does not occur in polymers, dislocations play a minor role. Vacancy formation also involves the breaking of primary bonds, and thus point defects of this type are not a major feature in polymers.

To understand the type of defect that can account for the behavior of crystalline polymers, consider the arrangement of the adjacent chains in a perfect region of the crystal (Fig. 4-24*a*). These chains are in a stable position

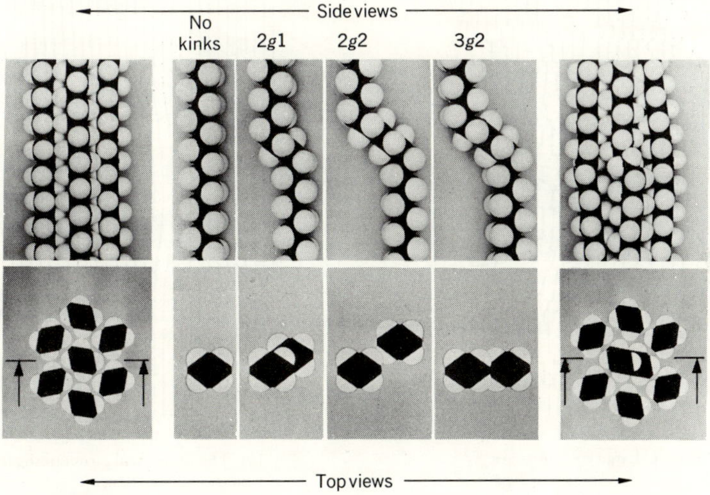

(*a*) Several chains in perfect registry in the crystal.

(*b*) Three types of kink in an originally straight chain.

(*c*) A 2gl-kink in one of the chains disturbs the perfection of the crystal.

FIGURE 4-24 Atomic models of chains in polyethylene illustrating the nature of kink-type defects. (*Courtesy W. Pechhold, University of Stuttgart, Germany.*)

and are not able to participate in structural changes. As discussed in Sec. 1-3, a chain can change from a straight configuration to a bent one by changing to an alternative (gauche) position of the carbon-carbon bond from the usual (trans) position. If at a certain position along the chain the carbon-carbon bonds take on the three successive positions, gauche, trans, and gauche (denoted $2g1$), the result is the first type of kink shown in Fig. 4-24*b*. Two other important kinks are also shown. Now, if a kink occurs somewhere along the length of a chain molecule (Fig. 4-24*c*) it has two important effects. First, in this region the molecule is less strongly bound to its neighbors and can "wriggle" past them a fraction of an atomic spacing by the passage of the kink along the length of the molecule. Second, the density of the crystal is lowered by the presence of the kink.

Kink-type defects can give a useful picture of a polymer both in the molten state and after crystallization. The concentration of kinks is high in the molten state (Fig. 4-25*a*), and accounts for the amorphous structure and

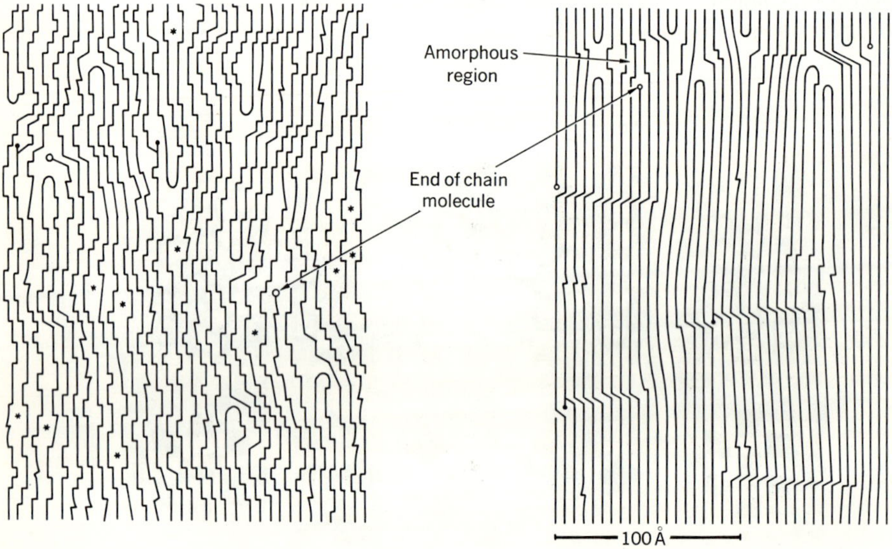

(*a*) The molten state. Large void areas are marked *.

(*b*) The solid state, containing regions where kinks are aligned.

FIGURE 4-25 Schematic illustration of the structure of polyethylene according to the kink-defect theory. (*After W. Pechhold.*)

the easy flow. After crystallization, the kinks are predominantly in two special configurations. Many of them are in the "amorphous regions" between the crystalline lamellae. For energetic reasons, the majority of the remaining kinks assume adjacent positions on neighboring chains within a lamella (Fig. 4-25*b*) and move cooperatively as a quasi-linear or -planar defect. Thus, kinks in polymers, although analogous to point defects, fulfill a function similar to a dislocation (line defect) or a boundary (planar defect). Chain ends (Fig. 4-25*b*), although significantly different from kinks, are also point defects.

PROBLEMS

4-1. Two types of quantities are needed to describe a transverse wave (Fig. 4-1*b*). First, the plane (perpendicular to the X, Y, or Z axis) in which it lies and second, the direction (X or Y, for the wave of Fig. 4-1*b*) in which it is propagating. Sketch the six modes of transverse vibrations.

4-2. (*a*) Show by a suitable sketch of a (one-dimensional) row of atoms that all the wave vectors $\mathbf{K}_q$ for a longitudinal wave in the positive direction can be described by vectors from the origin out to the points π/L, $2\pi/L$, $3\pi/L$, etc., along the X axis. What is the end point for the largest permissible wave vector? (*b*) Similarly, sketch the points for the wave vectors that describe all possible waves moving in all directions in the XY plane for a (two-dimensional) square lattice, including negative directions. Show that there are $(1/a)^2$ modes of vibration in a lattice 1 m on an edge. (*c*) In extending this type of representation to a (three-dimensional) simple-cubic lattice, sketch the region inside which all the wave vectors lie. This region is known as the *Brillouin zone* for the simple-cubic lattice. Show that there are N longitudinal modes of vibration in a lattice containing N atoms.

4-3. (*a*) The thermal energy of a crystal lattice containing N identical atoms is approximately $3Nn_{\text{av}}k\theta$ [Eqs. (4-5) and (4-7)]. Calculate the values of thermal energy for gold at 50°K, 100°K, 150°K, 200°K, 250°K, and 300°K, and make a plot of thermal energy versus temperature. (*b*) Show on the same plot the thermal energy according to the law of Dulong and Petit, $C_V = 3R \approx 6$cal/mole-°K, which is a good approximation for C_V above the Debye temperature θ. (*c*) Explain the cause of the difference between these two curves at low temperatures.

4-4. Show that the equilibrium fraction of vacancies, N_V°, in a metal at T (in degrees Kelvin) is given by

$$N_V^\circ = 3\left(\frac{\Delta L}{L} - \frac{\Delta a}{a}\right) \tag{4-79}$$

Δa and ΔL are the changes in lattice constant a and in length L measured between 0°K and T.

4-5. (*a*) Write the expression analogous to Eq. (4-18) for the rate R_d of destruction of vacancies at kinks on the surface. Note that ΔG_d^A can be written in terms of ΔG_f^A and ΔG_f. The expression must contain an additional factor, the number of vacancies per square meter of surface; write this factor in terms of N_V, Avogadro's number A, the molar volume V of the metal, and the interplanar spacing d of adjacent atomic layers.

(*b*) Since $R_f = R_d$ at equilibrium, show that K_d can be eliminated from your expression. (*c*) Derive the expression for the net rate of formation of vacancies [Eq. (4-19)].

4-6. For phosphorus in silicon, $E_d = 0.044$ ev. Calculate the temperature at which ionization is complete (that is, $N_e \approx N_d$) for a molar concentration $N_d = 10^{-6}$ of phosphorus in silicon.

4-7. Using an analysis similar to that employed in the text for a donor impurity, explain how an acceptor impurity can produce a bound energy state E_a just above the valence band. $E_a = 0.045$ ev for boron in silicon. A deficiency of n (nondefective) electrons in the valence band can alternatively be described as a full valence band plus n (nondefective) holes in another location.

4-8. Make a semiquantitative plot of the total carrier concentration $N_e + N_h$ in silicon doped with 10^{-6} mole fraction of phosphorus as a function of temperature from 0 to 1500°K. Note that calculations need be made only near the boundaries of the four regions of characteristic behavior: (1) incomplete ionization of $P_{\text{Si}}^{\times}$, (2) extrinsic region (near room temperature), (3) intrinsic region, (4) region dominated by vacancies (at very high temperatures).

4-9. Explain why the $V_{\text{S}}^{\times}$ vacancy undergoes the ionization reaction of Eq. (4-35). [HINT: Creation of this vacancy involves the removal of a S^{2-} ion and the *addition* of two electrons.]

4-10. (*a*) Using the value $K_p^{\circ} = 2.4 \times 10^{-2}$, calculate $[V'_{\text{Pb}}]$ in PbS at room temperature, 300°K. (*b*) Calculate the temperature at which $[V'_{\text{Pb}}] + [V_{\text{S}}^{\bullet}] = 2 \times 10^{-6}$.

4-11. Show that Eq. (4-17) can be obtained easily using the method of defect chemistry to write the equation for the formation of a vacancy in a metal. [Note: The activity of A_V of the vacancies in the metal can be expressed as $\gamma_V N_V^{\circ}$, where γ_V is the activity coefficient of the vacancies. A standard thermodynamic relation is $\gamma = \exp(\Delta G^{\text{xs}}/RT)$.]

4-12. Derive Eq. (4-40). State explicitly the necessary choice of standard state for each species.

4-13. (*a*) Using an argument similar to that employed in obtaining Eq. (4-52), show that Eqs. (4-48) and (4-49) give $(p_{\text{S}_2})^{1/2} = 1.76$ atm$^{1/2}$ for the boundary of regions I and II. (*b*) Why can this condition also be specified as $p_{\text{Pb}} = 1.5 \times 10^{-7}$ atm?

4-14. (*a*) Using the Brouwer diagram of Fig. 4-8, suggest a treatment of ordinary *n*-type PbS that would convert it to *p*-type. (*b*) Explain whether you could more conveniently control the pressure of lead or of sulfur in this treatment.

4-15. (*a*) Show that $\mathbf{h} \cdot \mathbf{R} = 0$ for the stacking fault of Fig. 4-11 and the (111) reflection. (*b*) Show by a sketch of the (111) planes that this particular stacking fault does not disrupt the periodicity. (*c*) Make a similar analysis for the $(11\bar{1})$ reflection.

4-16. The stacking fault illustrated in Fig. 4-12*b* is called an *intrinsic* (or *single*) stacking fault. During neutron irradiation of face-centered cubic metals, a disk of interstitial atoms can form as an extra plane of atoms. Using a sketch analogous to Fig. 4-12*a*, show the nature of an *extrinsic* (or *double*) stacking fault.

4-17. A Burgers circuit around the screw orientation of the dislocation line is shown in Fig. 4-13*a*. Construct the same circuit in the perfect lattice, and deduce the Burgers vector.

4-18. Consider a hypothetical element whose crystal structure consists of atoms on the points of a simple-cubic lattice. (*a*) Sketch a block of crystal structure about five atoms on a side showing a dislocation line in edge orientation. (*b*) Make a similar sketch for a left-hand screw orientation. (*c*) Devise a sketch to show effectively a curved dislocation line analogous to that in Fig. 4-13.

4-19. Consider the effect of a tensile stress in the X_1 direction (σ_{11}) on a dislocation in the X_2 direction whose Burgers vector is in the X_1 direction. (*a*) What is the orientation of the dislocation, edge or screw? (*b*) Use Eq. (4-63) to evaluate the components F_k of the force acting on the dislocation. You need consider only one term since most of the σ_{ij} and e_{ijk} components are zero. (*c*) Explain how the stress distribution around an edge dislocation leads to this behavior. (*d*) The following computer program can be used to evaluate Eq. (4-63) in more difficult cases. The subroutine, SUBCHK, evaluates e_{ijk}; recall that e_{123}, e_{231}, and e_{312} each equals +1 [see Eq. (1-7*a*)]. The symbol XI(I) represents ξ_i.

```
C  CALCULATION OF FORCE COMPONENTS, F(K)
      DIMENSION F(3), XI(3), SIGMA(3,3), B(3)
    1 FORMAT (9F7.3)
    2 FORMAT (E13.4)
C  READ DATA
      READ (5,1) XI
    3 READ (5,1) ((SIGMA(I,J), J = 1,3), I = 1,3)
    4 READ (5,1) B
C  CALCULATION
    5 DO 14 K = 1,3
    6 F(K) = 0
    7 DO 13 J = 1,3
    8 DO 12 I = 1,3
    9 CALL SUBCHK (I,J,K,E)
   10 DO 11 L = 1,3
   11 F(K) = F(K) + E * XI(I) * SIGMA(J,L) * B(L)
```

```
   12  CONTINUE
   13  CONTINUE
   14  CONTINUE
C  OUTPUT SECTION
   15  WRITE (6,4) F
   16  STOP
       END

           SUBROUTINE SUBCHK (I,J,K,E)
   C  SUBROUTINE TO EVALUATE E(I,J,K)
   DIMENSION COUNT(3), NCOMB(3)
   DATA NCOMB/123, 231, 312/
C  DETERMINE IF E(I,J,K) = 0
    1  DO 2 M = 1,3
    2  COUNT (M) = 0
    3  DO 6 M = 1,3
    4  IF (I. EQ. M) COUNT(M) = COUNT(M) + 1
    5  IF (J. EQ. M) COUNT(M) = COUNT(M) + 1
       IF (K. EQ. M) COUNT(M) = COUNT(M) + 1
    6  CONTINUE
    7  DO 11 M = 1,3
    8  IF (COUNT(M). NE. 0) GO TO 11
    9  E = 0
   10  GO TO 19
   11  CONTINUE
C  DETERMINE IF E(I,J,K) = -1
   12  NUMBER = (I * 100) + (J * 10) + K
   13  DO 15 M = 1,3
   14  IF (NUMBER. EQ. NCOMB(M)) GO TO 18
   15  CONTINUE
   16  E = -1
   17  GO TO 19
   18  E = 1
C  RETURN TO MAIN PROGRAM
   19  RETURN
       END
```

4-20. Use a sketch similar to Fig. 4-19 to picture constriction and cross slip of an extended screw dislocation. (*a*) Show the extended dislocation at the point when it has slipped on the (111) plane so that the δB partial is a distance $w/2$ beyond the line BC of cross slip, whereas the $C\delta$ partial is an equal distance in front of this line; w is the width of stacking fault separating the two partials. (*b*) Using the central portion of the line BC in the sketch of part *a*, show how the extended dislocation might be constricted in a suitable environment and form a section of perfect dislocation along *BC*. (*c*) Make a second sketch showing the extended dislocation moving along the $(1\bar{1}1)$ plane after cross slip has been completed.

4-21. The addition of two Burgers vectors is simply the addition of two lattice displacements and is performed by ordinary addition of the corresponding components of the two vectors. (*a*) Express *BA*, *DB*, and *DA* in the Thompson tetrahedron in terms of direction vectors in the lattice. (*b*) Add the components of *BA* and *DB* and show that they equal *DA*. (*c*) How is this particular addition indicated vectorially in the Thompson tetrahedron?

4-22. Displacements produced by intersection of two dislocation lines are easily visualized in a hypothetical simple-cubic metal. Considering only $\langle 100 \rangle$-type Burgers vectors and slip on $\{100\}$ type planes, show that all three types of displacements of the dislocation line occur in the two possible kinds of intersection of edge dislocations.

REFERENCES

Amelinckx, S.: "Direct Observation of Dislocations," Academic Press Inc., New York, 1964.

Cottrell, A. H.: "Mechanical Properties of Matter," John Wiley & Sons, Inc., New York, 1964.

Dienes, G. J.: Point Defects in Metals, in "Energetics in Metallurgical Phenomena," vol. 2, p. 139, Gordon and Breach, Science Publishers, Inc., New York, 1965.

Friedel, J.: "Dislocations," Pergamon Press, New York, 1964.

Gibbons, J. F.: "Semiconductor Electronics," McGraw-Hill Book Company, New York, 1966.

Hirsch, P. B., et al.: "Electron Microscopy of Thin Crystals," Butterworth & Co. (Publishers), Ltd., London, 1965.

Hirth, J. P., and J. Lothe: "Theory of Dislocations," McGraw-Hill Book Company, New York, 1968.

Hull, D.: "Introduction to Dislocations," Pergamon Press, New York, 1965.

Kittel, C.: "Introduction to Solid State Physics," 3d ed., John Wiley & Sons, Inc., New York, 1966.

Kröger, F. A.: "Chemistry of Imperfect Crystals," North Holland Publishing Company, Amsterdam, 1964.

Smallman, R. E., and K. H. G. Ashbee: "Modern Metallography," Pergamon Press, Oxford, 1966.

van Gool, W.: "Principles of Defect Chemistry of Crystalline Solids," Academic Press Inc., New York, 1966.

Wang, S.: "Solid State Electronics," McGraw-Hill Book Company, New York, 1966.

Weertman, J., and J. R. Weertman: "Elementary Dislocation Theory," The Macmillan Company, New York, 1964.

Chapter Five

Phases and Interfaces in Material Systems

In this chapter we examine in detail the nature of the phases that constitute material systems. Differences among phases are of little importance for certain theoretical generalizations; for example, the phase rule specifies merely the number of phases. For practical uses, however, a vapor phase differs immensely from a solid phase; in fact, one type of solid will be decisively different from another. In Chap. 1 we studied the major types of bonding in solids (ionic, covalent, and metallic), and now we will see that the character of a solid phase depends on how the component atoms are distributed in this phase. Although solid phases are of primary importance in materials science, gases and liquids are frequently of interest in their own right or in connection with reactions involving a solid phase.

NONCRYSTALLINE PHASES

Although a gas or an ordinary liquid is obviously noncrystalline, there is a gradual transition to crystallinity involving undercooled liquids, certain polymers, and amorphous solids. In the following sections we consider this entire range of phases.

5-1. Gases and Liquids

The gas phase is employed in many different ways in the production and treatment of materials. The doping of semiconductors and the creation of special surface layers can be accomplished efficiently by a vapor method. The vapor phase is one of the possible paths for transport of material in the process of changing the form of a solid and is especially important in geologic ice formations. Many calculations concerning gases merely involve the ideal gas law or (at pressures above $\sim 10^2$ atm) suitable corrections for nonideal behavior and transition to the liquid phase. At very high temperatures, however, two additional processes must be considered. The first process, characteristic of molecular gases such as H_2, is *dissociation* into neutral atoms. In calculations based on the gas laws, failure to consider dissociation leads to a large apparent increase in volume, above that for the ideal molecular gas (Prob. 5-1). At still higher temperatures the second process, *ionization*, occurs and converts the previously insulating gas into an electrically conducting *plasma* by virtue of the ions and electrons that are produced. In addition to their importance in nuclear energy research, plasmas are useful as heat sources because of their high heat content; also, the ionized atoms in a plasma undergo chemical reactions with special ease. Practical plasma devices, such as an induction plasma torch used to melt Al_2O_3, produce the ionization electrically.

The many ways in which liquids are employed in materials science—as raw material, intermediate product, or heat-transfer agent—are evident from the classification of the liquids given in Table 5-1. The variety of uses is

TABLE 5-1. CLASSIFICATION OF LIQUIDS

Type	Examples	Typical use
Monatomic	He, Ar	Liquid helium for cryogenic purposes
Molecular	C_6H_6, CCl_4	Benzene as solvent for organic substances
Associated molecular	NH_3, H_2O	Aqueous solutions of inorganic salts
Ionic (molten salts)	Li_2CO_3, cryolite	Fuel cells; electrolytic process for aluminum
Polymeric	Organic polymers, molten oxides (for example, SiO_2)	Polymer and glass technology
Metallic	Fe, Na	Steel making; nuclear power plants

a result of the range in essential character of the various liquids; this character can be judged qualitatively from the nature of the bonding of the corresponding solid. For example, metallic and ionic liquids are both electrical conductors (though by different mechanisms), whereas the remaining liquids are nonconductors; associated molecular liquids occupy an intermediate position. X-ray diffraction shows that the structure of a liquid, determined near its freezing point, has some resemblance to a crystalline structure in local regions. The number of nearest neighbor atoms is smaller by about 10 percent than in the solid, and a corresponding amount of "void space" exists.

Interesting phenomena may occur when one type of liquid is able to dissolve appreciably in another. Generally, liquids of one type tend to exhibit complete mutual solubility (liquid zinc in liquid copper, ammonia in water, carbon tetrachloride in benzene, and so on). Conversely, there is usually little mutual solubility between liquids of two different types. Because both metals and ionic crystals are useful at high temperatures, the interactions of their liquids has considerable technical importance. In general, liquids of these two types are essentially insoluble, but there is one notable exception. A given molten halide, KBr, for example, will take its own metal into solution. If the metal is an alkali metal, an alkaline earth, or a rare earth, the resulting solution has metallic properties since the electrons from the metal are mobile in the liquid. Other metals generally undergo a chemical reaction with the molten halide and lose their metallic character.

5-2. Glasses

When a liquid phase is cooled to its freezing temperature, usually the process of crystallization converts it into a crystalline solid (Fig. 5-1). Some liquids, for reasons of complex molecular configuration or of slow kinetics, do not crystallize on being cooled to low temperatures but instead form a rigid structure known as *glass*. In materials for which crystallization is possible, two major factors determine whether a given liquid will crystallize or form a glass during cooling. First, if the cooling rate is sufficiently high, in principle any liquid can be converted to a glass. Second, if the "basis" in the crystal structure (see Fig. 1-12) is difficult to produce from the liquid structure, crystallization will be slow, and glass formation will be favored. Metals, ceramics, and polymers illustrate qualitative differences in this regard. The basis in metallic crystals contains only a few atoms, often only one, and therefore easy crystallization is the rule. Metallic glasses can be produced only by special techniques, such as "splat cooling," which we considered below. Ceramic crystals are generally more complex, especially those such as SiO_2 that form three-dimensional networks. Glass formation is a common behavior, although crystallization also occurs in most ceramic materials. Finally, the long-chain polymers have two structural difficulties: (1) simple bases are rarely possible, and (2) suitable rearrangement of the existing chain segments is accomplished only by the slow diffusion of whole sections of a chain, without breaking and reforming bonds (in contrast to the behavior in metals and ceramics where a given atom can transfer a bond from one neighbor to another fairly easily). Glass formation is the rule in most polymers; crystallization plays a secondary role.

The *glass transition* point T_g is for glasses what the melting point T_f is for crystalline solids. Whether a material forms a glass depends on the variation of its viscosity with temperature, viscosity being a quantitative measure of resistance to continuous shearing. A regularity observed in almost all glass-forming materials is a change in the temperature dependence of volume at a

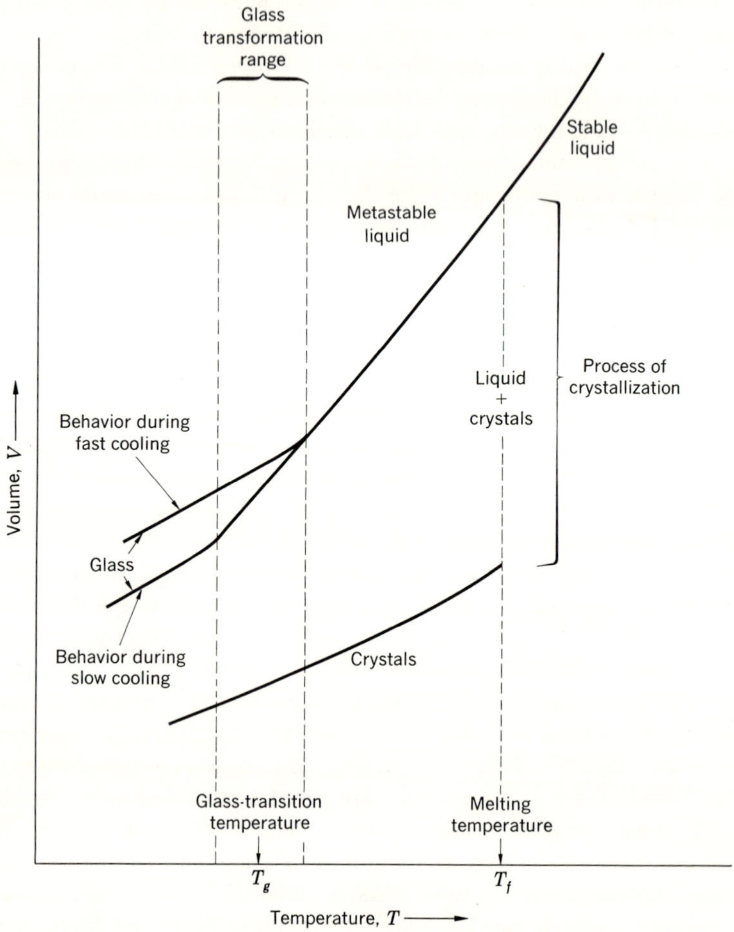

FIGURE 5-1 Volume changes in a liquid during crystallization or in the transition to a glass in the range of temperature near T_g.

temperature T_g, where the viscosity is in the range 10^4 to 10^6 N-sec/m^2 (Figs. 5-1 and 5-3). At higher temperatures (lower viscosities) the structural units are able to reorganize themselves quickly as a quasi-equilibrium liquid. At temperatures below T_g reorganization among the structural units almost ceases, and the resulting rigid material is known as a *glass*. The experimental value obtained for T_g depends on the time scale of the experiment (Fig. 5-1). During the long time of slow cooling, for example, the structural units have more time to reorganize, and thus T_g is lower.

An important thermodynamic change occurs at the glass transition temperature. At temperatures below T_g, effects due to changes in configuration are almost absent since the atoms are unable to change their relative positions. One of the properties affected by this change is the volume V, which is related to the entropy S (see Eq. 2-29) under equilibrium conditions by the standard relation

$$-\left[\frac{\partial V}{\partial T}\right]_P = \left[\frac{\partial S}{\partial P}\right]_T = \left[\frac{\partial S_v}{\partial P}\right]_T + \left[\frac{\partial S_c}{\partial P}\right]_T \tag{5-1}$$

where P is the pressure and T is the temperature. This equation describes the behavior at temperatures above the melting point T_f (Fig. 5-1), where the liquid is the thermodynamically stable phase. If crystallization is hindered for any reason, Eq. (5-1) continues to be obeyed since the liquid exists below T_f as a metastable phase and exhibits the usual equilibrium properties. For example, if the liquid is brought to a certain temperature in this range either by heating or cooling, a characteristic (metastable) structure is established. However, if the temperature is lowered to near T_g, the structural units of the liquid have significantly less capacity to adjust their configurations, and $[\partial S_c/\partial P]_T$ in Eq. (5-1) becomes negligibly small. Only S_v continues to change, and therefore the slope of the curve of volume versus temperature decreases.

METALLIC GLASSES. The existence of metallic glasses shows the generality of the glass-forming process; in addition, the physical picture in this case is especially simple. With the presently attainable cooling rates, no pure metal has yet been produced as a glass presumably because the liquid can be converted easily into the simple crystal structure. However, if the liquid is an alloy of two (or more) metals whose atoms differ greatly in size, the crystallization process is more difficult and can be prevented in some instances by extremely rapid cooling of the liquid (10^6 °C/sec). The two experimental techniques for rapid cooling are (1) propelling a liquid drop at high velocity and "splatting" it against a cooled copper surface and (2) flattening a liquid drop between a piston and an anvil. X-ray patterns from the solidified alloys are similar to those for liquid metals. This and other evidence suggest that the structure is a glass.

Another method for preparing metallic glasses is by vacuum deposition onto a substrate cooled to approximately 80°K. Pure metals again crystallize under these conditions, but suitable alloys do form glasses, some of which are stable when brought to room temperature. An example is the silver-copper system (Fig. 2-5), in which the two solid metals have only about 8 percent solubility for each other. Therefore, if a melt containing 50 $^{w}/_{0}$ silver and 50 $^{w}/_{0}$ copper is slowly cooled, two crystalline phases are produced. When this same alloy is obtained by vacuum deposition, a metastable metallic glass is formed. Initially the glass has metallic conductivity comparable to the extrapolation of the curve for the liquid alloy. When the glass is heated for $\frac{1}{2}$ hr at

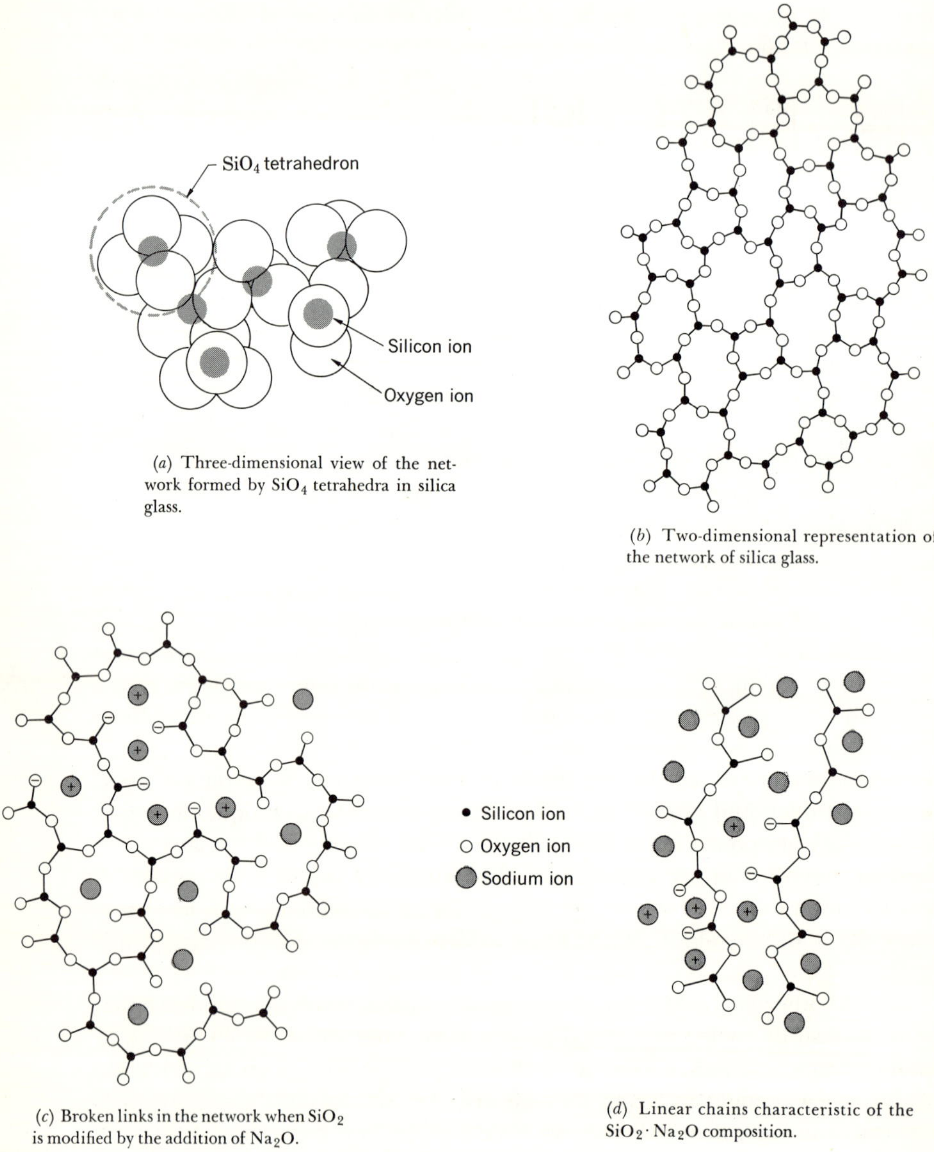

(*a*) Three-dimensional view of the network formed by SiO_4 tetrahedra in silica glass.

(*b*) Two-dimensional representation of the network of silica glass.

(*c*) Broken links in the network when SiO_2 is modified by the addition of Na_2O.

(*d*) Linear chains characteristic of the $SiO_2 \cdot Na_2O$ composition.

FIGURE 5-2 Schematic illustrations of various structures in glasses based on silica. (*After sketches by W. H. Zachariasen, by B. E. Warren, and by W. D. Kingery.*)

380°K, its structure converts to a crystalline, metastable face-centered cubic solid solution. Only on heating at about 500°K does the equilibrium, two-phase structure form. Many pure nonmetallic elements, such as S, Se, Ge, and Si, form a glass under suitable conditions.

OXIDE GLASSES. The same bonding forces responsible for the network structure in crystalline SiO_2 and silicates (see Sec. 1-6) produce similar networks in glasses based on SiO_2 (Fig. 5-2*a* and *b*). Other oxides, B_2O_3 and P_2O_5 in particular, can also act as the primary *network former*, but the present discussion will be limited to silica glasses. If a basic oxide, such as Na_2O, is added to silica glass, a bond in the network is broken by the following reaction:

$$-\mathrm{Si}-\mathrm{O}-\mathrm{Si}- + \mathrm{Na_2O} \rightarrow 2(-\mathrm{Si}-\mathrm{O^-Na^+}) \qquad (5\text{-}2)$$

and the relatively mobile sodium ion becomes a part of the structure (Fig. 5-2*c*). Examples of such network modifiers are Na_2O, K_2O, Li_2O, CaO, MgO, and PbO. With increase in the amount of modifier, the average number of oxygen-silicon bonds forming a bridge between two silicon atoms decreases as follows:

O/Si ratio in glass	**Bridging O–Si bonds/Si atom**
2	4
2.5	3
3	2

Two bridging bonds per silicon atom could correspond, for example, to a linear chain structure (Fig. 5-2*d*), and any further increase in modifier would reduce the length of the chain. The principal effect of a modifier is to lower the melting and working temperature by decreasing the viscosity. An excess of modifier can make the structural units in the melt sufficiently simple and mobile that crystallization occurs in preference to the formation of a glass.

The compositions and properties of typical glasses are listed in Table 5-2. Soda-lime glass has not only the advantage of being made from inexpensive raw materials, but also the practical convenience of low working temperatures (Fig. 5-3). Most glass-forming operations (pressing, drawing, etc.) begin at the temperature corresponding to the working point but can continue while the glass cools. At the softening point a glass can still deform because of its own weight. At the annealing point, stresses can be removed in a conveniently short time (several minutes). From the strain point (or a lower

TABLE 5-2. TYPICAL OXIDE GLASSES

Type	Composition, w/o								Properties or uses
	SiO_2	Na_2O	K_2O	CaO	MgO	B_2O_3	Al_2O_3	Other	
Soda-lime	72	14		9	4		1		Window glass
Silica glass (fused quartz)	99.5+								High-temperature applications; low coefficient of expansion
96% silica glass	96.3	<0.2	<0.2			2.9	0.4		Comparable to fused quartz
Borosilicate	80.5	3.8	0.5			12.9	2.2		Resistant to heat and to chemicals
Light flint optical	54	1	8					37PbO	High index of refraction
Surface-strengthened glass	55	16	2	2		2	19	$4TiO_2$	Cookware
Glass-ceramic	56				15		20	$9TiO_2$	Radomes

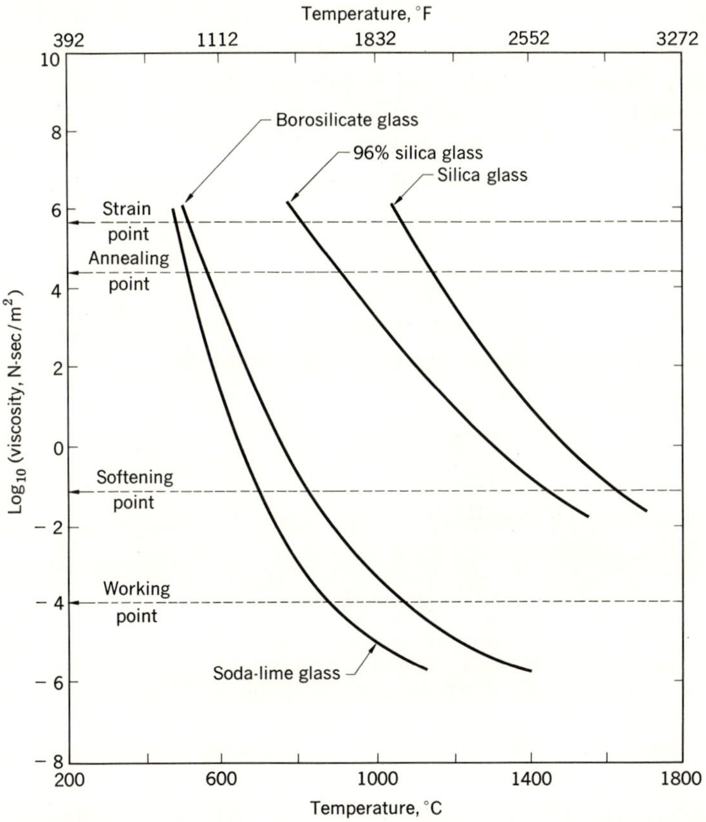

FIGURE 5-3 Important temperatures for the fabrication of silicate glasses, shown in relation to the curves of viscosity versus temperature (conversion of units: M N-s/m^2 $\times$ 10^9 = N poise).

temperature) a glass can be rapidly cooled without introducing permanent, dangerous residual stresses.

The strength of glass can be increased by *surface* strengthening, in which the composition of the surface is changed by ion exchange. The finished article is fabricated from a glass whose structure contains a small cation, such as lithium. Next, the article is placed in a molten salt bath containing a large cation (sodium or potassium) and treated for a few hours at a temperature below the strain point. During this time the large ion replaces the small one in a surface layer about 0.1 mm thick. Permanent compressive stresses are then generated in the surface of the glass during subsequent cooling. Since tensile stresses are the cause of failure in glasses, the surface-strengthened glass exhibits greatly increased tensile strength.

GLASSY POLYMERS. The noncrystalline aspect of polymers is of major importance for these materials. Several factors determine the extent to which a given polymer is crystalline at the temperature of use (usually room temperature). Molecular weight and tacticity are discussed here. During the synthesis of long polymer chains (Fig. 1-9) both the average chain length (molecular weight) and the stereoregularity of the molecule can often be controlled. The prefix *stereo-* means *space,* and Fig. 5-4 shows that a chain of

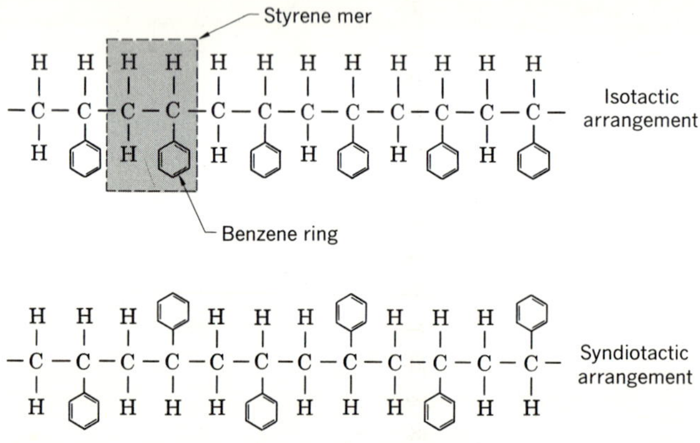

FIGURE 5-4 Two types of stereoregularity of the benzene side group in polystyrene. Commercial polystyrene has a random arrangement of the benzene rings, an arrangement termed *atactic.*

styrene monomers can form three different long-range arrangements in three dimensions, although each has the chemical formula $[-CH_2-CH(C_6H_5)-]_n$, where n is the number of mer units in the chain molecule. The most common arrangement, that of commercial polystyrene, is the random or *atactic* one. Suitable choice of catalyst and other conditions of manufacture permit the preparation of a chain having long-range regularity either of the isotactic or *syndiotactic* type.

The average molecular weight of the chains constituting a polymeric material strongly affects the melting temperature (Fig. 5-5) and therefore the properties of the polymer. When polyethylene has a molecular weight of about 200 ($n = 14$), T_f is far below room temperature (Fig. 5-5*a*), and this material is a viscous liquid. At a molecular weight of 400 ($n = 29$), the melting temperature is 62°C, and this polymer is fully crystallized at room temperature. T_f continues to increase with increase in molecular weight, reaching 137°C at $M = 10^4$. The material surrounding the crystalline lamellae is still flexible because the glass transition temperature for polyethylene (−110°C) is far below room temperature. Any other polymer, such as isotactic polystyrene, behaves in a similar manner with increase in molecular weight.

However, since T_g = 80°C for long-chain isotactic polystyrene, this polymer consists of a rigid glass phase surrounding the crystalline lamellae.

Styrene-butadiene rubber

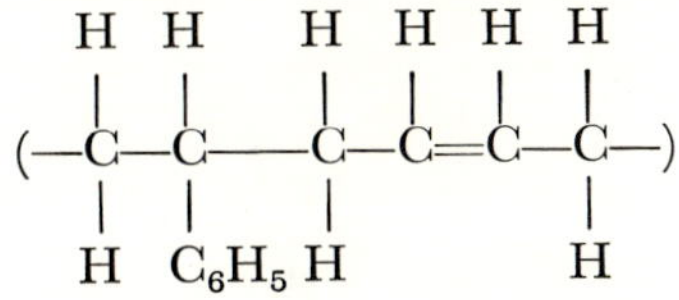

and atactic polystyrene are examples of chain molecules that have difficulty in crystallizing because of the complexity of their basis (Fig. 1-12) in a possible crystal lattice. When a high-molecular-weight polymer of this type cools from a temperature in the molten region (Fig. 5-5*b*), it passes through a diffuse transition zone analogous to a melting zone and enters a zone of rubbery or leathery properties. Styrene-butadiene rubber is in this condition at room temperature. When subjected to a tensile stress, such a material deforms elastically by the partial straightening of coiled chains, but also flows viscously by the slipping of one chain past another (see Sec. 9-7). The permanent deformation represented by this viscous flow can be largely prevented by *cross-linking* adjacent chains. If the polymer is reacted with a small amount of sulfur at a high temperature (the process of vulcanization), enough of the neighboring chains are cross-linked at the position of their double bonds to prevent viscous flow. The rubber then behaves in an elastic manner. Cross-linking of almost all of the double bonds results in the rigid mass known as *hard rubber*. Cross-link bonds in a thermosetting polymer are too strong to be broken by moderate heating, and therefore the typical polymer of this type does not soften or melt.

The examples of Fig. 5-5 show the importance of the T_g and T_f transitions (Fig. 5-1) for polymers. The effect of polymer composition and structure on T_f and T_g is often rather complex, but a simple example is given in Fig. 2-6. Those materials that are rubbers or are flexible plastics like polyethylene must have T_g below room temperature. Of course, if the operating temperature is lowered below their T_g transition, even these materials become rigid. Polymers like isotactic polystyrene, that have their T_g transition above room temperature, are rigid under usual conditions of testing. Commercial (atactic) polystyrene behaves in a rubbery manner at temperatures above its glass transition (100°C), but at room temperature it has a glass structure and is hard and brittle. If polystyrene is heated to above T_g, it reverts to the rubbery condition characteristic of the higher temperatures. Polymers that behave in this manner are termed *thermoplastic*. In contrast, *thermosetting polymers* lose this reversibility, because they cross-link during initial processing. These polymers do not become molten on being reheated; instead, they decompose like other polymers if heated to high temperatures (300 to 500°C). A useful type of cross-linking is furnished by ionic bonding in "ionomers." At high temperatures

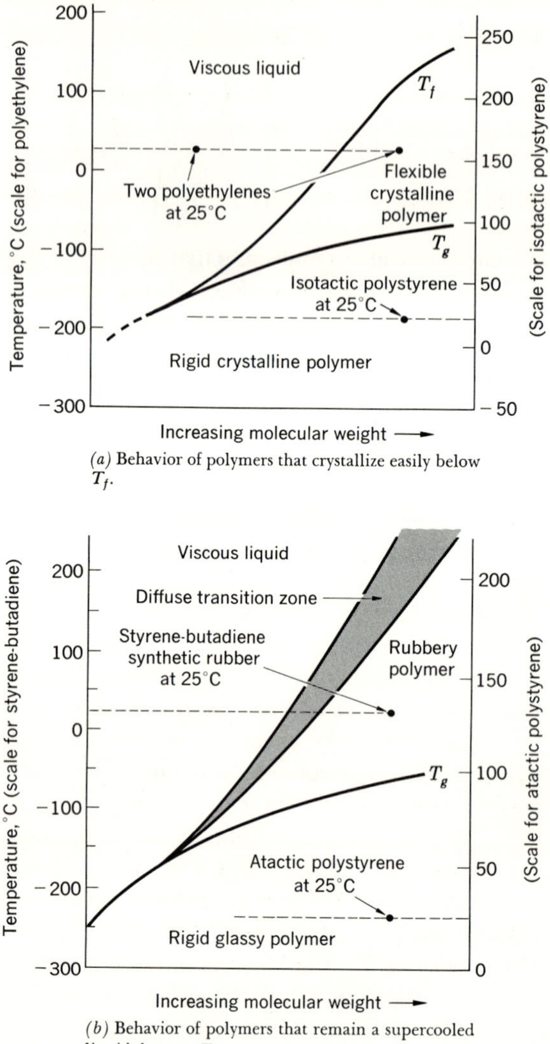

(*a*) Behavior of polymers that crystallize easily below T_f.

(*b*) Behavior of polymers that remain a supercooled liquid down to T_g.

FIGURE 5-5 Effect of molecular weight and of stereoregularity on the properties of typical polymers. (*After J. A. Brydson.*)

this type of cross-linking is appreciably weakened, and the polymer is fluid enough to be formed by the customary methods for thermoplastic materials; at room temperature the ionic bonding improves stiffness and toughness.

The variety of possible polymers is virtually unlimited, and the combinations of attainable properties is also extremely large. Over 100 different monomers are currently employed in the manufacture of commercial polymers. The chain molecules vary according to whether homopolymerization, copolymerization, block polymerization (Sec. 2-4), or graft polymerization have been employed singly or jointly. The latter technique permits the addition of long, bulky side groups to the main "backbone" of a polymer. In "ladder molecules" the chains are stiffened by the formation of a double chain with regular links between adjacent chains like the rungs of a ladder. Combining the various aspects of polymer structure in numerous combinations (Fig. 5-6) gives an impressive array of characteristics and of present and potential uses. The additional effects of mechanical treatments (orienting of fibers) or of heat treatments (changing the morphology) are considered in later chapters.

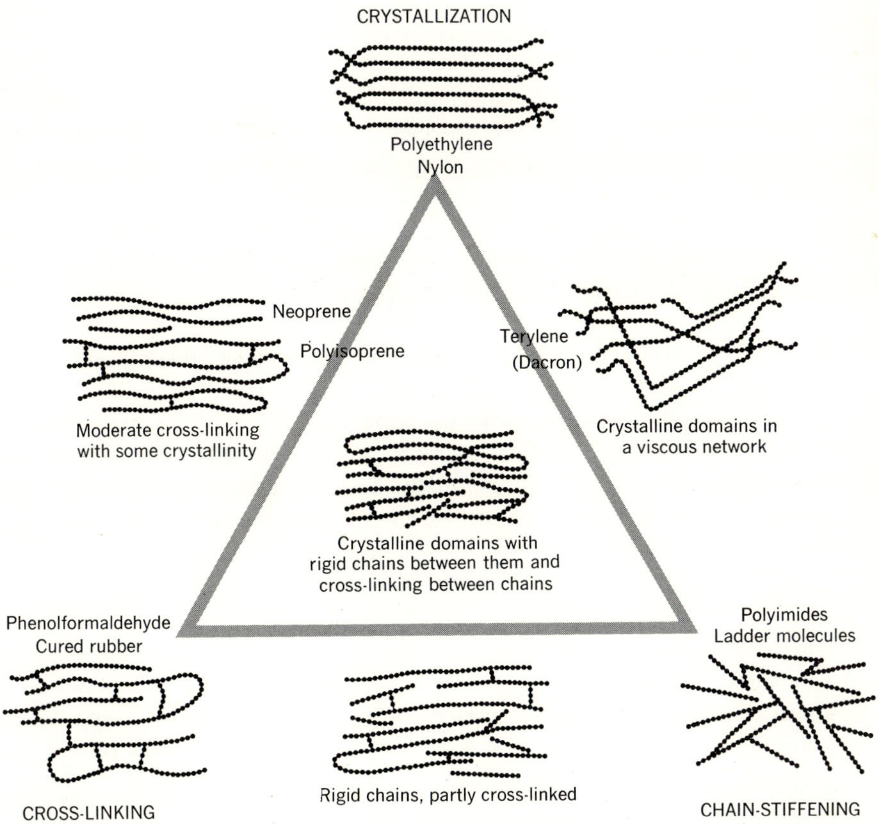

FIGURE 5-6 Schematic illustration of how the three principal strengthening mechanisms—crystallization, cross-linking, and chain-stiffening—are combined to give polymers with a variety of characteristics and uses. (*After H. F. Mark.*)

CRYSTALLINE PHASES

Crystalline phases constitute 100 percent of the structure of a typical metal, about 90 percent in many ceramics, and perhaps an average of 70 percent in high-strength polymers; the remaining phases are the glasses, previously discussed, and the "porosity phase" in ceramics (Fig. 6-21). Often two or more phases are present, each with characteristic properties and capacity for reaction or similar modification that can greatly improve the behavior of a material under given conditions of service. As a preparation for the study of phase reactions in the following chapters, the present section considers important aspects of crystalline phases not already treated in Chap. 1. The study of the many types of crystalline phases is simplified by using the somewhat arbitrary division into simple phases (Fe, MgO, etc.), solid solutions (C in Fe, Al_2O_3 in MgO, etc.), and intermediate phases (FeCr, $2MgO \cdot SiO_2$, etc.).

5-3. Simple Phases

Most elements and certain compounds form relatively simple crystalline phases. The crystal structures of the elements and various simple compounds were discussed in previous chapters and are summarized in the table on the inside back cover. Other important structural features of real crystals include (1) linear and point defects (Chap. 4), (2) boundaries between grains or within a grain, which are also classed as two-dimensional defects, and (3) external surfaces in contact with a vacuum or a vapor phase.

SURFACES. Although the surface layer is extremely small compared to the thickness of an ordinary crystal, several atomic layers contribute to the special properties associated with the surface (Fig. 5-7). The most energetic atoms are on the top layer; the atoms in several subsequent layers have successively lower energies and gradually become indistinguishable from the atoms in the bulk crystal. The surface energy γ has the general definition

$$\gamma = \frac{dW}{ds} \qquad \text{J/m}^2 \tag{5-3}$$

where dW is the reversible work in joules (J) necessary to create an area ds of new surface. When the temperature is sufficiently high, the effect of elastic strain is negligible. Often the surface energy can be pictured as being caused entirely by a force per unit length, F/l, acting in the plane of the surface and perpendicular to its boundary. Since $F\,dx = dW$, Eq. (5-3) can be rearranged in the useful form

$$dW = \frac{F}{l}(l \times dx) = \frac{F}{l}\,ds = \gamma\,ds \tag{5-4}$$

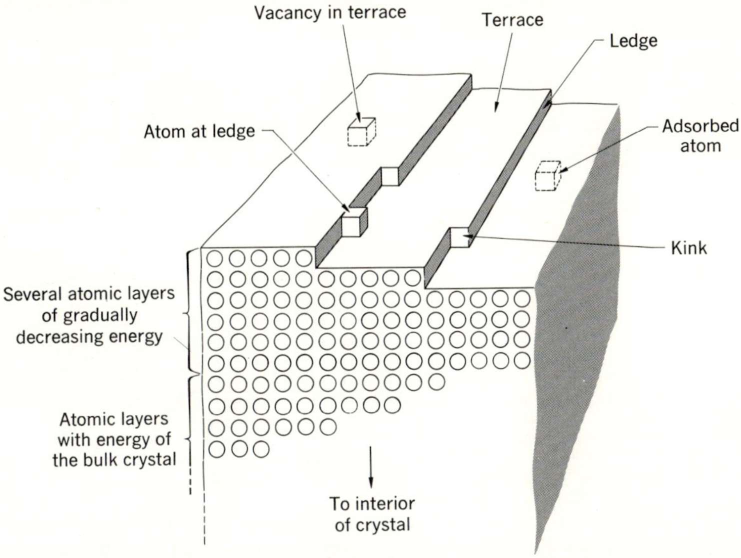

FIGURE 5-7 Schematic illustration of a vicinal surface of a crystal (when a low-index plane is almost parallel with the surface). The major structural features are the terraces, ledges, and kinks.

In this case, the surface energy can be treated as a surface tension, $\gamma = F/l$, with units of newtons per meter (Prob. 5-3).

In an *n*-component alloy the change in free energy,† dA, is the sum

$$dA = dW + \sum_{i=1}^{n} \mu_i \, dN_i \tag{5-5}$$

of the work, dW, to create an area ds of new surface and of the energy added by the adsorption of dN_i molecules on the area ds. The chemical potential μ_i was defined previously [see Eq. (2-50)]. After division by ds, Eq. (5-5) can be written

$$\gamma = \frac{dA}{ds} - \sum_{i=1}^{n} \mu_i \frac{dN_i}{ds} \tag{5-6}$$

In only a few cases (such as in a one-component system, where $dN/ds = 0$) is γ identical with the free energy per unit area, dA/ds. In most alloys adsorption also contributes to the value of γ.

† A is the Helmholtz free energy and is related to the Gibbs free energy G by the equation $G = A + PV$. Often the PV term is negligible, and then $G \approx A$.

Experimental values of γ are difficult to obtain even for a pure component since the surface must be extraordinarily clean to avoid extraneous effects. One of the few complete studies of γ as a function of crystallographic orientation of the surface is that for gold (Fig. 5-8). Because of the high symmetry of a cubic crystal, every type of crystal plane is included in the reference stereographic triangle (Fig. 5-8*a*), which has the points representing

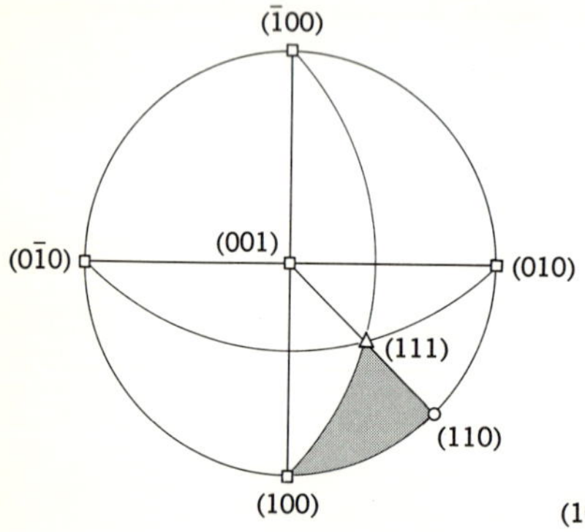

(*a*) Position of the reference stereographic triangle in the stereographic projection for a cubic crystal.

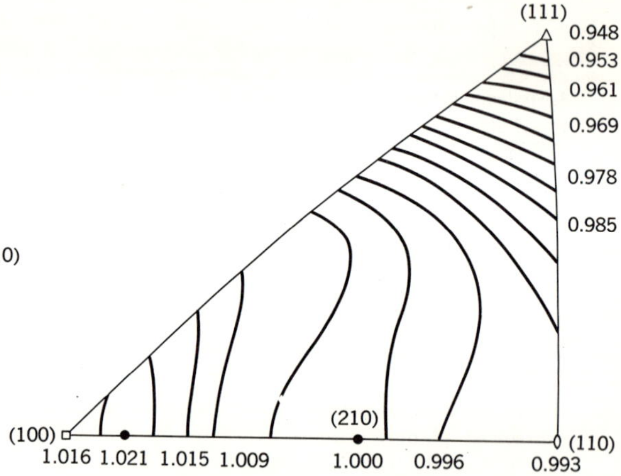

(*b*) Variation of the surface tension, γ, of gold with crystallographic orientation at 1030°C in a hydrogen atmosphere. The γ value of the (210) plane is used as the reference.

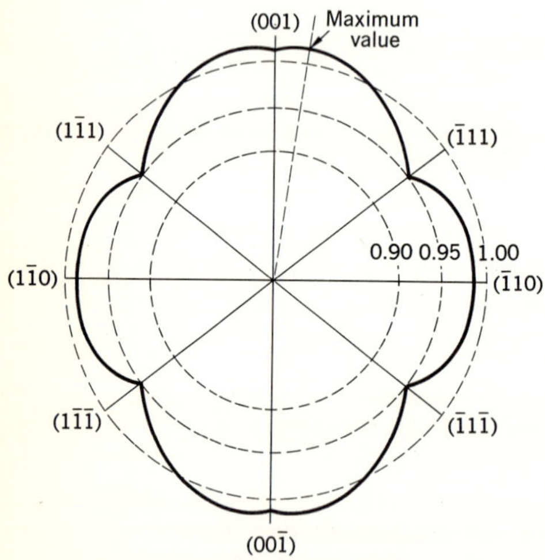

(*c*) Two-dimensional (polar) plot of γ values for the crystal planes whose pole is [110].

FIGURE 5-8 Surface energy of gold at 1030°C in a hydrogen atmosphere. (*Courtesy of W. L. Winterbottom and N. A. Gjostein, Ford Scientific Laboratory.*)

the (100), (110), and (111) planes as its three vertices. This triangle permits plotting the complete three-dimensional variation of γ in a crystal of gold (Fig. 5-8*b*). The values plotted are the ratios $\gamma_{(hkl)}/\gamma_{(210)}$, where the value for the (210) plane is used as the reference. The variation of γ can be shown more easily in a two-dimensional plot (Fig. 5-8*c*), and γ is then seen to be almost isotropic. The γ plot permits classifying surfaces among three types: (1) singular surfaces, like {111} or {100}, are characterized by a minimum (cusp) in the γ plot and could ideally exist as a smooth terrace; (2) vicinal surfaces (Fig. 5-7) have an orientation close to a singular surface and can be described by the terrace-ledge-kink model; (3) diffuse surfaces are "rough" on the atomic scale.

Ionic and covalent crystals are expected to exhibit much more anisotropy in their γ values. The schematic polar plot for NaCl in Fig. 5-9 shows

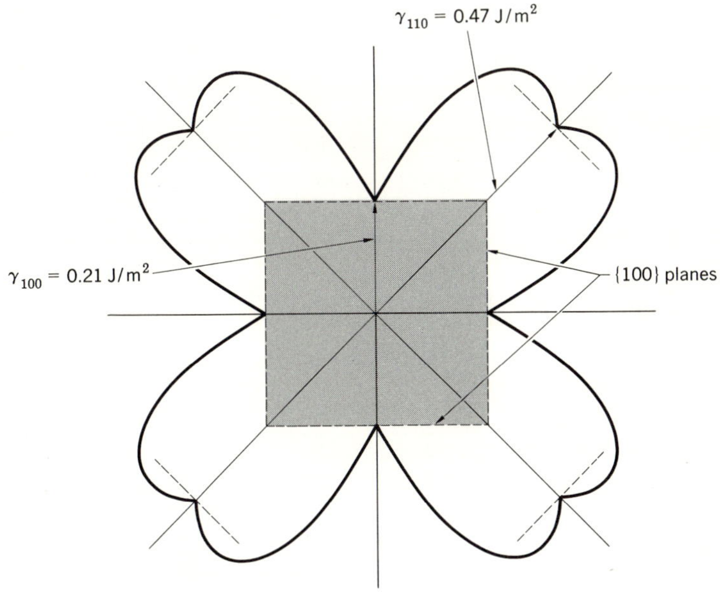

FIGURE 5-9 Schematic polar plot of the surface tension γ based on calculations for NaCl. The equilibrium form of the crystal should consist only of {100} type planes (conversion of units: $M\,\text{J/m}^2 \times 10^3 = N\,\text{erg/cm}^2$).

that the {110} planes have more than twice the surface energy of {100} planes; {111} planes should have even higher energy. The total surface energy depends both on the γ values and on the extent of each type of surface, and therefore, a special procedure is used to determine the exterior shape having the minimum surface energy. This procedure is the Wulff construction in which (*hkl*) planes (*lines* in two dimensions) are constructed at the end of the line representing $\gamma_{(hkl)}$ and perpendicular to this line. The equilibrium crystal shape is the geometric figure formed by those plane segments closest to the origin of the polar plot. Although pure NaCl in a clean atmosphere should consist of cubic crystals according to the construction in Fig. 5-9, kinetic factors or impurities (see Prob. 5-4) can result in other external forms.

INTERNAL BOUNDARIES. The surface separating two crystals of different orientations is called a *boundary.* In the previous chapter we saw that small-angle grain boundaries have a dislocation density that increases in proportion to the angle of misorientation, θ. Consequently, the energy of a small-angle boundary increases linearly with θ near $\theta = 0$ (see Prob. 5-5) and then more slowly as the stress fields of adjacent dislocations interact more strongly. This behavior is shown for copper in Fig. 5-10. An average surface tension γ_{gb} can

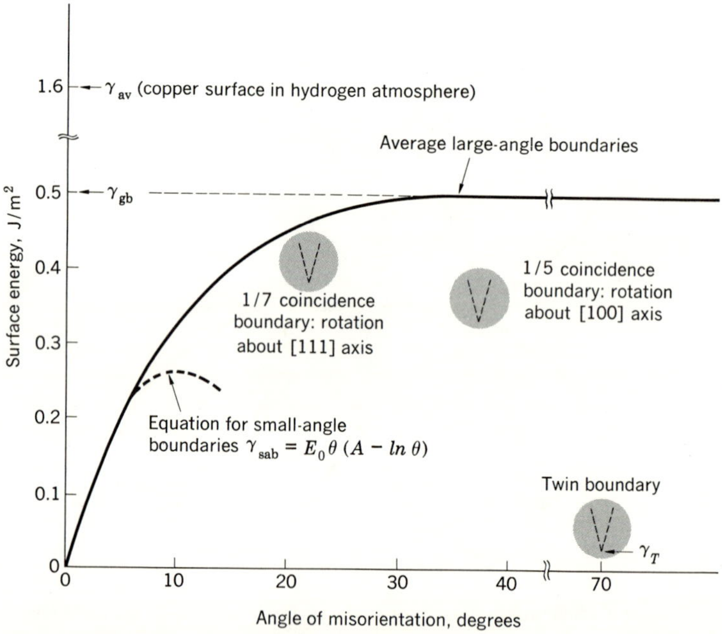

FIGURE 5-10 Typical relations among the surface tension values for various types of surfaces. Estimated values for copper are shown (conversion of units: M J/m² × 10³ = N ergs/cm²). (*Based in part on data of N. A. Gjostein and F. N. Rhines.*)

be associated with an ordinary (large-angle) grain boundary, which is composed of a mixture of dislocations of various types. Since the value of γ_{gb} is relatively high, it helps determine the stable shapes assumed by the grains of a given material. There are a few special boundaries for which a particular large-angle orientation of the two adjacent crystals produces a low value of γ. These special boundaries can be divided into two categories, *coincidence boundaries* and *twin boundaries*. A coincidence boundary (Fig. 5-11) is incoherent like an ordinary grain boundary; that is, most of the atoms at the boundary of one crystal do not match the lattice positions in the other crystal. On the average, however, mismatch in a coincidence boundary is less in proportion to the density of coincidence sites. For example, in Fig. 5-11, one atom in seven

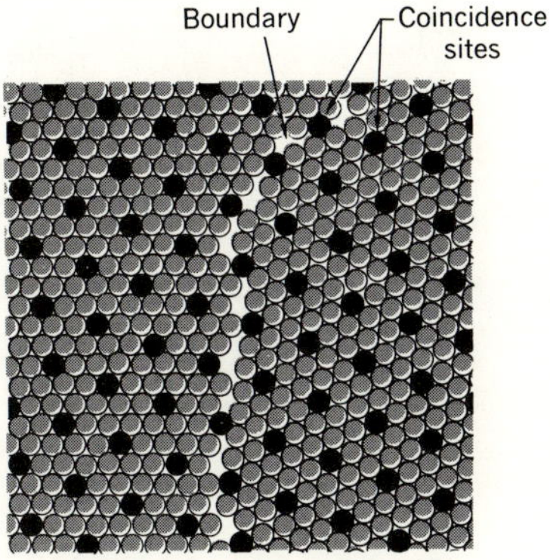

FIGURE 5-11 The coincidence lattice formed by every seventh atom in two grains misoriented 22° by rotation about a ⟨111⟩ axis. A typical coincidence boundary is indicated. (*After the late M. L. Kronberg, H. F. Wilson, and R. W. Cahn.*)

in an average boundary is a lattice site in both crystals. This boundary is called a $\frac{1}{7}$ *coincidence boundary*, and the atomic sites in question are said to form a *coincidence lattice* within the two grains. Coincidence lattices occur in all of the common crystal structures and have site densities varying from $\frac{1}{3}$ to $\frac{1}{19}$ and lower.

A twin boundary is often a type of coincidence boundary but is usefully treated separately because it is *coherent*; that is, the atomic positions in the two crystals match along the boundary, and no dislocation need be present. The energy γ_T of a twin boundary is usually only about 0.1 γ_{gb} (Fig. 5-10), whereas a coincidence boundary has an energy only slightly less than γ_{gb}. The two most common types of twin orientation are (1) rotation (coincidence) twins, produced by a suitable rotation about a direction [*hkl*] called the *twinning axis*, and (2) reflection twins, in which the lattice sites are mirror images across a plane (*hkl*) called the *twinning plane*. Twins in the more symmetrical lattices, such as the face-centered cubic, satisfy both definitions. The twinning plane K_1

in the FCC lattice is (111). Figure 5-12 shows how a displacement in the twinning direction (η_1), [11$\bar{2}$], on each successive twinning plane can produce a twinned region, separated from the original crystal by twin boundaries. Recalling that a Shockley partial dislocation produces a displacement in the [11$\bar{2}$] direction on the (111) plane, we see that twinning occurs when one partial dislocation moves along each (111) plane in a given region of the crystal.

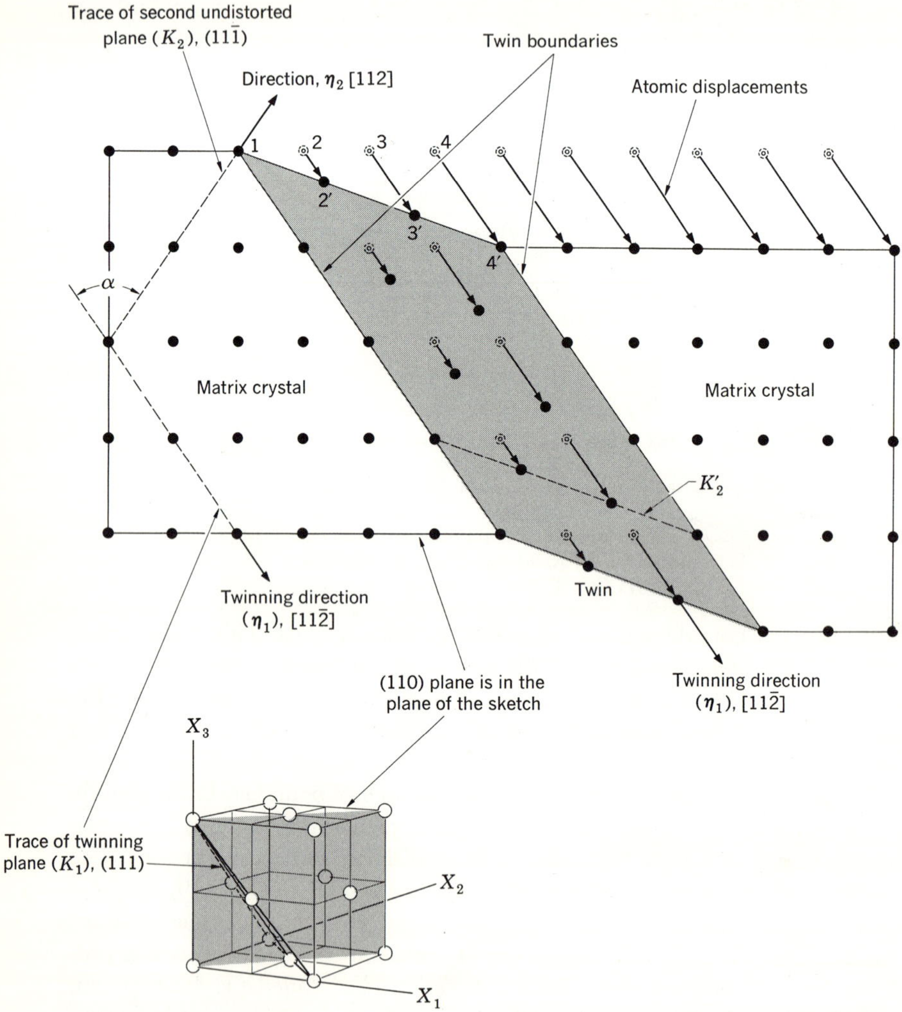

FIGURE 5-12 Sketch of a twin and twin boundaries in the face-centered cubic lattice. Because the plane of the sketch is the (110) plane, the (111) and (11$\bar{1}$) planes are perpendicular to the page, and the [11$\bar{2}$] and [112] directions lie in the plane of the sketch.

The shear strain accompanying the production of the twinned region from the original crystal causes changes in distances and angles in almost every crystal plane. For example, the row of atoms 1, 2, 3, and 4 in Fig. 5-12 changes to the row 1, 2′, 3′, and 4′, having a larger interatomic spacing. There are two special planes, however, on which such distortion does not occur. One is the twinning plane K_1; the other is called the *second undistorted plane*, K_2, and has a shearing direction η_2 associated with it. In the face-centered cubic lattice K_2 is $(11\bar{1})$, and η_2 is [112]. As a result of twin formation, K_2 changes from an orientation in which η_2 and $(-)\eta_1$ make an angle α to a new orientation in which η_2' and $(+)\eta_1$ make this same angle.

The surface tensions of the various types of boundaries explain many of the grain structures observed in polycrystalline substances. Consider first the junction of three grain-boundary surfaces (Fig. 5-13*a*). Each surface tries to

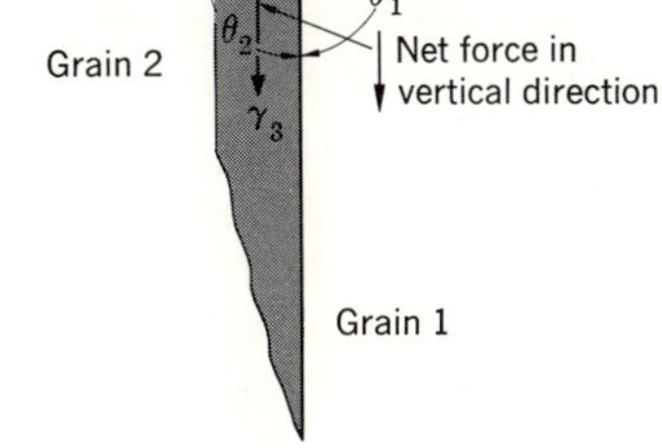

(*a*) A grain junction in the interior of a one-phase material. This junction is in an unstable configuration because the surface tensions are not balanced. All three angles would be 120° in a stable configuration.

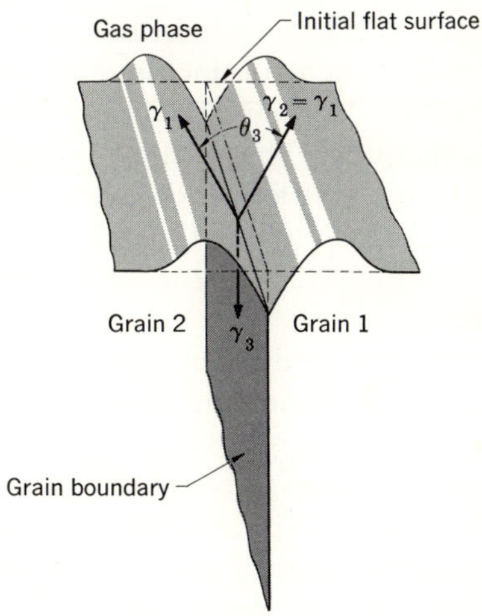

(*b*) A grain-boundary groove at the surface of a one-phase material. The groove is the stable configuration at the junction of a grain boundary with the surface.

FIGURE 5-13 Schematic illustrations of structural features in polycrystalline substances controlled by surface tension.

reduce its area. This tendency can be represented by a force γ_i per unit length acting on the grain junction, where γ_i is the surface tension of the *i*th grain-boundary surface. In this notation the surface tension of the boundary between grains 1 and 2 is denoted by γ_3, rather than by γ_{12}. If all three boundaries have the same surface tension, equilibrium of forces requires that the three angles around the junction be the same and equal to 120°. An unstable boundary configuration produces a net force on the grain junction; consequently, the grain structure tends to change toward the 120° junction angles. Problem 5-8 considers deviations from this normal stable configuration. If two similar surfaces meet a grain boundary at a junction (Fig. 5-13*b*), equilibrium of forces requires that

$$\gamma_3 = 2\gamma_1 \cos \frac{\theta_3}{2} \tag{5-7}$$

An initially flat region of surface, where a grain meets the surface perpendicularly, develops a groove in response to the unbalanced force. The groove angle θ_3 in a copper surface is the subject of Prob. 5-9.

Two experimental techniques used to study surface structures on the atomic level are LEED (low-energy electron diffraction; see Sec. 1-11) and field-ion microscopy. LEED patterns reveal a periodicity in many surface structures quite distinct from the three-dimensional periodicity of the crystal structure. In some cases the periodicity is that of minute amounts of impurities on the surface, but a clean vicinal surface is expected to show regularity of terracing, kink formation, or distribution of surface vacancies. At temperatures near the melting point, certain "atomically smooth" surfaces may change into "atomically rough" surfaces because of the extreme mobility of the surface atoms.

The field-ion microscope (Fig. 5-14*a*) employs a specimen in the form of a hemispherical tip having a radius of about 10^{-7} m. Through an intermediate process of ionization of an "imaging gas," such as helium, the image of the tip surface is radially projected by an accelerating potential onto a fluorescent screen at a magnification of about 10^6. The curvature of the tip brings the edges of successive atomic layers into view (Fig. 5-14*b*). Only the edge atoms of close-packed layers (the uppermost layer of Fig. 5-14*b*, for example) are imaged (Fig. 5-14*c*). In less closely packed layers, every surface atom is seen in the field-ion micrograph. Consequently, the presence of a lattice vacancy can be detected as shown in Fig. 5-14*c*. Since atomic positions are revealed in such detail, field-ion microscopy is useful for studying grain boundaries (Fig. 5-14*d*) and other structural features at a surface.

5-4. Solid Solutions

Foreign atoms that form a solid solution (see Sec. 2-3) with a simple phase may be deleterious in some respects, but they are often essential for properties of technical interest. The doping of silicon to produce a more useful semiconductor is one example. In ionic crystals the foreign atoms may be used

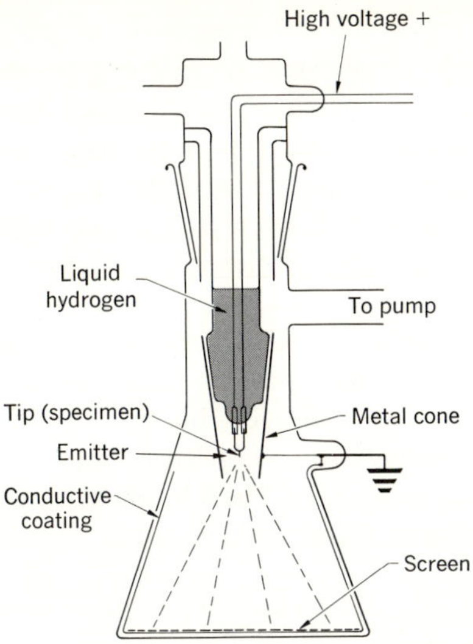

(a) Schematic illustration of the operation of a field-ion microscope.

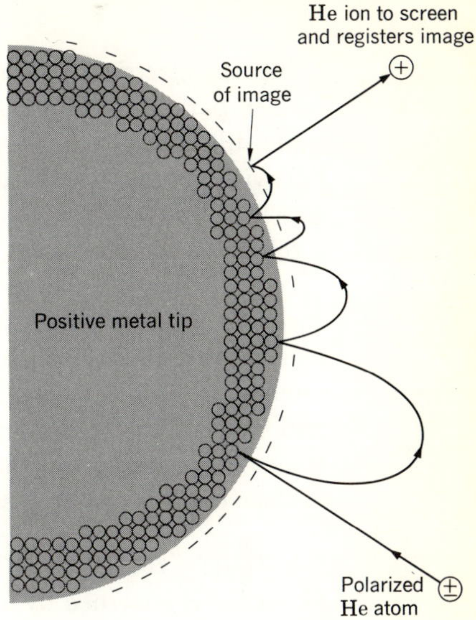

(b) Array of atomic planes at the tip and the mechanism of ionization.

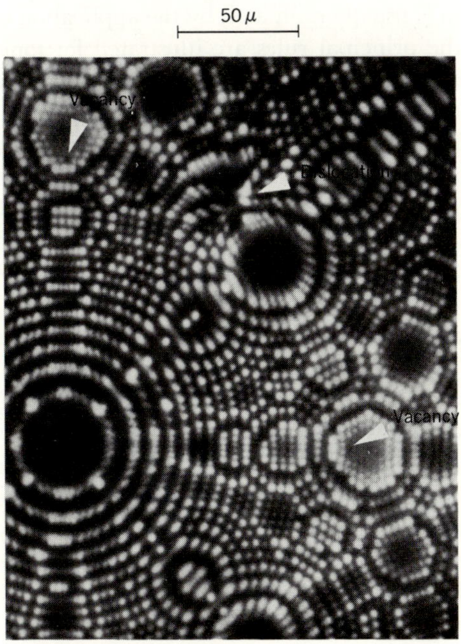

(c) Micrograph showing the presence of two vacancies and a dislocation in the surface of an indium specimen.

(d) Pattern from tungsten metal showing a grain boundary. The atomic positions are altered only in the immediate vicinity of the boundary.

FIGURE 5-14 Illustration of the field-ion microscope and its use in studying atomic arrangements in surfaces. (*Courtesy E. W. Mueller, Pennsylvania State University.*)

to increase electrical conductivity, whereas in metals their role in strengthening is often paramount. Foreign atoms present as a second crystal structure have quite different effects on properties. By the term *solid solution* we mean the following specific type of alloying: *the incorporation of an arbitrary, perhaps limited amount of substance B into the crystal lattice of substance A.* Even pure A may have a relatively complex array of "species" on its lattice. A pure MX compound (Sec. 4-5), for example, contains M_M, X_X, V_M, V_X, and also interstitial species (perhaps only interstitial vacancies V_i). If only the interstitial sites are occupied by substance B, the term *interstitial solid solution* is employed. The most common interstitial solutes in metals are the small atoms, hydrogen, boron, nitrogen, and oxygen. In the more usual case of a *substitutional solid solution*, both substance A and substance B occupy the normal lattice sites. The solution behavior of ionic crystals and the associated defects were discussed in Sec. 4-5.

FACTORS LIMITING SOLID SOLUBILITY. The extent of solid-solution formation is governed by free-energy curves in the manner described in Chap. 2. Only rarely are the actual thermodynamic data available, and adequate methods of calculation have not yet been developed. Consequently, the unknown behavior of a given system is usually estimated by the application of several empirical rules. Three of the principal rules are illustrated for ionic crystals by the phase diagrams of Fig. 5-15. One alkali halide, KCl, might be expected to incorporate easily into the crystal lattice of another, say RbCl, since both have the sodium chloride crystal structure. Figure 5-15*a* shows that in fact there is complete solid solubility in this case; the cation sites in the crystal are occupied, almost randomly, by K^+ and Rb^+ ions. The KCl–NaCl phase diagram (Fig. 5-15*b*) shows that solid solubility is more limited in this case; the cause is the dissimilarity in size of the K^+ and Na^+ ions. At low temperatures the resulting strain energy causes a mixture of the α_1 and α_2 phases to have lower energy than a single solid solution. When the difference in size is even greater (Fig. 5-15*c*), solid solubility is negligibly small, although in principle the entropy of mixing causes a finite (but perhaps unmeasurable) amount of NaCl to dissolve in RbCl, and vice versa.

The effect of atomic size on solid solubility is recognized for metallic systems and is expressed through the Hume-Rothery 15 percent rule. If the radius ratio is more than 1.15, the formation of solid solutions is generally limited to about 1 percent. In metals with the same crystal structure, complete solid solubility occurs if the radius ratio is less than about 1.08. In both ionic and covalent crystals the critical size ratios are at somewhat higher values than in metals. A probable explanation is the larger energy associated with space-filling requirements in metals, which are more closely packed and have higher coordination numbers. The large K^+ ion can be more easily accommodated among its six nearest neighbors in the NaCl lattice than can the Zr atom

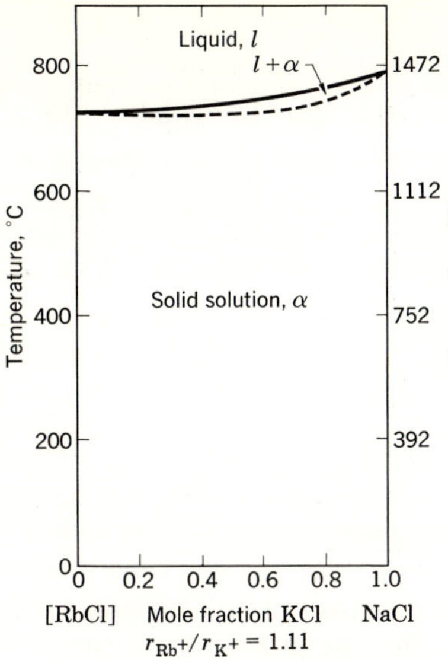

(*a*) Complete series of solid solutions between like atoms of similar size.

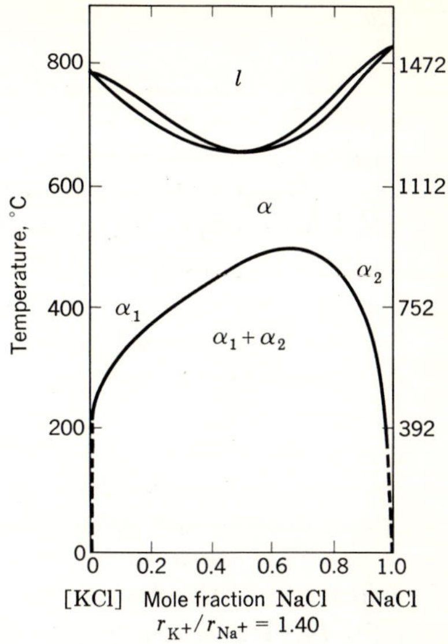

(*b*) Limited solid solubility at room temperature between like atoms of dissimilar size.

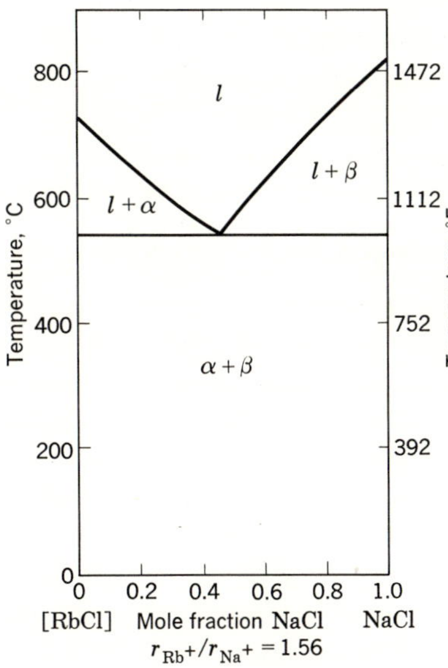

(*c*) Negligible solid solubility between like atoms when size difference is great.

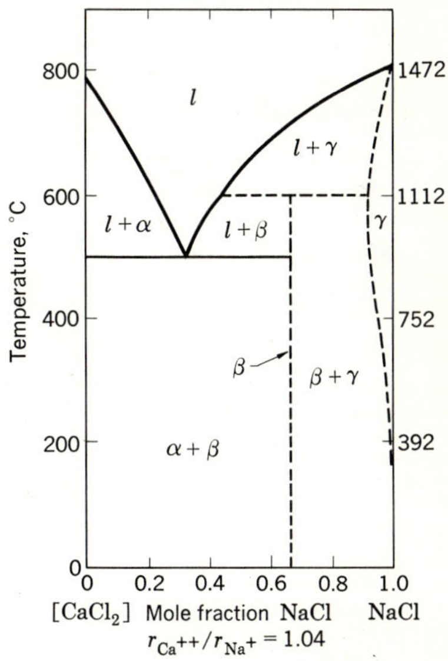

(*d*) Only limited solubility can occur between crystals having different crystal structures. The solubility is greater in the crystal of lower valence.

FIGURE 5-15 Effect of size factor, lattice type, and relative valence on the extent of solid solution between ionic crystals. (*Diagrams adapted from E. M. Levin, et al., "Phase Diagrams for Ceramists," American Ceramic Society, Columbus, Ohio 1964.*)

(r_{Zr}/r_{Cu} = 1.25) among 12 nearest neighbors in copper. In the latter case no solubility can be detected at room temperature.

The two other rules can be illustrated using the Ca^{+5} and Na^+ ions, which have nearly the same size (Fig. 5-15*d*). Since $CaCl_2$ and NaCl have different types of chemical formula, they also have different crystal structures ($CaCl_2$ has a structure similar to C4(TiO_2), shown on the inside back cover). The following *rule of lattice type* then applies: in general, a continuous series of solid solutions is possible only when the two components have the same type of crystal structure. The example of a metallic alloy of iron (body-centered cubic) and nickel (face-centered cubic) is considered in Prob. 5-10. Limited solid solubility at both ends of the diagram is still possible and occurs commonly in metallic systems. The following *rule of relative valence* then applies: the solubility is greater in the component of lower valence. For example, silver (valence +1) dissolves 40 A/o Zn although zinc (valence +2) dissolves only 5 A/o Ag. The reason for the operation of this rule in ionic (and covalent) crystals can be understood by considering the respective defect reactions that apply for the $CaCl_2$–NaCl system. The dissolution of a molecule of $CaCl_2$ in NaCl can be written

$$
CaCl_2 + V_V \begin{array}{l} \nearrow \; Ca_{Na}^{\bullet} + V_{Na}' + 2Cl_{Cl}^{\times} \qquad \text{(defect notation)} \\ \searrow \; Ca_{Na}^{++} + V_{Na} + 2Cl_{Cl}^{-} \qquad \text{(usual valence)} \end{array} \tag{5-8}
$$

Since a Ca^{++} ion on a Na^+ site is effectively positive, ionization occurs and should be included in the reaction

$$
CaCl_2 + V_V \begin{array}{l} \nearrow \; Ca_{Na}^{\times} + V_{Na}^{\times} + 2Cl_{Cl}^{\times} \\ \searrow \; Ca_{Na}^{+} + V_{Na}^{+} + 2Cl_{Cl}^{-} \end{array} \tag{5-9}
$$

Creation of a vacancy $V_{Na}^{\times}$ on a sodium-ion site in the lattice is required to preserve the crystal structure. At the other end of the phase diagram, dissolution of a molecule of NaCl in $CaCl_2$ requires the reaction

$$
NaCl + V_V \begin{array}{l} \nearrow \; Na_{Ca}' + V_{Cl}^{\bullet} + Cl_{Cl}^{\times} \\ \searrow \; Na_{Ca}^{+} + V_{Cl} + Cl_{Cl}^{-} \end{array} \tag{5-10}
$$

Ionization of Na′ to $Na^{\times}$ (Na^+ to Na^{++}) involves breaking the rare-gas electron configuration of neon and therefore does not occur. Thus, the process of dissolving NaCl in $CaCl_2$ requires the higher energy for two reasons: (1) a V_{Cl} is more difficult to form than a V_{Na} because of the large size of the chloride ion, and (2) unfavorable charge effects are associated with the reaction of Eq. (5-10). The disparity in solubilities at the two ends of the phase diagram (Fig. 5-15*d*) is a result of this difference in behavior. The technically important MgO–Al_2O_3 system is considered in Prob. 5-11 from this point of view.

A useful characteristic of solid solutions is their ability to exist in various metastable conditions. The technically important process of precipitation hardening (see Chap. 7) usually begins with a supercooled solid solution, and the "hardenability" of steel (Sec. 7-8) is essentially a measure of the metastability of austenite containing interstitially dissolved carbon and several substitutional alloying elements. An extreme type of metastability is produced by the same technique employed in making metallic glasses. For example, a complete series of solid solutions can be prepared by instantaneously cooling copper-silver alloys, even though the equilibrium solid solubility is only about 10 percent.

DEPARTURES FROM RANDOMNESS. Consider a solid solution of B atoms in element A, often pictured for simplicity as a *random* distribution of the two types of atoms. Actually, randomness occurs only in special cases, such as in dilute solutions. Several factors tend to make the average solid solution nonrandom. One is the relation of the bonding E_{AA} and E_{BB} between like atoms to the bonding E_{AB} between unlike atoms. The oversimplified, but still useful, quasi-chemical approach pictures these energies as determined only by the two atoms in question (in fact, many neighboring atoms are involved). If one A–A bond and one B–B bond are broken and the atoms rejoined to make two A–B bonds, the net energy E per A–B bond is

$$E = E_{AB} - \frac{E_{AA} + E_{BB}}{2} \tag{5-11}$$

If $E = 0$, the solid solution should be random since neither type of bonding is preferred. If AB bonds are more stable ($E < 0$), then a given atom tends to have unlike atoms as neighbors, a condition known as *ordering* (Fig. 5-16*a*). Clustering of like atoms (Fig. 5-16*b*) occurs when the bonding of like atoms is more stable ($E > 0$). A second factor contributing to nonrandomness is the strain produced by the difference in size of the A and B atoms. This factor influences both the energy and the entropy of the solid solution. Finally, changes in electronic structure within each of the atoms can favor one configuration over another by way of magnetic or electrical interactions. Although

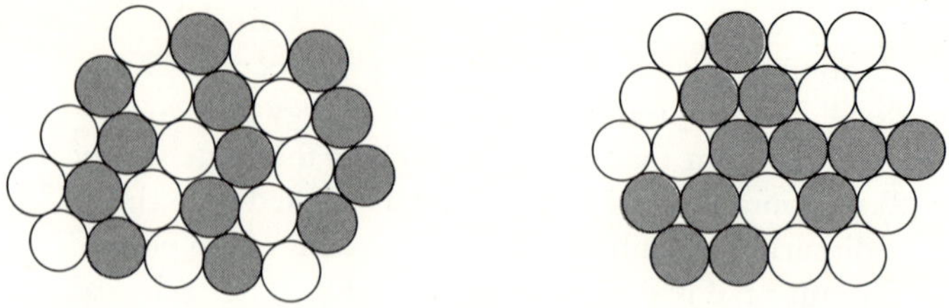

(a) Degree of ordering corresponding to $\alpha = -0.3$

(b) Degree of clustering corresponding to $\alpha = +0.3$

FIGURE 5-16 Schematic illustrations of a 50A/o solid solution of A and B atoms showing two types of short-range order—ordering and clustering.

no complete theory of solid-solution formation is available, the factor described by Eq. (5-11) is often the dominant one.

The size of an atom may vary with change in concentration of the solid solution. Atomic size is given directly by the partial molal volume $\bar{V}_i$ (Fig. 2-11), and in some systems a given atom changes its size by about 10 percent in going from a dilute solution to a concentrated solution. More commonly, $\bar{V}_A$ and $\bar{V}_B$ are almost constant (ideal-solution behavior), and thus both the molal volume and the lattice parameter plot as almost straight lines (Vegard's law) as a function of concentration. In a solid solution containing two atoms of different size, the position of an atom will not coincide exactly with a lattice point in the crystal lattice since many different interatomic distances occur. An average *static displacement* can be measured by x-ray techniques; typical values are given in Table 5-3. A second type of randomness in atomic locations results from

TABLE 5-3. REPRESENTATIVE VALUES OF STATIC AND DYNAMIC DISPLACEMENTS IN SOLID SOLUTIONS†

Material	Temperature, °K	Static displacements, Å	Dynamic displacements, Å
Cu–25 A/o Au	90	0.08	0.09
	295	0.08	0.14
Ni–50 A/o Au	180	0.11	0.13
	295	0.11	0.16
Mg–30 A/o Li	90	0.10	0.16
	295	0.10	0.25
Ti–20 A/o Mo	300	0.075	0.093

† Data courtesy B. L. Averbach, Massachusetts Institute of Technology.

thermal vibrations and is related to the phonon spectrum in the material. These *dynamic displacements* are relatively small at low temperatures but grow rapidly as the temperature rises. For a solid solution of any concentration, either ordering or clustering of the atoms can be described quantitatively by a single parameter, α. Consider a B atom surrounded by its first coordination "shell" of neighboring atoms, 12 atoms in the face-centered cubic lattice. If the solid solution contains N_A mole fraction of A atoms, then in the random arrangement the probability is N_A that a given neighbor is an A atom. In an actual solution this probability P_A is either greater or less than N_A; correspondingly, the quantity

$$\alpha = 1 - \frac{P_A}{N_A} \tag{5-12}$$

is negative if ordering occurs and positive if clustering occurs (Fig. 5-16). Order coefficients can also be determined for the second, third, and higher coordination shells. For comparison with long-range order, the degree of short-range order is described by the parameter

$$\sigma = \frac{P_A - N_A}{P_{max} - N_A} \tag{5-13}$$

P_{max} is the maximum probability of having an A atom as the nearest neighbor of a B atom in a given ordered structure.

At compositions corresponding to certain atomic ratios, AB_3, AB, or A_3B, for example, the two atoms in a solid solution may occupy specific positions in the crystal structure and form a phase with long-range order. The ordered phase may form from the disordered solid solution on cooling in two quite different ways (Fig. 5-17) in various alloys. In the copper-gold system (Fig. 2-23) the change from disorder (that is, only short-range order) to long-range order occurs almost completely at the critical temperature T_c of ordering. Hexagonal close-packed phases (Prob. 5-12) behave similarly. For some unknown reason, certain solid solutions (especially body-centered cubic) behave quite differently. Two common types of ordered phase in body-centered cubic solid solutions are shown in Fig. 5-18. When a Cu–50 $^A\!/_0$ Zn alloy is cooled to slightly below T_c, only a limited amount of ordering occurs (Fig. 5-17). Additional ordering occurs only when the temperature is again lowered. The degree of order, S, is defined as

$$S = \frac{P_A - N_A}{1 - N_A} \tag{5-14}$$

where P_A is the fraction of "A-type" sites actually occupied by A atoms. In a random solid solution $P_A = N_A$, and therefore $S = 0$. In the fully ordered structure $P_A = 1$, and $S = 1$; however, this structure develops only at a temperature considerably below T_c.

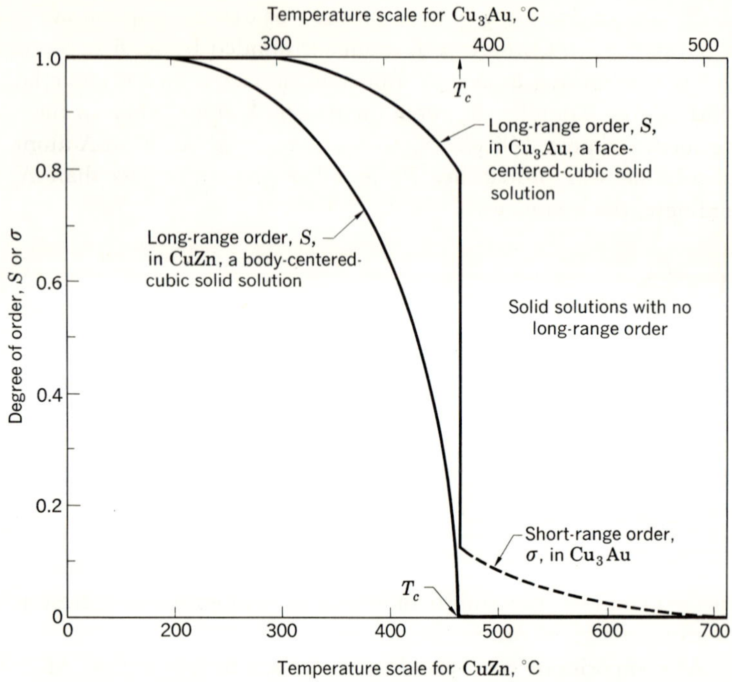

FIGURE 5-17 Variation in degree of equilibrium long-range and short-range order with temperature in two types of solid solutions.

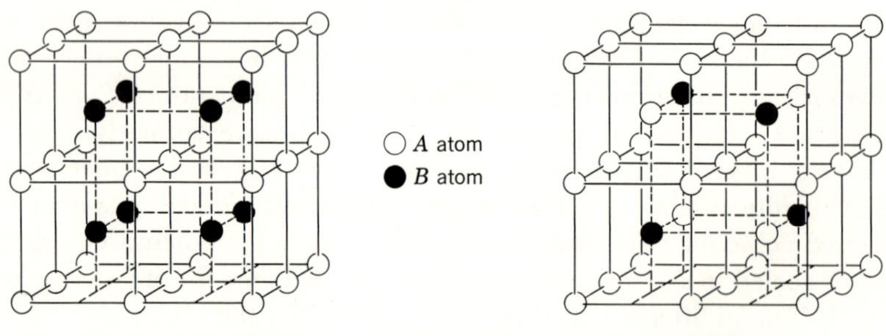

(*a*) The CsCl structure in an *AB* composition such as FeAl.

(*b*) The Fe_3Al structure in an A_3B composition.

FIGURE 5-18 Illustrations of two types of ordered structures that can develop from body-centered cubic solid solutions. (*Adapted from C. S. Barrett and T. B. Massalski, "Structure of Metals," 3d ed., McGraw-Hill Book Company, New York, 1966.*)

Two interesting structural aspects of ordered phases are domain structures and superdislocations. When a solid solution transforms to an ordered phase, the process usually involves the formation of many separate nuclei of the ordered structure and the growth of these ordered regions (domains) until they impinge on one another. Since four distinct "phases" (or configurations) of order can exist in a given Cu_3Au (or Mg_3Cd) crystal, a boundary forms at the junction of two domains that have different phases. These boundaries are usually called *antiphase domain boundaries*. Like grain boundaries, domain boundaries have energies and adjust themselves to produce a low-energy arrangement. Dislocations are intimately associated with domain boundaries (see Prob. 5-13). If a dislocation with $\mathbf{b} = \frac{1}{2}[110]$ moves through a previously perfect crystal of Cu_3Au, a domain boundary is formed separating two different domains. A second dislocation with the same Burgers vector would bring the two domains back into coincidence and remove the domain boundary; therefore, a single dislocation with $\mathbf{b} = [110]$ is the unit dislocation for the ordered Cu_3Au structure. Because the lattice of the ordered structure is called a *superlattice* (to distinguish it from the lattice of the random solid solution), the larger dislocation is known as a *superdislocation*. Such a dislocation is composed of four Shockley partial dislocations, which move cooperatively to produce slip in ordered Cu_3Au.

STRENGTH OF SOLID SOLUTIONS. A principal reason for producing a solid solution is to strengthen the initial material. As a preparation for Chaps. 9 and 10, the structural basis of the strength of solid solutions at normal and low temperatures is presented here. When the critical shear stress τ_c (Sec. 9-7) is reached, a material begins to deform plastically because the dislocations are able to move relatively freely. At temperatures low compared to the Debye temperature, little thermal activation occurs. The effect of solute atoms is then especially large, and generally increases τ_c (Fig. 5-19*a*).

Although many factors influence the mobility of a dislocation in a solid solution, an important component is the "friction force" created by the interaction of solute atoms with the dislocation line. The two primary contributions to the force are: (1) difference in size between the solvent and solute atoms, described quantitatively by the misfit parameter $\delta = (1/a)(da/dN_A)$, where a is the lattice parameter and N_A is the mole fraction of solute; and (2) other disturbances of electronic structure often conveniently approximated by the difference in shear modulus between the solvent and solute. In a dilute solid solution the flexible dislocation line is in contact with neighboring solute atoms, and therefore the friction force is proportional to $(N_A)^{1/2}$ (see Prob. 5-14). This type of behavior is observed experimentally in many materials (Fig. 5-19*a*, *b*, and *c*). At higher concentrations of solute the friction force is sometimes found to be proportional to N_A, as in the data of Fig. 5-19*a*.

The effectiveness of a given concentration of solute in impeding dislocations depends on whether the lattice is distorted symmetrically or asymmetrically. Symmetric distortion, produced in substitutional metallic solutions (Fig. 5-19*a*), is only moderately effective. The mechanism, *Cottrell locking*,

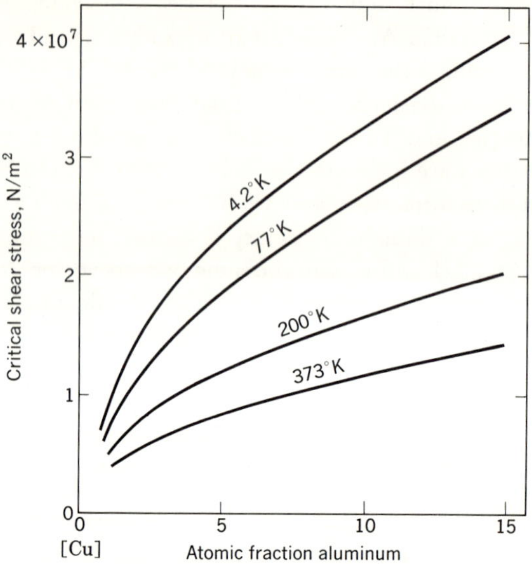

(a) Effect of aluminum in strengthening copper. The effect is greater at very low temperatures. (Data of T. J. Koppenaall and M. E. Fine.)

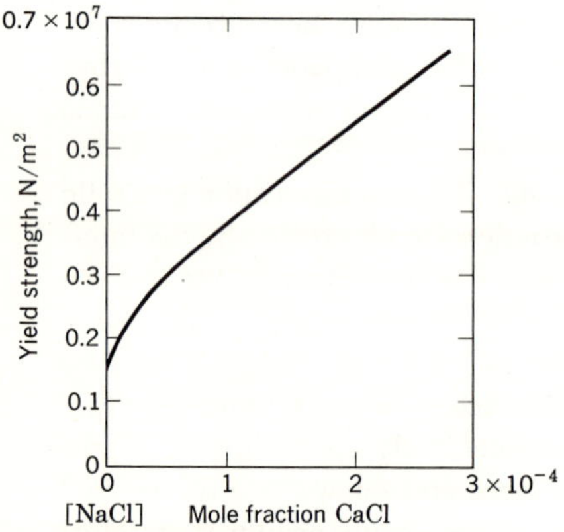

(b) Strengthening of NaCl by $CaCl_2$ in solution. Ion-vacancy pairs are present. (After W. D. Kingery.)

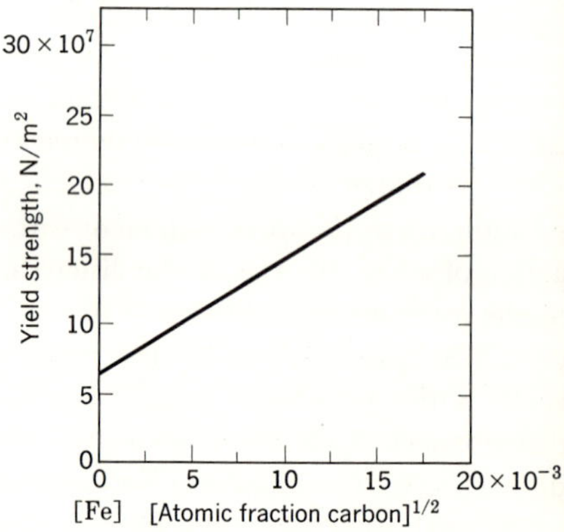

(c) Effect of interstitially dissolved carbon on the strength of iron. Tetragonal distortions are produced. (Data of C. A. Wert.)

FIGURE 5-19 Solid-solution strengthening in several representative systems [conversion of units: (M N/m²) $1.020 \times 10^{-7} = N$ kg/mm²; (M n/m²) $1.450 \times 10^{-4} = N$ lb/in²].

appears to be the following. If the A atom is larger than the solvent atom, the energy is lowered when A atoms occupy the expanded region near a dislocation. Similarly, smaller solute atoms tend to occupy the region of compressive stresses. In both cases an additional force is needed to move the dislocation through such an environment. The nature of asymmetric distortion, which is much more effective for strengthening, can be seen from the two examples in Fig. 5-20. The interstitial site in body-centered cubic iron chosen by a carbon

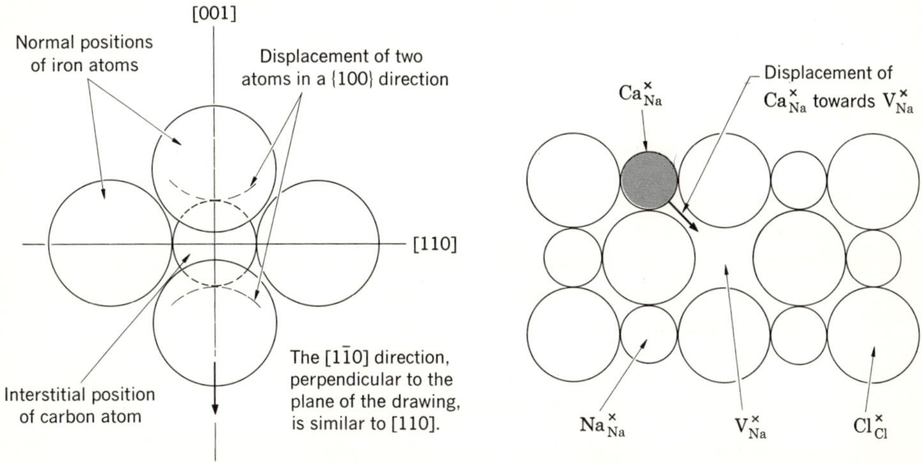

(a) Tetrahedral distortion of the body-centered cubic lattice by a carbon atom in an octahedral interstice.

(b) Asymmetric distortion in sodium chloride because of attractive forces binding the $Ca_{Na}^{\times} - V_{Na}^{\times}$ pair.

FIGURE 5-20 Two examples of asymmetric distortion of the solvent lattice by solid-solution formation.

atom (Fig. 5-20*a*) has a much smaller dimension in the direction of a $\langle 100 \rangle$ axis than in the $\langle 110 \rangle$ directions indicated. In the vicinity of a carbon atom the crystal symmetry is changed from cubic to tetragonal, with the result that only 10^{-4} atomic fraction of carbon causes the pronounced hardening shown in Fig. 5-19*c*. An ion-vacancy pair produces a more complex distortion in ionic crystals (Fig. 5-20*b*), and often leads to the type of strengthening illustrated in Fig. 5-19*b*. Negligible hardening occurs when solid-solution formation involves ions of the same charge (and comparable size), FeO in MgO or NaBr in NaCl, for example. The significance of electrostatic effects in ionic crystals is also illustrated by the charges present on jogs in dislocation lines (Prob. 5-15). The force to move the dislocation must then include a contribution for separating these additional electrical charges.

Some effects of solid solutions on dislocations are predominantly chemical. For example, the addition of a solute often lowers the stacking-fault energy of the matrix material. Since the partial dislocations then have a larger separation, screw dislocations are less able to cross-slip from one slip plane to another, and therefore the ductility is reduced. A related effect is *Suzuki locking* of extended dislocations. The region of stacking fault (which connects the two partial dislocations) has a different structure than the body of the crystal. Consequently, a different equilibrium concentration of solute can be established there by diffusion, producing a low-energy state that is an extra obstacle to movement of the dislocation.

Even in a pure material the yield strength σ is greatly increased by the presence of grain boundaries; the Hall-Petch equation is

$$\sigma = \sigma_i + \frac{k}{d^{1/2}} \tag{5-15}$$

where d is the grain diameter and σ_i and k are constants. In solid solutions the segregation of solute at the grain boundaries can further exaggerate the peculiarities of these regions. For example, a pronounced peak in hardness occurs near grain boundaries in some solid solutions, such as a dilute solution of MgO in Al_2O_3.

5-5. Intermediate Phases

In the broadest sense, an intermediate phase is any crystalline chemical compound. This definition encompasses such an immense number of phases that various subclassifications are employed in various specialties, for example, ionic compounds, covalent compounds, organic compounds, intermetallic compounds, and semiconducting compounds. Compounds are also classified by structural type, such as B1-type[1] or NaCl (crystal structure), III-V (valence electrons), Laves phases (space filling) and σ-type (complex phase). Because of the many properties of interest and the gradations in properties among the large number of intermediate phases, each system of classification is imprecise, especially if it covers a sizable fraction of the subject.

TYPES OF INTERMEDIATE PHASES. The classification of Table 5-4 is intended to give an overall view of the intermediate phases of primary interest in materials science and to serve as a framework for the discussion of several topics of special importance. The basis of the classification is the generalized chemical formula MX, which was employed in Sec. 4-5. This approach has been broadened still further, such that the intersection (M)(X) of row M and column X of Table 5-4 represents a class of phases formed between a normal metal M and a nonmetallic element X. These phases have various chemical formulas, MX, M_2X, M_2X_3, etc., or could even exhibit extensive solid solubility. If, besides M and X, we also employ in Table 5-4 the groups T (transition elements), S (semimetals), I (interstitial elements), and A_xB_y (com-

[1] This convenient method of designating various crystal structures is summarized on the inside back cover.

Component 1 / Component 2	M (normal metal)†	T (transition metal)	S (semimetal)	I (interstitial atom)	X (nonmetallic atom)	A_xB_y (stoichiometric compound)
M	Valence phases Al_2Ca $PtAl_3$ Electron phases β (~CuZn) γ (~Cu_5Zn_8) ε (~$CuZn_3$)	Valence phases Sn_2Pt β (~MnZn) γ (~Mn_5Zn_{21}) ζ (~$MnZn_{13}$) Laves phases $MgCu_2$ $MgZn_2$ $TiBe_2$	Valence phases Mg_2Si II–VI compounds PbTe CdTe	Dicarbides CaC_2 BaC_2	Valence phases Cu_2S CaF_2 MgO Al_2O_3 I–VII compounds AgBr AgI II–VI compounds ZnS CdS	Metallo-organic compounds CH_3Li methyl lithium
T		Laves phases (cont.) $MnBe_2$ $MgNi_2$ $TiCr_2$ $TaFe_2$ $HfCr_2$ Sigma-type phases σ, μ, χ, δ, P, R, E Superconducting phases Nb_3Pt Mo_3In	Valence phases MnSe NiAs structures MnTe NiSb Ni_2In Pd_5Sb_3 Nb_3Sn V_3Si	Simple interstitial phases ZrH TiC Cr_2N W_2C Ni_3N Complex interstitial phases Fe_3C $Cr_{23}C_6$ Ni_3C	Valence phases Cr_2O_3 FeO	Metallo-organic compounds $(C_2H_5)_2Fe$ ferrocene
S			Valence phases SiC III–V compounds GaP InSb	None known	Valence phases B_2O_5 P_2O_5	Inorganic compounds (rare) $Ce_2S\cdot Ce$ Metallo-organic compounds $(CH_3)_4Si$ tetramethylsilane
X					Compounds of organic chemistry CH_3Cl methyl chloride C_2H_4 ethylene	Inorganic compounds (rare) $2FeO\cdot O$
A_xB_y						Organic polymers Inorganic polymers Silicate chains Spinels $MgO\cdot Al_2O_3$ Silicates $CaO\cdot SiO_2$

† Classification of chemical elements in the periodic table (a few elements appear in more than one classification).
M—Elements of groups IA, IIA, IB, IIB, plus Al.
T—Elements of groups IIIB, IVB, VB, VIB, VIIB, VIII, plus the lanthanides and the actinides.
S—B, C, Si, Ge, P, As, Sb, Bi, Se, Te, Po.
I—H, B, C, N, Si.
X—Elements of group VIIA plus O, S, N, Si, C, and H.

pounds), then the various phases tend to fall into characteristic positions. For example, most of the phases of organic chemistry are in the class (X)(X). Phases with appreciable metallic character appear in several locations in the table and have many representatives since about 80 percent of the chemical elements are in the M, T, or S group. Phases with semiconducting properties, or which are otherwise useful in the electronics industry, are especially widely distributed among the various classifications.

The constituents of polymeric phases and of many ceramic phases appear consecutively in two locations in the table. The structural unit, C_2H_4 or MgO, for example, appears first as a simple phase. Then, in classification $(A_xB_y)(A_xB_y)$, each appears again, as an organic polymer (polyethylene) and in the spinel $MgO \cdot Al_2O_3$, respectively. Furthermore silica, as well as other network formers, can produce inorganic polymers, thus strengthening the analogy between ceramic materials and (organic) polymeric materials.

STRUCTURE OF IMPORTANT PHASES. In addition to the phases already discussed, several phases listed in Table 5-4 deserve special comment. The Laves phases, which have the general formula AB_2, have compactly packed structures determined by the relative size of the two atoms. One of the three Laves structures, $MgCu_2$, is shown in Fig. 5-21. The A atoms are at the sites

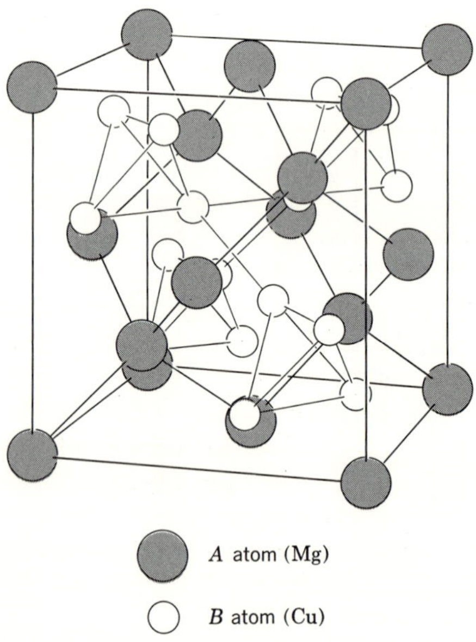

FIGURE 5-21 The $MgCu_2$ structure, one of three similar structures of the Laves phases, AB_2. (*After J. H. Wernick.*)

of a diamond-cubic structure (Fig. 1-8*b*). The small B atoms form tetrahedra, and a tetrahedron is located in each of the four interior sites that are unoccupied in the diamond-cubic structure. The unit cell contains a total of eight $MgCu_2$ units and is face-centered cubic. Laves phases have only small ranges of solid solubility. The NiAs-type structure ($B8_1$; see inside back cover) can be pictured as an hexagonal-close-packed array of semimetal atoms with the small transition-metal atoms in the octahedral openings formed between two adjacent hexagonal layers. The metallic character of this type of phase is increased if additional transition-metal atoms enter interstitial positions in the hexagonal layers. In Pd_5Sb_3 these positions are partially occupied, whereas in Ni_2In ($B8_2$) they are completely filled.

Compounds of interest for semiconductivity and related physical properties occur in the broad central area of Table 5-4. The structures and properties of many of the III-V, II-VI, and I-VII compounds can be understood with reference to silicon (or germanium). Silicon forms the diamond-cubic structure (Fig. 1-8*b*), and each silicon atom is tetrahedrally bonded; that is, its four valence electrons form covalent bonds with four symmetrically placed neighboring atoms. A III-V compound, such as GaP, has an analogous structure. If half of the silicon sites are occupied by gallium (valence 3) and half by phosphorus (valence 5), the full octet of electrons characteristic of covalent bonding can still be achieved, although some ionic nature is also present. The structure is now that of zinc blende (Fig. 1-18) since two types of sites occur, but the locations of the atomic sites are the same as in the diamond-cubic structure. Similar reasoning applies to the II-VI compounds, such as CdTe, which also has the zinc-blende structure. Ionic character is stronger in these compounds and becomes dominant in I-VII structures.

The principal interstitial phases (Table 5-4) are in the (T)(I) classification. If the radius ratio r_I/r_T is less than 0.59, then Hägg's rule predicts that a simple structure will form which is similar to an interstitial solid solution. Although pure titanium is not face-centered cubic, TiC can be pictured as a face-centered cubic arrangement of titanium atoms with all the octahedral interstices (Fig. 1-15*d*) occupied by carbon atoms. This structure corresponds to that of NaCl, but a range of solid solutions exists from 50 $^A/_0$ C down to 33 $^A/_0$ C. Complex interstitial phases form when r_I/r_T is greater than 0.59, for example in Fe_3C (Fig. 5-22). In this case the crystal structure is orthorhombic with 12 iron atoms and four carbon atoms in the unit cell.

An important type of phase that doesn't fit into Table 5-4 is the defect phase, which is intimately connected with departures from the stoichiometric composition. In a random solid solution, such as AgZn, or in an interstitial compound, such as TiC, the structure is not greatly changed by deviations from the exact chemical formula. The situation is quite different in a compound

FIGURE 5-22 A model of the structure of Fe_3C (cementite), a complex interstitial compound. The structure can be visualized as the stacking of layers *ABAB*, each a pleated HCP-type layer. (*Courtesy E. J. Fasiska, Fundamental Research Laboratory, U.S. Steel Corporation, and G. A. Jeffrey, Crystallography Laboratory, University of Pittsburgh.*)

such as PbS where the cation and anion sites on the lattice are distinct. If the concentration of lead is to be above the stoichiometric composition, either the excess lead atoms must occupy interstitial sites, or a corresponding number of sulfur sites must be vacant. As we have seen in Chap. 4, PbS is actually a vacancy-type nonstoichiometric compound. Other examples of vacancy-defective compounds are NiAl, CoAl, NiGa, and CoGa. Compounds of this type have unusual properties, such as fast diffusion, because of their high concentrations of vacancies.

TWO-PHASE SYSTEMS

Useful materials often consist of more than one phase. For example, a polymeric material may contain crystalline regions in a glassy matrix; a typical ceramic is composed of several crystalline phases, a glass phase, and porosity; and a metallic alloy is often an intimate union of one phase dispersed in another. The following chapters consider in detail these complex, multiphase materials from many aspects. In the present section we consider only a few equilibrium properties and limit our treatment to the simple case of two-phase systems.

5-6. Equilibrium of a Condensed Phase with its Vapor

Vaporization is significantly different for simple phases than for solid solutions, the simple phases being the easier to treat. When a simple phase (such as a metal) vaporizes at any temperature T, the reaction

$$M(s \text{ or } l) \rightarrow M(g) \tag{5-16}$$

is characterized by a change in free energy, $\Delta G(T)$, and in enthalpy, $\Delta H(T)$, often called the *heat of reaction*. As shown previously in Sec. 4-5, $\Delta G(T)$ permits calculation of the vapor pressure, whereas $\Delta H(T)$ can be used in the Clausius-Clapeyron equation [Eq. (2-1)] to determine the change of vapor pressure with temperature, dP/dT. A stoichiometric compound, such as KCl, vaporizes in a relatively simple manner because the bonding forces require that equal numbers of potassium and chlorine atoms leave the solid. In a solid solution no such requirement exists, and generally the equilibrium vapor has a different composition than the solid. If the thermodynamic activities a_i are available,[1] the partial pressure p_i of each component is easily determined using the relation

$$a_i = \frac{p_i}{P_i^{\circ}} \tag{5-17}$$

where P_i° is the vapor pressure of pure component i at the temperature in question. The cadmium-magnesium system can serve as an example in which complete solid solubility occurs. At 543°K the vapor pressure of the pure solid metals are $P_{\text{Cd}}^{\circ} = 1.5\ \text{N/m}^2$ and $P_{\text{Mg}}^{\circ} = 0.10\ \text{N/m}^2$. In an alloy of 10 A/0 Mg in cadmium, the activities are $a_{\text{Cd}} = 0.890$ and $a_{\text{Mg}} = 0.0044$. From Eq. (5-17) the ratio of partial pressures is found to be

$$\frac{p_{\text{Cd}}}{p_{\text{Mg}}} = \frac{1.3\ \text{N/m}^2}{0.44 \times 10^{-3}\ \text{N/m}^2} = 3.0 \times 10^3 \tag{5-18}$$

In an alloy of 90 A/0 Mg in cadmium, the activities are $a_{\text{Cd}} = 0.0042$ and $a_{\text{Mg}} = 0.885$, giving the ratio of partial pressures

$$\frac{p_{\text{Cd}}}{p_{\text{Mg}}} = \frac{0.61 \times 10^{-2}\ \text{N/m}^2}{0.89 \times 10^{-2}\ \text{N/m}^2} = 0.69 \tag{5-19}$$

Chemical effects play a major role in causing pronounced changes in the ratio of partial pressures with change in composition in many solid-solution systems.

[1] Ralph Hultgren, et al., "Selected Values of Thermodynamic Properties of Metals and Alloys," John Wiley & Sons, Inc., New York, 1963.

The possibility of various complexities in vaporization behavior must be recognized even for simple phases. Association of atoms [$2Bi(g) \rightarrow Bi_2(g)$] or of molecules [$n\, LiBr(g) \rightarrow (LiBr)_n(g)$] often occurs and affects both ΔH and gas-law calculations. A number of compounds vaporize to produce quite different species; for example,

$$SiO_2(s) \rightarrow SiO(g) + \tfrac{1}{2}O_2(g) \tag{5-20}$$

and

$$2TiBr_3(s) \rightarrow TiBr_2(s) + TiBr_4(g) \tag{5-21}$$

Equation (5-21) is an example of a disproportionation reaction. Compounds that exhibit appreciable variability in composition often have a wide range of vapor pressure at a given temperature. ZrC is an interstitial phase with the NaCl structure, but its composition at about 1800°C can range from 35 to 50 A/o C. For the 50 A/o C composition the ratio of carbon to zirconium in the vapor is 10^3 with a total pressure of 6×10^{-6} N/m^2; for the 35 A/o C composition the ratio is 10^{-6}, and the pressure is 5×10^{-4} N/m^2. PbS, discussed in Sec. 4-5, shows a similar, though less extreme, behavior.

The vapor pressure above a curved surface may be different from that above a plane surface for two reasons. First, the surface tension γ may vary appreciably with orientation in a crystalline phase (the behavior of gold was discussed in Sec. 5-3). Even if γ can be assumed to be constant, the Gibbs-Thomson equation,

$$\ln \frac{p_i}{p_i^\circ} = \frac{\bar{V}_i \gamma}{RT}\left(\frac{1}{r_1} + \frac{1}{r_2}\right) \tag{5-22}$$

specifies that above a curved surface the partial pressure p_i of component i is different from p_i°, the value above a plane surface. $\bar{V}_i$ is the partial molal volume of component i in the condensed phase (which may be a liquid); r_1 and r_2 are the principal radii of curvature and are considered negative when the surface is concave toward the vapor. Only for very small particles need Eq. (5-22) be taken into consideration. Typically, p_i is 1 percent larger than p_i° for particles 1 μ in diameter (so that $r_1 = r_2 = +0.5\ \mu$). One stage in the sintering of finely powdered ceramic or metallic materials is attributable to this factor.

5-7. Equilibrium between Solid and Liquid Phases

Many aspects of the behavior of a solid phase and a liquid phase in contact are governed by the surface tension γ of the solid-liquid interface. Except in a one-component system, this interface generally has a different composition than either of the two bulk phases. This behavior can be understood by considering the application of the Gibbs adsorption equation to a single phase containing two or more components. A useful form of the equation is

$$\Gamma_i = -\frac{a_i}{RT}\frac{d\gamma}{da_i} \tag{5-23}$$

where a_i is the thermodynamic activity of component i and Γ_i is the excess of this component in the surface layer over its composition in the bulk phase. In view of the minus sign, the accumulation of a component is associated with a decrease in surface tension. The application of Eq. (5-23) to a solid-liquid interface is considered in Prob. 5-16.

Only under special circumstances is the surface of a solid free of adsorbed substances and therefore capable of exhibiting its natural high value of surface tension. Although the surface tension of mica is known to be 5 J/m^2 (Table 5-5), almost any substance in the atmosphere surrounding a freshly

TABLE 5-5. INTERFACIAL TENSIONS OF MICA IN SEVERAL ENVIRONMENTS†

Environment	Interfacial tension, J/m^2
Freshly cleaved surface in a high vacuum	5.00
Surface containing adsorbed hexane vapor	0.271
Surface immersed in liquid hexane	0.255
Surface containing adsorbed water vapor	0.180
Surface immersed in water	0.110

† Adapted from F. P. Bowden and S. Kay.
Conversion of units: M J/m^2 × 10^3 = N erg/cm^2.

cleaved surface becomes (physically) adsorbed through the action of van der Waals forces. Even a nonpolar molecule like hexane, C_6H_{14}, causes a drastic decrease in surface tension. The physical adsorption of a polar molecule such as water is much stronger and causes a further decrease in surface tension. An example of the application of these ideas is to the lubrication of metals. If two clean metal surfaces are in sliding contact, the large surface tension causes welding at high points on the two surfaces. The resulting tearing and grooving of the surfaces is the phenomenon of wear. The surfaces can be prevented from contacting by the presence of various adsorbed substances. A nonpolar oil is only weakly attached to the surface of the metal by physical adsorption. The presence of a polar group gives a stronger physical adsorption and a greater ability to resist disruption of the layer of lubricant by the load that presses the two metals together.

Several important aspects of the surface tensions of solid and liquid phases play a role in the behavior of a cap of liquid on a solid substrate (Fig. 5-23). A common behavior is the occurrence of an equilibrium contact angle θ,

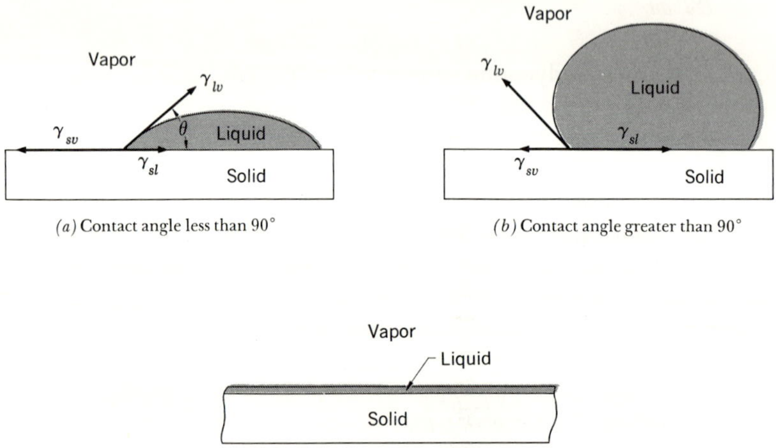

(*a*) Contact angle less than 90°

(*b*) Contact angle greater than 90°

(*c*) Liquid wets the solid; contact angle is zero.

FIGURE 5-23 Three types of equilibrium configurations after a cap of liquid is placed on a solid substrate.

considerably less than 90° (Fig. 5-23*a*). Since the equilibrium condition of the horizontal components of the three surface tensions is

$$\gamma_{sv} = \gamma_{sl} + \gamma_{lv} \cos \theta \tag{5-24}$$

$\theta < 90$ occurs for γ_{sv} values greater than γ_{sl} but less than $\gamma_{sl} + \gamma_{lv}$. If γ_{sl} is greater than $\gamma_{sl} + \gamma_{lv}$, the contact angle becomes zero, and the liquid spreads over the surface of the solid (Fig. 5-23*c*). Two factors can strongly influence the behavior of a given solid-liquid combination. Adsorption of impurities or formation of a surface film can decrease the value of γ_{sv} to only a fraction of the value characteristic of a clean surface. This factor tends to prevent wetting. If the solid and liquid phases are mutually soluble, composition changes may occur and lower the value of γ_{sl}, a factor acting in the opposite direction but less strongly.

The value of γ_{sl} is usually low for an interface between a solid and a liquid of similar character, for example, between solid silica and liquid sodium silicate, or between solid copper and liquid copper sulfide (Table 5-6). For this

TABLE 5-6. APPROXIMATE VALUES OF INTERPHASE SURFACE ENERGIES, IN JOULES PER SQUARE METER, FOR SEVERAL MATERIALS†

Material	γ_{sv}	γ_{sl}	γ_{lv}
Iron	1.900		1.700
Fused silica (*s*)	0.300		0.200
Copper	1.700	0.200	1.300
Copper (*s*)-sodium silicate (*l*)		1.500	
Copper (*s*)-copper sulfide (*l*)		0.100	
Fused silica (*s*)-sodium silicate (*l*)		0.020	
Fused silica (*s*)-copper (*l*)		1.400	

† Conversion of units: $M\ \text{J/m}^2 \times 10^3 = N\ \text{erg/cm}^2$.

reason a solid ceramic is easily wet by a ceramic glaze, as is a solid metal by a low-melting alloy of suitable character. On the other hand, the joining of a ceramic and a metal by a coating process is more difficult since γ_{sl} has high values in this case, as shown by the two pertinent examples in Table 5-6. Porcelain enameling of metals such as copper and iron is possible because their γ_{sv} values are high and the γ_{lv} values for liquid ceramics are usually very low. The wetting of solid ceramics by molten metals encounters the basic difficulty that γ_{sv} is low, whereas both γ_{sl} and γ_{lv} are high. The liquid metal cap then tends to assume the shape illustrated in Fig. 5-23*b*. Suitable alloy additions to the metal can reduce γ_{sl} by causing reactions between the metal and the ceramic.

The microstructures of solidified materials are strongly influenced by the interfacial tensions that existed at high temperatures between the solid and liquid phases. When solidification is nearly complete, the remaining liquid is in various configurations with respect to the solid phases. For the simple case of only one solid phase, the configuration of Fig. 5-24 is useful for visualizing

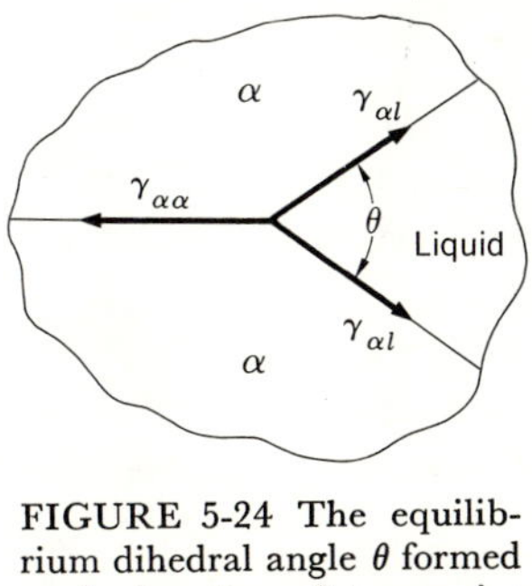

FIGURE 5-24 The equilibrium dihedral angle θ formed at the junction of two grains of the solidified α phase and a region of liquid phase.

the dihedral angle θ produced by the equilibrium of the horizontal components of the interfacial tensions,

$$\gamma_{\alpha\alpha} = 2\gamma_{\alpha l} \cos \frac{\theta}{2} \tag{5-25}$$

The configuration shown can exist for $\gamma_{\alpha l}/\gamma_{\alpha\alpha}$ ratios greater than 0.5, but below this value the high-energy α–α (grain) boundaries are replaced by α–l boundaries, and a continuous layer of liquid phase surrounds the solidified portion of the alloy. At a three-grain junction (Fig. 5-25*a*) the liquid assumes a characteristic form depending on the dihedral angle. The photomicrograph of Fig. 5-25*b* illustrates the usual behavior in metallic systems; the dihedral angle is zero, and liquation (formation of liquid in grain boundaries) is the

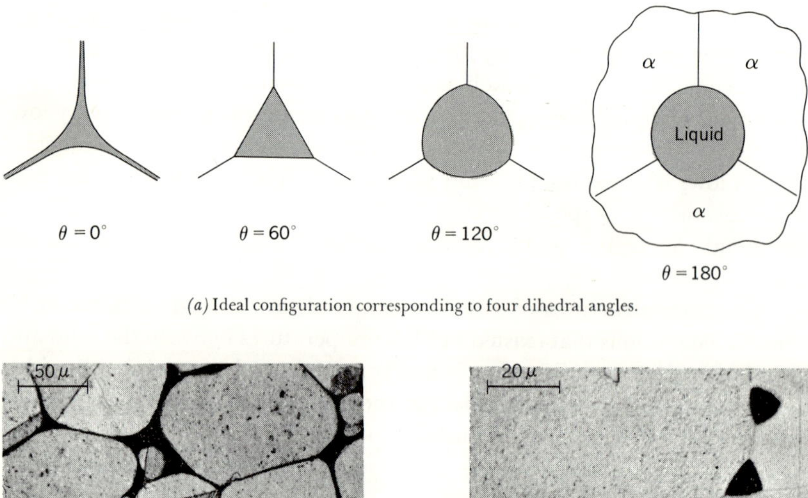

(*a*) Ideal configuration corresponding to four dihedral angles.

(*b*) In a Cu-15 w/o Ag alloy quenched from 850°C, the liquid phase "wets" the grain edges of the grains. $\theta \simeq 0°$.

(*c*) In a Cu-30 Zn-2 w/o Pb alloy quenched from 750°C, no "wetting" occurs. $\theta \simeq 80°$.

FIGURE 5-25 Dihedral angle produced by liquid phase at the junction of three grains of the solid α phase. (*After C. S. Smith.*)

first stage in melting. In a leaded-brass alloy (Fig. 5-25*c*) $\gamma_{\alpha l}$ has the high value of about 0.4 J/m^2 for the lead-rich liquid against the copper-rich solid. Liquation does not occur in this alloy since the dihedral angle is 80°.

5-8. Multiphase Solids

Many industrial materials contain two or more phases. In materials for critical applications, such as composites, the multiphase microstructure is purposely chosen because the resulting properties are distinctly superior to those of a one-phase substance. In some large-tonnage materials, such as the mullite[1] ceramic in Fig. 5-26, the initial, low-cost raw materials are largely

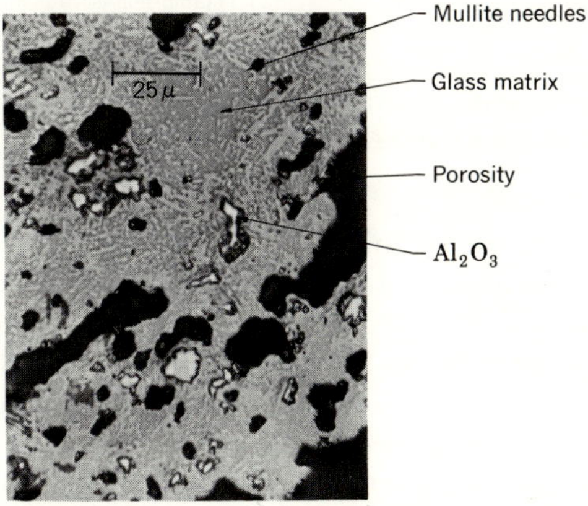

FIGURE 5-26 A typical multiphase microstructure, a mullite ceramic showing two crystalline phases, a glassy phase, and porosity. (*Courtesy R. W. Fulrath, Inorganic Materials Research Division, Lawrence Radiation Laboratory, University of California.*)

responsible for the multiphase structure. Usually one phase (the matrix) forms the main part of the structure and has its mechanical, physical, and chemical properties modified by the additional phases. The change in properties may be simply a result of dispersion of an "inert" second phase, but lattice "matching" at the interface between the matrix and the second phase often plays a role, as discussed in Sec. 7-9. Figure 5-27 shows an array of interfacial dislocations at a matching interface between the two phases in a metallic

[1] Mullite, $3Al_2O_3 \cdot 2SiO_2$, is one of the compounds in the Al_2O_3–SiO_2 binary diagram, but it also occupies a field of primary crystallization in several important ternary diagrams, such as Al_2O_3–SiO_2–MgO (Fig. 2-27).

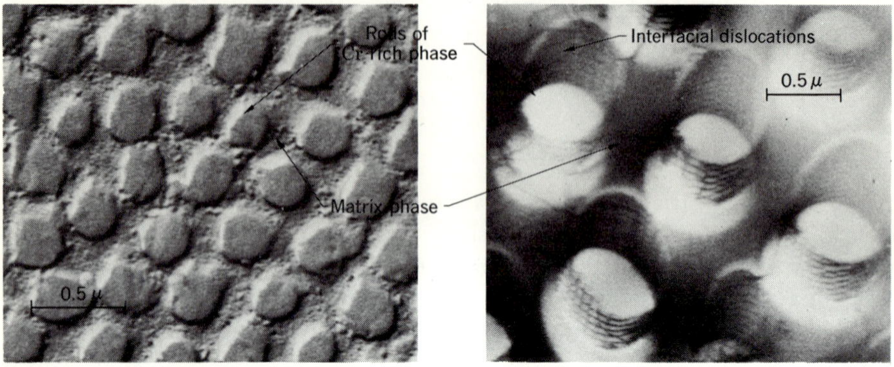

(a) Replica electron-micrograph showing the rod structure in the solidified alloy.

(b) Transmission electron-micrograph showing the dislocation pattern at the rod-matrix interface.

FIGURE 5-27 Rods of the Cr-rich phase in the NiAl-rich matrix phase produced by directional solidification of the NiAl–Cr eutectic liquid. (*Courtesy J. L. Walter and H. E. Cline, General Electric Research and Development Center.*)

eutectic, produced by the technique of directional solidification (Sec. 7-6). These dislocations are analogous to those at grain boundaries in a one-phase material (Sec. 4-12).

Multiphase materials can be divided into two categories: (1) those that belong to an ordinary *n*-component system, for example, a cast ceramic, a typical metallic alloy (even if directionally solidified), and a partially crystalline polymer, and (2) those produced "mechanically" from components not normally considered as forming a system. Most composite materials (Fig. 5-28) belong to this second category. A phase with high elastic properties and tensile strength is prepared in the form of fine fibers (Fig. 5-28*a*) which are either continuous or discontinuous. In some cases, such as boron filaments coated with SiC (Fig. 3-9), the fiber itself is a multiphase material, useful in minimizing chemical reactions between the fiber and the matrix. The high-strength fiber is then incorporated in a suitable matrix material such as a polymer (Fig. 5-28*b*) or a metal (Fig. 5-28*c*). The main role of the matrix is to transmit the load to the fiber and to carry the load locally if a fiber should fracture. The rubber in an automobile tire is also a complex composite, the main reinforcing phase being fine particles of carbon black (Fig. 5-28*d*). The fine particles increase the resistance of the rubbery polymer to tearing and to wear.

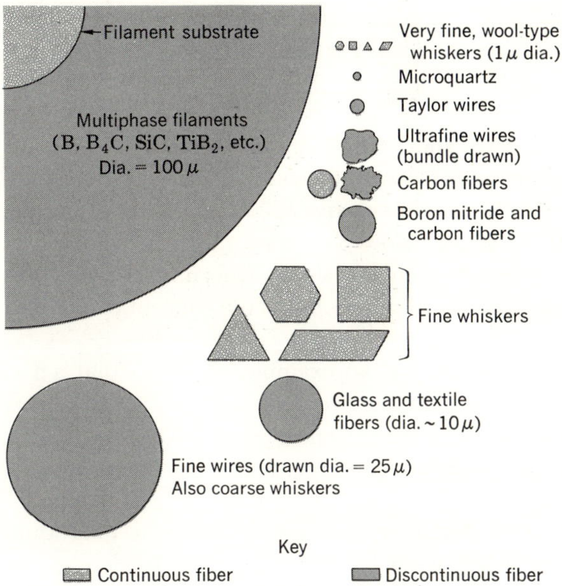

(a) Sizes and shapes of the cross-sections of typical reinforcing fibers. (After W. H. Sutton.)

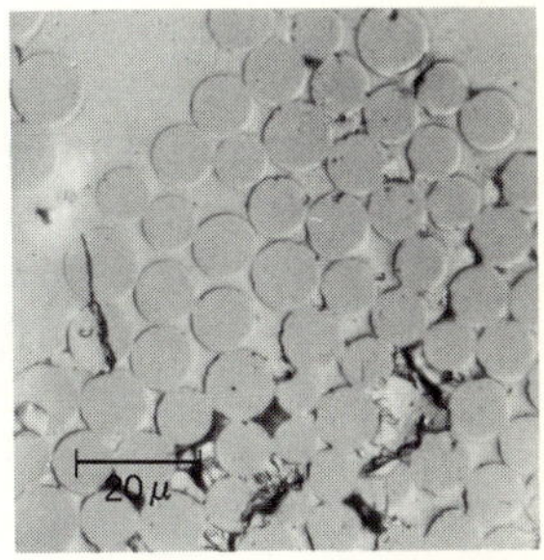

(b) Cross-sectional view of fiberglass (epoxy resin reinforced with glass fibers). (Courtesy L. J. Broutman, Illinois Institute of Technology.)

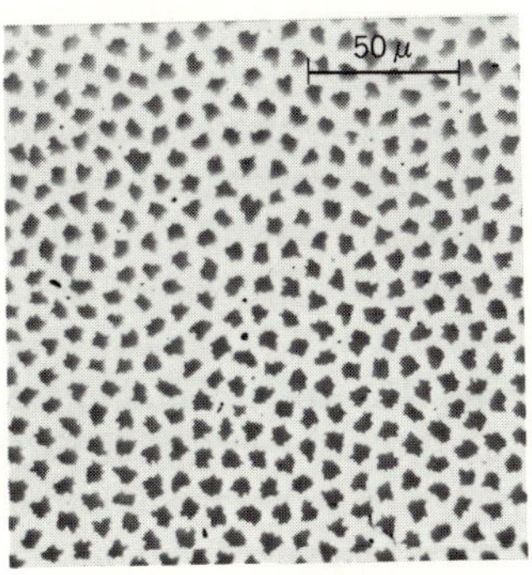

(c) Metallic nickel reinforced with 50 V/o of Thornel "50" carbon fiber. (Courtesy D. E. Niesz, Battelle Memorial Institute.)

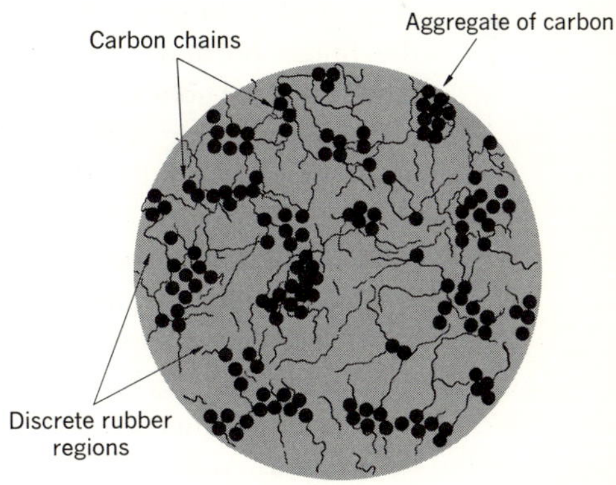

(d) Schematic structure of a rubbery polymer reinforced with 0.1μ particles of carbon black. (After Blanchard.)

FIGURE 5-28 The microstructures of representative composites, illustrating the reinforcing action of fibers and particles.

PROBLEMS

5-1. Assume that at 3000°K and 1 atm pressure 10 percent of the H_2 molecules are dissociated. Using the ideal gas law, calculate the volume of 10^{-1} moles of H_2.

5-2. Viscosity (in newton-seconds per square meter) ranges from 10^{-13} for air, through 10^{-7} for glycerin and 10^1 for pitch, to 10^5 for glacial ice. An appreciation of this huge range of viscosities can be obtained in terms of forces. Under its own weight, window glass at 1200°C flows like glycerine, whereas at about 700°C it only slowly deforms (10^{-1} N-sec/m²). (*a*) Estimate the force required to similarly slowly deform glass when the viscosity is 10^4 N-sec/m². (*b*) At 10^{11} N-sec/m² is the concept of viscosity still valid? Explain.

5-3. Using a soap film (two surfaces) as an example, show (*a*) that a force/meter, γ, of 2×0.073 N/m acting over 1 m produces 2 m² of new surface per unit length having a surface energy γ of 0.073 J/m², and (*b*) that the units newtons per meter and joules per square meter are actually identical.

5-4. A large number of small NaCl crystals are kept for a long time in a saturated aqueous solution for which $\gamma_{(100)} = 0.2$ J/m², $\gamma_{(110)} = 0.25$ J/m², and $\gamma_{(111)} = 0.4$ J/m². Use the Wulff construction to predict the equilibrium shape of the final crystals.

5-5. (*a*) Estimating the energy per unit length of dislocation line to be Gb^2 [Eq. (4-61)], calculate the surface tension of a small-angle boundary in the linear portion of Fig. 5-10; for example, at $\theta = 2^0$. The shear modulus G for gold is about 0.5×10^{11} N/m². (*b*) A digital computer is useful for working with equations containing two parameters, such as $\gamma_{sab} = E_0\theta(A - \ln \theta)$, Fig. 5-10. Given an experimental curve of γ_{sab} versus θ, the corresponding value of E_0 and of A can be determined by the following procedure. Use various combinations of values of E_0 and A, and calculate γ_{sab} as a function of θ in steps of 0.01 rad out to about $\theta = 0.2$. When a calculated curve matches the experimental curve, the desired values of E_0 and A have been found. Use the following program to determine which pair of (E_0, A) values best matches the curve in Fig. 5-10. Use E_0 values of 1.0, 1.5, and 2.0 and A values of -0.50, -0.75, and -1.00.

```
C  CALCULATION OF GAMMA FOR SMALL-ANGLE BOUNDARY
      DIMENSION EO(3), A(3)
    1 FORMAT (3F5.2, /, 3F5.2)
C  READ DATA
    2 READ (5,1) E0,A
```

```
C  CALCULATION
 3   DO 11 I = 1,3
 4   DO 10 J = 1,3
 5   DO 8 K = 1,20
 6   THETA = (K/100.0)
 7   GAMMA = EO(I) * THETA * (A(J) - ALOG(THETA))
 8   WRITE (6,9) EO(I), A(J), THETA, GAMMA
 9   FORMAT (1X, 'EO =', F6.3, 2X, 'A =', F6.3,
 *              2X, 'THETA =', F6.3, 2X, 'GAMMA =',
 *              F6.3)
10   CONTINUE
11   CONTINUE
12   STOP
     END
```

5-6. The deformation behavior of thin wires (10^{-4} m diameter) at elevated temperatures is largely controlled by surface tension. A typical wire has a "bamboo structure," consisting of many grains with their boundaries perpendicular to the axis of the wire. If a wire has n grains per meter of length, show that a mass m, given by

$$mg = \pi r \gamma - \pi r^2 n \gamma_{gb} \tag{5-27}$$

must be hung on the wire to keep it from contracting. [HINT: A material does not deform under hydrostatic compression; $\sigma_1 = \sigma_2 = \sigma_3$. Consider the radial stresses produced in a unit length of wire both by γ_{gb} and by γ. The longitudinal stress has contributions from γ and from the force mg, where g is the acceleration of gravity, 9.781 m/sec^2.

5-7. Describe the twin boundary of Fig. 5-12 as a coincidence boundary. Specify the rotation axis and the number of degrees of rotation.

5-8. A pure metal such as copper shows significant deviations from 120° in the angles at grain injunctions. Using the data of Fig. 5-10, calculate the largest deviation that could be produced by a coincidence boundary.

5-9. Use the data of Fig. 5-10 to calculate the angle of a grain-boundary groove in copper. Assume the geometry shown in Fig. 5-13*b*.

5-10. The atomic sizes of iron and nickel are almost the same. (*a*) Sketch a few unit cells of iron (body-centered cubic) with half of the sites occupied by nickel atoms. (*b*) Make a similar sketch for nickel (face-centered cubic) containing iron in solid solution. (*c*) Explain whether it is possible to have a single solid-solution phase extend completely across this phase diagram.

5-11. The two primary solid solutions, α and γ, in the MgO–Al_2O_3 system (Fig. 5-29) have greatly different solubilities. Use reactions of defect chemistry to explain this behavior.

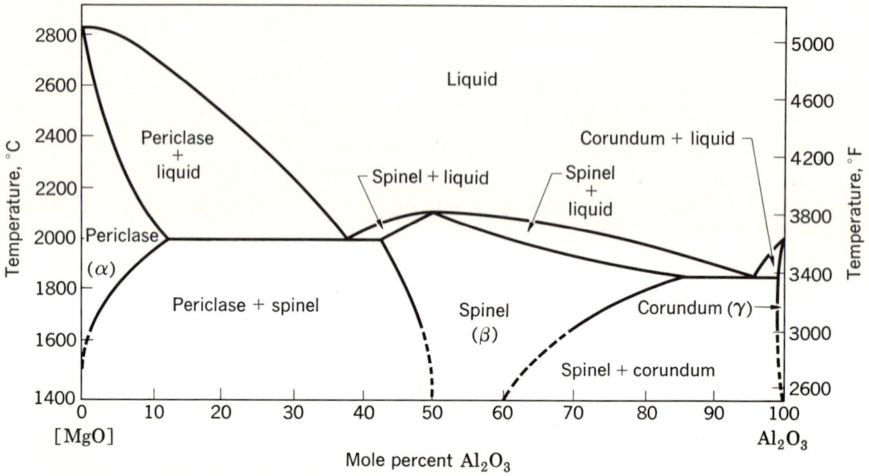

FIGURE 5-29 The phase diagram of the MgO–Al_2O_3 system. The mineralogical names for the three solid solutions are shown. The structure of the spinel phase is illustrated in Fig. 12-19*a*. (*Diagram adapted from E. M. Levin, et al., "Phase Diagrams for Ceramists," American Ceramic Society, Columbus, Ohio, 1964.*)

5-12. (*a*) Show that the Mg_3Cd-type of ordered phase, which forms from a hexagonal-close-packed solid solution with lattice spacings a_1 and a_3, can be pictured as four interpenetrating hexagonal close-packed lattices, each with $a_1' = 2a_1$ and $a_3' = a_3$. The cadmium atoms occupy one of the four lattices. (*b*) Similarly, sketch the MgCd ordered phase. Note that the proper choice of two of the four lattices gives higher symmetry. (*c*) By analogy with the cubic-to-tetragonal distortion exhibited by CuAu (Fig. 2-23), discuss possible distortion from the symmetry of the hexagonal crystal structure.

5-13. Consider a block of ordered Cu_3Au crystal structure, perfect except for a dislocation line with $\mathbf{b} = \frac{1}{2}[110]$ running from the front surface to the back surface. (*a*) Using a Burgers circuit, show that a circular path around the dislocation must cross a domain boundary. (*b*) Sketch a reasonable location for the boundary, labeling the regions on either side 1 and 1′ (Why not 1 and 2?).

5-14. (*a*) Make a simple sketch representing the random distribution in a dilute solid solution of about 20 A atoms (solute) on a slip plane. Draw a straight dislocation line in your sketch. If the dislocation line attempts to move perpendicular to its length but is impeded locally by an A atom, show that a wavy (but still fairly straight) line is produced. Show that the number of A atoms per unit length of this line is inversely proportional to the average spacing of the A atoms. (*b*) Using the method of stereometric microscopy (Sec. 3-7), show that the average spacing in three dimensions is inversely proportional to the square root of N_A, the mole fraction (more correctly, volume fraction) of solute. (*c*) Explain why the force to move the dislocation line should vary as $(N_A)^{1/2}$.

5-15. The edge orientation of the dislocation in sodium chloride (Fig. 4-13) is sketched with a row of Na^+ ions as the lower part. (*a*) Sketch a jog in this dislocation by beginning a row of Cl^- ions below the present row of Na^+ ions. (*b*) Demonstrate that there is a net charge on the jog.

5-16. Application of the Gibbs adsorption equation [Eq. (5-23)] to a solid-liquid interface is complicated by the fact that the mole fraction $N_i(l)$ of component *i* in the liquid phase is usually different from $N_i(s)$. Using the fact that $a_i(l) = a_i(s)$, show that both phases accumulate (or else both lose) component *i* in the vicinity of the interface.

5-17. In Fig. 5-25*c* the lead-rich phase is shown in three different locations: (1) within a single grain, (2) at a grain boundary, and (3) at the junction of three grains. On the basis of interfacial energies, explain as quantitatively as you can the geometrical features of the configurations of this phase in (1) and (2).

REFERENCES

Adamson, A. W.: "Physical Chemistry of Surfaces," Interscience Publishers, New York, 1967.

Alfrey, T., and E. F. Gurnee: "Organic Polymers," Prentice-Hall, Inc., Englewood Cliffs, N.J., 1967.

Barrett, C. S., and T. B. Massalski: "Structure of Metals," 3d ed., McGraw-Hill Book Company, New York, 1966.

Barrow, G. M.: "Physical Chemistry," McGraw-Hill Book Company, New York, 1966.

Brandup, J., and E. H. Immergut: "Polymer Handbook," John Wiley & Sons, Inc., New York, 1966.

Broutman, L. J.: "Modern Composite Materials," Addison-Wesley Publishing Company, Inc., Reading, Mass., 1967.

Burke, J. J., et al.: "Surfaces and Interfaces, I; Chemical and Physical Characteristics," Syracuse University Press, Syracuse, N.Y., 1967.

Cahn, R. W. (ed.:): "Physical Metallurgy," 2d ed., North Holland Publishing Company, Amsterdam, 1970.

Hughel, T. J. (ed.): "Liquids: Structure, Properties, Solid Interactions," American Elsevier Publishing Company, New York, 1965.

Kingery, W. D.: "Introduction to Ceramics," John Wiley & Sons, Inc., New York, 1960.

Meares, P.: "Polymers: Structure and Bulk Properties," D. Van Nostrand Company, Princeton, N.J., 1965.

"Metal Surfaces," ASM Seminar, American Society for Metals, Metals Park, Ohio, 1963.

Pask, J. A. (ed.): "An Atomistic Approach to the Nature and Properties of Materials," John Wiley & Sons, Inc., New York, 1967.

Rodriguez, F.: "Principles of Polymer Systems," McGraw-Hill Book Company, New York, 1970.

Sundheim, B. R. (ed.): "Fused Salts," McGraw-Hill Book Company, New York, 1964.

Ubbelohde, A. R.: "Melting and Crystal Structure," Clarendon Press, Oxford, 1965.

Chapter Six

Transport in Materials

The word *transport* is used here to mean the relative movement of a particle with respect to its neighbors. In its broadest sense, transport treats many types of particles (molecules, atoms, electrons, phonons, etc.) driven by such varied forces as electric fields, temperature gradients, and thermal vibrations. Diffusion of atoms in a gradient of concentration is an important transport process and is used here as the main example.

If the atoms in a solid were immobile in the framework of the crystal structure, clearly no reactions in solids would occur. Not only is each atom vibrating at a frequency of 10^{12} cycles/second, but at suitably high temperatures it "jumps" to new positions in the crystal lattice as often as 10^8 times each second. This is the phenomenon of diffusion and can occur in gases and liquids as well as in solids. Diffusion is especially important in solids because it is usually the only mechanism for atomic transport, whereas either convection or mechanical mixing are effective alternatives in gases or liquids.

INTRODUCTORY EXAMPLES OF DIFFUSION

Because of the immense number of jumps made by the particles in a system undergoing diffusion, statistical treatments are often useful. An especially general approach is that of *random walk*, which can be applied to gases, liquids, or solids. Consider a material in which a typical particle jumps Γ times/second and executes the displacement $\mathbf{r}_i$ during the ith jump. In time t the particle will make $\Gamma t = n$ diffusive jumps (Fig. 6-1). Since the net displacement can be written

$$\mathbf{R}_n = \mathbf{r}_1 + \mathbf{r}_2 + \cdots + \mathbf{r}_n \tag{6-1}$$

the square of $\mathbf{R}_n$ is the dot product (see Prob. 6-1)

$$R_n^2 = \mathbf{R}_n \cdot \mathbf{R}_n = \sum_{i=1}^{n} r_i^2 + \text{terms } \mathbf{r}_i \cdot \mathbf{r}_j \qquad i \neq j \tag{6-2}$$

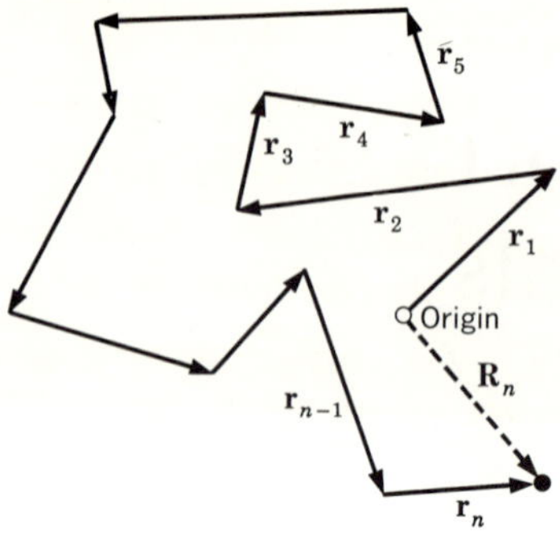

FIGURE 6-1 Schematic illustration of n random jumps made by a typical diffusing particle in time t. The net displacement from the origin is $\mathbf{R}_n$.

If this reasoning is applied to a large number of particles, an average value can be obtained as follows:

$$\bar{R}_n^2 = n\bar{r}^2 + \text{terms } \mathbf{r}_i \cdot \mathbf{r}_j \qquad i \neq j \tag{6-3}$$

where $\bar{r}$ is now the average displacement. Because the displacements are in random directions, the terms $\mathbf{r}_i \cdot \mathbf{r}_j$ can be collected in pairs having the same magnitude but opposite sign. Consequently, all the terms cancel in Eq. (6-3) except $n\bar{r}^2$, giving the simple result

$$\bar{R}_n^2 = n\bar{r}^2 \tag{6-4}$$

This equation applies to gases, liquids, and solids, but it has a particularly simple interpretation for solids since $\bar{r}$ is then the nearest neighbor distance d.

6-1. Fick's First Law

Consider now the effect of a gradient in concentration, dc/dx, on the diffusive motion of a collection of particles. An approximate derivation can be formulated in terms of two regions of width $\bar{R}_n$ on either side of a reference plane (Fig. 6-2). The cross section is chosen to be of unit area, 1 m^2. If c is the average number of particles per unit volume in region I, the total number of particles in this region is $\bar{R}_n c$. If one-third of the particles diffuse in each of the three coordinate directions, then $\frac{1}{6}$ will diffuse in the $-x$ direction and $\frac{1}{6}$ in the $+x$ direction. Therefore, the number crossing the reference plane in time t is

$$\vec{N} = \tfrac{1}{6}\bar{R}_n c \tag{6-5}$$

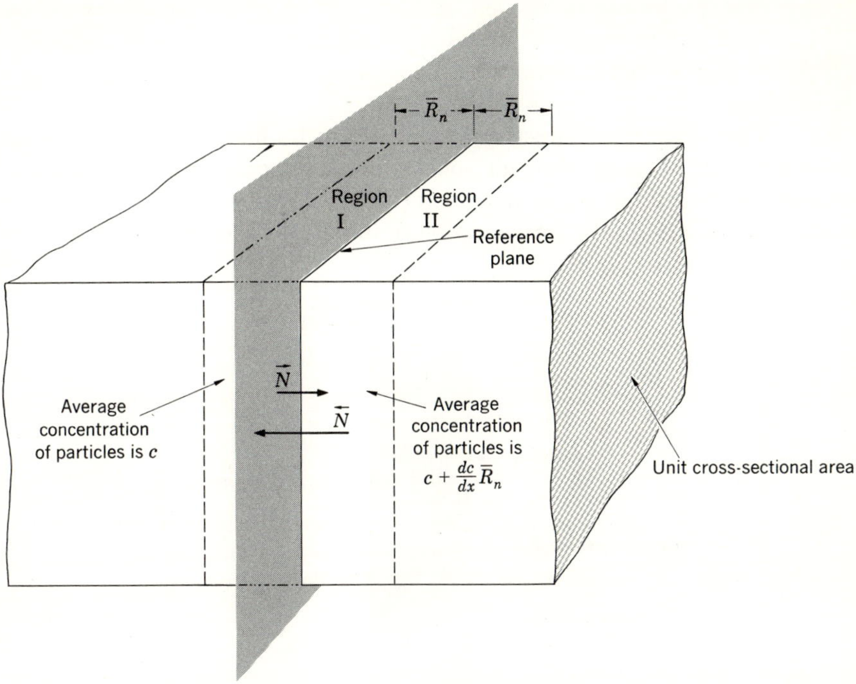

FIGURE 6-2 Schematic illustration of the numbers of particles crossing the reference plane in opposite directions in the presence of a concentration gradient dc/dx in a medium undergoing diffusion.

The average number of particles in region II is $c + (dc/dx)\bar{R}_n$, and thus the number crossing the reference plane in the opposite direction is

$$\overleftarrow{N} = \frac{1}{6}\bar{R}_n c + \frac{1}{6}\frac{dc}{dx}\bar{R}_n^2 \tag{6-6}$$

The net flux J of particles per second per unit area is then

$$J = \frac{N\,(\text{net})}{t} = -\frac{\bar{R}_n^2}{6t}\frac{dc}{dx} = -\frac{\Gamma\bar{r}^2}{6}\frac{dc}{dx} \tag{6-7}$$

where Γ is the jump frequency n/t. The substitution

$$D = \frac{\Gamma\bar{r}^2}{6} \tag{6-8}$$

then gives the usual form of Fick's first law,

$$J = -D\frac{dc}{dx} \qquad \text{particles/m}^2\text{-sec} \tag{6-9}$$

where D is the diffusion coefficient and has units of square meters per second. When appropriate empirical values are assigned to the diffusion coefficient, Eq. (6-9) can be used to describe diffusion phenomena under a wide variety of conditions.

One of the most direct applications of Fick's first law is to the passage (permeation) of gases through solids, including polymers and inorganic glasses. If the pressure P_2 on one side of a sheet of glass is higher than on the other side (Fig. 6-3), the concentration (solubility) of the gas at the surface of

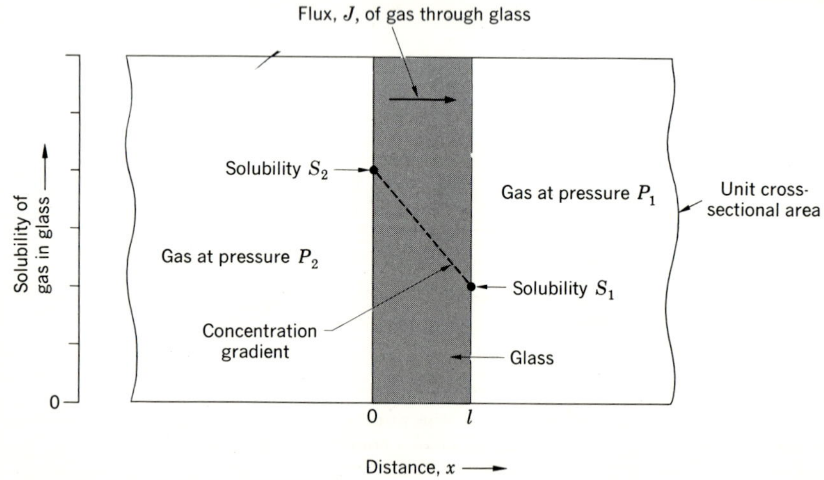

FIGURE 6-3 Schematic illustration of the permeation of a gas through a sheet of glass at the steady-state condition.

the glass will also be higher. Consequently, a concentration gradient exists, and diffusion can occur. The solution of Fick's law in this case merely involves separation of the variables and integration; thus

$$\int_{x=0}^{x=l} J\,dx = -\int_{c=S_2}^{c=S_1} D\,dc \tag{6-10}$$

The result is

$$J = D\,\frac{S_2 - S_1}{l} \tag{6-11}$$

for the steady-state condition shown in Fig. 6-3. In principle, an initial unsteady-state period should also be considered (see Prob. 6-2), but it can usually be neglected. Values of the coefficient D for the diffusion of helium in Pyrex glass at various temperatures are given by the upper curve in Fig. 6-4.

Often the pressures P_1 and P_2 on either side of a sheet of material are known, but not the solubility values necessary for use of Eq. (6-11). The flow rate F of a gas through a glass, plastic, or metallic container can then be determined from the equation

$$F = \frac{K(P_2 - P_1)A}{l} \tag{6-12}$$

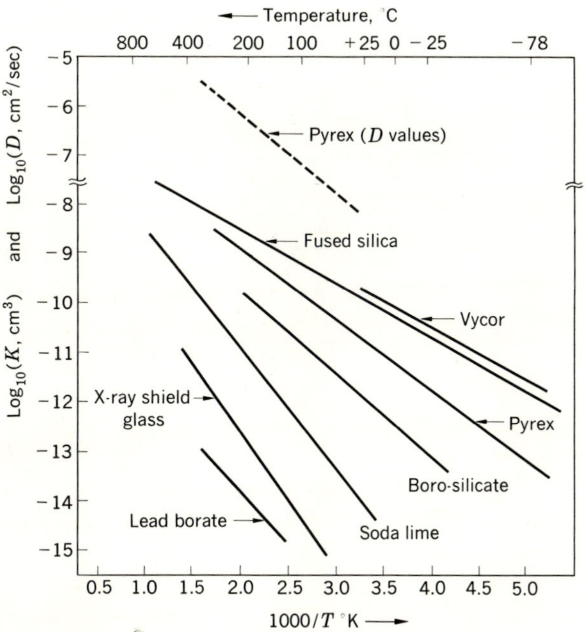

FIGURE 6-4 Permeation of various glasses by helium gas. For Pyrex glass the corresponding diffusion coefficient is given for comparison (conversion of units: M cm²/s $\times 10^{-4} = N$ m²/s). (*Data of F. H. Norton and others.*)

For the permeation coefficients K given in Fig. 6-4, the pressure is in atmospheres and the area A and thickness l are in centimeters. F is expressed in cubic centimeters of gas (at 0°C and 1 atm) per second. Problem 6-3 illustrates a typical use of Eq. (6-12).

6-2. Fick's Second Law

Most diffusion processes cause the concentration of the diffusing species to change with time. To handle this type of situation, a further deduction is made using Fick's first law. Consider an element of volume $(dx)(1\ m^2)$ defined by two reference planes in the manner shown in Fig. 6-5. The concentration gradient dc/dx that is responsible for the diffusion flux J_1 across the first reference plane also causes the flux J_2 across the second plane to have the value

$$J_2 = J_1 + \frac{dJ_1}{dx}\,dx \tag{6-13}$$

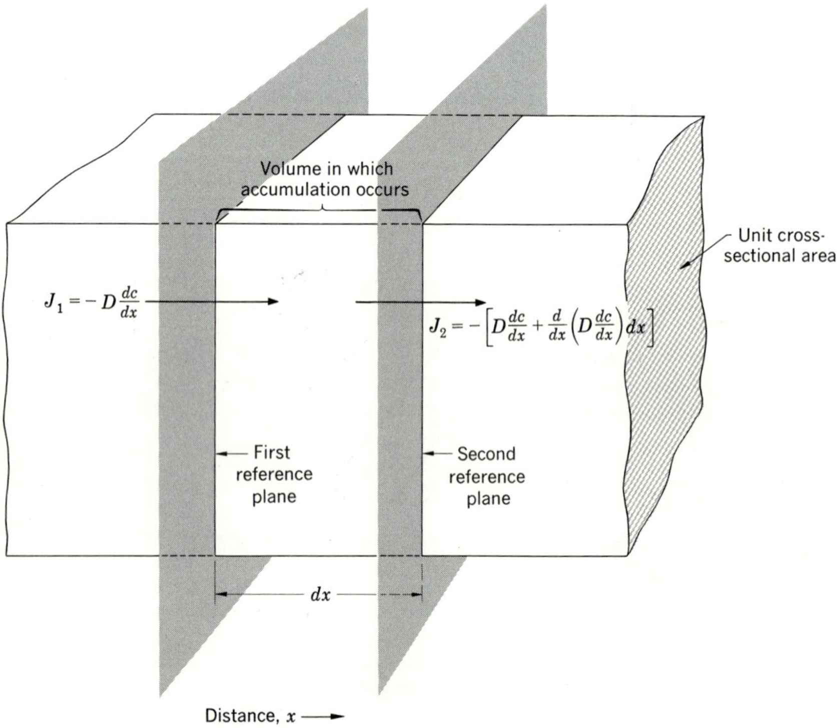

FIGURE 6-5 Schematic illustration of the derivation of Fick's second law. The diffusing particles accumulate in a volume element of unit cross section and of width dx.

The amount of material, dm, that accumulates in the volume element per second is the difference

$$dm = J_1 - J_2 = -\frac{dJ_1}{dx}\,dx = +\frac{d}{dx}\left(D\,\frac{dc}{dx}\right)dx \tag{6-14}$$

Since $dm/(dx \times 1\ m^2 \times 1\ sec)$ is the rate dc/dt at which the concentration in the volume element is changing with time, Eq. (6-14) can be written (changing to partial derivatives)

$$\frac{\partial c}{\partial t} = \frac{\partial}{\partial x}\left(D\frac{\partial c}{\partial x}\right) \tag{6-15}$$

This is the general form of Fick's second law for diffusion in one dimension that applies when the diffusion coefficient D is a function of concentration. Often D can be considered constant, permitting use of the simpler equation

$$\frac{\partial c}{\partial t} = D\frac{\partial^2 c}{\partial x^2} \tag{6-16}$$

Most of the following illustrative examples make use of Eq. (6-16). The analog of this equation in three dimensions is $\partial c/\partial t = D\nabla^2 c$, where $\nabla^2 c$ is the Laplacian of c.†

Equation (6-16) has various solutions, depending on the initial and boundary conditions in question.[1] An experimental arrangement widely employed for investigations of diffusion in gases, liquids, and solids is shown in Fig. 6-6. A typical example is the interdiffusion of H_2 and N_2 gases. The concentration of one of the components (say H_2) in region II is N_o. From the definition of mole fraction, the concentration of the other component is $1 - N_o$; that is, there is only one independent concentration variable in a two-component system. Because of the mathematical nature of Fick's second law, the usual solutions are functions of a single variable $\lambda = x/t^{1/2}$. For this reason, the *partial* differential equation [Eq. (6-16)] converts to the *ordinary* differential equation (see Prob. 6-4)

$$-\frac{dN}{d\lambda}\frac{\lambda}{2} = D\frac{d^2N}{d\lambda^2} \tag{6-17}$$

The substitution $p = dN/d\lambda$ leads to an equation in the variables p and λ. The variables can be separated, and the resulting equation can be integrated to give

$$\frac{-\lambda^2}{4} = D \ln p + \text{constant} \tag{6-18}$$

† Its use is discussed in P. G. Shewmon, "Diffusion in Solids," McGraw-Hill Book Company, New York, 1963.

[1] Many useful types of diffusion problems are considered in J. Crank, "The Mathematics of Diffusion," Oxford University Press, London, 1956.

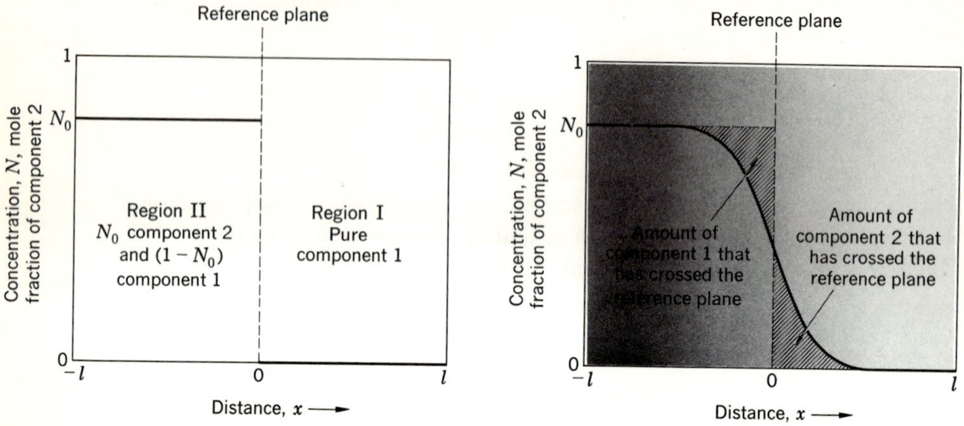

(*a*) Initial condition, time = 0. (*b*) Condition after diffusion for time *t*.

FIGURE 6-6 Conditions of interdiffusion between two regions of unit cross section and of effectively infinite length in the *x* direction. In a gaseous system, for example, a diaphragm at the location of the reference plane is removed to start the process of diffusion.

If the constant is set equal to $-D \ln A$, where A is a new constant, Eq. (6-18) can be written

$$\frac{dN}{d\lambda} = Ae^{-\lambda^2/4D} \tag{6-19}$$

After another integration and evaluation of the integration constants, the final form of the solution is obtained; thus

$$N = \frac{N_0}{2}\left(1 - \frac{2}{\sqrt{\pi}}\int_0^{x/2\sqrt{Dt}} e^{-z^2}\,dz\right) \tag{6-20}$$

where z is simply a variable needed in the integral. The quantity in parentheses is the complementary error function, $\operatorname{erfc}(x/2\sqrt{Dt}) = 1 - \operatorname{erf}(x/2\sqrt{Dt})$ and has the values shown in Fig. 6-7. This function determines the concentration profile produced by diffusion between two infinite regions, as shown in Fig. 6-6*b* (see Prob. 6-5).

An important aspect of interdiffusion experiments is illustrated by the interdiffusion of pure H_2 and N_2 gases under the conditions of Fig. 6-6. At 15°C and 1 atm, for example, this experiment gives only a single coefficient for the two gases, the interdiffusion coefficient $\tilde{D} = 0.7 \times 10^{-4}\ \text{m}^2/\text{sec}$. The self-diffusion coefficient (Sec. 6-3) in H_2 is about twice this value, however, whereas that in N_2 is only about half as great. Actually, in the interdiffusion experiment also, a hydrogen molecule diffuses about four times as rapidly as

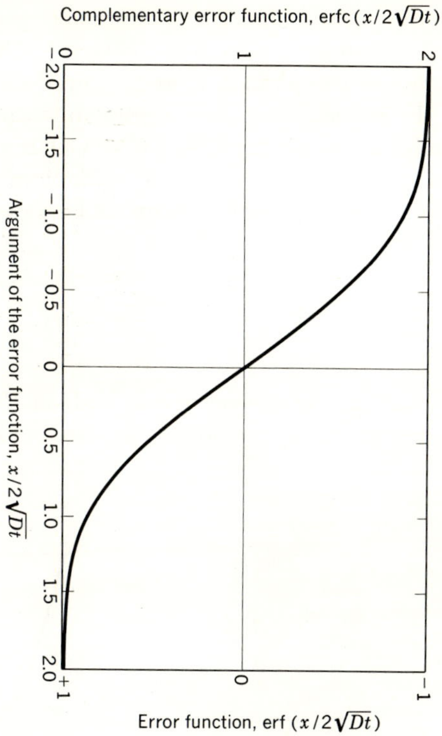

FIGURE 6-7 Values of the complementary error function erfc $(x/2\sqrt{Dt})$ for the significant range of its argument. The corresponding values of erf $(x/2\sqrt{Dt})$ are plotted on the right-hand ordinate.

does nitrogen. How can this fact be reconciled with the equal areas on the two sides of the initial plane separating regions I and II in Fig. 6-6*b*? In this type of experiment the number of molecules in a given region must remain constant. Therefore, if an excess of hydrogen molecules cross the interface by diffusion, an equal number of molecules of the gas mixture will be "pushed" across the interface in the opposite direction. Thus, the usual diffusion coefficient actually measures a combination of diffusive jumping and bulk flow of the two components.

The diffusion coefficients for gases[1] are uniformly in a small range near 10^{-4} m²/sec. Similarly, liquids have D values near 10^{-9} m²/sec. The uniformity of these values is in sharp contrast to the extremely wide range for solids, illustrated in the following examples.

[1] W. Jost, "Diffusion in Solids, Liquids and Gases," Academic Press Inc., New York, 1960.

A common experimental condition of diffusion (Fig. 6-8) is the diffusion of component 2 into an infinite region of component 1. Component 2 can be maintained at a constant composition c_0 at the surface of component 1 by various means. In the *carburizing* of iron, a constant carbon content can be provided by a gas mixture containing CH_4 and/or CO. Since carbon dissolves interstitially in iron, the carbon diffuses by jumping from one interstitial site to another (Prob. 6-6). The appropriate solution of Fick's second law is similar to Eq. (6-20); thus

$$c = c_0 \operatorname{erfc} \frac{x}{2\sqrt{Dt}} \tag{6-21}$$

Since the diffusion coefficient of carbon in iron is about 10^{-11} m²/sec at the temperatures used in carburizing (~950°C), the data of Fig. 6-7 predict that

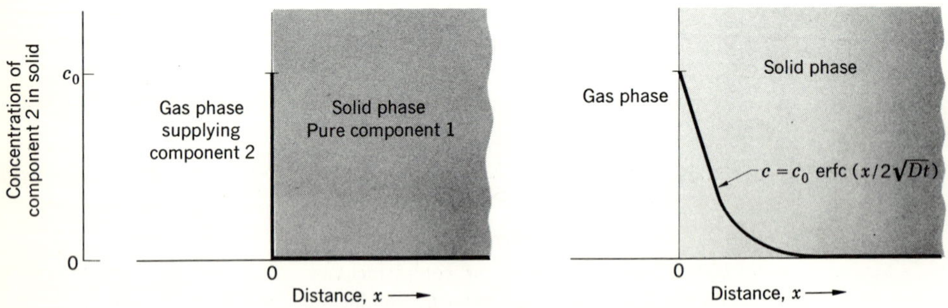

(*a*) Initial condition, time = 0.

(*b*) Condition after diffusion for time *t*.

FIGURE 6-8 Conditions for the diffusion of component 2 from the surface into an (effectively) infinite region of component 1.

several hours of carburizing are required to produce a 1-mm "case" or carbon-enriched surface layer (see Prob. 6-7).

Three types of diffusion behavior are important in applications of polymers:

1. Polymer films (used to wrap food, for example) should be essentially impermeable to water vapor and to gases that would taint the food. A low permeability constant K in Eq. (6-12) is obtained if the polymer is below its glass transition temperature T_g; K is further lowered by the presence of crystalline regions.

2. When solvent molecules diffuse into a glassy polymer, they may react chemically and cause the swelling shown in Fig. 11-1*b*. The polymer loses its original characteristics in the swollen region and may exhibit *environ-*

mental stress cracking. Since the overall process is a combination of diffusion, chemical reaction, and degradation of properties, Fick's laws are not obeyed.

3. The diffusion of ions (such as Na^+ and Cl^-) through certain polymeric films involves the phenomenon of *permselectivity*. For example, in a cation-selective membrane, the pores in the polymer film may be lined with such charged groups as COO^- or SO_4^{2-}. These charges repel the anions but permit the cations to enter and pass through the pores. Permselective films are used in desalinization equipment and in fuel cells.

INFLUENCE OF DEFECTS ON DIFFUSION

The examples of diffusion discussed above have a common feature—each diffusing particle has many possibilities for jumping. A molecule in the gas phase has much free space within which to move between collisions. An interstitial carbon atom has many neighboring interstices into which it can jump. This feature is lacking for the atoms in an ordinary crystal structure, gold atoms on their face-centered cubic lattice, for example. Each atom is closely packed within 12 neighboring atoms, and a prohibitively high energy would be required to cause two atoms, jumping simultaneously, to exchange positions. Diffusion would not occur in a perfect crystal, and only the presence of defects of various kinds accounts for the mobility of atoms in crystal lattices. The following treatment considers mainly defects related to lattice vacancies.

An overall picture of the defect structure of a polycrystalline material is shown in Fig. 6-9. Each type of defect can be associated with a characteristic diffusion coefficient D_s (surface), D_{gb} (grain boundary), D_p ("pipe" around a dislocation line), and D_v, or simply D (volume diffusion). Fundamentally, however, the lattice vacancy can be considered as the cause of diffusion of all types; the other defects simply modify the behavior of the vacancies in the neighborhood of a dislocation, grain boundary or surface. Since the behavior of vacancies can be understood most easily for volume diffusion, we begin with a consideration of this important type of diffusion in crystals.

6-3. Self-diffusion

The use of radioactive ("marked") atoms permits the study of diffusion in a pure element, such as gold. As the vacancies jump randomly in a crystal of gold, shown in Fig. 6-10*a* by two vacancies about to jump in opposite directions, the macroscopic condition of the crystal is not changed if all of the gold atoms are indistinguishable. When a gold tracer Au* is present, however, radiochemical analysis can show that tracer atoms have crossed a given interface by diffusion. A common experimental technique, that of the planar

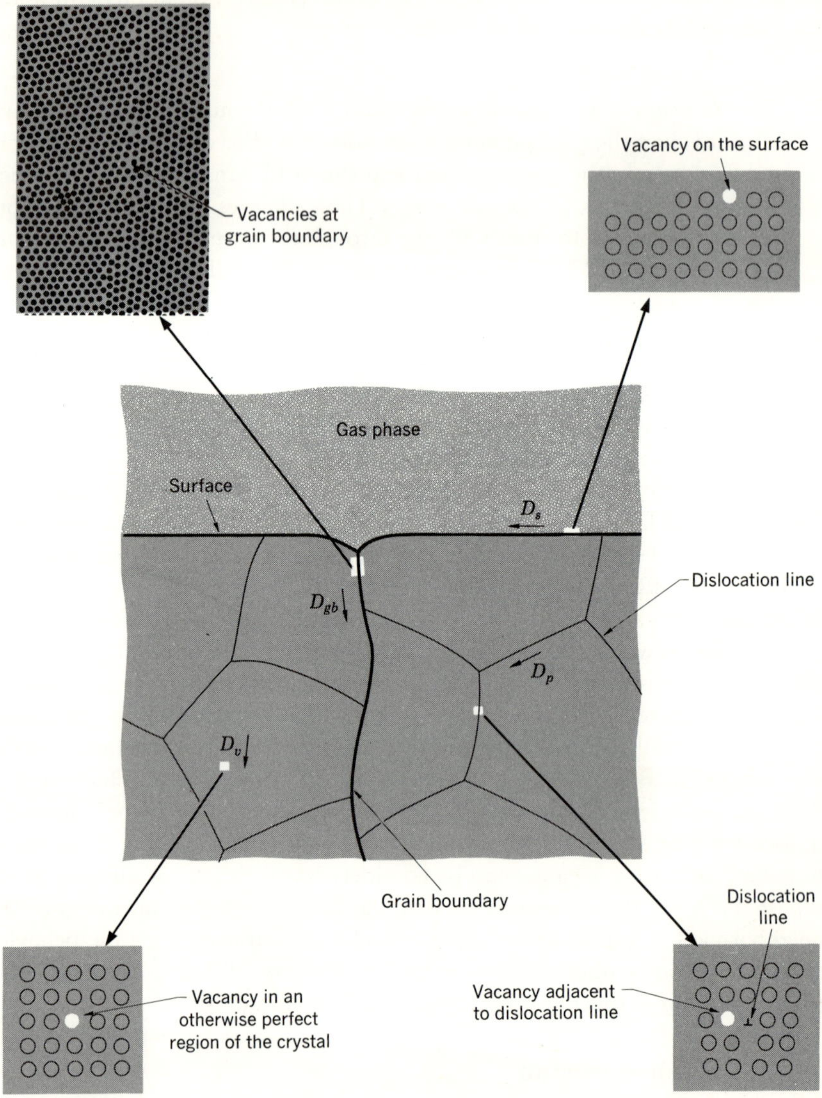

FIGURE 6-9 Schematic illustration of the types of diffusion coefficients associated with various regions of defect structure in a polycrystalline material.

source, shows how this principle can be implemented. A certain amount M of radioactive gold is deposited as a thin layer on the ends of two crystals of gold. The two crystals are then joined (Fig. 6-10*b*) to form a specimen having a planar source of Au* atoms at the origin $x = 0$. After this specimen has been heated for many hours, the distribution of tracer atoms satisfies the relation (see Prob. 6-8)

$$c = \frac{M}{2(\pi D t)^{1/2}} e^{-x^2/4Dt} \tag{6-22}$$

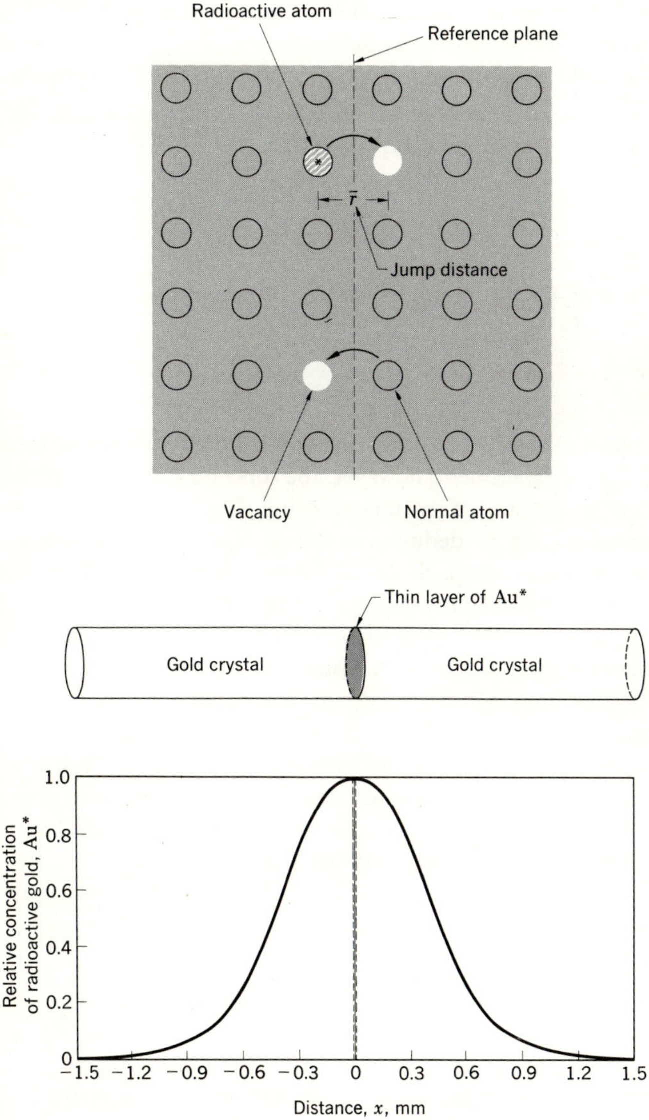

(a) A tracer atom provides experimental evidence that diffusion has occurred.

(b) Initial diffusion couple with planar source of radioactive gold, Au*.

(c) Distribution of Au* after diffusion for 100 hours at 920°C. (Data of H. M. Gilder and D. Lazarus.)

FIGURE 6-10 Illustration of the principle of tracer diffusion and of the planar-source method for determining the self-diffusion coefficient of gold.

where M is the total amount of tracer atoms in the experiment. Analysis of the experimental data of Fig. 6-10*c* (Prob. 6-9) gives $D = 2 \times 10^{-13}$ m²/sec as the self-diffusion coefficient of gold at 920°C.

The theory of self-diffusion can be built on the random-walk relation [Eq. (6-8)] because the vacancies (although not the atoms) move randomly. For example, one of the vacancies in Fig. 6-10*a* has an equal probability of exchanging with any one of its nearest neighbors on the first jump, and then it again moves in a random direction during successive jumps. The jump distance $\bar{r}$ is $a/\sqrt{2}$ in a face-centered cubic lattice, where a is the lattice constant. For the diffusion of a vacancy in gold, Eq. (6-8) becomes

$$D_V = \frac{\Gamma_V a^2}{12} \tag{6-23}$$

This diffusion coefficient is appropriate for use in the equation

$$J_V = -D_V \frac{dc_V}{dx} \tag{6-24}$$

and is employed when a significant concentration gradient of vacancies is present. In a self-diffusion experiment, however, the concentration of vacancies is constant, and thus no net flux of vacancies occurs. The average jump frequency of the gold atoms can be deduced as follows. Every time a vacancy jumps, a gold atom must also jump. Since all the gold atoms must share the small fraction N_V of vacancies, the average jump frequency of the gold atoms is only $\Gamma_V N_V$. A typical value of N_V (Fig. 4-4) is 10^{-4}.

The jumping of gold atoms is not completely random because of the phenomenon of correlation (Fig. 6-11). When a vacancy first becomes a neighbor of a given atom (step 1), this atom has only one chance in 12 of jumping (step 2) since there are 11 other neighbors of the vacancy. However, if the jump does occur, its effect in displacing the atom across a reference plane may be canceled if the reverse jump occurs (step 3′) rather than separation of the vacancy from the atom (step 3). A quantitative treatment shows that only 78 percent of the jumps of an atom in a face-centered cubic lattice contribute to diffusion, and thus the correlation factor f has the value 0.78. The desired expression for the self-diffusion coefficient is therefore

$$D = \frac{f N_V \Gamma_V a^2}{12} \tag{6-25}$$

This equation permits an experimental determination of Γ_V, the jump frequency of the vacancies. Using the value of D from Fig. 6-10 and the value of N_V from Fig. 4-4, we find that Γ_V is 5×10^{10} per second at 920°C in gold.

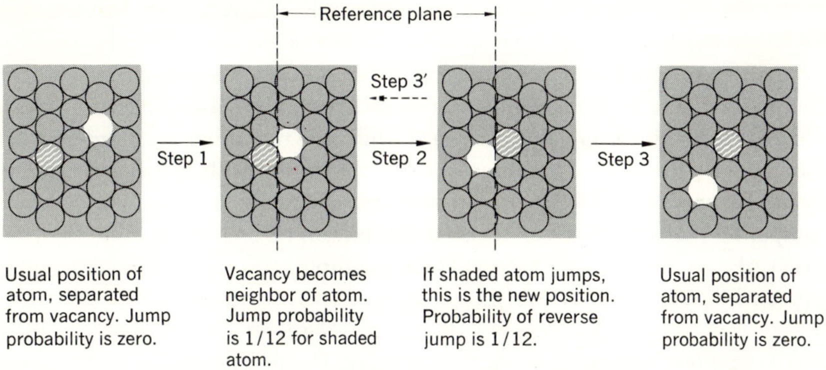

FIGURE 6-11 Simplified picture of the correlation phenomenon in self-diffusion in gold. Every atom behaves similarly, but the shaded atom is singled out for discussion.

Although compounds have a more complex defect structure than pure metals, self-diffusion investigations of ionic compounds can in principle be interpreted in a similar manner. At temperatures above 500°C, a crystal of NaCl contains about equal numbers of vacancies of the two types V_{Na} and V_{Cl} (see Sec. 4-5), but the jump frequencies of the vacancies are much different. A small sodium ion is able to exchange with a V_{Na} more easily than a large chloride ion with a C_{Cl}. Consequently, the self-diffusion of the sodium ion is found to be over 10^{-13} m^2/sec at 900°C, whereas the value for the chloride ion is about 10^{-14} m^2/sec. Interstitialcies contribute significantly to diffusion in many compounds. In AgCl, for example, the principal defects are V_{Ag} and Ag_i, the silver interstitialcy being easier to form than a vacancy on a chloride-ion site. The difference in self-diffusion coefficients is even larger than for NaCl, 10^{-10} m^2/sec at 450°C for the silver ion and less than 10^{-13} m^2/sec for the chloride ion. Some small fraction of V_{Cl} must be present to account for the diffusion of the chloride ion. The interstitialcy mechanism of diffusion employed by the silver ions is analogous to the vacancy mechanism discussed for gold. An interstitial silver ion displaces a neighboring silver ion from a lattice site, and the latter ion enters an adjacent interstitial site. Correlation in the case of an interstitialcy mechanism is the subject of Prob. 6-10.

Compounds generally have defect structures more complex than the Schottky pair (in NaCl) or the Frenkel pair (in AgCl) just described (see Sec. 4-5). Consider the effect of an impurity, say $CaCl_2$, in a stoichiometric compound such as NaCl at lower temperatures. Since one V_{Na} is produced for each calcium ion incorporated into the NaCl lattice [Eq. (5-9)], a supply of vacancies is available that is independent of thermal excitation. Below about 500°C the concentration of thermal vacancies (Schottky pairs) is so low that even "pure" NaCl contains a higher concentration of vacancies due to impurities. Consequently, diffusion rates below 500°C are higher than predicted by an extrapolation of high-temperature data.

Defect structure is especially significant for inherently nonstoichiometric compounds, such as PbS, which was considered in Sec. 4-5. Thermal production of defect pairs is only one of several equilibria that must be considered, and the major factor is often the interaction of the crystal with its gaseous environment. In particular, if the partial pressure of one of the components of the compound is changed, the effect on diffusion can be used to identify the defect structure. In the case of $Cd_{1+\delta}O$, for example, the diffusion coefficient of oxygen is found to vary as $(p_{O_2})^{1/6}$ when experiments are made at various pressures of oxygen. Two possible defect structures for CdO involve (1) interstitial cadmium ions or (2) oxygen vacancies. The first possibility is the subject of Prob. 6-11; the second is treated here as an example.

When oxygen is added to CdO, the effect is the filling of the oxygen vacancies; thus

$$\tfrac{1}{2}O_2(g) + V_O \rightarrow O_O \tag{6-26}$$

The corresponding chemical equilibrium is influenced, however, by the state of ionization of the vacancies. If the vacancies are doubly ionized according to the reaction

$$V_O \rightarrow V_O^{\bullet\bullet} + 2e' \tag{6-27}$$

then the appropriate form of Eq. (6-26) is

$$\tfrac{1}{2}O_2(g) + V_O^{\bullet\bullet} + 2e' \rightarrow O_O \tag{6-28}$$

By analogy with the development of Eq. (4-47), the equilibrium constant for this reaction can be written as

$$K = \frac{1}{(p_{O_2})^{1/2} N_V N_e^2} \tag{6-29}$$

since the oxygen ion O_O has unit activity. Charge neutrality requires that $N_e = 2N_v$, and therefore the variation of the concentration of oxygen vacancies with the pressure of oxygen is

$$N_V = \left(\frac{1}{4K}\right)^{1/3} (p_{O_2})^{-1/6} \tag{6-30}$$

This relation is consistent with the observed dependence of the diffusion coefficient of oxygen on the pressure of oxygen. Verification of a defect structure requires not only self-diffusion measurements, but also studies of other properties such as electrical conductivity and density.

6-4. Tracer diffusion

In Fig. 6-10, the radioactive gold might be diffused into silver, rather than into gold. This experiment is one type of tracer diffusion, but tracers can also diffuse in solid solutions, in two-phase systems, and under a special driving force, such as an electric field. In each case the goal is to take advantage of the relatively simple theoretical interpretation possible with marked atoms, usually without the complicating effects of a chemical concentration gradient.

The new factors involved are shown most clearly in the diffusion of a radioactive impurity in an ionic crystal, Ca* ions in calcium-doped NaCl, for example. For diffusion at high temperatures, we saw above that the vacancies created by the impurity atoms are negligible compared to thermal vacancies for self-diffusion of Na* ions. The same argument cannot be applied to tracer diffusion of Ca* ions since each ion is bound by electrical forces to a vacancy, and at any instant most of the Ca* ions have a vacancy as a neighbor. This factor is decisive in causing the tracer diffusion coefficient D^*_{Ca} to be much larger than the self-diffusion coefficient D^*_{Na}. The effect of correlation can be seen by considering the jumping of a vacancy near an impurity atom B in a matrix of A atoms. Three jump frequencies Γ can be distinguished as follows: (1) Γ_{B}, for exchanging with the impurity atom, (2) $\Gamma_{\text{A(B)}}$, for exchanging with a matrix atom that is a neighbor of the impurity, and (3) $\Gamma_{\text{A(A)}}$, for exchanging with a matrix atom that is not a neighbor of the impurity, thus causing the vacancy to become dissociated from the impurity atom. If Γ_{B} is comparable to or larger than $\Gamma_{\text{A(B)}}$, the correlation factor has a low value since there are many "return jumps" of the impurity. If Γ_{B} is small compared to the other two jump frequencies, the correlation factor is about unity, but again the overall result is to lower the diffusion coefficient of the impurity relative to that of the matrix.

In many systems, such as a metallic solid solution of B in matrix A, the B atom neither creates a vacancy nor forms a close impurity-vacancy association. An analysis based on the jump probabilities Γ_{B}, $\Gamma_{\text{A(B)}}$, and $\Gamma_{\text{A(A)}}$ then permits a wide range in the ratio $D^*_{\text{B}}/D^*_{\text{A}}$ in agreement with experimental data. For diffusion in silver at 1000°K, $D^*_{\text{B}}/D^*_{\text{Ag}}$ ranges from 11 for lead to 0.01 for ruthenium with intermediate values for other impurities. Theory suggests that the relative valence and relative size of the A and B atoms influence this ratio. When the second component is present in large amounts, changes in vacancy concentration have a pronounced effect on diffusion behavior. In a homogeneous alloy of 30 $^{\text{A}}\!/_{0}$ zinc in copper, for example, tracer diffusion studies at

700°C using Zn* and Cu* atoms show that both D^*_{Zn} and D^*_{Cu} are more than 20 times larger than in pure copper. Since the alloy is much closer to the melting temperature (Fig. 3-16), its higher vacancy concentration can account for most of this effect (Prob. 6-12).

6-5. Chemical Diffusion

Diffusion usually involves a gradient of chemically dissimilar atoms, such as copper and nickel, rather than a gradient of Cu* tracer in Cu. Because a normal copper atom and a radioactive copper atom are essentially identical, their solid solutions are ideal (Fig. 6-12*a*), and the thermodynamic activity

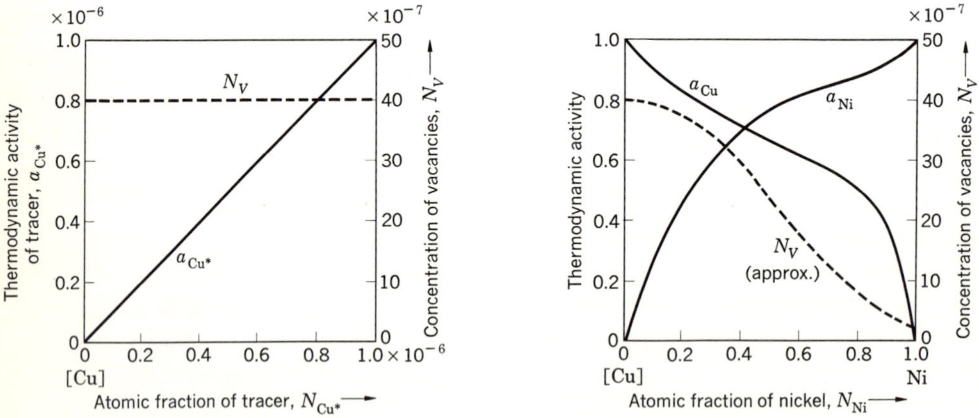

(*a*) The ideal solution of tracer atoms, Cu*, in ordinary copper. The N_V value is for 973°K.

(*b*) A typical real solution, the copper-nickel system. The data are for 973°K.

FIGURE 6-12 Comparison of an ideal solution and a real solution with respect to thermodynamic activities and vacancy concentrations.

a^*_{Cu} is equal to the atomic fraction N^*_{Cu}. Also, the vacancy concentration N_V is independent of the amount of Cu* in the solid solution. These two features greatly simplify the diffusion behavior in an ideal solution compared to that in a real solution, such as the copper-nickel system (Fig. 6-12*b*). Consider first the gradient dN_i/dx in Fick's laws. At any composition, $dN_{Ni}/dx = -dN_{Cu}/dx$ since $N_{Ni} + N_{Cu} = 1$. In a real solution, however, the activity gradient

$$\frac{da_i}{dx} = \frac{da_i}{dN_i}\frac{dN_i}{dx} \tag{6-31}$$

more accurately expresses the tendency for diffusion to occur. The "correction terms" da_i/dN_i vary appreciably with composition. At the nickel end of the diagram (Fig. 6-12*b*) $da_{\mathrm{Cu}}/dN_{\mathrm{Cu}} \approx 5$ whereas $da_{\mathrm{Ni}}/dN_{\mathrm{Ni}} = 1$. Thus, the thermodynamic factor favors the diffusion of copper. At the copper end of the diagram $da_{\mathrm{Cu}}/dN_{\mathrm{Cu}} = 1$, and $da_{\mathrm{Ni}}/dN_{\mathrm{Ni}} \approx 3$; thus the diffusion of nickel is favored. The vacancy concentration varies roughly inversely as the liquidus temperature of the alloy in question. In alloys near pure nickel (high liquidus temperature) less than one-tenth as many vacancies are available to participate in the diffusion process as in a copper-rich alloy; therefore, the rate of diffusion tends to be low. Other effects of vacancies are discussed later.

INTERDIFFUSION COEFFICIENT $\tilde{D}$. For mathematical convenience, diffusion in a two-component system is usually treated using the simple Fick laws and a single interdiffusion coefficient $\tilde{D}$. Since $\tilde{D}$ includes many of the complexities of the diffusion process, it often varies greatly with composition, as shown in Fig. 6-14 for the copper-nickel system. To understand how a single coefficient can be employed to describe the diffusion of two components, consider the curve of concentration versus distance (Fig. 6-13*a*) produced by the interdiffusion of pure copper and pure nickel. The argument employed with Fig. 6-6 shows that equal amounts of copper and nickel cross the interface that initially separated the copper and nickel. In the diffused specimen this plane is called the *Matano interface* and is located on the experimental concentration-distance curve to produce the two equal areas shown. The two fluxes J_{Cu} and J_{Ni} are then equal and opposite, as required by Fick's first law,[1]

$$J_{\mathrm{Cu}} = -\tilde{D}\,\frac{dC_{\mathrm{Cu}}}{dx} \tag{6-32}$$

$$J_{\mathrm{Ni}} = -\tilde{D}\,\frac{dC_{\mathrm{Ni}}}{dx} \tag{6-33}$$

since $dC_{\mathrm{Cu}}/dx = -dC_{\mathrm{Ni}}/dx$. ($C_i$ is the concentration in volume fraction: correspondingly, J_i is the number of cubic meters of component i crossing the interface per square meter per second.) Therefore, the Fick laws are able to treat interdiffusion in a mathematically consistent manner in terms of a single interdiffusion coefficient $\tilde{D}$.

INTRINSIC DIFFUSION COEFFICIENT, D_1. The copper atoms are actually jumping (diffusing) in the alloy appreciably faster than the nickel atoms, a

[1] The numerical value of the concentration unit *volume fraction* $C_i = N_i(\overline{V}_i/V)$ is roughly the same as atomic fraction N_i since the partial molal volumes $\overline{V}_i$ of the components of a solid solution are not greatly different from the molar volume V. Although N_i does not have the correct units for use in Fick's first law, C_i does; its units are amount (in cubic meters) of component i per cubic meter of solution. Because atomic fraction is widely employed in the literature, this concentration unit is used wherever possible in the present treatment.

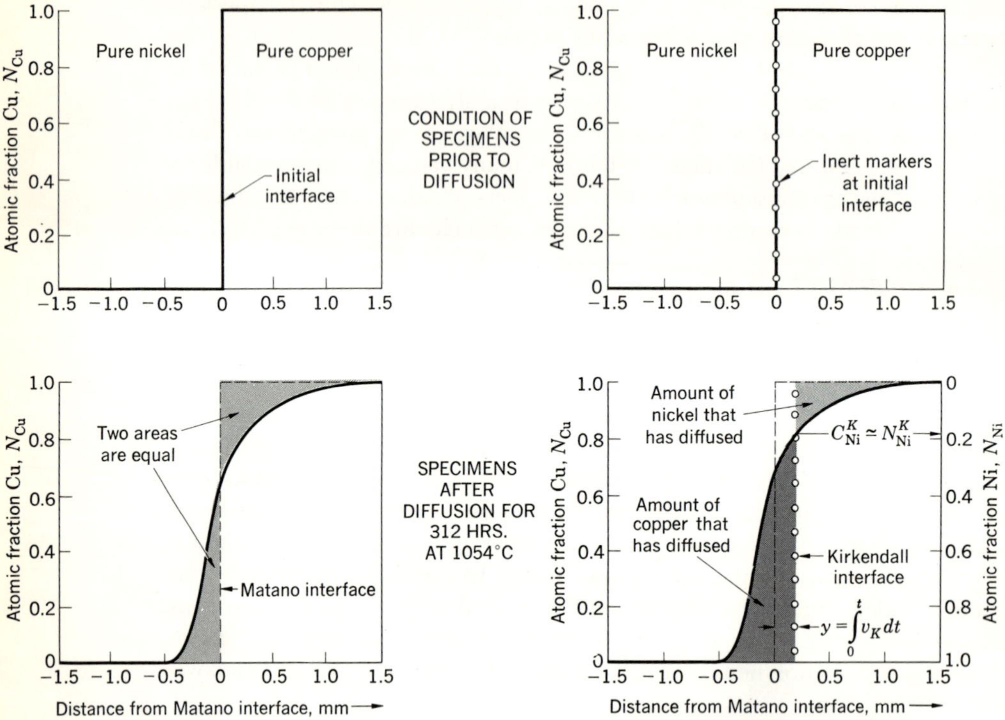

(*a*) The usual interpretation of diffusion between copper and nickel as interdiffusion.

(*b*) Use of a Kirkendall interface permits determination of the intrinsic diffusion behavior of both components.

FIGURE 6-13 Interpretations of diffusion behavior appropriate for an interdiffusion coefficient $\tilde{D}$ and for intrinsic diffusion coefficients D_{Cu} and D_{Ni}. (*Data of L. C. C. da Silva and R. F. Mehl.*)

fact revealed by the *Kirkendall effect*. The fluxes of Eqs. (6-32) and (6-33) are equal because each is the sum

$$J_{total} = J'_{diffusion} + J_{flow} \tag{6-34}$$

of an intrinsic diffusive flux J' and a flux attributable primarily to plastic flow of the alloy. This plastic flow occurs whenever one diffusive flux (J'_{Cu}) is greater than the other (J'_{Ni}). Measurement of intrinsic diffusion in a crystal merely requires that a suitable reference plane be fixed relative to the crystal lattice in the region of measurement. If small particles of some kind, such as fine oxide particles, are placed at the initial interface, they serve to locate the displaced position of this interface (*Kirkendall interface*) at the conclusion of the diffusion experiment. The equations describing diffusion in terms of intrinsic diffusion coefficients D_{Cu} and D_{Ni} are

$$J'_{Cu} = -D_{Cu} \frac{dC_{Cu}}{dx} \tag{6-35}$$

$$J'_{Ni} = -D_{Ni} \frac{dC_{Ni}}{dx} \tag{6-36}$$

where the prime (′) indicates that the reference plane is the moving Kirkendall interface. The experiment of Fig. 6-13*b* shows that D_{Cu} is three times as large as D_{Ni}; that is, $J'_{Cu} = -3J'_{Ni}$. A net increase in number of atoms has occurred in the volume to the left of the Kirkendall interface and has caused the displacement

$$y = \int_0^t v_K \, dt \tag{6-37}$$

of the interface. At any time during the diffusion process the velocity v_K of the Kirkendall interface is (see Prob. 6-13)

$$v_K = -(J'_{Cu} + J'_{Ni}) \tag{6-38}$$

For the experiment of Fig. 6-13*b*, $v_K = \frac{2}{3}J'_{Cu}$.

We now have a physical picture of the two terms in Eq. (6-34) whose sum is the flux across the Matano interface. In the case of nickel, the intrinsic diffusive flux J'_{Ni} is rather small. The larger contribution arises when the excess, $\frac{2}{3}J'_{Cu}$, of the intrinsic flux of copper atoms causes plastic flow to occur at a velocity v_K in the vicinity of the Kirkendall interface. The corresponding flux of nickel across the Matano interface is $v_K C_{Ni}^K$, a quantity of nickel merely carried along by the overall plastic flow of the alloy. Equation (6-34) can then be written

$$J_1 = J'_1 + v_K C_1^K \tag{6-39}$$

and applies to either component and to any plane for which the Kirkendall velocity can be measured.

The relation between $\tilde{D}$ and the two intrinsic diffusion coefficients follows from Eq. (6-39) (see Prob. 6-14) and has the form

$$\tilde{D} = C_1 D_2 + C_2 D_1 \approx N_1 D_2 + N_2 D_1 \tag{6-40}$$

The data for the copper-nickel system (Fig. 6-14) show that $\tilde{D}$ varies between $\tilde{D} = D_{Ni}$ at $N_{Ni} = 0$ ($N_{Cu} = 1$) to $\tilde{D} = D_{Cu}$ at $N_{Ni} = 1$, in agreement with

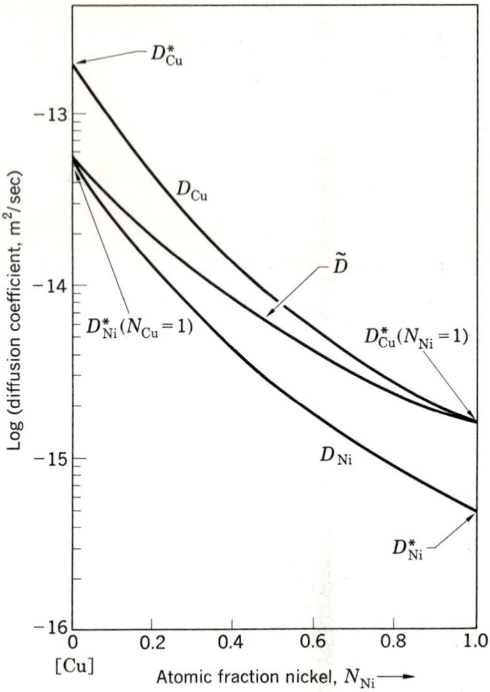

FIGURE 6-14 Relations among the various diffusion coefficients in the copper-nickel system at 1000°C (conversion of units: M m²/s × 10^4 = N cm²/s). (*Data mainly of J. Levasseur and J. Philibert.*)

Eq. (6-40). The decrease in all three diffusion coefficients with increasing nickel content in the alloy is due to the decrease in vacancy concentration (Fig. 6-12*b*). The terminal values of the three curves coincide with the self-diffusion coefficients D^*_{Cu} and D^*_{Ni} and the tracer-diffusion coefficients in dilute solution, $D^*_{Cu}(N_{Ni} = 1)$ and $D^*_{Ni}(N_{Cu} = 1)$. For a general concentration N_{Ni}, tracer-diffusion coefficients can be compared with intrinsic coefficients only if corrections are made for a thermodynamic factor (see Prob. 6-15) and for the effect of the flux of vacancies, $J'_V = -(J'_{Cu} + J'_{Ni})$. The existence of a flux of vacancies causes a further departure from random jumping of the atoms since a vacancy then approaches an atom more often from one side than the other. These vacancies also play a fundamental role in the Kirkendall effect. They form on that side of the interface where their concentration has been lowered below the equilibrium value, and they cause this region to increase in volume. In the course of the diffusion process they reach the other side of the interface, creating a supersaturation of vacancies. Most of the excess vacancies are annihilated at such locations as jogs on dislocation lines and cause shrinkage of this portion of the specimen. Some of the vacancies agglomerate and form small regions of empty space (voids) within the specimen, producing a zone of porosity on one side of the Kirkendall interface.

DOPING OF SILICON. The manufacture of semiconductor devices illustrates the application of many principles of diffusion. In microelectronics, for example, specific regions of the silicon matrix must be alloyed (doped) with a definite concentration of a suitable impurity, such as boron, to produce desirable electronic properties. The first step in the process is to form a "masking" layer of SiO_2 on the surface of the silicon and then to make small openings ("windows") in this layer (Fig. 6-15*a*) at the locations where doping

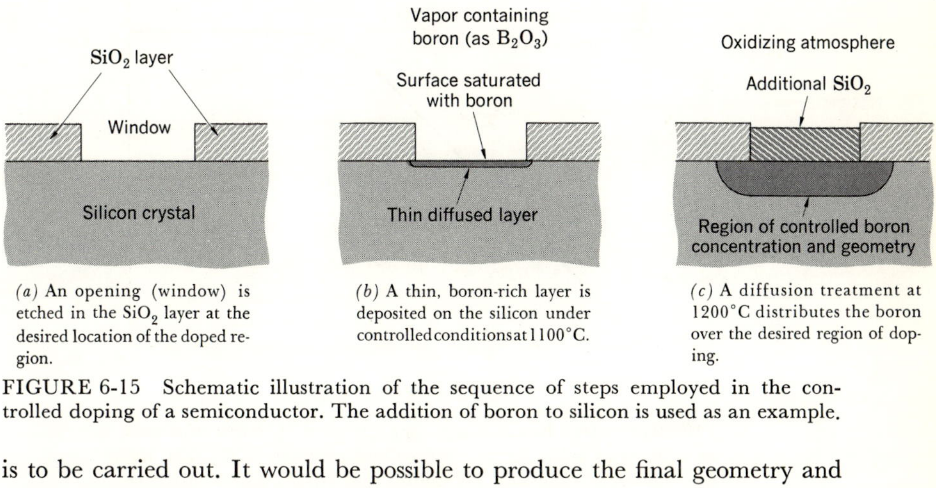

(*a*) An opening (window) is etched in the SiO_2 layer at the desired location of the doped region.

(*b*) A thin, boron-rich layer is deposited on the silicon under controlled conditions at 1100°C.

(*c*) A diffusion treatment at 1200°C distributes the boron over the desired region of doping.

FIGURE 6-15 Schematic illustration of the sequence of steps employed in the controlled doping of a semiconductor. The addition of boron to silicon is used as an example.

is to be carried out. It would be possible to produce the final geometry and concentration (Fig. 6-15*c*) in a single step, comparable to the carburizing of steel (Fig. 6-8). An alternative procedure, giving improved control of both surface concentration and depth of penetration, is the use of a deposition step (Fig. 6-15*b*) followed by a separate diffusion step (Fig. 6-15*c*).

The purpose of the deposition step is to add an accurately controlled amount M of boron to the surface of the silicon. Integration of Eq. (6-21) for a given time t of diffusion gives

$$M = \int_0^\infty c\,dx = 2c_0\left(\frac{Dt}{\pi}\right)^{1/2} \tag{6-41}$$

The surface concentration c_0 is evidently a crucial variable in determining the value of M. One method of adding boron to silicon is by a gas-solid reaction, using a compound such as BCl_3 or B_2O_3. For example, a given partial pressure of B_2O_3 in the gas phase (Fig. 6-16) corresponds to a certain equilibrium concentration of boron in the surface of the silicon. One particular concentration, the maximum solubility of boron in silicon, c_s, can be obtained precisely and reproducibly with only rough control of the partial pressure. Since the $\tilde{D}$ value for the diffusion of boron in silicon is 4×10^{-17} m²/sec at 1100°C, Eq. (6-21) becomes

$$c = 3 \times 10^{26} \operatorname{erfc} \frac{x}{2\sqrt{4 \times 10^{-17} t}}$$

for the deposition step. Only a few minutes is required to deposit sufficient dopant in a thin surface layer; the calculated curve for 6 min is shown in Fig.

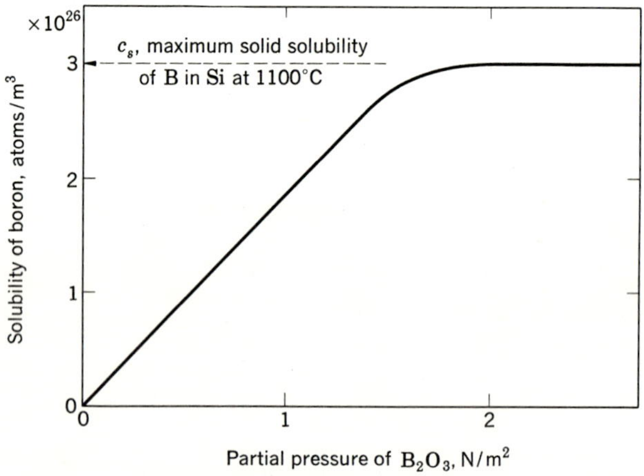

FIGURE 6-16 Equilibrium concentration of boron in silicon at 1100°C as a function of the partial pressure of B_2O_3 in the gas phase [conversion of units: (M N/m²) 0.75×10^{-2} = N torr; M atoms/m³ $\times 10^{-6}$ = N atoms/cm³]. (*Adapted from A. S. Grove, "Physics and Technology of Semiconductor Devices," John Wiley & Sons, Inc., New York, 1967.*)

6-17. Use of a logarithmic scale for the concentration c permits plotting the large range of concentrations of interest for applications to semiconductors. The usual linear scale was used in Fig. 6-8.

How is the actual behavior during deposition related to the calculated curve? The most important aspect is reproducibility. If the parameters controlling the calculated curve are the same for two different deposition treatments, the actual deposition curves are the same. The departures of the actual curve from the calculated one are caused by two factors. First, as explained in Sec. 6-8, the finite rate of reaction at the gas-solid interface prevents the surface concentration c_0 from reaching the saturation value c_s during a brief diffusion treatment. For this reason the actual curve in Fig. 6-17 begins at a concentration appreciably below c_s. The second factor arises because of the electronic carriers (holes) associated with the boron atoms. The holes are much more mobile and attempt to diffuse separately. As they move away from the ionized boron atoms, an electric field is established which aids the diffusion of the boron. For impurity concentrations which are large compared to the intrinsic carrier concentration for silicon (10^{25} at 1100 °C), the effective diffusion coefficient $\tilde{D}_{\text{eff}}$ is twice as large as the usual value $\tilde{D}$ measured in the absence of an aiding field. This causes the average depth of penetration to be greater than calculated and also decreases dc/dx for concentrations above 10^{25}. These two differences are shown by the curves of Fig. 6-17.

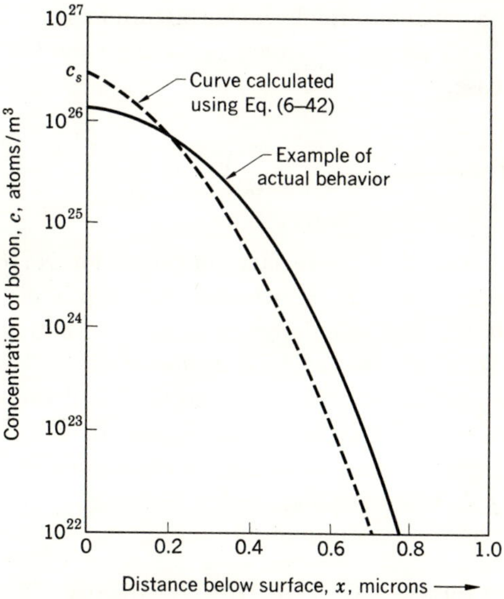

FIGURE 6-17 Comparison of calculated and experimental results for the deposition step in the doping of silicon with boron. Deposition for 6 min at 1100°C.

Following the short deposition process, a long diffusion treatment is employed to lower the surface concentration and to increase the depth of the doped region. A satisfactory calculation can be made on the assumption that the deposited layer is similar to the thin film used in tracer studies (Fig. 6-10). Since diffusion occurs in only the $+x$ direction in the present case, the solution differs from Eq. (6-22) by a factor of 2 and becomes

$$c = \frac{M}{(\pi D t)^{1/2}} e^{-x^2/4Dt} \tag{6-42}$$

To prevent loss of boron through the surface or the addition of impurities, the diffusion treatment is carried out in an oxidizing atmosphere. A layer of SiO_2 forms, blocking the window, and effectively impedes diffusion to or from the surface. The total amount M of boron in the initial deposited layer can be estimated from Eq. (6-41) with $c_o = c_s$. With this substitution, Eq. (6-42) becomes

$$c = \frac{2c_s}{\pi} \frac{(Dt)^{1/2}_{\text{dep}}}{(Dt)^{1/2}_{\text{dif}}} e^{-x^2/4(Dt)_{\text{dif}}} \tag{6-43}$$

This equation is valid only when the product $(Dt)_{\text{dif}}$ for the diffusion process is much larger than the corresponding product for the deposition step, $(Dt)_{\text{dep}}$. A typical diffusion treatment might be for 80 min at 1200°C. At this temperature $\tilde{D}$ is 3×10^{-16}, almost 10 times larger than for the deposition step at 1100°C. In this case,

$$\left[\frac{(Dt)_{\text{dep}}}{(Dt)_{\text{dif}}}\right]^{1/2} = \left(\frac{4 \times 10^{-17} \times 6 \times 60}{3 \times 10^{-16} \times 80 \times 60}\right)^{1/2} = \left(\frac{1}{100}\right)^{1/2} = 0.10$$

Equation (6-43) shows that the surface concentration is reduced by this ratio below the reference value $2c_s/\pi$, whereas the depth of the doped region increases as $2(Dt)^{1/2}_{\text{dif}}$. The calculated curve corresponding to these values is shown in Fig. 6-18. Reduction of the duration of the diffusion treatment by a factor of 10 changes the ratio from 0.10 to 0.32 and causes the pronounced changes in diffusion profile shown in the figure. Problem 6-16 considers the effect of longer diffusion treatments. Factors that cause departures in practice from the calculated curves are discussed in Sec. 6-8.

6-6. Alternatives to Volume Diffusion

Transport of atoms from one position in a crystal lattice to another occurs by several mechanisms acting simultaneously. Volume diffusion by the motion of individual defects is the main topic of the present chapter because it is the dominant mechanism in ordinary polycrystalline materials at high temperatures. Under other conditions, however, alternative mechanisms can be dominant. If the vapor pressure is sufficiently high, about 10 N/m^2, rapid transport can occur by evaporation in regions of high chemical potential and by condensation on other regions of the material. This mechanism is important for glacial ice and for a few other substances under special conditions. Plastic flow (as in the Kirkendall effect) can contribute appreciably to the overall process of material transport and is especially significant in the sintering process discussed below.

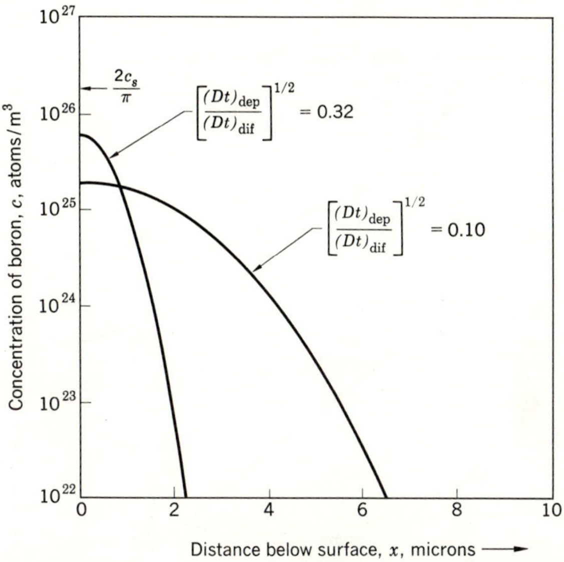

FIGURE 6-18 Calculated curves showing the effect of two different diffusion treatments on the deposited layer of Fig. 6-17. The temperature of diffusion is 1200°C. Times of 8 and 80 min give the two ratios shown.

SHORT-CIRCUIT DIFFUSION. Diffusion that is very fast compared to volume diffusion occurs in thin layers adjacent to surfaces and grain boundaries or around a dislocation line (Fig. 6-9). Experimental values of the corresponding diffusion coefficients are available for only a few metals, but the data for silver (Fig. 6-19) can serve as a basis for discussing a few important principles. Diffusion by more than one mechanism is often a complex process. As a useful simplification, each type of diffusive flux can be described by the following version of the first Fick law:

$$J = -DA_{\text{eff}} \frac{dc}{dx} \tag{6-44}$$

where A_{eff} is the fraction of the diffusion interface effectively employed for a given mechanism. For volume diffusion, A_{eff} is essentially unity. The values to be associated with the other mechanisms are illustrated schematically in Fig. 6-20 for a rod r m in radius having a concentration gradient along its length. If diffusion along the surface occurs in a region of thickness δ, the value of A_{eff} is $2\delta/r$. Since the value of δ is believed to be about 5 Å, A_{eff} is only 10^{-7} for diffusion in an ordinary rod ($r = 10^{-2}$ m). Although D_s is larger than D by

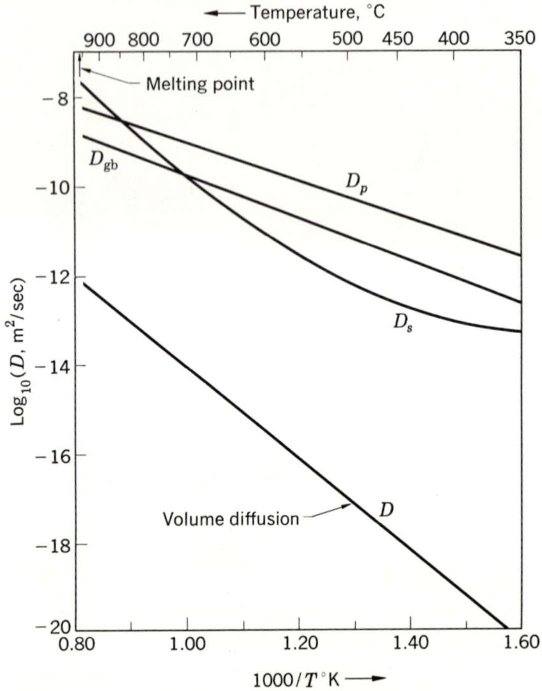

FIGURE 6-19 Variation with temperature of the usual diffusion coefficient for silver, D, in comparison with the other three types of diffusion coefficients D_{gb} (grain boundary), D_p (dislocation pipe), and D_s (surface layer). The data, especially for D_s, are approximate.

about 10^4 at high temperatures, only a negligible fraction (10^{-3}) of the diffusive flux is carried by the surface layer. In fine rods or small particles with a radius of 10 μ or less, surface diffusion can be of equal or greater importance than volume diffusion. The additional driving force caused by a curved surface [Eq. (5-22)] also becomes effective in this size range.

The effects of dislocations and of grain boundaries on diffusion are detectable only at relatively low temperatures. A single dislocation line in the rod of Fig. 6-20 has an A_{eff} value of $(\delta/r)^2$, or about 10^{-15}, for a rod 10^{-2} m in diameter. Even if the dislocation density is high (10^7 lines in the cross section of the rod), A_{eff} is only about 10^{-8}, and the total flux along dislocation lines is only a small fraction of volume diffusion at high temperatures. The ratio D_p/D increases with decreasing temperature and at 400°C reaches 10^8, compensating for the unfavorable value of A_{eff}. The dislocations then con-

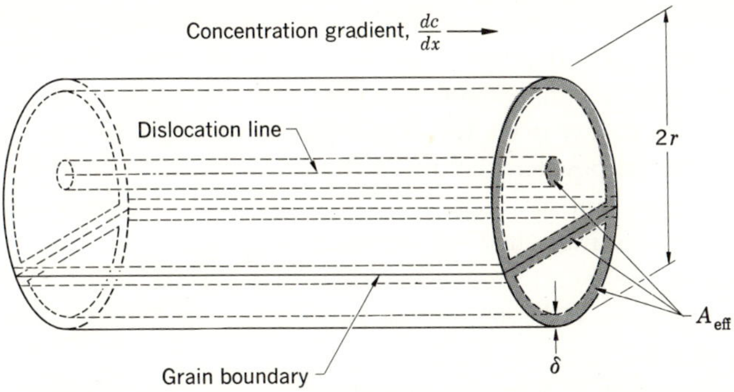

FIGURE 6-20 Schematic illustration of the effective areas associated with diffusion at the surface, a grain boundary, and a dislocation line. The same characteristic distance δ has been used in each case.

tribute equally with the volume of the material to the total diffusion flux. A number of processes, such as plastic deformation, can increase the dislocation density from the value of 10^{11} lines/m^2 characteristic of annealed crystals to values on the order of 10^{14} lines/m^2. The contribution of D_p is then greatly increased. Important structural changes in solid materials, such as precipitation from solid solution, involve diffusion distances of 10^{-7} m or even less. Since a measure of the completeness of diffusion is $x^2/4Dt = 1$ [Eq. (6-22)], a value $D_p A_{\text{eff}} = 10^{-18}$ is adequate to permit precipitation to occur in about an hour ($4t \approx 10^4$ sec). Such a low effective diffusion coefficient would be inadequate for a process such as carburizing in which the diffusion distance is about 10^{-4} m.

SINTERING PROCESSES. Important examples of short-circuit diffusion occur in the sintering of finely powdered materials. Although this process is also used with ceramics and with polymers, it is widely known by the name *powder metallurgy.* A fine powder is first poured into a mold and pressed to form a "compact" of the desired geometry; it is then given a heat treatment known as *sintering* to produce a strong, coherent mass. Sintering can be characterized as *solid-state* or *liquid-phase* depending on whether the mass transport processes that occur at elevated temperatures to bond the individual particles together are accomplished in the absence or presence of some liquid in the pressed compact. For example, cemented-carbide metal-cutting tools are a mixture of carbide and metal powders (WC–Co or TiC–Ni/Mo). During sintering, at about 1450°C for 1 hr, the metal powder melts and forms a cementing matrix that bonds the carbide particles into a strong mass which is essentially free of porosity. As is the case with many sintered products, the cemented carbides cannot be manufactured by conventional processes such as melting and casting.

Examples of solid-state sintering are sintered-iron products, widely used as gears. Their success is dependent on the ability to form a part to a precise shape and thereby eliminate expensive machining operations. Another feature of sintering is the ability to control porosity of the final product by appropriate choice of the initial powders, cold-pressing pressures, and sintering conditions (temperature, time, and atmosphere). Typical products are the porous bearings that are self-lubricating when impregnated with oil. Other examples of sintered materials are found among ceramics (e.g., bricks, oxide magnets, and translucent alumina) and even among polymers when Teflon is fabricated by a sintering process.

The sintering of a single component (powdered oxide or pure metal, for example) occurs in the stages illustrated in Fig. 6-21. At the point of contact between two particles (Fig. 6-21*a*), transport of material occurs and causes the growth of a neck joining the two particles. The surface energy is the principal driving force for this process, although it operates in several ways for various materials. In glass, for example, it produces a tensile stress in the surface of the neck, and the resulting viscous flow leads to further growth of the neck. In most materials the surface energy γ acts through the excess concentration of vacancies, ΔC, or the reduced vapor pressure $-\Delta p$ [see Eq. (5-22)]

$$\frac{\Delta C}{C_0} = \frac{-\Delta p}{p_0} = \frac{\gamma V_0}{RT\rho} \tag{6-45}$$

that exists below the concave surface of curvature ρ. Evaporation from the main surface and condensation in the neck region is the mechanism characteristic of sodium chloride. However, the most common mechanism is the diffusion of atoms along the surface and their deposition in the neck region by exchange with the excess vacancies. By the time the neck region has grown to about 0.2 of the cross-sectional area of the particle, the empty space surrounding each particle has been reduced to a network of channels joined at nodes (Fig. 6-21*b* and *c*). This structure is characteristic of the intermediate stage of sintering, during which most of the increase in density (and also appreciable grain growth) occurs. The volume of pore space decreases during this stage largely by diffusion of vacancies from the pore network into the lattice of the sinter body. Eventually, the pore channels close, changing into grain boundaries and leaving isolated pores at the grain junctions. This structure (Fig. 6-21*d*) characterizes the final stage of sintering and develops further by the gradual elimination of at least a portion of the remaining porosity. This final stage of sintering occurs by the diffusion of vacancies along the grain boundaries away from the pores. If, as a result of grain growth, the grain boundary migrates away from a pore, shrinkage of the pore virtually ceases since slow lattice diffusion is then the mechanism of densification. In the manufacture of

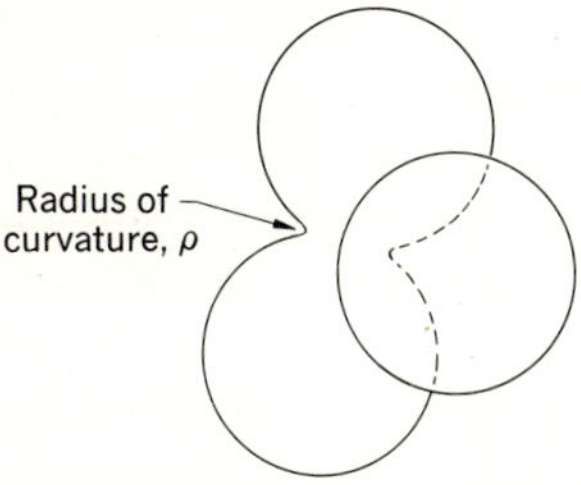

(*a*) Sphere-model showing growth of a neck at the point of contact between two particles.

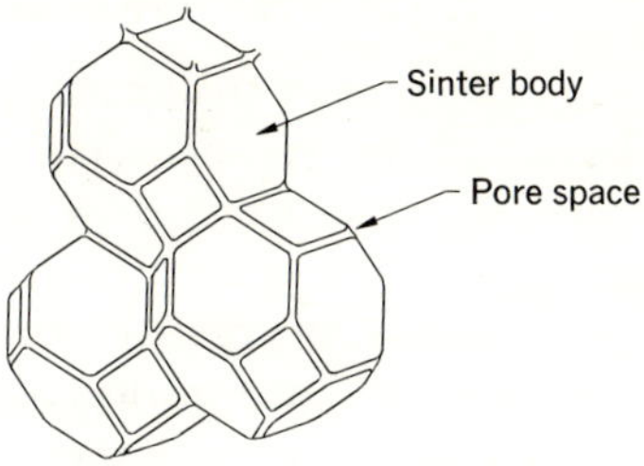

(*b*) Intermediate stage of sintering involving intermeshed networks of pore space and of sinter body.

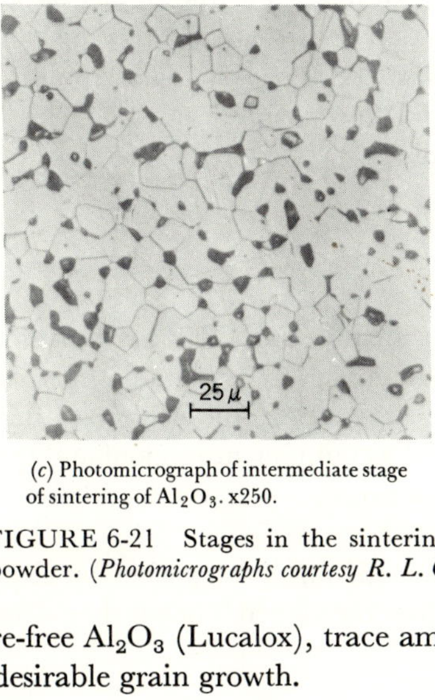

(*c*) Photomicrograph of intermediate stage of sintering of Al_2O_3. x250.

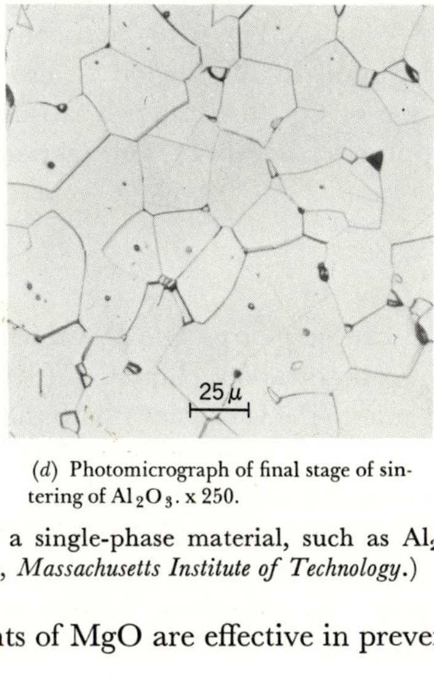

(*d*) Photomicrograph of final stage of sintering of Al_2O_3. x 250.

FIGURE 6-21 Stages in the sintering of a single-phase material, such as Al_2O_3 powder. (*Photomicrographs courtesy R. L. Coble, Massachusetts Institute of Technology.*)

pore-free Al_2O_3 (Lucalox), trace amounts of MgO are effective in preventing undesirable grain growth.

OTHER TRANSPORT PHENOMENA

Many technically important aspects of particle transport in materials require an amplification of the basic treatment of diffusion given in Secs. 6-1 and 6-2. Especially for electronic carriers, the driving force may be not only a concentration gradient but also an electric field or a temperature gradient. Diffusion may occur in series between two phases, a gas and a solid; the overall transport then involves an interfacial reaction at the gas-solid interface. Finally, if the diffusion system contains three or more components, the rules for multicomponent diffusion apply.

6-7. General Transport Equation

The present treatment applies to atoms as well as to defect particles (electrons, holes, vacancies, etc.) in a single phase. Electronic carriers are used as an example since they show especially clearly the phenomena of creation (at sources) and destruction (at sinks). In pure silicon at room temperature the concentration c_n of negative carriers (defect electrons e') is equal to c_p, the concentration of positive carriers (defect holes $h^\bullet$) (Fig. 4-7). In units of particles per cubic meter, each of these concentrations is 1.45×10^{16}; therefore, by Eq. (4-25) K_g is

$$K_g = c_n c_p = 2.1 \times 10^{32} \tag{6-46}$$

If a donor dopant, such as phosphorus, is diffused into the surface layer of silicon (compare Fig. 6-18) to a concentration of 10^{24} atoms/m^3, the electron concentration in this region is also $c_n = 10^{24}$ (Fig. 4-7). The equilibrium condition of Eq. (6-46) is still satisfied, and thus $c_p = 2.1 \times 10^8$. Fick's first law applies separately to both carriers; thus

$$J_i = -D_i \frac{dc_i}{dx} \tag{6-47}$$

where i can be n or p. The D_i values are shown in Fig. 6-22 as a function of dopant concentration. (Equation (6-47) also applies to both atomic species, Si and P, and to the vacancies.) Both the majority carriers (the electrons) and the minority carriers (the holes) can be pictured as jumping at random and therefore causing a diffusive flux J_i of particles per square meter per second provided that a concentration gradient dc_i/dx exists.

Quite a different transport phenomenon, *drift*, occurs when an electric field $E = -dV/dx$ volts/m, is applied across the doped silicon. The flux of particles is then

$$J_i' = -c_i M_i \frac{dV}{dx} \tag{6-48}$$

where M_i is the mobility[1] of the particle in question (Fig. 6-22). The particles can be pictured as drifting in response to the electric field, and thus the flux is proportional to the concentration of particles, c_i. Since M_p is opposite in sign to M_n, holes drift oppositely to electrons. The current density I_i' is simply

$$I_i' = J_i' z_i e \tag{6-49}$$

[1] The symbol μ is generally employed in the electronics literature for mobility. M_i is used here to avoid confusion with the present use of μ for chemical potential.

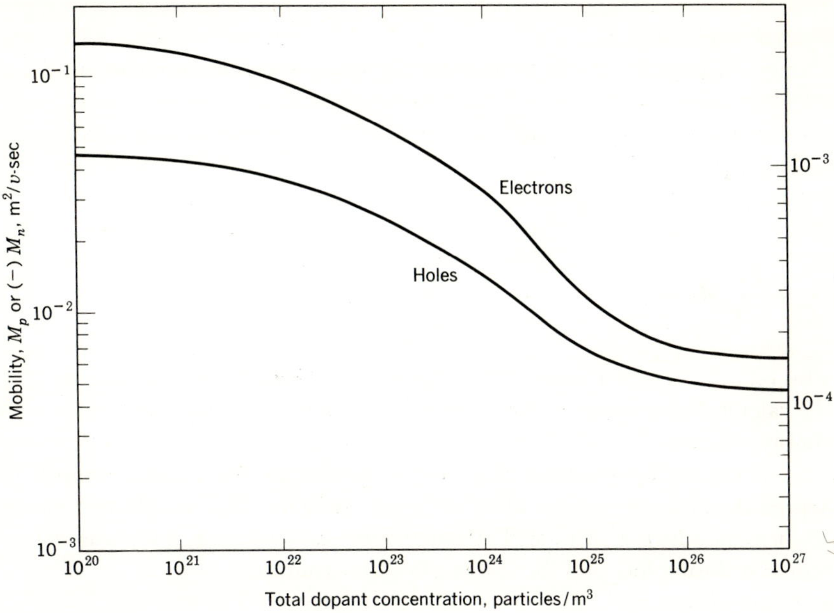

FIGURE 6-22 Mobility and diffusion coefficients of silicon as a function of (ionized) dopant concentration at room temperature.

where e is the electronic charge and z_i is the charge number. Since $z_p = +1$ and $z_n = -1$, I_n' and I_p' have the same sign and merely sum to give the total current density.

In spite of the quite different natures of a chemical gradient and an electric field, D_i and M_i for a given type of particle are related in a simple, useful manner. The fluxes of Eqs. (6-47) and (6-48) combine to give the total flux

$$J_i'' = -D_i \frac{dc_i}{dx} - c_i M_i \frac{dV}{dx} \tag{6-50}$$

Imagine that the transport of particles proceeds until the steady-state condition is reached, and thus $J_i'' = 0$. The "driving force" of the concentration gradient on a particular particle is then just equal and opposite to that due to the voltage gradient. The two driving forces can be converted into the same units (ev/m per migrating particle) and then canceled from the equation. The conversion of the concentration gradient involves the chemical potential $\mu_i = kT \ln \gamma_i c_i$ and the assumption that the activity coefficient γ_i is constant. The force of the electric field is $z_i^* eE = -z_i^* e\, dV/dx$; the *effective charge number*

z_i^* is identical with the actual charge for electrons and holes but is an experimental coefficient for particles such as substitutional atoms. With these changes, Eq. (6-50) can be written for the steady-state condition as follows:

$$0 = -\frac{D_i c_i}{kT}\frac{d\mu_i}{dx} - \frac{M_i c_i}{z_i^* e} z_i^* e \frac{dV}{dx} \tag{6-51}$$

or

$$D_i = \frac{kT}{z_i^* e} M_i \tag{6-52}$$

This relation between D_i and M_i, known as the *Nernst-Einstein equation*, permits the calculation of D_p or D_n from the measured mobility coefficients since z_i^* is simply $+1$ or -1.

For atomic particles, Eq. (6-52) is used in the reverse manner to express the mobility in terms of the diffusion coefficient D_i. The effective charge z_i^* then includes two factors: (1) the field force $z_i eE$, where z_i is the true charge on the diffusing ion ($+4$ for carbon, C^{4+}, for example), and (2) the effect of collisions of electronic carriers with the ion in question. The values in Table 6-1 show the wide variation in z_i^* in both magnitude and sign.

A significant aspect of transport of electronic carriers is the "creation-destruction" reaction that maintains the equilibrium expressed by Eq. (6-46). This reaction can be described mathematically by the equation

$$S_i = \frac{1}{\tau_i}(c_{io} - c_i) \qquad \text{particles/m}^3\text{-sec} \tag{6-53}$$

S_i can either be positive or negative and is the rate of creation (at sources) or destruction (at sinks) of the particle in question. c_{io} is the equilibrium concentration of particles; for example, c_{po} is 1.45×10^{16} in pure silicon at room temperature, but is 2.1×10^8 in the phosphorus-doped silicon discussed above. c_i is the actual concentration at any instant. If c_i is larger than c_{io}, the quantity $c_{io} - c_i$ is negative, and S_i then represents the destruction of particles at sinks. The quantity $1/\tau_i$ serves as a rate constant in Eq. (6-53); however, τ_i is also a useful measure of the lifetime of a transient excess of particles, as will now be explained.

The initial excess (or deficiency) of particles may result from a distinctly different mechanism than that involved in S_i. For this reason (among others) the symbol G is employed to represent the *rate of generation* (positive or negative) in units of particles per cubic meter per second. For example, light falling on the phosphorus-doped silicon can cause the generation of electron-hole pairs at a rate G_L. At any instant, the concentration of holes changes at a rate dc_p/dt, which is the sum of G_L and S_p [Eq. (6-53)]; thus

$$\frac{dc_p}{dt} = G_L + \frac{1}{\tau_p}(c_{po} - c_p) \tag{6-54}$$

After a time of about $5\tau_p$, c_p reaches the steady-state value $(c_p)_L$, and dc_p/dt becomes zero. From Eq. (6-54),

$$(c_p)_L = c_{po} + \tau_p G_L \tag{6-55}$$

Now, if the light is turned off, G_L becomes zero, and Eq. (6-54) is

$$\frac{dc_p}{dt} = \frac{1}{\tau_p}(c_{po} - c_p) \tag{6-56}$$

subject to the condition that $c_p = (c_p)_L$ at zero time. The solution of Eq. (6-56) is

$$c_p(t) = c_{po} + [(c_p)_L - c_{po}]e^{-t/\tau_p} \tag{6-57}$$

The decay of c_p, according to this equation, is shown in Fig. 6-23 for three values of τ_p. Typical experimental values of τ_i are given in Table 6-1.

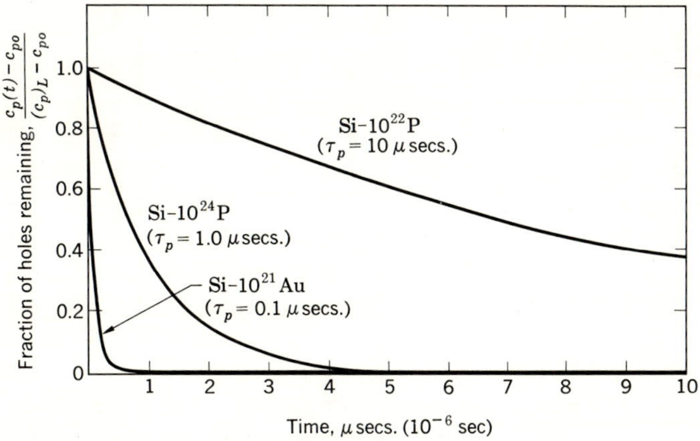

FIGURE 6-23 The decay of an excess of defect particles (holes) through the action of a sink. In semiconductors this process is called *recombination* since an electron-hole pair combines and disappears.

TABLE 6-1. TRANSPORT COEFFICIENTS FOR SEVERAL TYPICAL MATERIALS†

Semiconductors

Material	Temperature, °C	$\tilde{D}$, m²/sec	M_i, m²/volt-sec	Temperature, °C	D_n, m²/sec	D_p, m²/sec	M_n, m²/volt-sec	M_p, m²/volt-sec	D_n^T, electrons/m-°C-sec	D_p^T, holes/m-°C-sec	τ_n, sec	τ_p, sec
Silicon	1100	4×10^{-17}	—	27	3×10^{-3}	1×10^{-3}	-1.4×10^{-1}	5×10^{-2}	4×10^{12}	1.5×10^{12}	2×10^{-5}	3×10^{-5}
Si-10^{25}B	1100	4×10^{-17}	—	27	3×10^{-4}	2×10^{-4}	-1×10^{-2}	7×10^{-3}	~ 0	3.5×10^{19}	1×10^{-6}	4×10^{-7}
Si-10^{22}P	1050	6×10^{-17}	—	27	2×10^{-3}	1×10^{-3}	-1×10^{-1}	4×10^{-2}	1.2×10^{18}	~ 0	3×10^{-6}	4×10^{-6}
Si-10^{23}Au	727	2×10^{-17}	1×10^{-9} (at 1350°C)	27	1×10^{-3}	6×10^{-4}	-6×10^{-4}	2×10^{-2}	—	—	3×10^{-10}	1×10^{-9}

Metals and Alloys

Material	D_0 m²/sec	Q, kcal/mole	Temperature, °C	D^*, m²/sec	$\tilde{D}$, m²/sec	D_1, m²/sec	D_2, m²/sec	z^*	Q^*, kcal/mole	τ_V, sec
Gold	0.107×10^{-4}	42.26	920	2.0×10^{-13}	—	—	—	−9	+16	10^5 (at 60°C)
Silver	0.905×10^{-4}	45	850	1.7×10^{-13}	—	—	—	−25	~ 0	
Au–3 A/o Ag	4.7×10^{-6}	38.5	940	—	5.2×10^{-13}	2.8×10^{-13}	5.3×10^{-13}	—	−4	
Fe–2 A/o C	0.27×10^{-4}	34.5	927	—	1.6×10^{-11}	—	—	+5	+2	

† Conversion of units: M atoms/m³ × 10^{-6} = N atoms/cm³; M m²/sec × 10^4 = N cm²/sec.

$\tilde{D}$, interdiffusion coefficient [Eq. (6-32)]

D_0 and Q enter in the equation $D = D_0 e^{-Q/RT}$

D^*, self-diffusion coefficient [Eq. (6-16)]

D_1 (solvent) and D_2 (solute), intrinsic diffusion coefficients [Eq. (6-35)]

D_n and D_p, diffusion coefficients for electrons and holes [Eq. (6-47)]

M_n and M_p, mobility coefficients for electrons and holes [Eq. (6-48)]

M_i and z^*, mobility coefficient and effective charge number for atoms [Eq. (6-52)]

D_n^T and D_p^T, thermodiffusion coefficients for electrons and holes [Eq. (6-59)]

Q^*, index of thermodiffusion for atoms [Eq. (6-60)]

τ_n and τ_p, lifetime of electrons and of holes [Eq. (6-53)]

τ_V, lifetime of vacancies [Eq. (6-53)]

A temperature gradient produces two types of transport effects. One is the transport of heat,

$$H = -K \frac{dT}{dx} \qquad \text{cal/m}^2\text{-sec} \tag{6-58}$$

where K is the thermal conductivity. This equation is analogous to Fick's first law [Eq. (6-9)], and the methods of solution are similar[1] but are not of interest for the present treatment. A temperature gradient can also produce a flux of particles,

$$J_i^T = -D_i^T \frac{dT}{dx} \tag{6-59}$$

where D_i^T is the thermodiffusion coefficient. An argument similar to that used in deriving Eq. (6-52) shows that D_i^T is related to D_i by the equation

$$D_i = \frac{RT^2}{Q_i^* c_i} D_i^T \tag{6-60}$$

Q_i^*, the heat of transport, is the quantity of heat transported across the diffusion interface per mole of particles. As in the case of z_i^*, Q_i^* is essentially an empirical coefficient that determines the direction and amount of thermodiffusion. Typical values are given in Table 6-1.

In a given region of a specimen, all of the transport and generation processes for particles of type i can be considered to operate independently and to produce a net rate of change of concentration, dc_i/dt. The same reasoning that led from Fick's first law, $J_i = -D_i\, dc_i/dx$, to the second law, $\partial c_i/\partial t = D\, \partial^2 c_i/\partial x^2$ [Eq. (6-16)], can also be applied to Eqs. (6-48) and (6-59). When the resulting expressions are combined with Eq. (6-54), the general transport equation is obtained; thus

$$\frac{dc_i}{dt} = D_i \frac{\partial^2 c_i}{\partial x^2} + c_i M_i \frac{\partial^2 V}{\partial x^2} + D_i^T \frac{\partial^2 T}{\partial x^2} + G_L + \frac{1}{\tau_i}(c_{io} - c_i) \tag{6-61}$$

In principle, this equation can be used to describe the transport behavior of each atomic and defect particle in a given material. An adequate description of some electronic devices requires consideration of all five terms, but many simple situations can be treated using only one or two terms.

[1] H. S. Carslaw and J. C. Jaeger, "Conduction of Heat in Solids," Oxford University Press, London, 1959.

ELECTROTRANSPORT OF CARBON IN IRON. Because carbon is an interstitial solute in iron, its electrotransport behavior is relatively simple. The electric field (plus the transfer of momentum from collisions with electronic carriers) act on the C^{4+} ion and tend to move it within the rigid iron lattice toward the cathode. A gradient in carbon concentration then develops and opposes the motion caused by the field. The form of Eq. (6-61) suitable for this case is

$$\frac{\partial c_C}{\partial t} = D\frac{\partial^2 c_C}{\partial x^2} + M_C c_C \frac{\partial^2 V}{\partial x^2} \tag{6-62}$$

When a homogeneous 0.5 $^{w}/_{0}$ carbon steel is subjected to a relatively small field, $E \approx 10$ volt/m (Fig. 6-24*a*), a current density of about 10^7 amp/m^2 flows in the steel specimen and heats it to the temperature of the experiment, 927°C (1200°K). Initially the carbon concentration has the constant value 1.8×10^{27} atoms/m^3, but gradients in concentration develop at the two ends of the specimen in the early stages of electrotransport (Fig. 6-24*b*). Since carbon atoms cannot leave or enter the specimen, carbon atoms accumulate at the right and are depleted at the left, the corresponding concentration curves

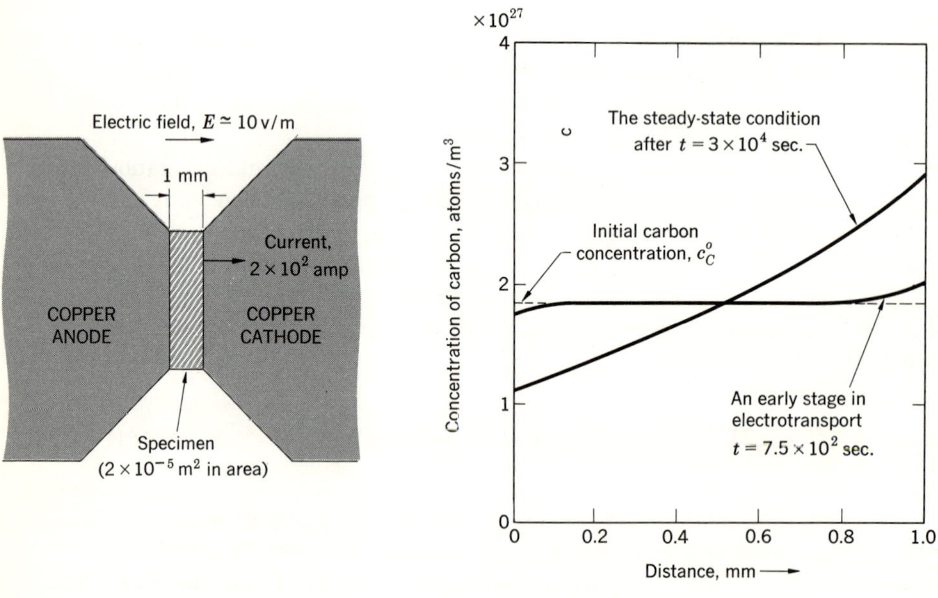

(*a*) An experimental arrangement for studying electrotransport.

(*b*) Redistribution of carbon is austenite produced by electrotransport at 927°C.

FIGURE 6-24 Illustration of the process of electrotransport in alloys, using carbon in iron as an example. (*Courtesy T. Okabe, Georgia Institute of Technology.*)

being given by the appropriate solution of Eq. (6-62). If electrotransport is continued for a sufficiently long time, $5\tau_D$, the steady-state condition is reached, in which the diffusion flux and the electrotransport flux balance each other. Since $\partial c_C/\partial t = 0$, Eq. (6-62) can then be solved (Prob. 6-19) to obtain the equation

$$c_C = \frac{c_C^{\circ} l}{(D_C/M_C E)(e^{M_C E l/D_C} - 1)} e^{M_C E x/D_C} \tag{6-63}$$

The corresponding curve is plotted in Fig. 6-24*b* and shows the large effects due to electrotransport. Possible applications include refining of metals, analogous to zone refining (Sec. 7-6).

Just as in chemical diffusion, lattice vacancies often play the major role in electrotransport. The behavior of a pure metal can be described by analogy with Fig. 6-24*b*. At one end of the specimen the concentration of vacancies, N_V, is forced below the equilibrium value N_V°. As a result, vacancies are generated in this region, and the volume of the crystal lattice increases locally. At the other end of the specimen, N_V is driven above the equilibrium value. Usually the excess vacancies are annihilated at sinks, such as dislocations, and cause a local shrinking of the specimen. In the electrical current leads (aluminum "stripes") in microelectronic circuits the current density may be 10^9 amp/m^2. Vacancies generated by electrotransport can then precipitate as voids and eventually cause failure of the stripe.

THERMODIFFUSION IN PURE GOLD. The effect of a temperature gradient on an initially homogeneous system is similar to that of an electric field. A driving force acts on all of the particles of the system and causes concentration gradients to appear. The mechanism of thermodiffusion depends on the particle being considered and the phase within which it is migrating. In solid phases the particles of interest are (1) electrons or holes (see Prob. 6-20), (2) interstitial solutes, which behave in a manner analogous to that shown for electrotransport of carbon in iron (Fig. 6-24), and (3) substitutional solutes or pure substances, such as the following example of pure gold. Although no concentration gradient of atomic particles can occur in this case, the gradient of vacancies is sometimes important. The form of Eq. (6-61) that applies is

$$\frac{\partial c_V}{\partial t} = D\frac{\partial^2 c_V}{\partial x^2} + D_V^T \frac{\partial^2 T}{\partial x^2} + \frac{1}{\tau_V}(c_{V_0} - c_V) \tag{6-64}$$

If a specimen containing a marked reference plane (Fig. 6-25*a*) is subjected to a steep temperature gradient (10^5 °C/m) for a long time at a high temperature, the markers move in the specimen (Fig. 6-25*c*).

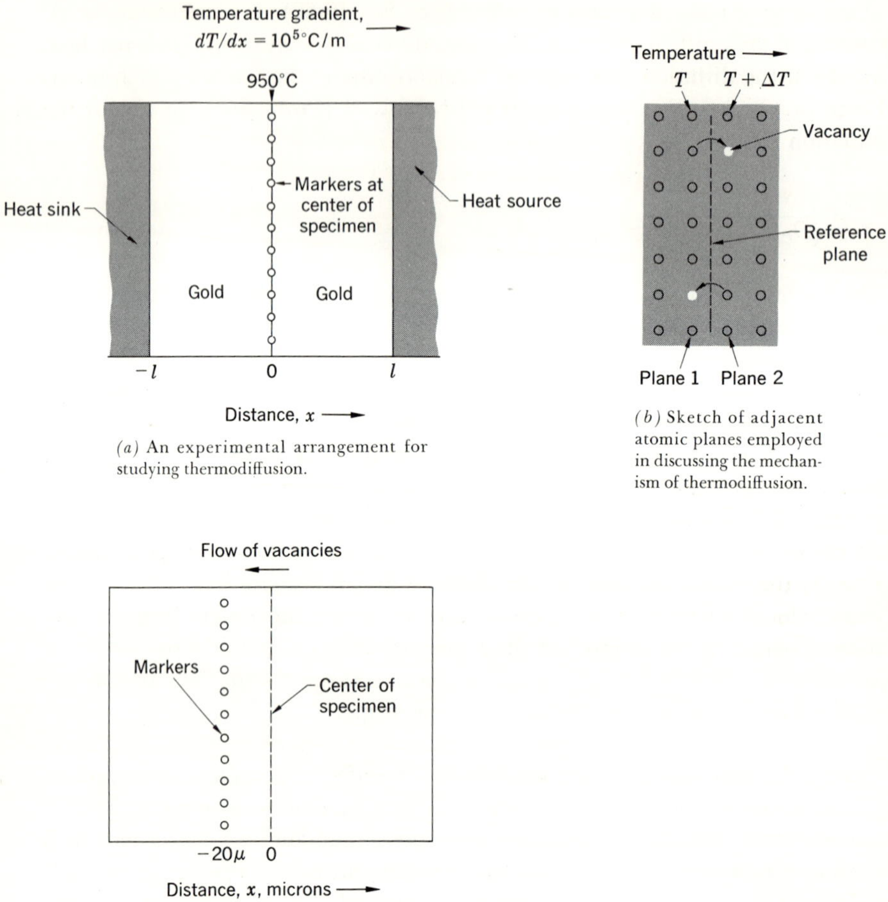

(a) An experimental arrangement for studying thermodiffusion.

(b) Sketch of adjacent atomic planes employed in discussing the mechanism of thermodiffusion.

(c) Migration of markers produced by thermodiffusion for 100 hrs.

FIGURE 6-25 Schematic illustration of the process of thermodiffusion in pure gold near 950°C. (*Data of C. J. Meechan and G. W. Lehman.*)

In some materials the movement is toward the hot end of the specimen; in gold it is toward the cold end. The two factors determining the direction of movement can be understood from the sketch of Fig. 6-25*b*. The net flux of vacancies to the right is proportional to the difference $N_V^1\Gamma_{Au}^2 - N_V^2\Gamma_{Au}^1$, where N_V^1 is the concentration of vacancies at plane 1 and Γ_{Au}^2 is the jump frequency of a gold atom located on plane 2. In gold, N_V increases more rapidly with temperature than does Γ_{Au}, and therefore the net flux of vacancies is to the left for this metal. As explained for the analogous case of electrotransport, the vacancies are created in one portion of the specimen and destroyed in another (although not necessarily at the end of the specimen) and cause the observed shifting of the marked interface. Thermodiffusion of atoms or of vacancies can be of technical importance only in the exceptional case when high temperatures and high heat fluxes are both present, for example, in special heat-exchange devices. Thermodiffusion of electrons and holes occurs more commonly and is discussed in Sec. 12-4 in connection with thermoelectric phenomena.

6-8. Two-Phase Diffusion

Consider two phases α and β (which may be solid, liquid, or gas) in a two-component system. Examples may be the gaseous and solid phases in the silicon-boron system or the α and β solid phases in the silver-copper system (Fig. 2-5). Within each phase the components diffuse in accordance with Fick's laws, as described in the previous sections. If the two phases are joined, a different process of atomic transport occurs at the interface between the two phases. For example, if pure silver (α) is joined to pure copper (β), a phase boundary α/β separates the two regions of the couple. The first step in atomic transport of either metal into the other must be the crossing of this phase boundary. If silver leaves the α phase and enters the β phase on the other side of the boundary, the process can be represented

$$\mathrm{Ag}(\alpha) \rightarrow \mathrm{Ag}(\beta) \tag{6-65}$$

Since the atomic motion involved may be quite different than a diffusive "jump," atomic transport across a phase boundary is described as the process of *interfacial reaction.*

Usually the phase boundary moves during two-phase diffusion, but if the movement is negligibly small, the coupling of interfacial reaction and diffusion can be described more simply. For example, in the initial stage of the deposition of boron on silicon (Fig. 6-26*a*) the net transfer of boron from the gas phase to the surface occurs at the rate

$$R = R_f - R_r = k[c_s - c_o(t)] \tag{6-66}$$

R_f is the rate of the forward reaction (gaseous phase to surface), R_r is the rate of the reverse reaction (surface to gaseous phase), k is a rate constant on the order of 10^{-9} m/sec, c_s is the saturation value (Fig. 6-16), and c_o is the actual surface concentration at a given time t. A diffusive flux $J = -D(dc/dx)$, which is equal to R, drains the boron atoms away from the surface. Consequently, the actual boundary condition in the early stage of deposition is

$$k[c_s - c_o(t)] = -D\frac{dc}{dx} \tag{6-67}$$

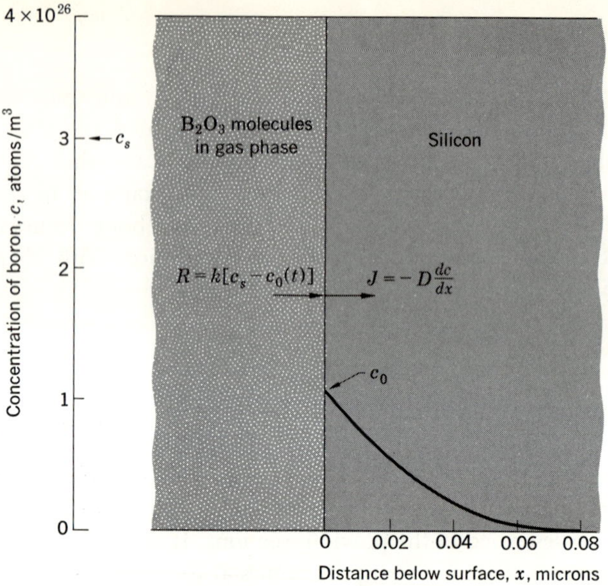

(*a*) Schematic illustration of the deposition process at an early stage.

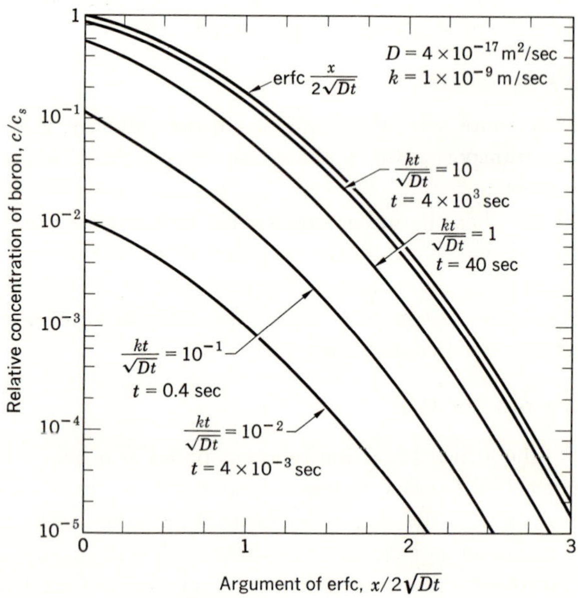

(*b*) Solutions of Eq. (6-73) for various stages in the deposition process. (After A. S. Grove.)

FIGURE 6-26 Influence of the gas-solid interfacial reaction on the deposition of boron on silicon at 1100°C.

The saturation value c_s is only gradually attained at the surface after an appreciable time of deposition. The solution of the second Fick law, employing Eq. (6-67), is

$$\frac{c}{c_s} = \operatorname{erfc}\frac{x}{2\sqrt{Dt}} - \left[\exp\left(\frac{kt}{\sqrt{Dt}}\right)^2 \exp\left(\frac{kt}{\sqrt{Dt}}\right)\left(\frac{x}{2\sqrt{Dt}}\right) \operatorname{erfc}\left(\frac{kt}{\sqrt{Dt}} + \frac{x}{2\sqrt{Dt}}\right)\right] \tag{6-68}$$

The first term is the approximation [Eq. (6-21)] neglecting the surface reaction, and the term in brackets is the appropriate correction. For values of $kt/\sqrt{Dt}$ of 10 or less the correction term is appreciable (Fig. 6-26*b*). For $D = 4 \times 10^{-17}$ and $k = 10^{-9}$, the time corresponding to $kt/\sqrt{Dt} = 1$ is

$$\frac{k^2t}{D} = 1^2 \qquad t = \frac{4 \times 10^{-17}}{10^{-18}} = 40 \text{ sec} \tag{6-69}$$

Similarly, the 6-min deposition of Fig. 6-17 corresponds to the ratio $kt/\sqrt{Dt} =$ 3 and is consistent with a surface concentration appreciably below c_s.

Two-phase diffusion and the accompanying interfacial reaction occur in most phase transformations in two- (or higher-) component systems, discussed in the next chapter. Often the reaction-rate constant k is relatively high, and the experimental behavior can be adequately interpreted by diffusion data alone. For example, the curve for $kt/\sqrt{Dt} > 10$ in Fig. 6-26*b* differs little from the curve erfc $[x/(2\sqrt{Dt})]$, obtained by neglecting the interfacial reaction. In some important transformations, however, interfacial reactions and interfacial structure dominate the experimental behavior (Secs. 7-6 to 7-9).

6-9. Multicomponent Diffusion

Diffusion often occurs in complex materials and involves more than the two components considered in the usual treatments. Examples are the doping of silicon with two impurities, such as boron and phosphorus, and the heat treatment of steels containing not only carbon but also one or more alloying elements. The necessary modification of the Fick laws to include n components (see Prob. 6-21) can be illustrated for the relatively simple case of a three-component system.

In general, the concentration gradient of any component can affect the diffusion flux of component 1, leading to the equation

$$J_1 = -\tilde{D}_{11}\frac{dC_1}{dx} - \tilde{D}_{12}\frac{dC_2}{dx} - \tilde{D}_{13}\frac{dC_3}{dx} \tag{6-70}$$

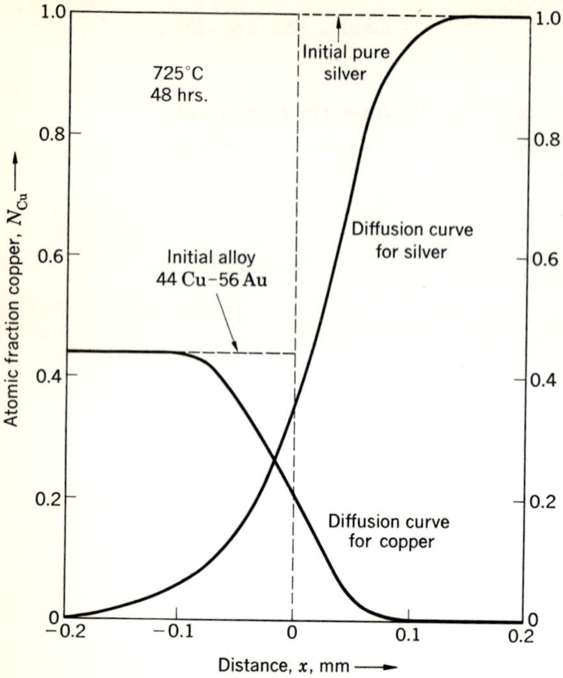

(*a*) Diffusion between a copper-gold alloy and pure silver gives two independent diffusion curves.

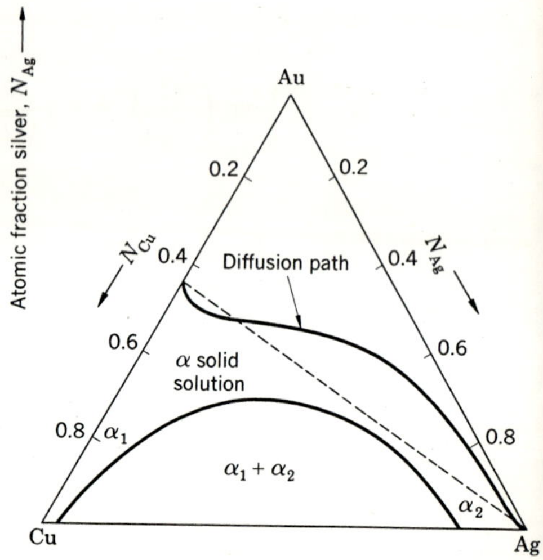

(*b*) The compositions across the specimen in (*a*) plot as a curved "diffusion path" in the ternary composition triangle.

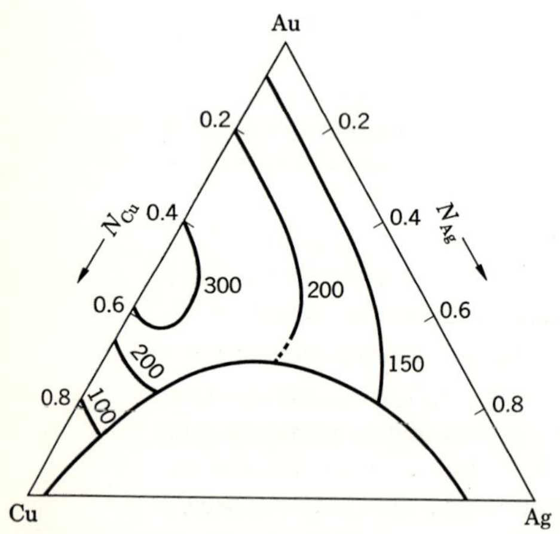

(*c*) Values of the ternary diffusion coefficient $\tilde{D}^{Au}_{CuCu}$ at 725°C as a function of composition. The units are 10^{-16} m²/sec.

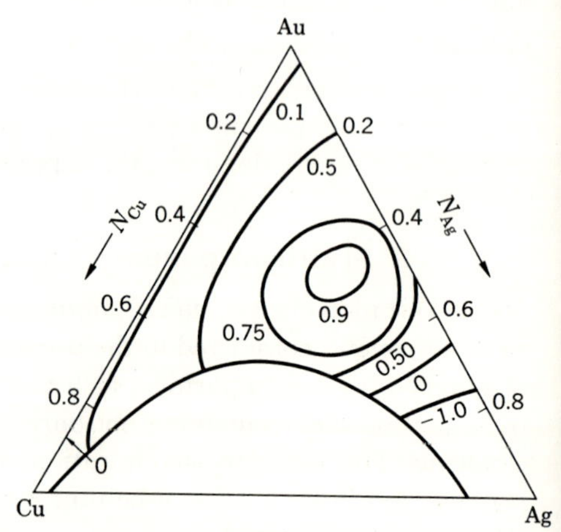

(*d*) Values of the ternary diffusion coefficient $\tilde{D}^{Au}_{AgCu}$, expressed as the ratio $\tilde{D}^{Au}_{AgCu} / \tilde{D}^{Au}_{AgAg}$. Note the large variation, including negative values.

FIGURE 6-27 Various aspects of three-component diffusion illustrated by data on diffusion in the copper-silver-gold system at 725°C. (***Data of T. O. Ziebold and R. E. Ogilvie.***)

where the $\tilde{D}$'s are suitable coefficients. Similar equations can be written for J_2 and J_3, the fluxes of the other two components. The first special feature of multicomponent diffusion is the necessity of choosing one of the three components to be the dependent concentration variable, say C_3. Since $C_1 + C_2 + C_3 = 1$, the expression for C_3 is

$$C_3 = 1 - (C_1 + C_2) \tag{6-71}$$

It follows that

$$\frac{dC_3}{dx} = -\left(\frac{dC_1}{dx} + \frac{dC_2}{dx}\right) \tag{6-72}$$

When dC_3/dx is eliminated from Eq. (6-70), the resulting equation can be written

$$J_1 = -\tilde{D}_{11}^3 \frac{dC_1}{dx} - \tilde{D}_{12}^3 \frac{dC_2}{dx} \tag{6-73}$$

where $\tilde{D}_{11}^3 = \tilde{D}_{11} - \tilde{D}_{13}$ and $\tilde{D}_{12}^3 = \tilde{D}_{12} - \tilde{D}_{13}$. In view of the definition of the Matano interface, $J_1 + J_2 + J_3 = 0$, and therefore the flux of component 3 is also a dependent quantity. The flux of component 2 is given by

$$J_2 = -\tilde{D}_{21}^3 \frac{dC_1}{dx} - \tilde{D}_{22}^3 \frac{dC_2}{dx} \tag{6-74}$$

and therefore a total of four diffusion coefficients is adequate to describe the diffusion behavior of a three-component system.

An example of three-component diffusion is shown in Fig. 6-27*a* for a couple consisting originally of a copper-gold alloy and a specimen of pure silver. When gold is chosen as the dependent concentration variable, the results of a diffusion treatment can be described by the two concentration-distance curves shown. A significant aspect of these curves is revealed when the range of concentration across the diffused specimen is plotted on a ternary concentration triangle (Fig. 6-27*b*). The concentrations plot as a curved line, the *diffusion path*, rather than following a straight line between the two initial compositions. The corresponding values of the four $\tilde{D}_{ij}^{\mathrm{Au}}$ coefficients, two of which are shown in Figs. 6-27*c* and *d*, differ appreciably among themselves and also vary with concentration in the three-component system.

Lack of experimental data on the four diffusion coefficients and their variation with composition seriously limits diffusion calculations in a typical three-component system. In principle, however, Eqs. (6-73) and (6-74) should be employed in place of the usual Fick law for a system containing appreciable amounts of three components. The details of the calculation are then conveniently handled by a digital computer.

PROBLEMS

6-1. Using Eq. (6-2), write out the expanded form of the dot product $\mathbf{R}_n \cdot \mathbf{R}_n$, and obtain a few representative terms. These terms fall into the two classifications shown in Eq. (6-3). Why are there n terms of the type r_i^2?

6-2. Usually a calculation of gas permeation is made to insure that the amount is not excessive. (*a*) Use a sketch similar to Fig. 6-3, but begin with a glass with zero concentration of the gas in question. Show by a sequence of four or five sketches the unsteady-state period of diffusion that precedes the steady-state. (*b*) If a calculation is made assuming that the steady-state condition exists from zero time onward, is the resulting estimate of permeation a conservative one? Explain.

6-3. (*a*) Using the value of K from Fig. 6-4, calculate the initial rate of loss of helium at 5 atm and 200°C from a Pyrex bulb 50 cm in diameter and with a wall thickness of 0.1 cm. (*b*) The same result must be given by the diffusion equation [Eq. (6-11)], and therefore the solubility of helium in Pyrex (at 5 atm and 200°C) can be determined. Make this calculation, expressing the solubility as cubic centimeters of helium at STP per cubic centimeter of Pyrex. (*c*) What would be the rate of loss if ordinary soda-lime glass were the material of the bulb?

6-4. Making use of relations of the form $\partial c/\partial t = (\partial c/\partial \lambda)(\partial \lambda/\partial t)$, show that Eq. (6-16) can be converted to Eq. (6-17).

6-5. An experiment corresponding to Fig. 6-6 is performed using a mixture of 0.8 H_2–0.2 N_2 in region II and pure N_2 in region I. If the diffusion coefficient is 0.7×10^{-4} m²/sec, at what distance from the interface in region I will the concentration of H_2 be 0.1 after 10^2 sec of diffusion?

6-6. (*a*) Which type of interstice (Fig. 1-15*d*) is occupied by most of the carbon atoms in γ-iron (face-centered cubic)? (*b*) Explain why some carbon atoms must be in the other interstice if diffusion is to occur.

6-7. Consider the carburizing of pure iron at 950°C. (*a*) If the carburizing medium, a hydrocarbon gas, saturates the surface with carbon in solution in γ-iron, what is the value of c_0 in Eq. (6-21)? See the iron-carbon phase diagram (Fig. 2-22). (*b*) If the "case" is considered as the surface region containing at least 0.25 w/o carbon, determine the time of carburizing required to produce a 1-mm case. (*c*) What time would be required to produce a 2-mm case? Recall the relation $\lambda = x/t^{1/2}$. (*d*) Since the complementary error function ERFC is one of the standard functions supplied with Fortran IV, a complete diffusion curve corresponding to Eq. (6-21) is easily obtained. The values of c_0 (CO), D, and $t(T)$ are first specified. A suitable initial value of x, XSTART, is used in calculating the first point on the curve. Additional points are calculated for 2(XSTART), 3(XSTART), etc. The calculation is stopped when the value of erfc drops to 10^{-3}.

```
C  CALCULATION OF A DIFFUSION CURVE
    1  FORMAT (F7.4, 3E10.3)
C  READ DATA
    2  READ (5,1) CO, D, T, XSTART
C  CALCULATION
    3  X = XSTART
    4  ARG = X/(2 * SQRT(D * T))
    5  ERROR = ERFC(ARG)
    6  C = CO * ERROR
    7  IF (ERROR. LT. 1.OE – 3) GO TO 11
    8  WRITE (6,1) C
    9  X = X + XSTART
   10  GO TO 4
   11  STOP
       END
```

6-8. (*a*) Prove by differentiation that Eq. (6-22) is a solution of Fick's second law. (*b*) What other conditions must this equation satisfy to be the proper solution for the experimental arrangement of Fig. 6-10*b*?

6-9. Determine the self-diffusion coefficient of gold by analyzing the experimental data of Fig. 6-10*c*. [NOTE: The value of M need not be known.]

6-10. (*a*) Explain the difference between an interstitial mechanism of diffusion (carbon in iron) and an interstitialcy mechanism. (*b*) Show by suitable sketches whether correlation will occur in either of these mechanisms.

6-11. (*a*) Using an analysis similar to that leading to Eq. (6-30), determine whether the data on self-diffusion of oxygen in CdO are consistent with a defect structure consisting of interstitial cadmium ions. (*b*) Explain how density measurements might be used to determine whether the defect in CdO is an oxygen vacancy or a cadmium interstitialcy.

6-12. (*a*) Using the data for copper in Fig. 4-4 and the Cu–Zn phase diagram (Fig. 3-16), estimate the ratio of the vacancy concentration in a 30 $^{A}/_{0}$ Zn alloy to that in pure copper. (*b*) What other factor, and of what magnitude, would you suggest to account for the factor of 20 in the tracer diffusion coefficients of the alloy?

6-13. Prove Eq. (6-38). Consider diffusion through an area of 1 m^2 for a time interval $\delta t = 1$ sec, and calculate the corresponding displacement δs.

6-14. Derive Eq. (6-40) from Eq. (6-39). Use the relations $J_1 = -J_2$ and $C_1 + C_2 = 0$.

6-15. Using Fick's first law in the form, $J_1 = -D_1 d(N_1/V)/dx$, where V is the molar volume (assumed constant), show that $D_1 = D_1^*(1 + d \ln \gamma_1/d \ln N_1)$ if both D_1 (intrinsic) and D_1^* (tracer) are measured at the same value of N_2. Use the fundamental equation, $J_1 = -B_1 N_1 (d\mu_1/dx)/V$, where B_1 is the *mobility*; $\mu_1 = \text{constant} + RT \ln (\gamma_1 N_1)$.

6-16. (*a*) What time of diffusion at 1200°C is required to obtain a boron concentration of 10^{24} at a depth of 8 μ? Begin with the deposited layer of Fig. 6-17. (*b*) What is the corresponding concentration at the surface?

6-17. Assuming that the grains are cubical for simplicity, estimate the grain size for which the diffusive flux along grain boundaries would be equal to the volume diffusion at 600°C in silver. In polycrystalline silver both contributions to diffusion must be considered in the range from about 500 to 700°C.

6-18. Using the method of separation of variables, solve Eq. (6-56), and obtain Eq. (6-57). [HINT: Multiply both sides of Eq. (6-56) by $dt/(c_{po} - c_p)$. Each side can now be integrated from $(c_p)_L$ at $t = 0$ to c_p at $t = t$.

6-19. Show that Eq. (6-63) is the steady-state solution of Eq. (6-62) for the conditions of Fig. 6-24. Since M_C, E, and D_C are constants, the two variables c_C and x can be separated. In the process of integration, if the integration constant is expressed as $\ln A$, the ratio c_C/A can be formed. After changing the equation from a logarithmic to an exponential form, you can evaluate the constant A by the integration

$$\int_0^l c_C \, dx = c_C^0 l$$

where c_C^0 is the initial uniform concentration and l is the length of the specimen.

6-20. Consider a semiconductor such as silicon in which the number of defect electrons increases with increasing temperature. When a temperature gradient is applied, the flux of phonons is from the hot to the cold end of the specimen. Show that a current of electrons also flows in this direction, as a combined result of diffusion in a concentration gradient and of thermodiffusion (assume that the mechanism of thermodiffusion involves collisions with phonons).

6-21. Using the index notation [see Eq. (1-7)], write the equivalent of Eqs. (6-73) and (6-74) for an n-component system. Only a single equation is needed.

REFERENCES

Adda, Y., and J. Philibert: "La Diffusion dans les Solides," 2 vols., Presses Universitaires, Paris, 1966.

Boltaks, B. I.: "Diffusion in Semiconductors," Academic Press Inc., New York, 1963.

Crank, J.: "The Mathematics of Diffusion," Oxford University Press, Fair Lawn, N.J., 1956.

——— and G. S. Park (eds.): "Diffusion in Polymers," Academic Press Inc., New York, 1968.

Doremus, R. H.: Diffusion in Non-crystalline Silicates, in "Modern Aspects of the Vitreous State," vol. 2, p. 1, 1962.

Grove, A. S.: "Physics and Technology of Semiconductor Devices," John Wiley & Sons, Inc., New York, 1967.

Jost, W.: "Diffusion in Solids, Liquids and Gases," Academic Press Inc., New York, 1960.

Kofstad, Per: "High-Temperature Oxidation of Metals," chap. 3, John Wiley & Sons, Inc., New York, 1966.

Kumins, C. A. (ed.): "Transport Phenomena in Polymeric Films," Interscience Publishers, New York, 1965.

Lazarus, D.: Intermetallic Diffusion, in "Energetics in Metallurgical Phenomena," vol. 1, p. 1, Gordon and Breach, Science Publishers, Inc., New York, 1965.

Manning, J. R.: "Diffusion Kinetics for Atoms in Crystals," D. Van Nostrand Company, Inc., Princeton, N.J., 1968.

Meares, P.: "Polymers: Structure and Bulk Properties," chap. 12, D. Van Nostrand Company, Inc., Princeton, N.J., 1965.

Shewmon, P. G.: "Diffusion in Solids," McGraw-Hill Book Company, New York, 1963.

Sundheim, B. S.: "Fused Salts," chap. 3, McGraw-Hill Book Company, New York, 1964.

Chapter Seven

Phase Transformations

Most materials are subjected to one or more phase transformations in the course of their manufacture. For example, steel is manufactured in the liquid state and undergoes a transformation to the solid state on cooling to room temperature. A later heat treatment, involving the formation of a different crystal structure, may be employed to increase strength as a result of a phase transformation in the solid state. The more crucial the properties of a material, the more likely it is that a phase transformation will be used in its preparation. A different aspect of phase transformations is their unwanted occurrence during some uses of materials. Few materials are employed in their equilibrium conditions, and therefore phase transformations of various kinds may occur at an appreciable rate. Undesirable phase transformations are especially likely to occur at high temperatures or in reactive environments.

Phase transformation is an extensive subject with many important technologies associated with it, for example, casting of metals, sintering of ceramics, heat treating of steels, and molding of polymers. A few common principles underlie the various technologies, and these principles are the focus of the present treatment. The development begins with vapors because of their simplicity and progresses to liquids and then to solids.

VAPOR → SOLID TRANSFORMATIONS

Transformations involving the vapor phase are interesting for two reasons. First, they are directly involved in many technological operations, in particular, in those of the semiconductor industry. Second, these transformations are more easily studied, both experimentally and theoretically, than are transformations in liquids and solids. Many of the approaches developed for the vapor reactions have been successfully adapted to treat the more involved behavior in condensed phases. The connection between a vapor-phase transformation and the corresponding equilibrium condition is especially evident and serves as a convenient starting point for this topic.

7-1. Condensation-Evaporation Equilibrium

The simple kinetic theory of gases gives a suitable picture of the equilibrium between a solid and its vapor at the (equilibrium) pressure P_e. Pressure is force F per unit area, and by Newton's second law,

$$F = \frac{d(mv)}{dt} \tag{7-1}$$

This equation permits us to determine the flux J (in molecules per square centimeter per second) corresponding to the pressure P_e (dynes per square centimeter). Using the average velocity given by simple kinetic theory, $v = (3kT/m)^{1/2}$ and noting that the change in momentum of a particle reflected from a wall is $+mv - (-mv) = 2mv$, we obtain the approximate relation

$$P_e = 2Jmv = 2J(3mkT)^{1/2} \tag{7-2}$$

A more rigorous derivation gives a slightly different numerical value, and thus the final expression for the flux of molecules is

$$J = \frac{P_e}{(2\pi mkT)^{1/2}} \tag{7-3}$$

Typical values of J are on the order of 10^{18} molecules/cm^2-sec. Under equilibrium conditions one flux of this magnitude J_v is leaving the surface of the solid (vaporizing), and a second flux J_c is condensing on the solid. Each molecule carries out its individual reaction unimpeded (see Prob. 7-1) since interference between the opposing fluxes becomes important only at very high fluxes (on the order of 10^{25}).

A difference ΔP between the actual pressure P and the equilibrium pressure P_e causes a net condensation or evaporation of atoms to occur. Under ideal conditions the rate is given directly by the analog of Eq. (7-3), but to allow for various factors that can decrease the rate under actual conditions, a coefficient α is added; thus

$$J_v = \frac{\alpha_v(P_e - P)}{(2\pi mkT)^{1/2}} \tag{7-4}$$

$$J_c = \frac{\alpha_c(P - P_e)}{(2\pi mkT)^{1/2}} \tag{7-5}$$

The evaporation coefficient α_v and the condensation coefficient α_c are usually determined experimentally, but they also have useful theoretical interpretations in terms of the mechanism of the corresponding transformation.

The same steps are involved in evaporation and in condensation, but let us first consider them in the order they occur during evaporation. The terrace-ledge-kink model (Fig. 5-7) suggests that a typical, clean surface has the structure shown in Fig. 7-1. An atom need not evaporate from a position within a terrace (Fig. 7-1*a*), or even from the less tightly bound position within a ledge (Fig. 7-1*b*). Instead, it can attain the position of an adsorbed atom on the surface (Fig. 7-1*c*) and then evaporate by the following four-stage process: (1) an atom at a kink position breaks free and begins moving along the ledge; (2) this atom breaks its connection to the ledge and becomes an adatom (adsorbed atom) on the surface; (3) the adatom diffuses on the surface; and (4) the adatom enters the vapor phase by the process of desorption.

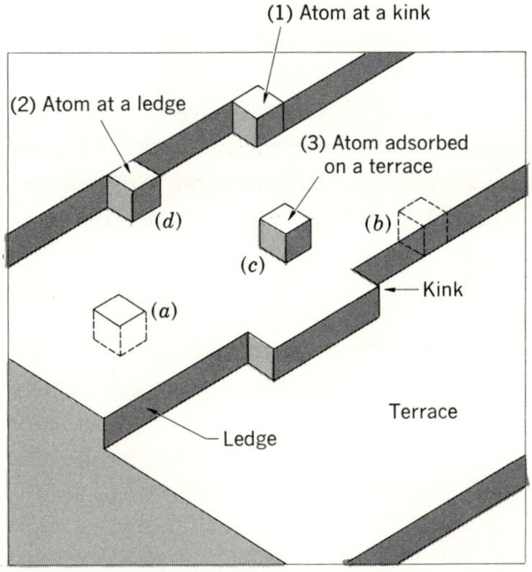

FIGURE 7-1 Schematic view of the terrace-ledge-kink surface of a typical solid. An atom progresses through stages 1 to 3 in the process of evaporation. (*After J. P. Hirth and G. M. Pound.*)

Except in the special case of a perfect surface without ledges, stage 4 is the rate-determining step. This fact can be appreciated more easily for the condensation process in which the stages occur in the reverse order. As the vapor flux strikes the surface, each atom is adsorbed and quickly becomes involved in the other stages, each of which is driven by only a small fraction of the overall driving force ΔP. A change in P, and thus a change in the vapor flux, directly influences the rate of adsorption and therefore the rate of condensation. This is a commonly observed behavior, and the coefficient α_c is then unity. In the unusual case in which no ledges exist, α_c can be essentially zero for small values of ΔP since no mechanism for attachment is then available and the adatoms must eventually desorb.

7-2. Vaporization

A major industrial use of evaporation of solid or liquid materials is to obtain a flux of atoms or molecules for deposition on some solid substrate. Almost any metal can be treated in this way, but silicon is an especially interesting example because of its importance in semiconductor technology. For reasonable rates of evaporation into a vacuum, the vapor pressure must be about 10^{-5} atm. Like most materials, silicon attains this vapor pressure only above its melting point (1410°C) and is usually heated to about 1550°C for efficient evaporation. The evaporation rate is then 7×10^{-5} g/cm²-sec. The substrate on which condensation occurs can be any kind of material and may be hot or cold.

A result equivalent to ordinary (physical) evaporation can be produced by chemical evaporation, in which a chemical vapor removes atoms from the surface of the material by a chemical reaction. An important example is the evaporation of molybdenum at high temperatures by the reaction

$$Mo(s) + \tfrac{3}{2}O_2(g) \rightarrow MoO_3(g) \qquad (7\text{-}6)$$

In this case, chemical evaporation causes serious loss of metal at temperatures where physical evaporation is negligible. Chemical evaporation of silicon can occur by the reaction

$$SiCl_4(g) + Si(s) \rightarrow 2SiCl_2(g) \qquad (7\text{-}7)$$

The rate of physical evaporation can be controlled in several ways. A high rate is easily obtained by using a high temperature and keeping α_v near unity, as discussed below. Often the more difficult problem is to minimize vaporization in a device that must operate at high temperatures. Usually the solution is to employ an impurity of some kind to lower α_v (by many orders of magnitude in some cases). The impurity may affect any of the four stages in the overall process of vaporization (Sec. 7-1) or may impose a slow diffusion step at the solid-vapor interface. An example is the use of nitrogen gas instead of a vacuum in an ordinary electric light bulb to impede vaporization of the tungsten filament. By this means α_v is changed from unity to about 10^{-3}. An unusual suppression of vaporization occurs in a few materials, such as phosphorus, of whose vapor consists polymers, P_4. The desorption step is then difficult, and α_v may be as low as 10^{-4}. Dissociation may cause a similar effect, as for Al_2O_3, whose vapor includes Al, O, AlO, Al_2O and Al_2O_3, and for which α_v is about 10^{-4}.

7-3. Condensation

Transformation of a vapor to a solid can have two different aspects. If the driving force ΔP is sufficiently high, the condensation coefficient α_c is unity in Eq. (7-5), and the transformation appears to be simply the addition of atoms to the surface of the solid. For small values of ΔP, however, an important new process becomes evident, that is, the nucleation of the new phase on a substrate. When gold vapor is only slightly supersaturated, its initial rate of deposition on a flat crystal of MoS_2 is very slow (Fig. 7-2*a*).

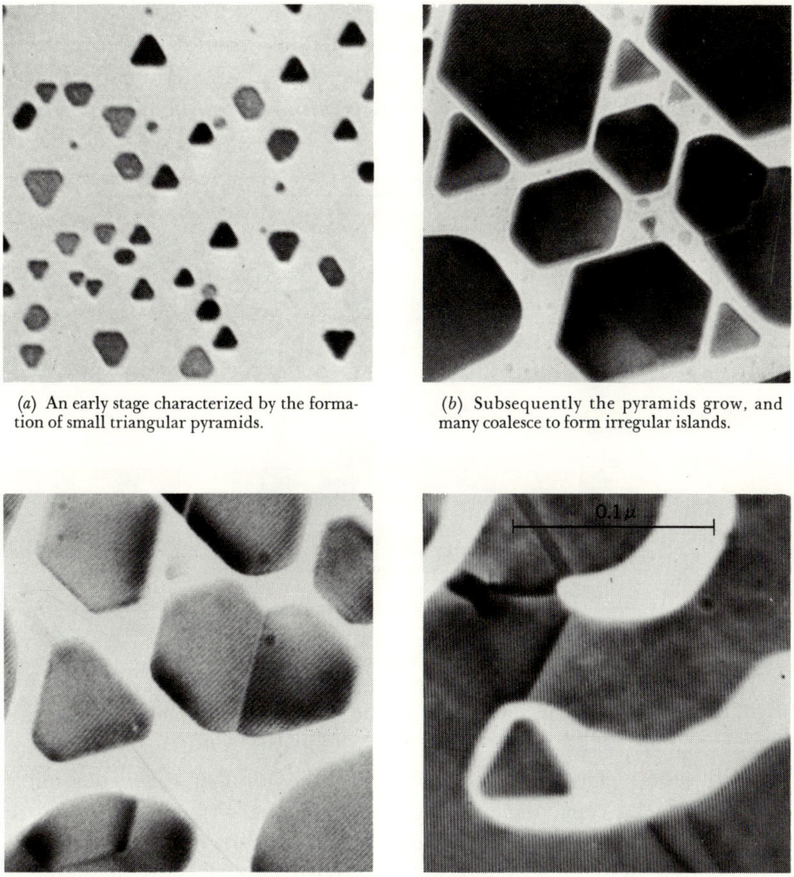

(*a*) An early stage characterized by the formation of small triangular pyramids.

(*b*) Subsequently the pyramids grow, and many coalesce to form irregular islands.

(*c*) The continuing process of coalescence leaves channels between the islands.

(*d*) The final stage is the closure of a few remaining channels.

FIGURE 7-2 Stages in the formation of a continuous film of gold on an MoS_2 substrate by condensation of gold vapor. (*Courtesy D. W. Pashley, Tube Investments Research Laboratory.*)

Individual atoms reevaporate, and small aggregates tend to dissociate. Eventually an aggregate of a critical size forms and serves as the nucleus to which additional atoms can attach themselves. At an intermediate stage (Fig. 7-2*c*), the substrate is partially covered by "islands" of solidified gold, but only at a later time has further deposition led to a continuous film of gold covering the substrate. This film has the useful characteristic of being a single crystal with the same orientation as the single crystal of MoS_2 on which it was deposited. This behavior, *epitaxial growth*, occurs because each initial nucleus forms in a low-energy configuration with a (111) plane parallel to the substrate and with a quite regular orientation in this plane. Consequently, when two islands coalesce, the new island will be entirely one orientation, though usually containing dislocations.

With increase in the driving force for the vapor → solid transformation of a pure substance, the character of the nucleation step changes, as shown schematically in Fig. 7-3. The factors determining these changes can be under-

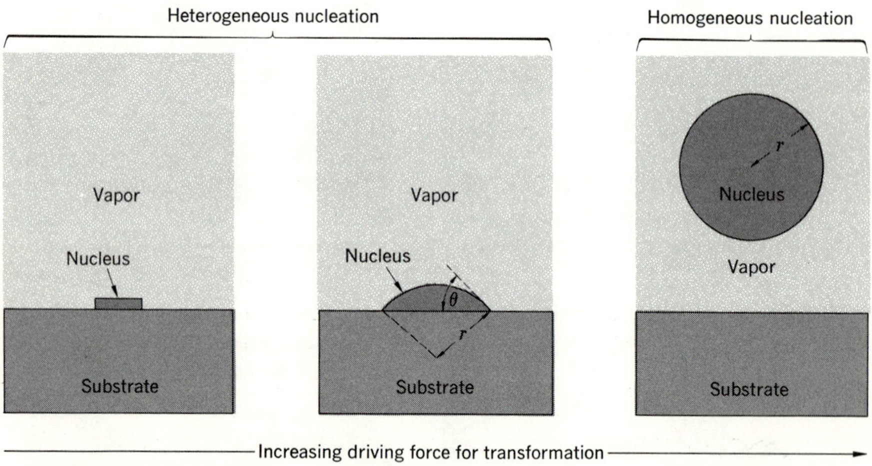

(*a*) Formation of two-dimensional nucleus.
(*b*) Formation of cap-shaped nucleus.
(*c*) Formation of spherical nucleus.

FIGURE 7-3 Schematic illustration of the change in character of nucleation with increase in the driving force for vapor → solid transformation.

stood by considering the most common behavior, heterogeneous nucleation involving a cap-type nucleus (Fig. 7-3*b*). A quantitative expression for the driving force can be obtained from Eq. (2-20) since pressure can be substituted for activity. The change in molar free energy accompanying the change (at temperature T) from a vapor at pressure P to a solid of vapor pressure P_e is

$$\Delta G = RT \ln P_e - RT \ln P = RT \ln \frac{P_e}{P} \tag{7-8}$$

To emphasize the essentially negative character of the change in free energy, this equation can be written[1]

$$\Delta G_v = -\frac{RT}{V} \ln \frac{P}{P_e} \tag{7-9}$$

ΔG_v is the driving force per unit volume, V is the molar volume of the solid, and P/P_e is the supersaturation ratio. Before nucleation has occurred, the atoms exist on the substrate not only as isolated atoms, but also as aggregates of various sizes known as embryos. The larger the volume of the embryo, the greater is the bulk driving force ΔG_b tending to cause transformation to the solid phase. In the case of a hemispherical cap (Fig. 7-3*b*), the expression for the driving force is

$$\Delta G_b = \tfrac{4}{3}\pi r^3 S \, \Delta G_v \tag{7-10}$$

S is a shape factor that depends on the equilibrium contact angle θ,

$$S = \frac{(2 + \cos\theta)(1 - \cos\theta)^2}{4} \tag{7-11}$$

A second type of energy that must be considered in the formation of a nucleus of the solid phase is the interfacial energy γ of unit area of the vapor-solid interface. In the case of homogeneous nucleation (Fig. 7-3*c*) a simple calculation shows the variation with radius r of the relative contributions of γ and ΔG_v to the total change in free energy for the formation of embryos of various sizes (Prob. 7-2). Above a critical size r^*, the contribution from ΔG_v is larger, and thus embryos of the critical size can grow spontaneously and serve as nuclei for the phase transformation. A large number of atoms must agglomerate to form the critical nucleus for homogeneous nucleation, and therefore this process requires a large supersaturation ratio. Except in a few special situations, heterogeneous nucleation occurs before the supersaturation can reach the high value needed for homogeneous nucleation. The crucial factor is the interfacial energy γ_{sv} between the substrate and the vapor. By analogy with Fig. 5-23, a large value of γ_{sv} corresponds to a small value of θ and partially compensates for the energy used to form the interface between the embryo and the vapor. The surface energy of a cap-shaped embryo can be written as

$$\Delta G_s = 4\pi r^2 S \gamma \tag{7-12}$$

[1] Strictly, the value of P_e depends on the radius of the particle of solid [see Eq. (6-45)]. A treatment considering this factor is given by M. E. Fine, "Introduction to Phase Transformations in Condensed Systems," p. 7, The Macmillan Company, New York, 1964.

where the shape factor is given by Eq. (7-11). The total energy of an embryo, ΔG, is then the sum of Eqs. (7-10) and (7-12) and varies with r in the manner shown by the middle curve in Fig. 7-4. At a critical size given by

$$r^* = -\frac{2\gamma}{\Delta G_v} \tag{7-13}$$

ΔG passes through a maximum value

$$\Delta G^* = \frac{16\pi\gamma^3 S}{3\,\Delta G_v^2} \tag{7-14}$$

The addition of atoms to an embryo of critical size causes a decrease in total free energy; therefore such an embryo is a nucleus that can grow spontaneously.

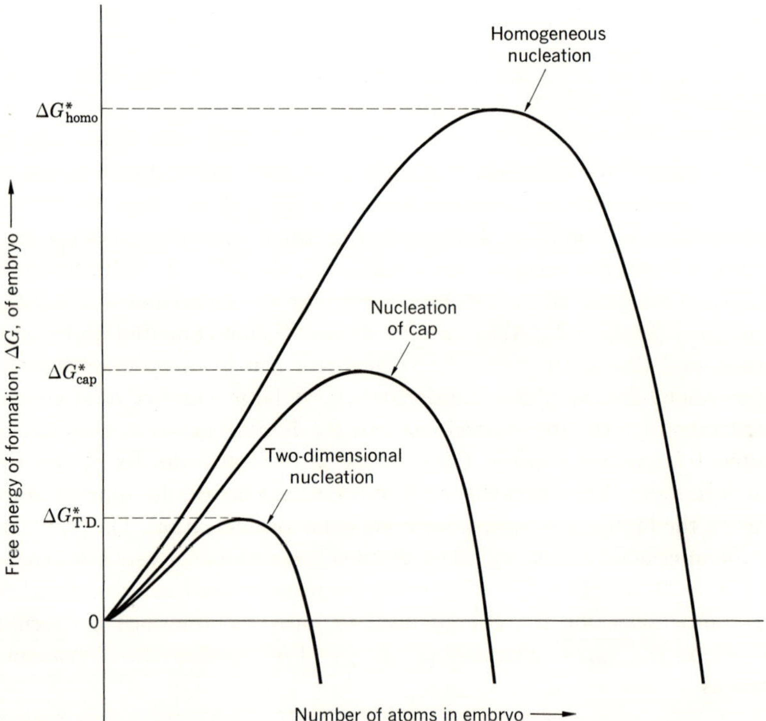

FIGURE 7-4 Schematic illustration of the free energy ΔG^* for the formation of a critical embryo (nucleus) for three types of nucleation in vapor → solid transformations.

The rate of nucleus formation, J^*, is given by the product

$$J^* = \omega n^* \tag{7-15}$$

where n^* is the equilibrium concentration of critical embryos and ω is the number of atoms per second added to a given embryo, usually by surface diffusion. n^* is given by the exponential function

$$n^* = ne^{-\Delta G^*/kT} \tag{7-16}$$

where n is the concentration of atoms in the vapor phase. Because of the presence of the factor ΔG_v^2 in the expression for ΔG^*, n^* varies extremely sharply with change in ΔG_v at a given temperature. Consequently, slightly below a critical supersaturation ratio $(P/P_e)^*$ nucleation does not occur, whereas slightly above this ratio nucleation occurs instantaneously. Although ΔG^* is lowest for two-dimensional nucleation, the rate of this type of transformation is slow because of the necessity for achieving special orientation relations between the embryos and the substrate. The temperature of the substrate must be relatively high to give the necessary atomic mobility, and P/P_e must be between the critical values for two-dimensional nucleation and for nucleation of a cap. When the value $(P/P_e)^*_{\text{cap}}$ is exceeded, the rapid process of forming cap-shaped nuclei becomes dominant.

The above treatment of condensation of physical vapors also applies in large part to the related technique of *chemical vapor deposition*, the deposition of a material from a chemical compound. The essential difference between the two procedures can be illustrated for the deposition of silicon. The physical evaporation of silicon requires that the heat of evaporation be supplied. The resulting silicon vapor is then very reactive and must be deposited immediately, often on a substrate at room temperature. Chemical evaporation of silicon as a stable compound, such as $SiCl_4$, is a spontaneous reaction that occurs at relatively low temperatures. The product of evaporation can be stored and conveniently manipulated in the vapor phase. The deposition reaction occurs when a mixture of the compound with hydrogen is heated to a sufficiently high temperature to activate the reaction

$$SiCl_4(g) + 2H_2(g) \rightarrow Si(s) + 4HCl(g) \tag{7-17}$$

The rate of reaction is determined, not only by the temperature, but also by the chemical composition of the vapor (Fig. 7-5). Increasing the concentration of the compound in the vapor phase causes an initial increase in the rate of deposition, but at large concentrations chemical evaporation can occur in this system because of the reaction of Eq. (7-7). Chemical vapor deposition is an especially flexible process because several compounds can be employed simultaneously to produce an alloy layer (doped silicon, for example); alternatively, a compound such as a carbide or oxide can be deposited directly.

Both the crystal structure and the morphology of vapor-deposited materials vary over a wide range. Low-temperature deposits may be amorphous or may consist of small, imperfect grains. At higher temperatures the deposits often consist of oriented crystals, sometimes forming a columnar pattern. Especially at low temperatures, the growing crystals may assume a filamentary form, often in the shape of *whiskers* that may be single crystals. Only a few of the factors determining the morphology of vapor deposits are understood at present. In some instances dislocations are responsible for the growth of whiskers, but other processes have also been identified. An interesting one is the vapor-liquid-solid mechanism involving a small drop of liquid on the surface of the substrate. In the case of silicon (Fig. 7-6) a small amount of gold forms a liquid alloy with silicon. Silicon atoms from the $SiCl_4$ vapor enter the liquid region at a high rate (Fig. 7-6*a*) and build up a high supersaturation. Deposition on the substrate can then occur by a mechanism

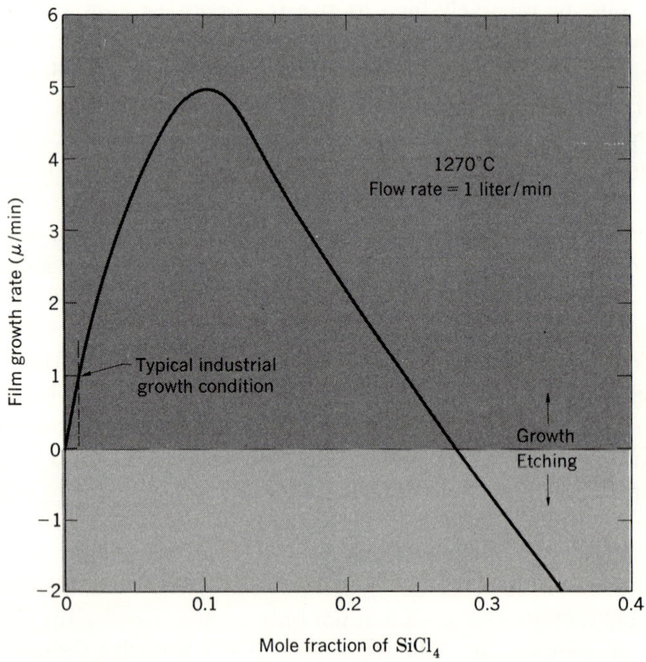

FIGURE 7-5 The effect of concentration of $SiCl_4$ on the rate of deposition or chemical vaporization of silicon. (*By permission from A. S. Grove, "Physics and Technology of Semiconductor Devices," John Wiley & Sons, Inc., New York, 1967.*)

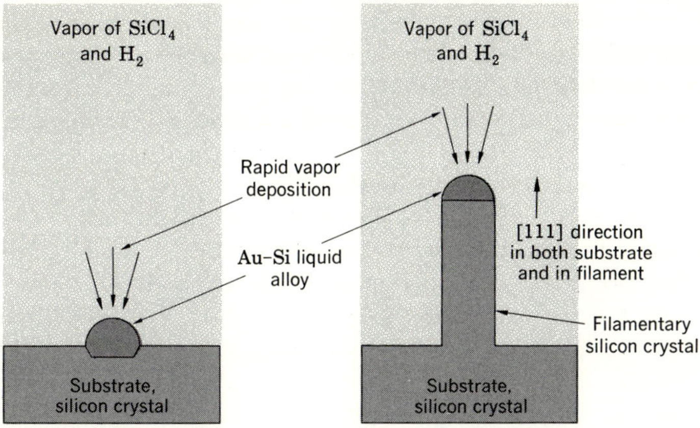

(*a*) Initial formation of liquid-alloy layer on substrate crystal.

(*b*) Typical stage in the growth of a filamentary crystal.

FIGURE 7-6 Schematic illustration of the V-L-S mechanism of filamentary growth. This example shows the growth of silicon by chemical vapor deposition. (*After R. S. Wagner and W. C. Ellis.*)

(uniform attachment) that does not require the presence of a dislocation. The liquid drop is lifted by the growing whisker (Fig. 7-6*b*), and thus the process is able to continue.

LIQUID → SOLID TRANSFORMATIONS

Many materials, but especially metals and alloys, undergo a liquid → solid transformation in the course of their manufacture. Ordinarily the aim is to produce the material in a finished or in a semifinished form. A significant fraction of the production of metals and alloys is in the form of finished castings produced by solidification of the liquid alloy in molds usually made of sand or metal. Other finished products contain "cast" structures where molten welding or brazing alloys have solidified. Most metallic materials are used in the wrought condition, produced by rolling, extrusion, etc., but generally a large casting (ingot) is an intermediate stage in the process. Each of the commercial processes involving a liquid → solid transformation has a specialized technology and certain factors of special importance, but the following treatment considers those aspects of the transformation common to a wide variety of applications.

7-4. Nucleation

Solidification of a liquid, like condensation of a vapor, requires the formation of nuclei and their subsequent growth. We begin with the simple case of homogeneous nucleation of a pure substance and consider the process of forming spherical solid embryos within the body of the liquid. If the temperature T is below the equilibrium freezing temperature T_e, then the change in bulk free energy accompanying the solidification of 1 cm^3 of liquid is

$$\Delta G_v = G^s - G^l = (H^l - H^s) - T(S^s - S^l) \tag{7-18}$$

G^s and G^l are the free energies per cubic centimeter of the solid and liquid phases. Since $H^s - H^l \approx -\Delta H_f$, the heat of fusion, and

$$S^s - S^l \approx -\Delta S_f \approx -\frac{\Delta H_f}{T_e} \tag{7-19}$$

Eq. (7-18) can be written

$$\Delta G_v = -\Delta H_f \frac{\Delta T}{T_e} \tag{7-20}$$

ΔT is the undercooling in degrees Celsius. An interfacial free energy γ is associated with 1 cm^2 of interface separating the embryo from the liquid, and therefore the total change in free energy accompanying the formation of an embryo of radius r is

$$\Delta G = -\Delta H_f \frac{\Delta T}{T_e} \frac{4}{3} \pi r^3 + 4\pi r^2 \gamma \tag{7-21}$$

The critical values r^* and ΔG^*, corresponding to the formation of a nucleus that can grow spontaneously, are given by Eqs. (7-13) and (7-14) with $S = 1$ for a spherical shape. By considering the rate of transfer of atoms from the liquid to the solid across the interface of the nucleus, a process characterized by the activation energy ΔG^A, the following type of expression is obtained for the rate of homogeneous nucleation:

$$J^* = K_v e^{-\Delta G^A/kT} e^{-\Delta G^*/kT} \tag{7-22}$$

Although the factor K_v can be estimated, its complete determination is still the subject of research.

As in the corresponding case of Eq. (7-15), J^* is an extremely sensitive function of ΔG_v [Eq. (7-20)], and homogeneous nucleation can be expected to occur when a specific, large value of supercooling, ΔT, is achieved. Even in pure materials, however, many sites for heterogeneous nucleation exist in the liquid and produce solidification at only slight supercooling. If liquid mercury is finely dispersed in a suitable medium, the sources of heterogeneous nucleation are present in only a small fraction of the droplets, which then solidify first; the remaining portion of the mercury solidifies at a much greater supercooling, $\Delta T = 77°C$. The observed supercooling ΔT for homogeneous nucleation leads to a value of the interfacial energy γ, in agreement with other methods. If this value is used in Eq. (7-13), the radius r^* of the critical nucleus can be calculated as a function of the degree of supercooling. The data for copper (Fig. 7-7) are typical in showing that the critical radius increases rapidly in size as ΔT decreases.

Nucleation of liquid → solid transformations is always heterogeneous in practice. Epitaxial nucleation on a substrate can occur in special cases, such

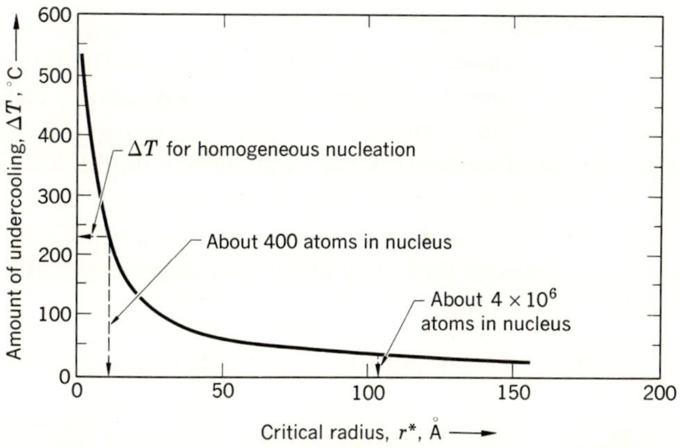

FIGURE 7-7 Variation of the critical radius for homogeneous nucleation of copper with degree of supercooling ΔT. (*Adapted from Bruce Chalmers, "Principles of Solidification," John Wiley & Sons, Inc., New York, 1964.*)

as a solute precipitating from solution, but is of minor importance. The usual behavior, as in the solidification of metallic castings, involves random formation of cap-shaped nuclei. The theory is analogous to that for heterogeneous vapor → solid transformations. The effect of a substrate (mold walls, impurity particles, etc.) is added to the theory of homogeneous nucleation, using Eqs. (7-10) to (7-14). The final theory then predicts that only a degree or two of supercooling is required for nucleation. An important consequence of heterogeneous nucleation is in providing a method for control of the grain size of the solidified casting. A fine grain size is usually desirable and can be obtained by supplying many sites for heterogeneous nucleation. For example, if only 0.01 percent titanium is present in a molten aluminum alloy, many small particles of TiC form prior to the main solidification and serve as the substrate on which the embryos can form. The result is a casting with a much finer grain size than would be possible without the addition of titanium.

7-5. Growth of Pure Substances

The nucleation step is followed by growth of the nuclei in the complete process of solidification of a liquid. Alloying elements play a crucial role in actual growth mechanisms, but their effects can be better appreciated when the essential features of growth in a pure substance are first understood. The case of the pure substance is simpler because only a single solid phase forms, and no changes in composition can occur.

The entropy of fusion [Eq. (7-19)] is a basic factor in determining the manner of attachment of atoms to the solid phase at the liquid-solid interface. The ratio $\Delta S_f/k = \Delta H_f/kT_e$ is the quantity usually considered. If the entropy change is small enough to allow $\Delta H_f/kT_e$ to be about unity, as in the case of metals and some other substances, the interface between the liquid and solid is not a sharp demarcation but is spread over a few atomic distances (Fig. 7-8*a*). Growth occurs continuously at all positions along the interface and produces a planar interface between the solid and the liquid. The interface in the transparent compound carbon tetrabromide, for which $\Delta H_f/kT_e$ is 0.8, is shown in Fig. 7-9*a*. Presumably the liquid-solid interface in a pure metal is similar. In this case, theory predicts that the rate of growth, g, is

$$g = k\,\Delta T \tag{7-23}$$

with a large reaction-rate constant k (Fig. 7-10). The rate of growth is so high

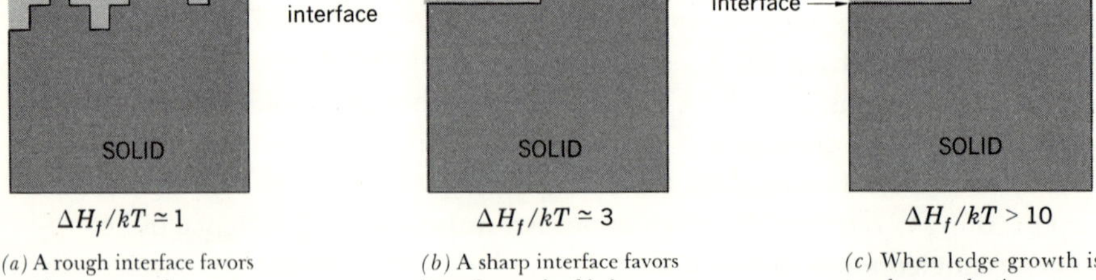

$\Delta H_f/kT \simeq 1$

(*a*) A rough interface favors continuous growth.

$\Delta H_f/kT \simeq 3$

(*b*) A sharp interface favors lateral growth of ledges.

$\Delta H_f/kT > 10$

(*c*) When ledge growth is too slow, nucleation can occur in advance of the growing interface.

FIGURE 7-8 Schematic illustration of the division of growth into three classes on the basis of an entropy of fusion factor.

(*a*) Planar interface in carbon tetrabromide. $\Delta H_f/kT_e = 0.8$.

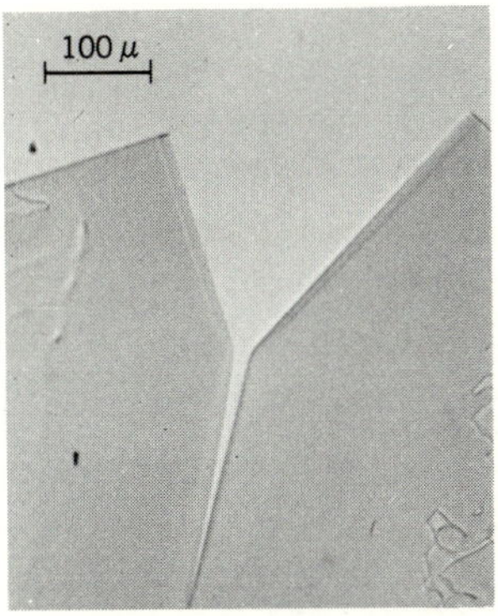

(*b*) Faceted interface in *t*-butyl alcohol. $\Delta H_f/kT_e = 2.6$.

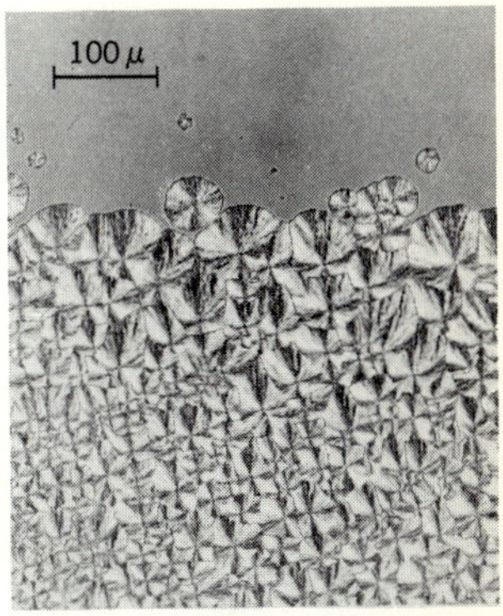

(*c*) Spherulitic growth with repeated nucleation in tristearin. $\Delta H_f/kT_e = 63$.

FIGURE 7-9 Photomicrographs illustrating the characteristic growth of the three classes of substances described in Fig. 7-8. The three organic compounds shown here form transparent crystals that can be conveniently photographed during solidification. (*Courtesy K. A. Jackson, Bell Telephone Laboratories.*)

for metallic systems that the rate of solidification of a casting is controlled by by the rate at which heat is removed from the system.

Substances having intermediate values of $\Delta H_f/kT_e$ include the semimetals, such as silicon and germanium, and many inorganic and organic compounds. Because of the relatively high entropy, a small number of atoms participates in the attachment process at a given location, and consequently the interface is sharp (Fig. 7-8*b*). Ledges and kinks in the surface account for most of the attachment, and thus growth consists largely of the lateral motion of monatomic ledges across surfaces with low Miller indices, (*hkl*). Since these crystallographic surfaces do not coincide, in general, with the plane favored for crystallization by the temperature gradient, the actual liquid-solid interface is facetted (Fig. 7-9*b*) along a few low-index faces. The theoretical growth law in this case (assuming the presence of screw dislocations) is

$$g = k'(\Delta T)^2 \tag{7-24}$$

The reaction-rate constant k' is distinctly lower than the constant for continuous growth (Fig. 7-10). For the crystallization of polymers and other complex substances, $\Delta H_f/kT_e$ has such high values that not only facetted growth but also competitive nucleation occur (Fig. 7-8*c*). The reaction-rate constant k' is low enough to allow supercooling to attain values in front of the liquid-solid interface that result in repeated nucleation (Fig. 7-9*c*). The overall solidification process is then largely determined by the rate of nucleation rather than by the rate of growth.

Experimental observations of growth behavior suggest that the degree of undercooling ΔT may affect the mode of growth. At low values of ΔT the interface should be sharp and grow at the rate given by Eq. (7-24) provided that defects are present. If they are not, the growth rate is much slower. With increase in ΔT, the rate passes through a transition stage and then follows Eq. (7-23) at suitably high values. At large values of ΔT, molecules can be added readily at any site on the interface; growth is not confined to ledges and kinks.

A significant factor, especially in the solidification of metals, is whether the liquid-solid interface advances as a smooth plane (Fig. 7-9*a*) or becomes

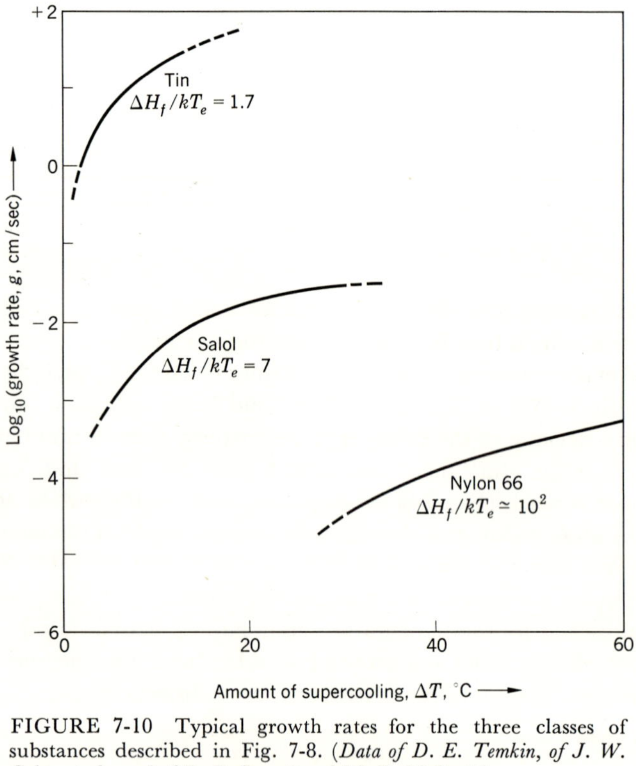

FIGURE 7-10 Typical growth rates for the three classes of substances described in Fig. 7-8. (*Data of D. E. Temkin, of J. W. Cahn, et al., and of B. B. Burnett and W. F. McDevit.*)

unstable and grows as a collection of filaments (Fig. 7-12*a*). Although alloying elements play a dominant role in determining this behavior (Sec. 7-6), the variation in amount of undercooling ΔT is the essential criterion. Let us first consider the effect of temperature gradients[1] in the solid phase, G^s, and in the liquid phase, G^l (Fig. 7-11). If heat is continuously supplied to the liquid

[1] The symbol G, for temperature gradient, widely used in the literature, should not be confused with the free energy.

phase, the equilibrium melting temperature T_e could be maintained at the liquid-solid interface (Fig. 7-11*a*), and the interface would be stationary. The temperature gradients would simply maintain equal heat flows

$$K^sG^s = K^lG^l \tag{7-25}$$

through the two phases; K^s and K^l are the thermal conductivities. Under the

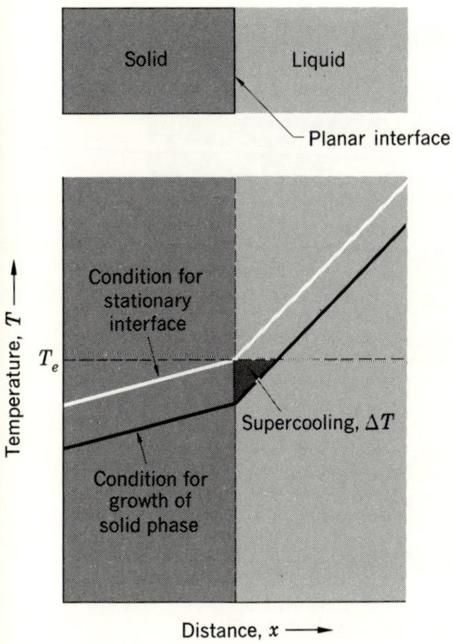

(*a*) A positive temperature gradient, the type encountered in most metal-casting operations.

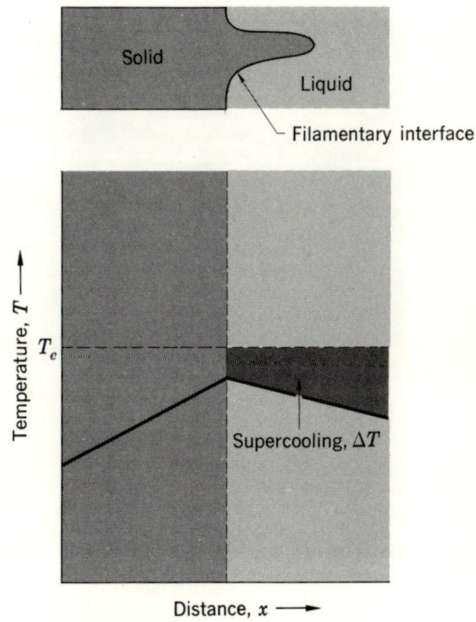

(*b*) A negative temperature gradient, which occurs only under special conditions.

FIGURE 7-11 Two types of temperature gradients G^l in the liquid phase during the solidification of a pure substance. (*After W. C. Winegard, "An Introduction to the Solidification of Metals," Institute of Metals, London, 1964.*)

usual conditions of casting, heat flow through the solid is greater, and solidification can occur at a rate R such that

$$K^sG^s = K^lG^l + R\,\Delta H_f \tag{7-26}$$

That is, the rate of solidification is controlled by the rate at which the latent heat of fusion can be conducted away. Only a small part δT of the total amount of supercooling ΔT is available for driving the liquid → solid transformation. The major part is employed in the mechanism of heat transfer. When random fluctuations produce a small protuberance on the interface into the liquid phase, ΔT is decreased at this location, as shown in Fig. 7-11*a*. The protuberance grows more slowly than the balance of the interface and soon disappears. Thus, a planar interface is stable in the presence of a positive gradient G^l.

Under some circumstances the temperature gradient in the liquid is negative (Fig. 7-11*b*). When a random protuberance arises in this case, it extends into a portion of the liquid where ΔT is larger, facilitating both dissipation of heat and crystal growth. A filamentary growth of some kind can then occur, often in the form of a branched dendrite (Fig. 7-12*a*) that propagates

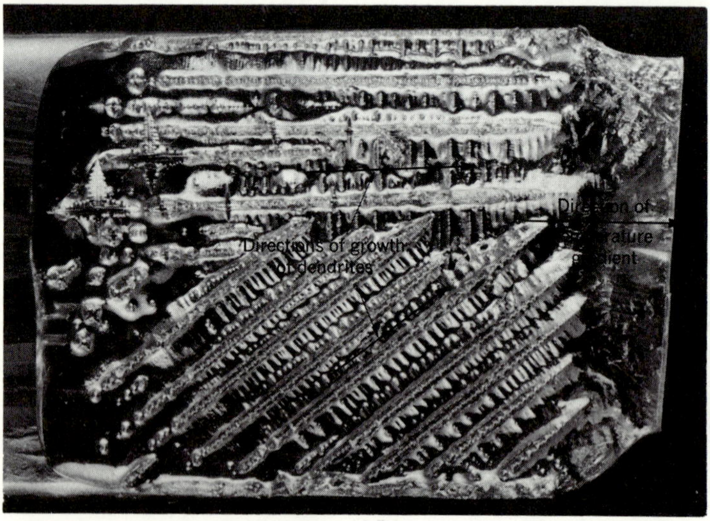

(*a*) Dendritic structure formed during the solidification of molten lead. (By permission from Bruce Chalmers, *Principles of Solidification,* John Wiley and Sons, New York, 1964.)

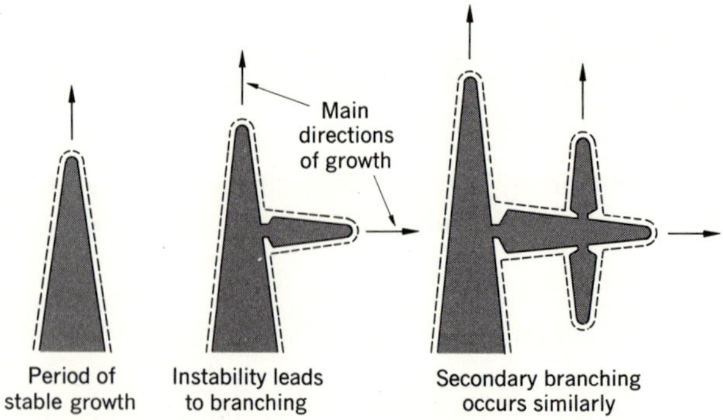

(*b*) The occurrence of periodic branching during dendritic growth. The constriction at the base of a branch facilitates detachment of branches. (after G. S. Cole) .

FIGURE 7-12 Illustration of the dendritic mode of solidification.

rapidly in advance of the usual liquid-solid interface. Although dendrites grow preferentially in the direction of the temperature gradient, their growth axis is a crystallographic axis [(100) for cubic metals, for example]. Branching of dendrites (Fig. 7-12*b*) is a result of the instability of a relatively flat interface. When a spherical protuberance is small, its sharply curving surface is stable against secondary fluctuations. As growth continues and portions of the surface become less curved, side branches can form and grow.

7-6. Growth from Liquid Solutions

Most commercial materials are alloys rather than pure substances, and thus new factors enter into their liquid → solid transformations. If only a single solid phase is produced in the transformation, then the principal factor is the distribution of the solute (alloying element) between the liquid and solid solutions. When two solid phases are being formed, the morphologies adopted depend on additional parameters including surface energies and rates of diffusion.

FORMATION OF A SOLID SOLUTION. During equilibrium solidification, considered in Sec. 2-3, a liquid solution of initial composition C_0 would follow

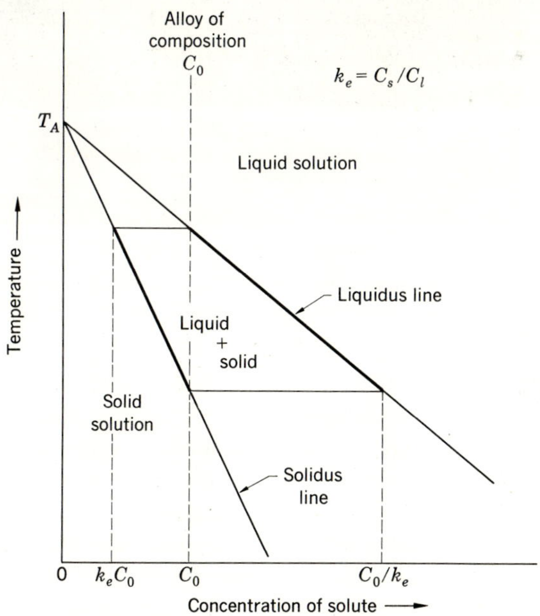

FIGURE 7-13 Portion of a phase diagram for which $k_e < 1$ for equilibrium solidification. The liquidus and solidus curves are approximated by straight lines.

the liquidus line (Fig. 7-13), and the composition of the solid phase would follow the solidus line. The final result would be a uniform solid of composition C_0 (Fig. 7-14) since sufficient time is allowed for the necessary diffusion in the solid phase. Under actual conditions of solidification, diffusion in the solid is often negligibly small, but the liquid phase may be kept uniform by suitable mixing. This case is considered in Sec. 2-12; Eq. (2-65) leads to the concentration distribution shown in Fig. 7-14. The most common conditions of solidification are approximated by the following assumptions: (1) negligible diffusion in the solid, (2) partial diffusion in the liquid (but no other type of mixing), and (3) equilibrium maintained at the liquid-solid interface. Consider first the major, steady-state portion of the solidification curve shown in Fig. 7-14. Since the net result of solidification during this stage is the disappearance of liquid of composition C_0 and the appearance of an equal amount of solid of the same composition, the composition of the liquid at the interface must be $C^l = C_0/k_e$. This composition is the maximum point on a steady-state

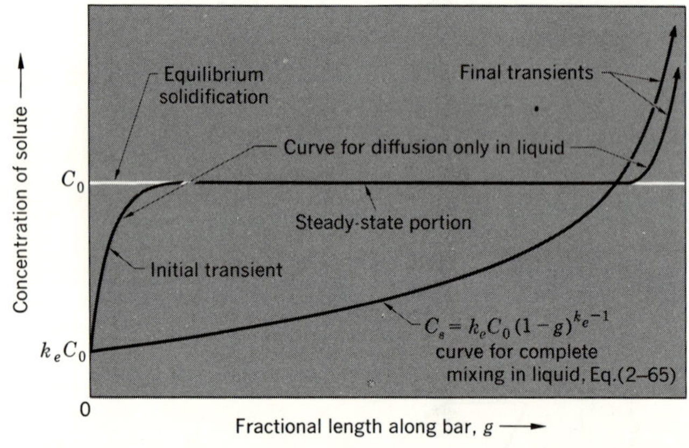

FIGURE 7-14 Concentration distribution along the length of a bar solidified from a liquid of composition C_0 under three different conditions.

concentration profile (Fig. 7-15) moving with the liquid-solid interface. The equation describing this profile can be derived using a coordinate system x' moving with the profile. The net accumulation in any volume element dx' must be zero at the steady state. One portion of the net accumulation, the diffusive flux, is given by Fick's second law as $D(d^2C/dx'^2)$. The second portion arises as follows. If the liquid-solid interface is advancing at the rate R cm/sec, a flux of solute for solidification $-RC$, can be pictured as flowing past an interface whose composition is C. A derivation similar to that employed for Fick's second law (Prob. 7-4) shows that the accumulation in dx' is $R(dC/dx')$. The net accumulation is then

$$D\frac{d^2C}{dx'^2} + R\frac{dC}{dx'} = 0 \tag{7-27}$$

The solution of this equation (Prob. 7-5) is the desired expression for the solute concentration C_l in the liquid phase in front of the interface; thus

$$C^l = C_o\left(1 + \frac{1 - k_e}{k_e} e^{-Rx'/D}\right) \tag{7-28}$$

The ratio D/R determines the length of the enriched zone since C_l decreases by $1/e$ below its peak value at $x' = D/R$. The final transient (Fig. 7-14) occurs over a similar length, but the initial transient is longer by the factor $1/k_e$. The above analysis is valid also for $k_e > 1$ (Prob. 7-6).

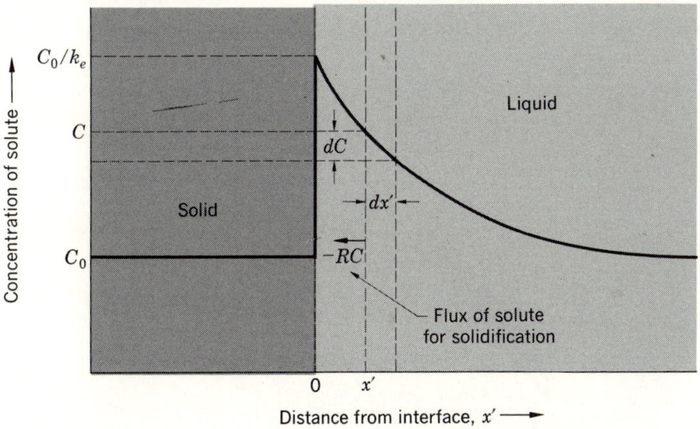

FIGURE 7-15 Steady-state concentration profile extending into the liquid phase from the solid-liquid interface. Solidification is occurring at the rate R cm/s.

The zone of enriched solute can cause the liquid phase to be supercooled even though the temperature gradient is positive, a condition known as *constitutional supercooling*. A quantitative expression for the increase in freezing temperature with increase in x' (Fig. 7-16) is obtained from the linear approximation to the liquidus curve of Fig. 7-13,

$$T^l = T_A + mC_l \tag{7-29}$$

Here m is the slope of the liquidus curve and is negative for $k_e < 1$. Substitution of the expression for C_l [Eq. (7-28)] gives the curve

$$T^l = T_A + mC_o\left(1 + \frac{1 - k_e}{k_e} e^{-Rx'/D}\right) \tag{7-30}$$

an example of which is sketched in Fig. 7-16. Provided that the actual temperature gradient G_l is less than a critical value

$$\frac{G^l}{R} = -\frac{mC_0}{D}\left(\frac{1-k_e}{k_e}\right) \tag{7-31}$$

a region of constitutional supercooling exists adjacent to the liquid-solid interface.

Unlike a pure metal, which solidifies with a planar interface in a positive temperature gradient, a solid solution develops two types of instabilities as a result of constitutional supercooling—cells and dendrites. When the amount of supercooling is small, random projections on the surface can develop only a short distance ahead of the main interface (Fig. 7-17*a*). A zone enriched in solute develops not only in front of the growing filaments but also

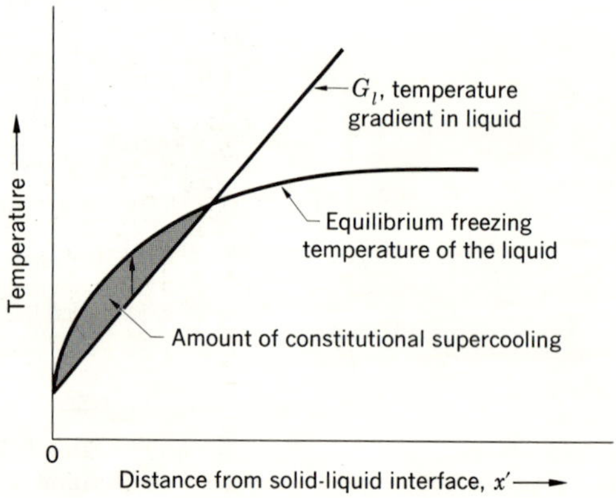

FIGURE 7-16 Relation of actual temperature in the liquid to its equilibrium freezing temperature, giving rise to a region in which constitutional supercooling exists.

at their sides, creating a layer rich in solute in the boundary between neighboring cells. Since the cells are usually hexagonal in cross section, cellular solidification results in a segregation of the solute in a grid pattern, extended in the direction of solidification. With increase in amount of constitutional supercooling, cell formation gradually changes into dendritic growth (Fig. 7-17*b*) roughly similar to that for pure metals in a negative temperature gradient. Significant differences are the enrichment of the liquid phase, the occurrence of diffusion, and the limited region of supercooling; these factors tend to limit the amount of dendrite formation. The liquid enriched in solute accumulates in the interdendritic areas and solidifies there by addition to the dendritic branches. As a consequence, the segregated regions are distributed throughout the dendrite, except for the main axis and branches.

One of the factors neglected in the above treatment is thermal convection in the liquid phase, which is known to affect the distribution of solute. Thermal convection also contributes to the formation of an equiaxed grain structure, the type of structure desired in most castings. The flow of liquid produced by thermal convection can sweep suitable embryos into the region of supercooling, where growth can then occur. Fluid flow may also dislodge portions of existing dendrites by partial melting and make them the nuclei for new grains. Fluid motion can also be caused by such mechanical means as

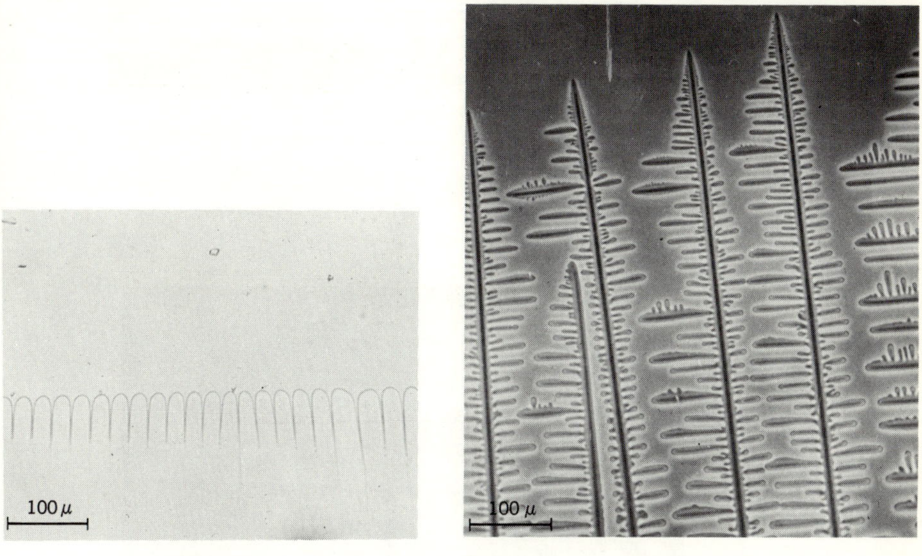

(a) Cellular growth. *(b)* Dendritic growth.

FIGURE 7-17 Illustrations of cellular and dendritic growth in carbon tetrabromide containing an impurity. (*Courtesy K. A. Jackson, Bell Telephone Laboratories.*)

electromagnetic stirring or simple mechanical oscillation of a solidifying casting. The resulting effective grain-refining action is attributable largely to fragmentation of existing filamentary grains.

ZONE MELTING. In this method, only a small zone of length l need be molten at any one time (Fig. 7-18*a*). This simple concept of zone melting has led to important advances in such areas as purification, uniform distribution of solute, preparation of eutectic and peritectic alloys, determination of portions of phase diagrams, and preparation of single crystals. The essential features of the method can be understood by considering its application to purification, the subject of *zone refining*. The effective distribution coefficient k is defined as

$$k = \frac{C^s}{C^l} \tag{7-32}$$

where C^s is the composition of the solid being formed, and C^l is the *average* composition of all of the liquid in the molten zone. Enrichment of solute at the

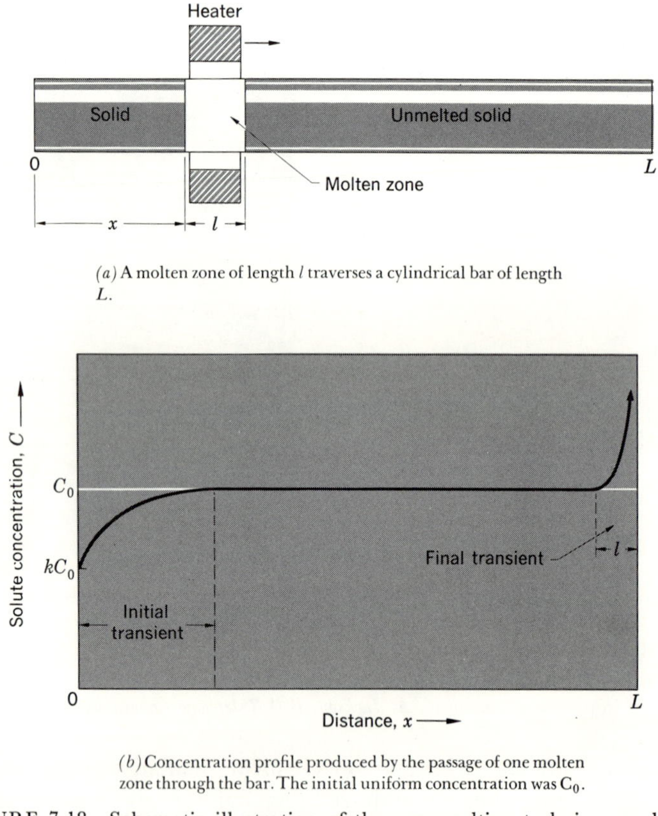

(*a*) A molten zone of length *l* traverses a cylindrical bar of length *L*.

(*b*) Concentration profile produced by the passage of one molten zone through the bar. The initial uniform concentration was C_0.

FIGURE 7-18 Schematic illustration of the zone-melting technique and of the change in concentration produced by its use. (*By permission from W. G. Pfann, "Zone Melting, 2nd ed., John Wiley & Sons, Inc., New York, 1966.*)

liquid-solid interface, although still occurring, is now accounted for in the empirical determination of k. Numerical values of k lie somewhere in the range between the equilibrium value k_e and unity. The faster the motion of the zone and the less the mixing of the liquid, the more nearly k approaches unity and the less efficient is the purification. The method of Sec. 2-12 can be used with Eq. (7-32) and the assumption of negligible diffusion in the solid to obtain the following equation for the concentration profile produced by one passage of a molten zone of length l:

$$C^s = C_0[1 - (1 - k)e^{-kx/l}] \tag{7-33}$$

This equation describes the initial transient (Fig. 7-18*b*) with characteristic length l/k and also the subsequent stage of steady-state solidification. It does not include the final transient, which occurs during solidification of the last zone length of the specimen. An interesting feature of the concentration profile of Fig. 7-18*b* is the long region of constant composition. Its application to the production of uniform compositions, the technique of *zone leveling*, is the subject of Prob. 7-7.

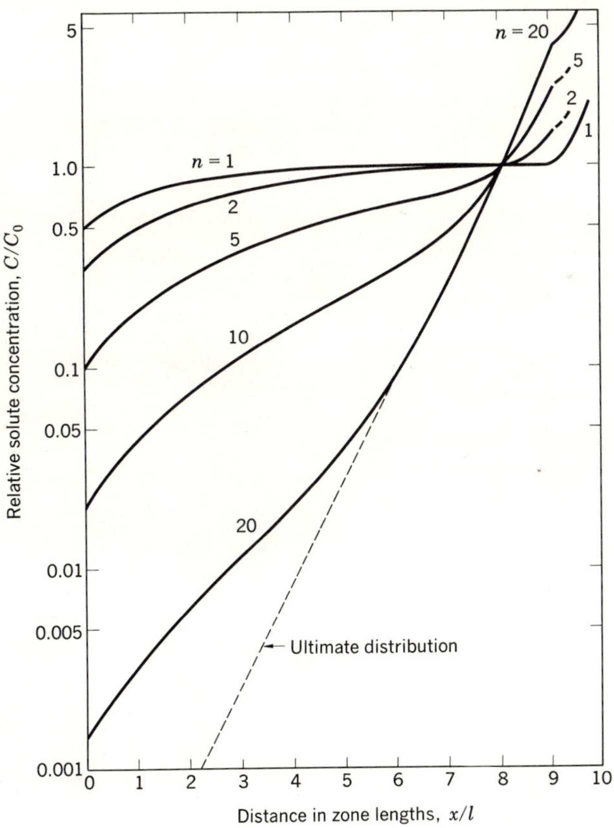

FIGURE 7-19 The effect of repeated zone melting when $k = 0.5$ and l is one-tenth of the length of the specimen. (*After W. G. Pfann.*)

An advantage of zone melting over conventional solidification is that the effect of successive zone-melting operations is cumulative. For example, if $k = 0.5$, the solute concentration at $x = 0$ is decreased to $0.5\,C_0$ by the first pass of the molten zone (Fig. 7-19). In a second pass the average composition of the first zone length is about $0.6\,C_0$, and therefore a solid with about $0.5 \times 0.6\,C_0 = 0.3\,C_0$ is formed at $x = 0$. Similarly, during successive passes the entire concentration profile in the main portion of the specimen is lowered, and only the final zone length or two becomes enriched in solute. Purification of the first few zone lengths, by factors ranging up to many powers of 10, is achieved in a number of passes that depends sensitively on the values of k and of L/l. In every case an ultimate distribution exists that sets a limit to the purification possible by zone refining.

Contamination from the mold used in ordinary zone melting is avoided in the technique of floating-zone melting (Fig. 7-20). If only a limited volume of metal is molten, surface tension is sufficient to hold the molten zone in place

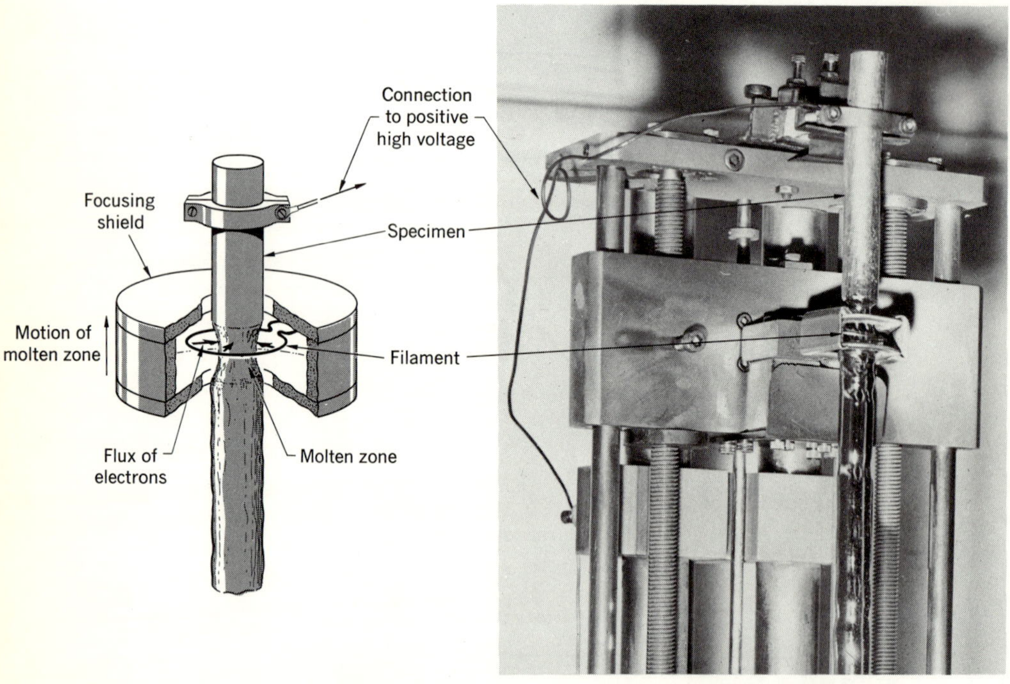

(*a*) Schematic illustration of the use of an electron beam to heat the molten zone.

(*b*) Commercial device for floating-zone melting, showing a specimen near the end of a pass.

FIGURE 7-20 Illustration of floating-zone melting. The device operates inside a vacuum jar. (*Courtesy Materials Research Corporation.*)

if the specimen is vertical. Electron-beam melting is done in a vacuum, and thus an additional source of purification is evaporation of impurities from the well-mixed molten zone. Floating-zone melting often results in a specimen that is a single crystal over most of its length, and therefore it is one of the standard procedures for preparing single crystals.

FORMATION OF A TWO-PHASE SOLID. The microstructures in Chaps. 2 and 3 illustrate some of the morphologies produced when two solid phases are formed during solidification. The eutectic structure (Fig. 2-5) is especially interesting for two reasons: (1) the fine mixture of two different phases often gives a material with a useful combination of strength and ductility; and (2) by suitable control of the conditions of solidification, the two phases of some eutectics can be prepared as a composite material (Fig. 5-27*a*) with controlled orientation and spacing of the reinforcing phase. Gradual solidification of a bar along its length, the technique of unidirectional solidification, is essential for the production of aligned composites. Although the science and technology of controlled solidification are only now developing, some fundamental ideas have been proposed and are presented here.

Not all eutectic systems form the parallel lamellae or cylindrical rod morphology usually found in metallic systems. A useful classification of eutectic structures can be based on the value of $\Delta H_f/kT_e$ for each component, following the pattern previously used for the solidification of a pure component (Fig. 7-8). The general pattern of solidification of a liquid of eutectic composition (Fig. 7-21*a*) begins with the heterogeneous nucleation of one of the phases, say α, at the bottom of the crucible when the supercooling ΔT has reached the critical value. Since α is rich in component A, component B must be rejected into the liquid, both laterally and in front of the α nucleus. If both phases have low values of $\Delta H_f/kT_e$, characteristic of metals, the β phase nucleates in a low-energy orientation relationship on the α phase in regions of liquid that are supercooled and also enriched in component B. If the volume fraction of β phase is less than $1/\pi$, the amount of interface is lower if the β phase forms as cylindrical rods. Interfacial energy favors the lamellar morphology if the volume fraction of the minor phase is greater than $1/\pi$ (and if the interfacial free energy is isotropic). Once the β phase has formed, its rejection of the A component encourages the growth of α phase along the far side of the β lamella. This process occurs by bridging over from the initial particle of α. In this manner a complete eutectic grain of α and β lamellae (Fig. 7-21*b*) is built up. A eutectic grain is composed of a single grain of α phase intermeshed with a single orientation of β phase, which is also a single, interconnected grain.

If the two phases composing the eutectic structure both have high values of $\Delta H_f/kT_e$ so that each grows with faceted interfaces (Fig. 7-21*c*), coupled growth of the type just described is unlikely to occur. One phase

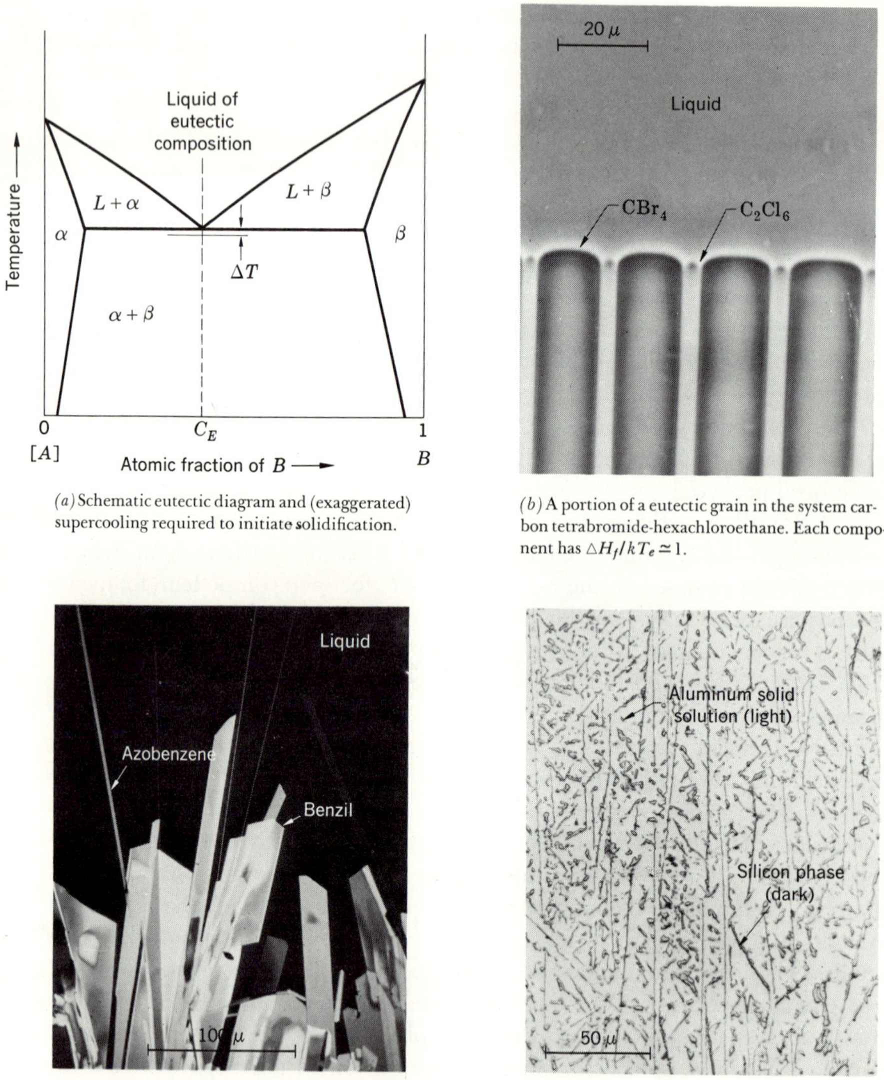

(*a*) Schematic eutectic diagram and (exaggerated) supercooling required to initiate solidification.

(*b*) A portion of a eutectic grain in the system carbon tetrabromide-hexachloroethane. Each component has $\Delta H_f/kT_e \simeq 1$.

(*c*) Faceted crystals of the two phases form the eutectic structure in the system azobenzene-benzil. Each component has $\Delta H_f/kT_e \simeq 5$.

(*d*) Irregular coupled eutectic growth produced by unidirectional solidification of the aluminum-silicon eutectic. Aluminum has $\Delta H_f/kT_e \simeq 1$ and silicon has $\Delta H_f/kT_e \simeq 3$.

FIGURE 7-21 Illustration of three types of eutectic morphologies, determined by the character of the components A and B. (*Courtesy K. A. Jackson, Bell Telephone Laboratories.*)

nucleates first, and its rejection of solute encourages the nucleation of the second phase at an adjacent location. This second phase is likely to nucleate separately on the substrate and to grow with no crystallographic relation to the existing phase. The final result of eutectic solidification is then an irregular mixture of faceted crystals of the two phases. When only one of the phases facets (Fig. 7-21*d*), coupled growth usually occurs, but the morphology tends to be irregular. Suitable control of the solidification process may give a rod structure, as illustrated in Fig. 5-27*a*, but the rods are faceted to some degree.

Impurities can affect dendritic growth in two different ways. Even small concentrations of certain impurities can modify the interfacial tensions

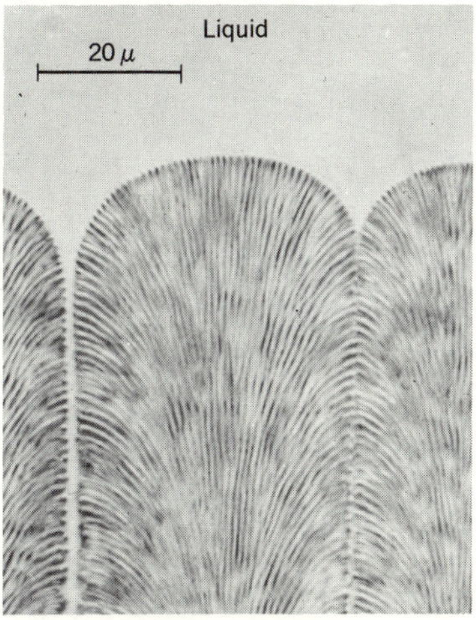

(*a*) Three lamellar colonies growing in a cellular manner as a result of an impurity.

(*b*) A pure eutectic growing with a planar liquid-solid interface. Only a single eutectic grain is shown.

FIGURE 7-22 Cellular growth fronts produced by an impurity compared to the planar growth front in a pure eutectic. Transparent organic materials are illustrated. (*Courtesy K. A. Jackson, Bell Telephone Laboratories.*)

and their anisotropies; the attachment mechanism or other aspects of the growth process can also be strongly affected. The second effect of impurities is to cause cellular growth (Fig. 7-22*a*). Parallel growth of lamellae (or rods) occurs in a pure eutectic during unidirectional solidification (Fig. 7-22*b*) because the growth front is planar. If sufficient impurity is present, its rejection by a growing colony can cause each colony to grow with a cellular interface and a consequent bending of the lamellae. The production of aligned lamellae or rods requires adequate purity to avoid constitutional supercooling by impurities. Constitutional supercooling by the two components of the eutectic will now be considered.

A eutectic structure can be produced from liquids having compositions C_0 far removed from the eutectic composition C_E (Fig. 7-23). The resulting structure may have more useful properties because of a change from a lamellar to a rod morphology (or vice versa) or because of the increased amount of the reinforcing phase in the composite structure. The tin-lead system, which is used here as an example, forms a lamellar structure on solidification of the eutectic composition C_E. A composition such as C_0, however, begins its solidification with a region of α phase (Fig. 7-24; see Prob. 7-8). Only after a second

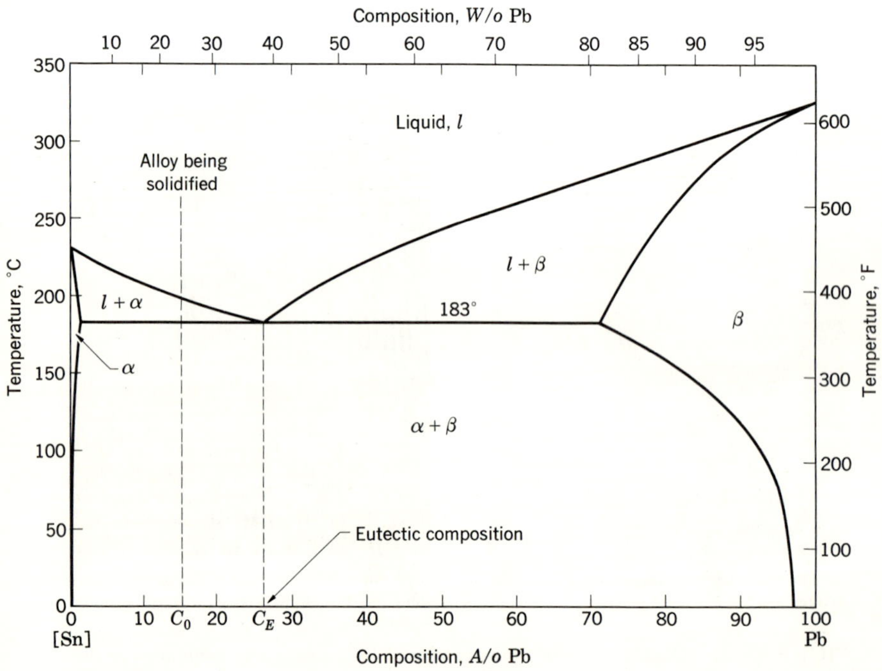

FIGURE 7-23 The tin-lead phase diagram used as an example of two-phase solidification. (*Adapted from M. Hansen and K. Anderko, "Constitution of Binary Alloys," McGraw-Hill Book Company, New York, 1958.*)

transient stage and under suitable conditions (including negligible convection) is a lamellar or rod structure produced at the steady state. The equation for solute distribution at the steady state is similar to Eq. 7-27 if the lamellar spacing is small compared to D/R, diffusion parallel to the interface[1] then having little effect on the solute concentration out in the liquid. The resulting

[1] This type of diffusion determines the spacing λ of the lamellae. If the rate of solidification, R, is low, diffusion can produce thicker lamellae. A general relation is $\lambda^2 R$ = constant.

equation for the composition C of the liquid at a distance x' in front of the liquid-solid interface is

$$\frac{C - C_0}{C_E - C_0} = e^{-Rx'/D} \tag{7-34}$$

An argument similar to that used to obtain Eq. (7-31) shows that for values less than

$$\frac{G^l}{R} = -\frac{m(C_E - C_0)}{D} \tag{7-35}$$

a region of constitutional supercooling exists in front of the liquid-solid interface.

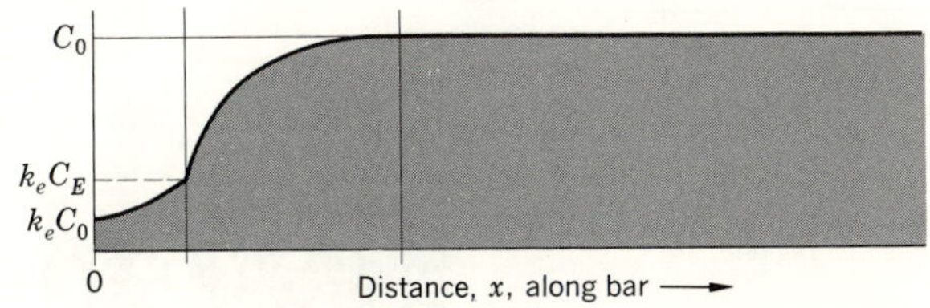

(*a*) Variation of average composition of the solid in the course of solidification.

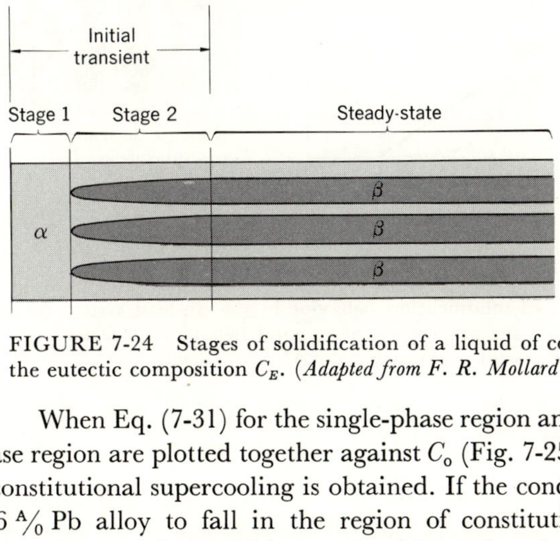

(*b*) Changes in the microstructure at various stages of solidification.

FIGURE 7-24 Stages of solidification of a liquid of composition C_0 different from the eutectic composition C_E. (*Adapted from F. R. Mollard and M. C. Flemings.*)

When Eq. (7-31) for the single-phase region and Eq. (7-35) for the two-phase region are plotted together against C_0 (Fig. 7-25*c*), an overall prediction of constitutional supercooling is obtained. If the conditions of growth cause a 12.6 A/$_0$ Pb alloy to fall in the region of constitutional supercooling (Fig. 7-25*d*), the result is dendritic growth of the α phase into the supercooled liquid and subsequent formation of a lamellar eutectic structure in the interdendritic spaces. The same alloy solidified at a higher value of G/R (Fig. 7-25*a*) is above the region of constitutional supercooling and forms a rod structure. An alloy closer to the eutectic composition (Fig. 7-25*b*) forms a lamellar structure.

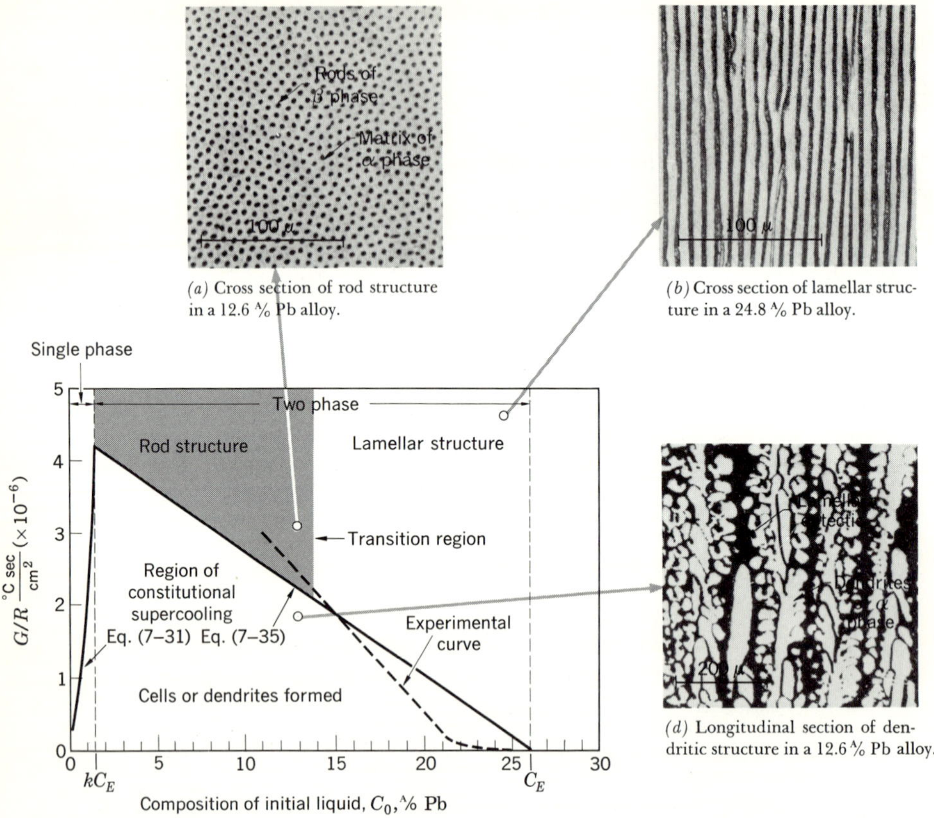

(a) Cross section of rod structure in a 12.6 % Pb alloy.

(b) Cross section of lamellar structure in a 24.8 % Pb alloy.

(c) Predicted boundaries of region of constitutional supercooling and a portion of a boundary determined experimentally.

(d) Longitudinal section of dendritic structure in a 12.6 % Pb alloy.

FIGURE 7-25 Comparison of solidification behavior in the tin-lead system predicted by Eqs. (7-31) and (7-35) and the observed experimental behavior. (*Adapted from F. R. Mollard and M. C. Flemings. Photomicrographs courtesy of M. C. Flemings.*)

TRANSFORMATIONS IN SOLIDS

Phase transformations in the solid state have many analogies to transformations involving the vapor and liquid states, but important novel features are also present. A pervading influence is the crystalline nature of the initial (matrix) phase and also of the product phase. The reaction path is usually influenced by geometric matching between the two crystal lattices, and one result is the frequent production of nonequilibrium transition phases. Transformations involving rigid phases also lead to straining of both matrix and product phases. The corresponding stresses help determine the reaction path and in extreme cases can cause plastic deformation or fracture. Imperfections are essential for the occurrence of all solid-state transformations. Vacancies or other point defects permit phase changes by a diffusion mechanism, whereas dislocations play either a dominant or a secondary role in all types of transformation.

In spite of the technological importance of solid-state transformation and much research effort, only modest progress has been made toward a basic theory. As a consequence, the development of complex, useful alloys (martensitic-hardening steels or precipitation-hardening aluminum alloys) has been accomplished largely empirically. An extensive literature is concerned with the technology of complex multicomponent materials of this type. On the other hand, effective theoretical studies ordinarily begin with the simplest systems (usually two-component) and establish a sound basis for treating more complex systems. Although much is known about polymorphic transformations in pure substances and a limited knowledge exists on three-component systems, the present treatment will be limited to the transformation behavior of two-component systems.

7-7. Formal Theories of Transformation

As a background for the examples of solid-state transformations considered in the following sections, some relevant aspects of the theory are summarized here. In addition to the topics of nucleation and growth, equations describing the overall process of transformation are discussed.

NUCLEATION. If effects due to strain are neglected, a simplified (classical) theory of homogeneous nucleation is obtained from the corresponding theory for the liquid → solid transformation of a pure substance. The essential modification is to determine an expression for the free-energy change ΔG_v driving the reaction, taking account of the chemical compositions of the matrix and product phases. A precipitation reaction is a convenient example. In general, a composition C_o (Fig. 7-26*a*) remains a solid solution on rapid quenching from T_q and undergoes a precipitation reaction only on being reheated to a suitable temperature. If the temperature of precipitation is far below T_q, interpretations of the free-energy curves may reveal large driving forces for transition products as well as for the equilibrium precipitate. At temperature T_1 the schematic free-energy curves of Fig. 7-26*b* are applicable. In the vicinity of a particle of precipitate the initial solid solution can be pictured as having decomposed according to the reaction

$$\alpha(C_o) \rightarrow \alpha(C_1) + \text{precipitate } (C) \qquad (7\text{-}36)$$

where each phase is of uniform composition. The composition of the precipitate is purposely left undetermined for the moment. The composition C_1 must

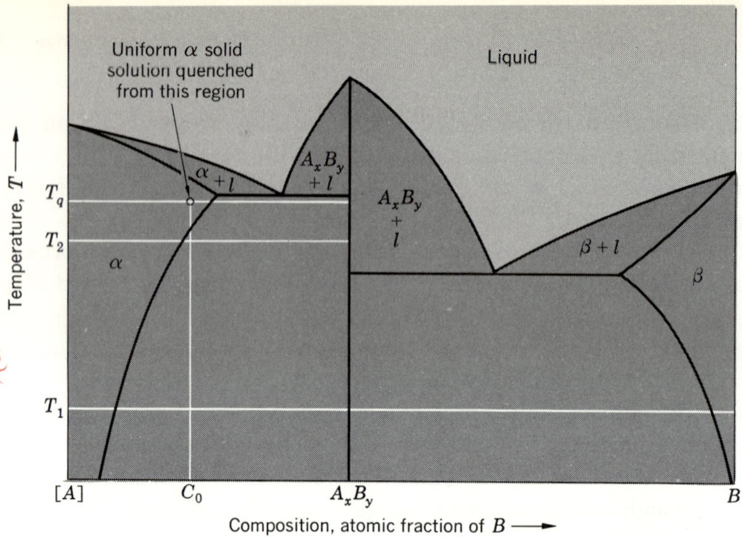

(*a*) A phase diagram in which the composition C_0 can undergo precipitation reactions at various temperatures, such as T_1 and T_2.

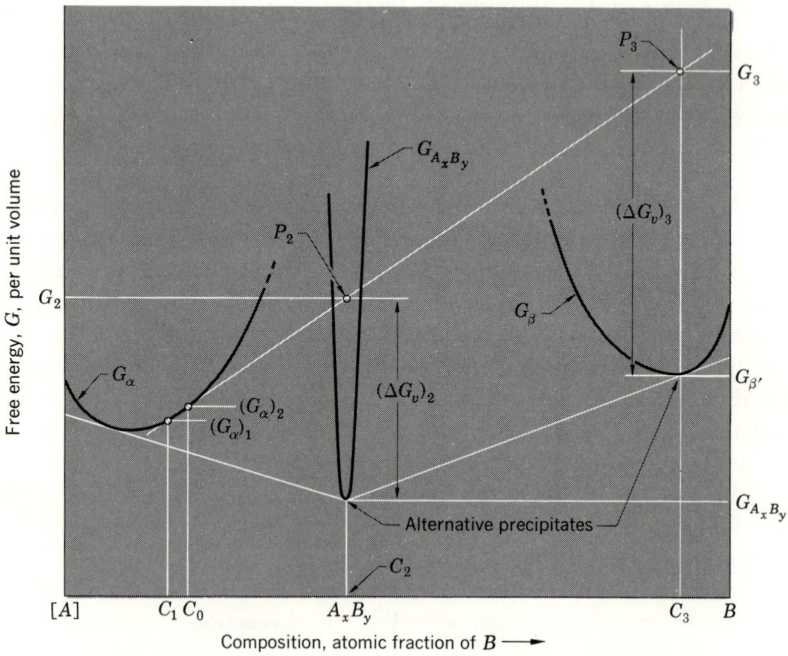

(*b*) Schematic free-energy curves at T_1 for the α, β, and A_xB_y phases. The graphical determination of ΔG_v for alternative products of the precipitation reaction is illustrated.

FIGURE 7-26 Schematic illustration of the driving force ΔG_v for alternative nucleation processes according to the classical theory. (*After M. E. Fine.*)

lie on the free-energy curve of the α phase and is close to C_o (by the lever rule) since the amount of precipitate in the nucleus is very small. Consider the case in which the precipitate is the intermediate phase A_xB_y. By analogy with the tangent-line construction for free-energy curves (Fig. 2-11), the initial solid solution can be pictured as dividing its free energy along the tangent line through C_o; thus

$$(G_\alpha)_o = \frac{C_2 - C_o}{C_2 - C_1}(G_\alpha)_1 + \frac{C_o - C_1}{C_2 - C_1}G_2 \tag{7-37}$$

Because $(G_\alpha)_1$ is less than $(G_\alpha)_o$, G_2 must be greater and is located at point P_2 at the composition C_2. G_2 represents the sum of the free energy $G_{A_xB_y}$, of the intermediate phase and of the free energy $(\Delta G_v)_2$ available to drive the corresponding precipitation reaction.

In the present example two compositions of precipitate can be considered as especially likely to form (1) the intermediate phase A_xB_y just considered and (2) the β phase. In the latter case, the tangent line in Fig. 7-26*b* extends to point P_3 at composition C_3, near the minimum on the free-energy curve of the β phase. The higher value of G_3 compared to G_2 is possible because the amount of precipitate is smaller in this case. Even though $G_{\beta'}$ is higher than $G_{A_xB_y}$, the driving force $(\Delta G_v)_3$ has a larger value than $(\Delta G_v)_2$. Kinetic factors determine the extent to which each type of reaction occurs. Typically, the transition phase β nucleates first and is the major product in the early stage of precipitation. Later the equilibrium precipitate A_xB_y may nucleate adjacent to particles of the β phase and during subsequent growth may cause the β phase to disappear.

The slope of the tangent at C_o on the free-energy curve (Fig. 7-26*b*) largely determines the magnitude of the free energy ΔG_v available for driving various precipitation reactions. When precipitation occurs at temperature T_2, the composition C_o is near the minimum on the free-energy curve of the α phase (see Prob. 7-9). The above construction then shows that a thermodynamic driving force exists only for precipitation of the equilibrium precipitate; nucleation of the β phase is thermodynamically impossible.

Lattice strain affects the total free-energy change ΔG for homogeneous nucleation in two ways, conveniently described by the following equation:

$$\Delta G = V(\Delta G_v + \sigma) + fV^{2/3}\gamma \tag{7-38}$$

where V is the volume of an embryo. First, the strain energy σ accompanying the formation of a unit volume of embryo is a positive quantity and therefore lowers the (negative) bulk free energy ΔG_v given by the classical theory. Second, the total surface energy is affected both by changes in the interfacial tension γ and by increases in surface area (described by the factor f) when the embryo is not spherical. The value of f is about 5 for a sphere, but can be many times larger if the embryo has the shape of a disk or a needle. Even if the interface between the embryo and the matrix is assumed to be incoherent (as is an ordinary grain boundary), the strain energy σ calculated for a disk-shaped embryo is only a small fraction of the corresponding value for a sphere. For most materials, however, the coherency strains associated with coherent or semicoherent interfaces (Fig. 7-27) probably exert a major influence on the shape of the embryo. An interface between the matrix α and a precipitate β can form especially easily along a plane of the matrix on which atom-to-atom matching of crystal planes in the two lattices can occur. If the disregistry of the

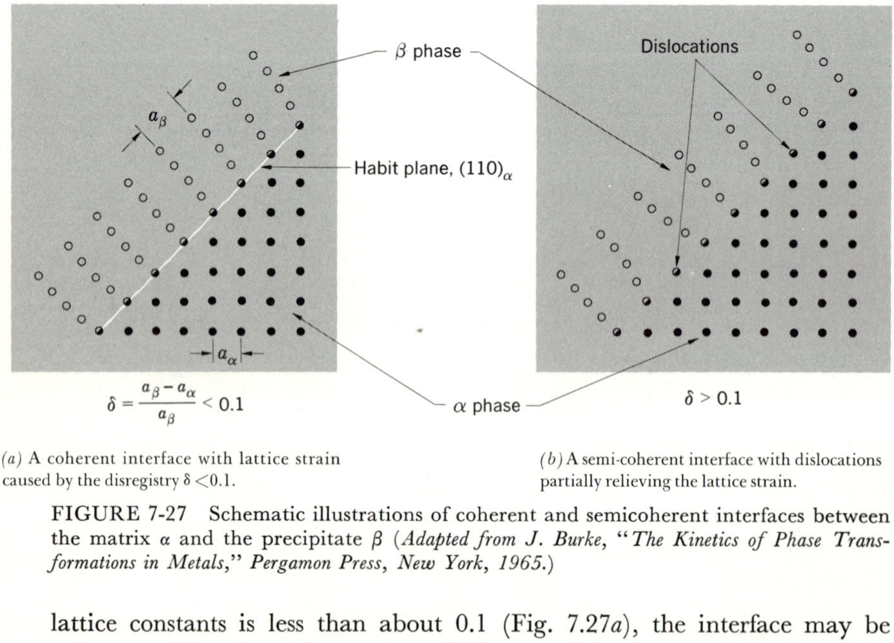

(*a*) A coherent interface with lattice strain caused by the disregistry $\delta < 0.1$.

(*b*) A semi-coherent interface with dislocations partially relieving the lattice strain.

FIGURE 7-27 Schematic illustrations of coherent and semicoherent interfaces between the matrix α and the precipitate β (*Adapted from J. Burke, "The Kinetics of Phase Transformations in Metals," Pergamon Press, New York, 1965.*)

lattice constants is less than about 0.1 (Fig. 7.27*a*), the interface may be coherent, although strain energy will be stored in the crystal lattices. The interfacial tension γ for a coherent boundary is on the order of only 100 ergs/cm^2, and consequently encourages embryos to form as platelets on the habit plane; elastic anisotropy is also a factor. When the disregistry is larger than about 0.1 (Fig. 7-27*b*), the total energy is lower if the interface consists of a regular array of interfacial dislocations, thus decreasing the amount of strain energy. This semicoherent interface has an interfacial tension γ ranging between about 200 and 500 ergs/cm^2, depending on the dislocation density. For higher values of disregistry the interface is incoherent, and γ has a value on the order of 1,000 ergs/cm^2.

Nucleation of solid-state transformations is almost always heterogeneous (GP zones, considered below, may nucleate homogeneously). A foreign particle can supply a substrate that acts to reduce the interfacial energy in a manner similar to that described for solidification in Sec. 7-4. Grain boundaries are effective sites for nucleation since one of the interfaces of an embryo can replace a corresponding area of grain boundary with little expenditure of energy (or even with a gain in energy). An even greater fraction of the surface of the embryo can be exchanged for existing boundary when nucleation occurs at a junction of three or four grains. When nucleation occurs randomly within the volume of a grain, the embryos often form at dislocation lines and nodes or at stacking faults. A principal factor in this case appears to be the release of the strain energy of the imperfection in the process of nucleation of the embryo.

GROWTH. The quantitative details of the growth of nuclei vary widely depending on the specific conditions, but a few qualitative features are almost universal. Growth processes are conveniently divided into two classes (Fig. 7-28), reconstructive and displacive. *Reconstructive growth* involves the reorganization of the initial phase into a distinctly different phase or phases. Breaking of bonds and diffusion of atoms must occur. *Displacive growth* is discussed under martensitic reactions (Sec. 7-9), but its significant feature is cooperative movement of large numbers of atoms. When a given phase (e.g., austenite in steel) can transform either reconstructively (to pearlite) or displacively (to martensite), the displacive transformation can occur only if cooling from T_e to below T_0 is adequately fast to prevent appreciable reconstructive growth. As explained later in connection with Fig. 7-42, appreciable supercooling below the equilibrium temperature T_0 is required to initiate a displacive reaction. A common feature of reconstructive growth is the maximum in growth rate at an intermediate amount of supercooling (Fig. 7-28). Initially the growth rate increases from zero at the equilibrium temperature T_e as the amount of supercooling (and therefore the volume free energy of transformation) increases. However, thermal activation is required to cause the changes in atomic configuration involved in a reconstructive transformation, and at lower temperatures activation is less effective. Consequently, the rate of reconstructive growth decreases to negligibly small values at large amounts of supercooling. When no displacive transformation is possible at low temperatures, the initial phase can then exist indefinitely in a metastable condition.

To produce the radical change in crystal structure involved in reconstructive growth, an individual atom must cross the boundary separating the initial and product phases; this process is the interfacial reaction considered in Sec. 6-8. A diffusion process of some kind is also involved whether it is boundary

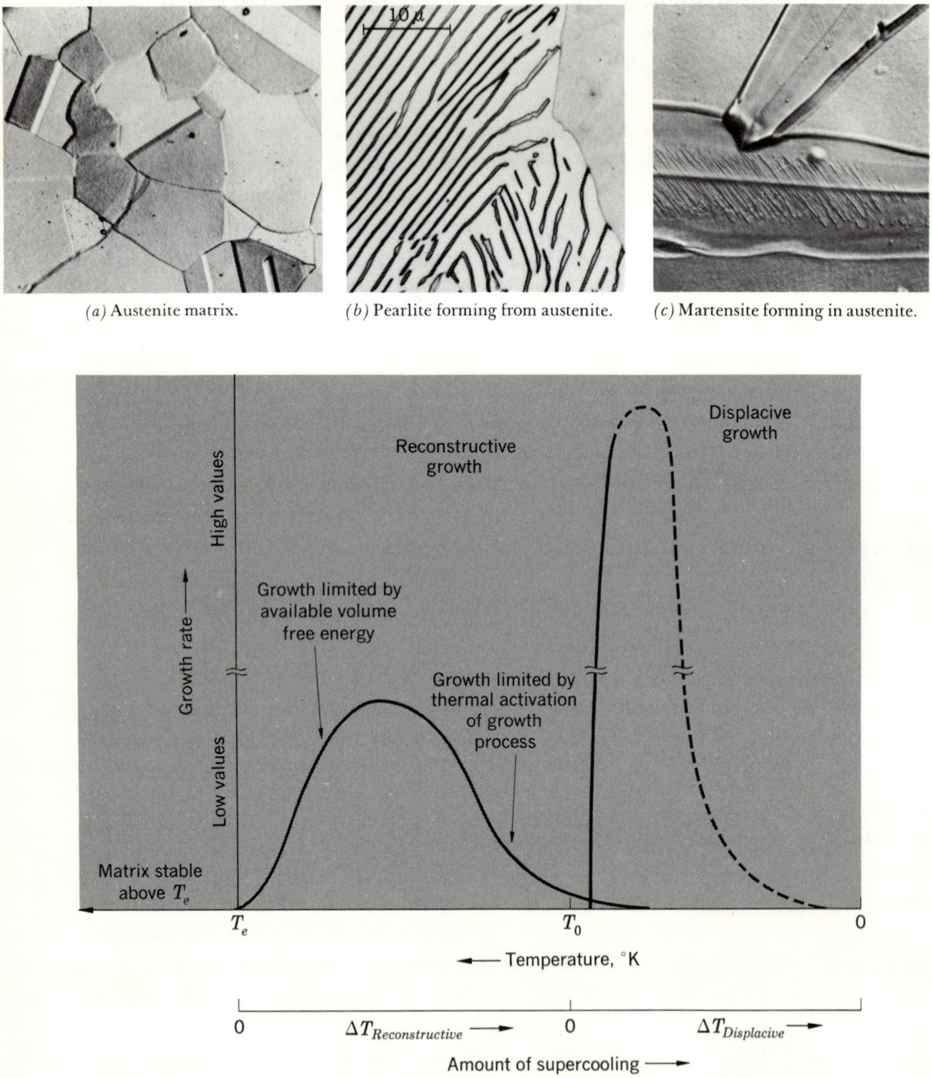

(*a*) Austenite matrix. (*b*) Pearlite forming from austenite. (*c*) Martensite forming in austenite.

FIGURE 7-28 Schematic illustration of the variation of growth rate with amount of supercooling for a reconstructive transformation and for a displacive transformation. The micrographs show how the austenite matrix in steel (*a*) transforms reconstructively to pearlite (*b*) or displacively to martensite (*c*). (*Micrographs* (*a*) *and* (*b*) *courtesy of L. S. Darken;* (*c*) *Courtesy C. M. Wayman.*)

diffusion along the interface in massive transformations (Sec. 7-9) or long-range volume diffusion in most cases, such as precipitation or cellular reactions (Sec. 7-8). Often an adequate treatment of the process of growth requires an analysis of the coupled action of diffusion and interfacial reaction. However, many important transformations can be considered to be controlled only by diffusion or only by the interfacial reaction (Fig. 7-29). If the interfacial reaction is controlling (Fig. 7-29*a*), rapid diffusion maintains an almost constant solute concentration C^α in the matrix. By the method of Sec. 6-8, the rate of transport, R, of solute across the interface is (see Prob. 7-11)

$$R = k(C^\alpha - C_{eq}) \qquad g/(cm^2\text{-}sec) \tag{7-39}$$

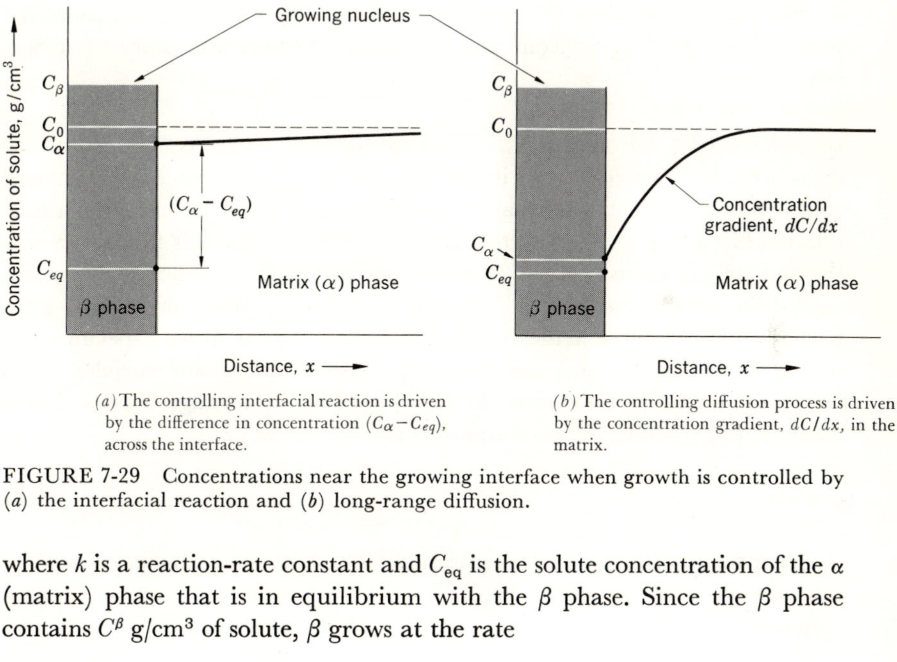

FIGURE 7-29 Concentrations near the growing interface when growth is controlled by (*a*) the interfacial reaction and (*b*) long-range diffusion.

where k is a reaction-rate constant and C_{eq} is the solute concentration of the α (matrix) phase that is in equilibrium with the β phase. Since the β phase contains C^β g/cm³ of solute, β grows at the rate

$$g = \frac{R}{C^\beta} \qquad cm/sec \tag{7-40}$$

Let y be the fraction of the total solute that has become incorporated in the β phase. Then the value of C^α varies approximately as

$$1 - y = \frac{C^\alpha - C_{eq}}{C_o - C_{eq}} \tag{7-41}$$

When the expression for C^α given by Eq. (7-41) is substituted in Eq. (7-39), the final expression for the growth rate is

$$g = \frac{k(C_o - C_{eq})(1 - y)}{C^\beta} \tag{7-42}$$

In the initial stages of growth, when y is small compared to unity, the growth rate is almost constant. Later a gradual decrease sets in.

The corresponding treatment of diffusion-controlled growth (Fig. 7-29*b*) is somewhat more involved, but the result shows the expected $t^{0.5}$ dependence on the linear dimension of the growing particle during the initial stages. Since interface control leads to a $t^{1.0}$ dependence, a growth process controlled jointly by interfacial reaction and diffusion can be expected to show an intermediate behavior. A factor not considered in this simple treatment is a possible change in type of control in the course of the growth process (see Sec. 6-8).

DESCRIPTION OF OVERALL TRANSFORMATION. As a given phase transformation proceeds from its initial state to its final state, the fractional completion y of the transformation at any time is given by some functional relationship $y = f(t)$. Among the factors that determine this function are n, the rate of nucleation, g, the rate of growth, the interference with growth caused by the impingement of adjacent particles, the shape of particles after partial impingement, and the limited dimensions of plate or wire specimens. A simple example is the diffusion-limited growth of precipitate particles, all of which nucleated at time zero. At time t, when the average particle has a radius r, the volume-fractional transformation has the form

$$y = Ar^3 \tag{7-43}$$

A depends on the initial concentration of nuclei, the subsequent rate of nucleation being negligibly small. In the early stages of growth (simple diffusion control)

$$r = Bt^{0.5} \tag{7-44}$$

Therefore the growth rate $g = dr/dt$ can be taken as proportional to $t^{-0.5}$ with a rough correction for depletion of the matrix given by $1 - y$, as described by Eq. (7-41). The resulting expression is

$$\frac{dr}{dt} = Ct^{-0.5}(1 - y) \tag{7-45}$$

When Eq. (7-43) is differentiated with respect to t and combined with Eqs. (7-44) and (7-45), the desired differential equation is obtained; thus

$$\frac{dy}{dt} = A' t^{0.5}(1 - y) \tag{7-46}$$

Separation of variables and integration gives the final result as follows:

$$y = 1 - e^{-kt^{1.5}} \tag{7-47}$$

often called an *Avrami equation*.

This type of equation describes many kinds of phase transformations depending on the exponent n in the factor t^n (Fig. 7-30). Usually the overall

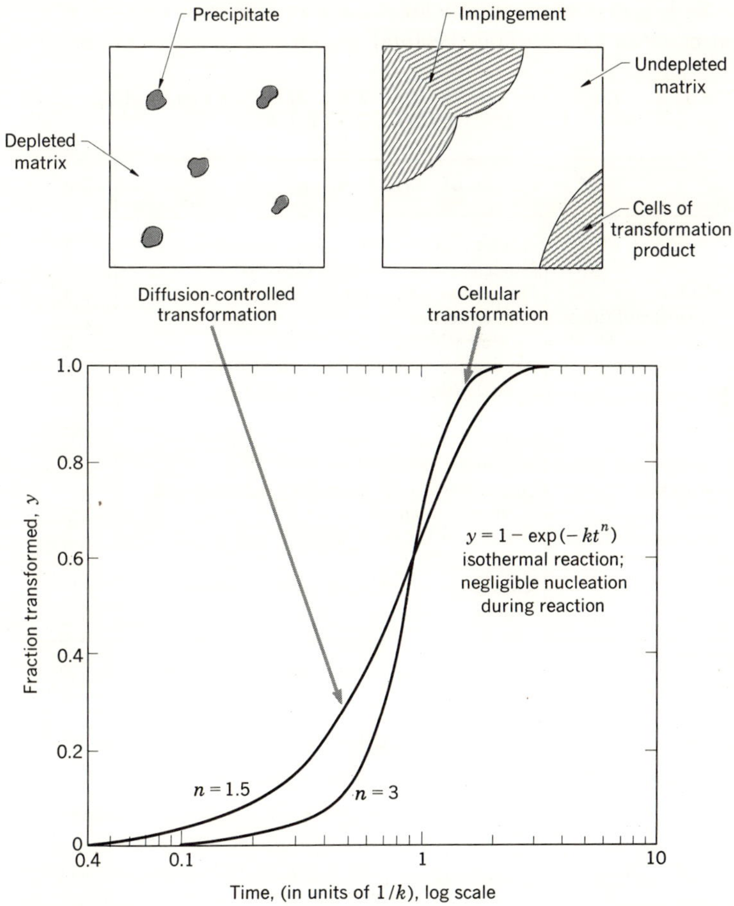

FIGURE 7-30 Two Avrami-type equations describing the overall course of phase transformations. Table 7-1 lists the values of n predicted for various conditions of transformation.

curve has an S shape with a central region where the rate of reaction is highest. Diffusion-controlled transformations, such as the precipitation reaction just considered, differ significantly from a group of transformations (eutectoid reactions, polymorphic changes, discontinuous precipitation, and others) known collectively as *cellular transformations*. In diffusion-controlled transformations a small amount of product phase forms within the matrix, which is gradually depleted of solute. Because of approximations in its derivations, the Avrami equation holds only for the initial stages of the transformation. In cellular transformations, regions of product phase grow into undepleted matrix, and therefore the rate of advance of a cell boundary is constant. When adjacent cells impinge, both stop growing along the boundary of impingement. The factor $1 - y$, used previously for depletion, now takes account of impingement in equations describing the total amount of transformation. Table 7-1

TABLE 7-1. CONDITIONS OF TRANSFORMATION AND THE VALUE OF n IN THE AVRAMI EQUATION.†

Conditions of transformation		n
Cellular transformations	Nucleation only at start of transformation	3
	Nucleation at constant rate	4
	Nucleation at an increasing rate	>4
	Nucleation at start plus continuing nucleation at grain edges	2
	Nucleation at start plus continuing nucleation at grain boundaries	1
Diffusion-controlled transformations	Initial growth of particles nucleated only at start of transformation	1.5
	Initial growth of particles nucleated at a constant rate	2.5
	Growth of isolated plates or needles of finite size	1
	Thickening of plates after their edges have impinged	½

† Adapted from J. W. Christian, "The Theory of Transformations in Metals and Alloys," Pergamon Press, New York, 1965.

summarizes the effects of various conditions of transformation on the value of n in the Avrami equation. If the finite size of the specimen affects the transformation, n can be reduced by as much as 1 for a sheet or 2 for a wire.

CRYSTALLIZATION IN ORGANIC POLYMERS. If a crystallizable polymer melt is cooled below its melting temperature T_f, it will crystallize according to an Avrami type equation (Fig. 7-31). However, the exponent n generally has a different interpretation than it does for the transformation of the simpler

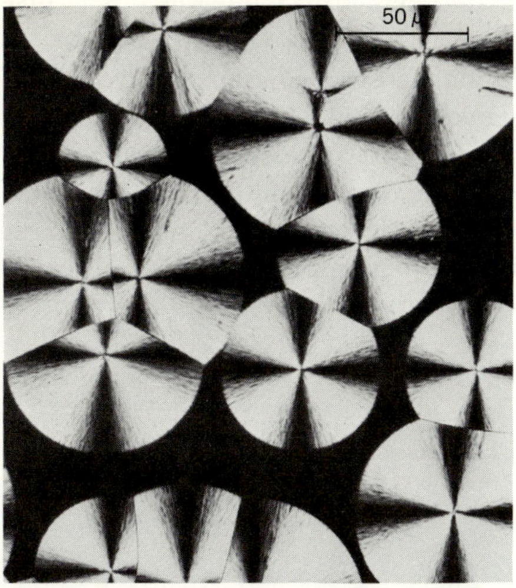

Spherulites growing at constant temperature into supercooled melt (dark areas). Polarized-light photomicrograph.

Spherulites at completion of isothermal crystallization. Radial arrangement of lamellae is evident.

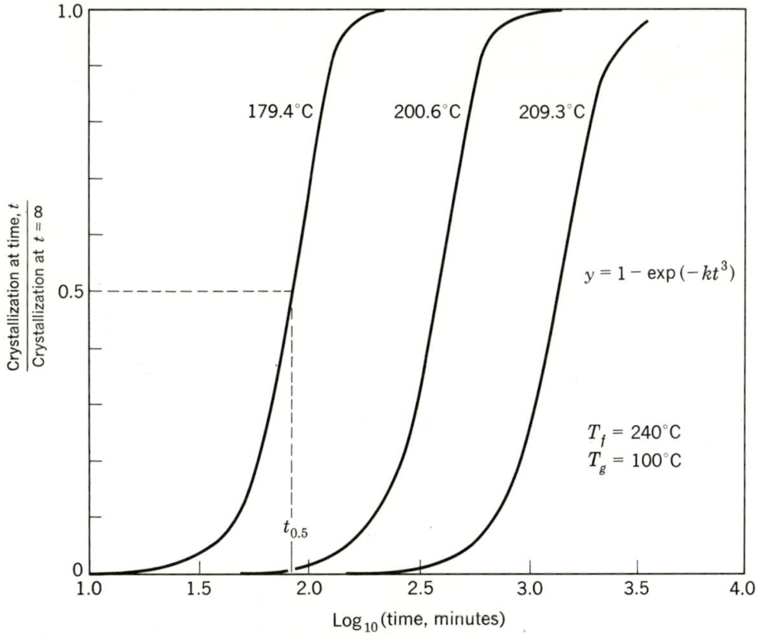

FIGURE 7-31 The rate of crystallization of isotactic polystyrene at temperatures close to the melting point. At a given temperature T, the amount of crystallization at equilibrium (crystallization at $t = \infty$) is greater the larger the difference $T_f - T$. (*Data of J. N. Hay. Photomicrographs courtesy H. D. Keith, Bell Telephone Laboratories.*)

inorganic materials. The primary nucleation step is the folding of a molecular chain (Fig. 1-9*f*) to initiate the growth of a chain-folded lamellar crystal. At temperatures near T_f the rate of nucleation increases with the degree of supercooling $(T_f - T)$. Following primary nucleation, the lamellar crystal undergoes a complex branching development into a spherulitic structure. The rate of growth of the lamellae forming the spherulite depends on the mobility of the segments of the molecular chain and therefore decreases as the temperature is lowered. The overall rate of crystallization, determined by both nucleation

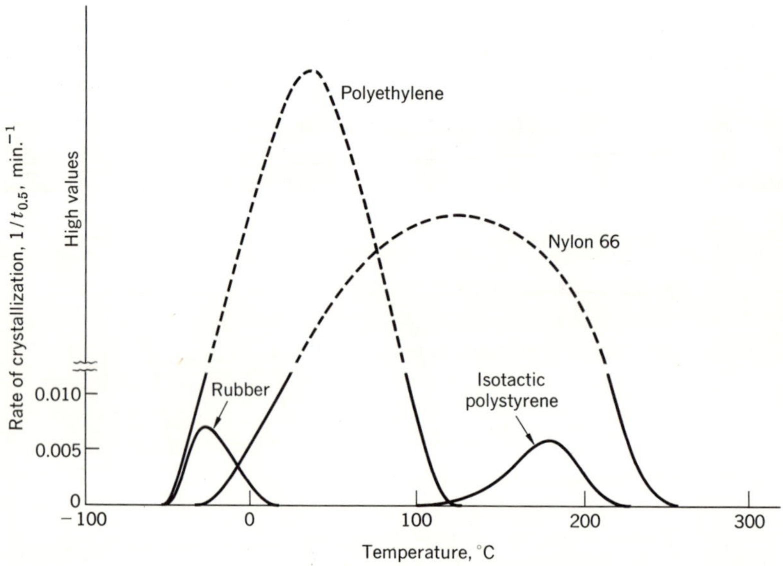

FIGURE 7-32 Variation of the rate of crystallization with temperature for several polymers. (*Adapted from A. R. Ubbelohde, "Melting and Crystal Structure," Clarendon Press, Oxford, 1965.*)

and growth, passes through a maximum (Fig. 7-32) and approaches zero at temperatures near the glass transition T_g.

The extent to which crystallization can be prevented by rapid cooling of a molten polymer varies considerably. The rate constant *k* for polyethylene reaches such high values during cooling that even the quenched material has a high degree of crystallinity. Nylon is an example of a polymer that does not crystallize if cooled quickly, but which undergoes crystallization on being reheated. Crystallization is also promoted by plastic deformation. In the process of manufacturing nylon fibers the liquid polymer is extruded through fine openings called *spinnerettes*; the resulting fiber cools quickly and usually with little crystallization. Because of its rubbery character, the fiber can be elongated to about four times its original length in a simple drawing operation. Crystallization occurs during this process, and the crystallites become preferentially oriented in the direction of drawing. The strength-to-weight ratio of such fibers is comparable to that of steel wires.

7-8. Transformations to Stable Phases

The distinction between the formation of stable phases or of transition phases, already discussed in connection with Fig. 7-26 for precipitation reactions, is especially evident in the TTT (time-temperature-transformation) diagram (Fig. 7-33) describing the transformation of metastable austenite in a given type of steel. The TTT diagram consists of two areas, corresponding to distinctly different behaviors of austenite cooled instantly[1] to a temperature where the austenite is metastable. Above the M_s temperature the austenite undergoes a reconstructive transformation to one or both of the stable phases, ferrite and a carbide similar to Fe_3C. Different ranges of transformation are distinguished, as shown in Fig. 7-33*c*, but a common characteristic is an isothermal reaction curve, a simple example of which is given in Fig. 7-33*a*. These curves are qualitatively of the Avrami type, but the exponent n is difficult to interpret. Below the M_s temperature austenite transforms displacively to form martensite. The amount of reaction increases with supercooling below M_s (Fig. 7-33*b*), but at a given temperature the characteristic amount of reaction is completed in about 10^{-6} sec. For example, when austenite is quenched from 850°C to 300°C, 50 percent martensite forms during cooling between the M_s temperature and 300°C. If the specimen is then held at this temperature, virtually no additional martensite forms. The martensite reaction is discussed in Sec. 7-9, but the remaining reactions in the TTT diagram are now considered among examples of transformations to stable phases.

ISOTHERMAL TRANSFORMATIONS IN STEEL. If the carbon content of a steel is not the eutectoid value, ~0.8 w/o C, a *proeutectoid reaction* (prior to the eutectoid reaction) may occur when austenite is cooled from about 900°C. For carbon contents above eutectoid (*hypereutectoid* steels) proeutectoid cementite can form during slow cooling as an objectionable brittle layer at the original austenite grain boundaries. When high-carbon tools (drills, knives, etc.) are quenched during heat treatment, this proeutectoid reaction is avoided. For the heat treatment of *hypoeutectoid* (below eutectoid) steels, the reactions involving proeutectoid ferrite are of paramount interest. These reactions can result in several morphologies of the ferrite phase. At moderate amounts of supercooling, the ferrite nucleates at grain boundaries and normally grows smoothly along

[1] A suitably small specimen of (stable) austenite can be cooled from an initial (austenizing) temperature of about 850°C to the temperature of reaction in a fraction of a second.

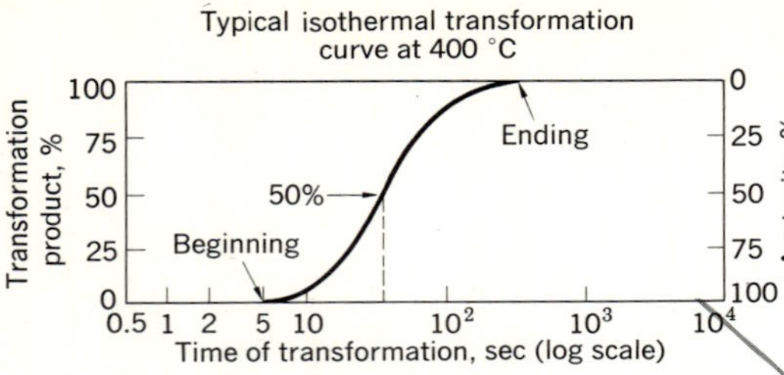

(*a*) Isothermal transformation curve for metastable austenite at 400°C. Each point on the curve is determined by microscopic examination of a specimen reacted for the indicated time.

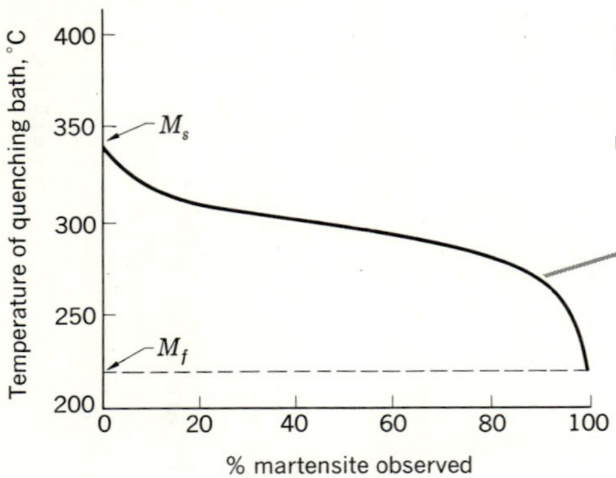

(*b*) Experimental data on the transformation of austenite to martensite on cooling. Transformation begins at M_s (340°C) and ends at M_f (220°C).

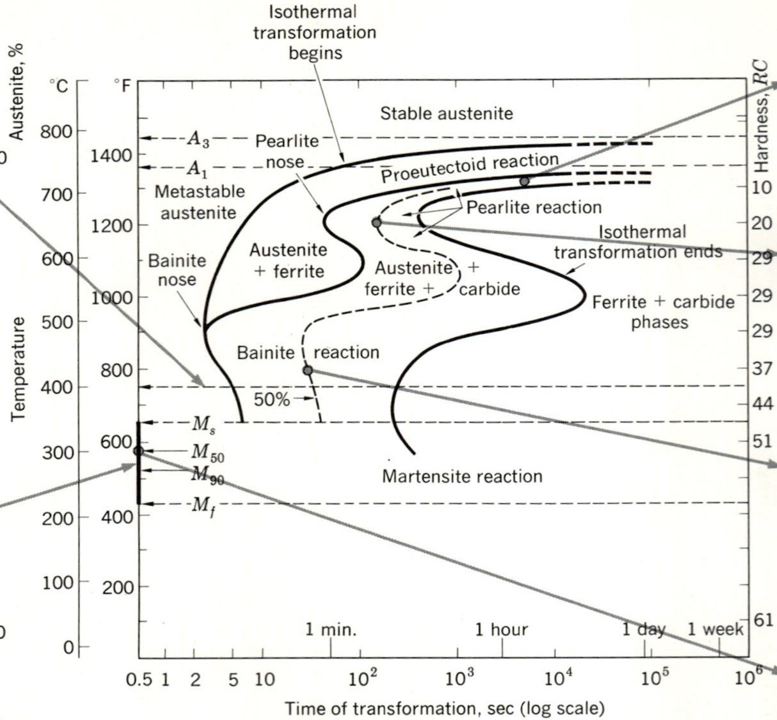

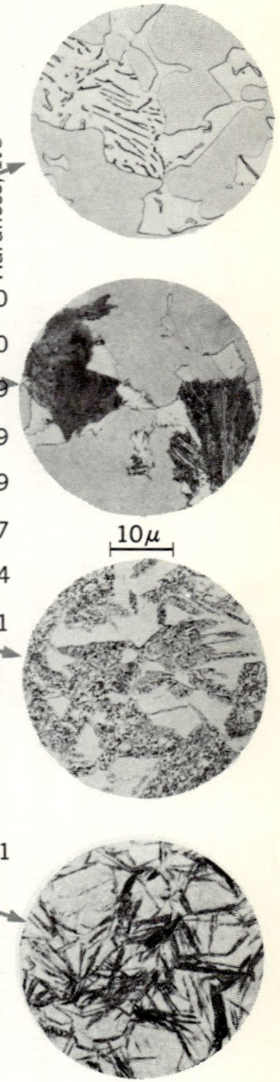

(*c*) The time-temperature-transformation (TTT) diagram for 4140 steel. Typical transformation products are illustrated by photomicrographs. The hardness values of the completely transformed specimens are listed.

FIGURE 7-33 Typical experimental transformation curves, (*a*) isothermal transformation, and (*b*) transformation on cooling, employed in constructing the TTT diagram for a 4140 steel. The composition of this steel is 0.4 w/o C, 0.8 w/o Mn, 1.0 w/o Cr, and 0.2 w/o Mo. (*Courtesy United States Steel Corporation.*)

them as grain-boundary allotriomorphs.[1] The initial, isolated grain-boundary allotriomorphs grow very rapidly along the grain boundary and soon produce a complete boundary network. The slow growth in the direction normal to the boundary is limited in pure iron-carbon alloys by the diffusion of carbon away from the incoherent interface. When alloying elements are present, an interfacial reaction may also become important and greatly decrease the rate of boundary movement.

In the pearlite (eutectoid) reaction (Fig. 7-34) the significant new feature is the cooperative growth of *two* phases: a high-carbon product, Fe_3C, as well as the low-carbon ferrite phase. Thus we have

austenite $\rightarrow$ Fe_3C + ferrite

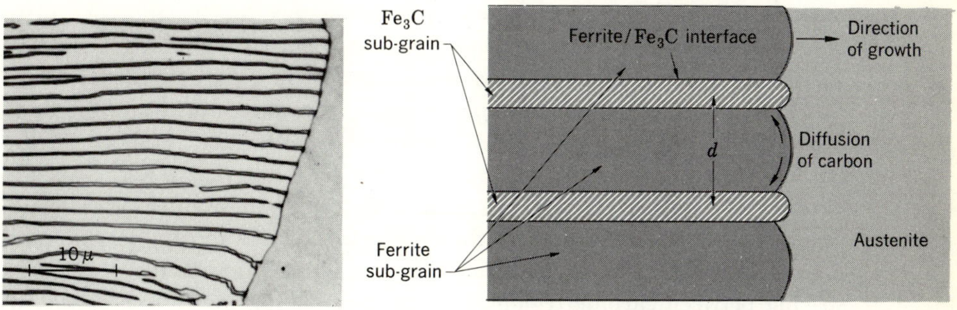

(*a*) Photomicrograph of a pearlite grain growing into prior austenite.

(*b*) Schematic illustration of the relations among the austenite, ferrite, and cementite phases during the growth of pearlite.

FIGURE 7-34 Illustrations of the transformation of austenite to pearlite. (*Photomicrograph courtesy U. S. Steel Research Laboratories.*)

The lamellar arrangement of these two phases in the pearlite morphology permits a convenient redistribution of the carbon in the austenite between the ferrite and the Fe_3C. As with the proeutectoid reaction, (1) nucleation occurs primarily at grain boundaries, (2) the orientation relation between the ferrite (α) and austenite (γ) is the same in both cases, $(111)_\gamma \| (110)_\alpha$, $[110]_\gamma \| [111]_\alpha$, (3) the growth interface is essentially planar and incoherent, and (4) imperfections and dislocation arrays are generated in both matrix and product phases because of an increase in volume accompanying the phase transformation. The lamellar spacing d (Fig. 7-34*b*) is important because the strength of a steel is inversely proportional to $d^{\frac{1}{2}}$ (Sec. 9-12). Although a small spacing facilitates diffusional redistribution of the carbon, the closeness of the lamellar spacing is limited by the energy required to form the corresponding amount of

[1] An *idiomorph* has its own, characteristic shape. An *allotriomorph* adapts its shape to the environment.

α/Fe_3C interface, which is proportional to $1/d$. To a first approximation, a fixed fraction of the total driving force for the austenite $\rightarrow$ pearlite transformation is converted to interfacial energy, and thus d decreases roughly in proportion to the amount of supercooling. For pearlite in an Fe–0.8 $^{\mathrm{W}}\!/_{0}$ C steel, d is halved by each additional 40°C of supercooling. A decrease in d results in a faster redistribution of carbon by diffusion (Fig. 7-34*b*) since the distance for diffusion is smaller. This factor is a major contributor to the increased rate of formation of pearlite as the reaction temperature is lowered from A_1 to the temperature of the pearlite nose of the TTT diagram (Fig. 7-33).

The isothermal transformation of austenite at temperatures in the range of about 350 to 550°C produces the bainite microconstituent (Fig. 7-33*c*). As with pearlite, bainite is a mixture of ferrite and Fe_3C, but instead of a cooperative growth of the two phases, growth is initiated by the ferrite phase in steels of moderate carbon content. The first step is the formation of a plate of ferrite, supersaturated to some degree in carbon. At high temperatures in the bainite range, the Fe_3C phase forms in austenite parallel to the ferrite plates in a morphology called upper bainite. At lower temperatures the Fe_3C forms also within the ferrite plates to produce lower bainite. As shown by the Rockwell hardness C scale (RC) values in the TTT diagram of Fig. 7-33*c*, the hardness of a given steel increases as the transformation product forms at successively lower temperatures, in the order proeutectoid ferrite, pearlite, and bainite. An important cause of this behavior, in addition to the decrease in spacing between carbide regions, is the corresponding increase in density of dislocations.

CELLULAR PRECIPITATION. Precipitation from a supersaturated solid solution can occur in several ways. Reactions that lead to precipitation hardening will be considered in Sec. 7-9. A significantly different reaction (sometimes called *discontinuous precipitation*) produces a reaction product resembling pearlitic cells, composed of the two stable phases. Because of the similarities in their reactions, both eutectoid and precipitation reactions of this type are often classified as cellular transformations. An example of cellular precipitation is the decomposition of supersaturated β phase in the lead-tin system (Fig. 7-23). If an alloy containing about 10 $^{\mathrm{A}}\!/_{0}$ tin is quenched from a temperature in the β-phase region and then allowed to transform, the tin-rich α phase precipitates in a cellular morphology together with β phase, which is much depleted in tin. The reaction curve is fit by an Avrami equation with $n \approx 3$ as expected (Table 7-1) since nucleation occurs at the grain boundaries near the beginning of the reaction.

SPINODAL DECOMPOSITION. The thermodynamic character of a supersaturated solution lying between the spinodal curves was discussed previously using Fig. 2-15. Unlike the usual *metastable* phase, this phase is *unstable* relative to even a small fluctuation in composition. This difference in behavior accounts for the two different modes of decomposition of a supersaturated solution, illustrated for the aluminum-zinc system in Fig. 7-35*a*. If composition c_0 is quenched from T_q to T_2, it lies outside of the region of spinodal decomposition, and therefore a zinc-rich nucleus must form (Fig. 7-35*b*) to permit a reduction in free energy. An interface exists between the nucleus and the depleted matrix, and growth occurs by the motion of this interface. The solute needed for growth is supplied by diffusion down the concentration gradient. If the

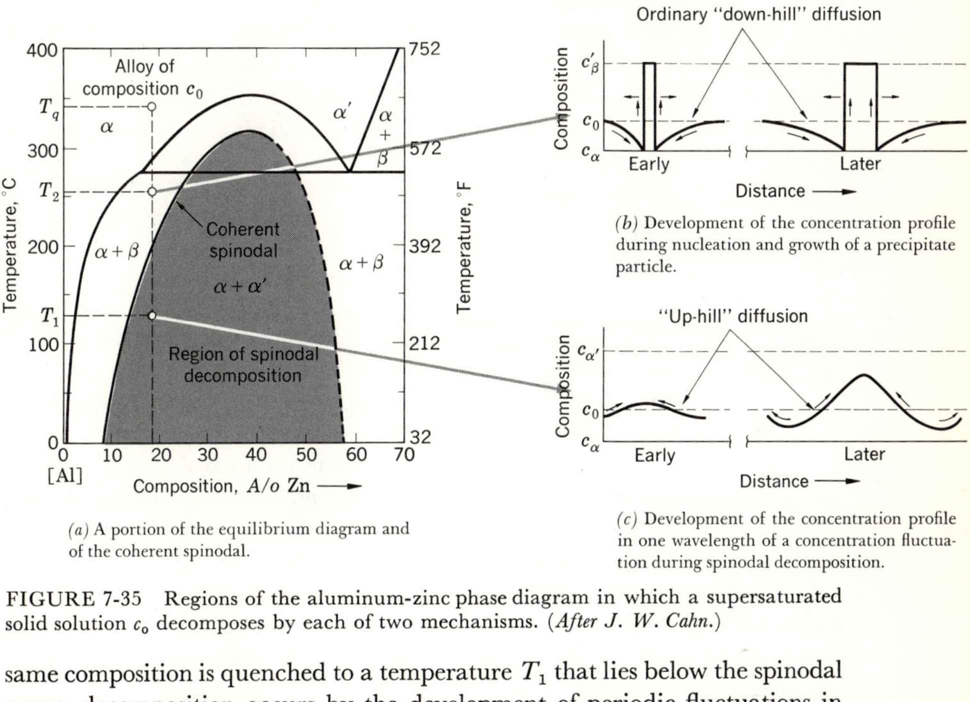

(a) A portion of the equilibrium diagram and of the coherent spinodal.

(b) Development of the concentration profile during nucleation and growth of a precipitate particle.

(c) Development of the concentration profile in one wavelength of a concentration fluctuation during spinodal decomposition.

FIGURE 7-35 Regions of the aluminum-zinc phase diagram in which a supersaturated solid solution c_0 decomposes by each of two mechanisms. (*After J. W. Cahn.*)

same composition is quenched to a temperature T_1 that lies below the spinodal curve, decomposition occurs by the development of periodic fluctuations in composition (Fig. 7-35*c*), each fluctuation having a wavelength of about 100 Å. No interface in the usual sense exists between adjoining regions of high and low concentration.

The theory of spinodal decomposition involves two new factors, negative diffusion coefficients and the energy of steep concentration gradients. If diffusion of the solute is first described in terms of chemical potential μ and a suitable coefficient B,

$$J = -B \frac{d\mu}{dx} \tag{7-48}$$

a conversion can easily be made to the free energy G of the solution (see Prob. 7-16); thus

$$J = -B\frac{d^2G}{dc^2}\frac{dc}{dx} \tag{7-49}$$

Diffusion usually occurs in solutions for which d^2G/dc^2 is positive. The diffusion coefficient $B(d^2G/dc^2)$ is then positive, and diffusion occurs down the concentration gradient, as in Fig. 7-35*b*. In the region of spinodal decomposition, however, d^2G/dc^2 is negative, the diffusion coefficient is negative, and diffusion must occur up the concentration gradient (uphill diffusion). The solution of the diffusion equation in the presence of a steep concentration gradient dc/dx involves an additional free energy $K(dc/dx)^2$, where K is a suitable coefficient. The solution of Eq. (7-49), after addition of a gradient-energy term, has the form

$$c = c_o + e^{\alpha t}\cos\frac{2\pi x}{\lambda} \tag{7-50}$$

That is, uphill diffusion during spinodal decomposition produces fluctuations in composition around the initial composition c_0. The fluctuations can have various wavelengths λ, the amplitude of each type of fluctuation growing and decaying with time according to its characteristic value of α. Wavelengths in a certain range develop preferentially and create the final three-dimensional pattern of fluctuations. Compositions near c_o in Fig. 7-35*a*, after decomposition at 65°C, develop fluctuations having wavelengths near 30 Å; the time for 50% reaction is about 1 hr.

Spinodal decomposition in inorganic glasses was described in connection with Fig. 2-16; a typical microstructure involving two glass phases is shown in Fig. 7-36*a*. An important strengthening process involves the conversion of a glass article to a predominantly crystalline structure by suitable heat treatment. For this purpose two phases are required. A minor phase with a strong tendency for crystallization must exist as a fine dispersion in the major phase, which is a relatively stable glass. Phase separation of this kind can occur in a number of ways, but spinodal decomposition appears to be an especially effective mechanism. In the glass-ceramic listed in Table 5-2, the minor phase is a glass rich in TiO_2. After being formed in the usual manner, the glass article is heated to cause nucleation in the minor phase but not in the major phase. A further increase in temperature produces crystallization of the major phase at the multitude of existing nuclei and results in a final structure that is fine-grained and almost completely crystalline.

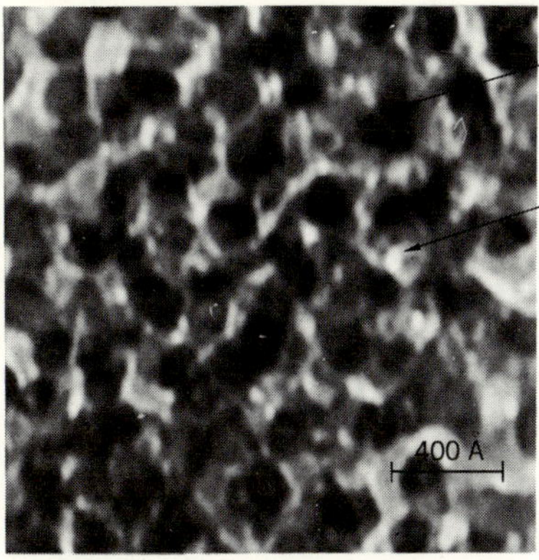

(*a*) Uniform distribution of transformation products resulting from the spinodal decomposition of a sodium borosilicate glass. (Courtesy A. M. Turkalo, F. E. Wagstaff and R. J. Charles, General Electric Research and Development Center.)

(*b*) Two-phase structure originating from the spinodal decomposition of a copper-nickel-iron solid solution. (Courtesy R. B. Nicholson, University of Manchester).

FIGURE 7-36 Electron micrographs showing the fine-scale phase separation produced by spinodal decomposition.

Strain energy, although not localized at a sharp interface, can play a dominant role in spinodal decompositions in solid solutions. The ordinary spinodal curve (of Fig. 2-15, for example) considers only chemical interactions. The coherent spinodal curve (Fig. 7-35*a*) also includes the strain energy and may lie far below the ordinary curve. Anisotropy of elastic constants (Chap. 9) may result in strongly directional precipitates. In cubic alloys (Fig. 7-36*b*) the preferred direction is often along ⟨100⟩.

ORDER-DISORDER TRANSFORMATIONS. The two types of ordered phases that were distinguished in Fig. 5-17 on the basis of their equilibrium behavior also transform quite differently. Consider first the composition Mg_3Cd (Cu_3Au-type) for which the critical temperature is about 160°C. If the alloy is quenched from 180°C, the solid solution (with only short-range order) is obtained at room temperature. Two processes occur when the alloy is then heated at an intermediate temperature (Fig. 7-37): (1) nucleation and growth of domains of nearly perfect order, and (2) coalescence of existing domains with resulting increase in domain size. In the early stages of transformation the alloy consists mostly of disordered regions. Ordinary dislocations can move easily in these regions, and therefore the material is relatively soft. As the ordered domains develop and occupy most of the volume of the specimen, the hardness increases because the ordinary dislocations must create antiphase boundary on passing through an ordered domain. Finally, when almost no disordered material remains, superdislocations account for essentially all of the

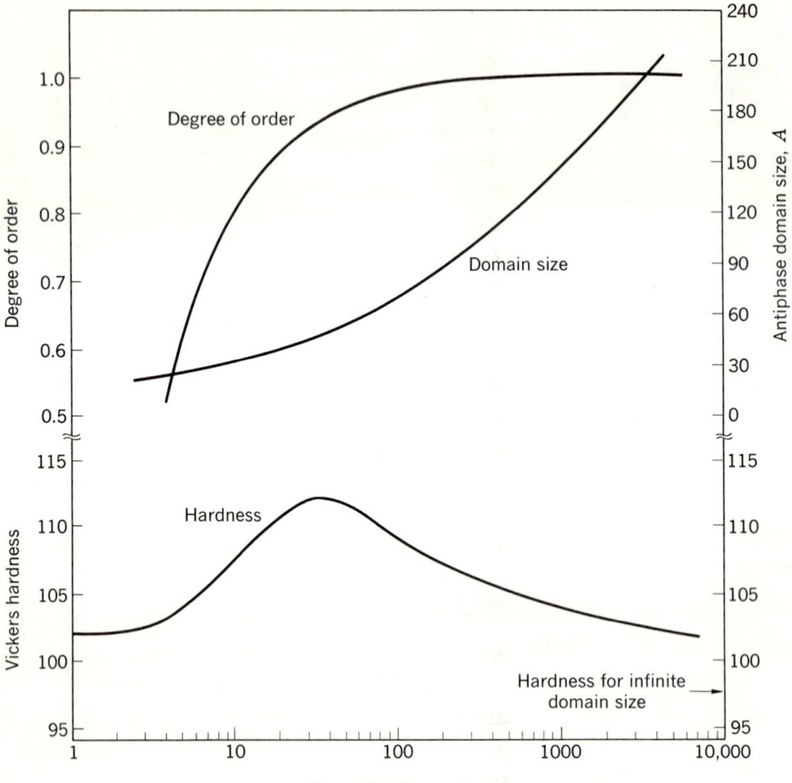

FIGURE 7-37 Development of long-range order and corresponding change in hardness during transformation of the Mg_3Cd solid solution at 68°C. (*Data of R. G. Davies and N. S. Stoloff.*)

plastic deformation. As the domain size increases, the superdislocations experience less boundary restraint, and the hardness again decreases.

Solid solutions that form a CsCl-type ordered phase tend to undergo both ordering and domain coalescence extremely rapidly. The composition FeCo, for example, orders completely during water quenching from above the critical temperature. As generally observed for ordering transformations, the rate of reaction can be decreased markedly by displacing the composition from the stoichiometric value. In this case, if 1 percent of both iron and cobalt is replaced by vanadium, the resulting Fe–Co–2%V solid solution can be retained by quenching and its superior fabrication properties employed in forming this otherwise brittle material for use in magnetic devices. A subsequent heat treatment then permits development of long-range order and coalescence of domains in a manner similar to that shown in Fig. 7-37.

7-9. Transformations to Transition Phases

Useful improvements in the strength and hardness of many materials occur as a result of phase transformations that produce transition phases, rather than stable phases. A large driving energy, as shown for one case by Fig. 7-26, is always necessary since a transition phase has higher energy. Structural relations among the phases in a given material are of decisive importance, however, both for the actual formation of a transition phase and for its effectiveness in strengthening the material. In this section we consider several instances of the formation of transition phases, including examples of precipitation hardening in aluminum alloys and martensitic hardening in steel.

PRECIPITATION REACTIONS. Precipitation from a supersaturated solid solution occurs in any alloy system that has a suitably sloping solvus line, Fig. 2-24. However, the effect of this precipitation on properties often depends on a sequence of reactions of the type

$$\begin{array}{lll} & \nearrow & \alpha(\text{I}) + \text{transition phase (I)} \\ & & \qquad\qquad\qquad\quad \searrow \\ \alpha(\text{supersaturated}) & \longrightarrow & \qquad \alpha(\text{II}) + \text{transition phase (II)} \\ & & \qquad\qquad\qquad\qquad\qquad \searrow \\ & \searrow & \qquad\quad \alpha(\text{equilibrium}) + \text{equilibrium precipitate} \end{array} \qquad (7\text{-}51)$$

These reactions produce useful increases of hardness or strength in many materials. Two instructive examples are the aluminum-copper and aluminum-silver systems (Fig. 7-38*a* and *b*). The aluminum matrix is alloyed with a few atomic percent of solute so that a solid solution α is produced by a high-temperature *solution heat treatment*. Next, the solid solution is quenched to room temperature to prevent transformation to the equilibrium precipitate during cooling. However, transformations of other kinds do occur. Vacancies, which are supersaturated by a factor of about 10^7, partially precipitate as dislocation loops and on dislocations and other imperfections. The first stage in the precipitation of solute also occurs during quenching, and regions containing predominantly solute atoms form in the matrix lattice. These GP (Guinier-Preston) zones (Fig. 7-38*c*) take various shapes depending on the relative sizes of the matrix and solute atoms. In an Al–4.4 $^{A}\!/_{0}$ Ag alloy the two atoms are almost the same size, and thus the zones can form as spheres without developing much strain energy. In an Al–2 $^{A}\!/_{0}$ Cu alloy, however, the difference in atomic size is 13 percent, and the zones develop as thin plates, only about 2 atoms thick, on the $\{100\}$ planes of the matrix.

Most alloys are given a *precipitation heat treatment* to develop the optimum properties. The corresponding structural changes depend on the temperature

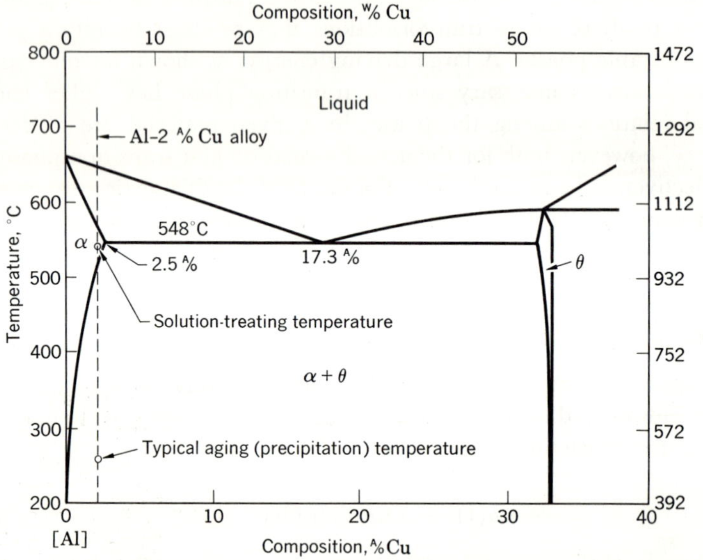

(*a*) Location of the Al-2 w/o Cu alloy in the aluminum end of the aluminum-copper phase diagram. The theta (θ) phase is C16 type structure, illustrated on inside back cover.

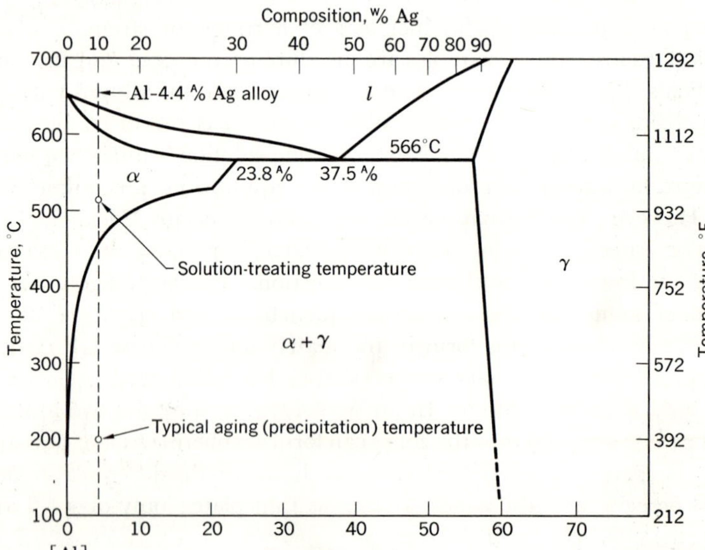

(*b*) Location of the Al-4.4 w/o Ag alloy in the aluminum end of the aluminum-silver phase diagram. The gamma (γ) phase has an HCP structure.

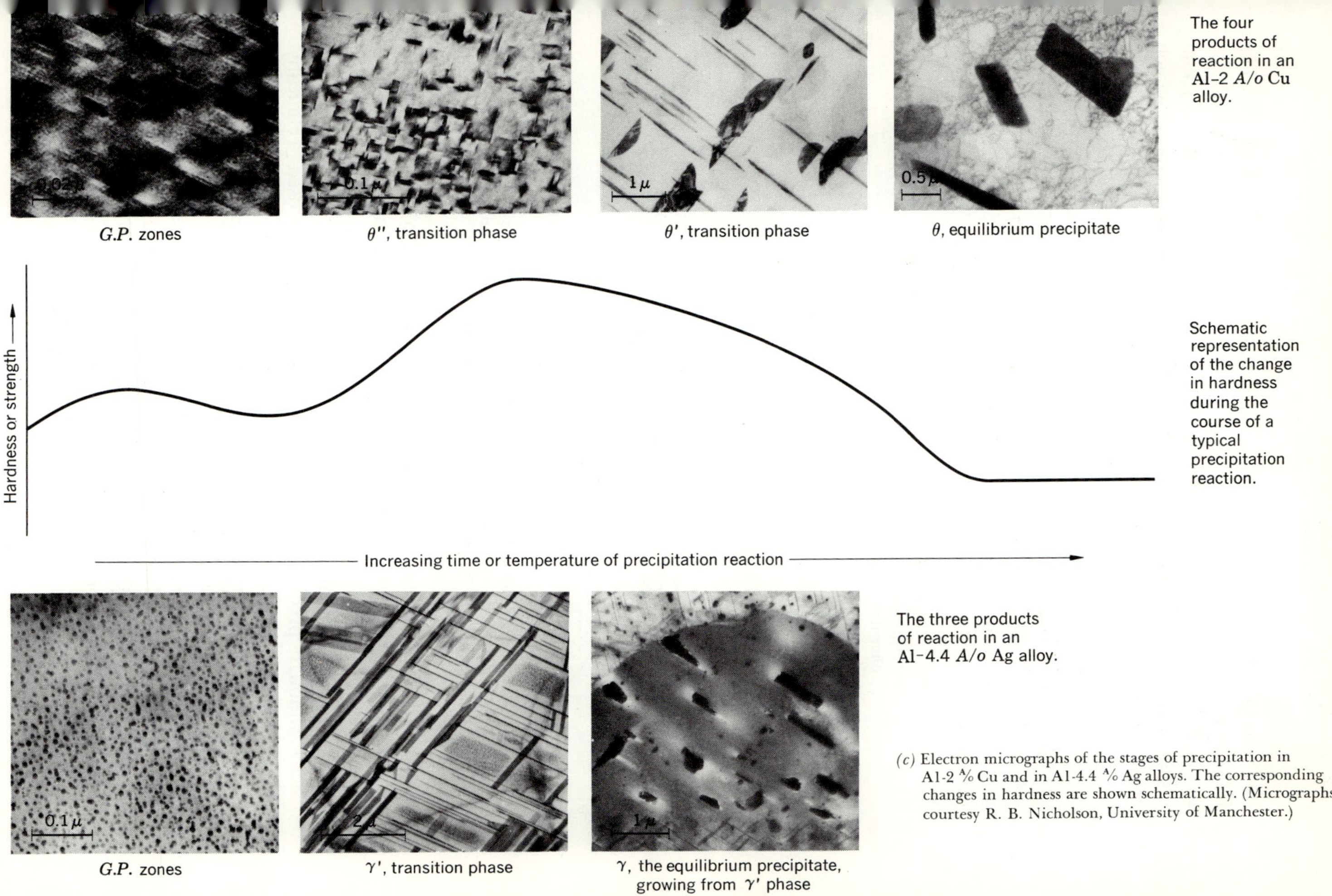

(*c*) Electron micrographs of the stages of precipitation in Al-2 A/o Cu and in Al-4.4 A/o Ag alloys. The corresponding changes in hardness are shown schematically. (Micrographs courtesy R. B. Nicholson, University of Manchester.)

FIGURE 7-38 Examples of precipitation reactions (age hardening) in two metallic systems, aluminum-copper and aluminum-silver. (*Diagrams adapted from M. Hansen and K. Anderko, "Constitution of Binary Alloys," McGraw-Hill Book Company, New York, 1958.*)

and time of the treatment, but typical changes are illustrated in Fig. 7-38. The GP zones continue to develop during the early stages and are responsible for appreciable hardening. A transition phase, θ'' or γ', appears fairly early and grows to produce a condition of maximum hardening, the stage at which commercial heat treatments are ended. Distortion of the matrix around the coherent transition phase (Fig. 7-39) is one of the factors contributing to hardening. Another factor is long-range ordering in the precipitate phase, for example, in the Ni_3Al phase (Fig. 2-19) that occurs in many alloys used for service at high temperatures. If the precipitation reaction is allowed to continue, phases with less coherency are produced, θ' (semicoherent) and θ (incoherent) in Al–2 $^A/_0$ Cu and γ (semicoherent) in Al–4.4 $^A/_0$ Ag. The size of

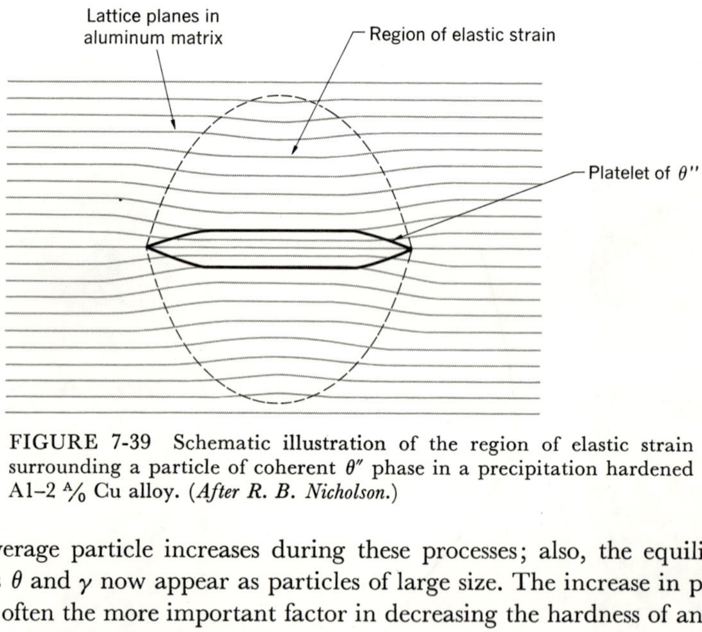

FIGURE 7-39 Schematic illustration of the region of elastic strain surrounding a particle of coherent θ'' phase in a precipitation hardened Al–2 $^A/_0$ Cu alloy. (*After R. B. Nicholson.*)

the average particle increases during these processes; also, the equilibrium phases θ and γ now appear as particles of large size. The increase in particle size is often the more important factor in decreasing the hardness of an alloy, but loss of coherency also plays a role.

Although precipitation reactions have been widely used technically only in metallic systems, they can occur also in semiconductors and in ceramics. In the MgO–Fe_2O_3 system, precipitation is greatly different under equilibrium conditions (Fig. 2-24) and under nonequilibrium conditions. If a MgO–1 $^W/_0$ Fe_2O_3 alloy is solution treated, quenched, and aged (Fig. 7-40*a*), coherent particles of $MgFe_2O_4$ having the spinel structure $H1_1$ precipitate within the MgO-rich solid solution. The particles are octahedrons only 150 Å in size and are coherent with the MgO matrix along {111} planes. Another interesting example is the composition MgO–75 $^M/_0$ Al_2O_3, which is a solid solution (spinel) at high temperatures (Fig. 5-29). The solid solution can be retained by rapid cooling to room temperature. On aging at 1000°C, a transition phase precipitates within the spinel matrix, increasing the hardness and especially the abrasion resistance. As in metallic systems, the equilibrium corundum (Al_2O_3-rich) phase is produced by prolonged heating. Quite a different pattern of precipitation (Fig. 7-40*b*) is shown by a crystal of sapphire (Al_2O_3) in which 0.5 percent of the aluminum atoms have been replaced by titanium. This structure is produced by a 10-hr precipitation treatment at 1500°C and results in appreciable increase in hardness and fracture strength.

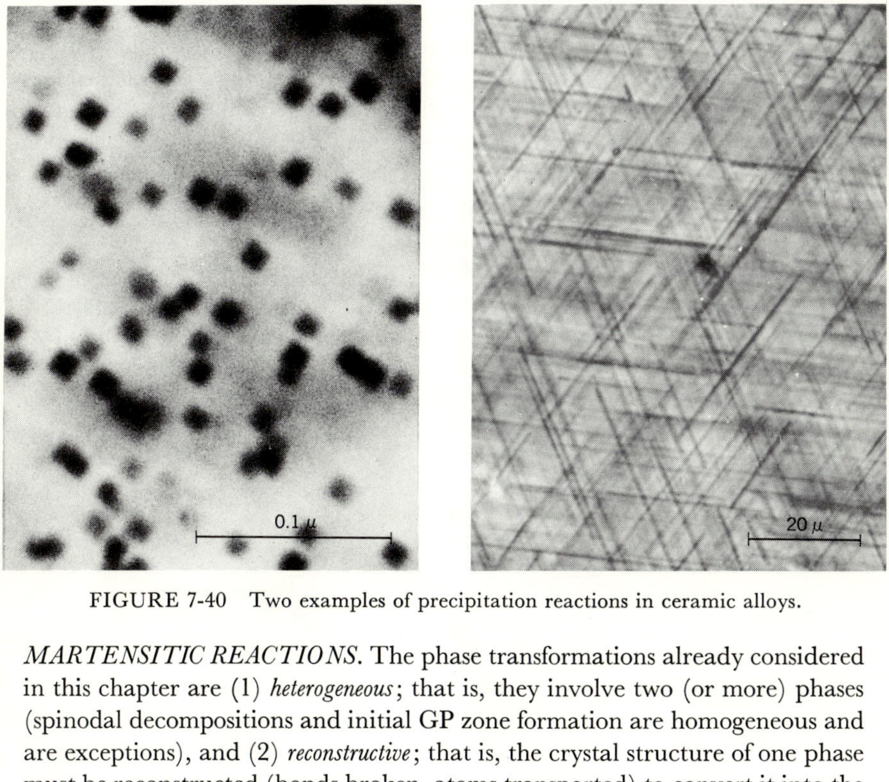

FIGURE 7-40 Two examples of precipitation reactions in ceramic alloys.

MARTENSITIC REACTIONS. The phase transformations already considered in this chapter are (1) *heterogeneous*; that is, they involve two (or more) phases (spinodal decompositions and initial GP zone formation are homogeneous and are exceptions), and (2) *reconstructive*; that is, the crystal structure of one phase must be reconstructed (bonds broken, atoms transported) to convert it into the new phase (Fig. 7-28). Heterogeneous transformations may alternatively occur by quite a different mechanism, provided that the phase in question can be cooled far below the range of temperature in which it is stable. The thermodynamic driving force then becomes great enough to allow each atom to shift a slight distance relative to its neighbors and produce a different phase (different crystal structure). This is the process of displacive transformation, and usually no thermal activation is necessary. Typically, the reaction occurs at very high rates, the interface between the two phases sometimes moving at almost the velocity of sound. A transformed crystal has a different shape than the original region from which it formed since the slight deformation is repeated regularly by all of the atoms in the crystal. This characteristic feature of a displacive transformation produces unevenness at surfaces and may even cause fracture of the material.

The name *martensitic reaction* used in the metallurgical literature comes from the well-known martensite phase in steel, produced by a displacive transformation. The familiar needle-like appearance of martensite in a photomicrograph (Fig. 2-1*c*) reflects the plate structure of this transformation product, Fig. 7-41. A characteristic feature is the surface tilting associated with each plate of martensitic transformation product. The faces of the plate lie along a characteristic habit plane of the initial solid solution. Unlike the atom-by-atom movements involved in diffusional transformations, martensitic reactions occur by the coordinated movement of many atoms, each atom moving relative to its neighbors by only a fraction of an atomic distance.

The essential thermodynamic aspects of martensitic reactions are illustrated by the simplified free-energy diagram of Fig. 7-42. The initial phase γ (usually a solid solution) is stable above a certain temperature T_e. In a range of temperature beginning slightly below T_e the equilibrium phase (or phases, such as $\alpha + Fe_3C$ in iron-carbon alloys) can nucleate and grow by a diffusional mechanism. If this mode of transformation of the γ phase is prevented by suitably rapid cooling (missing the nose of the TTT diagram, Fig. 7-33), then the martensitic transformation will begin when the temperature

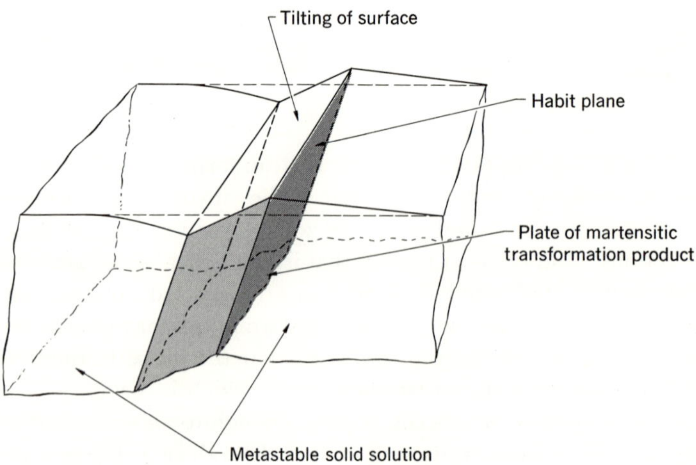

FIGURE 7-41 Schematic illustration of the surface tilting produced by the formation of a plate of martensitic transformation product.

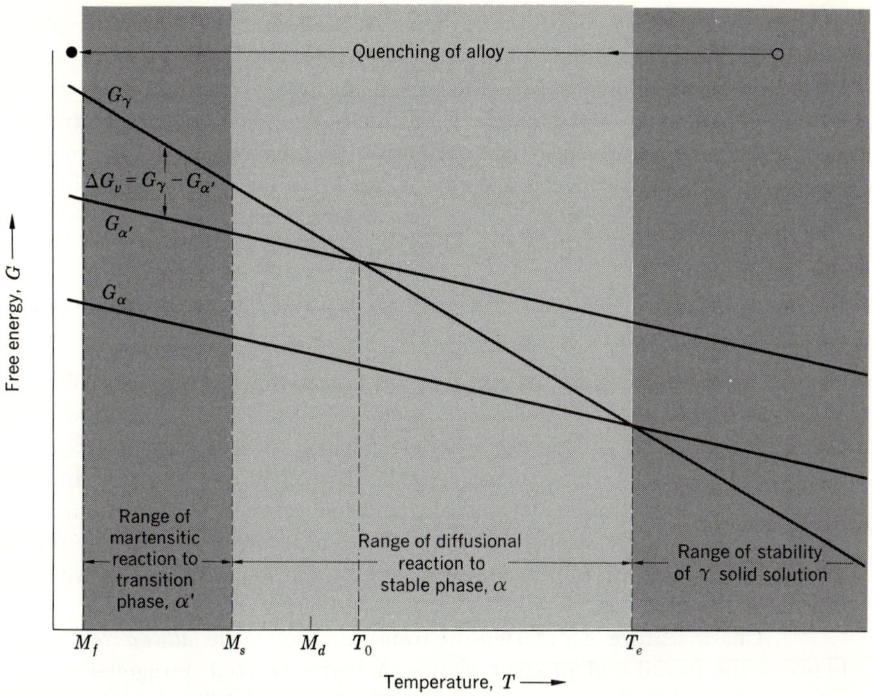

FIGURE 7-42 Stability of the initial solid solution γ relative to the stable phase α and to the transition phase α'. Quenching of the γ phase from a temperature above T_e may prevent diffusional reactions; the martensitic reaction then occurs during cooling in the temperature range between M_s and M_f. (*Adapted from J. Burke, "The Kinetics of Phase Transformations in Metals and Alloys," Pergamon Press, New York, 1965.*)

reaches M_s, the "martensite start" temperature. Although the γ phase becomes metastable with respect to the martensitic phase α' at a higher temperature T_0, the formation of the critical nucleus requires appreciable elastic energy, which must be supplied by ΔG_v, the volume free energy. Consequently, M_s is lower than T_0 by 20 to 400°, depending on the alloy. A portion of this elastic energy can be supplied by an externally applied force, and thus there is a temperature M_d at which γ begins to transform to α' when plastic deformation is performed on the quenched alloy. Because of the supercooling required to produce nucleation, a large driving force is usually available for the growth process. The nuclei then grow at a very high velocity, growth being stopped by impingement against a grain boundary, another martensite plate, or a similar impediment. Elastic strain energy is also generated during the growth process and is one of the factors limiting growth at a given temperature. If the temperature is lowered, ΔG_v increases, and additional energy becomes available for further nucleation or growth. If the temperature is then raised, the strain energy drives the reaction in the reverse direction, and the martensite plates change back into the initial γ phase. In a few systems, including iron-carbon alloys, this reversibility is prevented by a secondary reaction, such as the formation of a compound (carbide) by diffusion below the M_s temperature.

An important aspect of martensitic transformations is the overall deformation P_{ij} that converts a vector R_j in the initial solid solution into its new position R_i' in the martensitic product. In index notation [Eq. (1-7*a*)] this overall deformation is written for cartesian coordinates as follows:

$$R_i' = P_{ij}R_j \tag{7-52}$$

The overall deformation matrix P_{ij}, must account for the experimental observation that the habit plane is essentially undistorted and unrotated. Consequently, P_{ij} is not merely a homogeneous lattice deformation B_{ij}, but must also contain a heterogeneous shear S_{ij} (corresponding to slip, twinning, or faulting on an atomic scale) and also a lattice rotation Ω_{ij}. An example of these crystallographic aspects of martensitic transformations is the transformation of austenite to martensite in a 1 w/o carbon steel, which we now consider.

A basic feature of a martensitic transformation is the *lattice correspondence* between the initial and product phases. A body-centered tetragonal lattice, related to that of martensite, can be derived from the face-centered cubic austenite in the manner shown in Fig. 7-43. A vector $\mathbf{r}$ [with components r_i, see Eq. (1-2)] out to a given point in the austenite lattice corresponds to a vector $\bar{\mathbf{r}}$ (with components $\bar{r}_i$) to the same point, but referred to the body-centered tetragonal lattice. For the relative orientations of the two lattices shown in Fig. 7-43, the relation between the r_i and $\bar{r}_i$ components is (Prob. 7-19)

$$\begin{aligned} \bar{r}_1 &= 1r_1 - 1r_2 + 0r_3 \\ \bar{r}_2 &= 1r_1 + 1r_2 + 0r_3 \\ \bar{r}_3 &= 0r_1 + 0r_2 + 1r_3 \end{aligned} \tag{7-53}$$

In index notation [Eq. (1-7*a*)] this equation is

$$\bar{r}_i = \bar{C}_{ij}r_j \tag{7-54}$$

where $\bar{C}_{ij}$ is the lattice-correspondence matrix having the components shown in Fig. 7-43.

A lattice deformation B_{ij} is required to convert the tetragonal lattice of Fig. 7-43 with $\bar{a}_3/\bar{a}_1 = \sqrt{2}$ to the tetragonal lattice of martensite, for which the

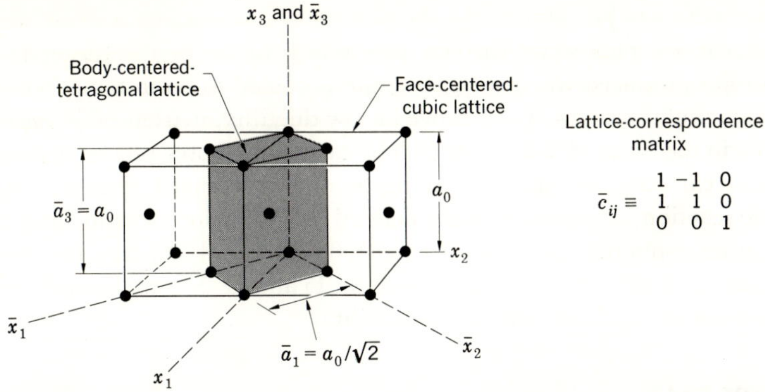

FIGURE 7-43 Illustration of the lattice correspondence between a face-centered cubic lattice (austenite) and a body-centered tetragonal lattice (related to martensite). This lattice correspondence is described by the matrix $\bar{C}_{ij}$. (*After C. M. Wayman.*)

a_3/a_1 ratio is only about 1.05. This lattice deformation consists of a contraction along the $\bar{x}_3$ direction and extensions along the $\bar{x}_1$ and $\bar{x}_2$ directions. If $\bar{X}_i$ are the orthonormal coordinates of a vector in the undeformed state, then $\bar{Y}_i$, given by

$$\bar{Y}_i = B_{ij}\bar{X}_j \tag{7-55}$$

are the coordinates of this vector after the lattice deformation.

A second type of deformation, the simple shear S_{ij}, is able to account for the occurrence of a planar interface. This shear occurs on an atomic scale and may be caused by slip, twinning, or the formation of stacking faults. Thus, the matrix P_{ij} [Eq. (7-52)] that adequately describes the overall deformation of the martensite transformation is a suitable product of the three simpler matrices B_{ij}, S_{ij}, and Ω_{ij}. In a 1.0 $^w/_0$ carbon steel, electron microscopy has verified that the shear S_{ij} is caused by twins about 10 Å wide within the martensite plates.

The strengthening of steel by a martensitic reaction is an important and extensive subject. Some essential aspects are listed here and are described in detail in the references. Asymmetric distortion of the iron matrix by the carbon atoms is the essential source of strengthening, and therefore a carbon content of at least 0.4 $^w/_0$ is usually employed. Since the formation of pearlite must be avoided during quenching of the austenite, alloying elements such as molybdenum, manganese, nickel, and chromium are added to the steel to shift the nose of the TTT diagram to longer times. The *hardenability* of the steel is then increased, and a bar of larger size can be successfully quenched to produce martensite. Arrays of dislocations at the surface of the martensite plates also contribute to strengthening. If the metastable austenite is plastically deformed at a temperature just above M_d, the process of Ausforming, a dense network of dislocations plus some carbide precipitate forms in the austenite. The martensite produced when such austenite is cooled to room temperature has especially high strength. To improve its low ductility, martensite is *tempered* by heating in the range of 200 to 600°C. The carbon then precipitates as transition or stable carbides, and the dislocations become rearranged. One type of martensitic reaction, *Maraging*, does not depend on carbon for strengthening. When martensite containing 18 w/o Ni, 0.4 w/o Ti, 8 w/o Co, and 5 w/o Mo, for example, is heated at 500°C, fine precipitates of Ni_3Ti and Fe_2Mo are formed and give appreciable strength and adequate ductility.

PROBLEMS

7-1. Using a reasonable estimate of molecular velocity, 10^5 cm/s, make a rough calculation of the separation of a molecule from adjacent molecules in the gas phase at the instant of evaporation, (*a*) for $J = 10^{18}$ and (*b*) for $J = 10^{25}$.

7-2. (*a*) Write the expression for $\Delta G(r)$, the change in free energy accompanying the (homogeneous) formation of a spherical embryo of radius r (Fig. 7-3*c*). Include one term involving ΔG_v and one involving γ. (*b*) Take the derivative $d[\Delta G(r)]/dr$, and set it equal to zero to determine the critical value r^* at which $\Delta G(r)$ has its maximum value ΔG^*. (*c*) Explain why the behavior of an embryo of radius r^* justifies the special designation *nucleus*.

7-3. (*a*) Consider the hypothetical experiment in which the center of a perfect crystal is heated above the melting point while the exterior is maintained below the melting point. Explain why the solid could exist in this metastable condition. (*b*) When an actual crystal is heated uniformly, melting begins at the surface with negligible superheating. Why?

7-4. Using the sketch of Fig. 7-15, show that solute accumulates in a volume element at the rate $R(dC/dx')$ because of the advance of the liquid-solid interface at the rate R cm/sec.

7-5. (*a*) Using one of the standard methods of differential equations (such as the substitution $C^l = Ae^{mx'}$), show that the general solution of Eq. (7-27) is $C^l = Ae^{-Rx'/D} + B$. (*b*) Evaluate the constants A and B by fitting this solution to the boundary conditions at $x' = 0$ and $x' = \infty$, and put the resulting equation in the form of Eq. (7-28).

7-6. Describe solidification in a system for which $k_e = 2$, using suitable modifications of Figs. 7-13, 7-14, and 7-15. Note especially changes in the initial and final transients.

7-7. Consider the use of zone leveling to produce a uniform concentration in a bar containing local segregation of solute but averaging $C_0 = 2$ A/o. (*a*) If $k = 0.1$, why would one pass of ordinary zone melting be unsatisfactory? (*b*) An alternative procedure is to prepare a special specimen in which the first l cm has the composition C_0/k or 20 A/o; l is the zone length. Show that zone leveling will be produced by one pass in all but the last zone length of such a bar.

7-8. (*a*) Explain the occurrence of stage 1 in Fig. 7-24 by analogy with the initial transient in Fig. 7-14. (*b*) At the end of stage 1 the composition of the solid being formed is $k_e C_E$. Why? (*c*) From the beginning of stage 2 and throughout the steady-state process, the liquid at the solidifying interface has the composition C_E. Use suitable sketches to explain the solidification behavior (1) near the beginning of stage 2 and (2) near the end of stage 2.

7-9. (*a*) Make a sketch analogous to Fig. 7-26*b* but for temperature T_2. (*b*) Determine ΔG_v for the nucleation of the A_xB_y phase by a graphical construction. (*c*) According to your sketch, is it possible for the β phase to nucleate? Explain.

7-10. Explain under what conditions a precipitate particle will take the form of (*a*) a platelet and (*b*) a spheroid.

7-11. Using the method of Sec. 6-8, show that the rate R in Eq. (7-39) is the net result of a competition between a reaction in the forward and in the backward direction.

7-12. Explain in terms of the boundary conditions of the corresponding diffusion problem in Chap. 6 why a $t^{0.5}$ dependence exists only in the early stage of diffusion-controlled growth.

7-13. Carry out the steps in the derivation of Eq. (7-47), beginning with Eq. (7-43).

7-14. Explain why the rate of transformation of austenite to pearlite goes through a maximum (the nose of the TTT diagram) as the reaction temperature is lowered below A_1.

7-15. Diffusion of carbon at the austenite/pearlite interface can occur by three alternative paths. Sketch them, and estimate which one should be preferred. The diffusion coefficient of carbon in ferrite is about 10^2 larger than in austenite.

7-16. Using the fact that $\mu = dG/dc$, show that Eq. (7-49) can be derived from Eq. (7-48).

7-17. Some ordered phases such as CuAu (Fig. 2-23) have a different crystal structure than the corresponding disordered solid solution. (*a*) Use a sketch to explain why elastic straining would occur in the domain structure in this case. (*b*) By analogy with Fig. 7-39, explain the fact that order hardening in a few cases is comparable in magnitude to precipitation hardening.

7-18. Not all alloy systems with a sloping solvus line are able to develop appreciable precipitation hardening. Contrast the behavior of a Pb–10 A/o Sn alloy (Fig. 7-23) and of an Al–2 A/o Cu alloy when each is given a solution heat treatment followed by precipitation at a lower temperature.

7-19. (*a*) By enlarging the body-centered tetragonal lattice in Fig. 7-43, show that the vector **r** out to the [1,1,1] point in the austenite lattice corresponds to the vector $\bar{\mathbf{r}}$ out to the [0,2,1] point in the body-centered tetragonal lattice. (*b*) Verify that Eq. (7-54) also gives this same correspondence. (*c*) Computer programs can conveniently perform calculations involving matrices. Two simple examples are the following:

```
C  MATRIX MULTIPLICATION R̄(I) = C̄(I,J)R(J)
C  DIMENSION
      DIMENSION RBAR(3), CBAR(3,3), R(3)
    1 FORMAT (3F7.3)
C  READ DATA
    2 READ (5,1) CBAR, R
C  CALCULATION
    3 DO 4 I = 1,3
    4 RBAR(I)=CBAR(I,1) * R(1)+CBAR(I,2) * R(2)+CBAR(I,3) * R(3)
C  OUTPUT SECTION
    5 WRITE (6,1) RBAR
    6 STOP
      END
```

```
C  MATRIX MULTIPLICATION R(I,J) = B(I,K)S(K,J)
C  DIMENSION
      DIMENSION R(3,3), B(3,3), S(3,3)
    1 FORMAT (3F9.3)
C  READ DATA
    2 READ (5,1) B, S
C  CALCULATION
    3 DO 5 I = 1,3
    4 DO 5 J = 1,3
    5 R(I,J) = B(I,1) * S(1,J) + B(I,2) * S(2,J) + B(I,3) * S(3,J)
C  OUTPUT SECTION
    6 WRITE (6,1) ((R(I,J), J = 1,3), I = 1,3)
    7 STOP
      END
```

REFERENCES

Bain, E. C., and H. W. Paxton: "Alloying Elements in Steel," American Society for Metals, Metals Park, Ohio, 1961.

Burke, J.: "The Kinetics of Phase Transformations in Metals," Pergamon Press, New York, 1965.

Christian, J. W.: "The Theory of Transformations in Metals and Alloys," Pergamon Press, New York, 1965.

"Mechanisms of Phase Transformations in Crystalline Solids," Metals and Metallurgy Trust, London, 1969.

Peckner, D. (ed.): "The Strengthening of Metals," Reinhold Publishing Corporation, New York, 1964.

Pfann, W. G.: "Zone Melting," 2d ed., John Wiley & Sons, Inc., New York, 1966.

Powell, C. F., et al. (eds.): "Vapor Deposition," John Wiley & Sons, Inc., New York, 1966.

Rutner, E., et al. (eds.): "Condensation and Evaporation of Solids," Gordon and Breach, Science Publishers, Inc., New York, 1964.

Shewmon, P. G.: "Transformations in Metals," McGraw-Hill Book Company, New York, 1969.

Tiller, W. A.: Solidification, in R. W. Cahn (ed.), "Physical Metallurgy," 2d ed., p. 403, North Holland Publishing Company, Amsterdam, 1970.

Wayman, C. M.: "Introduction to the Crystallography of Martensitic Transformations," The Macmillan Company, New York, 1964.

Zackay, V. F. (ed.): "High-Strength Materials," John Wiley & Sons, Inc., New York, 1965.

Chapter Eight

Annealing Processes in Materials

The term *annealing* broadly covers the heating of materials for a certain period to produce desired changes in properties. In contrast to phase transformations, which also involve heating the material, annealing processes ordinarily occur within the existing phase or phases. The material simply changes from a higher-energy (metastable) state to a more stable one. The initial state may be that of a pressed powder, of a solidified casting, or of a rapidly transformed solid. However, the most important initial state is that produced by prior deformation, especially the cold working discussed in Sec. 8-1. The successive stages in the annealing of cold-worked materials are *recovery* (of certain properties), *recrystallization* (of the grain structure), and *continued growth* of the relatively stable grains. Since most ceramics are too brittle to be cold worked appreciably, metallic materials (and also polymers) are the principal subjects in this chapter.

INITIAL STATE OF MATERIAL

8-1. Cold-worked Materials

A significant fraction of commercial metals are used in a cold-rolled (or other cold-worked) condition and owe much of their strength to the effect of cold working. Chapter 9 considers the subject of plastic deformation more broadly, and here we wish only to understand the changes in properties and structure produced by cold working. The effect of cold working on mechanical properties is roughly the same whether the method of working is rolling, elongation in tension, wire drawing, etc. The amount of working is usually expressed as the percent elongation of the specimen in a tension test or as the percent reduction in cross-sectional area in rolling or in wire drawing (see Prob. 8-1). Figure 8-1 shows how the tensile strength of copper and of three copper-base alloys gradually increases with amount of cold working. Doubling of the initial strength is typical for large amounts of cold work. The presence

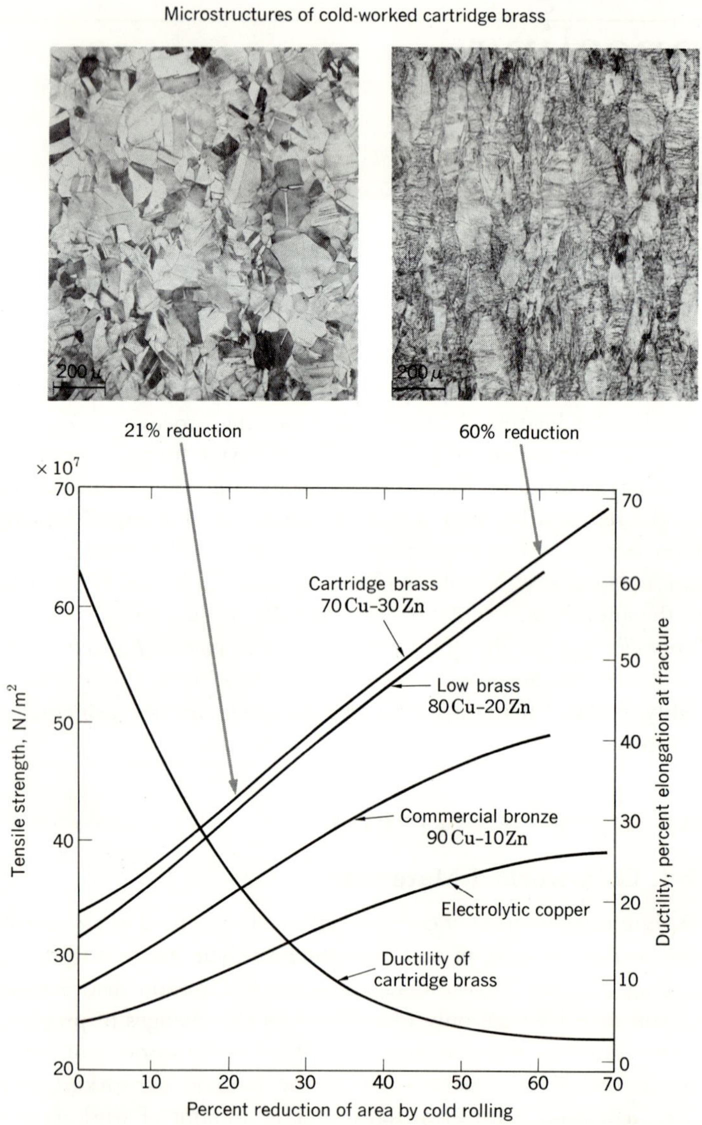

FIGURE 8-1 Tensile strength determined on sheet specimens of copper and copper-base alloys that had been previously subjected to various amounts of cold rolling [conversion of units: (M N/m^2) $1.450 \times 10^{-4} =$ N lb/in^2; (M N/m^2 $1.020 \times 10^{-7} = N$ kg/mm^2]. (*Photomicrographs courtesy Anaconda American Brass Company.*)

of alloying elements not only increases the strength of the unworked specimen but also increases the strengthening effect of a given degree of cold work—the rate of *strain hardening* (or work hardening) of the material (see Prob. 8-2). Refining the grain size also increases the rate of strain hardening. The microstructures of Fig. 8-1 show that distortion of the grain shapes becomes pronounced only at high degrees of cold work. A typical loss in ductility accompanying cold working is illustrated for cartridge brass.

Imperfections such as dislocations, point defects, stacking faults, and twins are known to be responsible for the changes in properties accompanying cold working. Prior to cold working the density of dislocations is relatively low, but even a few percent of cold working (Fig. 8-2) produces the high density of dislocations shown in the first electron micrograph. With increasing amounts of cold working, the dislocations tend to segregate into cell walls, leaving the center of the cells relatively free of dislocations. The increase in the yield strength of iron is seen to parallel the increase in density of dislocations. A

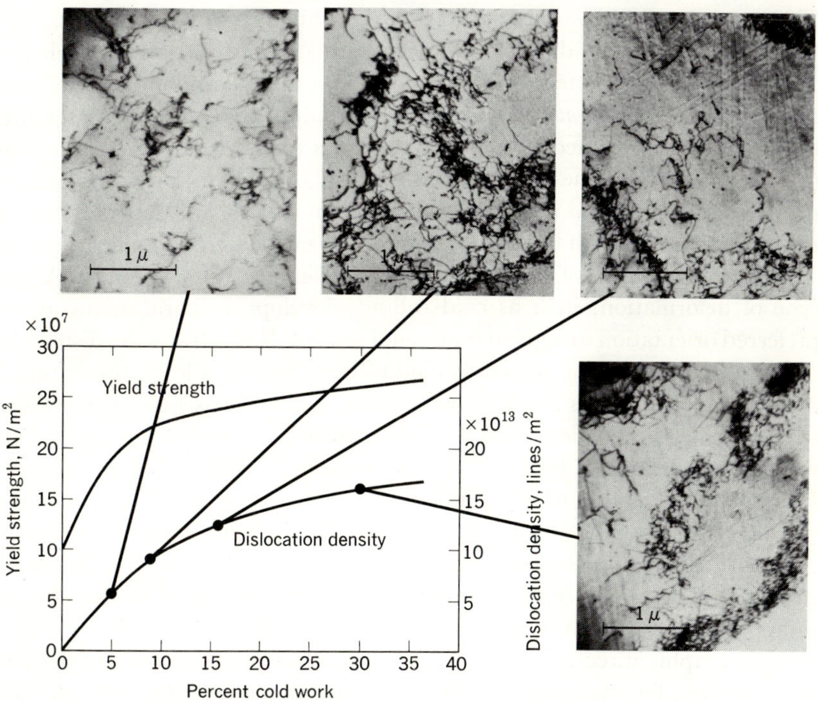

FIGURE 8-2 Effect of cold working on the yield strength of pure iron. The electron micrographs show the corresponding changes in number and distribution of dislocations [conversion of units: M lines m^2 $\times$ 10^{-4} = N lines cm^2]. (*Research of the late A. S. Keh, Research Center, U. S. Steel Corporation. By permission from Direct Observation of Imperfections in Crystals, Interscience Publisher, New York, 1962.*)

useful measure of total quantity of imperfections produced by cold working is the resulting stored energy in the deformed material. The stored energy is the driving force for the processes of recovery and recrystallization of the cold-worked structure. Structural imperfections can also be produced by neutron irradiation, by quenching from high temperatures, or by some solid-state transformations. These imperfections are also eliminated by annealing.

A useful concept for estimating the properties of various materials is that of homologous (or corresponding) temperature, T/T_f. Two substances can be expected to behave similarly with respect to a given property (such as cold-working behavior) when their temperatures are the same fraction of their absolute melting points (in degrees Kelvin). Because annealing removes the effects of cold working, if working is performed at a sufficiently high temperature, the structural imperfections are immediately eliminated. Deformation under these conditions is called *hot working* and corresponds to a homologous temperature of about 0.6 T_f. Tin, whose melting temperature is 505°K, can be hot worked at room temperature. On the other hand, tungsten (T_f = 3683°K) can be cold worked up to 1700°C, while it is at a white heat.

Because plastic deformation occurs in certain directions in each grain of a polycrystalline material (see Chap. 9), an initially random array of grains develops a *preferred orientation* or *texture* as a result of cold working. A texture is conveniently represented by plotting the poles (or normals) of certain planes (such as the {200} planes, determined by x-ray diffraction) on a stereographic projection. If the grains are oriented at random (Fig. 8-3*a*), the poles from the multitude of grains will be spread uniformly over the stereographic projection. This uniform density of poles is assigned a reference value of unity. A given type of deformation, such as cold rolling, develops a characteristic type of preferred orientation in a given material. Figure 8-3*b* illustrates the distribution of grain orientations produced by cold rolling silver at a low temperature. In certain preferred areas the relative density of poles is as high as 6, whereas over large areas it is less than 1. This texture is simple enough to be very roughly approximated by the "ideal orientation" (110) [$1\bar{1}2$], in which the (110) plane is parallel to the plane of cold rolling and the [$1\bar{1}2$] direction is parallel to the rolling direction (see Prob. 8-3). The texture of cold-rolled copper (Fig. 8-3*c*), although roughly similar to that for silver, is sufficiently more complex to require that a mixture of two ideal orientations must be used to approximate its distribution of poles. In the cold drawing of wire a given crystallographic direction tends to become aligned along the axis of the wire; an iron wire, for example, has the [110] direction as its *fiber axis* (see Prob. 8-4).

Textures can be described and controlled, but a basic theory of their origin is still lacking. In Chap. 9 we will see that the orientation of a single crystal changes in a definite manner during certain types of deformation. This

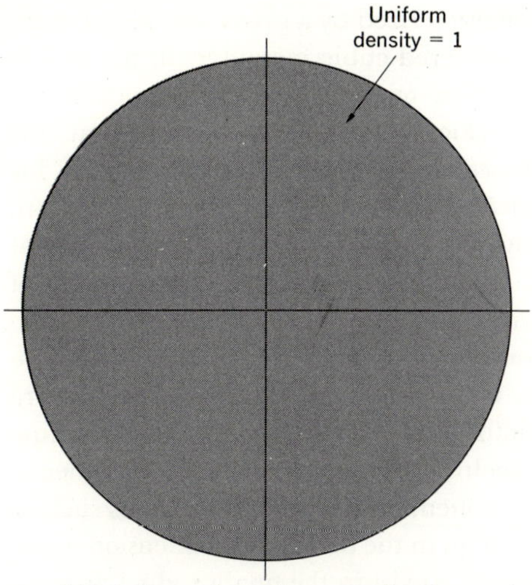

(a) Pole figure of a random array of grains, ideally present in a material before plastic deformation.

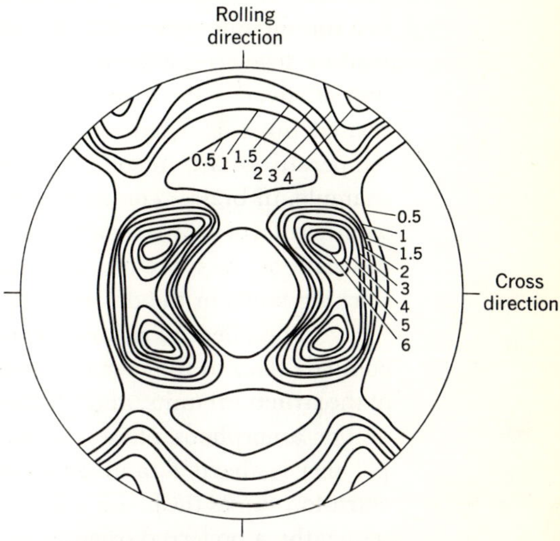

(b) Pole figure of texture present in silver sheet cold rolled 91% at 0°C. (Data of H. Hu and R. S. Cline.)

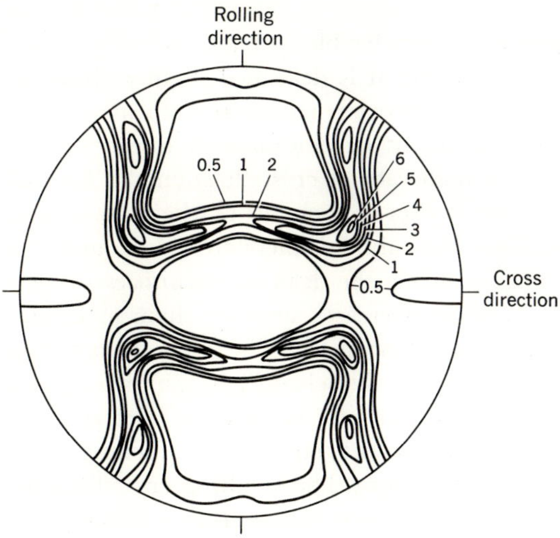

(c) Pole figure of texture present in copper sheet cold rolled 97%. (Data of H. Hu and S. R. Goodman.)

FIGURE 8-3 Stereographic projections of the {200} poles of the grains in polycrystalline materials to compare a random array (*a*) with two types of preferred orientations (*b*) and (*c*).

simple picture becomes greatly complicated, however, when the multitude of mutually constraining grains in a sheet of metal are subjected to deformation in a rolling mill. Nevertheless, metals characterized by a given crystal structure tend to develop similar textures. Face-centered cubic metals (with the exception of silver) have a rolling texture similar to that of copper (Fig. 8-3*c*). A related but somewhat simpler texture (Fig. 8-3*b*) develops in silver at low temperatures and in solid solutions based on face-centered cubic metals, for example, in brasses containing at least 5 percent zinc. The brass-type texture seems to form in materials with a low stacking-fault energy. Slip of partial dislocations then occurs more easily, leading to a relatively high density of stacking faults in the deformed material.

The effect of prior deformation on the structure of polymers is more complex than in the case of metals for two reasons: (1) the long-chain nature of the structural units, and (2) the various states of the polymer (as a crystalline and/or amorphous phase) below its melting temperature T_f. Even when a polymer is above T_f and therefore is "molten," a deformation process such as extrusion causes the chain molecules to align in the direction of extension. Consequently, a preferred orientation is found to exist in the final product at room temperature, whether or not crystallization occurs during final cooling. The orientational effects are greatly enhanced, however, by crystallization of a partially aligned melt.

An amorphous, but potentially crystallizable, polymer behaves in two quite different ways depending on whether it is deformed slightly above or below T_f. The equilibrium value of T_f increases markedly with increase in tensile stress. Also, as the molecular chains become aligned during tensile deformation, kinetic factors favor a high rate of crystallization. T_f for undeformed natural rubber is 30°C, but it rises to about 70°C when a specimen is extended by 300 percent (Fig. 8-4). The rubber is therefore able to undergo crystallization (and become stronger) during the deformation process. When the stress is removed, however, T_f again decreases, and the chain molecules tend to return to their more random arrangement. The other behavior is shown by nylon 6-6 (Fig. 8-4), which is already below T_f in the undeformed condition but fails to crystallize because the rate is too slow. Deformation encourages crystallization both by raising the equilibrium melting temperature and by improving the kinetic factors. The resulting crystallization persists after the stress has been removed since the temperature is below T_f.

Polymers containing a large fraction of crystalline regions deform differently depending on whether the temperature of deformation is above or below T_g. Below T_g, molecular motion is so limited that most polymers (including amorphous ones) can deform only slightly before they fracture. Above T_g, even a macrospherulitic structure (Fig. 3-7*b*) is able to convert into a fibrillar one (Fig. 3-8*b*) by a sliding motion of chain molecules in the crystal

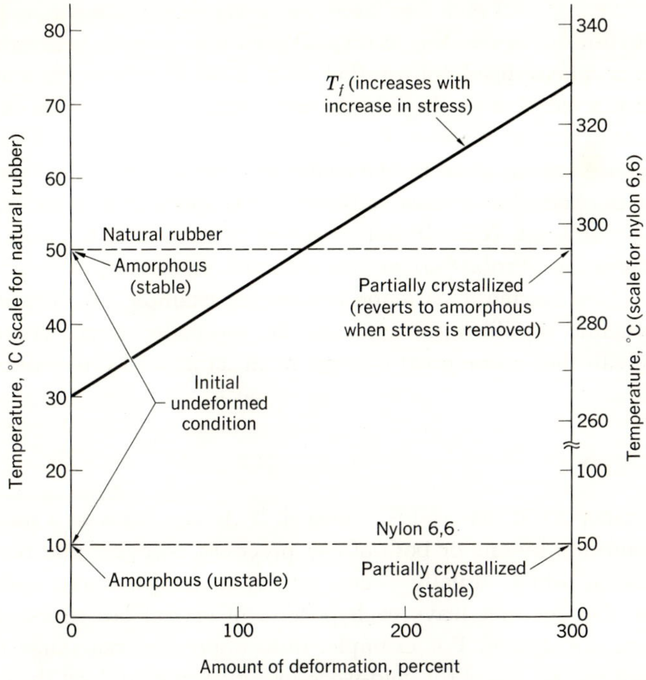

FIGURE 8-4 Effect of tensile deformation at 50°C on the crystallization of two initially amorphous polymers, natural rubber (initially above its T_f) and Nylon 6,6 (below its T_f).

lattice. A possible mechanism for this conversion is shown in Fig. 3-8*c*. The thickness ("long period") of the lamellae composing the fibrils depends on the temperature of deformation.

8-2. Other Metastable States

A given material may be in a metastable condition for many reasons besides cold working. The driving force for sintering of compacted fine powders is discussed in Sec. 6-6. Drastic departures from a stable condition may be shown by materials subjected to irradiation by such particles as neutrons, electrons, and ions. Major changes in chemical composition may occur and are especially important for polymers. The effect of ordinary light (photons) on polymers is discussed in Chap. 11. If the main effect of irradiation is only a local disruption of the normal crystal structure, then an annealing process is often effective in returning the material to its original condition.

A quite general condition of metastability exists in every material to the extent that it departs from its condition of thermodynamic equilibrium. For example, even a steel that has been relatively slowly cooled to produce a coarse pearlitic structure (Fig. 2-1*a*) will gradually change because of surface energy to a spheroidized constitution (Fig. 2-1*b*) if heated for a sufficiently long time at a temperature high in the ferrite-plus-carbide region of the phase diagram. In a similar manner, the lamellae in a crystalline polymer have an initial thickness determined by the conditions under which they were formed. A greater thickness has greater stability and is approached during a suitable annealing treatment. If the initial condition of a material has been produced by rapid cooling (Prob. 8-5) or has involved certain types of phase transformations (the martensite reaction in steel, for example), the departure from thermodynamic equilibrium tends to be especially large. An annealing treatment can then cause great changes in structure and properties.

RECOVERY

When a cold-worked material is annealed, it changes back to a stable, lower-energy condition by one or both of two processes, recovery and recrystallization. *Recovery*, which usually occurs at lower temperatures, includes any structural changes that leave each portion of crystal lattice essentially unchanged in orientation. For example, dislocations can rearrange themselves and cause a partial loss of the cold-work strengthening, and yet the orientation of the lattice is almost unaffected. If the degree of cold work is small or if a single crystal deforms by slipping on only one set of slip planes, recovery alone may be sufficient to return the plastically deformed material to a stable state equivalent to that existing prior to cold working. Usually, however, a second process, *recrystallization*, also occurs, in which the boundary separating two regions having different orientations, A and B, moves across area A and causes it to change to orientation B. Although recovery and recrystallization both occur during a typical annealing treatment, often competitively, each process is considered separately in the following section.

8-3. Nature of the Recovery Process

The nature of the recovery process in a given material depends on the manner in which the "disturbed" structure is produced—by cold working, by neutron irradiation, or by quenching. All of these processes cause an increase in the electrical resistivity of the material, and thus the progress of recovery can be followed by measuring the gradual disappearance of this extra resistivity. An especially simple example is the recovery of a metal that has been quenched from a temperature near its melting point into liquid nitrogen (80°K) (curve *a* in Fig. 8-5). With increase in annealing temperature the resistivity of quenched copper remains almost unchanged until room temperature has been reached. Over a short interval of temperature the resistivity drops almost to the value of normal copper, i.e., to the value before quenching. The main process causing this recovery phenomenon is known to be the migration (and eventual elimination at dislocations) of excess vacancies (see Prob. 8-6). This process, conveniently denoted as stage IV recovery, also occurs in cold-worked metals (curve *b* in Fig. 8-5) but in this case is preceded by stage III recovery, probably caused by the migration of vacancies as bound pairs (divacancies). Prior to recrystallization of a cold-worked metal, stage V recovery is observed and is attributed to some type of reaction involving the

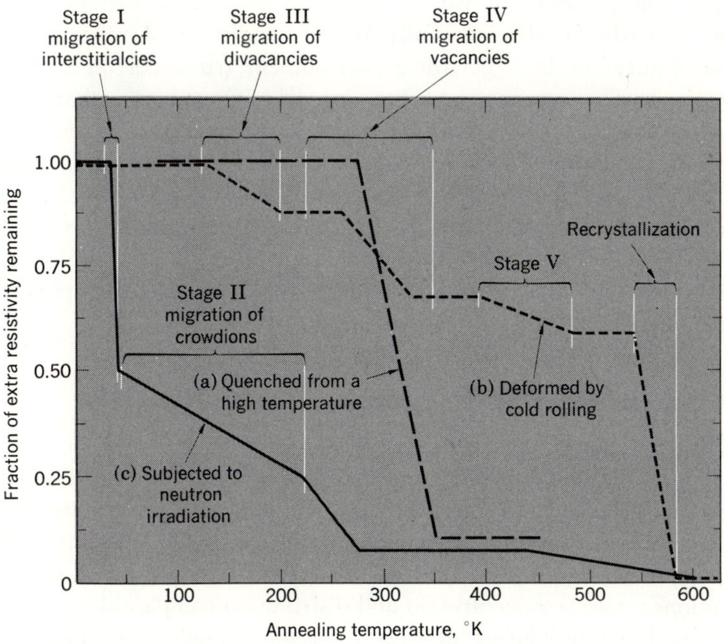

FIGURE 8-5 Schematic isochronal annealing curves showing the recovery of electrical resistivity in copper that had been distorted by various treatments. (*After J. S. Koehler, et al.*)

dislocations. Curve *c* in Fig. 8-5 shows that the radiation damage produced by neutron irradiation is distinctly different from the structural distortions after quenching or cold working. Although excess vacancies are present, a new feature is the occurrence of many interstitialcies (atoms squeezed into interstitial sites). The first two stages of recovery then involve the migration of single interstitialcies and of bound linear groups of interstitialcies (crowdions).

The observed effect of a given recovery anneal depends on the property being measured. Electrical resistivity, for example, changes greatly during recovery, but hardness and strength are usually only slightly affected. An important example of this difference in behavior is shown by the two types of residual stresses introduced into a specimen during deformation. *Body stresses* are those stresses that vary over macroscopic dimensions. For example, a cold-drawn bar (Prob. 8-7) has tensile stresses in the surface region, but these gradually change to compressive stresses at the center. *Textural stresses* are associated with such microstructural features as a particle of precipitate, a plate of martensite, or inhomogeneous deformation from one grain to another. Tensile and compressive stresses alternate within a microscopic region, creating a balanced texture of stresses. Textural stresses are an essential part of useful strengthening processes in various materials, but body stresses often contribute to early failure, especially under conditions of repeated loading (fatigue failure) or in a corrosive environment (stress-corrosion cracking). *Compressive* surface stresses, however, are widely used to enhance fatigue resistance (see Sec. 10-7). Fortunately, a low-temperature recovery anneal permits the small amount of plastic deformation of each local region adequate to reduce the long-range body stresses to a safe level; in most cases the textural stresses, and therefore the strength, are reduced only slightly or not at all. Usually the added strength of cold-worked metals is lost only slightly even at temperatures high in the recovery range but disappears completely after recrystallization.

8-4. Behavior of Dislocations During Recovery

Before examining ordinary cold-worked metals, we consider two special cases. If a single crystal has been deformed by the gliding of dislocations along a single set of planes and without any bending, annealing of the deformed crystal permits the residual dislocations to move on their slip planes. Dislocations of opposite sign attract and annihilate each other, and soon the residual dislocations attain a concentration and distribution comparable to that of the undeformed crystal. The second case also involves single glide (Fig. 8-6*a*) but is complicated by bending and the resulting slight misorientation of adjacent regions of the deformed crystal. An excess of edge dislocations of one sign remains in the slip planes to accommodate the bending. Since annealing removes the bent shape only partially, it cannot remove the extra dislocations needed to compensate for the bending. Instead, the diffusion-controlled recovery process consists of the movement of individual dislocations by glide and climb to form walls (Fig. 8-6*b*) perpendicular to the initial alignment of the dislocations along slip planes. This process is shown schematically in Fig. 8-6*c* and results in a lower interaction energy for the group of dislocations. At higher temperatures of recovery (Fig. 8-6*d*) all of the dislocations are in the polygonized boundaries, and coalescence of adjacent boundaries occurs with a further lowering of energy. Recrystallization does not occur in either of these two special cases.

The recovery of a typical cold-worked, polycrystalline material also involves the rearrangement of dislocations. The process of rearrangement actually begins during the cold-working operation. Dislocations then arrange

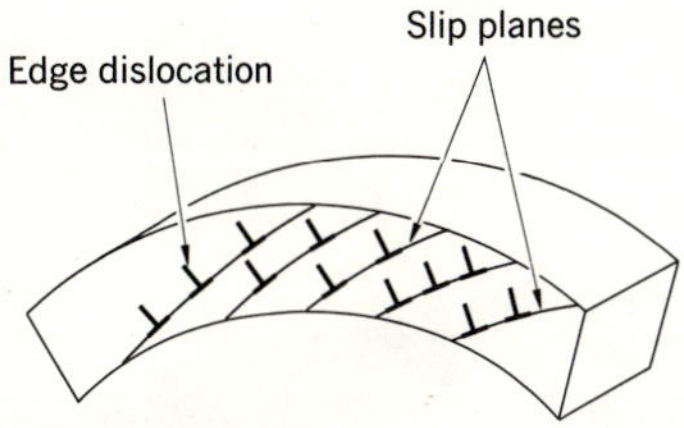

(*a*) Dislocations on the (110) slip planes that permit the plastic bending shown.

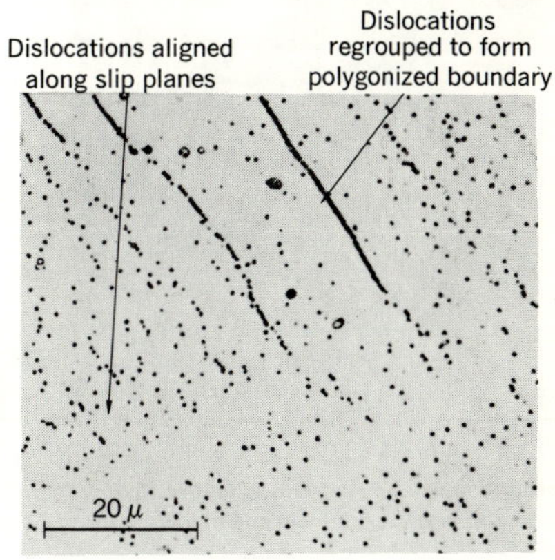

(*b*) After 1 hr. recovery anneal at 700°C the initial alignment of dislocations along slip planes is only partially changed to polygonized boundaries.

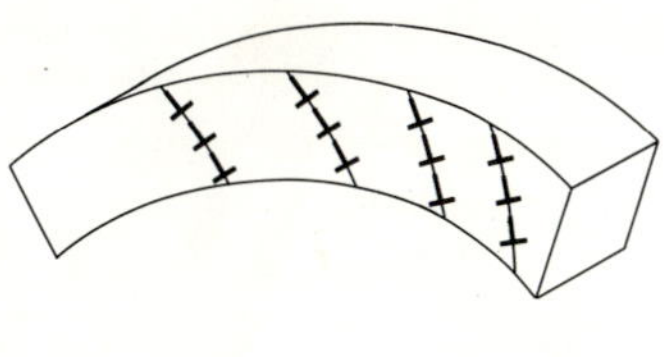

(*c*) Edge dislocations align themselves vertically to lower their interaction energies, thus forming polygonized boundaries.

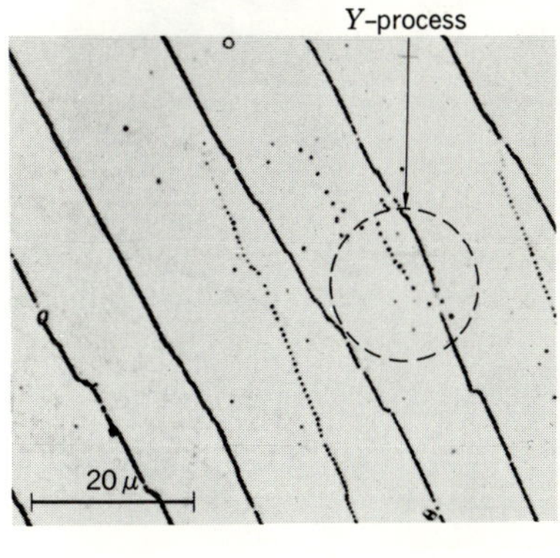

(*d*) After 1 hr. recovery anneal at 925°C almost all dislocations are in polygonized boundaries. The coalescence of adjacent boundaries (*Y*-process) is progressing.

FIGURE 8-6 Illustrations of the plastic bending of a single crystal of body-centered cubic iron-silicon alloy and of the polygonized structure that gradually forms during recovery annealing. (*Courtesy C. G. Dunn, Research and Development Center, General Electric Company.*)

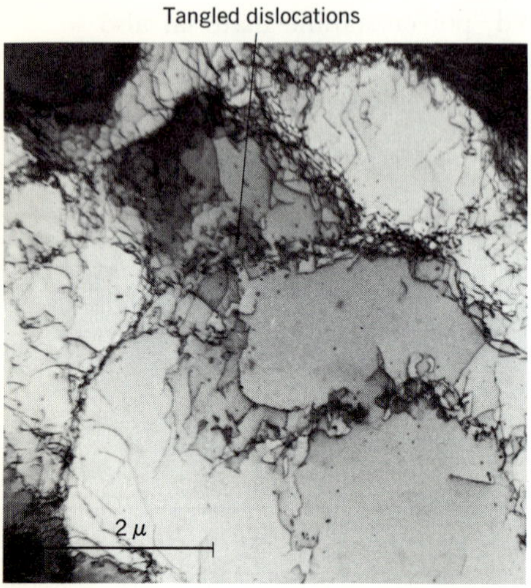

(*a*) Initial structure of cold-worked aluminum; cell boundaries are dislocation tangles with loops and dipoles.

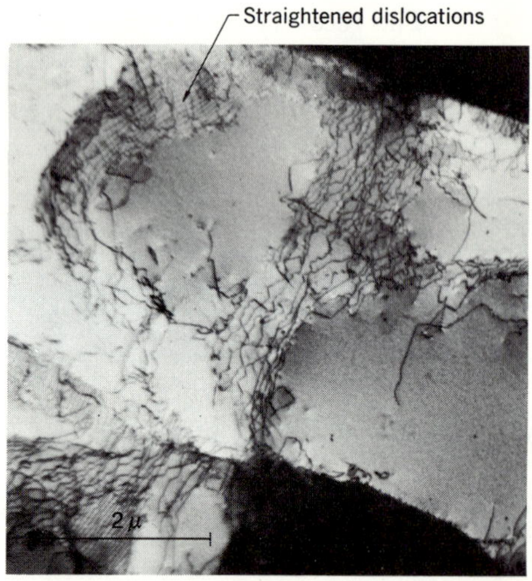

(*b*) After recovery for 0.1 hr. the dislocations in some of the initial tangles have become fairly straight.

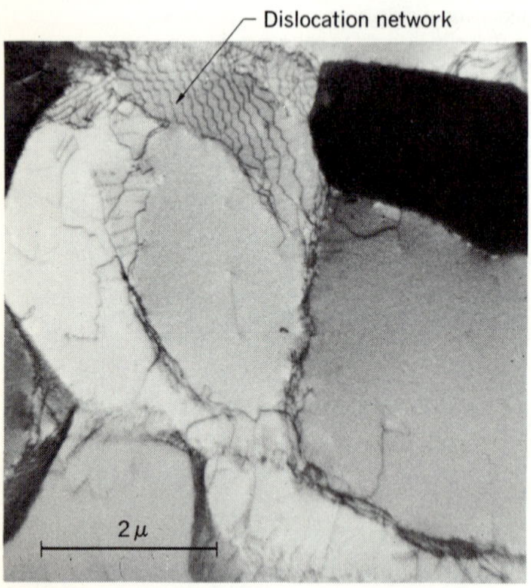

(*c*) After recovery for 50 hrs. the dislocations in each cell wall have formed a network; the subgrains have elongated.

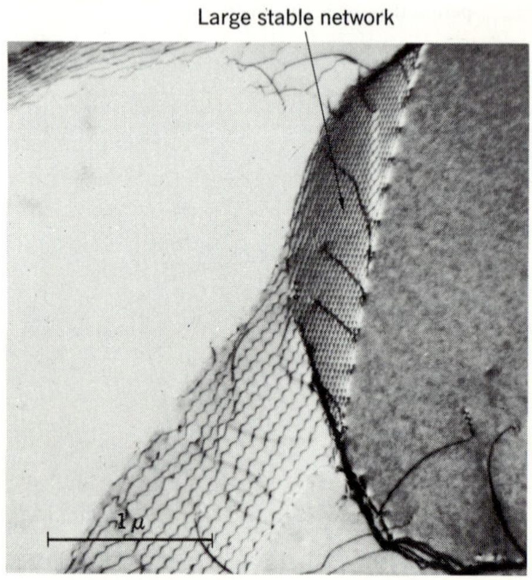

(*d*) After recovery for 300 hrs. some networks have broken up and joined more stable networks.

FIGURE 8-7 Electron micrographs illustrating the development of cell boundaries during recovery at 200°C of pure aluminum, initially cold worked 5 percent at room temperature. (*Courtesy C. T. Young and J. L. Lytton, Virginia Polytechnic Institute.*)

themselves in tangles that form the diffuse cell boundaries in the cold-worked structure (Fig. 8-7*a*). After the earliest recovery processes, such as the disappearance of excess vacancies, the least stable dislocation arrangements in the tangles begin to change (Fig. 8-7*b*). Dislocations in the body of the cell tend to migrate to the cell boundaries; prismatic dislocation loops, which were generated near the tangles during cold working, begin to shrink and disappear. The tangled dislocations in the cell boundaries form a straighter, less confused arrangement. As recovery progresses (Fig. 8-7*c*), the cell proper becomes almost free of dislocations and assumes a uniform orientation. The term *subgrain* is now more descriptive than cell. The dislocations in the boundaries take on the low-energy configuration of networks, providing a system of sharp, distinct subboundaries. Coalescence of subgrains occurs when the intervening wall of dislocations dissociates and its dislocations join more stable arrays in adjoining walls. The subgrains then increase in size and elongate in preferred crystallographic directions. Sometimes the total energy is lowered when adjacent dislocation networks join to form a subboundary of more closely spaced dislocations (Fig. 8-7*d*). This process of subboundary coalescence also contributes to the growth of subgrains during prolonged recovery treatments. The subgrain size tends to be smaller the lower the temperature of recovery and the higher the amount of prior cold work.

The behavior of aluminum described here involves the principal features exhibited in the recovery of most cold-worked materials, although cell formation may be less pronounced. Rearrangement of dislocations, rather than their disappearance, is the dominant process. Usually only a slight decrease in strength accompanies recovery, the decrease being greater the larger the final subgrain size. Removal of the bulk of the dislocations produced by cold working occurs only during recrystallization, by the process described in the following section.

RECRYSTALLIZATION

A pronounced change in configuration of the grains characterizes recrystallization both in inorganic materials and in organic polymers. In polymers the "grains" in question are the lamellae and the (supergrains) macrospherulites. The mechanisms of recrystallization are sufficiently different for the two classes of materials to warrant separate consideration.

8-5. Recrystallization of Inorganic Materials

When the process of recrystallization is observed microscopically, a new set of grains seems to replace the original, cold-worked set. Recrystallization is easily observed with the optical microscope in metals (Fig. 8-8) and in those ionic crystals capable of sufficient plastic deformation. Although new grains

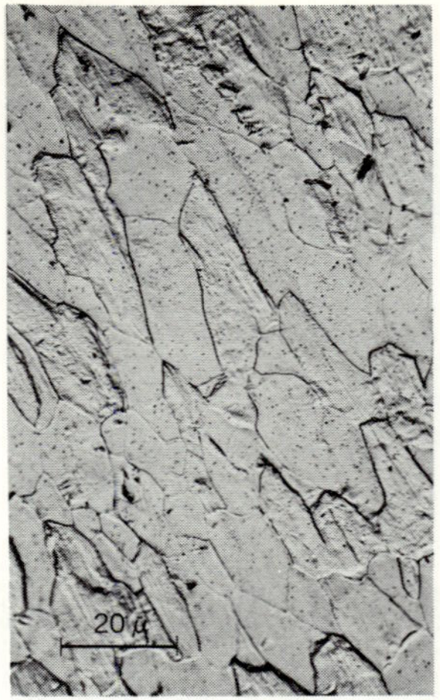

(*a*) Microstructure of cold-worked alloy.

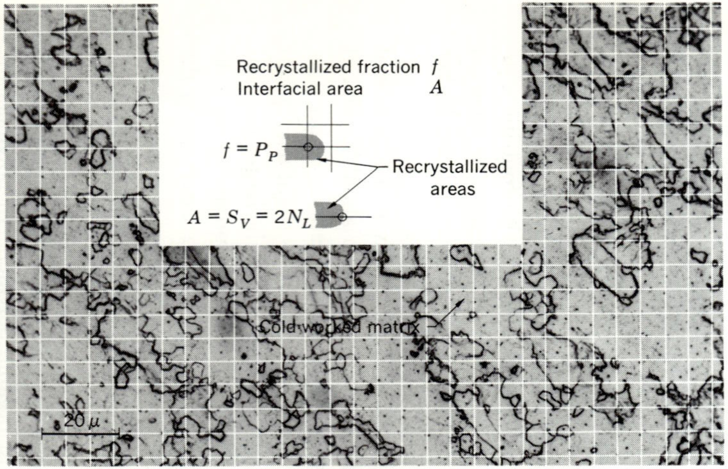

(*b*) Measurement of f and A on partially recrystallized specimen using a grid and point-counting methods.

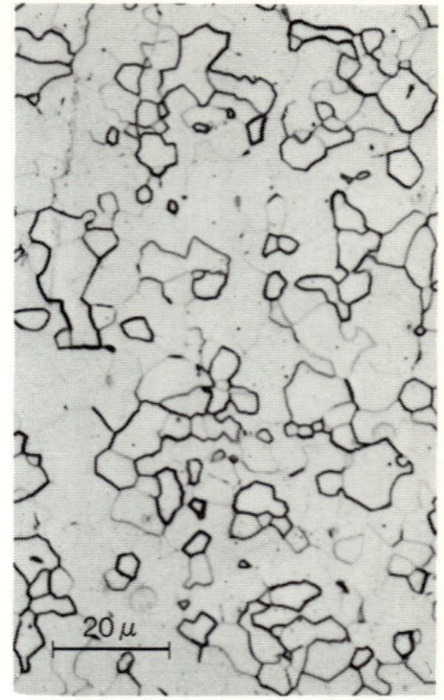

(*c*) Microstructure after complete recrystallization (100 sec. at 700°C).

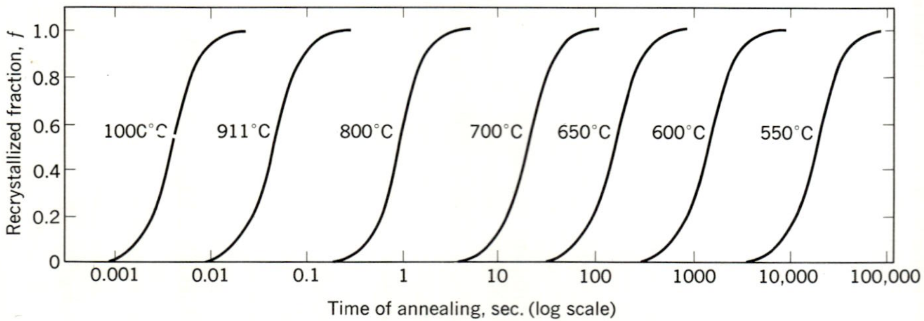

(*d*) Isothermal recrystallization curves for temperatures in the range 550° to 1000°C.

FIGURE 8-8 Recrystallization produced by annealing a 3.25 w/o silicon-steel that had been previously cold-rolled 60 percent. (*Courtesy* appear to nucleate and then grow, the actual mechanism consists of the migration of existing boundaries. When a cold-worked metal is first heated into the appropriate range of temperature for recrystallization, a few of the millions of subboundaries undergo some modification during an initial induction period and become mobile. Boundaries with high misorientation or in a highly deformed area of the specimen are usually activated first. A boundary advances by taking in atoms from the cold-worked matrix and adding atoms to the almost dislocation-free region behind the boundary. The principal source of energy driving the advancing boundary is the strain energy per unit volume, F_v, of the cold-worked matrix. If unit area of boundary advances a distance dx, the energy liberated, $F_v\,dx$, is approximately equal to the work done on the boundary,

$$F_v\,dx = p\,dx \tag{8-1}$$

Therefore, F_v is equal to p, the effective pressure acting on the boundary. The rate of migration, G, produced by a given value of F_v depends, among other factors, on the relative orientation of the two crystals and on the orientation of the boundary separating the two crystals. A tilt boundary migrates faster than a twist boundary or a boundary of mixed character. Relative orientations approximating those of coincidence boundaries tend to predominate.

The progress of recrystallization in a cold-worked metal (Fig. 8-8) is determined by microscopic examination of a series of specimens heated for various lengths of time. The point-counting technique of stereometric microscopy (Sec. 3-7) can be used to determine the fraction f of recrystallization in each specimen (Fig. 8-8*b*; Prob. 8-8) and also the amount of interfacial area per unit volume, $A = S_V$ (Prob. 8-9). When the measured values of f are plotted against the logarithm of the time of heating at a given temperature (Fig. 8-8*d*), a characteristic S-shaped curve is obtained for each temperature. At 700°C, for example, recrystallization begins after about 4 sec of heating, proceeds at an increasing rate up to $f = 0.5$, and then slows down during the approach to 100 percent recrystallization at 100 sec. At lower temperatures the process of recrystallization proceeds more slowly, and at higher temperatures it proceeds more rapidly.

The process of isothermal recrystallization can be described mathematically in two ways. By analogy with solidification, the recrystallized regions can be pictured as "nucleating" at the rate N and increasing in radius at the rate G (Prob. 8-10). Although this approach is adequate in many cases, a more satisfactory analysis is based on the area A of migrating boundary per unit volume of material and the average velocity G of the boundary. In a time dt the volume V of recrystallized material increases by the amount

$$AG\,dt = dV/m^3 = df \tag{8-2}$$

where f is the fraction of recrystallization completed at a given time. Since both A and f can be determined experimentally (Fig. 8-8), G can be determined as a function of time at a given temperature. The data corresponding to the curves of Fig. 8-8 are plotted in Fig. 8-9 and obey the relation

$$G = \frac{K_G}{t} \tag{8-3}$$

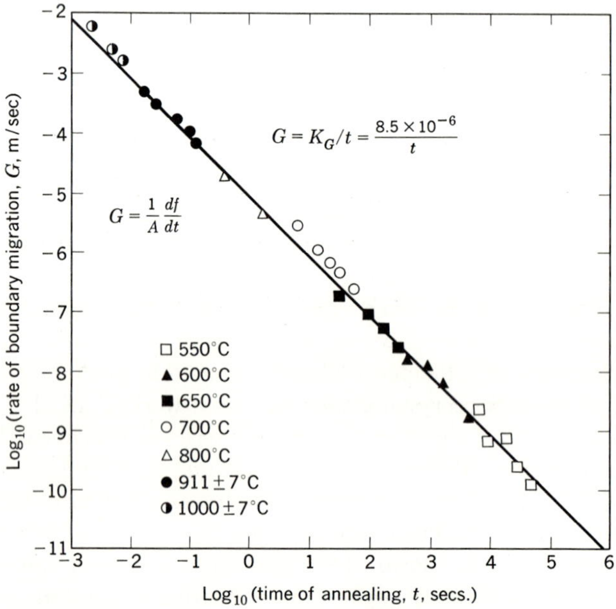

FIGURE 8-9 Rate of boundary migration, G, as a function of the time of annealing of 3.25 w/o silicon-steel after 60 percent cold working. (*After G. R. Speich and R. M. Fisher.*)

K_G is a constant, and t is the time of annealing at which G is measured. The amount of boundary area A also obeys an empirical relation

$$A = K_A f(1 - f) \tag{8-4}$$

where the constant K_A is the same for the entire range of temperatures. With an increase in the fraction of recrystallization, f, A increases from zero, passes through a maximum at $f = 0.5$, and then decreases to zero at $f = 1$.

Many engineering devices depend on the added strength produced in metals by cold working. Since this added strength is lost when recrystallization occurs, a method is needed for estimating recrystallization behavior during the many years of useful life of typical devices. A relation analogous to Eq. (4-17) permits the necessary extrapolation from laboratory studies of reasonable duration. If at temperature T_1 the time for 50 percent recrystallization is t_1, then

$$\text{Rate of reaction} = \frac{1}{t_1} = Be^{-Q/RT_1} \tag{8-5}$$

A linear relation between log $(1/t)$ and $1/T$ is obtained by taking the natural logarithm of both sides of Eq. (8-5); thus

$$\log\frac{1}{t} = \frac{\ln(1/t)}{2.303} = \frac{-Q}{2.303RT} + \text{constant} \tag{8-6}$$

The experimental data t_1 and t_2 obtained at two different temperatures T_1 and T_2 determine a straight line that can be extrapolated to the long time or low temperature of practical interest (see Prob. 8-11). In view of the relation between t and G (Fig. 8-9), Q is the activation energy for migration of the boundary.

The grains in an annealed material exhibit some type of preferred orientation. One factor determining the final texture is the preferred orientation of the initial cold-worked structure, a factor which depends on alloy composition and on the type of deformation. If annealing results only in recovery, the cold-work texture is almost unchanged. Recrystallization, however, produces new crystalline orientations and almost always results in a different texture. Generally a recrystallization texture consists of several components, but in certain face-centered cubic metals rolled into sheets the texture is simply (100)[001], called the *cube texture*. Problem 8-12 considers this texture in copper with reference to the corresponding deformation texture (Fig. 8-3). As explained in the following section, further changes in grain structure may follow (primary) recrystallization. Characteristic textures develop during these processes, often influenced by surface energies and the furnace atmosphere in which the annealing is done.

In commercial practice cold-worked alloys are annealed to increase their ductility, decrease the strength, or effect other desired changes in properties. Isochronal curves (corresponding to an annealing time of 1 hr; for example) are often more useful for designing practical heat treatments than are isothermal curves of the type illustrated in Fig. 8-8*d*. The isochronal curves of Fig. 8-10 show the effect of annealing for 1 hr at various temperatures on three properties of cold-drawn yellow brass (65 Cu–35 w/o Zn) wire. Abrupt changes in tensile strength and elongation occur because of recrystallization near 260°C (Fig. 8-10*a*), but less pronounced changes also occur during heating at lower or higher temperatures. Figure 8-10*b* shows that the grain size increases if the annealing temperature is raised above the minimum temperature required to produce recrystallization in one hour. The latter temperature is defined as the *recrystallization temperature,* even though Fig. 8-8*d* demonstrates that recrystallization can occur over a wide range of temperature (provided that the time is appropriately chosen).

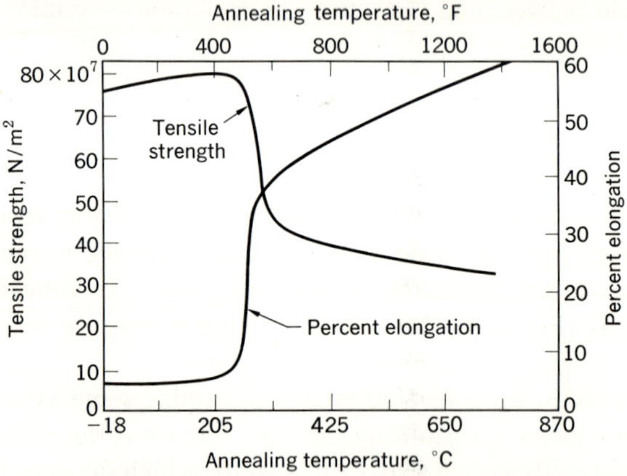

(*a*) Changes in tensile strength and elongation as determined in a tension test.

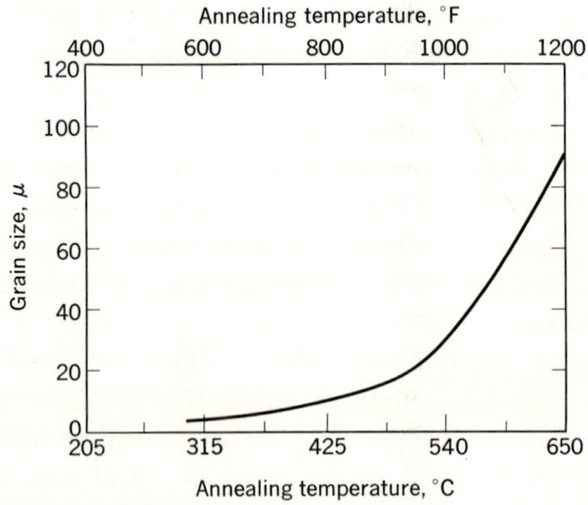

(*b*) Changes in grain size of the recrystallized alloy as determined by microscopic examination.

FIGURE 8-10 Changes in properties of yellow brass wire (previously cold-drawn 63 percent) as a result of annealing for 1 hr at various temperatures [conversion of units: (M N/m^2) 1.450 × 10^{-4} = N lb/in^2; (M N/m^2) 1.020 × 10^{-7} = N kg/mm^2]. (*Adapted from the "Metals Handbook," 8th ed., vol. 2.*)

Several factors besides the time and temperature of heating determine the response of a material to annealing. The amount of cold work is of basic importance (Fig. 8-11), and below some minimum amount of deformation (usually in the range of 2 to 20 percent) recrystallization does not occur even at temperatures near the melting point, although processes such as recovery are possible. The large grains produced by annealing after the "critical" deformation are probably a result of migration of existing boundaries, rather than a result of the nucleation and growth of new grains. Commercial working operations avoid deformation in this critical range since a fine grain size is ordinarily desired. The recrystallization temperature falls rapidly with increase in amount of deformation and tends to level off at about 50 percent cold work. A given degree of cold work produces a greater tendency for recrystallization the finer the grain size and the lower the temperature at which the cold work is performed. For typical commercial alloys, deformation performed at a homologous temperature of about 0.7 or higher is hot working and is ineffective in producing work hardening.

The composition of a material can have a pronounced effect on annealing behavior. High-purity metals typically recrystallize at a relatively low temperature (about 0.3 T_f), but the addition of even small amounts of impurities markedly raises the recrystallization temperature. Figure 8-12 illustrates this behavior for nickel and shows the marked difference between high-purity metal and the usual commercial material. Significant amounts of a suitable alloying element can further raise the recrystallization temperature, as shown by the curves for the nickel-molybdenum and nickel-chromium alloys. The presence of a second phase in a given matrix metal does not necessarily raise the recrystallization temperature, but two types of phases are found to be especially effective. A few percent of finely dispersed oxide can make a cold-worked metal stable at temperatures approaching its melting point. An outstanding example is SAP aluminum (sintered aluminum powder) which has a recrystallization temperature of about 500°C with 5 $^w\!/_0$ Al_2O_3; for comparison, commercially pure aluminum recrystallizes at 150°C. The corresponding nickel alloy, TD (thoria, ThO_2, dispersed) nickel, recrystallizes at a *lower* temperature than does commercially pure nickel after ordinary cold working. However, after a special thermomechanical treatment (repeated deformation and low-temperature annealing), TD nickel develops resistance against recrystallization at high temperatures. The rearrangement of dislocations that underlie this type of behavior is discussed in Sec. 8-9. A fine dispersion of pores (voids) is the other type of second phase that can raise the recrystallization temperature. The sintering of powders is one of the techniques for producing a suitable array of pores.

Several other practical aspects of annealing can only be mentioned here. Residual stresses resulting from plastic deformation (or previous heat

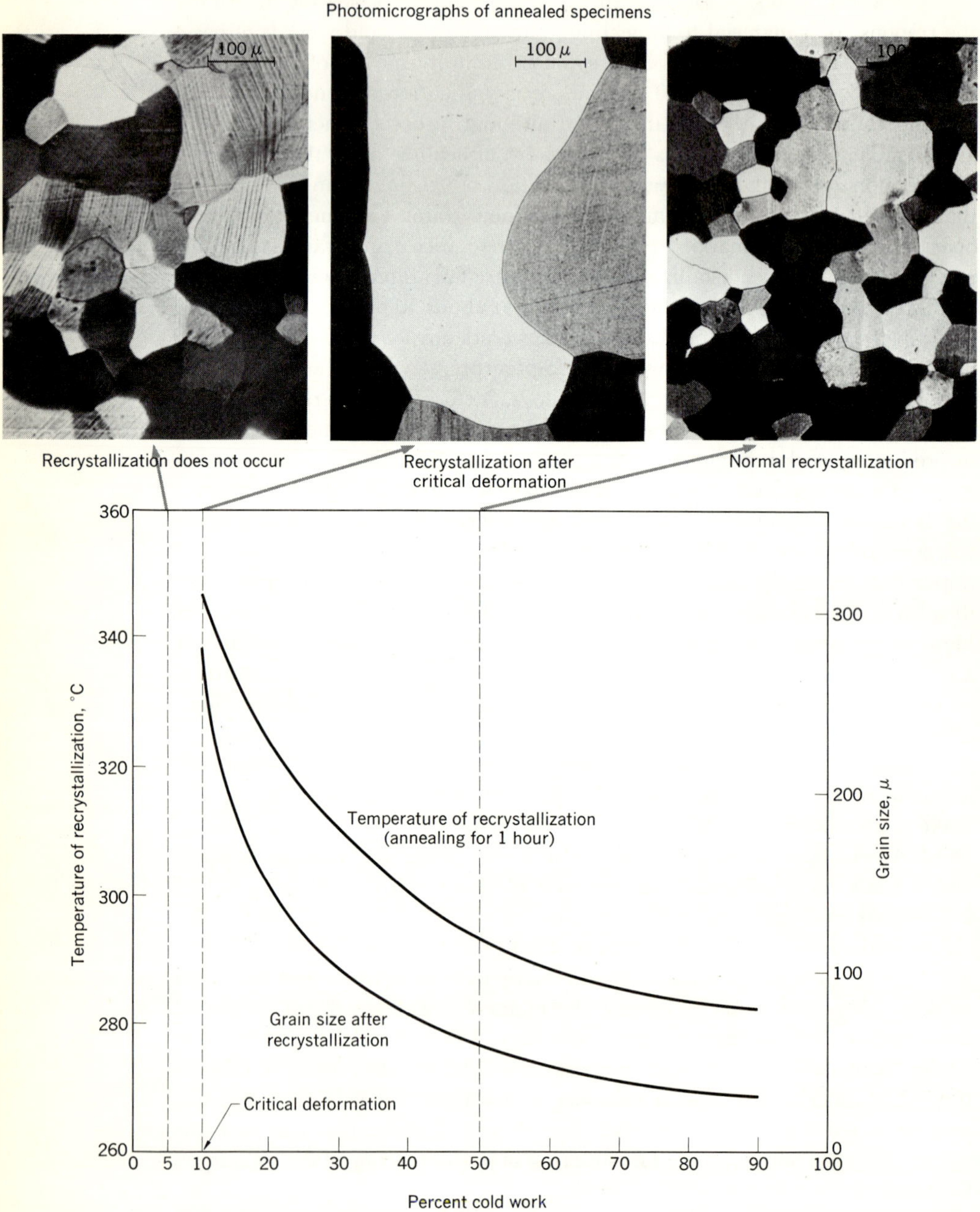

FIGURE 8-11 Effect of amount of cold work on the annealing behavior of 99.99% pure aluminum sheet. (*Courtesy W. A. Anderson, Research Laboratory, Aluminum Company of America.*

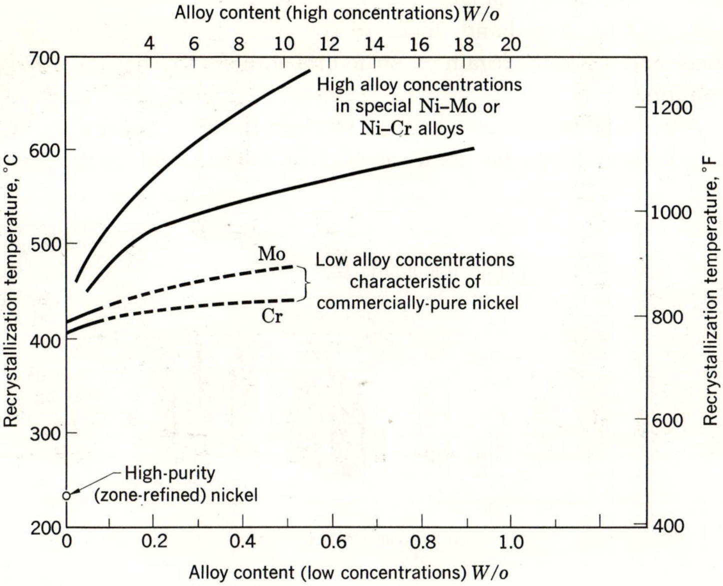

FIGURE 8-12 The effect of typical alloying elements on the recrystallization temperature of nickel. (*Courtesy K. Detert, AEG-Telefunken, Frankfurt, Germany.*)

treatment) can be effectively removed by low-temperature annealing, which does not destroy the beneficial mechanical properties conferred by the prior deformation. Annealing processes are employed in various ways to produce a desired grain size. For example, an ultrafine-grained steel (ASTM grain size 13.5; see Prob. 3-12) can be obtained by (1) heating the steel to the minimum temperature for formation of austenite, (2) deforming severely, preferably in a single operation, and (3) allowing recrystallization of the austenite to occur during a brief isothermal anneal or during suitable slow cooling. The resulting ultrafine grain size improves the strength and toughness of the steel in both the unhardened and the hardened condition. Ordinarily a fine dispersion of a second phase (Al_2O_3 in steel) is effective in maintaining a fine grain size ($\sim$ASTM 8) during annealing at ordinary heat-treating temperatures (see Sec. 8-9). For certain materials, such as tungsten lamp filaments, large elongated grains are obtained by using second-phase particles and causing secondary recrystallization.

8-6. Recrystallization of Organic Polymers

A significant difference between polymers and other crystalline materials in annealing behavior is the change in external shape. A metal rod, cold-drawn to form a wire, retains the form of a wire on recrystallization but experiences extensive changes in its crystalline structure. A cold-drawn polymer, on the other hand, tends to return to its initial length on being annealed and also to regain a structure similar to that existing before deformation.

Several factors cause polymers to react in this manner: (1) the long-chain molecules, (2) the phenomenon of chain folding, and (3) the presence of

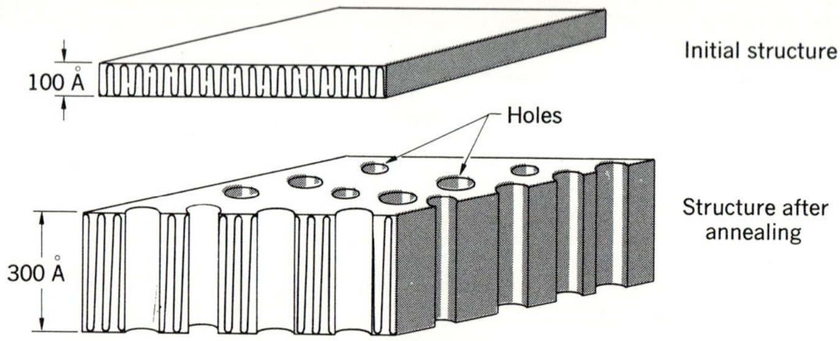

(*a*) An isolated lamella.

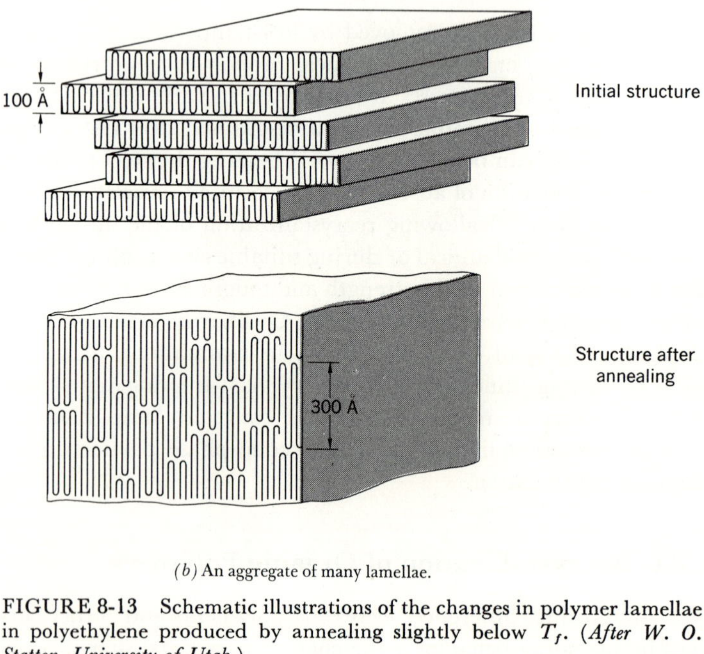

(*b*) An aggregate of many lamellae.

FIGURE 8-13 Schematic illustrations of the changes in polymer lamellae in polyethylene produced by annealing slightly below T_f. (*After W. O. Statton, University of Utah.*)

both amorphous and crystalline regions in a typical specimen. Consider first the simple case of an (undeformed) single-crystal lamella (Fig. 3-7*a*) annealed slightly below T_f (Fig. 8-13*a*). By a sliding motion the chain molecules are able to increase the long period in the crystal from 100 Å to 300 Å; holes are produced during the rearrangement, and the overall dimensions of the lamella change markedly. Similar changes occur in an aggregate of lamellae (Fig. 8-13*b*), except that interpenetration of neighboring lamellae occurs rather than the formation of holes.

The behavior during annealing becomes more complex if the initial material, originally containing macrospherulites (Fig. 3-7*b*) or microspherulites (Fig. 3-8*a*) has been elongated manyfold by drawing (Fig. 3-8*c*). Many chain molecules are then in an extended configuration and tend to cause an overall contraction of the specimen. Consequently, the results produced by annealing depend on whether the specimen is clamped or is free to contract. A clamped specimen not only retains the deformed shape, but also preserves or even increases the preferred orientation conferred by the drawing operation. However, a free specimen shrinks back towards its original dimensions and loses most of its preferred orientation. The processes that occur during annealing include (1) incorporation of an appreciable fraction of the amorphous chains into the crystalline regions, (2) growth of crystalline regions in their

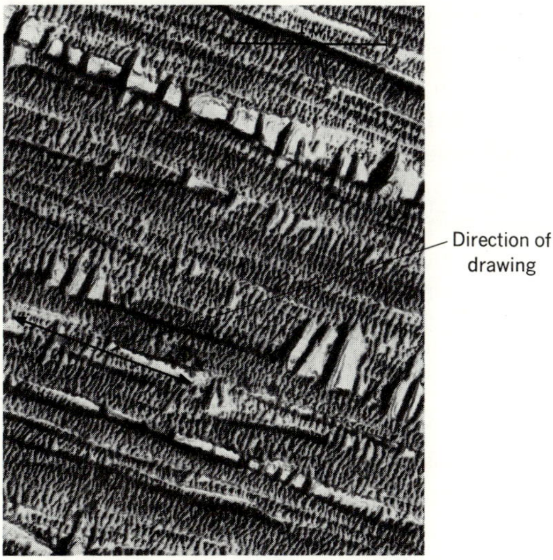

FIGURE 8-14 Electron micrograph of polyethylene, cold-drawn to 16 times its original length and then annealed at 120°C for 2hr. (*Courtesy K. Sakaoku and A. Peterlin, Camille Dreyfus Laboratory.*)

long period, and interpenetration, and (3) reduction in the concentration of defects. A typical annealed structure of a clamped specimen (Fig. 8-14) continues to show fibrils, and the orientation of the lamellae is more uniformly perpendicular to the direction of drawing than in the unannealed specimen.

GRAIN GROWTH

8-7. Migration of Grain Boundaries

As with recrystallization, grain growth occurs by the migration of boundaries, but there is a difference in the driving force. In a cold-worked metal, boundary migration occurs primarily because of the strain energy. The "stable" grain structure after recrystallization has negligible strain energy, but it has a certain amount of free energy associated with its grain boundaries. At a suitably high temperature the grain-boundary free energy can cause boundary migration and therefore growth of grains. Grain growth decreases the total free energy by decreasing the amount of grain-boundary area, but an important factor may also be the replacement of high-energy boundaries by those of lower energy. Grain growth can occur in virtually all materials including cast metals, two-phase systems, and sintered powders of metals, ceramics, and polymers. Figure 8-15 shows that the characteristic increase in grain size proceeds even in certain crystalline organic compounds. If a macroscopic stress is present, it will usually act as a driving force on a grain boundary. Since a stress causes motion of dislocations [Eq. (4-63)], this driving force can be studied easily in a bicrystal containing a single tilt boundary; in actual microstructures the situation is more complex.

The essential role of surface tension in grain growth is shown by experiments on wedge-shaped bicrystals (Fig. 8-16). For a wedge of unit thickness the surface free energy E of a curved section of boundary is

$$E = \gamma R \alpha \tag{8-7}$$

where γ is the surface tension, R is the radius of the cylindrical surface, and α is the wedge angle. Migration of the boundary towards the center of curvature causes a decrease in area and therefore in energy of the boundary. By analogy to the relation force $\times$ d(distance) $=$ d(energy), a virtual force F,

$$F = -\frac{dE}{dR} = \gamma\alpha \tag{8-8}$$

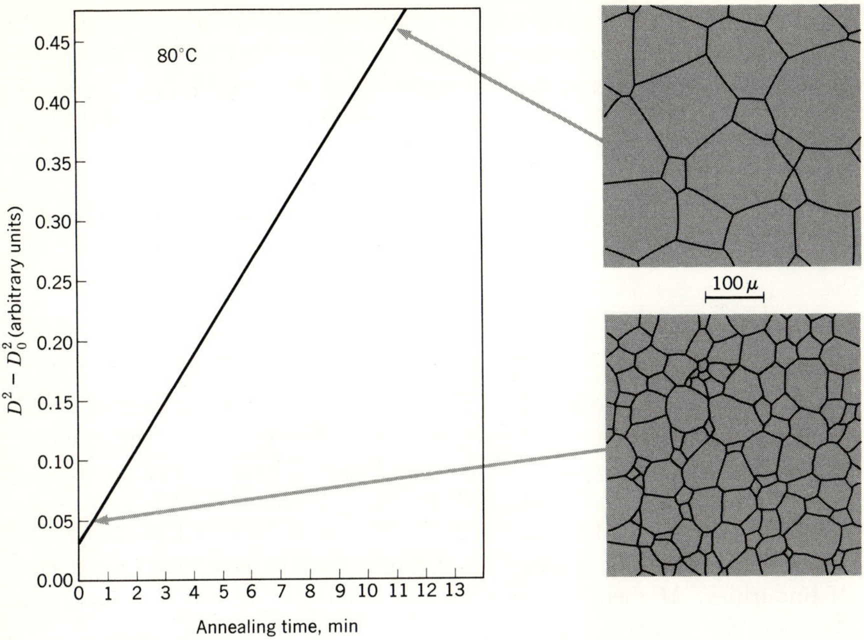

FIGURE 8-15 Increase in grain size with annealing time at 80°C in a thin-film specimen of carbon tetrabromide. Two typical microstructures are shown. (*Courtesy C. J. Simpson, C. J. Beingessner, and W. C. Winegard.*)

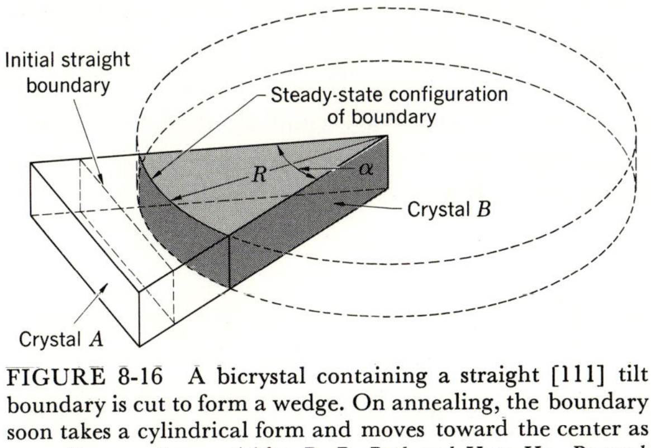

FIGURE 8-16 A bicrystal containing a straight [111] tilt boundary is cut to form a wedge. On annealing, the boundary soon takes a cylindrical form and moves toward the center as annealing continues. (*After B. B. Rath and Hsun Hu, Research Center, U.S. Steel Corporation.*)

can be pictured as acting on the boundary. The corresponding driving force P per unit area is

$$P = \frac{\gamma}{R} \tag{8-9}$$

Experimental studies show that the velocity v of the boundary is related to γ/R by the power relation

$$v = M\left(\frac{\gamma}{R}\right)^n \tag{8-10}$$

where M and n are constants for a given type of boundary in a given material. The driving force P increases directly with the surface tension γ of the boundary and inversely with the radius of curvature. Figure 5-10 shows that γ is small at small boundary misorientations, rises to a fairly constant value for most high-angle boundaries, and has low values for coincidence or twin boundaries. The lowest value of n is about unity, but it can increase to over 10 in the presence of impurities; M then decreases (in the usual case in which an impurity decreases the rate of grain growth).

The geometry of grain growth in a polycrystalline material is more complex than the migration of a single cylindrical boundary, but Eq. (8-10) leads to a useful relation in many instances. Curvature in three dimensions is described by two principal radii of curvature. If these are equal, then by analogy with Eq. (5-22) the appropriate form of Eq. (8-9) is

$$P = \frac{2\gamma}{R} \tag{8-11}$$

The curvature R is now some average value for the whole array of grains, but this value can be assumed to vary in proportion to the average grain diameter D. A grain boundary should then migrate towards its center of curvature at the rate

$$v = \frac{dD}{dt} = M\left(\frac{\gamma K}{D}\right)^n \tag{8-12}$$

If the average value of γ remains constant during grain growth, this equation simplifies to

$$\frac{dD}{dt} = \frac{K'}{D^n} \tag{8-13}$$

Although some grains shrink and disappear, the average grain size increases according to the relation obtained by integration of Eq. (8-13); that is

$$D^{n+1} - D_0^{n+1} = kt \tag{8-14}$$

where D_0 is the initial grain size at $t = 0$, the beginning of the anneal. If D_0 is negligibly small, Eq. (8-14) can be put in the form

$$D = k't^N \tag{8-15}$$

that describes grain growth in most materials. For many pure substances (see Fig. 8-15, for example) $N = 0.5$, corresponding to $n = 1$ in Eq. (8-10). For usual materials, however, a typical value of N is 0.3, and values below 0.1 are possible.

8-8. Grain Encounters

Several features of actual grain-growth processes must be considered in addition to the basic phenomenon of boundary migration. The topology of three-dimensional grain structures, discussed in Chap. 3, strongly influences the rate of grain growth. The two-dimensional analogy illustrated in Fig. 8-17*a* shows that a six-sided grain can have straight sides and still provide the 120° boundary junctions required for equilibrium. Since the radius of curvature, R, approaches infinity for a straight boundary, the driving force is negligibly small. If one of the grains in an array has only four sides (Fig.

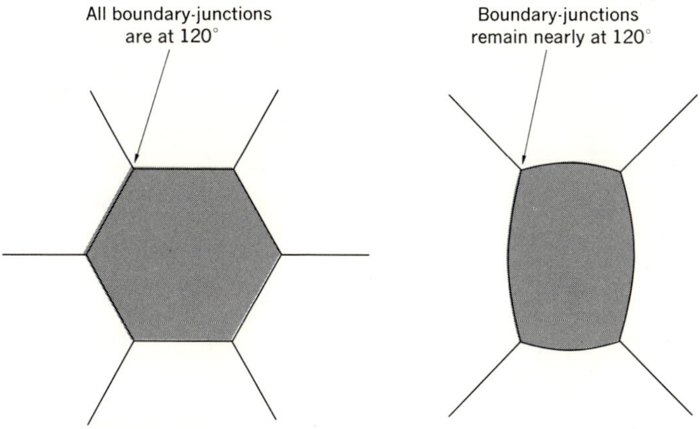

(*a*) A six-sided grain surrounded by other six-sided grains; all sides can be straight.

(*b*) A four-sided grain surrounded by six-sided grains; curvature must exist.

FIGURE 8-17 A two-dimensional analog showing why curved faces exist on grains having fewer than the stable number of faces.

8-17*b*), the boundary junctions can remain nearly at 120° only if the four sides become convex as shown. Each of the four boundaries moves toward its center of curvature since the driving force $P = \gamma/R$ has an appreciable value. As a grain gradually shrinks, one or more *grain encounters* usually occur; two grains, not previously neighbors, meet and form a new grain boundary. Figure 8-18*a* illustrates in two dimensions the grain encounter that occurs when a four-sided grain loses one of its sides. Figure 8-18*b* shows a simple type of encounter in three dimensions; more complex types can account for the loss of grain edges and faces. A face with only three edges gradually disappears without causing further encounters. Similarly, a grain with only four faces causes no en-

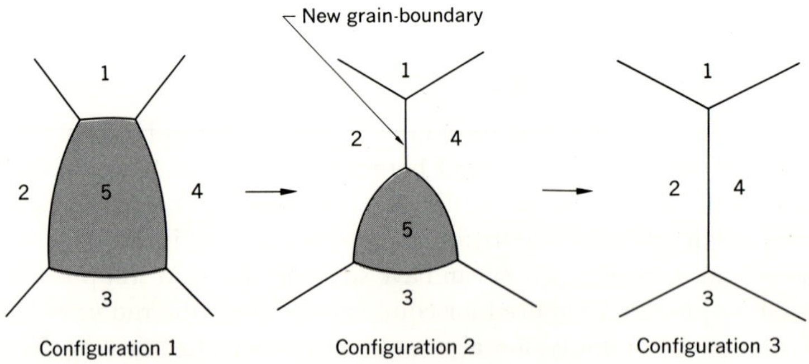

(a) A two-dimensional analog showing an encounter of grains 2 and 4 as grain 5 loses a side. Subsequently grain 5 disappears without causing any additional encounter.

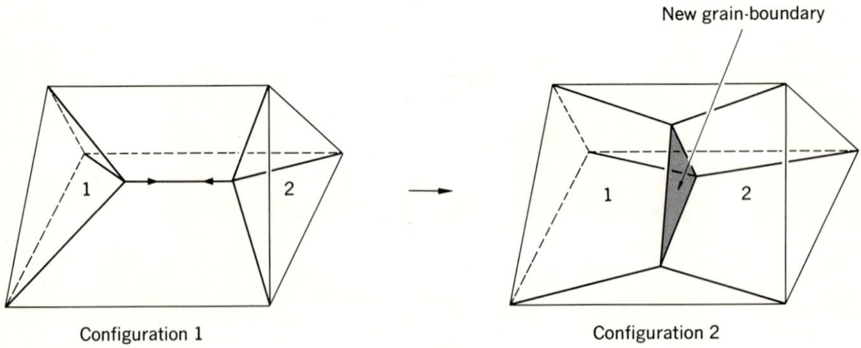

(*b*) A simple type of encounter in three dimensions. A new grain-boundary forms between grains 1 and 2.

FIGURE 8-18 The nature of a grain encounter, in which a new grain-boundary forms. (*Courtesy J. P. Nielsen, New York University.*)

counters as it disappears. Problem 8-14 shows that billions of encounters per cubic millimeter occur during grain growth in an initially fine-grained material.

Encounters are of two kinds, normal and special. Normally the orientations of the two grains involved in an encounter are such that a typical high-angle grain boundary is produced. In rare instances, however, one of the three special encounters occurs: (1) coalescence to form a single grain, when the two grains have virtually the same orientation, (2) formation of a twin boundary, when the two grains are in twin orientation, or (3) formation of a coincidence boundary, when the two grains are in a suitable orientation. Because twin and coincidence boundaries have low energies, they tend to stabilize a grain structure and impede further grain growth.

8-9. Impediments to Grain Growth

Second-phase particles (and pores) may also play an important role in impeding the migration of grain boundaries. As a boundary moves across a spherical particle (Fig. 8-19*a*), the energy of the boundary reaches a minimum at the midpoint of the particle since an area πr^2 of grain boundary has been "blanked off" by the particle. The corresponding energy $\pi r^2 \gamma$ is regained as the boundary migrates the distance r to the surface of the particle because a drag force F (compare Eq. 8-8)

$$F_{\max} = \frac{\Delta E}{\Delta x} = \frac{\pi r^2 \gamma}{r} = \pi r \gamma \tag{8-16}$$

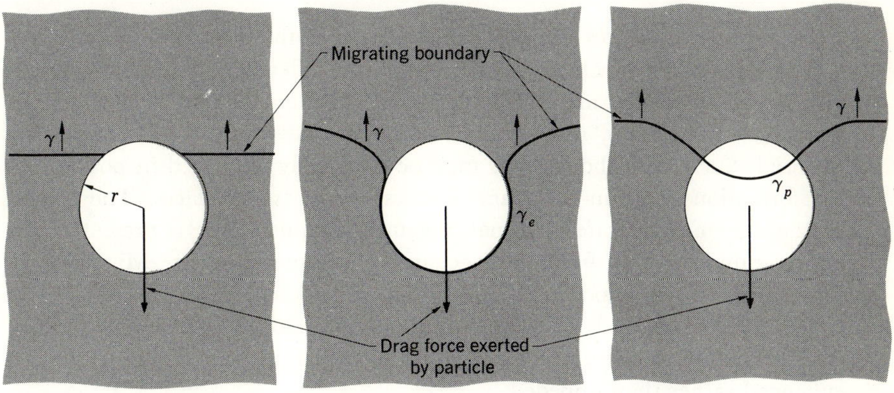

(*a*) Simple picture of "blanking-off" of grain boundary by a spherical particle.

(*b*) Grain boundary envelops particle forming a new boundary of energy γ_e.

(*c*) Grain boundary passes through particle forming a new boundary of energy γ_p.

FIGURE 8-19 Three mechanisms by which a drag force is exerted by a second-phase particle on a migrating grain boundary. (*After M. F. Ashby, et al., Harvard University.*)

is exerted on the boundary by the particle. Actually the grain boundary either envelops the particle (Fig. 8-19*b*) or may even pass through a coherent particle (Fig. 8-19*c*). If the corresponding surface tension, γ_e or γ_p, is small compared to γ, Eq. (8-16) continues to be a good estimate of the drag force. Since a unit area of a plane boundary will touch $2rN_v$ particles, where N_v is the number of particles per unit volume, the drag force P_D per unit area of boundary is

$$P_D = 2\pi r^2 \gamma N_v \tag{8-17}$$

When this force just balances the driving force for grain growth [Eq. (8-11)], we have the relation

$$\frac{2\gamma}{R} = 2\pi r^2 \gamma N_v \tag{8-18}$$

Consequently,

$$R = (\pi r^2 N_v)^{-1} \tag{8-19}$$

Substitution of the volume fraction of particles, $f = \frac{4}{3}\pi r^3 N_v$, in Eq. (8-19) gives the more useful relation

$$R = \frac{4r}{3f} \tag{8-20}$$

Problem 8-15 illustrates the use of this relation in the control of grain size.

Two other mechanisms are sometimes effective in impeding grain growth. In thin strip material with grain diameters two or three times the thickness of the strip, most grain boundaries pass completely through the material. Such grain boundaries may be effectively anchored in position by the formation of grain-boundary grooves at the two surfaces. The second mechanism may operate if a small amount of liquid phase is present at the grain boundaries, as frequently occurs in ceramic systems. Migration of boundaries will be impeded if the liquid-solid interfacial energy is lower than that of the grain boundary. If the liquid-solid interface has a high energy, then the situation is more complex and grain growth may, in some instances, be enhanced rather than impeded.

Some materials exhibit a type of grain growth in which a few grains grow at a greatly exaggerated rate, a phenomenon termed *abnormal grain growth* or *secondary recrystallization*. The characteristic microstructure is illustrated in Fig. 8-20 for a sintered oxide ceramic. The boundaries of the matrix

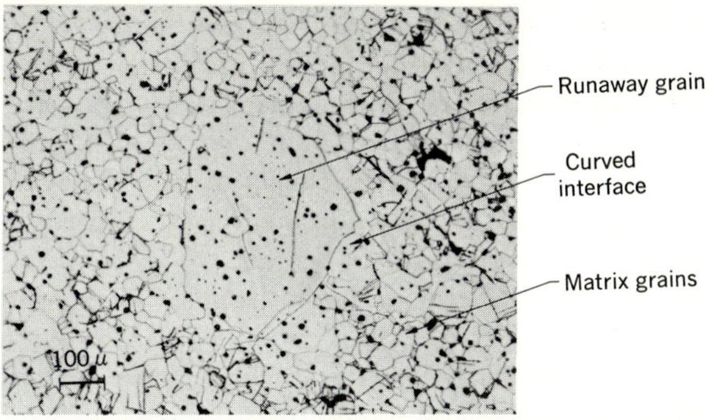

FIGURE 8-20 Microstructure of a sintered commercial (ceramic) ferrite illustrating the process of secondary recrystallization. (*Courtesy J. E. Burke, Research and Development Center, General Electric Company.*)

grains bow into the "runaway" grain (see Prob. 8-16), and therefore these boundaries migrate into the matrix grains. The minimum size D of a runaway grain, determined by a simple estimate of boundary energies (Prob. 8-17), is

$$D = \frac{2\sigma_{Dd}}{\sigma_{dd}} d \tag{8-21}$$

where d is the diameter of the average matrix grain, σ_{dd} is the interfacial energy between matrix grains, and σ_{Dd} is the interfacial energy between the runaway grain and a matrix grain. If $\sigma_{dd} = \sigma_{Dd}$, a grain need be only twice the average size to serve as a center for secondary recrystallization. In sintered ceramic materials made from very fine powder, the required difference in size exists and causes exaggerated grain growth. The runaway grains grow so rapidly that they incorporate the pores into their structure. The resulting porosity decreases strength and ductility significantly. In most materials, runaway grains can form only if normal grain growth is impeded by one of the mechanisms described above or by the existence of a pronounced texture.

PROBLEMS

8-1. The volume of a metal specimen remains almost constant during severe plastic deformation such as rolling or wire drawing. (*a*) Use this fact by using the fractional elongation of the specimen, $(l_f - l_o)/l_o$, or the fractional reduction in its cross-sectional area $(A_o - A_f)/A_f$; the subscripts ($_o$) and ($_f$) denote the initial and final dimensions. (*b*) The percent reduction in area is usually defined relative to the initial area $(A_0 - A_f)/A_0$. Use a rough plot to show its relation to percent elongation in the range of 0 to 300 percent elongation.

8-2. Using the data of Fig. 8-1, determine the rate of strain hardening of cartridge brass and of electrolytic copper in units of newtons per square meter per one percent reduction in area (*a*) at 10 percent reduction in area, (*b*) at 30 percent, (*c*), at 60 percent.

8-3. (*a*) Show where the {200} poles fall in the pole figure of Fig. 8-3*b* for the ideal orientation $(110)[1\bar{1}2]$. (*b*) Explain why a pole figure describes a texture more satisfactorily than does such an ideal orientation, especially for small amounts of cold working.

8-4. Construct the {110} pole figure for an iron wire having the ideal fiber texture [110]. Let the wire axis be vertical in the stereographic projection. Take account of the fact that a wire-drawing die produces cylindrical symmetry.

8-5. If a bar of metal is quenched from a high temperature, the surface layers are found to contain compressive stresses and the center portion tensile stresses. Explain this phenomenon. [HINT: The center remains in the "hot-working" range briefly after the surface has appreciably cooled and contracted.]

8-6. The curves of Fig. 8-5 are *isochronal*; that is, they are based on a constant time of annealing (10 min, for example) at each temperature. (*a*) Sketch possible *isothermal* curves of the recovery of quenched copper at 300 and 325°K. Make sure that your curves are consistent with the isochronal data. (*b*) Determine the rate of recovery (at 50 percent recovery) from each of the curves of part *a*. (*c*) Show that the activation energy of the recovery process can be determined from these two rates. (*d*) If this activation energy is the same as that found for the migration of vacancies in copper by diffusion studies, why can recovery be assumed to occur by this mechanism?

8-7. In the process of cold drawing, a bar is reduced in cross section by being drawn through a die. Tensile stresses develop in the surface layers but are balanced by compressive stresses in the center of the bar. If a flat surface is produced by machining on one side of such a bar, show that the bar must bow to produce a new equilibrium distribution of body stresses.

8-8. (*a*) Make the appropriate point count on the photomicrograph of Fig. 8-8*b* to estimate A_A for the recrystallized structure [see Eq. (3-7)]. (*b*) How is the fraction f of recrystallization in the volume of the specimen related to A_A?

8-9. (*a*) Make the appropriate count on the photomicrograph of Fig. 8-8*b* to estimate the number of intercepts per unit length of grid line, N_L, made by the outlines of the recrystallized grains. (*b*) How is $A = S_V$ in three dimensions related to the two-dimensional quantity N_L?

8-10. In view of the actual mechanism of recrystallization, explain why neither the concept of "nucleation of new grains" nor the assumption of uniform spherical growth is appropriate for a general analysis of recrystallization.

8-11. A 3.25 $^{\mathrm{w}}\!/_{0}$ silicon steel that has been cold rolled 60 percent is to be used in a machine rated for high-temperature operation for 10,000 hr. Assuming that the strength of the steel would be seriously reduced when 50 percent recrystallization had occurred, determine the highest temperature at which the machine could be operated for 10,000 hr. (*a*) From the curves of Fig. 8-8*d*, estimate the time for 50 percent recrystallization at 523 and 1273°K. Use these two data points to construct a linear plot that can be extrapolated to 10,000 hr. (*b*) Show that Q in Eq. (8-5) is the activation energy for the process of boundary migration. (*c*) The method of least squares[1] permits the use of data from several curves in determining the straight line in part *a*; thus

$$y = a + bx$$

$$\log \frac{1}{t} = a + \frac{-Q}{2.303R}\frac{1}{T} \qquad (8\text{-}22)$$

The coefficients a and b of the best straight line are determined by the following operations on N pairs (x_i, y_i) of experimental data points.

$$\text{Corrected sum of squares (CSS)} = \sum_{i=1}^{N} x_i^2 - \text{XBAR} \qquad (8\text{-}23)$$

where XBAR is given by Eq. (3-10).

$$\text{Corrected sum of cross products (CSCP)} = \sum_{i=1}^{N} x_i y_i - \frac{\left(\sum_{i=1}^{N} x_i\right)\left(\sum_{i=1}^{N} y_i\right)}{N} \qquad (8\text{-}24)$$

$$\text{Slope} = b = \frac{\text{CSCP}}{\text{CSS}} \qquad (8\text{-}25)$$

$$\text{Intercept} = a = \text{YBAR} - m\,\text{XBAR} \qquad (8\text{-}26)$$

Use the following program to determine the least-squares line [Eq. (8-22)] for the seven experimental curves of Fig. 8-8*d*. Your initial data are t (TIME) and T (TEMP).

[1] The basis of the method is described in the book by T. R. Dickson, "The Computer and Chemistry," p. 154 ff, W. H. Freeman and Company, San Francisco, 1968.

```
C  CALCULATION OF LEAST-SQUARES LINEAR FIT
 1  FORMAT (I2)
 2  FORMAT (2E9.3)
C  READ NUMBER OF DATA POINTS, N
 3  READ (5,1) N
C  MAKE N A REAL NUMBER
 4  REALN = N
 5  SUMX = 0
 6  SUMY = 0
 7  SUMXY = 0
 8  SUMX2 = 0
C  CALCULATION
 9  DO 16 I = 1,N
10  READ (5,2) TIME, TEMP
11  Y = ALOG10(1/TIME)
12  X = 1/TEMP
13  SUMX = SUMX + X
14  SUMY = SUMY + Y
15  SUMXY = SUMXY + X * Y
16  SUMX2 = SUMX2 + X ** 2
17  XBAR = SUMX/REALN
18  YBAR = SUMY/REALN
19  CSCP = SUMXY - SUMX * SUMY/REALN
20  CSS = SUMX2 - SUMX ** 2/REALN
21  B = CSCP/CSS
22  A = YBAR - B * XBAR
23  WRITE (6,2) A,B
24  STOP
    END
```

8-12. (*a*) Sketch the {200} pole figure of the cube texture of high-purity copper, which has a spread of about 5° about the ideal (100)[001] orientation. (*b*) Compare the orientations of the recrystallized grains with those of the initial cold-rolled grains, and explain qualitatively in terms of the mechanism of recrystallization how such pronounced changes in orientation might have occurred.

8-13. Important phenomena governed by recrystallization range in homologous temperature from 0.3 (recrystallization temperature of high-purity metals) to 0.7 (minimum temperature for hot working of typical alloys). Describe the two factors that account for this difference in homologous temperature.

8-14. Grain growth in many materials changes a grain size of about 10^{-3} mm into grains about 1 mm in diameter. (*a*) Show by a rough calculation that about 10^9 grains/mm^3 disappear during grain growth. (*b*) Recall from Chap. 3 that an average grain has 13 faces. Imagine that one grain disappears from an average array of grains, its neighbors rearranging themselves to reestablish an average stable arrangement. Show that an average of 6.5 grain encounters must occur.

8-15. Only a small fraction, about 0.01 percent by volume, of second-phase particles can be used in the practical control of grain size. If a stable grain size of 0.01 mm is desired, what must be the size of the second-phase particles?

8-16. By analogy with the cause of curvature in Fig. 8-17, explain with the aid of a sketch why a large grain (necessarily having *many* boundaries with grains of the matrix) produces the curvature shown in Fig. 8-20. Assume that $\sigma_{Dd} = \sigma_{dd}$.

8-17. Derive Eq. (8-21). (*a*) Begin with an array of grains of uniform size. By considering one single face of an average many-faced grain, show by a rough estimation that an increment *dr* of preferential growth in radius of this grain would result in an increase in interfacial energy of its own faces twice as great as the decrease in interfacial energy of the shrinking neighboring grains. (*b*) If $\sigma_{Dd} = \sigma_{dd}$, show that the factor of two in part *a* can be compensated if $D = 2d$. (*c*) Finally, show that Eq. (8-21) is obtained if σ_{dd} is different from (usually less than) σ_{Dd}.

REFERENCES

Barrett, C. S., and T. B. Massalski: "Structure of Metals," 3d ed., McGraw-Hill Book Company, New York, 1966.

Bonis, L. J., and H. H. Hausner (eds.): Fundamental Phenomena in the Material Sciences: vol. 1, "Sintering and Plastic Deformation," Plenum Press, New York, 1964.

Burgers, W. G.: Principles of Recrystallization, in J. J. Gilman (ed.): "The Art and Science of Growing Crystals," John Wiley & Sons, Inc., New York, 1963.

Byrne, J. G.: "Recovery, Recrystallization and Grain Growth," The Macmillan Company, New York, 1965.

Cahn, R. W. (ed.): "Physical Metallurgy," 2d ed., North Holland Publishing Company, Amsterdam, 1970.

Christian, J. W.: "The Theory of Transformations in Metals and Alloys," Pergamon Press, New York, 1965.

Fulrath, R. M., and J. A. Pask (eds.): "Ceramic Microstructures," John Wiley & Sons, Inc., New York, 1968.

Gordon, P.: Annealing Mechanisms in Deformed Metals, in W. M. Mueller (ed.), "Energetics in Metallurgical Phenomena," vol. 1, p. 207, Gordon and Breach, Science Publishers, Inc., New York, 1965.

Himmel, L. (ed.): "Recovery and Recrystallization of Metals," Gordon and Breach, Science Publishers, Inc., New York, 1963.

Kingery, W. D.: "Introduction to Ceramics," John Wiley & Sons, Inc., New York, 1960.

Shewmon, P. G.: "Transformations in Metals," McGraw-Hill Book Company, New York, 1969.

"Recrystallization, Grain Growth and Textures," ASM seminar, American Society for Metals, Metals Park, Ohio, 1966.

Chapter Nine

Deformation of Materials

INTRODUCTION

The various kinds of materials respond in quite different ways to an applied force. Consider first a tensile force that produces only *elastic deformation.* When the force is removed the specimen returns to its initial size and shape. In a metallic or ceramic material adjacent atoms are pulled only slightly apart, a process that generally requires a large force. Elastic deformation may be followed directly by fracture (treated in Chap. 10), as shown by the stress-strain curve for Al_2O_3 (Fig. 9-1*a*), representative of ceramic materials at room temperature. In a typical metal (Fig. 9-1*b*) elastic deformation is followed by an important stage of *plastic deformation* prior to fracture. The mechanisms of plastic deformation produce permanent deformation, which remains after the deforming force has been removed.

In those polymeric materials consisting of coiled long-chain molecules, elastic deformation can be extremely large (Fig. 9-1*c*). The term *elastomeric deformation* is often employed in this case to emphasize the characteristic behavior and deformation mechanisms. The coils straighten partially and can lengthen to many times their original size. In the absence of cross-linking, neighboring chains can slip relative to one another in a manner analogous to slip in a crystal lattice and produce plastic deformation. The data for poly-methyl-methacrylate (Fig. 9-1*d*) illustrate typical deformation of an amorphous polymer. The stress-strain curve would be significantly different, however, if the time of testing were appreciably shorter or longer. This is a feature of *viscoelastic deformation* (Sec. 9-13), which involves several mechanisms but whose characteristic is a pronounced dependence on the duration of loading and also on the temperature.

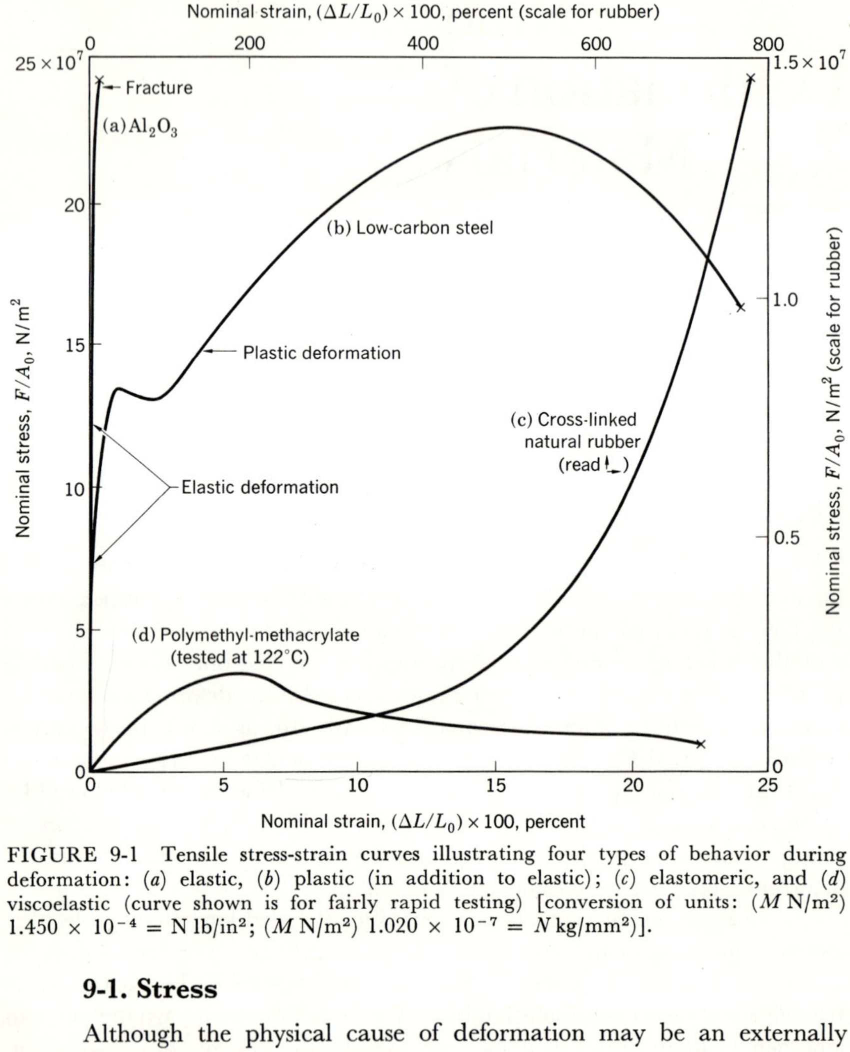

FIGURE 9-1 Tensile stress-strain curves illustrating four types of behavior during deformation: (*a*) elastic, (*b*) plastic (in addition to elastic); (*c*) elastomeric, and (*d*) viscoelastic (curve shown is for fairly rapid testing) [conversion of units: (M N/m^2) 1.450×10^{-4} = N lb/in^2; (M N/m^2) 1.020×10^{-7} = N kg/mm^2)].

9-1. Stress

Although the physical cause of deformation may be an externally applied force F, quantitative treatments of deformation usually employ the concept of stress,

$$\sigma = \frac{F}{A} \tag{9-1}$$

Not only is the area A of the specimen taken into account, but the stresses can adequately represent the "force condition" that exists at various locations within the material (Fig. 9-2). The method for achieving this representation is illustrated for point P. A small cube is constructed about P with its faces perpendicular to the three axes X_i of a cartesian coordinate system. Across any face, such as A_1, a force $\mathbf{F}_1$ is exerted on the cube by the adjoining material. This force has components F_{11}, F_{12}, and F_{13} along the three coordinate axes. Since each component acts on the area A_1, the stress components acting on face A_1 are

$$\sigma_{11} = \frac{F_{11}}{A_1} \qquad \sigma_{12} = \frac{F_{12}}{A_1} \qquad \sigma_{13} = \frac{F_{13}}{A_1} \tag{9-2}$$

A similar analysis for the other three faces gives a total of nine stress components σ_{ij}; here i and j can each have the values 1, 2, or 3. Because the small cube must be in static equilibrium with respect to rotation (Prob. 9-1),

$$\sigma_{12} = \sigma_{21} \qquad \sigma_{23} = \sigma_{32} \qquad \sigma_{31} = \sigma_{13} \tag{9-3}$$

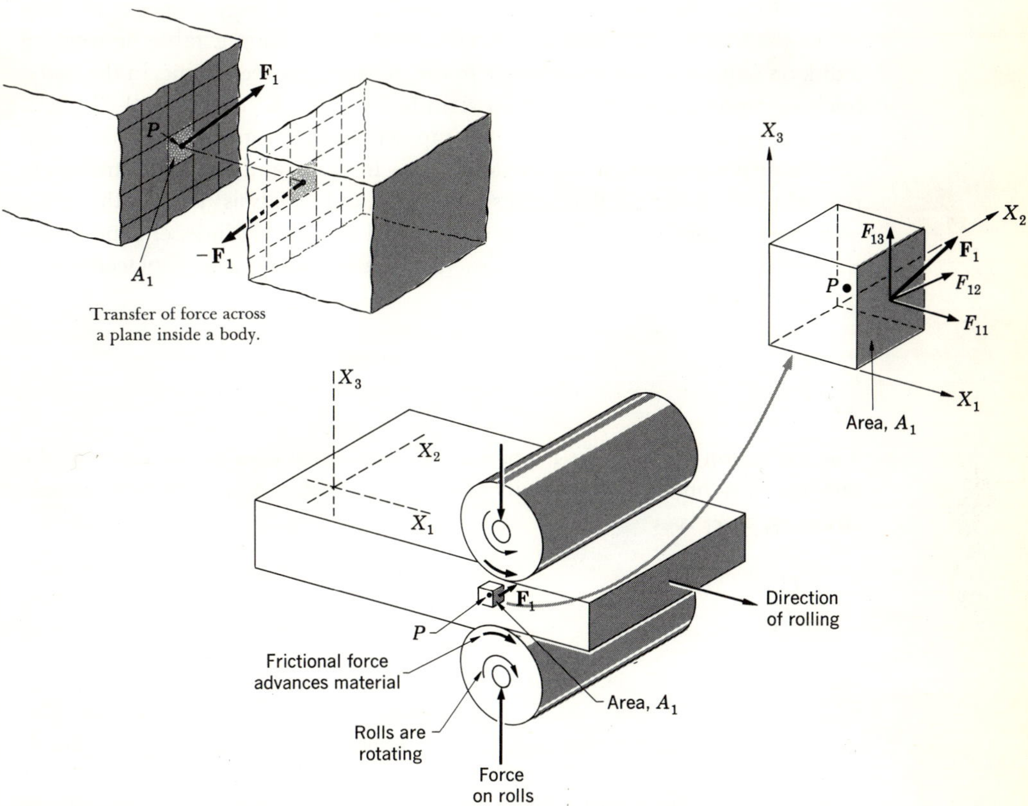

FIGURE 9-2 Schematic illustration of a material being rolled, showing the relation between the external forces and the stresses within the material at some point P.

Therefore, the state of stress at a point is described by only six independent components, often numbered for convenience with a single subscript; thus

$$\sigma_{11} = \sigma_1 \qquad \sigma_{22} = \sigma_2 \qquad \sigma_{33} = \sigma_3 \qquad \sigma_{23} = \sigma_4 \qquad \sigma_{31} = \sigma_5 \qquad \sigma_{12} = \sigma_6 \tag{9-4}$$

A normal stress (σ_1, σ_2, or σ_3) causes elongation of a material or contraction if the stress is negative. A shear stress (σ_4, σ_5, or σ_6) produces the shearing type of deformation shown in Fig. 9-3*b*. The definition of stress is general and is independent of the nature of the material. The response of the material to a given stress, however, may be elastic, elastomeric, plastic, or viscoelastic, as discussed in later sections. Often the nominal stress F/A_0 employing the initial area A_0 is used in place of the true stress defined by Eq. (9-1).

9-2. Strain

Strain measures the displacement of points of a body *relative to neighboring points*. Moving (displacing) an entire specimen along a table or rotating a disk does not produce any change in the distance between points in the body. Such displacements are known as rigid-body motions. The overall displacements of points of a body can be decomposed into (1) rigid-body motions and (2) the displacements giving rise to changes in shape and size. The strains are related to the latter displacements only. A suitable analysis[1] isolates the strain components from the total displacements and leads to six independent strains ϵ_i, analogous to the six stresses [Eq. (9-4)]. Each strain is a displacement U per unit length L; thus

$$\epsilon = \frac{U}{L} \tag{9-5}$$

The definition of strain at a point P, as with stress at a point, involves the analysis of a small cube with its center at P (Fig. 9-3*a*). A typical normal strain $\epsilon_{11} \equiv \epsilon_1$ is

$$\epsilon_1 = \frac{U_{11}}{L_1} \tag{9-6}$$

where U_{11} is the displacement in the X_1 direction of the end of the line L_1, which was initially along the X_1 axis and with one end fixed at the origin. The shear strains are sometimes defined analogously to shear stresses [Eq. (9-2)],

[1] This analysis is well explained in F. A. McClintock and A. S. Argon (eds.), "Mechanical Behavior of Materials," pp. 55–58, Addison-Wesley Publishing Company, Inc., Reading, Massachusetts, 1966.

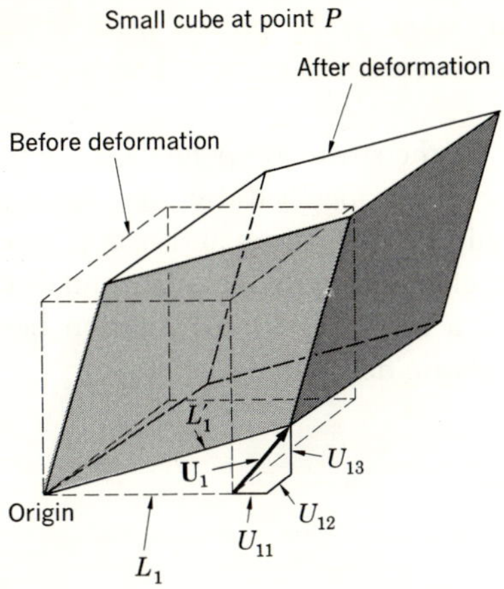

(*a*) The displacement, $\mathbf{U}_1$, resulting from the deformation of line L_1 to L_1'.

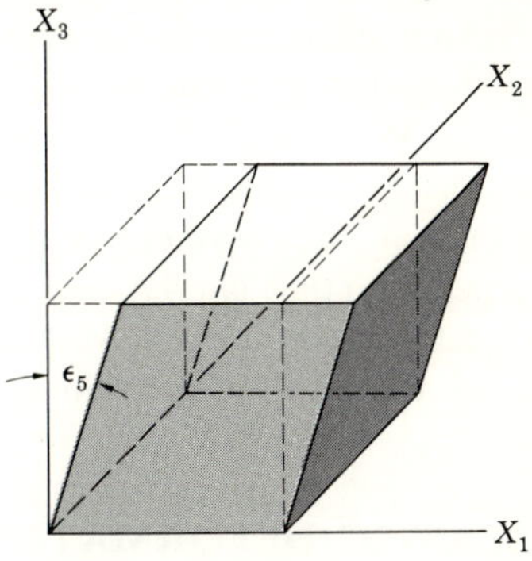

(*b*) Shear deformation associated with ϵ_5.

FIGURE 9-3 Schematic illustration of (*a*) the displacements U_{ij} used to define the strains and (*b*) a typical shear deformation ϵ_5.

but usually they are defined to describe the type of shear deformation shown in Fig. 9-3*b*; thus

$$\epsilon_5 = \frac{U_{31}}{L_3} + \frac{U_{13}}{L_1} \tag{9-7}$$

Like stress at a point, strain at a point is a general description that is valid regardless of the nature of the material.

Two methods for calculating strain are commonly used, depending on the magnitude of the strain. When a strain is small, the reference axes remain virtually unchanged, and the procedure is simple. In the case of normal strain ϵ_1, for example, $U_{11} = \Delta L_1$, and L_1 has essentially its initial length L_o. The strain is then simply

$$\epsilon_1^o = \frac{\Delta L_1}{L_o} \tag{9-8}$$

and is known as the *nominal strain*. For convenience, as in Fig. 9-1, nominal strain is often employed even when L_1 changes appreciably from its initial value. More correctly, the *true strain* should be used in such cases; thus

$$\epsilon_1 = \int_{L_o}^{L_1} \frac{dL}{L} = \ln \frac{L_1}{L_o} \tag{9-9}$$

This equation takes into account the change in the reference length L from its initial value L_0 to its value L_1 at some stage in the process of deformation. Problem 9-2 considers the replotting of a stress-strain curve of Fig. 9-1 in terms of true stress and true strain.

When a given force (described by the stress components σ_i) acts on a material, the resulting deformation can be specified by the strain components ϵ_i. The mathematical relationship between stress and strain, however, depends on the material in question and often on the temperature and duration of deformation as well. We begin with a relatively simple relation between stress and strain, that for infinitesimal elastic deformation.

ELASTIC DEFORMATION

9-3. Hooke's Law

For many technically important materials at ordinary temperatures (metals, ceramics, crystalline polymers, and wood) the response to relatively small stresses is quite simple (elastic) deformation. The time dependence is negligibly small, and the temperature aspect is partially included in the constants (S_{ij} or C_{ij}, discussed below) that describe the behavior of a given material. Thermal strains, if any, can be calculated separately as

$$\epsilon_T = \alpha \, \Delta T \tag{9-10}$$

where α is the coefficient of linear expansion (see Prob. 9-3). The term *Hookean behavior* is used to describe the usual case in which a given strain component varies linearly with certain of the stress components, as illustrated by curve *a* in Fig. 9-1. The principle of superposition ensures that the strain due to one stress can be simply added to that due to another stress. Therefore, the total strain component ϵ_i is [see Eq. (1-7*a*) for the index summation convention]

$$\epsilon_i = S_{ij}\sigma_j \qquad i,j = 1, 2, \ldots, 6 \tag{9-11}$$

Six equations of this type describe all six of the strain components and require a total of 36 *elastic compliances* S_{ij}. Alternatively, the stresses can be determined from the strains

$$\sigma_i = C_{ij}\epsilon_j \tag{9-12}$$

where C_{ij} are the *elastic constants.*

Several circumstances reduce the number of independent S_{ij} (or C_{ij}) values needed to describe the behavior of various materials: (1) An argument involving strain energy[1] shows that the S_{ij}, when arranged in the form of a matrix (Fig. 9-4*a*), occur as pairs of identical values on either side of the main diagonal. (2) As the crystal symmetry increases, the number of independent, nonzero compliances decreases from the maximum of 21, characteristic of triclinic crystals[1] to the following numbers: monoclinic, 13; orthorhombic, 9; tetragonal and rhombohedral, 7 or 6; hexagonal, 5; and cubic, 3 (Fig. 9-4*b*). An isotropic substance, such as glass or a fine-grained crystalline aggregate, has only two independent, nonzero compliances. (3) Often only a single strain component is of interest, so only the pertinent S_{ij} values need be considered.

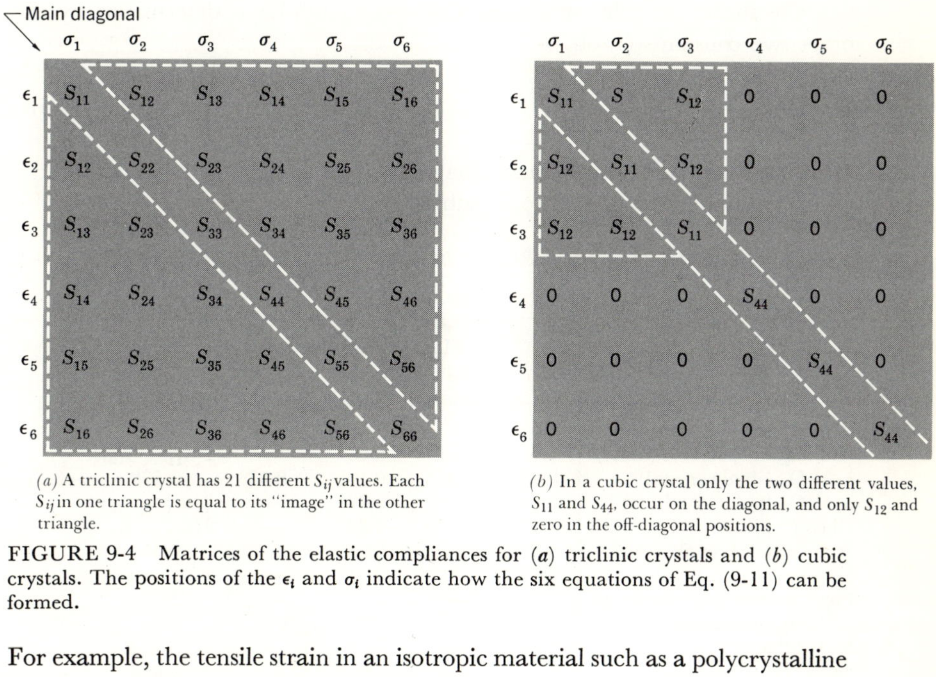

(*a*) A triclinic crystal has 21 different S_{ij} values. Each S_{ij} in one triangle is equal to its "image" in the other triangle.

(*b*) In a cubic crystal only the two different values, S_{11} and S_{44}, occur on the diagonal, and only S_{12} and zero in the off-diagonal positions.

FIGURE 9-4 Matrices of the elastic compliances for (*a*) triclinic crystals and (*b*) cubic crystals. The positions of the ϵ_i and σ_i indicate how the six equations of Eq. (9-11) can be formed.

For example, the tensile strain in an isotropic material such as a polycrystalline steel bar is fully described by the equation,

$$\epsilon_1 = S_{11}\sigma_1 + S_{12}(\sigma_2 + \sigma_3) \tag{9-13}$$

even if all six stresses have finite values. With the usual engineering notation, Table 9-1, this equation becomes,

$$\epsilon_1 = \frac{\sigma_1}{E} - \frac{\nu}{E}(\sigma_2 + \sigma_3) \tag{9-14}$$

[1] Further information can be found in J. F. Nye, "Physical Properties of Crystals," p. 136ff. Oxford University Press, London, 1960.

E is Young's modulus and ν is Poisson's ratio. (4) Finally, if only a single stress is acting on a material, Hooke's takes law is one of its simplest forms,

$$\epsilon_1 = \frac{\sigma_1}{E} \qquad \text{tensile deformation} \tag{9-15}$$

$$\epsilon_4 = \frac{\sigma_4}{G} \qquad \text{shear deformation} \tag{9-16}$$

$$\epsilon_2 = \frac{-\sigma_1 \nu}{E} \qquad \text{Poisson contraction} \tag{9-17}$$

G is the shear (or bulk) modulus. The shear modulus is determined by the other two constants (Prob. 9-3),

$$G = \frac{E}{2(1 + \nu)} \tag{9-18}$$

consistent with the requirement that only two independent values of the elastic constants exist for isotropic substances.

TABLE 9-1. ELASTIC COMPLIANCES AT 25°C FOR REPRESENTATIVE MATERIALS†

Material	Elastic compliances, 10^{-11} m²/N					
	S_{11}	S_{33}	S_{12}	S_{13}	S_{44}	S_{14}
Aluminum (single crystal)	1.59	—	−0.58	—	3.52	
Aluminum (polycrystalline)	1.6	—	−0.53	—	$2(S_{11} - S_{12})$	
α-Al_2O_3 (single crystal)	0.290	0.194	−0.105	−0.038	0.578	−0.171
α-Al_2O_3 (polycrystalline)	0.26	—	−0.071	—	$2(S_{11} - S_{12})$	
Glass (typical)	1.6	—	−0.38	—	$2(S_{11} - S_{12})$	
Iron (single crystal)	0.80	—	−0.28	—	0.86	
Iron (polycrystalline)	0.48	—	−0.14	—	$2(S_{11} - S_{12})$	
Polyethylene (high-density)	170	—	—	—	260	
SiO_2 (quartz crystal)	1.27	0.97	−0.17	−0.15	2.01	−0.43
SiO_2 (glass)	1.38	—	−0.28	—	$2(S_{11} - S_{12})$	
Tungsten (single crystal)	0.257	—	−0.073	—	0.660	
Tungsten (polycrystalline)	0.29	—	−0.082	—	$2(S_{11} - S_{12})$	
Engineering notation for isotropic materials	$\frac{1}{E}$	—	$-\frac{\nu}{E}$	—	$\frac{1}{G} = \frac{2(1+\nu)}{E}$	—

† Conversion of units: (M m²/N) × 10^{-1} = N cm²/dyne; (M m²/N) (6.90 × 10^3) = N in²/lb

The deformation of single crystals may be of interest because of either, (1) the anisotropic behavior of individual grains in a polycrystalline aggregate, or (2) some property such as piezoelectricity, employed technically with single crystals of inorganic substances. In either case, the complete set of elastic compliances S_{ij}, Table 9-1, or the corresponding elastic constants C_{ij} may be involved. An interesting example is the following method for detecting the presence of preferred orientation (see Fig. 8-3) in a polycrystalline material by measuring the velocity of high-frequency ultrasonic waves, v_s, in several directions. If the grains are randomly oriented, the velocity of a longitudinal wave is

$$v_s = \left[\frac{E}{\rho}\frac{1-\nu}{(1+\nu)(1-2\nu)}\right]^{1/2} \tag{9-19}$$

ν is Poisson's ratio, and ρ is the density. On the other hand, a high degree of preferred orientation would produce behavior approaching that of a single crystal. For longitudinal (e.g., compressional) waves along the [100], [110], and [111] directions in a cubic crystal, for example, the velocities are

$$[100]: \qquad v_s = \left(\frac{C_{11}}{\rho}\right)^{1/2} \tag{9-20}$$

$$[110]: \qquad v_s = \left(\frac{C_{11} + C_{12} + 2C_{44}}{2\rho}\right)^{1/2} = \left(\frac{C_{11} + C_a}{\rho}\right)^{1/2} \tag{9-21}$$

$$[111]: \qquad v_s = \left(\frac{C_{11} + 2C_{12} + 4C_{44}}{3\rho}\right)^{1\ 2} = \left(\frac{C_{11} + 4C_a/3}{\rho}\right)^{1/2} \tag{9-22}$$

The differences among these values depend on the anisotropy factor

$$C_a = C_{44} - \frac{C_{11} - C_{12}}{2} \tag{9-23}$$

In an isotropic material $C_a = 0$. [The corresponding factor for compliances (Table 9-1) is $S_a = S_{44} - 2(S_{11} - S_{12})$.] If preferred orientation is present in a sheet material, for example, then v_s varies with the direction of testing in the sheet (see Prob. 9-5).

9-4. Variation of Elastic Constants

A useful correlation among various substances with respect to elastic properties has been devised by J. J. Gilman.[1] Elastic stiffness is proportional to the stress necessary to increase the separation of the atoms or ions in a material. For ionic, covalent, and certain metallic materials, this stress is proportional to

$$\frac{F}{A} = \frac{\text{force of electrostatic attraction}}{\text{appropriate area}} \approx \frac{q^2/d^2}{d^2}$$

[1] J. J. Gilman, "Micromechanics of Flow in Solids," pp. 29–42, McGraw-Hill Book Company, New York, 1969.

where d is the interatomic separation. Therefore, the elastic stiffness should be proportional to q^2/d^4 and should be greater the larger the charge q and the smaller the interatomic spacing. This relation predicts a slope of -4 in a logarithmic plot of elastic constants versus interatomic spacings for several comparable substances. Figure 9-5 shows that materials having the same type of bonding (ionic, for example) and the same crystal structure fall on the same straight line.

Although various materials differ greatly in their elastic properties, a given substance usually has relatively constant elastic properties under the influence of moderate changes in temperature or composition. Small changes in temperature have little effect on Young's modulus (Fig. 9-6a). In critical

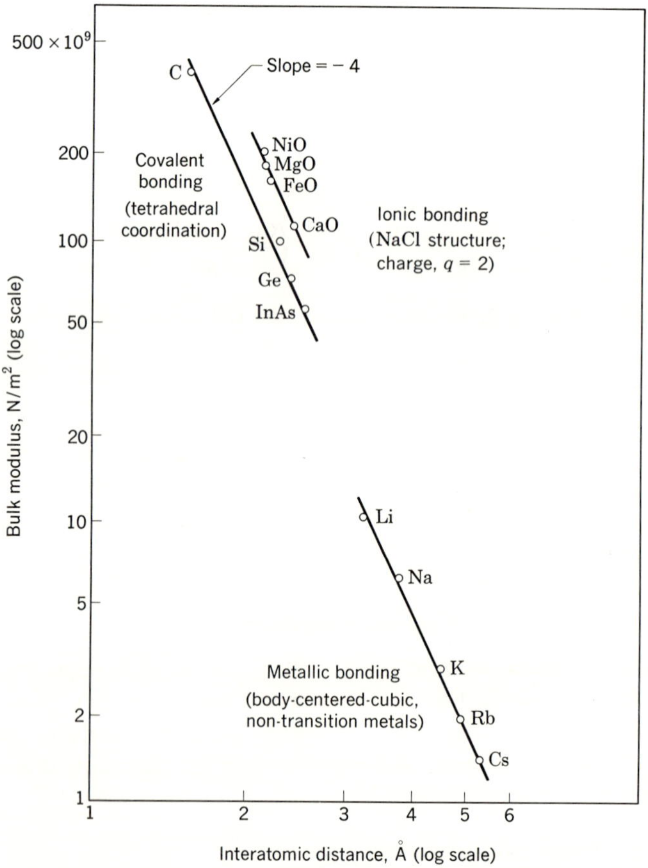

FIGURE 9-5 Dependence of the bulk modulus $E/[3(1-2\nu)]$ of several types of materials on their interatomic spacing [conversion of units: (M N/m²) $1.450 \times 10^{-4} = N$ lb/in²; M N/m² $\times 10 = N$ dynes/cm²]. (*After J. J. Gilman.*)

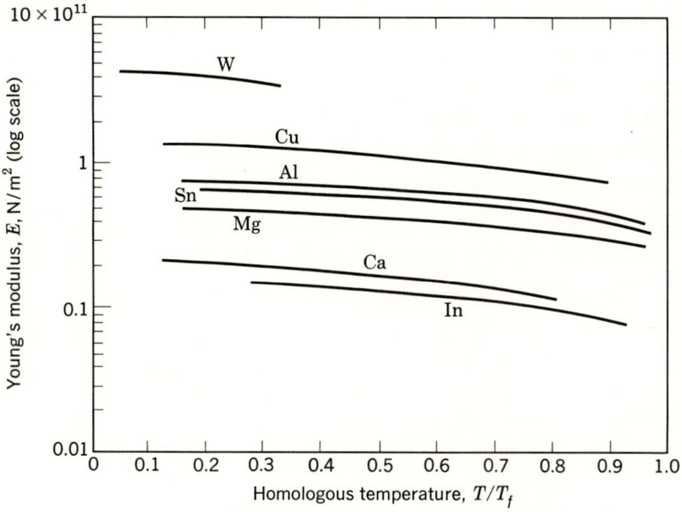

(*a*) Effect of temperature. (Data of W. Köster.)

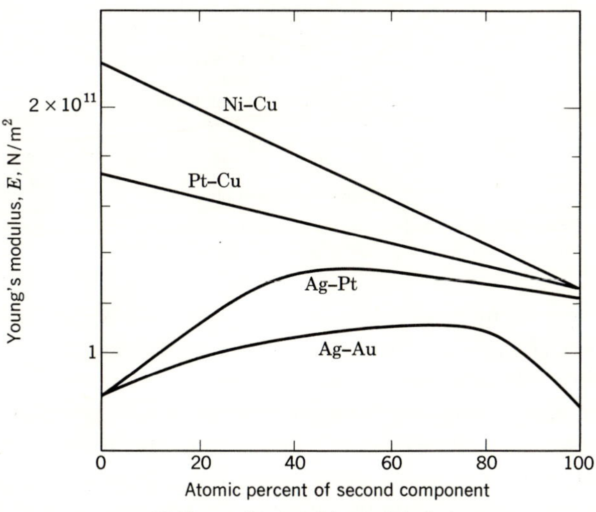

(*b*) Change of *E* across binary solid-solution systems.

FIGURE 9-6 The effect of temperature and of solid-solution formation on Young's modulus for several metals. (*Adapted from O. D. Sherby.*)

applications, such as precision springs, even this small effect can be eliminated by the use of special alloys (Elinvar, for example) for which $dE/dT \approx 0$ near room temperature. Small additions of an alloying element or impurity to a base material usually cause little change in elastic properties (Fig. 9-6*b*), especially if the alloying element forms a solid solution. Alterations in the structure of a material by cold working, precipitation hardening, or other strengthening procedure, generally reduce Young's modulus slightly, even though the yield strength of the material is increased.

An important feature of two-phase materials, such as composites, is the fact that the elastic properties vary in a roughly linear manner between those

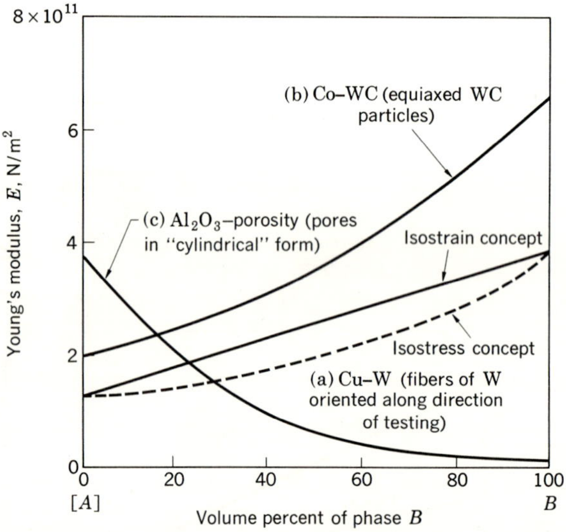

FIGURE 9-7 Effect of change in composition on Young's modulus for a two-phase (A-B) composite material.

of the two component phases. For example, when copper is reinforced with tungsten fibers (Fig. 9-7*a*), Young's modulus is doubled by the addition of 40 v/o of fibers. The observed linear dependence on volume percent of the second phase agrees with the isostrain concept, which postulates that both phases are strained the same amount when a composite deforms under a stress. Tests of this composite at right angles to the long fibers, however, give much lower values for Young's modulus, closer to the isostress concept (Prob. 9-6). Many composite materials consist of roughly equiaxed particles of a reinforcing phase in a matrix (Fig. 9-7*b*). The variation of Young's modulus is then found to lie somewhere between the predictions of the isostrain and isostress concepts. An important second phase is porosity, especially in sintered polycrystalline ceramics (Fig. 9-7*c*). The sharp drop of Young's modulus to about 50 percent of its initial value in the presence of only 20 percent porosity suggests that the porosity is present in an especially deleterious form, such as interconnected, roughly cylindrical volumes.

9-5. Anelasticity

The foregoing treatment of elastic deformation, which neglected any effect of time, is generally correct even though the several time-dependent processes shown in Fig. 9-8 are occurring in a typical material. The key to understanding this paradox is the concept of *relaxation time* τ. Even for complex

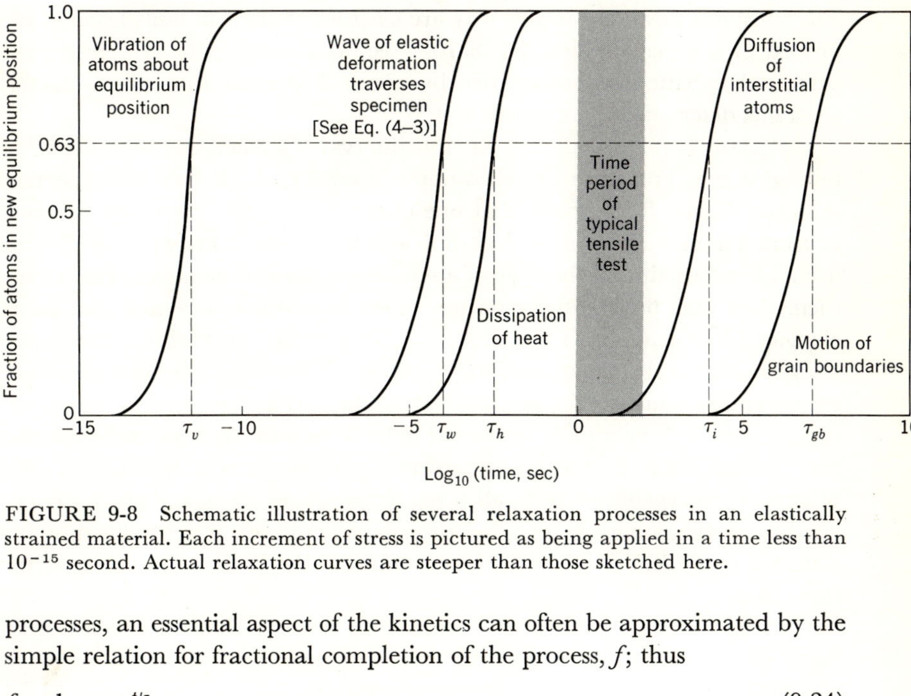

FIGURE 9-8 Schematic illustration of several relaxation processes in an elastically strained material. Each increment of stress is pictured as being applied in a time less than 10^{-15} second. Actual relaxation curves are steeper than those sketched here.

processes, an essential aspect of the kinetics can often be approximated by the simple relation for fractional completion of the process, f; thus

$$f = 1 - e^{-t/\tau} \tag{9-24}$$

At time $t = \tau$ after the beginning of the process, $f = (1 - e^{-1}) = (1 - 0.37) = 0.63$; that is, 63 percent of the system concerned has undergone the process in question. At $t = \tau/3$, however, the reaction has barely started ($f = 0.05$), and at $t = 3\tau$ it is almost complete ($f = 0.95$). For those processes that obey Eq. (9-24), the relaxation time indicates the relatively narrow interval of time in which a major part of a given process takes place. Consider the time (from about 10^0 to 10^2 sec) during which the data for a typical tensile test in the elastic range are recorded (Fig. 9-8). For the spectrum of relaxation times shown three processes have already relaxed (atomic vibration, wave of deformation, and dissipation of heat). Other slower processes are not detectable in the time range of testing. Consequently, for the period in which tensile-test data are being taken in this case, the various time-dependent processes can be neglected.

Time-dependent processes can appear in "elastic" measurements in several ways. An increase in the sensitivity of an experiment may reveal the occurrence of a relaxation process. A change in the temperature or the time of testing drastically shifts the position in the relaxation spectrum. For example the sonic method for determining elastic constants measures the time (about 10^{-5} sec) for an elastic wave to traverse a typical specimen [see Eq. (4-3)]. The constants measured in this way are *adiabatic* elastic constants because the process of heat conduction ($\tau_h \approx 10^{-4}$ sec) can occur only to a negligible extent. Adiabatic constants differ by about 1 percent from the *isothermal* constants determined by a tensile test.

Many materials fall into the time interval of a relaxation process during testing or use. The anelastic relaxation of iron-carbon alloys is an important example. Figure 5-20*a* shows that one of the three $\langle 100 \rangle$ directions becomes elongated in the vicinity of each carbon atom dissolved in body-centered cubic iron. Therefore, during the time of a tensile test the carbon atoms redistribute themselves and preferentially occupy those interstitial sites that are being elongated by the applied uniaxial stress σ. In so doing, the carbon atoms cause a slight additional elongation of the specimen. If the stress is quickly removed, the specimen contracts the amount σ/E of the elastic deformation. However, the anelastic deformation is recovered only at the rate that the carbon atoms can "jump" from the previously preferred sites into a neighboring site. This relaxation process occurs until all three types of sites are again equally populated by carbon atoms. In this case the relaxation time τ is determined by the diffusion coefficient D of the carbon atoms and the lattice constant a_0 of the iron; thus

$$\tau = \frac{a_0^2}{36\,D} \qquad (9\text{-}25)$$

At slightly elevated temperatures τ is of the order of 10^2 sec, corresponding to a D value of 10^{-24} m^2/sec. One use of anelastic studies is to determine kinetic parameters, such as diffusion coefficients, in low ranges not accessible by the usual experimental methods.

The *damping* of vibrations is another aspect of anelastic behavior. In only a few instances is a low damping capacity advantageous, for example, in bronze bells, in piezoelectric devices, and in delay lines in electronic circuits.

A high damping capacity is usually desirable in engineering materials to prevent the occurrence of large-amplitude vibrations. Vibrational energy may be constantly supplied (for example, in an airplane from its engine), but it must be dissipated by damping. If ΔU is the amount of vibrational energy dissipated per cycle, its ratio $\Delta U/U$ to the total vibrational energy U is one of the standard measures of damping capacity. Typical values for several materials are listed in Table 9-2. Damping usually increases with an increase

TABLE 9-2. DAMPING CAPACITY AT LOW STRESSES FOR TYPICAL MATERIALS

Material	Temperature, °C	$(\Delta U/U)$ per cycle
Aluminum	25	1×10^{-4}
Gray cast iron	25	60×10^{-4}
Polycrystalline Al_2O_3	25	3×10^{-3}
	1000	3×10^{-2}
	1200	2×10^{-1}
Soda-lime glass	−50	6×10^{-3}
	200	2×10^{-2}
	400	6×10^{-2}
Isotactic polypropylene	−25	0.2
	25	1
	75	2

in the stresses produced by the vibration and may vary significantly with the vibrational frequency. The high damping capacity of polymers makes them especially useful for damping of vibrations (as mounting pads or washers). Gray cast iron, widely used for the frame of machine tools, owes its relatively good damping capacity to the graphite flakes in its microstructure.

9-6. Elastomeric Deformation

The deformation of a slightly cross-linked, amorphous (rubbery) polymer differs significantly from that of metals and ceramics. The region of elastic behavior (Fig. 9-1*c*) extends over strains of a few hundred percent, compared to about 1 percent in metals. Over this large range of strains, significant departures from Hookean behavior ($\epsilon = \sigma/E$) occur. The mechanism of elastic deformation involves the partial straightening of randomly coiled chain molecules. This partial ordering causes the change in entropy, ΔS, to be negative. A standard thermodynamic equation gives the following relation involving the work W_d of elastic deformation:

$$W_d = \Delta E - T\,\Delta S \tag{9-26}$$

The change in internal energy, ΔE, is negligibly small in natural rubber for the range of deformation being considered. Therefore, $W_d \approx -T\,\Delta S$, and the work to produce a given degree of deformation (corresponding to a given change in entropy, ΔS) is proportional to the temperature of deformation (see Prob. 9-7). A statistical treatment of the change in entropy leads to the following equation for uniaxial elastic deformation:

$$\sigma = \frac{\rho RT}{M_c}\left(\lambda^2 - \frac{1}{\lambda}\right) \tag{9-27}$$

M_c is the molecular weight of a section of polymer chain between adjacent cross-links, ρ is the density, R is the gas constant, and $\lambda \equiv L/L_o = 1 + \epsilon^o$ is known as the *stretch ratio*. For strains of only a few percent the true stress σ is proportional to the nominal strain ϵ^o. At $\epsilon^o = 0.1$, the departure from linearity is 10 percent, at $\epsilon^o = 0.3$ it is 40 percent, and it increases rapidly at still higher values.

PLASTIC DEFORMATION

In contrast to elastic deformation, which depends primarily on stress in the simplest cases, plastic deformation is typically a function of stress (σ), temperature (T), and the rate of straining ($d\epsilon/dt \equiv \dot{\epsilon}$). The microstructure of a given material is often important also. Depending on the values of σ, T, and $\dot{\epsilon}$, a material may exhibit various mechanisms of plastic deformation. Although a given mechanism may be relatively simple, the simultaneous operation of several mechanisms often produces a complex behavior. In addition, the "stress path" may be significant. For example, if a steel specimen is both elongated and twisted in the plastic range, its strength is higher if the elongation is performed first, rather than if it is performed simultaneously or last. The mechanisms that account for the various aspects of plastic deformation are usefully treated in two categories: those that permit the relative displacement of adjacent portions of the lattice over large distances, and those that allow displacement of only one lattice spacing or less.

9-7. Atomic Displacements Over Multiple Lattice Spacings

Technical processes of plastic deformation, such as the drawing of polymers or the hot rolling of metals, may lengthen the object tenfold within several seconds. Plastic deformation of this magnitude generally occurs by the relative "slipping" of adjacent parts of the material by large distances relative to the interatomic spacing. In the next two sections we consider the nature of this process in (1) inorganic crystals, such as metals, and (2) organic polymers.

CRYSTAL SLIP. When a dislocation having Burgers vector b moves on its slip plane through a region of a crystal (Fig. 4-13), it causes relative movement of the two parts of the crystal on either side of the slip plane. Several factors determine the relation between this elementary step and the overall plastic deformation of polycrystalline materials. First, slip occurs on a number of *slip systems* in a given material (Table 9-3). In face-centered cubic metals, for example, slip occurs on one of four {111}-type planes and in any of three ⟨101⟩-type directions in a given plane, a total of 12 slip systems. Provided that at least five independent slip systems can operate, a grain in a polycrystalline material is able to deform into any arbitrary shape. This requirement seriously limits the deformation of Al_2O_3, for example, which slips only on the (0001) plane, even at high temperatures and in only two independent directions on this plane. In some materials, such as MgO, the number of active slip systems increases to five or more at sufficiently high temperatures; these polycrystalline materials tend to be brittle at low temperatures but become ductile at high temperatures. The shear stress† τ_c necessary to cause plastic deformation on a given slip system can be measured in single crystals (Prob. 9-8). Typical values are given in Table 9-3.

Consider now a typical grain within a polycrystalline specimen. A tensile stress σ applied to the specimen imposes a shear stress τ_i on the ith slip system. Its magnitude is

$$\tau_i = \sigma \cos \theta_i \cos \varphi_i = \sigma m_i \tag{9-28}$$

The Schmid factor m_i is determined by the angles the direction of the applied stress makes with the slip direction (θ_i) and with the normal to the slip plane (φ_i) (Prob. 9-8). The dislocations begin to move first on the *primary slip system* the one for which m_i (and therefore τ_i) is highest (Fig. 9-9*a*). The contribution $\dot{\epsilon}_i$ of a given slip system to the rate of tensile elongation can be expressed as

$$\dot{\epsilon}_i \equiv \frac{d\epsilon_i}{dt} = \rho_i b_i v_i m_i \tag{9-29}$$

ρ_i is the density of favorably oriented dislocations with Burgers vector b_i. Under the action of the stress σ acting at a given time, these dislocations move with an average velocity v_i. The Schmid factor is required to resolve the strain in the slip direction into strain in the direction of testing. In addition to the primary slip plane, other {111} planes are active in deformation (Fig. 9-9*a*), even in single crystals. Because the crystal rotates during deformation, the conjugate plane becomes an active slip plane. Usually it acts cooperatively

† A widely used notation for shear stress is τ, and it is conveniently employed here (rather than σ_4, σ_5, or σ_6). Similarly, γ denotes a shear strain.

TABLE 9-3. DATA ON SLIP AND TWINNING IN TYPICAL MATERIALS†

Material	Slip systems		Critical resolved shear stress	Twinning elements‡				Twinning shear strain
	Plane	Direction	τ_c, N/m²	K_1	η_1	K_2	η_2	
Copper (FCC)	{111}	⟨101⟩	1 × 10^6 (25°C)	{111}	⟨112⟩	{111}	⟨112⟩	0.707
Molybdenum (BCC) below 650°C	{112}	⟨111⟩	7 × 10^7 (25°C)	{112}	⟨111⟩	{112}	⟨111⟩	0.707
above 650°C	{110}	⟨111⟩						
Cadmium (HCP)	{0001}	$\langle 11\bar{2}0\rangle$	6 × 10^5 (25°C)	$\{10\bar{1}2\}$	$\langle\bar{1}011\rangle$	$\{\bar{1}012\}$	$\langle 10\bar{1}1\rangle$	0.175
Silicon (diamond cubic)	{110}	⟨111⟩		{111}	⟨112⟩	{111}	⟨112⟩	0.707
TiC (NaCl cubic)	{111}	⟨101⟩	1 × 10^8 (1100°C)					
Al_2O_3 (hexagonal)	{0001}	$\langle 11\bar{2}0\rangle$	4 × 10^7 (1500°C)	$\{10\bar{1}1\}$	$\langle\bar{1}012\rangle$	$\{\bar{1}012\}$	$\langle 10\bar{1}1\rangle$	0.202
Polyethylene	{hk0}	⟨001⟩	—	{110}	$\langle 1\bar{1}0\rangle$	—	—	—

† Conversion of units: (M N/m²) 1.020 × 10^{-7} = N kg/mm²; (M N/m²) 1.450 × 10^{-4} = N lb/in².
‡ See Fig. 5-12.

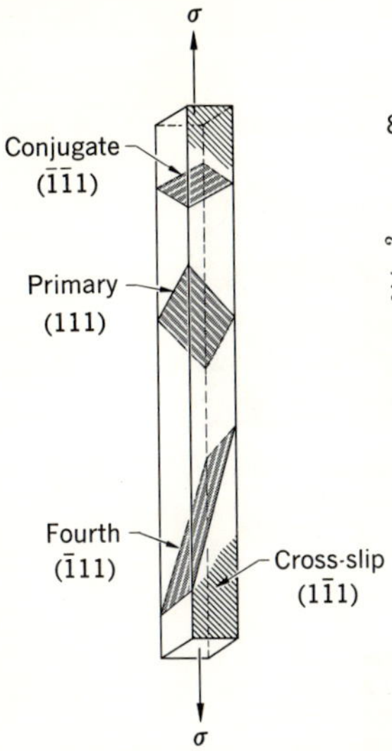

(a) An example of the geometry of the four types of slip planes, shown in a particular single crystal tested in tension.

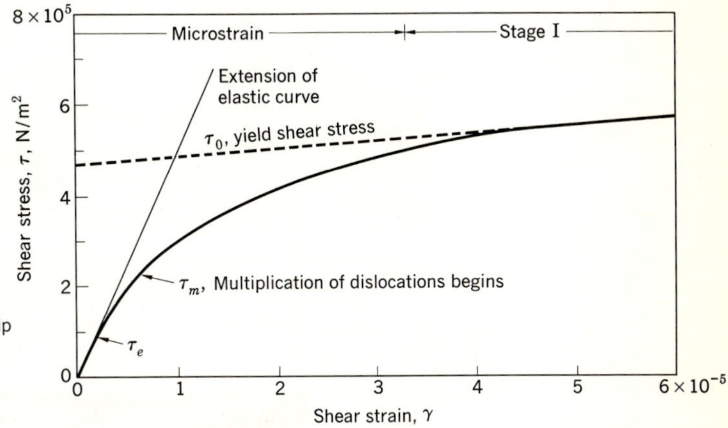

(b) The shear-stress *vs* shear-strain curve of the crystal, showing the pre-yield behavior and also the meaning of τ_0, the yield shear stress.

FIGURE 9-9 Deformation of a single crystal of copper by slip in the region of microstrain and in stage I (easy glide). (*After A. S. Argon and W. T. Brydges.*)

with the primary slip plane and produces double slip. A dislocation traverses a cross-slip plane in moving from one active slip plane to an adjacent, parallel plane.

The beginning of plastic deformation can be specified with various degrees of refinement (Fig. 9-9*b*). A sensitive detector will show the shear stress τ_e at which ordinary, linear elastic behavior ends. Above τ_e the existing dislocations move and produce a microstrain in addition to the elastic strain. At a higher stress τ_m, additional dislocations begin to be created by the various means of *dislocation multiplication*, such as a Frank-Read source. Plastic strain then quickly increases with increase in shear stress, and the stress-strain curve becomes linear at higher strains. This portion of the curve is called *stage I* or *easy glide*; the dislocations are moving principally along the primary slip plane. Extrapolation of the linear stage I curve back to the stress axis is a common procedure for determining the yield shear stress τ_0. Microscopic evidence of slip at this stage can be seen in electron micrographs and usually can also be observed on the surface of the specimen (Fig. 9-10). Individual slip lines tend to concentrate in groups to form a slip band.

Materials deformed appreciable amounts develop more complex dislocation arrangements. If a material has a high stacking-fault energy (undis-

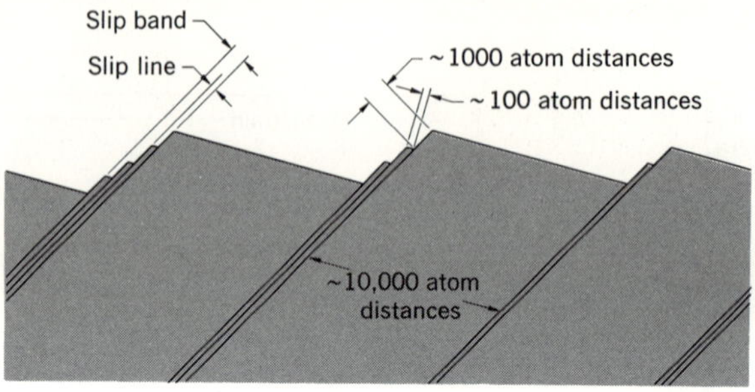

(a) Schematic illustration of slip bands within a crystal and the corresponding slip markings at the surface. (Adapted from S. Kalpakjian, *Mechanical Processing of materials,* D. Van Nostrand Company, Princeton, New Jersey, 1967.)

(b) Scanning-electron micrograph showing the slip markings on a deformed single crystal of cobalt. (Courtesy Eric Lifshin, Research and Development Center, General Electric Company.)

FIGURE 9-10 Illustrations of the characteristic markings produced by the intersection of active slip planes with the surface of a specimen.

sociated) dislocations are able to cross-slip onto secondary slip systems in stage II of deformation (Fig. 4-19). A result is the production of tangles of dislocations that form cell walls, outlining cellular volumes that are relatively free of dislocations (Fig. 9-11*a*). This cell structure persists through the final deformations (stage III) that precede fracture, but the misorientation of adjacent cells becomes greater. In materials whose low stacking-fault energy permits appreciable dissociation of dislocations, cross-slip is much reduced. The dislocation structure (Fig. 9-11*b*) tends to be more uniform and shows stacking

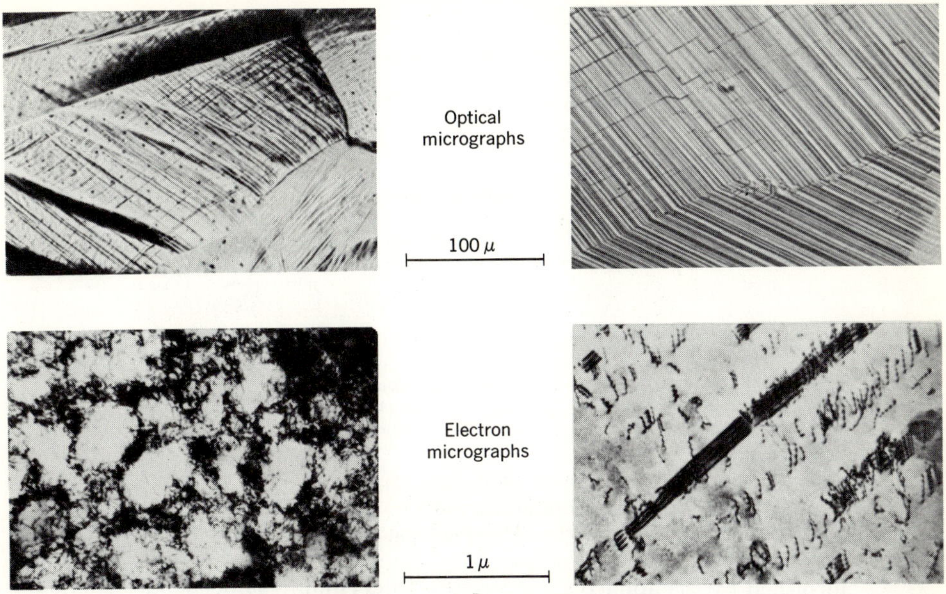

(*a*) Copper, having a high stacking-fault energy and therefore essentially undissociated dislocations. Cross-slip leads to the wavy slip bands and tangled dislocations.

(*b*) A Cu-7.5 w/o Al alloy, having a low stacking-fault energy and dissociated dislocations. The absence of cross-slip leads to straight slip traces and uniform dislocation lines.

FIGURE 9-11 Comparison of the deformation of two types of uniform polycrystalline materials in stage II deformation. (*Courtesy T. L. Johnston and C. E. Feltner, Ford Scientific Laboratory. By permission of Metallurgical Transactions*).

faults on various types of slip planes. The motion of dislocations can be impeded not only by interaction with other dislocations, but also by obstacles such as particles of a second phase, twin boundaries, or grain boundaries. A typical effect of such obstacles on the dislocation structure is the production of pile-ups (Fig. 9-12) during plastic deformation.

CHAIN MOBILITY. Deformation over multiple atomic spacings in polymers differs fundamentally from that in inorganic crystals because the structural units (such as the —CH_2— mer in polyethylene) cannot shift their main bonds but instead are permanent members of a long polymer chain. Consider first the effect of shear stresses (Fig. 9-13). In a glassy polymer two requirements must be met to permit deformation of a portion of chain, assumed to be straight initially. First, a certain "free volume" of empty space must become available within the bulk of the polymer (analogous to a vacancy for atomic diffusion) to accommodate the chain in its new position. Second, some bond angles must flex from low-energy positions to high-energy positions (Fig. 9-13*a*) to permit displacement of a portion of the chain. Energy for both of these processes is

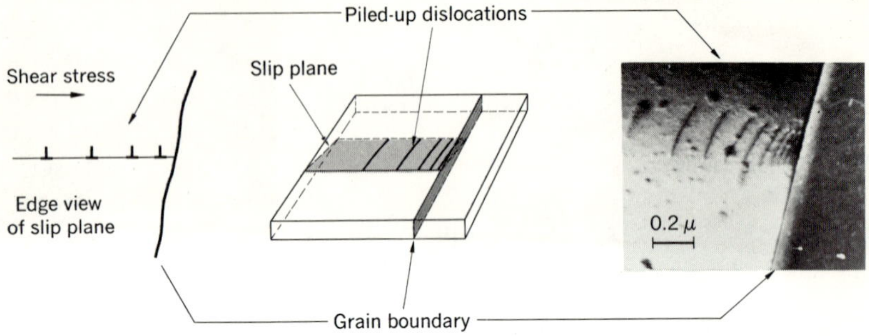

(*a*) Shear stress causes dislocations to pile up at an obstacle, a grain boundary in this case.

(*b*) Sketch of a slip plane passing at an angle through a thin foil used for transmission electron microscopy.

(*c*) Electron micrograph corresponding to (*b*), revealing piled-up dislocations in an aluminum alloy.

FIGURE 9-12 Schematic drawings and a transmission micrograph of the piling up of dislocations at an obstacle. (*Micrograph courtesy of J. J. Hren, University of Florida.*)

furnished by the applied shear stresses. In crystalline lamellae, deformation is restricted by the requirement for maintaining spatial regularity (Fig. 9-13*b*). An individual kink within a lamella (Fig. 4-24) moves under the influence of a shear stress. When the kink has traversed the entire length of a chain segment, this segment has been displaced relative to the rest of the lamella. At temperatures close to the melting point an entire segment may move half a lattice

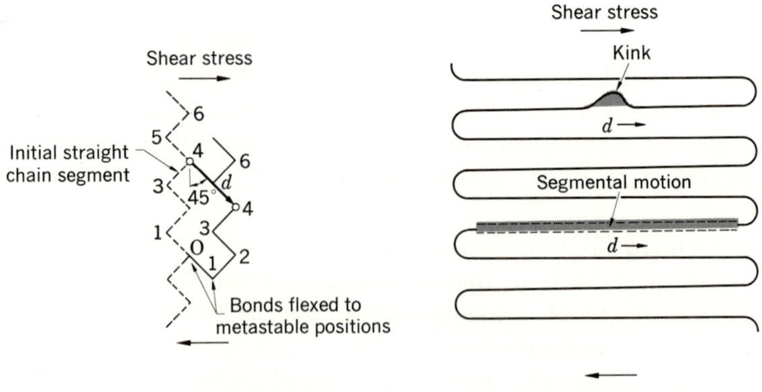

(*a*) In a glassy polymer, flexing of bonds permits the displacement *d* into an available free space. (After R. E. Robertson and C. W. Joynson.)

(*b*) Within a crystalline lamella a displacement *d* occurs by the motion of individual kinks or by the coordinated motion of an entire segment.

FIGURE 9-13 Schematic illustrations of mechanisms for motion of chain segments under the action of a shear stress in (*a*) a glassy polymer and (*b*) a lamella of a crystalline polymer.

distance in unison, provided that it rotates 180° to fit in properly at the new location. Other aspects of the deformation of crystalline polymers (twinning, rotation, and the formation of kink bands) are considered in Sec. 9-8.

Sufficiently large tensile stresses can cause polymer chains to slip and become aligned. In glassy polymers, slip bands form in the direction of maximum shear stress. At a later stage *crazes* develop perpendicular to the applied tensile stress; they are the precursors of the cracks that lead to eventual fracture. Crazes consist of bundles of highly aligned molecular chains alternating with submicroscopic cracks. In crystalline polymers, the lamellae first slip past one another and then, depending on their orientation, become compressed or physically separated. In the latter case, tie molecules in the amorphous region continue to connect adjacent lamellae.

9-8. Atomic Displacements Over Fractional Lattice Spacings

Under some circumstances the large local displacements of slipping are energetically unfavorable compared to small displacements. For example, at suitably low temperatures atoms can "shift" slightly under stress to produce a new arrangement such as a twin (Fig. 5-12); the process of slip would require a much higher stress. On the other hand, at high temperatures the motion of vacancies results in atom-by-atom displacements in response to a low stress; slip would be possible only at a high stress. We now consider the nature of these small displacements and the manner in which they can cause appreciable plastic deformation.

TWINNING. The mechanism of twinning (Fig. 5-12) results in an appreciable macroscopic shear even though the individual atoms shift less than an interatomic distance. Table 9-3 lists the twinning elements and the shear strain for several materials. Typically, twinning has the following characteristics: (1) the stress is large compared to that for slip, (2) nucleation is necessary, (3) growth occurs by a dislocation mechanism, and (4) the rate of formation of twins may be extremely high even at low temperatures.

Even though the stress for twinning is higher than that for slip, special circumstances can favor twinning. An important instance is the deformation of hexagonal metals at low temperature; slip then occurs only on the (0001) plane. If a tensile stress σ is almost perpendicular to this plane, a large value of σ would be necessary before the shear stress could reach the critical value τ_c necessary for slip on the (0001) plane. In fact, twinning occurs at a lower value of σ. The twinned region has an orientation more favorable for slip, and thus deformation can continue by slip in this region. Both twinning and slip occur as deformation of the metal proceeds. In many instances twinning is favored by high strain rates and/or low temperatures. Most metals form twins when deformed at high rates (for example, when formed by explosive charges), especially at subzero temperatures.

LOCAL ROTATION OF LATTICE. Deformation, especially of polycrystalline materials, is often locally heterogeneous due to grain boundaries, surface irregularities, or other inhomogeneities. An especially clear example is a kink band (Fig. 9-14*a*), which can form in crystals of ceramics, metals, or polymers. Although the crystal lattice in the kink band is appreciably rotated relative to the rest of the lattice, most of the atoms shift only slightly with respect to their neighbors. Figure 9-14*b* shows a kink band produced in polyethylene by a special shearing experiment; such bands form also in the course of industrial deformation processes. The general term *deformation bands* is applied to similar bands in deformed FCC metals. These bands may be caused by kinking or by slip on secondary planes. A final example is the rotation of lattice planes that occurs locally in many technical forming operations. A simple example is shown in Fig. 9-14*c*, in which slip plus lattice rotation permits a sapphire crystal to bend, even though only the (0001) slip plane is active.

DISPLACEMENTS AT HIGH TEMPERATURES. Even if the applied stresses are small, at sufficiently high temperatures two types of "stepwise" displacements can produce major amounts of deformation. The climb of edge dislocations (Fig. 4-17) occurs within the volume of a grain and combines with slip to produce creep at high temperatures. As discussed in Sec. 4-10, the mechanism depends on a sufficient number and mobility of vacancies. The second type of displacement occurs at boundaries, in particular at grain boundaries. Although a grain boundary interferes with deformation (especially with slip) at ordinary temperatures, its special defect structure makes it an active location for atom-by-atom rearrangement of crystal structure. At temperatures approaching the melting point, metals and ceramics may undergo one-third of their total deformation in the immediate vicinity of their grain boundaries by the sliding of one grain relative to another. Fracture at grain junctions accompanies this process.

TECHNICAL ASPECTS OF PLASTIC FLOW

Under the conditions existing in typical applications, many useful materials behave elastically at low stresses and plastically at high stresses. The specific details of the behavior vary considerably, even within one class of material, such as the steels. Most machines and structures must operate in the elastic range, and thus the stress at which plastic deformation first begins (the yield strength) is an important factor in design. Another factor is ductility, the

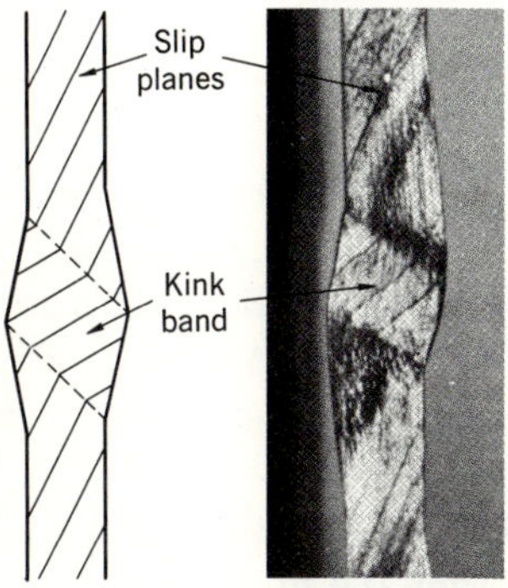

(*a*) Sketch and photograph of a kink band produced when a zinc crystal is deformed in tension.

(*b*) Electron micrograph of a kink band produced by shearing polyetheylene in which the fibrils were initially oriented in one direction. (Courtesy R. E. Robertson, Research and Development Center, General Electric Company.)

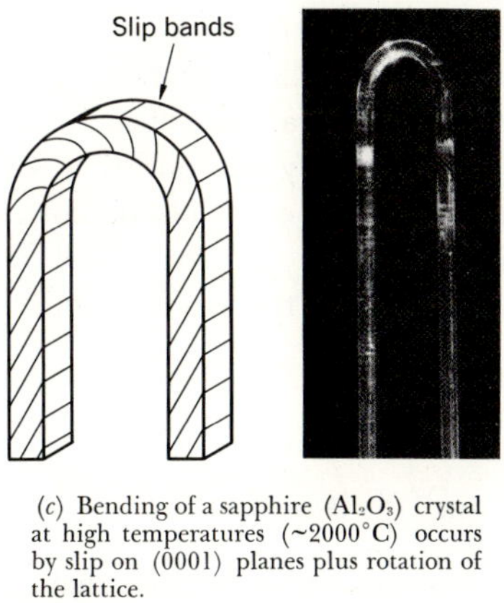

(*c*) Bending of a sapphire (Al_2O_3) crystal at high temperatures (~2000°C) occurs by slip on (0001) planes plus rotation of the lattice.

FIGURE 9-14 Examples of three types of local rotation of the lattice. (*Parts* (*a*) *and* (*c*) *adapted from J. J. Gilman, "Micromechanics of Flow in Solids," McGraw-Hill Book Company, New York, 1969.*)

amount of plastic deformation that occurs prior to fracture. Ductility often determines the margin of safety during "misuse" of a given object; a slight mishap only bends an automobile fender made of steel, since this material can withstand about 25 percent elongation before it fractures (Fig. 9-1*b*). We begin our study of several aspects of plastic flow by considering the general equations, analogous to Hooke's law for elastic deformation, describing plastic deformation in terms of a few constants for each material.

9-9. Equations for Plastic Flow

The most general description of plastic deformation of polycrystalline materials must account for various phenomena, for example, the dependence of yielding on the state and rate of stressing, the effect of prior deformation, and the interaction among the several stresses and strains. Although various detailed analyses can be found in the literature,[1] simpler approximate treatments are often preferable. A useful procedure, illustrated in Fig. 9-15, describes plastic flow by two equations. The first gives a smooth curve from the yield stress σ_o to the beginning of necking; the second is a straight-line portion out to the occurrence of fracture. The first region, corresponding to relatively uniform plastic deformation, is satisfactorily represented by the effective-stress–effective-strain curve,

$$\bar{\sigma} = K\bar{\epsilon}^n \tag{9-30}$$

K is the strength coefficient, and n is the strain-hardening exponent; Table 9-4 gives approximate values for several representative materials. The effective stress $\bar{\sigma}$ and the effective strain $\bar{\epsilon}$ are defined to accomplish a number of purposes. (1) They are adequate for a general state of stress (six σ_i values) and the corresponding general strains ϵ_i. (2) They are consistent with the experimental fact that a hydrostatic pressure does not cause plastic flow. (3) They reduce to $\bar{\sigma} = \sigma_1$ and $\bar{\epsilon} = \epsilon_1$ for a simple tension test, for which σ_1 is the single stress acting and ϵ_1 is the plastic elongation of the tensile specimen. The appropriate definitions are

$$\bar{\sigma} = \{\tfrac{1}{2}[(\sigma_{\mathrm{I}} - \sigma_{\mathrm{II}})^2 + (\sigma_{\mathrm{II}} - \sigma_{\mathrm{III}})^2 + (\sigma_{\mathrm{III}} - \sigma_{\mathrm{I}})^2]\}^{1/2} \tag{9-31}$$

$$\bar{\epsilon} = \{\tfrac{2}{9}[(\epsilon_{\mathrm{I}} - \epsilon_{\mathrm{II}})^2 + (\epsilon_{\mathrm{II}} - \epsilon_{\mathrm{III}})^2 + (\epsilon_{\mathrm{III}} - \epsilon_{\mathrm{I}})^2]\}^{1/2} \tag{9-32}$$

Only three stresses and three strains occur because they are referred to principal axes (indicated by the roman-numeral subscripts); by definition, shear stresses are zero for principal axes. Equation (9-30) has the important advan-

[1] See, for example, T. H. Lin, "Theory of Inelastic Structures," John Wiley & Sons, Inc., New York, 1968.

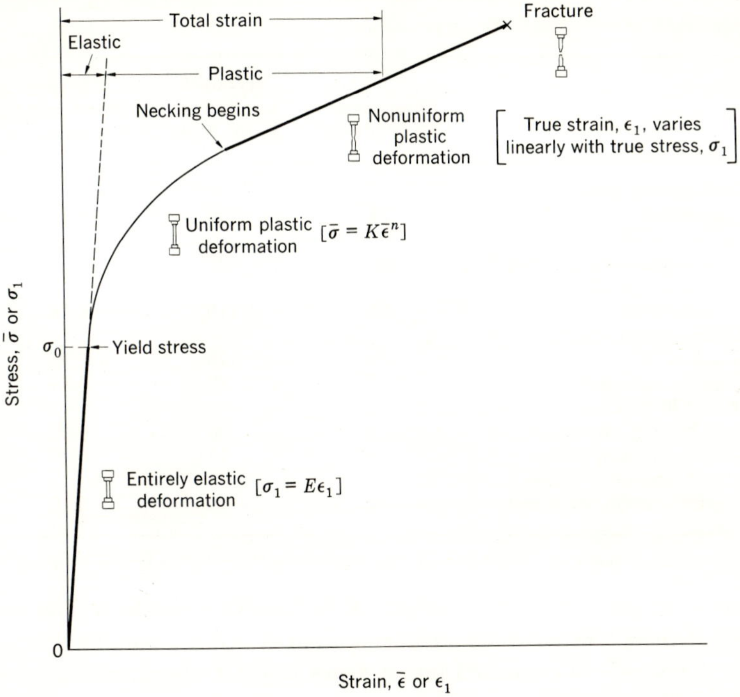

FIGURE 9-15 Schematic illustration of features in a useful treatment of plastic flow. Elastic deformation (exaggerated for clarity) is also produced by stresses in the ranges of plastic deformation, as illustrated for a typical total strain.

tage that the constants K and n can be determined for a given material from the data of an ordinary tensile test, $\sigma_1 = K\epsilon_2^n$ (Prob. 9-9). The behavior of the same material under conditions of complex loading can then be estimated by use of Eq. (9-30) in its more general form (Prob. 9-10).

9-10. Yielding and Strain Hardening

A significant phenomenon that occurs only in a few metallic materials but quite commonly in both glassy and crystalline polymers is a drop in stress at the yield point. Once deformation has been initiated, it continues at constant stress along a planar front, which advances until the entire body has been deformed. An increase in stress is then required to continue the deformation process. Unlike the necking process in Fig. 9-15 which represents an instability and leads quickly to fracture (see Sec. 10-1), the yield-point phenomena in both metals and polymers give a product of increased strength.

The mechanisms of yielding are completely different for metals and for

TABLE 9-4. VALUES OF σ_0, K, AND n FOR REPRESENTATIVE METALS AND ALLOYS†

Material	Flow stress σ_0, 10^7 N/m^2	Strength coefficient K, 10^7 N/m^2	Strain-hardening exponent, n
Low-carbon steel, annealed	21	50	0.28
0.6 w/o carbon steel, quenched and tempered	52	127	0.15
Alloy steel (4135), cold-rolled	65	110	0.14
Stainless steel (304), annealed	59	128	0.45
Copper, annealed	6	32	0.54
Brass (70 w/o Cu, 30 w/o Zn), annealed	8	90	0.49
Aluminum, annealed	4	18	0.20
Aluminum alloy (2024), precipitation hardened	31	70	0.16

† Conversion of units: (M N/m^2) 1.450×10^{-4} = N lb/in^2; (M N/m^2) 1.020×10^{-7} = N kg/mm^2

polymers. In polymers a realignment of the lamellae (or chain molecules in glassy polymers) occurs at the advancing front of deformation. In this process a typical polymer is converted into a *drawn condition* characterized by fibrils oriented along the direction of deformation and by increased strength. If the initial specimen already consists of fibrils oriented in the direction of stressing, no yield point occurs during a tensile test. In low-carbon steel, the principal metal exhibiting a yield point, the yield-point phenomenon is undesirable because it causes surface roughening. The mechanism involves the attraction of carbon and nitrogen atoms to dislocations and the resulting immobilization ("anchoring" or "pinning") of the dislocations. When a sufficiently high stress eventually frees some dislocations from their surrounding "atmospheres" of solute atoms, these dislocations can move at a lower stress and can also generate other mobile dislocations. The resulting front of plastic deformation, called a *Lüders band*, propagates across the specimen. The yield phenomenon can be eliminated either (1) by adding titanium to the steel, which produces compounds with carbon and nitrogen and thereby removes them from solution, or (2) by giving the sheet steel a small amount of deformation (skin rolling) shortly before the forming process. Skin rolling frees the dislocations from their atmospheres and permits yielding to occur at a slightly lower stress uniformly throughout the steel.

We now consider two factors, in addition to stress, that can influence the yield behavior of materials—strain rate and temperature. In view of the effect of slightly elevated temperatures in changing the structure of polymers, the deformation of polymers depends strongly on temperature. Also, if the strain rate is such that molecular processes have time to occur, an otherwise brittle polymer can deform appreciably. Another aspect of the strain rate is responsible for the useful phenomenon of *superplasticity* in certain alloys. These alloys can be elongated almost indefinitely under suitable conditions without local necking and consequent fracturing. The explanation lies in an abnormally high sensitivity of the stress to the strain rate $\dot{\epsilon}$, expressed quantitatively by the exponent m in the following equation:

$$\sigma = A\dot{\epsilon}^m \tag{9-33}$$

A is a constant. The index of strain-rate sensitivity, m, has a value of about 0.5 for superplastic alloys; it is less than 0.2 for normal alloys. Typically, m varies with temperature, with structure, and with strain rate; it has its maximum value for small grain sizes and usually at an intermediate strain rate. If a neck begins to form during the tensile deformation of a superplastic alloy, the local strain rate increases. The stress σ for continued deformation in this region then increases according to Eq. (9-33), and thus deformation shifts to another location. Among the superplastic alloys are a special stainless steel and several aluminum-base and titanium-base alloys.

The critical stress σ required for deformation at some rate $\dot{\epsilon}$ decreases greatly with increase in temperature in ordinary polycrystalline materials; the decrease of τ_c in single crystals is a similar phenomenon. This strong dependence of the motion of dislocations on temperature suggests that the shear stress has two parts;

$$\tau = \tau^* \text{ (thermal stress)} + \tau_a \text{ (athermal stress)} \tag{9-34}$$

Except at very low temperatures, the larger part of the stress, τ^*, is acting against obstacles which the dislocations can overcome with the help of thermal energy. These obstacles, such as the "forest dislocations" intersecting the active slip plane, effectively obstruct a dislocation only over a few atomic distances. Therefore, the coherent thermal fluctuations of a relatively small number of atoms can assist the dislocation past such an obstacle. The higher the temperature, the lower the value of τ^* required to cause the dislocation velocity corresponding to a given rate of strain. Some obstacles, such as particles of a second phase, act on a dislocation over a region large enough to make thermal activation ineffective. These obstacles account for the athermal stress τ_a, which is independent of temperature except for the slight decrease in the shear modulus with increase in temperature.

The activated complex method (Fig. 4-5) gives the following expression for the average velocity v of a dislocation:

$$v = Be^{-\Delta G^A/RT} \tag{9-35}$$

ΔG^A is the free energy of activation, and B is a constant determined by the vibrational frequency of the dislocation and by the distance moved. Substitution of Eq. (9-35) into Eq. (9-29) gives the following expression for the strain rate:

$$\dot{\epsilon} = Ce^{-\Delta G^A/RT} \tag{9-36}$$

where C is a constant. The following procedure, based on Eq. (9-36), identifies the "thermal" obstacle that is determining the rate of deformation at a given temperature. Measurements of $\dot{\epsilon}$ at various temperatures and stresses determine ΔH^A, the enthalpy of activation; see Eq. (4-17) for a similar analysis. The value of ΔH^A varies with temperature (Fig. 9-16). By comparison with values of ΔH^A obtained from other types of experiment on the same material, the controlling dislocation reaction in a given range of temperature can be identified, for example, the intersection of forest dislocations near 300°K in polycrystalline sodium chloride (Fig. 9-16).

Strain hardening (or work hardening) is the increase in flow stress of a material as it continues to be deformed. One quantitative measure of this basic property is the strain-hardening exponent (Table 9-4). Although the exact mechanisms of strain hardening are still in doubt, dislocation reactions of various types are certainly involved. Deformation increases the density of dislocations, of jogs, and of such causes of sessile dislocations as Lomer-Cottrell blocks (Fig. 4-20). At high temperatures thermal activation eliminates or overcomes most of these impediments to the motion of dislocations, and thus the rate of strain hardening is much reduced.

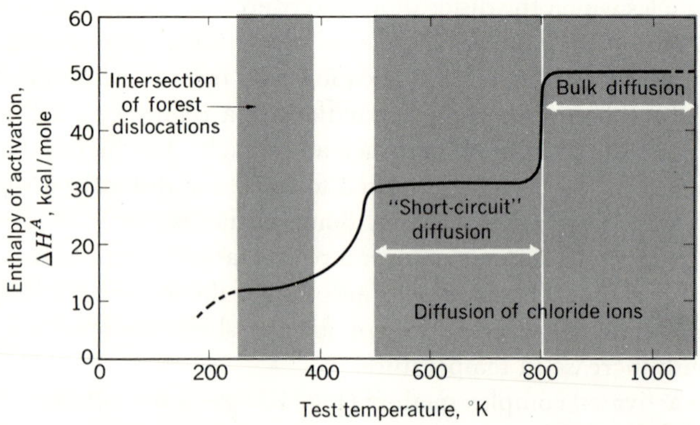

FIGURE 9-16 Data for deducing the rate-determining dislocation reactions for creep in polycrystalline sodium chloride. (*Adapted from P. M. Burke and O. D. Sherby.*)

9-11. Test Methods and Applications

Various aspects of plastic deformation are important for the fabrication and use of many materials. We first consider the range of temperature in which deformation is not affected by concurrent structural changes (such as recrystallization); later we will treat high-temperature deformation. The various processes—rolling, extrusion, pressing, forging, etc.—can be analyzed by the methods of continuum mechanics in the manner described in the references by Avitzur, D'Isa, and Mendelson, listed at the end of the chapter. The quantitative parameters, such as yield strength, needed to predict the behavior of a given material can be obtained from appropriate laboratory tests. Tensile tests of cylindrical or strip specimens are especially useful and common; the corresponding testing of a sheet specimen involves use of a punch (or of hydraulic pressure) to produce bulging and eventual fracture. Once a given type of material has been adequately characterized by testing and actual performance in a forming process, hardness testing can usually serve as a convenient, inexpensive control for routine production. A typical instrument (Fig. 9-17) uses a known load to force an indenter into a smooth area of the material to be tested. When the load is removed after a fixed time, about 10 sec, the dial gauge indicates the hardness number corresponding to the depth of penetration of the indenter. By the use of an appropriate load and a suitable type of indenter, a sensitive measure of hardness can be obtained for materials ranging from soft polymers to hard ceramics. The technique of *microhardness testing* permits a similar testing of individual microconstituents viewed in a microscope.

A common service condition (in an automobile axle, for example) involves many repetitions or reversals of the applied stress. A new design parameter, the fatigue (or endurance) strength, is required and is obtained by tests simulating the conditions of service. In a typical fatigue-testing machine (Fig. 9-18*a*), a specially prepared specimen is subjected to repetitions of a prescribed cycle of stressing, the maximum stress having a fixed value S. Ordinarily, the specimen fractures after some number N of cycles. When pairs of (S, N) values are plotted (Fig. 9-18*b*), the curve for most materials approaches an asymptote, the fatigue limit. Below this stress a specimen can withstand an indefinitely large number of cycles without breaking. The fatigue limit determined by laboratory tests is generally lower than the yield strength. A still lower value of the stress must be used for design purposes because of the adverse effect of ordinary surface roughness, scratches, or slight corrosion of the surface. The mechanism of fatigue fracture is considered in Sec. 10-7.

Deformation at high temperatures often has special importance. Many materials develop the capacity for plastic deformation only at elevated temperatures (Fig. 9-19); at lower temperatures they fracture under high applied stresses. Those useful metals and alloys that are capable of deforming over a

FIGURE 9-17 Measurement of hardness with a Rockwell hardness tester. The hardness test is a convenient method for determining the deformation characteristics of materials. (*Courtesy Wilson Instrument Division, American Chain & Cable Company.*)

wide range of temperatures exhibit a significant decrease in the stress required for deformation as the temperature increases, especially above a homologous temperature of about 0.6 T_f. This decrease is important for commercial forming processes (extrusion, rolling, etc.) because it reduces the energy needed to accomplish a given change in shape. Hot deformation also permits changes in microstructure that often improve the properties of the material. On the other hand, in service at high temperatures the decreased flow stress permits a gradual deformation (creep) of structural members and requires adequate control through proper design.

The flow stress is determined not only by the temperature, but also by the rate of straining, $d\epsilon/dt \equiv \dot{\epsilon}$. This relation is shown in Fig. 9-20*a* for commercially pure aluminum extruded at various combinations of temperature and strain rate. The experimental data fit the empirical relation

$$\dot{\epsilon} = A[\sinh(\alpha\sigma)]^n e^{-Q/RT} \tag{9-37}$$

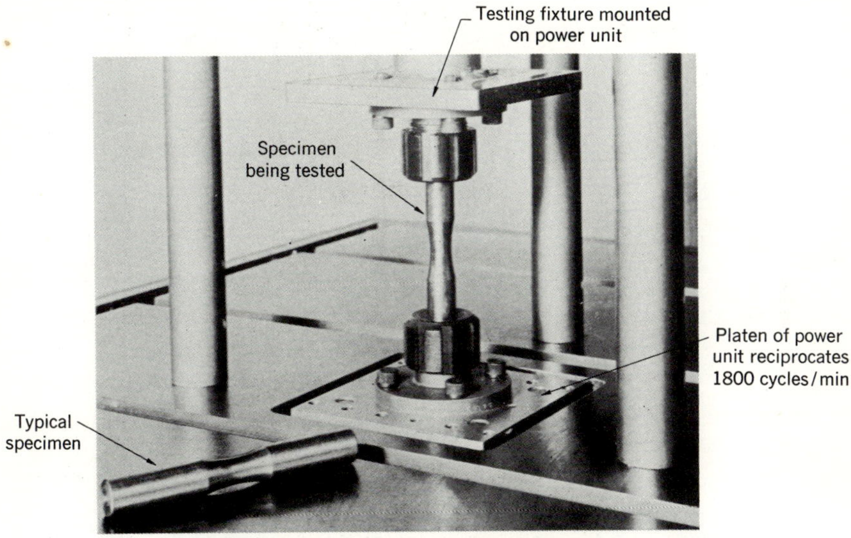

(*a*) Arrangement for testing a specimen in tension and/or compression. (Courtesy SATEC Systems, Inc.)

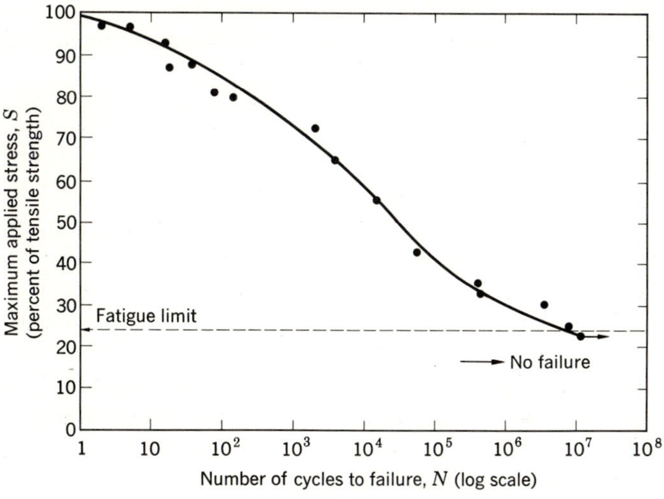

(*b*) Fatigue data for a composite (fiberglass embedded in phenolic resin) plotted as stress (S) *vs* number (N) of cycles to failure. Data of G. H. Stevens.)

FIGURE 9-18 A machine for fatigue testing and a typical S-N curve.

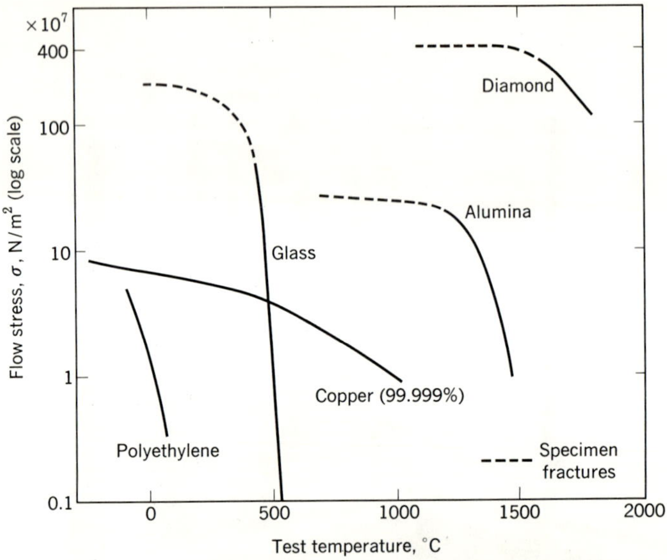

FIGURE 9-19 Variation of flow stress with temperature of testing for various materials. The flow stress of metals varies relatively little in the high-temperature region. (*After D. McLean.*)

where A, α, n, and Q are constants. A log-log plot then gives a straight line for each value of T, the slope being n (see Prob. 9-11). Rearrangement of Eq. (9-37) to the form

$$\frac{Z}{A} = [\sinh(\alpha\sigma)]^n \tag{9-38}$$

allows the stress to be pictured as producing a single effect of magnitude Z/A. Comparison of Eqs. (9-37) and (9-38) shows that

$$Z = \dot{\epsilon} e^{Q/RT} \tag{9-39}$$

Z is thus a strain rate compensated for the effect of temperature. The data of Fig. 9-20*a* plot as a single line in terms of Z/A (Fig. 9-20*b*). Furthermore, data on slow rates of deformation (creep) and on other types of metal forming, such as compression, also fit this same line quite well.

The parameter Z has an important microstructural significance. During hot deformation of a metal under a given stress σ, a typical substructure such as that of Fig. 9-21*a* develops early in the process and is then maintained as the "steady-state" structure. This substructure depends on the value of Z rather than on either temperature or strain rate individually. During hot

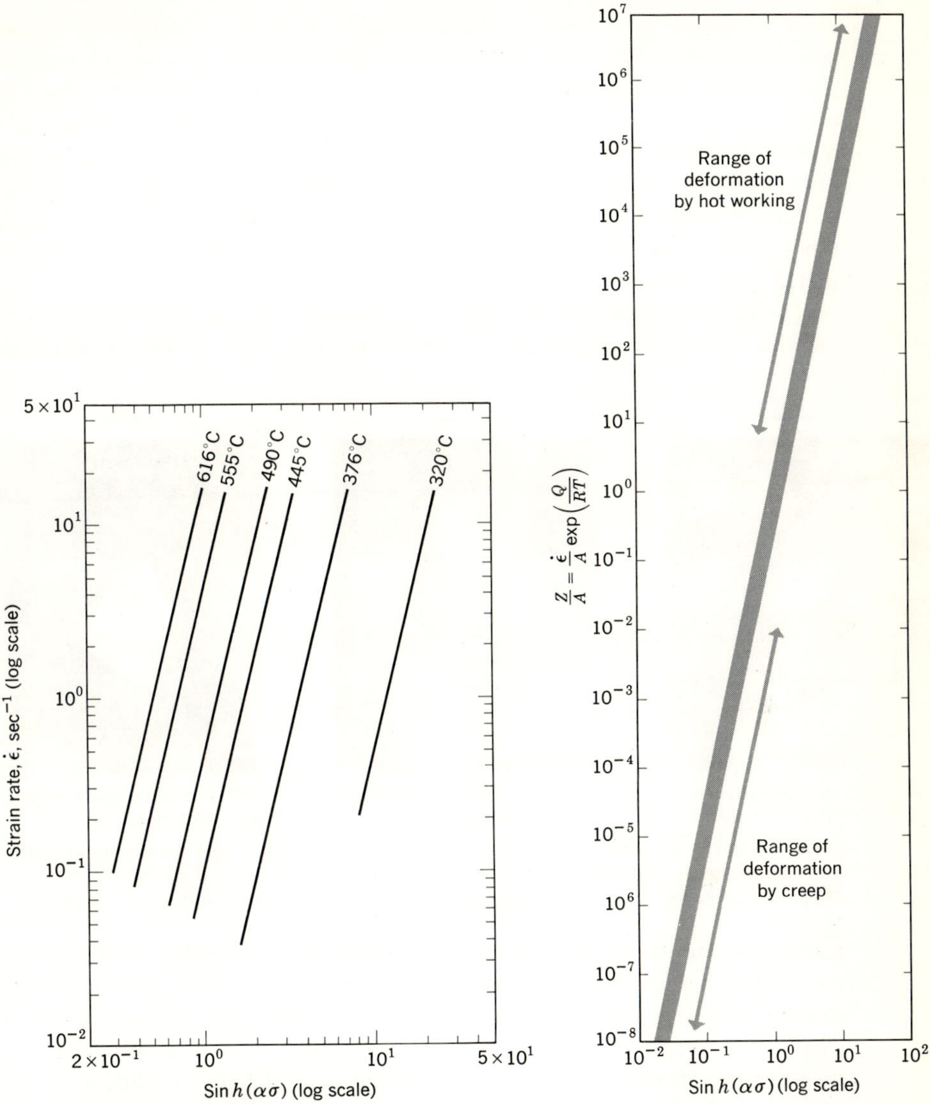

(*a*) A plot in which the separate effects of temperature and strain rate can be seen.

(*b*) A more useful plot uses the temperature-compensated strain-rate parameter, *Z*. Data from several types of deformation on aluminum of varying purity fall largely within the band indicated.

FIGURE 9-20 Data on the high-temperature extrusion of aluminum showing that the flow stress σ depends on both temperature and strain rate. (*Data of W. A. Wong and J. J. Jonas.*)

working a dynamic equilibrium is established between the generation of dislocations (tending to cause strain hardening) and their annihilation by the processes of recovery and, in some cases, recrystallization. With increase in temperature at a constant strain rate (Fig. 9-21) the substructure increases in perfection and in subgrain size; both changes reduce the hardness and strength of the hot-worked metal. Recovery and recrystallization that may occur during final cooling from the temperature of hot working are termed *static* to distinguish them from the *dynamic* processes occurring during deformation. Both types of processes influence the structure and properties of typical hot-worked metals and alloys.

Two laboratory tests, *creep* and *creep-rupture*, are used to evaluate materials for use as stress-bearing members at high temperatures. In both tests

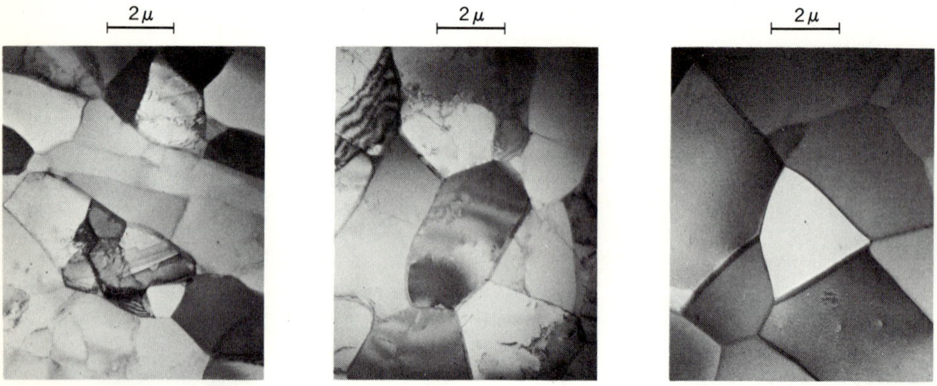

(*a*) At 250°C the subboundaries consist of ragged tangles of dislocations.

(*b*) At 350°C the boundaries are narrower and less ragged.

(*c*) At 450°C the subgrains are larger and more perfect.

FIGURE 9-21 Electron micrographs of longitudinal sections from aluminum extruded at three different temperatures at a strain rate of 1 per second. (*Courtesy W. A. Wong, H. J. McQueen, and J. J. Jonas. By permission from the Journal of the Institute of Metals.*)

a specimen similar to that employed in tensile testing is subjected to a fixed stress while being maintained at a constant high temperature. Deformation is initially relatively rapid (stage I) but constantly decreases and soon attains a steady rate (stage II). A creep test uses especially sensitive strain gauges to measure this steady rate. A typical creep strength of Udimet 700†, a nickel-base alloy for high-temperature service, is 4×10^7 N/m^2 for 10^{-3} percent elongation/hr at 982°C (1800°F). For applications in which relatively large deformations can be tolerated, rupture (fracture) may be the limiting factor. Creep-rupture tests also measure elongation, although less precisely, but permit the test to continue until fracture occurs. A typical rupture strength of

† The chemical composition is 18.5 w/o Co, 15 Cr, 5.2 Mo, 4.3 Al, 3.5 Ti, 0.8 Fe, 0.15 C, 0.04 B, and the balance nickel.

Udimet 700 is 5.2×10^7 N/m^2 to produce rupture in 1000 hr at 982°C (1800°F). In this case the elongation at rupture is about 8 percent.

9-12. Strengthening Mechanisms

For each of the thousands of industrially important materials—polymers, metals, and ceramics—strength is usually a significant property, often the property of paramount interest. The strength may be desired at high temperature, for repeated loading, for prolonged service, with specified ductility, or with other requirements. Many different approaches are employed to obtain useful strengthening for the various materials and conditions of

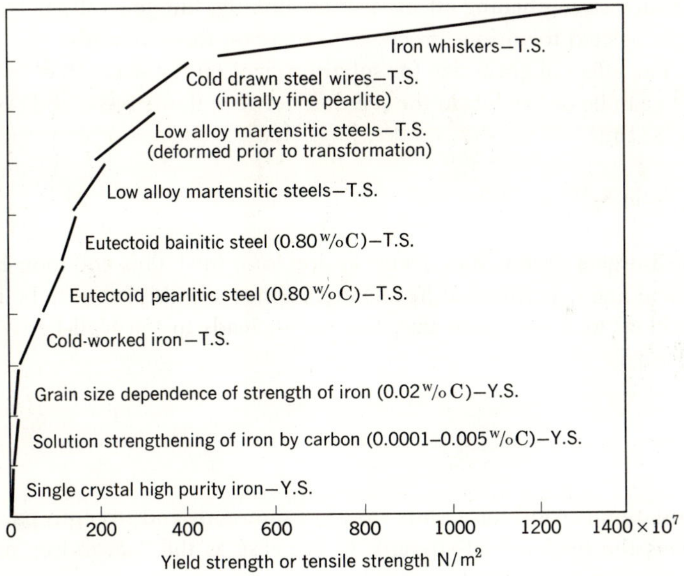

FIGURE 9-22 An example of typical strengthening of materials, the range of strength in iron and steels produced by various means. (*After V. F. Zackay.*)

service. The strengthening of steel (Fig. 9-22) can serve as an example. Both the weakest and the strongest forms of iron are single crystals. In large single crystals the dislocations move very easily; in microscopic whiskers, dislocations are immobile or perhaps absent. The strength of iron can be increased by any one of the following processes: decreasing the grain size, cold working, forming a solid solution, adding a second phase (Fe_3C), and producing martensite. More effective strengthening occurs when two or more of these processes are combined, for example, cold working a steel containing small particles of Fe_3C, or producing martensite from deformed austenite. The latter process involves prompt plastic deformation of a steel at about 500°C immediately after rapid cooling from the austenitic range. Martensite of unusually high strength and toughness is produced by the subsequent transformation of the deformed austenite. *Thermomechanical treatment* is the term used to describe such combinations of heat treatment and mechanical working.

The intrinsic strength of the base material (aluminum, MgO, or a given polymer, for example) tends to set a limit to the strength properties that can be achieved. The low melting point of aluminum, 660°C, prohibits the consideration of aluminum alloys for use under creep conditions, at say 600°C; MgO, on the other hand, can be used at much higher temperatures since its melting point is 2800°C. The "intrinsic" condition of a polymer may already include such strengthening effects as cross-linking, but generally polymers can also be subjected to various processes to increase their strength.

The effect of grain size (or subgrain size) on the strength of metals and ceramics can be derived from the relation between flow stress σ and dislocation density ρ; thus

$$\sigma = \sigma_i + \alpha\mu b\rho^{1/2} \tag{9-40}$$

b is the Burgers vector, and α and σ_i are constants. This equation has both theoretical and experimental justification. Since ρ is also found to be inversely proportional to d, the grain size, Eq. (9-40) leads to the Hall-Petch relation [Eq. (5-15)],

$$\sigma = \sigma_i + \frac{k}{d^{1/2}} \tag{9-41}$$

This equation is valid for a wide range of materials and strength parameters. It reflects the tendency of strength to increase as the "mean-free path" for movement of dislocations decreases. An important exception to the Hall-Petch relation occurs in certain high-strength materials at high temperatures, for example, in MgO and in complex high-temperature alloys. In these materials deformation in the grain boundaries may become a major component of the deformation process; rupture strength then increases with increase in grain size and may be a maximum for a single crystal of the matrix phase. In a complex alloy such a "single crystal" actually contains several finely dispersed phases.

Many commercial metallic and polymeric materials owe much of their strength to cold deformation during manufacture. The amount of cold rolling of brass sheet (Fig. 8-1) is referred to as its *temper* and is specified by such terms as *half-hard*, *hard*, and *spring tempers* (corresponding to 21, 37, and 60 percent reduction, respectively, by rolling). The hard temper increases the yield strength by about a factor of 3 while leaving a ductility of almost 10 percent elongation. Strengthening by cold work has the further advantage of being economical on a production scale. Cold drawing of polymers is a standard operation in the production of fibers, tubes, and films; it can increase the strength tenfold. Because the fibrils produced by cold drawing have superior properties in the direction in which they were drawn, various methods are employed to draw a given product in the directions requiring strength during service.

The term *texture strengthening* refers to improvements in the strength and general behavior of metals resulting from preferred orientations (see Sec. 8-1). For example, titanium-alloy sheets are especially strong under the biaxial stresses existing in a tank under pressure if the sheets have a basal-plane texture. The basal plane (which contains the $\langle 11\bar{2}0 \rangle$ slip direction common to all active slip systems) then lies in the plane of the sheet, and thus thinning of the sheet is impeded. Plastic deformation can produce (in addition to crystallographic textures) *mechanical fibering*, caused by the elongation of inclusions, special phases, or voids in the direction of working.

The most dramatic changes in strength of a given material result from suitable changes in its composition—*alloying*, in the most general sense. Moderate strengthening occurs if the alloying element forms a solid solution (Sec. 5-4), but pronounced strengthening, especially at high temperatures, can accompany the formation of a suitable two-phase structure. The techniques used to create desirable two-phase structures include conventional alloying and heat treating, powder metallurgy, directional solidification, and fabrication of special composites. Second-phase particles in polymers include fillers (such as carbon black), used in elastomers to limit the propagation of tears, and crystalline regions to strengthen an amorphous matrix.

The most effective strengthening of metals by a moderate amount ($\sim$10 percent) of a second phase occurs when the particles of the second phase are so small that their interparticle spacing d is only about 10^2 Å. A competition then occurs between two mechanisms for movement of a dislocation along a slip plane containing second-phase particles (Fig. 9-23). When d is large, the dislocation can bow easily between neighboring particles (Fig. 9-23*a*), reform as a line, and leave a dislocation loop around each particle. The shear stress needed for this process acts against the line tension of the dislocation [Eq. (4-62)] and is proportional to $1/d$. When d becomes sufficiently small, the stress required to bypass the particles in this way becomes greater than that required to cut through them, Fig. 9-23*b*. The resistance offered by the particles depends on their nature. In the case of precipitation hardening (Sec. 7-9), the GP zones are too similar to the matrix to have maximum effectiveness; only with the formation of the transition lattice does the yield strength rise to a high value. Maintenance of good creep strength at high temperatures is difficult

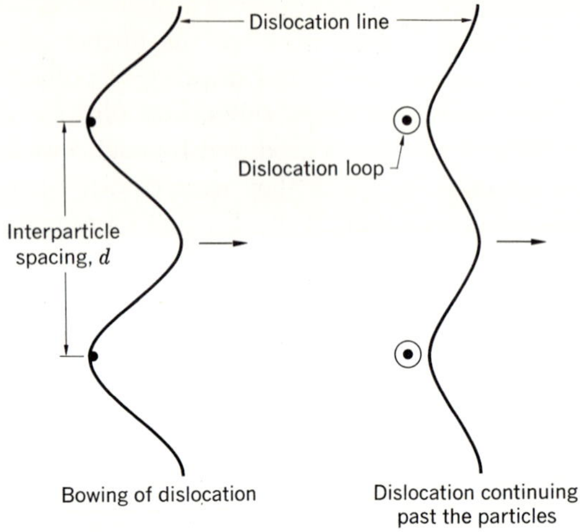

(a) Top view of slip plane showing a dislocation bowing between adjacent particles and leaving a loop of dislocation around each particle.

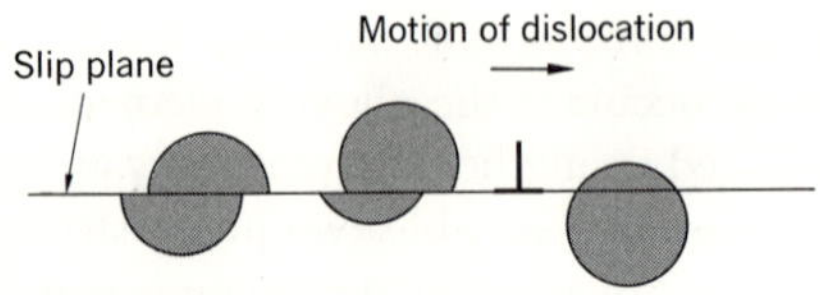

(b) Edge view of slip plane showing a dislocation cutting through second-phase particles as it moves. (After R. W. K. Honeycombe.)

FIGURE 9-23 Two mechanisms for the propagation of a dislocation along a slip plane containing second-phase particles.

because initially fine particles tend to grow during service. An exception is TD nickel, nickel strengthened by a fine dispersion of thoria, ThO_2. Thoria is virtually insoluble in nickel, and thus coarsening cannot occur even within 100°C of the melting point of nickel.

Among the numerous types of composite materials (briefly summarized in Sec. 5-8) the highest strengths are attained by a strong (usually brittle) fiber embedded in a ductile matrix. Borsic-aluminum composites (Fig. 3-9) can serve to illustrate the nature of two-phase strengthening in the case of continuous fibers. Since the fibers are firmly gripped by the matrix, elastic deformation occurs equally in both phases (the isostrain concept of Fig. 9-7). Young's modulus E of the composite then increases linearly with the volume fraction of fiber, V_f; thus

$$E = E_f V_f + E_m(1 - V_f) \tag{9-42}$$

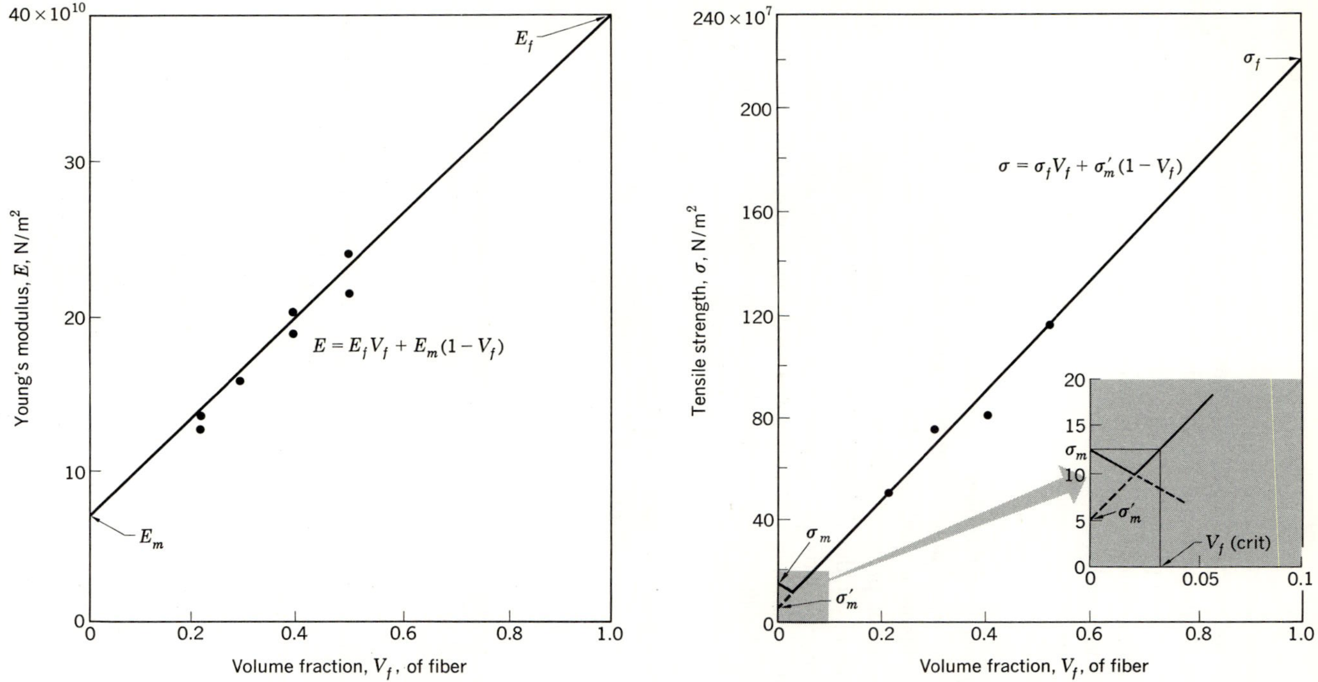

(*a*) Values of Young's modulus for the composite follow the linear relationship given by the isostrain concept.

(*b*) Since the V_f values for actual composites are above V_f (crit), tensile strength also follows a linear relation.

FIGURE 9-24 **The variation of Young's modulus and tensile strength with volume fraction of a continuous fiber in a composite. The data are for Borsic-aluminum composites. [conversion of units: (M N/m²) 1.450 × 10⁻⁴ = N lb/in²; (M N/m²) 1.020 × 10⁻⁷ = kg/mm²]. (*Data of K. Kreider and M. Marciano.*)**

Useful fibers have a Young's modulus E_f several times larger than that of the matrix, E_m. Consequently, at the maximum values of V_f (about 0.5) consistent with adequate toughness, a composite has good elastic stiffness (Fig. 9-24*a*).

The tensile strength σ of a composite is influenced by interaction of the stress fields of the two phases, by flaws, and by residual stresses. Nevertheless, the following simple linear relation approximates the variation of σ in the range of V_f values of interest:

$$\sigma = \sigma_f V_f + \sigma'_m(1 - V_f) \qquad (9\text{-}43)$$

The tensile strength of the fiber, σ_f, is much higher than that of the matrix, σ'_m, and thus the usual amounts of fiber give composites with high strengths (Fig. 9-24*b*). This figure indicates that a critical amount of fiber ($V_f = 0.03$ in this case) must be present before a composite has a strength larger than the tensile strength σ_m of the unreinforced matrix. In the usual composite, the matrix contributes a strength σ'_m somewhat greater than its yield strength.

The strengthening of composites made with discontinuous fibers is diminished by the low stress in the ends of relatively short lengths of fiber (Fig. 9-25*a*). The transfer of load between the two phases is a complex process, but a useful approximation pictures the stress as being zero at the end of the fiber and as reaching the maximum stress σ_{max} in the fiber at a distance $l_c/2$ given by the following relation (see Prob. 9-12):

$$\frac{l_c}{d} = \frac{\sigma_{max}}{2\tau_m} \qquad (9\text{-}44)$$

τ_m is the shear strength of the matrix, and d is the diameter of the fiber. The ratio l_c/d, known as the *critical aspect ratio* of the fiber, is about 10 for a Borsic-

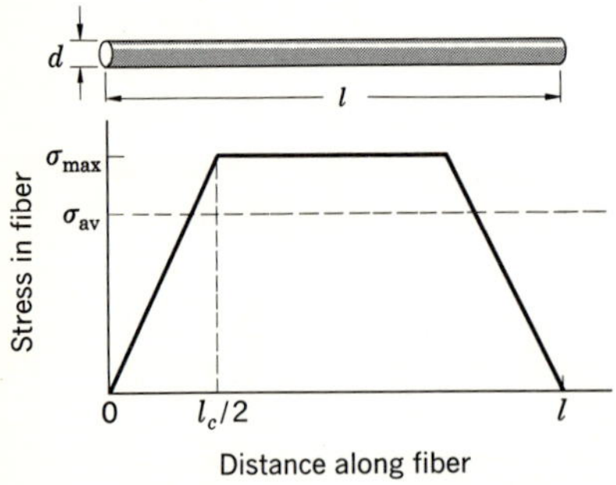

(*a*) A total length, l_c, at the two ends of a fiber carries less than the maximum stress.

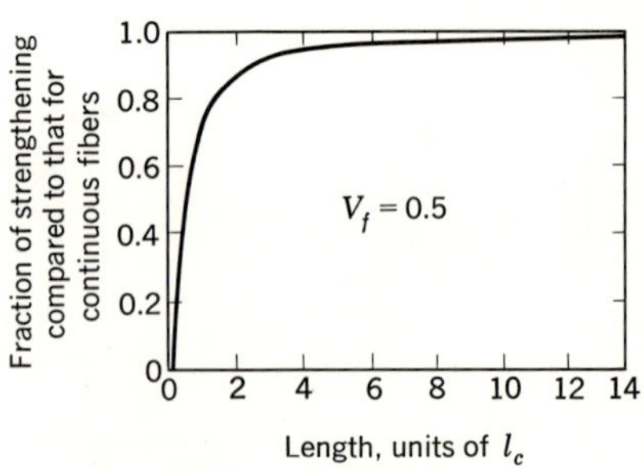

(*b*) Increasing the length of fibers increases their strengthening effect.

FIGURE 9-25 Factors affecting the tensile strength of composites reinforced with discontinuous fibers.

aluminum composite. If $d = 0.1$ mm, a fiber must be 1 mm in length to develop σ_{max} at its midpoint. Longer fibers carry a higher average stress, and therefore the strengthening effect is greater (Fig. 9-25*b*). As the length becomes much greater than l_c, the strength approaches that obtained from continuous fibers.

VISCOELASTICITY

The deformation of most materials is actually in a wide transition region between perfectly elastic behavior and ideally viscous behavior. Sufficiently precise measurements of the "elastic" deformation of metals at room temperature, for example, reveal that small viscous effects also occur. As the temperature of any material is raised, the viscous component becomes more important, and in a certain range of temperature the term *viscoelastic deformation* aptly describes the combined effect of the two mechanisms in determining the observed behavior. Viscoelastic behavior is particularly evident in glassy polymers, and therefore the following treatment focuses mainly on these materials.

9-13. Viscoelastic Deformation

If an elastic material is given an instantaneous deformation ϵ and (after any arbitrary time) the stress is measured, the ratio is Young's modulus,

$$E = \frac{\sigma}{\epsilon} \tag{9-45}$$

This result is independent of time. When the same type of experiment is done with a glassy polymer slightly above its glass transition temperature, the corresponding modulus $M(t)$ must be written

$$M(t) = \frac{\sigma(t)}{\epsilon} \tag{9-46}$$

The stress σ decreases with time t as the polymer gradually changes its internal structure in response to the constant strain ϵ. Shear tests (rather than tensile tests) are usually employed to measure this property, and thus the shear-relaxation modulus $G(t)$ is the quantity usually determined; $3G(t) = M(t)$. The data for atactic polystyrene at 100°C (Fig. 9-26) show a typical time dependence of the shear-relaxation modulus.

A helpful comparison of the behavior of a glassy polymer with that of related materials (Fig. 9-27) can be made using values of the shear-relaxation modulus determined for a time of 10 sec, $G(10)$. At low temperatures, $3G(10)$ is identical with Young's modulus. At intermediate temperatures crystalline

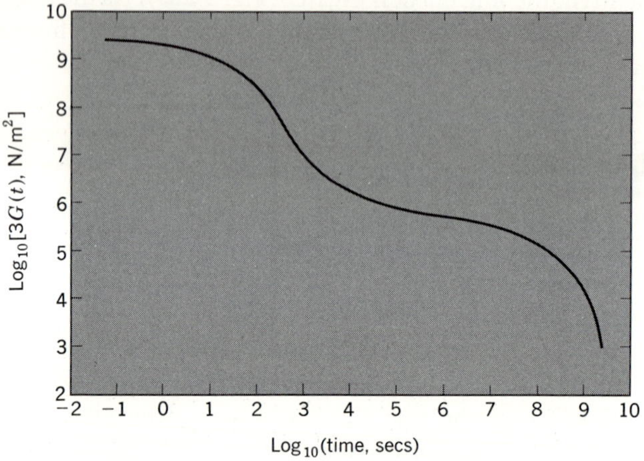

FIGURE 9-26 Time dependence of the shear-relaxation modulus $G(t)$ for atactic polystyrene at 100°C. (*After A. V. Tobolsky.*)

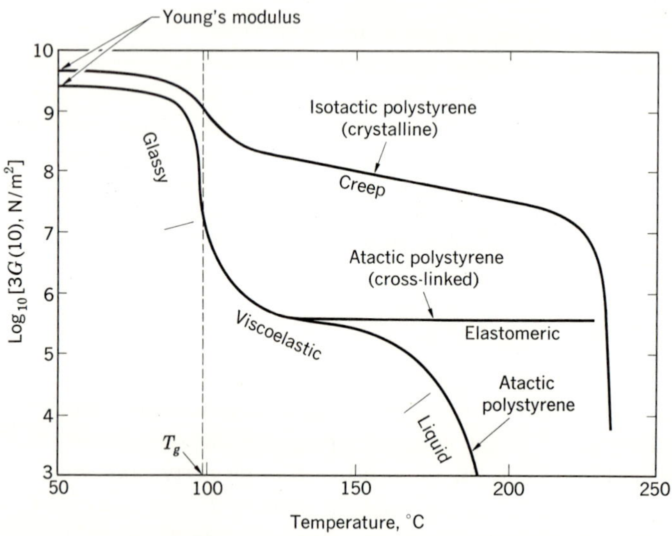

FIGURE 9-27 Comparison of the deformation behavior of atactic polystyrene with the 10-second value of $G(t)$, the shear-relaxation modulus. The behavior of isotactic (crystalline) and cross-linked atactic (rubber) polystyrene are also shown. (*After A. V. Tobolsky.*)

isotactic polystyrene exhibits a gradual decrease in $G(10)$ with increasing temperature. A metal or a ceramic behaves in a similar manner at temperatures above about 0.6 T_f. In this range these materials are said to show *creep deformation*, characterized by a relatively small dependence of deformation behavior on temperature (or on strain rate). The corresponding temperature (or strain-rate) dependence of atactic polystyrene is much greater and leads to the special phenomena encompassed by viscoelastic deformation, which will be discussed below. The addition of a relatively small fraction of cross-links (see Sec. 5-2) to atactic polystyrene interconnects the molecular chains into a network polymer with elastomeric (rubber) deformation behavior. The elastic stiffness of an elastomer is a factor of 10^4 lower than that of the glassy form of the same polymer.

The "memory" of a viscoelastic material for its past history can significantly affect its behavior. For example, the $G(t)$ curve of Fig. 9-26 must be determined on a specimen that has been at rest for a time long enough for practically complete relaxation. The consequence of previous history is shown in recovery following a period of deformation under constant stress (creep) (Fig. 9-28). The initial creep strain $\epsilon(t)$ after a time t under a stress σ can be expressed as

$$\epsilon(t) = \frac{J(t)\sigma}{3} \tag{9-47}$$

$J(t)$ is the shear-creep compliance and is roughly the inverse of $G(t)$. Values of $J(t)$ for polystyrene are given in Fig. 9-29. Even though the deformation after a given time τ of stressing is partly viscous and partly elastic, it can be essentially completely recovered with time (Fig. 9-28*a*) under suitable conditions. Cross-linking, crystallinity, or glassiness encourage this behavior, known

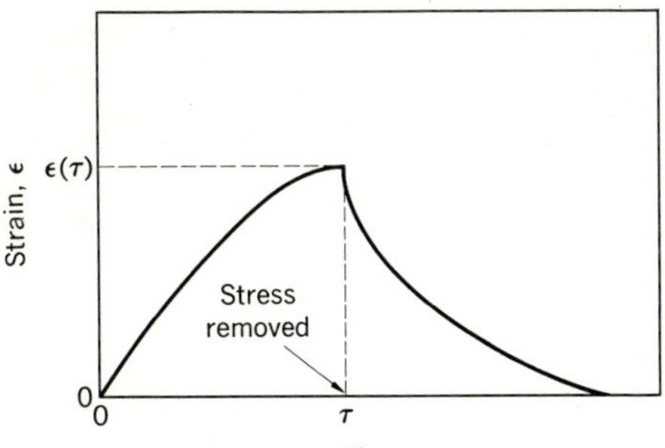

(*a*) Complete recovery is favored by low stress and low temperature of deformation.

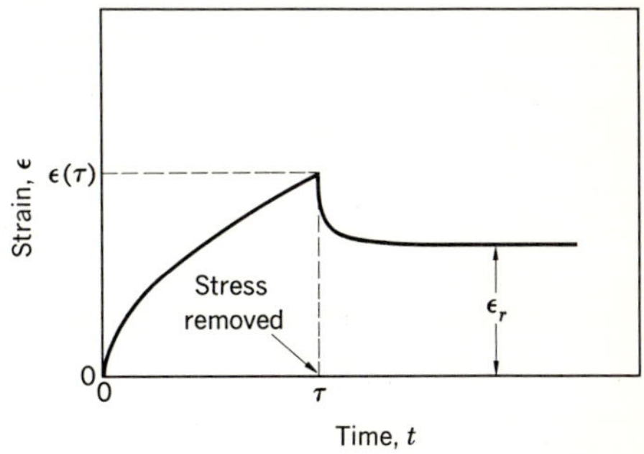

(*b*) A residual deformation, ϵ_r, remains when the conditions favor a large viscous component of the deformation.

FIGURE 9-28 Schematic illustrations of (*a*) complete and (*b*) partial recovery of an initial deformation $\epsilon(\tau)$ after removal of the stress at time τ.

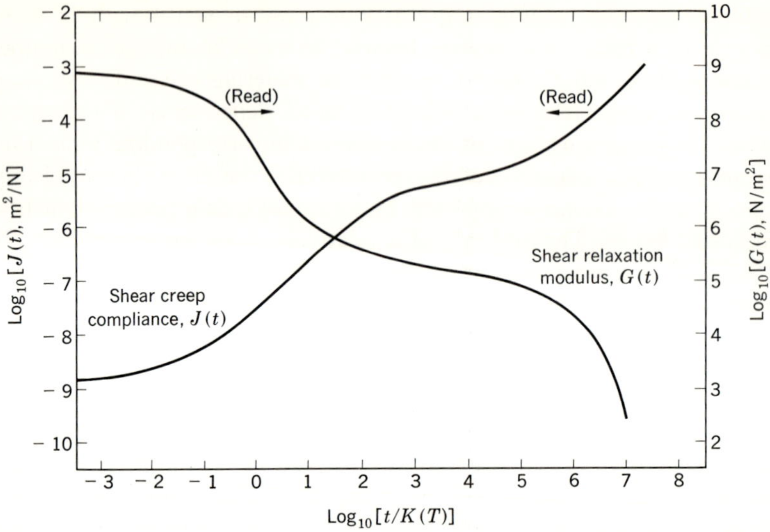

FIGURE 9-29 Temperature-time master curves of the shear-relaxation modulus $G(t)$ and the shear-creep compliance $J(t)$ for atactic polystyrene. (*After A. V. Tobolsky.*)

as the *elastic aftereffect.* Recovery is usually only partial, and a residual permanent deformation remains (Fig. 9-28*b*). Recovery is important in polymers that have been shaped by calendering (rolling), extrusion, or compression. The finished polymer "remembers" the deformation history to which it has been subjected, even if it is immediately taken to a lower temperature. It may then recover slowly over a long period, or it may recover rapidly if it is reheated.

Comparison of the deformation behavior of an amorphous viscoelastic material at various temperatures is greatly facilitated by the fact that a single parameter, $t/K(T)$, expresses the joint effect of time and temperature. $K(T)$ is a characteristic relaxation time and is defined for each temperature T as the time for $3G(t)$ to reach a value of 10^8 N/m^2. At 100°C, for example, the data of Fig. 9-26 show that $K(373°K)$ is about 240 sec for atactic polystyrene. Plots of $G(t)$ and $J(t)$ versus $t/K(T)$ (Fig. 9-29) give master curves that can be interpreted for any temperature from a knowledge of the value of $K(T)$ for that temperature.

9-14. Models of Viscoelastic Behavior

A useful insight into the nature of viscoelastic deformation can be obtained from a consideration of relatively simple models based on the spring and dash-pot. For example, if a spring and dash-pot are connected in series

(Fig. 9-30), application of a stress causes an immediate elastic strain ϵ_e, followed by uniform viscous flow during the period of stressing. Only elastic recovery occurs when the stress is removed from this model. A somewhat closer approximation to actual creep-recovery behavior is given by the model of Fig. 9-31*a*. A spring and dash-pot are connected in parallel in such a manner that both experience identical strains. Initially all of the stress σ is on the dash-pot, and thus the rate of strain is a maximum. As the spring elongates, its stress (σ_s) increases, thus decreasing the net stress ($\sigma - \sigma_s$) acting on the dash-pot. The rate of straining decreases toward zero as the strain approaches ϵ_{max}, the value at which $\sigma_s = \sigma$. When the stress is removed at time τ, the elastic stress in the spring causes the dash-pot to flow in the opposite direction, tracing an analogous exponential curve.

Various modifications of the simple Maxwell and Voigt elements can simulate the viscoelastic behavior of certain types of materials under particular conditions of service. If a spring is added in series with a Voigt element (Fig. 9-32*a*), the resulting model gives a good approximation to the creep-recovery behavior of a network (rubber) polymer (Fig. 9-28*a*). The necessary assumptions include strains of only a few percent and deformation dominated by a single mechanism. The addition of a dash-pot as a fourth parameter (Fig. 9-32*b*), produces a model that includes permanent flow as an additional aspect

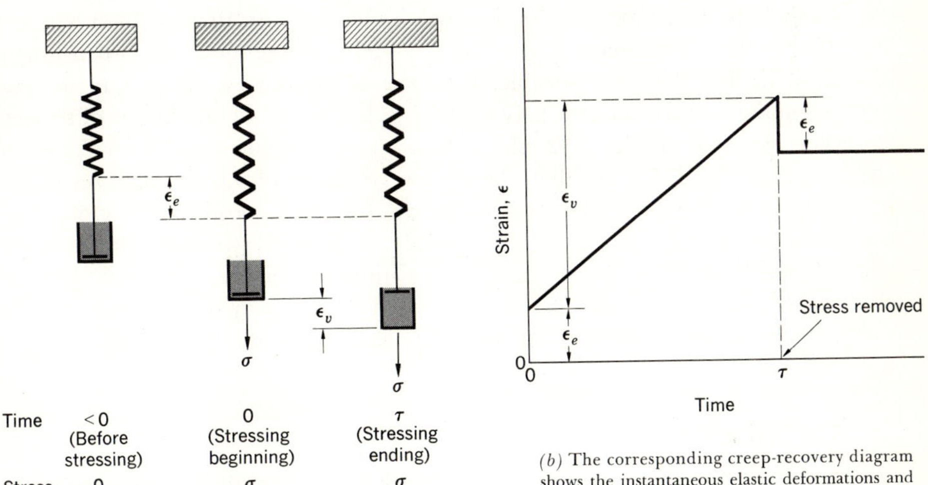

(*a*) Sketches showing the separate elastic deformation, ϵ_e, and viscous deformation, ϵ_v, under a creep stress σ.

(*b*) The corresponding creep-recovery diagram shows the instantaneous elastic deformations and linear viscous flow.

FIGURE 9-30 Behavior of the Maxwell element, a spring and dash-pot in series, during creep and recovery.

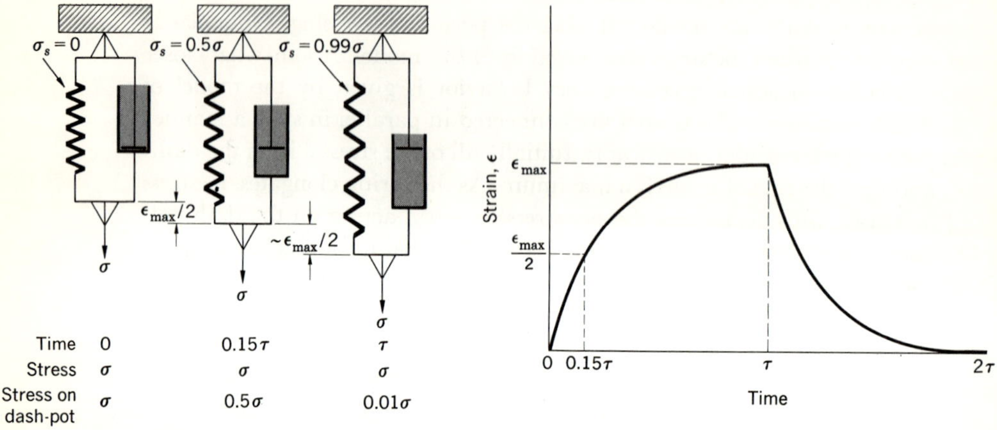

(a) Sketches showing the distribution of constant stress, σ, between the dash-pot and the spring, σ_s.

(b) The corresponding creep-recovery diagram shows exponential behavior in both parts of the curve.

FIGURE 9-31 Behavior of the Voigt element, a spring and dash-pot in parallel, during creep and recovery.

of viscoelastic deformation. This model approximates certain aspects of the behavior of a crystalline polymer, for example.

The moduli $G(t)$ and $J(t)$, which determine the entire spectrum of viscoelastic behavior, vary in a complex manner with time and temperature (Fig. 9-29). Consequently, no simple model can reproduce the broad ranges of behavior that are possible. However, a sufficiently complex model can adequately represent real polymers subjected to actual conditions of service. The example of Fig. 9-32*c* shows how Voigt elements can be combined to simulate rapid, intermediate, and slow responses of various features of the polymeric structure. Once such a model has been constructed for one condition of service, the values of its parameters E_i and η_i permit calculations of expected behavior under various conditions of service, provided that the deformations are small. The strain for each element or parameter can then be calculated separately and superposed linearly to predict the overall effect.

PROBLEMS

9-1. (*a*) Because the small cube in Fig. 9-2 is in equilibrium with respect to translation, each normal stress component, say $\sigma_{11} = F_{11}/A_1$, acting on a positive face must be balanced by an equal component acting on the corresponding negative face. Explain why this balancing component is positive (as with σ_{11}) in the case illustrated, making use of the concept

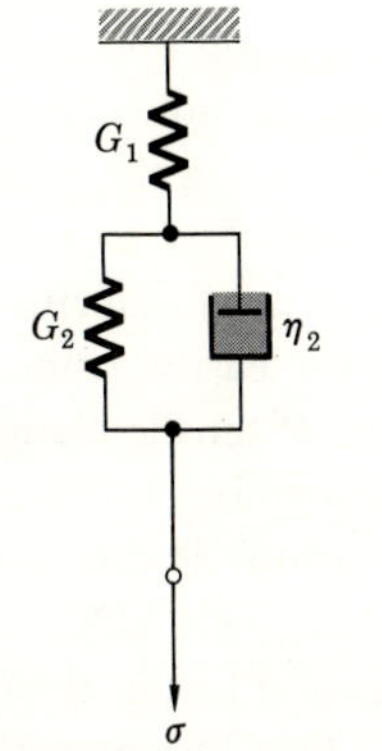

(a) Three-parameter model for a network polymer capable of complete recovery.

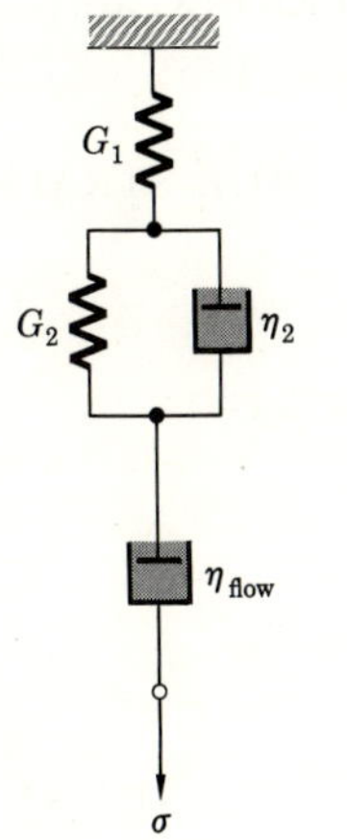

(b) Four-parameter model for a polymer exhibiting residual deformation.

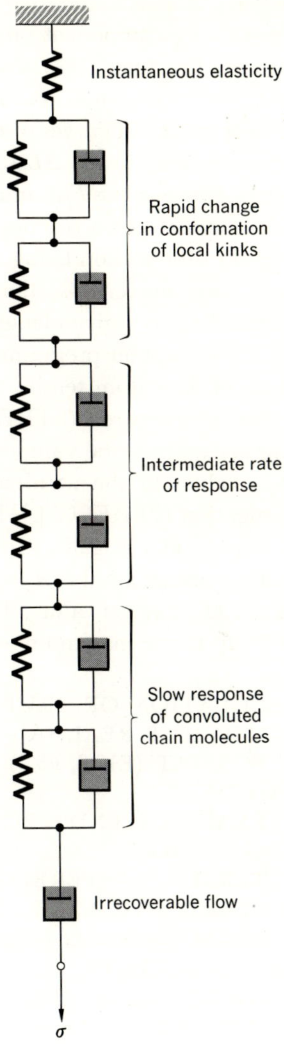

(c) Typical model for describing a broad spectrum of viscoelastic behavior.

FIGURE 9-32 More complex spring-and-dash-pot models that approximate the experimental behavior of several viscoelastic materials. (*Adapted from T. Alfrey and E. F. Gurnee, "Organic Polymers," Prentice-Hall, Inc., Englewood Cliffs, N.J., 1967.*)

of a *directed* unit normal for a given face. (*b*) A shear stress component, say σ_{12}, acting on positive face 1 must similarly be balanced by a corresponding component on the negative face. This pair of stresses constitutes a torque. Show that a matching pair σ_{21} must act on face 2 of the cube to maintain rotational equilibrium.

9-2. The stress-strain curves of Fig. 9-1 are plotted using nominal stress F/A_o and nominal strain $\Delta L/L_o$. Replot curve *b*, for low-carbon steel, in terms of true stress F/A and true strain $\ln L/L_o$ out to the point at which the nominal stress reaches its maximum value. Up to this point the metal in the section being tested (the gauge length) deforms uniformly, and thus the cross-sectional area A can be easily calculated, corresponding to a given length L in the process of tensile elongation.

Computer programs are useful for performing routine manipulations of data from tensile tests. A simple example is the calculation of values of true stress (STRESS) and true strain (STRAIN), corresponding to experimental values of load and elongation [or, in the case of Fig. 9-1, values of the nominal stress F/A_o (NOMSTS) and the percent elongation (PERELG)]. The following relations are employed: $L/L_o = 1 +$ percent elongation/100; $A_o/A = L/L_o$ since the volume is essentially constant; and $F/A = (F/A_o)(A_o/A) = (F/A_o)(1 +$ percent elongation/100). Pairs of values of NOMSTS and PERELG are read from Fig. 9-1*b* up to the maximum value of the nominal stress.

```
C  CALCULATION OF TRUE STRESS AND TRUE STRAIN
       IMPLICIT REAL (A – Z)
    1  FORMAT (E10.3, F5.2)
C  READ DATA
    3  READ (5, 1, END = 10) NOMSTS, PERELG
C  CALCULATION
    4  TERM = 1 + (PERELG/100.0)
    5  STRESS = NOMSTS * TERM
    6  STRAIN = ALOG (TERM)
C  OUTPUT SECTION
    7  WRITE (6,8) STRESS, STRAIN
    8  FORMAT (1X, 'STRESS =', E10.3, 1X, 'STRAIN =', E10.3)
    9  GO TO 3
   10  STOP
       END
```

9-3. An appreciation of the magnitude of possible thermal strains (and the accompanying stresses) can be obtained from the following example. Picture a cylindrical steel bar, 10^{-3} m^2 in cross section and 0.2 m long, mounted in a massive assembly that prevents any motion of the ends of the bar. During a gas-welding operation on the bar its average temperature is increased by 500°C. (*a*) Using Eq. (9-10), calculate the thermal strain (if the bar were free to elongate); α for steel is 11×10^{-6} °C. (*b*) According to Eq. (9-15), what magnitude of (compressive) stress tends to develop in the *constrained* bar? (*c*) Using the stress-strain curve of Fig. 9-1*b*, predict the actual elastic and plastic deformation of the bar.

9-4. Consider a unit cube of material acted upon by a normal tensile stress σ_1 (along the X_1 axis) and a compressive stress $\sigma_2 = -\sigma_1$ (along the X_2 axis). (*a*) Using Eq. (9-14), obtain expressions for ϵ_1 and ϵ_2. (*b*) These two stresses produce a maximum shear stress $\sigma_6' = (\sigma_1 - \sigma_2)/2 = \sigma_1$, referred to axes X_1' and X_2' making an angle of 45° with respect to the X_1 and X_2 axes; X_3' and X_3 are identical. Using the analog of Eq. (9-16), obtain an expression for ϵ_6' referred to the X_i' coordinate axes. By relating ϵ_6' to the vector sum of ϵ_1 and ϵ_2, prove Eq. (9-18). [HINT: The reference line for the shear strain in your sketch may be longer than that for the normal strains by a factor of $\sqrt{2}$.]

9-5. (*a*) Use the data of Table 9-1 to determine whether an ultrasonic method could be used more sensitively with a polycrystalline sheet of (1) aluminum or (2) iron for detecting the presence of preferred orientation. What is the maximum ratio of $v_s(\max)/v_s(\min)$ possible for each of these two metals? C_{11} (al) $= 1.07 \times 10^{11}$ N/m^2; C_{11} (Fe) $= 2.01 \times 10^{11}$ N/m^2.

9-6. If two phases having different values of E are to sustain the same stress, their respective regions in the material must deform different amounts. Using sketches of the two different conditions of loading, explain why the isostress concept might apply to long fine fibers *perpendicular* to the direction of stress, but not to such fibers parallel to this direction.

9-7. Two identical specimens of rubber are subjected to a stress of 5×10^4 N/m^2, but specimen A is tested at 275°K and specimen B at 325°K. (*a*) Sketch the (essentially linear) stress-strain curve for 300°K; use the data for the initial curve in Fig. 9-1*c*. (*b*) Making use of the relation $W_d = -T\,\Delta S$, plot the linear stress-strain curves for specimens A and B. [HINT: For a material having Young's modulus E, the work of deformation is as follows:

$$W_d = \int F\,ds = \int_0^{\epsilon_f} E\epsilon\,d\epsilon = \tfrac{1}{2}E\epsilon_f^2 = \tfrac{1}{2}\sigma\epsilon_f \qquad (9\text{-}48)$$

where ϵ_f is the final value of strain produced by the (slowly) applied stress σ.]

9-8. (*a*) Sketch a cylindrical single crystal of cross-sectional area A_0 subjected to a tensile force F; the corresponding tensile stress is then $\sigma =$

F/A_o. Show on your sketch a slip plane whose normal makes an angle φ with the axis of the cylinder. Sketch a slip direction in the slip plane and denote by θ the angle it makes with the axis of the cylinder. (*b*) Show that the two parts of the Schmid factor m in Eq. (9-28) account for (1) resolution of the force into a shear component in the direction of slip and (2) increase in the area over which the component force is acting. (*c*) Explain why slip occurs at a critical value τ_c of the resolved shear stress, rather than at a critical value of the tensile stress F/A_o.

9-9. (*a*) Show that $\bar{\sigma}$ is identical with the single tensile stress σ_1 applied in an ordinary tensile test. [HINT: $\sigma_1 = \sigma_I$, and $\sigma_{II} = \sigma_{III} = 0$.] (*b*) Similarly, show that $\bar{\epsilon}$ is identical with the strain ϵ_1 in the direction of the tensile stress. [HINT: Since the volume of the specimen is constant, show that $\epsilon_{II} = \epsilon_{III} = -\epsilon_I/2$.]

9-10. Relatively brittle materials can often be successfully deformed by special processes in which two (or three) principal stresses are applied simultaneously, one a tensile stress and the other a compressive stress of comparable magnitude. Using Eq. (9-31), show that plastic flow can occur even if each of these stresses is appreciably below the yield stress σ_o at which plastic flow begins when only a single tensile stress is applied.

9-11. (*a*) By taking the natural logarithm of both sides of Eq. (9-37), show the condition for plotting data involving $\dot{\epsilon}$ and σ to obtain a straight line. (*b*) Explain why a decrease in temperature should cause the line to shift to the right.

9-12. (*a*) The matrix metal exerts a force on the cylindrical surface of a short length of fiber equal to $\tau_m A$, where τ_m is the shear strength of the metal and A is the surface area. Show that only a small tensile stress $\sigma_\lambda = 4\tau_m\lambda/d$ is developed in a fiber of diameter d at a short distance λ from one end. (*b*) The value of σ_λ increases and reaches σ_{max} as λ increases to a value defined as $l_c/2$ Eq. (9-44). Show that the force due to shear balances that due to the tensile stress at the critical distance $l_c/2$.

9-13. (*a*) Use the model of Fig. 9-32*a* to sketch a creep-recovery curve similar to Fig. 9-28*a*. (*b*) Use the model of Fig. 9-32*b* to sketch a creep-recovery curve similar to Fig. 9-28*b*.

REFERENCES

Avitzur, B.: "Metal Forming: Processes and Analysis," McGraw-Hill Book Company, New York, 1968.

Barrett, C. B., and T. B. Massalski: "Structure of Metals," 3d ed., McGraw-Hill Book Company, New York, 1966.

Cottrell, A. H.: "Mechanical Properties of Matter," John Wiley & Sons, Inc., New York, 1964.

D'Isa, F. A.: "Mechanics of Metals," Addison-Wesley Publishing Company, Inc., Reading, Mass., 1968.

"Ductility," ASM seminar, American Society for Metals, Metals Park, Ohio, 1968.

Ferry, J. D.: "Viscoelastic Properties of Polymers," John Wiley & Sons, Inc., New York, 1961.

Honeycombe, R. W. K.: "Plastic Deformation of Metals," Edward Arnold (Publishers) Ltd., London, 1968.

Kalpakjian, S.: "Mechanical Processing of Materials," D. Van Nostrand Company, Inc., Princeton, N.J., 1967.

Lin, T. H.: "Theory of Inelastic Structures," John Wiley & Sons, Inc., New York, 1968.

McClintock, F. A., and A. S. Argon (eds.): "Mechanical Behavior of Materials," Addison-Wesley Publishing Company, Inc., Reading, Mass., 1966.

Mendelson, A.: "Plasticity: Theory and Application," The Macmillan Company, New York, 1968.

Nye, J. F.: "Physical Properties of Crystals," Oxford University Press, Fair Lawn, N.J., 1960.

Palmour, H., and W. Kriegel: "Mechanical Behavior of Ceramics," Interscience Publishers, New York, 1961.

Parkins, R. N.: "Mechanical Treatment of Metals," American Elsevier Publishing Company, New York, 1968.

Tegart, W. M. M.: "Elements of Mechanical Metallurgy," The Macmillan Company, New York, 1966.

Chapter Ten

Fracture of Materials

INTRODUCTION

Deformation, the topic of the previous chapter, is one aspect of the response of materials to stress. Another is *fracture*, the failure of a material by stress-induced separation into two or more parts. Experience has often taught us how to avoid fracture, and thus today the failure of most materials in service is caused by other processes such as corrosion, excessive wear, or excessive plastic deformation. Of course, fractures still do occur, and often with disastrous consequences. A broken axle may cause an automobile accident; a fractured turbine shaft may leave a community without electricity. A knowledge of the various phenomena associated with fracture can suggest means for minimizing the probability of such fracture.

A useful distinction is drawn between ductile fractures, those preceded by appreciable plastic deformation, and brittle fractures. Ductile fractures, although important for commercial forming operations, are of less concern in the design of load-carrying materials. An understanding of brittle fracture, however, is essential in the use of high-strength materials, including metals, ceramics, and composites. The Griffith theory and its modifications offer a unifying approach for rationalizing the behavior of these various materials. As a preparation for studying fracture under tension, the topic of principal interest, we first survey various aspects of this kind of fracture.

10-1. Characteristics of Tensile Fracture

Just as materials undergo several types of plastic deformation, so they exhibit a variety of modes of fracture depending upon the material, the stress condition, the temperature, the rate of straining, and the environment. Study of the modes of fracture is termed *fractography* and establishes the relationships

[1] This chapter was written by Professor A. J. McEvily, Head of the Department of Metallurgy, University of Connecticut.

between features of the fracture surface and the processes involved in stress-induced separation. Fractographic methods include visual observation, the use of binocular microscopes for low magnification, light microscopy, electron microscopy, and scanning-electron microscopy. The high resolution and great depth of focus available in both types of electron microscopes are especially valuable in fractographic examination since most fracture surfaces are rough even on a fine scale.

BRITTLE FRACTURE. The final stage in tensile testing (Fig. 10-1*a*) is separation of the specimen into two pieces. This process of fracture is termed *ductile* if it is preceded by appreciable plastic deformation and localized reduction in area. On the other hand, *brittle* or *semibrittle* fracture involves little prior inelastic deformation; failure occurs abruptly without localized reduction in area (Fig. 10-1*a*). Below the glass transition temperature, polymers and inorganic glasses both behave in this brittle manner. Polycrystalline materials can fracture in a brittle manner, either along grain boundaries (*intergranularly*) or through the grains (*transgranularly*). Transgranular fracture usually occurs by cleavage, that is, by separation along certain crystallographic planes having low Miller indices. Representative cleavage planes are {100} in sodium chloride and in body-centered cubic iron and (0001) in zinc. Certain titanium alloys and high-strength steels exhibit a type of brittle fracture called *quasi-cleavage*; it is not crystallographic but is similar in appearance to cleavage fracture.

Typical features of brittle-fracture surfaces are illustrated in Fig. 10-1. In an inorganic glass the origin of fracture is usually a surface defect (Fig. 10-1*b*). The crack spreads gradually through the mirror area and then more rapidly. The markings beyond the mirror area point backward toward the origin of fracture and are therefore useful in failure analysis. These markings are called *tear lines* and arise because various portions of the front of the crack propagate on slightly different levels. The shell-like appearance of the tear lines in an inorganic glass or in a brittle polymer gives rise to the term *conchoidal fracture*. Tear lines are also present on the bright cleavage facets of brittle metals (Fig. 10-1*c*) and often join to form a "river" pattern. As a crack front crosses a grain boundary, numerous tear lines initiate because of the change in crystallographic orientation. Macroscopic tear lines can also occur, for example, in steel plates that have fractured in a brittle manner (Fig. 10-1*d*). The lines are often in a chevron pattern with the apex of the pattern pointing toward the origin of fracture.

Fracture of brittle materials can occur under compressive loading also. For such materials the compressive strength is usually an order of magnitude greater than the tensile strength, and separation takes place by a shearing process. Concrete and cast iron are examples of materials that exhibit this type of failure.

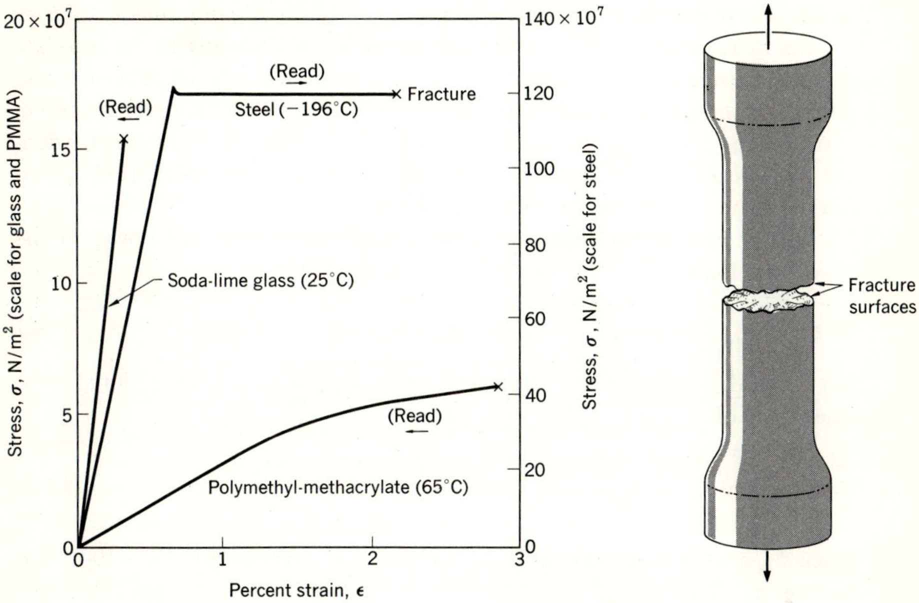

(*a*) Typical stress-strain curves and type of fracture for brittle materials.

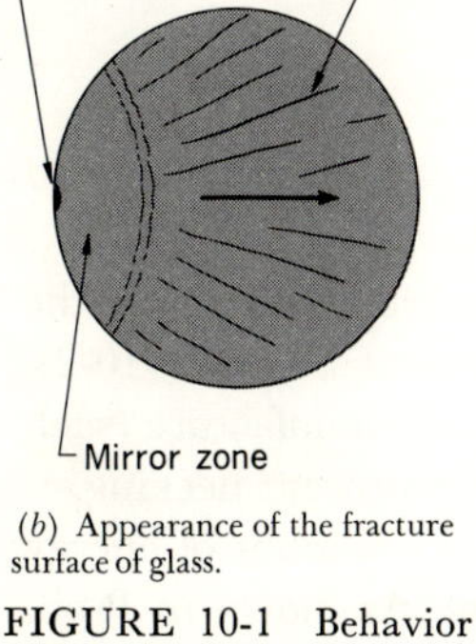

(*b*) Appearance of the fracture surface of glass.

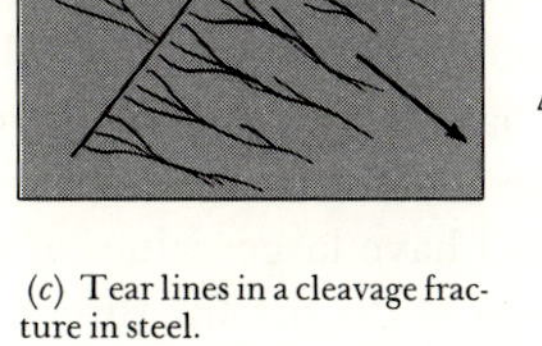

(*c*) Tear lines in a cleavage fracture in steel.

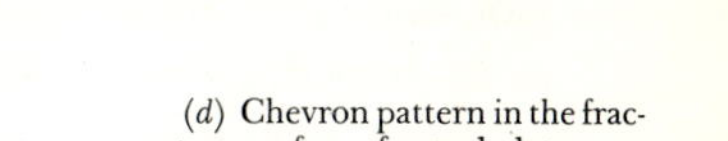

(*d*) Chevron pattern in the fracture surface of a steel plate.

FIGURE 10-1 Behavior of representative brittle materials and examples of fracture surfaces [conversion of units: (M N/m²) $1.450 \times 10^{-4} = N$ lb/in²; (M N/m²) $1.020 \times 10^{-7} = N$ kg/mm²].

DUCTILE FRACTURE. Fracture of ductile materials occurs in the necked region, which has undergone appreciable plastic deformation. The process of neck formation is most simply described in terms of the load F acting on a tensile specimen of initial cross-sectional area A_0. As the nominal stress F/A_0 reaches its maximum value (the tensile strength), uniform plastic deformation of the specimen can no longer continue. An unstable condition arises because a local decrease in cross-sectional area diminishes the load-bearing capacity of that area more than can be compensated by the accompanying strain hardening. Consequently, necking begins at the maximum load (Fig. 10-2*a*), and further reduction in area proceeds in a localized region.

The critical condition for necking can be obtained as follows. At the maximum load

$$dF = 0 \qquad (10\text{-}1)$$

But $F = \sigma A$, where σ is the true stress and A is the actual cross-sectional area. Therefore, $dF = \sigma\, dA + A\, d\sigma = 0$, and

$$\sigma = -\frac{A}{dA}\, d\sigma \qquad (10\text{-}2)$$

Since $dA/A = -d\epsilon$ (see Prob. 10-1), where ϵ is the true strain, Eq. (10-2) can be written

$$\frac{d\sigma}{d\epsilon} = \sigma \qquad (10\text{-}3)$$

This condition for the beginning of necking can be applied easily to those materials whose uniform plastic deformation can be approximated by the relation $\sigma = K\epsilon^n$ (Fig. 9-15). In this case,

$$\frac{d\sigma}{d\epsilon} = nK\epsilon^{n-1} = \frac{n\sigma}{\epsilon} \qquad (10\text{-}4)$$

Comparison of Eqs. (10-3) and (10-4) shows that the true strain ϵ at the beginning of necking equals the strain-hardening exponent n (Prob. 10-2). Consequently, alloys subjected to extensive deformation in manufacture (such as in deep drawing) should have large values of n to help prevent necking.

A significant feature of ductile fracture is the large amount of energy required compared to that necessary for brittle fracture. As shown in Prob. 10-4, the energy per unit volume is

$$\frac{\int F\, ds}{V} = \int_0^{\epsilon_f} \sigma\, d\epsilon \qquad (10\text{-}5)$$

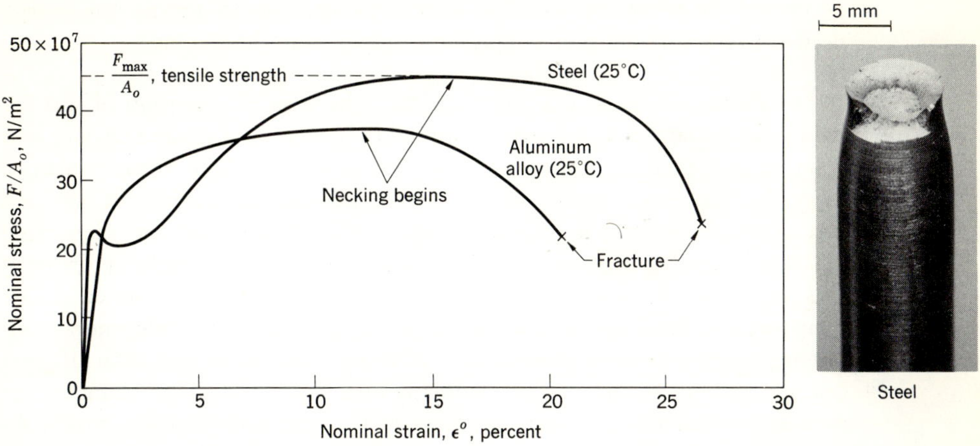

(a) Typical stress-strain curves and types of fracture for ductile materials.

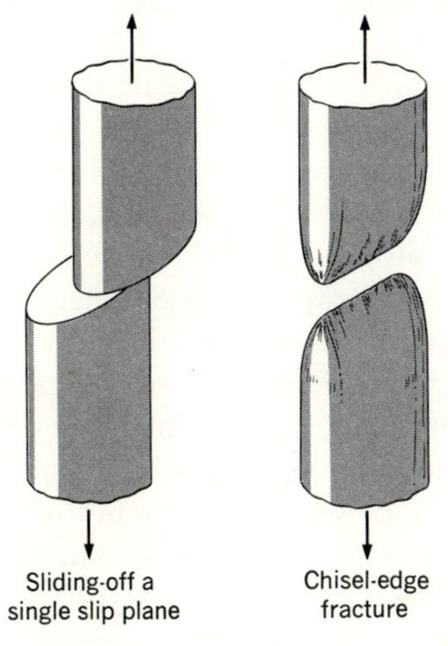

(b) Two modes of separation in single-crystal tensile specimens.

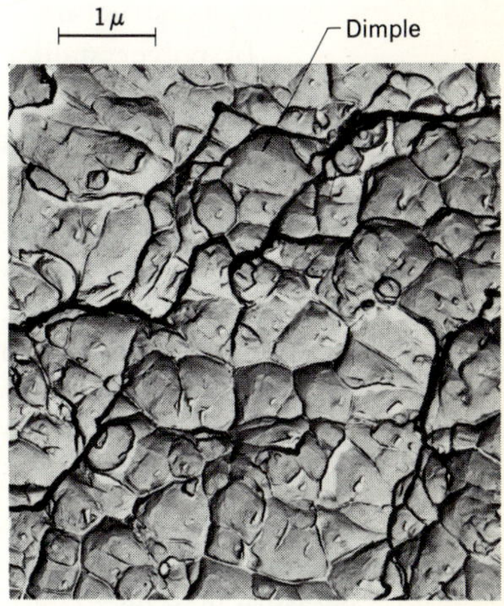

(c) Electron micrograph of fracture surface of steel. (Courtesy C. B. Beachem, U.S. Naval Research Laboratory.)

FIGURE 10-2 **Several characteristics of ductile fracture.**

where ϵ_f is the true strain at fracture and V is the volume. The second integral of Eq. (10-5) is the area under the true-stress–true-strain curve and is much greater for ductile materials than for brittle materials. The pertinent volume V varies during the course of a tensile test. It is the entire test section during the stage of uniform plastic deformation but later becomes restricted to a portion of the necked region. The volume factor tends to increase the energy needed to fracture alloys having a high rate of strain hardening.

INFLUENCE OF SLIP ON FRACTURE. The ease of operation of slip (or twinning) systems in a crystalline material influences its fracture behavior and determines the nature of the final separation process. If slip is virtually absent, as in Al_2O_3 at temperatures below 1000°C, the material is brittle. If only one slip system is operative in a single crystal, tensile deformation may then cause separation by "sliding off" on a slip plane (Fig. 10-2*b*). Suitably oriented crystals of zinc or niobium experience this type of failure when tested at low temperatures. In single crystals of the NaCl type, only two independent slip systems operate at room temperature. Although specimens can attain 10 percent elongation, fracture occurs suddenly and without localized reduction in area as a result of nucleation of cracks at the intersection of slip bands. At elevated temperatures where the slip systems are more easily made operative, single crystals of NaCl are ductile, and failure is of a chisel-edge type (Fig. 10-2*b*). This kind of failure is characteristic of ductile single crystals having a cubic crystal structure.

In polycrystalline materials the requirements for plastic deformation are more severe since continuity must be preserved across the boundary between grains having different orientations. Nevertheless, provided that a material has the necessary five independent slip systems (Sec. 9-7), as does an FCC metal, for example, it can undergo arbitrary changes in shape, including the formation of a neck. Randomly oriented polycrystalline materials containing less than the requisite five deformation modes are inherently brittle, however. The extent of necking in suitable polycrystalline, single-phase materials depends on the purity. In high-purity materials necking may proceed until almost 100 percent reduction in area has been attained, with the specimen drawing down to a point. In most materials, however, the presence of second phases and impurities results in a reduction in area less than this maximum. A typical value of reduction in area for a vacuum-melted steel at room temperature may be 70 percent, whereas the value for a less pure air-melted steel of the same nominal composition may be 50 percent.

The formation of tiny internal voids is a principal cause of decreased reduction in area. Since an impurity particle deforms differently than the matrix, the strains may become incompatible at the particle-matrix interface and give rise to a void at this location. This phenomenon is especially likely to occur within a neck because of the triaxial tensile stresses in this region. On the other hand, large hydrostatic compressive stresses suppress the formation of voids and increase the reduction in area. Once the voids have formed in a tensile test, they continue to grow with increasing strain and eventually coalesce to form a crack in the center of the neck. Shear strains now tend to concentrate near the tip of the crack; they occur predominantly in two thin sheets at an angle of approximately 30° to the axis. Numerous small voids form in these deformed regions and weaken them. One of the deformed regions eventually fails, causing the crack to extend a short distance both radially and out of the plane of maximum tensile stress. The extended crack again concentrates strain at its tip, voids form, and failure occurs. This time, however, the failure propagates the crack back through the plane of maximum tensile stress, thus starting a zigzag pattern of crack growth. The coalescence of voids produces a dimpled fracture surface (Fig. 10-2*c*); often an impurity particle can be associated with each dimple.

When this irregular but roughly planar crack spreads over a sufficiently large central area of the specimen, the high stresses in the remaining outer "shell" cause final separation to occur by one or another mechanism. In some alloys (steel and brass, for example) final separation also occurs along the plane of maximum shear strain by void formation. This final separation forms the cone portion of a cup-and-cone fracture (Fig. 10-2*a*). In metals such as gold, silver, copper, aluminum, and nickel the final separation takes place by an alternating slip process similar to the process of chisel-edge formation in single crystals. The resulting fracture surface has a "double-cup" appearance and contains striations perpendicular to the direction of advance of the crack.

Two-phase materials sometimes fail by a combination of brittle and ductile rupture processes. For example, Borsic-aluminum composites (Fig. 3-9), consisting of an aluminum matrix reinforced with continuous boron fibers, fail by brittle fracture of the fibers and ductile rupture of the matrix (Fig. 10-3).

An important phenomenon in certain materials is a transition from ductile to brittle behavior as the temperature of testing is lowered (Fig. 10-4). In a given type of material, such as steel, the transition is less marked in

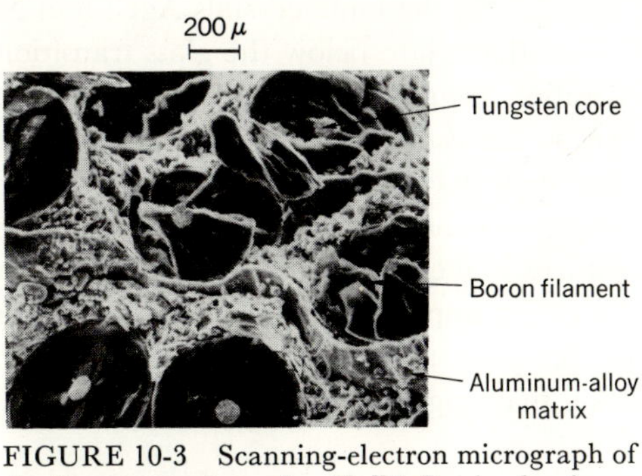

FIGURE 10-3 Scanning-electron micrograph of the fracture surface of a 50 $^{v}/_{o}$ Borsic-aluminum composite. (*Courtesy K. J. Kreider, United Aircraft Research Laboratories.*)

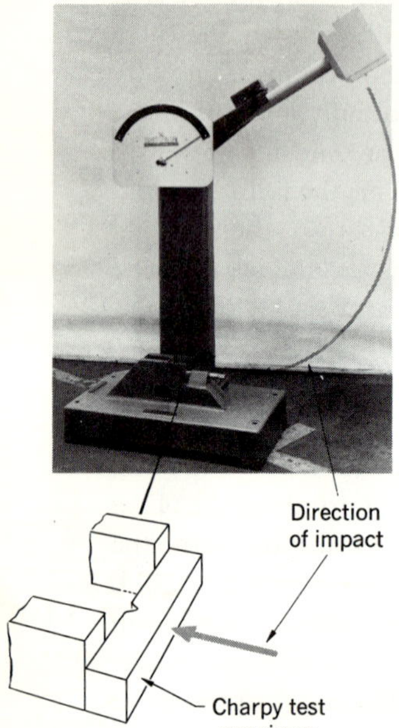

(a) A multi-purpose impact tester. The procedure for a notched charpy test is shown. (Courtesy Tinius Olsen Testing Machine Co., Inc.)

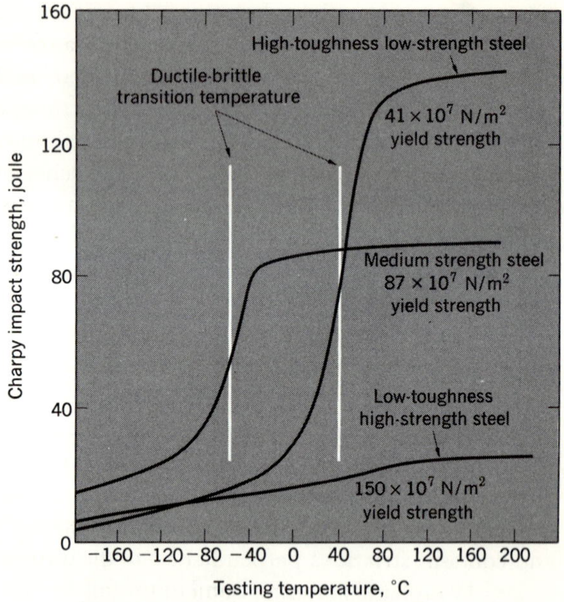

(b) Impact test data for various temperatures can reveal significant changes in toughness and fracture behavior. (After A. S. Tetelman and A. J. McEvily.)

FIGURE 10-4 Illustration of the notch-impact resistance of typical commercial steels as a function of temperature (conversion of units: M J × 0.737 = N ft-lb).

specimens that have been strengthened in various ways, such as by heat treatment. Materials exhibiting this type of transition include alloys of Fe, Ta, Nb, Sn, Mo, W, Be, Cr, and Si, the ionic crystals AgCl and NaCl, and polymers such as polymethyl-methacrylate below the glass transition temperature. The temperature of transition from a cleavage mode to a ductile mode in crystalline materials is known as the *ductile-brittle transition temperature*. It is not a fixed value, however, but depends on the type of specimen and the conditions of loading. For example, notched specimens and high rates of straining give transition temperatures higher than that determined by an ordinary tensile test. Because of this effect, impact tests of notched specimens are used to determine the transition temperature in an effort to simulate the most unfavorable conditions of loading that might be encountered in service.

CREEP FRACTURE. Prolonged application of a constant load [as in creep-rupture testing (Sec. 9-11)] may produce significantly different behavior than does an ordinary tensile test. The stress for fracture is lower the longer the time of loading. For polymers the loss of strength with time is appreciable even at room temperature. For example, the data of Fig. 10-5 show that the biaxial stress required to fracture polyethylene decreases by about 50 percent as the time of loading increases by a factor of 10^2. Furthermore, a change in fracture mechanism occurs as the stress is decreased. In the region of high stresses (corresponding to short times to fracture) the spherulitic structure deforms, and the polymer behaves in a ductile manner; the amount of deformation at fracture is about 100 percent. The lower stresses do not cause appreciable deformation of the spherulites but instead encourage the growth and coalescence of voids and cracks, resulting in brittle fracture at deformations of less than 10 percent.

Creep in most structural metals is a high-temperature phenomenon, significant above about one-half of the absolute melting temperature. The curve of elongation versus time for a typical creep-rupture test has the three stages shown in Fig. 10-6*a* and is terminated by fracture. Knowledge of creep behavior is important in the design of many structures, for example, nuclear power plants, which may have a service life of the order of 40 years. Such components as heat exchangers and superheaters utilize metals in the creep range, and thus safe working stresses must be determined, corresponding to the desired rates of elongation and rupture life. For many materials the product of rupture life t_f and minimum strain rate $\dot{\epsilon}_{min}$ is a constant, and thus estimates of long-time behavior can be made by extrapolation from short-time tests. Of course, environmental factors and the stability of the metallurgical structure must be considered in making such predictions.

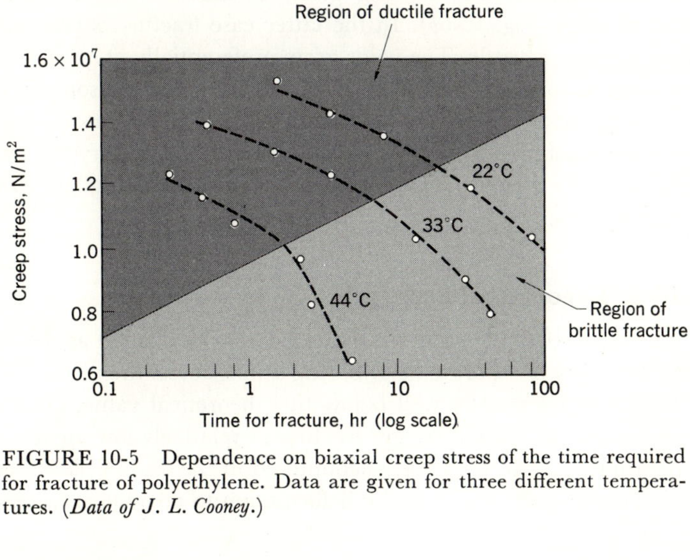

FIGURE 10-5 Dependence on biaxial creep stress of the time required for fracture of polyethylene. Data are given for three different temperatures. (*Data of J. L. Cooney.*)

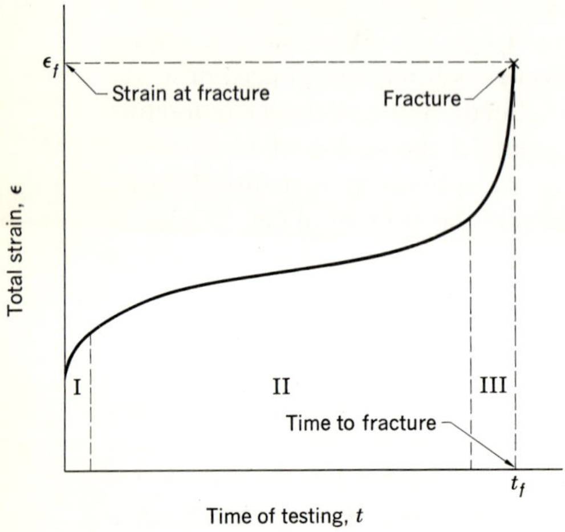

(*a*) Schematic creep-rupture curve showing the three stages of creep. Stage III terminates in fracture.

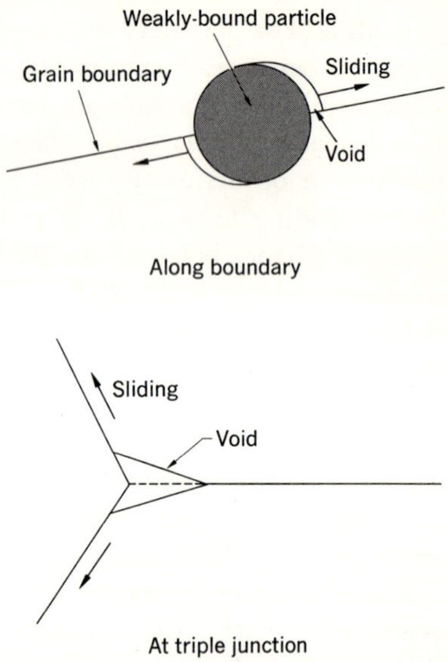

(*b*) Void formation at grain boundaries resulting from grain-boundary sliding during creep.

FIGURE 10-6 Aspects of fracture in creep, including the prior deformation and the formation of voids at grain boundaries.

Creep failure can occur simply because of decrease in cross-sectional area as the material elongates, but in many polycrystalline metals rupture results from localized sliding along grain boundaries, which leads to the formation of voids (Fig. 10-6*b*). In the latter case fracture is generally intergranular and rather brittle. The voids grow preferentially at grain boundaries that are normal to the principal tensile stress. For this reason directionally solidified alloys with boundaries oriented parallel to the tensile axis exhibit superior creep resistance. Still better are high-strength single crystals, which are being evaluated for use as turbine blades at the high operating temperatures of jet aircraft engines.

10-2. Theoretical Strength

The theoretical strength of a material can be defined as the force per unit area necessary to separate neighboring atoms across the plane of fracture. Actual strengths are usually well below this theoretical value, often because flaws act as stress raisers and trigger fracture at relatively low applied stresses. The situation is analogous to that encountered in the case of plastic deformation, where dislocations allow plastic deformation to take place at stress levels that are but a small fraction of the theoretical shear strength. To obtain an estimate of the theoretical tensile strength σ_T, we consider two planes of atoms (Fig. 10-7) separated at equilibrium by the distance a. The stress σ required to produce large displacements x of one plane relative to the other can be approximated by a sine curve as follows:

$$\sigma = \sigma_T \sin \frac{2\pi x}{\lambda} \tag{10-6}$$

The maximum of the sine curve is the theoretical strength σ_T. Taking $x_f = a/10$ as the average of various estimates of the theoretical displacement at fracture and assuming that Hooke's law remains applicable, we have

$$\sigma_T = E\epsilon_f = E\frac{x_f}{a} = \frac{E}{10} \tag{10-7}$$

This rough estimate suggests that high values of strength are theoretically possible. A comparison of Young's modulus and the actual strength of several materials (Table 10-1) shows that actual strengths are typically far below their potential values. These tabulated data are for ordinary bulk specimens.

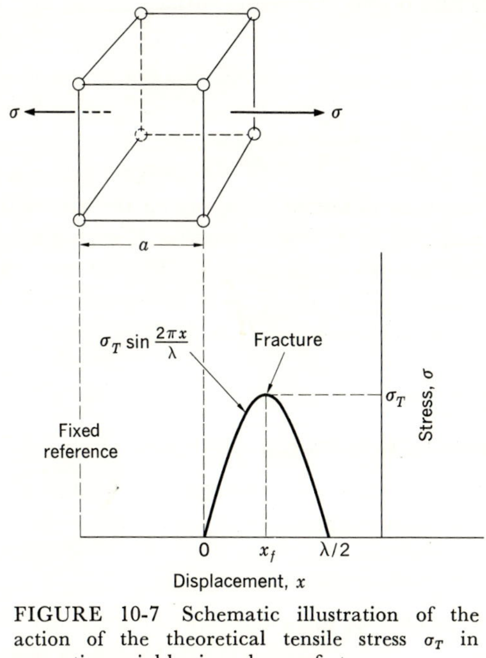

FIGURE 10-7 Schematic illustration of the action of the theoretical tensile stress σ_T in separating neighboring planes of atoms.

Strengths much closer to the theoretical values are obtained on tests of fiber specimens, carefully prepared and free of flaws.

TABLE 10-1. COMPARISON OF ACTUAL STRENGTH WITH THE INDICATOR OF THEORETICAL STRENGTH, YOUNG'S MODULUS†

Material	Young's modulus, N/m^2	Tensile strength, N/m^2
High-strength steel	21×10^{10}	3×10^9
High-strength aluminum alloy	7×10^{10}	7×10^8
Soda-lime glass	7×10^{10}	6×10^7
Phenolic-glass fabric	2×10^{10}	4×10^8
Polymethyl-methacrylate	3×10^9	5×10^7
Polystyrene	3×10^9	6×10^7

† Conversion of units: $(M\ N/m^2)\ 1.450 \times 10^{-4} = N\ lb/in^2$; $(M\ N/m^2)\ 1.020 \times 10^{-7} = N\ kg/mm^2$.

An important quantity in fracture analysis is the so-called *true surface energy* γ_s, defined as the work done in creating new surface area by the breaking of bonds. If we consider a unit volume of material, then the area under the stress-displacement curve (Fig. 10-7) out to $x = \lambda/2$ is the energy needed to complete the process of fracture. Two new surfaces, each of unit area, are created when fracture occurs. Equating surface energy and strain energy, we obtain the relation

$$2\gamma_s = \int_0^{\lambda/2} \sigma_T \sin \frac{2\pi x}{\lambda}\, dx = \frac{\lambda}{\pi}\, \sigma_T \tag{10-8}$$

An estimate for λ can be obtained by applying the approximation (for small angles) $\sin x \approx x$ at (the large angle) $\sin x = 1$, giving $x_f = \lambda/2\pi$. Equation (10-7) can then be written

$$\sigma_T = \frac{\lambda}{2\pi}\frac{E}{a} \tag{10-9}$$

When λ/π is eliminated between Eqs. (10-8) and (10-9), the result is

$$\sigma_T = \left(\frac{E\gamma_s}{a}\right)^{1/2} \tag{10-10}$$

The true surface energies γ_s for some representative materials are listed in Table 10-2.

TABLE 10-2. TRUE SURFACE ENERGIES†

Material	γ_s, J/m²
NaCl	0.1
Gold	1.4
Copper	1.7
Iron	2.3
Mica	4.5
Glass	0.9
Polymethyl-methacrylate	0.5

† Conversion of units: M J/m² × 10³ = N erg/cm².

FACTORS DETERMINING FRACTURE BEHAVIOR

Of the various aspects of fracture behavior, three factors are of special importance for many technical materials. Flaws, initially present or produced during service, are decisive in the fracture process. Plastic deformation can strongly influence the processes of crack nucleation and growth. Finally, the fracture of high-strength materials requires special attention.

10-3. Effect of Flaws

Griffith in 1920 recognized the importance of flaws in reducing the fracture strength of brittle materials. An existing flaw in a brittle material propagates to failure when the stress increases to a critical level; the driving force for this process is a reduction of the overall energy of the system. This energy consists of the elastic strain energy stored in the specimen and also the surface energy of the advancing crack, both of which change as a crack advances. The relationship between them determines the condition for propagation of a preexisting flaw. Consider a wide sheet of brittle material containing a crack of length $2c$ through its thickness (Fig. 10-8*a*). Fixed external restraints maintain a displacement ΔL in the sheet; correspondingly, a force F is exerted on the sheet. This force decreases by some amount ΔF as the crack advances by Δc since a smaller force is then sufficient to maintain the fixed displacement ΔL. The decrease in stored elastic energy, ΔW_e (Fig. 10-8*b*), can be calculated as the difference between the two values shown and is equal to $\Delta L\ \Delta F/2$. Griffith showed that this decrease is given by

$$\Delta W_e = \frac{2\pi\sigma^2}{E} c\ \Delta c \tag{10-11}$$

for the case of an elliptical crack of vanishingly small tip radius.

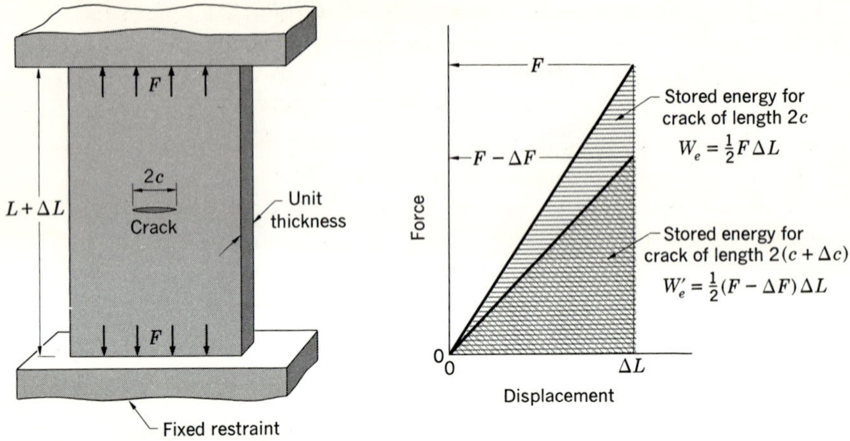

(*a*) External restraints produce a force *F* in a sheet specimen containing a crack of length 2c.

(*b*) Force-displacement diagrams give the stored energy for two values of crack length.

FIGURE 10-8 Explanation of the decrease in stored elastic energy accompanying the growth of a crack in a specimen subjected to a fixed displacement.

The creation of additional fracture surface requires an expenditure of energy to raise atoms to positions of higher potential energy at the new surface of the crack. This increase in surface energy, ΔW_s, is given by

$$\Delta W_s = 4\gamma_s\,\Delta c \tag{10-12}$$

where γ_s is the surface energy per unit of area. The numerical factor 4 appears because two increments of surface area are created at each end of the crack. The Griffith condition for fracture is obtained by equating the rate at which strain energy is released to the rate at which energy is required to create the new surfaces. This condition can be expressed as

$$\lim_{\Delta c \to 0} \frac{\Delta W_e}{\Delta c} = \lim_{\Delta c \to 0} \frac{\Delta W_s}{\Delta c}$$

or

$$\frac{dW_e}{dc} = \frac{dW_s}{dc} \tag{10-13}$$

The values given by Eqs. (10-11) and (10-12) then lead to the final expression

$$\frac{2\pi\sigma_c^2 c}{E} = 4\gamma_s$$

or

$$\sigma_c = \left(\frac{2E\gamma_s}{\pi c}\right)^{1/2} \tag{10-14}$$

Here σ_c is the critical value of stress required for propagation of the flaw. This is the Griffith equation, a cornerstone of modern fracture theory. It is a go/no-go condition; that is, the flaw will not propagate until the critical value of stress is reached. Once the crack starts to propagate in a brittle material, its rate of growth accelerates since the strain energy released as it lengthens is increasingly in excess of that required for the creation of new surface. This acceleration continues until the crack reaches a terminal velocity v_t on the order of one-half the velocity of a longitudinal sound wave in the material; that is,

$$v_t \approx \frac{1}{2}\left(\frac{E}{\rho'}\right)^{1/2} \tag{10-15}$$

E is Young's modulus, and ρ' is the density of the material.

The effect of flaws on the fracture strength of brittle materials can also be treated by considering the flaw to be a stress raiser. This approach considers the fracture process at the tip of the flaw. The stress-concentration factor for an elliptical flaw, the type in Fig. 10-8*a*, is given by $\sigma_{\max} = \sigma[1 + 2(c/\rho)^{1/2}]$. For a tip radius ρ very small with respect to c,

$$\sigma_{\max} = 2\sigma\left(\frac{c}{\rho}\right)^{1/2} \tag{10-16}$$

Provided that $\sigma_{\max}$ reaches a value equal to σ_T [Eq. (10-10)], fracture occurs by the rupturing of interatomic bonds. Since $\sigma_T = (E\gamma_s/a)^{1/2}$, the critical value of applied stress σ_c for fracture is given by

$$2\sigma_c\left(\frac{c}{\rho}\right)^{1/2} = \left(\frac{E\gamma_s}{a}\right)^{1/2}$$

or

$$\sigma_c = \left(\frac{E\gamma_s\rho}{4ca}\right)^{1/2} \tag{10-17}$$

A reasonable value, $\rho = 8a/\pi$, for the radius of the tip of the crack makes this value of σ_c the same as that given by the previous method [Eq. (10-14)].

10-4. Role of Plastic Deformation in Fracture

Although Griffith's original experimental investigation on glass specimens substantiated his theory for a brittle material, later work with less brittle materials demonstrated the need for considering the effects of plastic deformation. This modification is especially needed for ordinary structural materials, which are much more ductile than the glass modeled and tested by Griffith. We now consider the importance of the plastic deformation that usually accompanies nominally brittle fracture.

CRACK PROPAGATION. When steels containing cracks of a known size are tested in the brittle range below the ductile-brittle transition temperature, the experimental values of γ are much larger than γ_s, the surface energy. Localized plastic deformation occurs during fracture, and x-ray studies confirm that a heavily distorted layer is present immediately beneath the fracture surface. The effect of such plastic deformation on the fracture stress σ_c is represented by γ_p in the following modified equation:

$$\sigma_c = \left(\frac{2E(\gamma_s + \gamma_p)}{\pi c}\right)^{1/2} \tag{10-18}$$

Since the work of plastic deformation, γ_p, is usually much larger than γ_s, Eq. (10-18) can be written

$$\sigma_c = \left(\frac{2E\gamma_p}{\pi c}\right)^{1/2} \tag{10-19}$$

This equation retains the Griffith formalism but with a new interpretation placed on γ. The modification is important because γ_p (and therefore fracture toughness) can be increased by suitable changes in the microstructure of a material. If the value of γ_p is increased, a higher value of stress can be applied with flaws of a given size, or flaws of a larger size can be tolerated at a given applied stress.

The discrepancy between γ_s and the measured value of γ is not restricted to metals. For example, Fig. 10-9 shows the variation in strength of a glassy polymer, polymethyl-methacrylate, as a function of flaw size. The value of γ_p calculated from these data is 2×10^2 J/m^2, which is almost 10^3 times larger than γ_s (Table 10-2). Inelastic deformation involving rearrangement of the polymer chains along the fracture surface increases the effective surface energy during fracture. Even in many inorganic glasses inelastic processes can cause the value of γ_p to be much greater than γ_s.

CRACK INITIATION. The above discussion has emphasized the importance of plastic deformation in the propagation of preexisting flaws. Plastic deformation can also play a role in the initiation of cracks. Especially clear evidence is provided by tensile tests at −195°C on specimens of low-carbon steel having various grain sizes (Fig. 10-10). In the range of small grain sizes the steel is

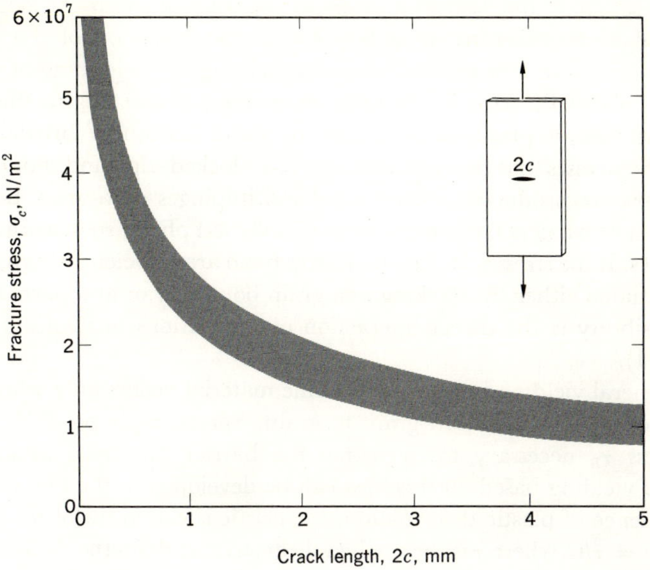

FIGURE 10-9 Fracture strength of samples of polymethyl-methacrylate containing a prior crack of length $2c$. (*Data of J. P. Berry.*)

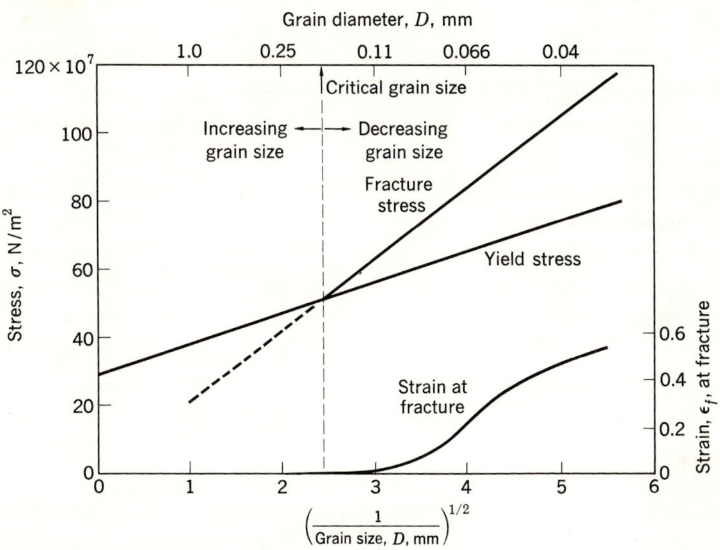

FIGURE 10-10 Variation of fracture stress and yield stress with grain size in low-carbon steels tested at $-195°C$. (*After J. R. Low, Jr.*)

fairly ductile, and thus curves can be plotted for both yield stress and fracture stress. These curves intersect at a grain size of about 0.2 mm. At larger grain sizes the data for fracture stress fall, *not* on the extension of the previous fracture-stress curve, but on the curve representing the beginning of yielding. Evidently, plastic deformation initiates the process of cracking in this case.

The role of plastic deformation in the initiation of brittle fracture involves the stresses that develop at the tip of a blocked slip band. For example, high stresses are produced when a slip band impinges on a grain boundary. Other types of barriers include particles of a second phase, such as carbides or precipitates. If the stresses at the tip of a slip band are sufficiently high, fracture can be initiated either by cracking at a grain boundary or at a particle. A less likely possibility is the direct interaction of dislocations to nucleate a crack (Fig. 10-11).

General yielding in a polycrystalline material occurs only when plastic deformation can spread from grain to grain. Spreading is possible once the shear stress τ_B necessary to overcome the barrier has been attained. An analysis of yielding based on this idea can be developed in the following way. In the absence of plastic deformation, the elastic shear strain of the region is given by $\gamma = \tau/G$, where τ is the applied shear stress and G is the shear modulus. The associated elastic displacement Δ is given by (see Prob. 10-8)

$$\Delta = \frac{\tau d}{G} \tag{10-20}$$

where d is the grain size. When yielding occurs, the elastic displacement relaxes and is replaced by a plastic displacement.

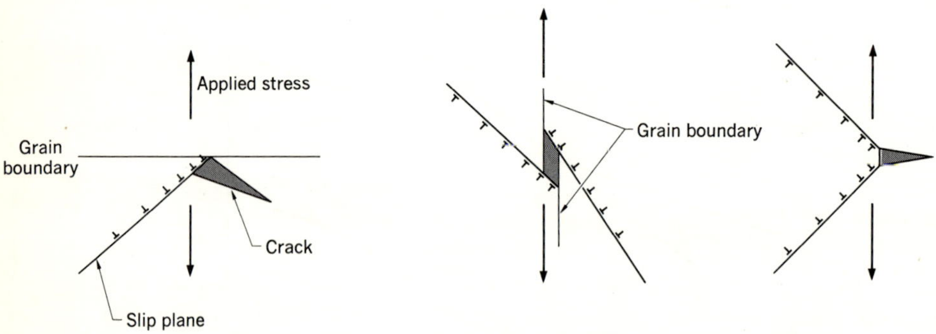

(*a*) Dislocations piled up at a grain boundary.

(*b*) Dislocations piled up at both sides of a grain boundary.

(*c*) Interaction of dislocations on intersecting slip planes.

FIGURE 10-11 Proposed mechanisms for the nucleation of cracks by the interaction of dislocations. (*Adapted from H. Conrad and H. L. Sujata.*)

The number n of dislocations in the pile-up at the end of a blocked slip plane can be estimated since $nb \approx \Delta$; b is the Burgers vector of each dislocation. Therefore,

$$n = \frac{\tau d}{Gb} \tag{10-21}$$

If the stress τ_A at the leading dislocation of the pile-up is assumed to be

$$\tau_A = \tau n \tag{10-22}$$

then,

$$\tau_A = \frac{\tau^2 d}{Gb} \tag{10-23}$$

Since moving dislocations must also overcome a general frictional stress τ_i, Eq. (10-23) should be written

$$\tau_A = \frac{(\tau - \tau_i)^2 d}{Gb} \tag{10-24}$$

General yielding can occur when the applied shear stress τ reaches the value τ_y (the yield stress) at which τ_A equals the stress τ_B needed to overcome the barrier. Equation (10-24) can then be written in the form

$$\tau_y = \tau_i + (\tau_B Gb)^{1/2} d^{-1/2} \tag{10-25}$$

The corresponding expression in terms of tensile stresses is

$$\sigma_y = \sigma_i + k_y d^{-1/2} \tag{10-26}$$

Values for σ_i and k_y are usually determined experimentally. The yield strength values in Fig. 10-10 obey this relation and plot as a straight line as a function of $d^{-1/2}$. The fact that for large grain sizes the fracture stress lies on the extension of this line suggests the following picture. In the range of large grain sizes yielding can create flaws of sufficient size to satisfy the Griffith criterion for propagation. At grain sizes smaller than the transition grain size, any flaws created on yielding are too small to propagate, and hence a higher stress is needed before fracture will occur.

The constant k_y in Eq. (10-26) is a useful indicator of the tendency of a material to behave in a ductile manner. Since k_y is proportional to the stress

τ_B, low values indicate ease of plastic release of stress concentrations. If cross-slip is possible, stress concentrations are held to a low level as the wavy slip lines follow a relatively low-energy path (Fig. 9-11). Experimental data on copper and copper-zinc alloys (Fig. 10-12) show that low values of k_y are associated with materials displaying wavy slip and high values with planar slip.

Boundary effects of several kinds can affect fracture behavior. The presence of a free surface impedes the initiation of fracture since the stresses at pile-ups are more easily relieved in this region. Grain boundaries are preferred

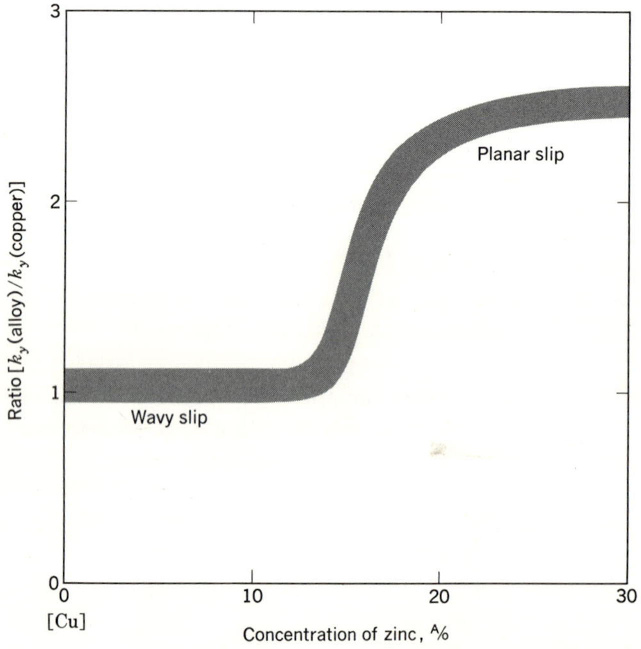

FIGURE 10-12 Data showing the increase in k_y caused by the addition of zinc to copper. Zinc decreases the stacking-fault energy and causes the mode of slip to change from wavy to planar. (*Data of T. L. Johnston, R. Ku, and A. J. McEvily.*)

sites for the segregation or precipitation of solutes. The segregation of certain elements, such as antimony, to grain boundaries during heat treatment of steel can lead to the embrittlement known as *temper brittleness*, the mechanism involved being a lowering of the bond strength. Brittle fracture along grain boundaries can also occur when slip has difficulty in propagating from one grain to another. As with other barriers, this impediment leads to the development of high stresses. In cases where the boundary itself acts as the principal barrier, impurities that have segregated to the boundary may further facilitate fracture.

The initiation of fracture in polymers is also generally associated with the prior presence of flaws since polymers contain many structural inhomogeneities which aid in the nucleation of fracture processes. Examples are the transition zones between crystalline and amorphous regions, low-molecular-weight fractions rejected during crystallization, dust particles, catalysts, scratches, and voids. Polymeric materials lose toughness below the glass transition temperature because of chain immobility. To improve toughness in this region, second phases such as butadiene rubber can be introduced to interfere with the propagation of cracks, Fig. 2-7. In some glassy polymers, absorbed water will act as a plasticizer, decrease the glass transition temperature, and increase toughness, but at the expense of the tensile strength.

Inorganic glasses, although containing some bulk flaws, are especially susceptible to surface damage such as fine scratches which may be introduced in the manufacturing process. Careful surface preparation can therefore greatly increase the strength of glass. Processing to impart residual compressive stresses at the surface may also be beneficial (Sec. 5-2).

The transitions in fracture behavior that occur as the temperature of testing is lowered can be studied in various ways, for example, using impact-test data (Fig. 10-4). When tensile tests are employed for this purpose (Fig. 10-13), a significant change in fracture stress is observed at T_{DB}, the ductile-brittle transition temperature. T_{DB} is analogous to the critical grain size in Fig. 10-10. Below T_{DB} the stress for fracture is equal to the yield stress because the flaws produced during yielding are of sufficient size to satisfy the Griffith criterion. Above this temperature a stress higher than the yield stress is required for fracture. Cleavage continues to be the mode of fracture in a limited range of temperature, but eventually necking begins, and ductile fracture becomes the dominant mode.

10-5. Fracture of High-strength Materials

A high-strength material is defined as one whose yield strength is in excess of $E/150$, a category which includes certain alloys of iron, titanium, aluminum, and magnesium. Unlike low-strength materials, which have great toughness provided that they are used above the ductile-brittle transition temperature determined by impact tests, high-strength materials have low toughness even when their mode of fracture is nominally ductile. Design of these materials employs an approach known as *fracture mechanics*, based on the modified Griffith formalism [Eq. (10-19)]. The fracture stress and the crack length are combined as a new parameter K_c, the *fracture toughness*; thus

$$K_c = \sigma_c(\pi c)^{1/2} \qquad (10\text{-}27)$$

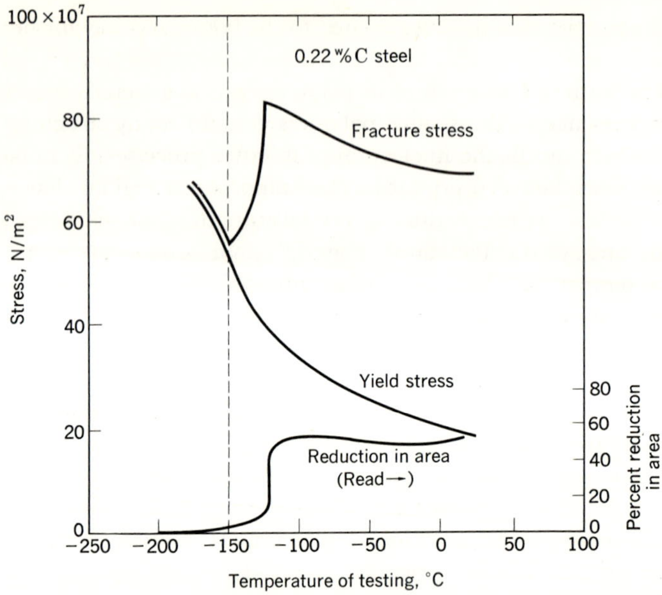

FIGURE 10-13 The ductile-brittle transition determined by tensile test data for a low-carbon steel. (*Data of G. T. Hahn, B. L. Averbach, W. S. Owen, and M. Cohen.*)

Rearrangement of Eq. (10-19) in the form

$$K_c = (2E\gamma_p)^{1/2} \tag{10-28}$$

shows that the fracture toughness depends on the theoretical strength (through Young's modulus E) and on the work of plastic deformation, γ_p. Furthermore, Eqs. (10-12) and (10-13) lead to the relation

$$2\gamma_p = \frac{1}{2}\frac{dW_e}{dc} \equiv G_c \tag{10-29}$$

The parameter G_c is a measure of the rate at which the strain energy is released upon propagation of a crack of critical size. Substitution of Eq. (10-29) in Eq. (10-28) gives the useful relation

$$K_c = (EG_c)^{1/2} \tag{10-30}$$

Fracture toughness K_c is a structure-sensitive property of the material and is responsive to treatments of various kinds.

The actual values of stress produced by given conditions of loading depend on the character of the flaws that are present and on their geometry relative to the specimen and the applied loads. The stress intensity parameter K is a useful quantity for calculations of stress distributions. Its value for a Griffith crack, for example, is $\sigma(\pi c)^{1/2}$. K and K_c are closely related in that fracture occurs when the stress intensity reaches the critical value $K = K_c$.

The fracture toughness of a high-strength material tends to decrease as its strength increases (Fig. 10-14). Therefore, the higher the strength, the greater is the susceptibility to flaws. The fracture toughness K_c decreases with increase in strength because of a decrease in the plastic work done in initiating fracture at the tip of a flaw. Fracture of the interface between an impurity particle and the matrix is often the critical step in a fracture process. If this interface is in the plastic zone at the tip of a flaw, fracture occurs when the stress-strain relation (Fig. 10-15) has been traced out to the fracture stress σ_f. The total strain produced by a given stress σ in the plastic range is given by

$$\epsilon = \frac{\sigma_y}{E} + \frac{(\sigma - \sigma_y)}{E_T} \tag{10-31}$$

The tangent modulus E_T is the slope of the plastic portion of the stress-strain curve. The strain at failure, ϵ_f, which is sensitive to the magnitude of the yield stress σ_y, is given by

$$\epsilon_f = \frac{\sigma_y}{E} + \frac{(\sigma_f - \sigma_y)}{E_T} \tag{10-32}$$

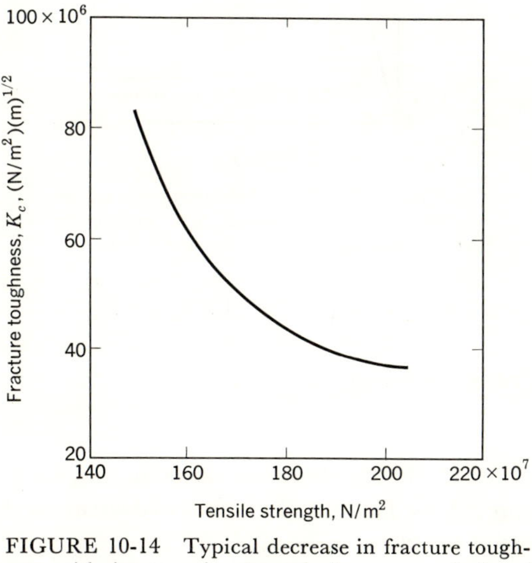

FIGURE 10-14 Typical decrease in fracture toughness with increase in strength for a quenched and tempered alloy steel [conversion of units: $M(\text{N/m}^2)(\text{m})^{1/2}\ 0.910 \times 10^{-3} = N(\text{lb/in}^2)(\text{in})^{1/2}$]

A suitable analysis, based on stress intensity factors for both the elastic and plastic portions of the fracture process, gives the following estimate for the two parts of the fracture strain when the tip radius ρ is extremely small:

$$\frac{\sigma_y}{E} + \frac{\sigma_f - \sigma_y}{E_T} = \frac{4K_c^2}{\pi E \sigma_f \rho} \tag{10-33}$$

Since E is large compared to E_T, the term σ_y/E can be neglected, and Eq. (10-33) gives the following expression for fracture toughness:

$$K_c = \left(\frac{(\sigma_f - \sigma_y)\pi E \sigma_f \rho}{4E_T}\right)^{1/2} \tag{10-34}$$

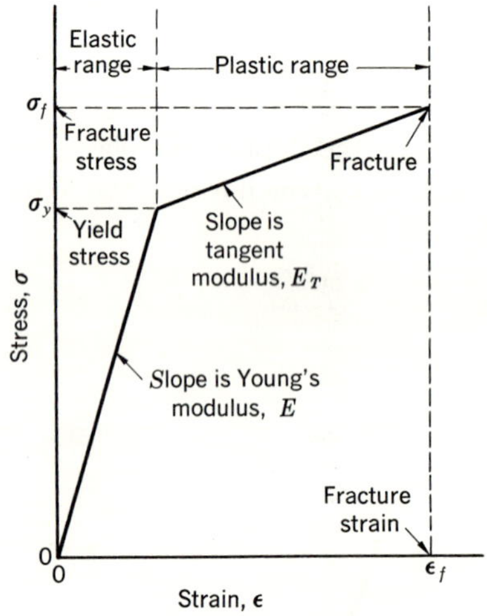

FIGURE 10-15 Schematic stress-strain curve useful for explaining fracture behavior at a sharp notch.

In view of the relation of G_c to K_c [Eq. (10-30)],

$$G_c = \frac{(\sigma_f - \sigma_y)\pi \sigma_f \rho}{4E_T} \tag{10-35}$$

Equations (10-34) and (10-35) show that fracture toughness is promoted by high strength (through the factor σ_f) but tends to decrease as the yield strength is raised (because of the factor $\sigma_f - \sigma_y$).

An important parameter affecting the fracture behavior of materials tested in sheet or plate form is the thickness of the specimen. This effect of thickness occurs for both low- and high-strength materials, but it is of particular importance in the latter case. Figure 10-16 shows the variation of fracture toughness with thickness for a high-strength aluminum alloy that had been precracked in fatigue to create sharp flaws. At a sufficient thickness, the fracture toughness reaches a minimum value K_{Ic}. Above the critical thickness a specimen containing a precrack is tested under conditions of plane strain; that is, under stress there is no appreciable strain in the thickness direction at the crack tip. A lateral contraction along the precrack front, which can occur in the case of a thinner specimen, is prevented by the mass of material adjacent to the precrack front. As in the case of elastic deformation at a notch, these

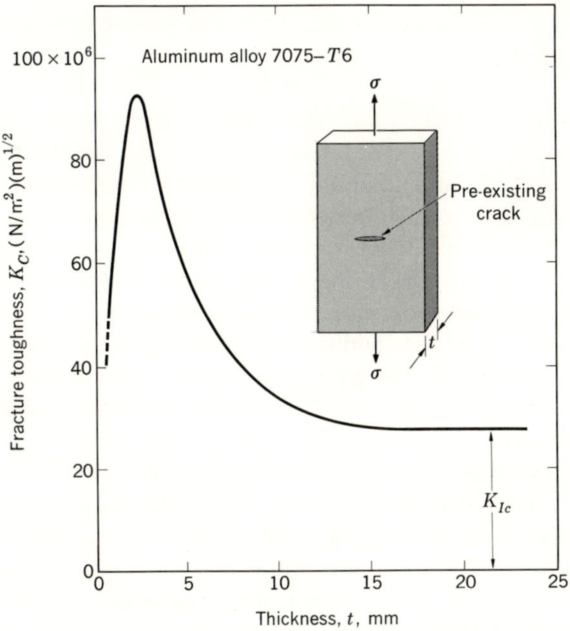

FIGURE 10-16 Variation of fracture toughness with thickness for a high-strength aluminum alloy.

constraints decrease the ability to deform plastically and reduce toughness to the K_{Ic} level. The quantity K_{Ic} is known as the *plane-strain fracture toughness* and is important in the design of heavy structures.

EFFECTS OF ENVIRONMENT AND REPEATED LOADING

10-6. Environmental Effects

The fracture behavior of materials is often influenced by environmental effects, sometimes to a marked degree. For example, certain normally ductile metals, when strained in liquid metals such as mercury, fail in a brittle manner after initial plastic deformation. In polycrystalline metals such liquid-metal embrittlement occurs along grain boundaries, but single crystals can be embrittled as well. Beneficial effects are also possible. For example, ionic crystals can undergo more plastic deformation in water than in air because of the blunting of surface flaws by dissolution. When the environment has an effect, the interaction tends to be specific for a particular combination of environment and material; for example, mercury can embrittle zinc but has little effect on magnesium. Although no general prediction of deleterious environment-material combinations is possible, known examples illustrate typical mechanisms and methods of control. Principal types of embrittlement include liquid-metal embrittlement, hydrogen embrittlement, and stress-corrosion cracking.

LIQUID-METAL EMBRITTLEMENT. Liquid-metal embrittlement occurs by a direct interaction between highly strained bonds at the tip of a stress raiser (such as a crack) and adsorbed metal atoms from the liquid. For brittle fracture to occur in a normally ductile material, the stress for bond rupture must be reduced to a level below that required for local plastic deformation. The ratio of these two stresses is thus a measure of the tendency for embrittlement. An increase in flow stress can contribute to an increase in the tendency for embrittlement. Once the condition for bond rupture has been met, crack propagation occurs rapidly during tensile loading and is limited only by the rate of flow of the liquid metal to the crack tip. A similar type of adsorption embrittlement can occur with nonmetals; for example, AgCl can be embrittled by the adsorption of $AgCl_4^{3-}$ ions. These adsorption-embrittlement processes are consistent with the Griffith treatment; the surface-energy term γ_c is greatly reduced as a result of weakening of bonds. In certain cases adsorption effects can even be beneficial. For example, in the drilling of quartz rock, if $AlCl_3$ is added to the water used as a lubricant, the speed of drilling can be doubled without increase in wear of the drill.

HYDROGEN EMBRITTLEMENT. Adsorption of hydrogen into high-strength steels can also cause embrittlement by reducing the bond strength. In this case the hydrogen diffuses within the steel to regions of high triaxial stress and causes nucleation of cracks and premature fracture. A steel specimen, not exposed in service to a hydrogen environment, may become embrittled by hydrogen introduced during previous processing. Such hydrogen can be removed by heating for a suitable time at about 200°C. Because bonds can be more highly strained the higher the strength level, the danger of hydrogen embrittlement in a given class of material increases as the overall strength becomes greater.

STRESS-CORROSION CRACKING. This type of cracking is a time-dependent phenomenon requiring the presence of a particular environment and also a tensile stress (often a residual stress). The time required for failure decreases as the stress increases (Fig. 10-17). As in the case of liquid-metal embrittlement, fracture is initiated by crack formation at the surface, followed by crack penetration into the material. Eventually the crack reaches the critical size for static failure of the remaining section. This type of fracture occurs both in crystalline and in noncrystalline materials; in metals the fracture path may be either intergranular or transgranular. There are many manifestations of this type of failure. When inorganic glass is stressed in tension in the presence of water vapor, cracking can initiate through chemical reactions at the surface (Sec. 11-1). The fracture behavior of polymeric materials is also strongly influenced by such environments as moisture, solvents, lubricants, and soap.

Stress-corrosion cracking differs from liquid-metal embrittlement and hydrogen embrittlement in that electrochemical dissolution is involved (Sec. 11-4). A crack may initiate at a surface pit, and its tip may then act as a local anode. Metal ions go into solution at the anode, and electrons flow from the

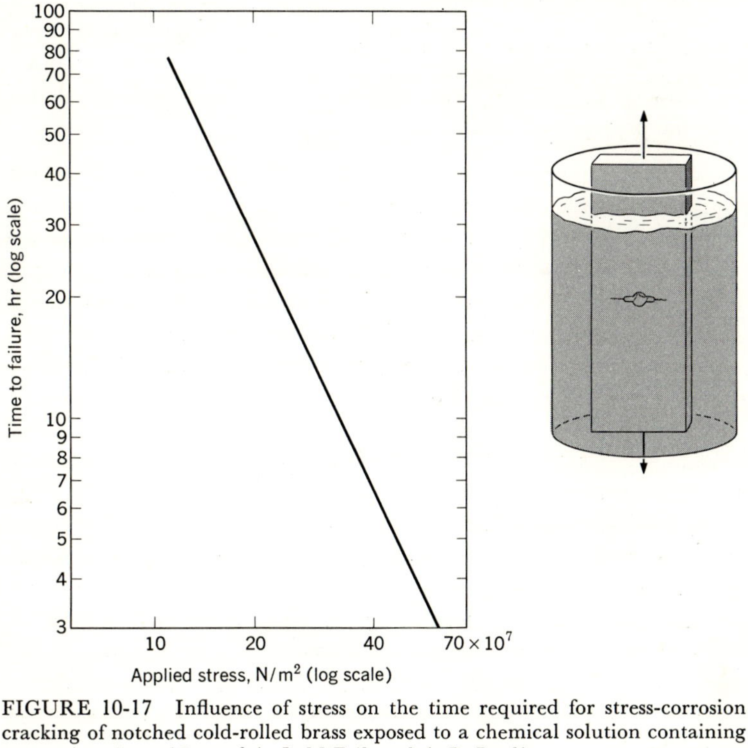

FIGURE 10-17 Influence of stress on the time required for stress-corrosion cracking of notched cold-rolled brass exposed to a chemical solution containing ammonium ions. (*Data of A. J. McEvily and A. P. Bond.*)

anode to a nearby region on the specimen surface which serves as a cathode. Stress probably acts to break down the protective surface films that are present initially. Once the crack has started to grow, plastic deformation at the tip of the crack either prevents the film from reforming or breaks it down again should it reform. In addition, local tearing of the material by the stress aids in advancing the crack. For a variety of metals the reaction at the crack-tip anode is of the type

$$Me^+ + H_2O \rightarrow MeOH + H^+ \tag{10-36}$$

Subsequent adsorption of the H^+ ion may also contribute to embrittlement and crack advance.

When a material must be used in an adverse environment, the fracture-mechanics approach can be used to minimize the danger of failure. Precracked high-strength steels tested in a salt solution (such as sea water) are subject to stress-corrosion cracking only if the stress intensity K is above a minimum value K_{SCC}. For stress intensities below K_{SCC} no further growth of cracks occurs (Fig. 10-18). If the initial value of K is above K_{SCC}, the existing flaws grow with time and cause the stress intensity to increase. Fracture eventually occurs when K equals the fracture toughness K_c. For safe design of structures that may contain flaws, the maximum initial value of K must be less than K_{SCC}. A suitable method of inspection can eliminate material containing detectable flaws, but the presence of smaller flaws (below the limit of detection) must be

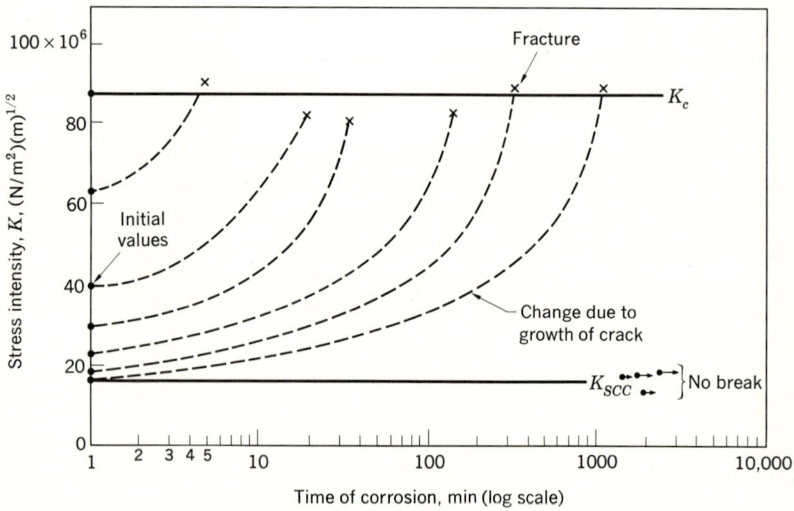

FIGURE 10-18 Effect of initial stress intensity K on crack growth and eventual fracture in precracked samples of an alloy steel exposed to a $3\frac{1}{2}\%$ solution of sodium chloride. (*Data of F. L. Brown and C. D. Beachem.*)

taken into account. The maximum size of such flaws is used in calculating the stress intensity K corresponding to a given applied stress. The value of K_{SCC} is found to decrease with increase in yield strength, but it can be substantially increased in steels by suitable alloying, such as the addition of 10 to 20 w/o nickel.

10-7. Fatigue

The failure of a material due to repeated stressing is known as *fatigue*. A common example is the failure of a paper clip after to-and-fro bending. Of course, fatigue is important in the design and operation of motor vehicles, power-generating equipment, and aircraft. For this reason it has been the subject of intensive study, and much is known about the failure process.

Fatigue-testing procedures (Fig. 9-18*a*) determine the number of cycles, N, required to produce failure when a specimen is subjected to a specified amplitude of either stress or strain. For stress amplitudes close to the tensile strength, failure occurs in a few cycles (as with the paper clip). As the stress amplitude is decreased, the number of cycles to failure increases markedly (Fig. 9-18*b*), and as many as 10^8 cycles may be required. For still lower stress amplitudes many materials exhibit a fatigue limit and are then in a safe range of cyclic load. A useful distinction is made between low-cycle fatigue (N less than about 10^4) and high-cycle fatigue. Plastic strain is of major importance in low-cycle fatigue, and the mechanism of fracture is related to that in static tension. The results of low-cycle fatigue tests on various metals correlate very well when analyzed in terms of the applied plastic strain (Fig. 10-19). The fatigue lifetime in this range can be expressed as

$$N = C\epsilon_p^{-n} \tag{10-37}$$

where ϵ_p is the plastic strain amplitude and the exponent n has a value close to 2. The constant C increases with increase in the ductility of the material.

In both high-cycle and low-cycle fatigue, the failure process involves plastic deformation. In high-cycle fatigue, the amount of plastic deformation required for fatigue fracture need not be large, and in many engineering materials fatigue failures can occur well below the nominal yield strength of the material (for example, in aluminum alloys). Heterogeneities within the alloy structure permit plastic deformation to occur in areas of local weakness. Such areas thus contribute to a decrease in fatigue resistance although they do not significantly affect static strength. Fatigue is therefore a particularly structure-sensitive property. The deleterious effects of inhomogeneities are lessened by procedures such as shot-peening or nitriding, which strengthen the crucial surface layers. The elimination of corrosive environmental constituents can also have a beneficial effect on fatigue behavior.

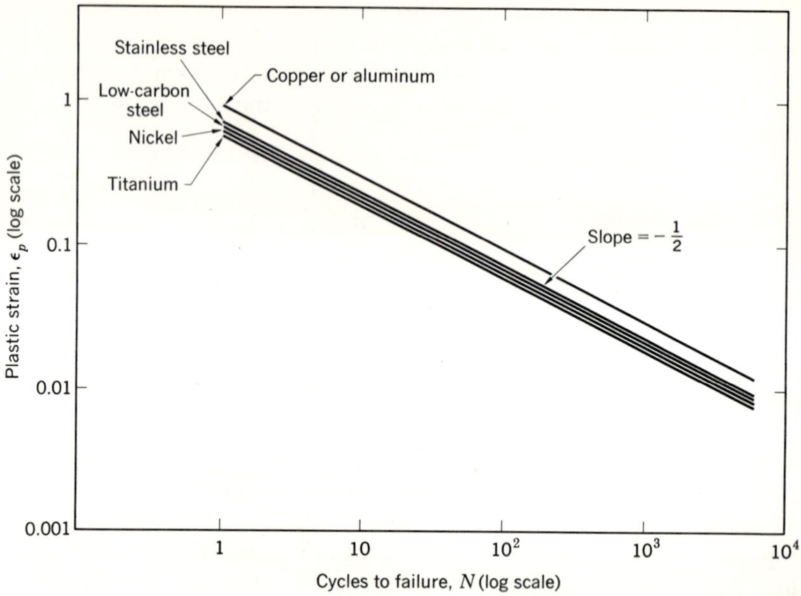

FIGURE 10-19 Relationship between plastic strain and failure in low-cycle fatigue. (*Data of L. F. Coffin and J. F. Tavernelli.*)

The fatigue process in a clean ductile metal can be divided into three stages: crack initiation, crack propagation, and tensile failure when the crack reaches a critical size. Crack initiation occurs because plastic deformation is not completely reversible during cyclic loading. Consequently, a notch-peak topography develops in the slip bands which serves to concentrate further deformation (Fig. 10-20). Cracks form at the base of the slip-band notches and propagate into the material parallel to the operative slip plane. This slip-plane cracking is known as *stage I* of the crack growth process. As the crack lengthens, its growth is controlled more by tensile than by shear stresses, and it turns and grows in a plane perpendicular to the maximum tensile stress. This stage of growth, known as *stage II*, terminates in tensile fracture.

In crack growth in stage II, the tip of the crack becomes blunt during each cycle of loading, and thus the amount of advance of the crack is limited. The crack is resharpened during subsequent unloading and acts to concentrate the stress for the next cycle of loading. As the crack advances an incremental amount per cycle, a characteristic fatigue striation (Fig. 10-21) is formed on the fracture surface. These markings are useful in failure analysis. The rate at which the crack advances is dependent upon the applied stress intensity factor K. Rates of growth vary from as low as 10^{-10} m/cycle up to 10^{-4} m/cycle at the onset of terminal failure.

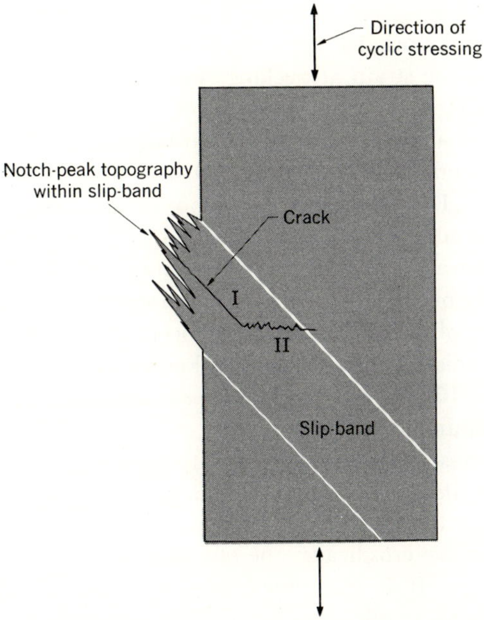

FIGURE 10-20 Schematic illustration of the two stages of growth of a fatigue crack following its initiation at the surface.

FIGURE 10-21 Example of fatigue striations in 2024 aluminum alloy. (*Courtesy of C. D. Beachem, U.S. Naval Research Laboratory.*)

PROBLEMS

10-1. The incremental true strain $d\epsilon$ is dL/L, where L is the actual length [Eq. (9-9)]. Show that $dL/L = -dA/A$ provided that the volume can be assumed to remain constant, that is, provided $A_0L_0 = AL$, where A_0 and L_0 are the original area and length.

10-2. Show that the strain at necking (for a material obeying the relation $\sigma = K\epsilon^n$) is given by $\epsilon = n$.

10-3. (*a*) Show that the condition for the beginning of necking, $d\sigma/d\epsilon = \sigma$ (where σ and ϵ are true stress and true strain), can be expressed as $d\sigma/d\epsilon^0 = \sigma/(1 + \epsilon^0)$, where ϵ^0 is the nominal strain $\Delta L/L_0$. (*b*) Show, using a suitable sketch, that the graphical interpretation of this condition is a straight line through the point ($\sigma = 0$, $\epsilon^0 = -1$) and tangent to the curve of true stress versus nominal strain. At the point of tangency the values of σ and ϵ^0 are those corresponding to the beginning of necking. This graphical method is known as *Considere's construction.*

10-4. Prove Eq. (10-5). Assume that $ds = dL$, and use the relations $\sigma = F/A$, $d\epsilon = dL/L$, and $V = AL$.

10-5. Typical stress-strain curves, such as those of Fig. 10-1*a*, are determined using a tensile tester that strains (elongates) the specimen at a constant rate (about 10^{-3}/sec). The two ends of the specimen are firmly held in two massive crossheads, one of which moves slowly with respect to the other. (*a*) If, instead of a tensile tester, heavy direct weights were used to load a tensile specimen (called *deadweight* loading), show that the stress-strain curve so obtained would terminate at F_{max}. (*b*) A "stiff" tensile tester is one that experiences little deflection of the crossheads under load (and therefore stores little elastic energy). Explain why the load F at which fracture occurs is lower the greater the stiffness of the testing machine employed. (*c*) Fracture of a container holding gas under pressure is much more violent than fracture of a similar container holding a pressurized liquid. Explain.

10-6. In brittle fracture the total change in energy during extension of a crack is the algebraic sum of Eqs. (10-11) and (10-12). For a glass containing a crack of length $2c = 0.5 \times 10^{-3}$ m, calculate the stress required to produce failure, assuming that the only barrier to growth of a crack is the creation of new surface.

10-7. Consider the growth of a crack under a fixed load F, rather than under a fixed restraint (Fig. 10-8). (*a*) Using a sketch similar to Fig. 10-8*b*, show that the stored elastic energy increases by $Fd(\Delta L)$ as the crack increases in length by dc and the specimen elongates by $d(\Delta L)$. (*b*) Show that the elastic energy available for creation of new fracture surface resulting from growth dc of a crack is the same for a fixed load as for a fixed restraint. [HINT: Use the relation $\sigma = E\epsilon$ or its equivalent $F = K\,\Delta L$, where K is a spring constant.]

10-8. Using Eq. (9-5), show that Eq. (10-20) gives the elastic displacement U (or Δ) in a distance L equal to the diameter d of a grain.

10-9. As shown in Fig. 10-16, the thickness of a specimen affects its fracture toughness K_c and therefore the critical stress for fracture σ_c. Making use of Eq. (10-27), calculate and plot the critical stress as a function of thickness in the range 5 to 20 mm. Assume that $c = 3$ mm.

10-10. A metallic material containing initial flaws no larger than 10^{-4} m in radius is stressed at 60×10^7 N/m^2 in a corrosive environment. Determine the minimum value of K_{SCC} needed for this material to avoid stress-corrosion cracking.

10-11. *S-N* curves of fatigue behavior are usually determined experimentally. The following empirical equation (valid for wide aluminum sheet, cycled over a range of stress between zero and σ) permits the calculation of *S-N* curves for flaws of various initial size $c_{\min}$ (CMIN). The rate of crack growth per cycle, dc/dN, is given by

$$\frac{dc}{dN} = \frac{0.08}{E\sigma_y}(\sigma^2 c - 6.4 \times 10^{12}) \tag{10-38}$$

E is Young's modulus (7×10^{10} N/m^2 for aluminum), and σ_y is the yield strength (YS). Equation (10-38) can be integrated between $c_{\min}$ at zero cycles and the final crack size, c_f (FINALC) at the number N of cycles at which failure occurs. The resulting equation can be written

$$N = \frac{E\sigma_y}{0.08\,\sigma^2} \ln \frac{\sigma^2 c_f - 6.4 \times 10^{12}}{\sigma^2 c_{\min} - 6.4 \times 10^{12}} \tag{10-39}$$

Use the following computer program to calculate the number of cycles, N, at a given stress σ for a crack to grow from an initial size CMIN $= 2 \times 10^{-3}$ m to the size at which failure occurs, assumed to be FINALC $= 2.5 \times 10^{-2}$ m. Use values of SIGMA from 36×10^7 N/m^2 down to 6×10^7 N/m^2 in steps of 6×10^7 N/m^2. Assume that YS $= 41 \times 10^7$ N/m^2. Plot the data as an *S-N* curve, and estimate the fatigue strength.

```
C  CALCULATION OF S-N CURVE
      REAL N
   2  YS = 41.0 E + 7
   3  E = 7.0 E + 10
   4  CONST = 6.4 E + 12
   5  CMIN = 2.0 E - 3
   6  FINALC = 2.5 E - 2
   7  GAMMA = (E * YS)/0.08
```

```
C  CALCULATION
    8  DO 14 I = 6,36,6
    9  SIGMA = I * 10 ** 7
   10  SIGMA2 = SIGMA * SIGMA
   11  N = (GAMMA/SIGMA2) * ALOG((SIGMA2 * FINALC-CONST)/
                                          (SIGMA2 * CMIN-CONST))
C  OUTPUT SECTION
   12  WRITE (6,13) SIGMA, N
   13  FORMAT (1X, 'SIGMA =', E9.2, 2X, 'N =', E9.2)
   14  CONTINUE
   15  STOP
       END
```

REFERENCES

Andrews, E. H.: "Fracture in Polymers," American Elsevier Publishing Company, New York, 1968.

Bonis, L. J., et al. (eds.): "Fundamental Phenomena in the Material Sciences: vol. 4, Fracture of Metals, Polymers, and Glasses," Plenum Press, New York, 1967.

Gassner, E., and W. Schutz (eds.): "Fatigue Design Procedures," Pergamon Press, Oxford, 1969.

Liebowitz, H. (ed.): "Fracture," vols. VI and VII, Academic Press Inc., New York, 1968–1969.

Madayag, A. F. (ed.): "Metal Fatigue: Theory and Design," John Wiley & Sons, Inc., New York, 1969.

Pratt, P. (ed.): "Fracture," Chapman & Hall, Ltd., London, 1969.

Tetelman, A. S., and A. J. McEvily: "Fracture of Structural Materials," John Wiley & Sons, Inc., New York, 1967.

Chapter Eleven

Deterioration of Materials

A significant factor in the use of a given material is its resistance to deterioration under ordinary conditions of service. The economic loss because of deterioration is enormous—estimated at more than \$10 billion per year in the United States—and therefore methods of combating deterioration are of great importance. Typical methods employed to reduce the rusting of automobile bodies, for example, are more effective protection of the surface (painting, etc.), modification of the composition of the sheet steel (alloying), replacement of the steel by a corrosion-resistant material such as aluminum or a plastic, or modification of the corroding environment (using noncorrosive agents on icy streets). Effective use of the various methods for reducing deterioration requires a knowledge of the underlying mechanisms by which deterioration occurs. These are primarily chemical in nature and may be considered under the following three types of reactions: (1) local chemical reactions, characteristic of polymers and ceramics, (2) electrochemical reactions in a liquid environment (e.g., ordinary corrosion), and (3) electrochemical reactions in a dry environment (e.g., high-temperature oxidation). An electrochemical reaction differs from an ordinary chemical reaction in that an electric current flows an appreciable distance in the system undergoing the reaction.

CHEMICAL REACTIONS OF POLYMERS AND OF CERAMICS

Compared to the most common metal, iron, nonmetallic materials are generally resistant to deterioration. In particular, they are immune to electrochemical corrosion, which causes rusting of iron. Nevertheless, under certain conditions nonmetals experience deterioration of other kinds. The examples in Fig. 11-1 show how the serviceability of concrete and of a plastic can be impaired by chemical reactions. The many types of deteriorative reactions will be treated

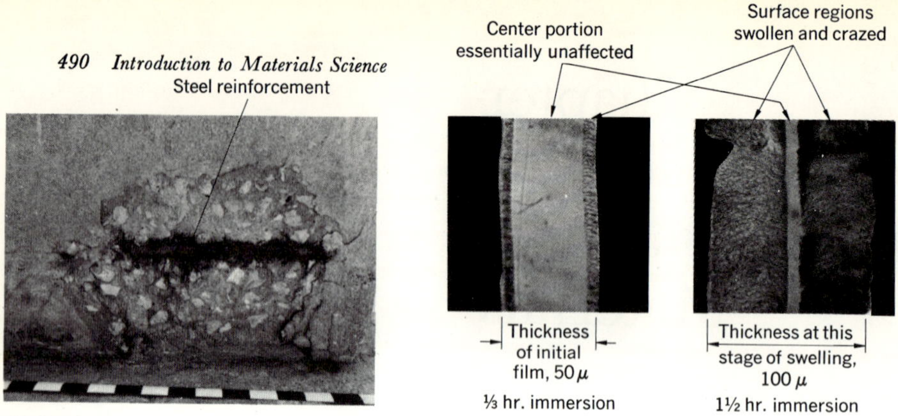

(*a*) Chemical attack of concrete and its steel reinforcement. (Courtesy American Concrete Institute).

(*b*) Swelling and resulting cracking (crazing) of a polystyrene film immersed in *n*-heptane at 35°C. (Courtesy H. B. Hopfenberg, North Carolina State University).

FIGURE 11-1 Examples of deterioration of nonmetallic materials.

under the following three headings: (1) reactions of components, (2) breaking of polymer chains, and (3) dissolution and swelling.

11-1. Reactions of Components

Ordinary glass is an example of a single-phase, homogeneous material, but some of its components are especially susceptible to chemical attack. The alkali oxides Na_2O and K_2O present in low-cost glasses (see Table 5-2) are not fully incorporated into the silicate network; instead, an alkali ion creates a "hole" in the network by forming a single bond with an oxygen ion, as shown schematically in Fig. 5-2*c*. When water is in contact with the surface of the glass, the alkali atoms adjacent to the surface undergo the following reaction:

$$\boxed{\text{Network}}\text{—Si—O—Na} + H_2O \rightarrow \boxed{\text{Network}}\text{—Si—OH} + Na^+ + OH^- \tag{11-1}$$

Under many conditions of service the OH^- ion is removed from the reacting system, for example, by rain against a window pane or through neutralization by an acid beverage inside a bottle. Further reaction then requires the slow diffusion of alkali ions from interior locations to the surface of the glass, and thus the rate of attack of the glass gradually decreases. Additions of CaO, MgO, and Al_2O_3 to a glass increase the resistance of the alkali ions to solution.

Under some conditions dissolution of the alkali ions is accompanied by chemical attack of the silicate network, a more serious type of deterioration. The OH^- ions produced by the reaction in Eq. (11-1) may remain in contact with the glass because of dirt on the surface or because a pocket of moisture has formed between two pieces of glass stored in a poorly ventilated warehouse. The silicon-oxygen bond can then be broken by the reaction

$$\boxed{\text{Network}}\text{—Si—O—Si— } + OH^- \rightarrow \boxed{\text{Network}}\text{—Si—OH} + \text{Si—O}^- \tag{11-2}$$

This type of reaction eventually forms a cloudy surface on the glass beneath a layer of siliceous gel. An alkaline solution can also furnish the OH^- ions needed for this reaction. Ordinary glass bottles are attacked during long storage by even moderately alkaline solutions, and eventually flakes of siliceous gel appear in the solution. Suitable additions, such as ZrO_2, can improve the resistance of glass to this type of deterioration. Glass is stable in the presence of most acids, but the silicate network is strongly attacked by hydrofluoric acid and by hot concentrated phosphoric acid.

Most ceramic materials consist of more than one phase, and each phase has its own chemical properties. In phase-separated glasses, for example, one phase may be prone to chemical attack. Concrete is a multiphase ceramic whose deterioration is of special interest in view of its exposure to weathering in highways, to the sea in harbor facilities, and to chemicals in large manufacturing or sanitary installations. Either external attack or internal reactions can cause deterioration of concrete. In both cases durability depends on the chemical composition of the aggregate (sand, gravel, etc.) and of the cement (portland, high-alumina, etc.) To resist external attack, a concrete must be impermeable to reactive solutions. This property depends on proper formulation, mixing, compaction, and curing of the concrete. Typical of chemicals that attack the surface of good-quality concrete are sulfate ions present in the soil, in ground water, or in sea water. The sulfate first reacts with tri-calcium aluminate in concrete made with portland cement; then a subsequent reaction with free lime produces a bulky product which cracks the concrete and promotes additional attack. Concrete made with cement containing less than 6 percent tri-calcium aluminate has good resistance to sulfate solutions.

Internal reactions sometimes occur between the aggregate and the cement or between a reinforcement and the concrete. Aggregates such as flint or volcanic rock can react with Na_2O in the cement to form a bulky silicate that cracks the concrete. If reactive aggregates must be used, the alkali content of the cement should be less than 0.6 percent. The steel rods employed as reinforcement in concrete are initially protected from corrosion by an alkaline environment (pH about 12), provided that the surrounding concrete has been made with portland or other alkaline cement. In the course of time, however, moisture is able to permeate the protective layer of concrete, usually a few centimeters in thickness. Carbon dioxide dissolved in the moisture reacts with the free lime in the concrete and lowers the pH. When the steel corrodes, the bulky oxides fracture the concrete and accelerate the deteriorative processes.

11-2. Breaking of Polymer Chains

Polymeric materials are like ceramics in being susceptible both to chemical reaction of their separate components and to disintegration of the main framework, the long-chain molecules in this case. Because of the chemical complexity of commercial polymers, their deterioration is only partially understood, but scission (breakage) of polymer chains is known to be of major importance. Deterioration is usually investigated by observing the changes in mechanical or dielectric properties during exposure to a specific environment. For example, Fig. 11-2 shows that prolonged outdoor exposure has quite different effects on the mechanical properties of two common plastics.

A convenient starting place for the study of scission is thermal degradation, produced by heating. The relative bond energies (see Table 1-1) C–F >

TABLE 11-1. THERMAL DEGRADATION OF POLYMERS

Polymer	Monomer unit	Percent of monomer in volatile products	T_h†, °C [°F]	K_{350}‡, %/min
Polytetrafluoro-ethylene (e.g., Teflon)	F F | | C=C | | F F	100	509 [948]	5.2×10^{-6}
Polyethylene	H H | | C=C | | H H	Negligible	414 [777]	0.004
Polymethyl-methacrylate (e.g., Lucite)	H CH_3 | | C=C | | H $COOCH_3$	100	327 [617]	5.2
Polyvinyl-chloride	H Cl | | C=C | | H H	Negligible	260 [500]	170

† Temperature at which a polymer loses half of its weight during 30 min heating in vacuum.
‡ Rate of volatilization at 350°C [662°F], percent loss of weight per minute.

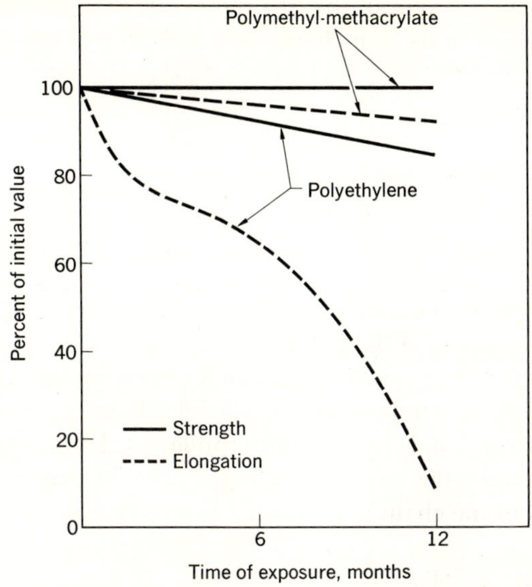

FIGURE 11-2 Reduction in strength and ductility of high-density polyethylene and of injection-molded polymethyl-methacrylate during prolonged outdoor exposure. (*Data of H. M. Quackenbos and H. Samuels, and of R. B. Barrett, respectively.*)

C–H > C–Cl explain qualitatively the thermal stability of the polymers listed in Table 11-1. One type of thermal degradation, illustrated by the behavior of polymethyl-methacrylate, involves the production of the monomer M_1. The reaction begins at the end of a polymer chain of length n with the production of two free radicals $P^\cdot$; thus

$$M_n \rightarrow P^\cdot_{n-1} + P^\cdot_1 \tag{11-3}$$

The monomer radical $P^\cdot_1$ eventually converts to the monomer. The long-chain radical undergoes a so-called *depropagation reaction* (the reverse of the initial polymerization reaction); thus

$$P^\cdot_{n-1} \rightarrow P^\cdot_{n-2} + M_1 \tag{11-4}$$

to produce a molecule M_1 of the monomer. This reaction can repeat until the entire polymer chain is consumed. Polyethylene behaves quite differently; scission of a chain occurs randomly along the length of the polymer. The initiation of scission is now random, giving the reaction

$$M_n \rightarrow P^\cdot_j + P^\cdot_{n-j} \tag{11-5}$$

and the subsequent reaction of the free radicals is at a random location in a neighboring chain or in its own chain; thus

$$P_j^{\cdot} + M_n \rightarrow M_j + P_n^{\cdot} \tag{11-6}$$

or

$$P_n^{\cdot} \rightarrow P_i^{\cdot} + M_{n-i} \tag{11-7}$$

The result is a rapid reduction in the average chain length and corresponding changes in the properties of the polymer.[1]

Two factors acting jointly, oxygen and ultraviolet light, cause polymers to deteriorate at room temperature. Ultraviolet light (or other high-energy radiation) can break bonds in a manner similar to thermal energy. In polyethylene, for example, the terminal hydrogen atom H may be removed from the remainder R of the chain to produce a free radical $R^{\cdot}$; thus

$$R\text{—}H \xrightarrow{\text{(radiation)}} R^{\cdot} + (H^{\cdot}) \tag{11-8}$$

In the absence of oxygen (or other reactants) this type of reaction leads to relatively slow degradation of the polymer, but the presence of oxygen permits continued degradation by such reactions as the following:

$$R^{\cdot} + O_2 \rightarrow ROO^{\cdot} \tag{11-9}$$

$$ROO^{\cdot} + RH \rightarrow ROOH + R^{\cdot} \tag{11-10}$$

The rate of oxidative deterioration of polymers can be greatly reduced by incorporating antioxidants (amines or phenols, for example) in the polymer. Let AH represent the antioxidant molecule including a reactive hydrogen atom. Free radicals produced by the reaction in Eq. (11-9) are eliminated as follows:

$$ROO^{\cdot} + AH \rightarrow ROOH + A^{\cdot} \tag{11-11}$$

The $A^{\cdot}$ free radical has the property that it does not propagate the deteriorative reactions.

Some polymers (synthetic or natural rubber, in particular) contain regularly spaced doubly bonded carbons that are especially susceptible to

[1] The fumes produced by thermal degradation of polymers are sometimes disagreeable and in rare instances may be dangerous. In particular, polyvinyl-chloride, polyvinylidene-chloride, and neoprene, which are often used in technical laboratories, can give off small amounts of the poisonous gas phosgene, $COCl_2$.

attack by ozone, O_3. Scission of the main chain can be caused by a reaction such as the following:

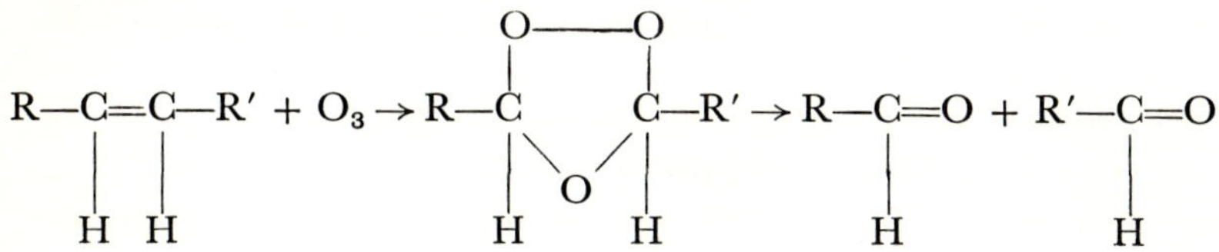

(11-12)

R and R′ represent portions of the polymer chain. Although the scission products can cause additional reactions, if the polymer specimen is unstressed a reacted layer forms on the surface and protects the bulk of the specimen from deterioration. However, most rubber-type polymers (elastomers) are subjected to stress in use, and under these conditions the deteriorative reactions cause the formation of deep crevices and eventual rupture. Various additions to the chemical composition of elastomers effectively decrease the rate of attack by ozone; antiozonants, such as primary diamines, act similarly to antioxidants.

Suitable treatments of the surface of ceramics or polymers can improve such properties as their resistance to deterioration without affecting their bulk mechanical or physical behavior. For example, ceramics used as dielectrics may be given a glaze rich in barium ions to reduce the atmospheric leaching of sodium ions, thus preventing the formation of a conducting electrolyte on the surface. The surface of polymers may be given a variety of chemical treatments (with oxygen, acids, alkalies, or halogens, for example) to improve numerous properties. Oxidation of some polymers improves their ability to retain printed words or designs on their surfaces; that is, their adhesive characteristics are improved. A typical effect of treatment of polyethylene with fluorine gas is to improve resistance to chemical attack.

11-3. Dissolution and Swelling

Nonmetallic materials are commonly degraded by a gaseous or liquid environment because of either incorporation of a gas or liquid into the material or solution of the material in a solvent. Ceramics, for example, may be slowly dissolved at high temperatures by molten salts or molten metals. Some clays possess the useful property of taking up water, swelling, and becoming more pliable. The behavior of polymers is especially significant because exposure to deleterious solvents commonly occurs in service. The polymers most susceptible to dissolution by organic liquids are glassy, not cross-linked, and nonpolar. Complete dissolution occurs in some cases (such as polystyrene in benzene), but more often the liquid has limited solubility in the polymer (acetone in polystyrene) and causes the changes in dimensions and properties denoted by the term *swelling* (Fig. 11-1*b*). The glass transition temperature of the polymer decreases with increase in amount of solvent, and thus an initially strong plastic becomes rubbery as swelling progresses.

The extent of solution of an ordinary glassy polymer in a given solvent can be estimated from the solubility parameter $(\Delta E/V)^{1/2}$ [Eq. (2-54)]. If the solubility parameters of the polymer and solvent are nearly the same, the polymer will dissolve completely. For differences in solubility parameters greater than about one unit (see Prob. 11-1) the solubility decreases progressively. A cross-linked polymer will not dissolve completely but is susceptible to swelling. A crystalline polymer is almost inert even in a solvent with the same solubility parameter. However, amorphous regions in highly crystalline polymers can experience a limited amount of swelling. When polar bonds are present in a polymer, then polar organic liquids such as phenol are especially effective solvents and may even attack crystalline structures.

The ability of a polymer to resist deterioration is determined by many factors; temperature, applied stresses, irradiation, and chemical environment are especially important. Also, a given designation such as *polyethylene* includes a wide range of materials with correspondingly varying properties. Furthermore, suitable additions can greatly change specific properties of a given polymer; for example, carbon black imparts excellent resistance against sunlight to high-density polyethylene. In view of these complexities, quantitative data on the deterioration of polymeric materials must usually be obtained by controlled testing under conditions similar to those encountered in service. Qualitative comparisons of various polymers, such as Table 11-2, may be useful as a general guide in choosing a suitable material, provided that the limitations of such comparisons are recognized.

ELECTROCHEMICAL FUNDAMENTALS

The corrosion of metals and semiconductors involves the flow of an electric current within the material. Since deterioration depends on electrical as well as chemical factors, its study requires the methods of electrochemistry. The simplified model of Fig. 11-3 shows the analogy between corrosion and the operation of an electrochemical cell. The essential corrosion process ($Fe \rightarrow Fe^{++}$, in the case of iron) corresponds to the anode in the cell. At the surface of the anode, electrical charge (in the form of the Fe^{++} ion) is transferred from the metal to the solution. Since the initially neutral anode has lost two electronic units of positive charge, it tends to be negatively charged by the presence of two excess electrons. This accumulation of charge soon prevents further

Condition of service	Type of polymer														
	ABS (acrylonitrile-butadiene-styrene), high-impact	Polymethyl-methacrylate	Cellulose acetate	Polystyrene, atactic	Polyvinyl-chloride	Epoxy resins, aluminum-filled	Polycarbonate, unfilled	Polyester, thermoplastic	Polyhexamethylene-adipamide (e.g., Nylon 6/6)	Silicones, cast resin	Polyethylene, high-density	Melamine-formaldehyde, alpha-cellulose-filled	Ionomers	Polytetrafluoro-ethylene (e.g., Teflon)	Polypropylene, unmodified
Temperature limit for continuous service, °C	70–100	70–100	60–105	65–75	65–80	95–260	120	80–120	80–150	200	120	100	70–105	290	120–160
Water absorption, %, 3-mm sheet in 24 hr	0.2–0.45	0.3–0.4	2–7	0.03–0.1	0.07–0.4	0.1–4	0.15	0.02	1.5	0.1	< 0.01	0.1–0.6	0.1–1.4	0	< 0.01
Resistance to attack by: Sunlight	G‡	G	F	F	F	G	F	F	F	G	P	F	F	G	P
Weak acids	G	G	F	G	G	G	G	G	G	G	G	G	F	G	G
Strong, oxidizing acids	P	P	P	P	G	F	P	F	P	P	F	—	P	G	F
Weak alkalies	G	G	F	G	G	G	F	G	G	G	G	G	G	G	G
Strong alkalies	G	P	F	G	G	F	P	P	G	P	G	—	G	G	G
Organic solvents	Attacked by ketones, esters, aromatic and chlorinated hydrocarbons					Attacked by chlorinated hydrocarbons			Dissolves in certain solvents		G	G	G	G	G

† Adapted from more detailed tables in "Modern Plastics Encyclopedia," pp. 84–101, October, 1968.
‡ G = good resistance; F = fair resistance; P = poor resistance.

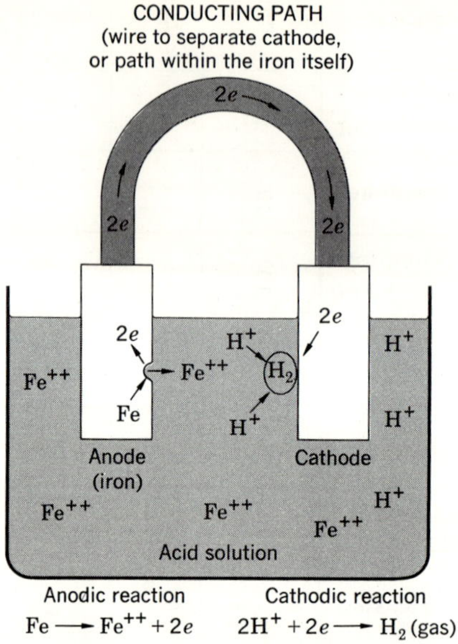

FIGURE 11-3 Schematic illustration of the two electrode processes involved in the corrosion of iron in an acid solution.

reaction unless the anode can act as a sink for electrons and remove the negative charge. The electrons removed from the region of the anodic reaction flow to the cathode, which can then serve as a source of electrons for the cathodic reaction (the reduction of H^+ ions to hydrogen atoms in this case). Corrosion is thus analogous to the action of an electrochemical cell composed of two electrodes; the current flowing between the two electrodes is a measure of the rate of corrosion. Depending on the nature of the corroding material (metal or semiconductor) and of the corrosive environment (dry air, aqueous solution, etc.), the electrode reactions vary greatly, and transfer of charge may involve not ions, but electrons, holes, or even protons.

11-4. Equilibrium Aspects

The corrosion process of Fig. 11-3 can be pictured as two electrode processes as follows:

Anodic process: $$Fe \rightarrow Fe^{++} + 2e \tag{11-13}$$

Cathodic process: $$2H^+ + 2e \rightarrow H_2 \tag{11-14}$$

or as the overall reaction [obtained by adding Eqs. (11-13) and (11-14)]

$$Fe + 2H^+ \rightarrow Fe^{++} + H_2 \qquad \Delta G^o = -20.3 \text{ kcal} \tag{11-15}$$

The standard free-energy change ΔG^o for the overall reaction at 25°C is obtained by the method of Sec. 2-5. Because Eq. (11-15) is an ordinary chemical reaction, it does not include the effects of applied electrical potentials or of external sources of current. However, these important aspects of electrochemical reactions can be investigated for the separate electrode reactions.

SINGLE-ELECTRODE REACTIONS. Imagine the hypothetical experiment in which a piece of iron is placed in water and only the reaction of Eq. (11-13) occurs. Since the electrons cannot be conducted away, an electrical potential φ called the *Galvani potential* is quickly established between the interior of the metal and the bulk of the solution. Consequently, an Fe^{++} ion can be added to the solution only so long as the following combined process is energetically possible: (1) formation of Fe^{++} at the surface, releasing chemical free energy ΔG_s, where the subscript s identifies a quantity referring to a single-electrode reaction, and (2) traversal by the Fe^{++} ion of the region of electrical potential, requiring electrical work equal to $2e\varphi$, where e is the charge on an electron. At equilibrium,

$$\Delta G_s = 2e\varphi \tag{11-16}$$

since the chemical free energy must supply the energy for the electrical work. This relation can be put in a more useful form by employing the standard molar free energy ΔG_s^o [see Eq. (2-21)], correspondingly replacing e by the Faraday constant F [23,060 cal/(volt) (g equiv)], and using z to indicate the ionic charge (2 for Fe^{++}). The result is

$$\Delta G^o + RT \ln K_s = zF\varphi \tag{11-17}$$

where K_s is the equilibrium constant for the single-electrode process; $K_s = a_{Fe^{++}}$ for the reaction in Eq. (11-13), for example. Since ΔG_s^o corresponds to a potential φ^o by Eq. (11-16), Eq. (11-17) can be written

$$\varphi = \varphi^o + \frac{RT}{zF} \ln K_s \tag{11-18}$$

The Galvani potentials φ are useful theoretical quantities but cannot yet be evaluated individually. However, a conveniently measurable quantity is the difference $E = \varphi - \varphi_{\text{ref}}$ between the potential of interest and that of a reference electrode; E is called the *electrode potential.* If the standard hydrogen electrode[1] is used as a reference and φ_{ref}^o is arbitrarily set equal to zero, then

$$E = \varphi - \varphi_{\text{ref}} = E^o + \frac{RT}{zF} \ln K \tag{11-19}$$

[1] In the standard hydrogen electrode, hydrogen gas at 1 atm pressure is in equilibrium with hydrogen ions at unit activity. K_s for this single electrode is unity. Equilibrium is maintained at the especially reactive (platinized) surface of a platinum electrode, which does not participate in the chemical reaction.

E is the electrode potential for a given concentration of the solution, and E° is the *standard* electrode potential for unit activity (that is, about unit molarity) of the ions in solutions. Use of a potentiometer permits a determination of E under conditions of equilibrium when no net current is flowing in the cell. The standard value E° for the $Fe|Fe^{++}$ electrode is -0.440 volt. Standard values for other electrodes are tabulated in the electromotive series (Table 11-3).

Before considering the quantitative application of electrode reactions to corrosion behavior, let us see how the electromotive series can be interpreted in a rough, qualitative manner. When the potential of iron is maintained at -0.440 volt, the Fe^{++} ions in an aqueous solution at unit activity are at equilibrium (Fig. 11-4). If this piece of iron is electrically connected to a Cu/Cu^{++} electrode, the potential of the iron will tend to rise (the actual amount of rise is considered later). The Fe^{++} ions can then be pictured as

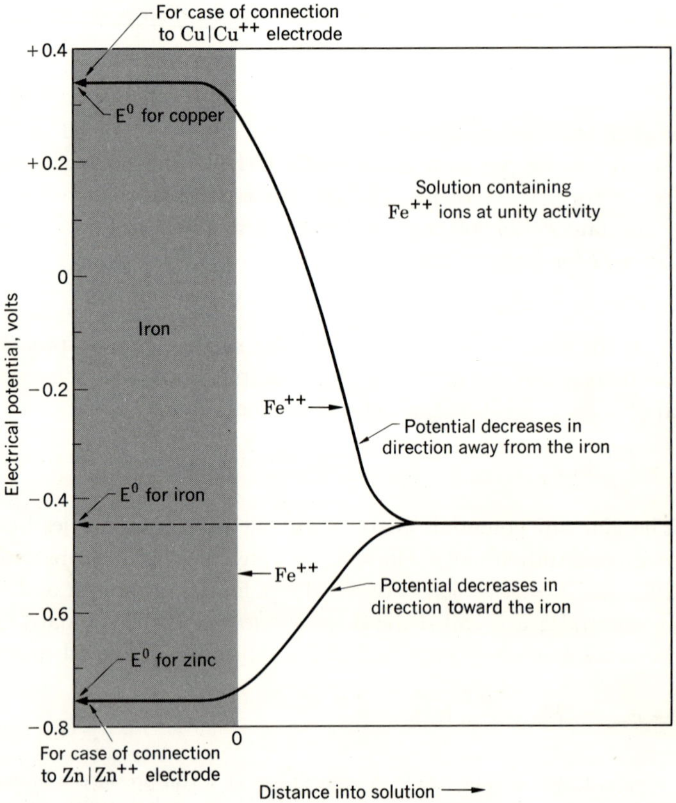

FIGURE 11-4 Simplified method for interpreting the electromotive series; illustrated for iron with either copper or zinc.

moving down the potential gradient, out of the metal, and into the solution, which is the process of oxidation or corrosion. Conversely, if the potential of the iron is lowered by connection to a $Zn|Zn^{++}$ electrode, the existing Fe^{++} ions tend to be reduced to iron, and the iron is protected. The same kind of reasoning can be applied to metals and alloys in a galvanic series (Table 11-3)

TABLE 11-3. REPRESENTATIVE MEMBERS OF THE EMF SERIES AND A GALVANIC SERIES†

Electromotive series			
Metal|metal-ion electrode	**Standard electrode potential at 25°C, $E°$, volts**		**Galvanic series in seawater**
$Au\|Au^{3+}$	+1.498	↑	Platinum
$Pt\|Pt^{++}$	+1.2		Gold
$Ag\|Ag^{+}$	+0.987	Increasingly	Titanium
$Cu\|Cu^{++}$	+0.337	cathodic	Silver
		(protected)	18-8 stainless steel (passive)
$H_2\|H^{+}$	0.000		Chromium stainless steel |11 to 30^{w}% Cr| (passive)
			Inconel (passive) (80 Ni,
$Pb\|Pb^{++}$	−0.126		13 Cr, 7 Fe)
			Nickel (passive)
			Monel (70 Ni, 30 Cu)
$Sn\|Sn^{++}$	−0.136		Bronze (~90 Cu, 10 Sn)
			Copper
$Ni\|Ni^{++}$	−0.250		Nickel or Inconel (active)
$Co\|Co^{++}$	−0.277		Tin
$Cd\|Cd^{++}$	−0.403		Lead
$Fe\|Fe^{++}$	−0.440		18-8 stainless steel (active)
$Cr\|Cr^{3+}$	−0.744		Steel
$Zn\|Zn^{++}$	−0.763		2024 aluminum (4.5 Cu,
		Increasingly	1.5 Mg, 0.6 Mn)
$Al\|Al^{3+}$	−1.662	anodic	Cadmium
$Mg\|Mg^{++}$	−2.363	(corroded)	1100 aluminum (99% Al)
$Na\|Na^{+}$	−2.714		Zinc
$K\|K^{+}$	−2.925	↓	Magnesium alloys

† Adapted from the more complete tables in M. G. Fontana and N. D. Greene, "Corrosion Engineering," McGraw-Hill Book Company, New York, 1967.

based on practically important electrolytes such as sea water. A high member in this series usually tends to establish a potential gradient that favors corrosion at the surface of a lower member.

Quantitative data, such as the electromotive series, permit calculation of the theoretical limit of oxidation (or reduction). For example, the equilibrium extent of oxidation of iron to Fe^{++} when in contact with a standard

$H_2|H^+$ electrode can be calculated using Eq. (11-19), which at 25°C can be written (Prob. 11-2)

$$E = E^\circ + \frac{0.0592}{z} \log K \tag{11-20}$$

Since $E = 0$ at equilibrium,

$$E^\circ = -0.440 = -\frac{0.0592}{2} \log a_{Fe^{++}} \tag{11-21}$$

and therefore,

$$\log a_{Fe^{++}} = \frac{0.88}{0.0592} = 14.9 \tag{11-22}$$

Thus, an impossibly large activity of Fe^{++} ions (about 10^{15}) would be required to establish equilibrium between an $Fe|Fe^{++}$ electrode and a standard hydrogen electrode. Problem 11-3 considers the $Cu|Cu^{++}$ electrode in an analogous manner and shows that reduction of Cu^{++} to copper tends to occur in this case since E° is positive.

POTENTIAL-pH DIAGRAMS. If the equilibrium conditions for a given electrode reaction are represented graphically, then the several electrode reactions of interest for a given material can be shown compactly in a single diagram, often called the *Pourbaix diagram* after its originator. Electrode potential is plotted against the pH of the solution in the manner shown in Fig. 11-5 for the Fe–H_2O system. Consider first the horizontal line 1 at $E = -0.62$ volt, which represents the significant equilibrium aspect of the reaction

$$Fe \rightleftarrows Fe^{++} + 2e \tag{11-23}$$

Equation (11-20) shows that $E = -0.62$ corresponds to an equilibrium concentration of Fe^{++} ions of 10^{-6} mole/liter (Prob. 11-4). This low concentration has been chosen as the basis for constructing the diagram because the corresponding *rate* of solution of the metal must then be negligibly low. Consequently, at electrode potentials below line 1, iron is immune from corrosion for practical purposes. At higher potentials, however, the equilibrium concentration of Fe^{++} ions (and therefore the tendency towards corrosion) increases

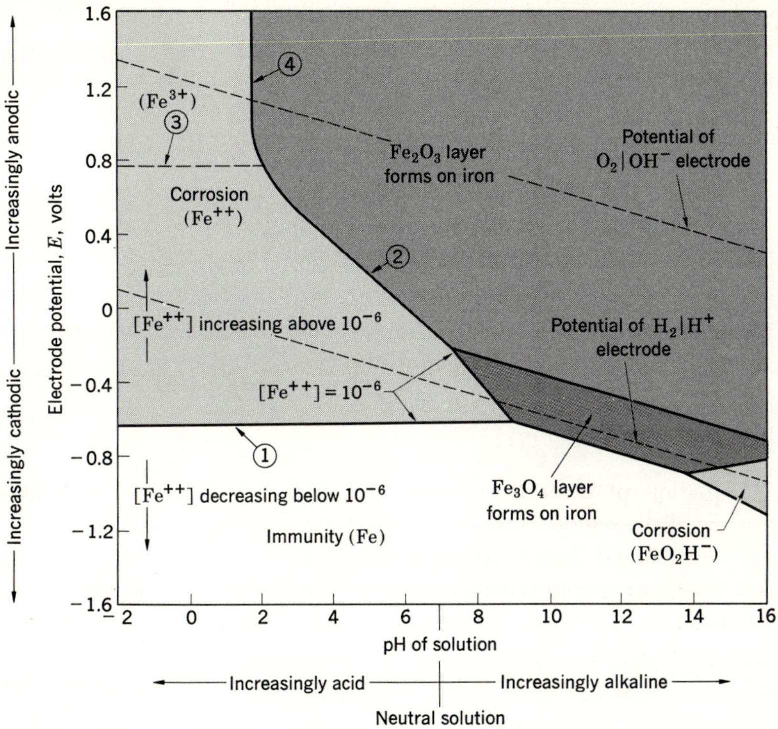

FIGURE 11-5 Simplified potential-pH diagram for the Fe–H_2O system at 25°C. (*After M. Pourbaix.*)

rapidly. At E° (-0.44 volt), for example, [Fe^{++}] is at unit activity or about 1 mole/liter.

The sloping line 2, separating the region of corrosion from that of layer formation in Fig. 11-5, refers to the electrode reaction (Prob. 11-5)

$$2Fe^{++} + 3H_2O \rightleftarrows Fe_2O_3 + 6H^+ + 2e \qquad \Delta G^\circ = 33.6 \text{ kcal} \tag{11-24}$$

The Fe^{++} ion is oxidized to Fe^{3+} and is then deposited on the iron as a layer of Fe_2O_3 that tends to protect the iron from further corrosion. The standard free-energy change ΔG° is obtained from data on the free energy of each species at unit activity (Table 2-1). Since a net ionic charge of $+2$ moves from the electrode surface into the solution,

$$E^\circ = \frac{\Delta G^\circ}{2F} = \frac{33{,}600}{46{,}120} = 0.73 \text{ volt} \tag{11-25}$$

The equilibrium constant for the reaction in Eq. (11-24) is

$$K = \frac{(a_{H^+})^6}{(a_{Fe^{++}})^2} \tag{11-26}$$

and therefore Eq. (11-20) becomes

$$E = 0.73 + \frac{0.0592}{2}(6 \log a_{H^+} - 2 \log a_{Fe^{++}}) \tag{11-27}$$

If 10^{-6} mole/liter of Fe^{++} ion is again taken as the criterion of corrosion and since $pH = -\log [H^+] \approx -\log a_{H^+}$, Eq. (11-27) becomes

$$E = 1.08 - 0.178 \text{ pH} \tag{11-28}$$

This equation plots as the sloping line that forms the upper right-hand boundary of the region of corrosion involving Fe^{++}.

The dashed line 3, separating the region of Fe^{++} ions from that in which Fe^{3+} ions predominate, corresponds to the *redox* (*red*uction-*ox*idation) reaction

$$Fe^{++} \rightarrow Fe^{3+} + e \qquad \Delta G^\circ = 17.8 \text{ kcal} \tag{11-29}$$

In a redox reaction the electrode surface serves merely as a sink (or source) for electrons. In contrast, the corresponding charge transfer in a typical corrosion reaction involves formation and migration of an ion, as in Eq. (11-13). The electrode potential at which the Fe^{++} and Fe^{3+} ions have equal activities is $E = +0.77$ (see Prob. 11-6), indicated by line 3 in Fig. 11-5. As the potential increases above this value, Fe^{3+} quickly becomes the predominant ion in solution.

Line 4 in Fig. 11-5 differs from the boundary lines previously considered because its position is determined only by the pH of the solution (not by the electrode potential). This line refers to the reaction

$$2Fe^{3+} + 3H_2O \rightleftarrows Fe_2O_3 + 6H^+ \qquad \Delta G^\circ = -1.9 \text{ kcal} \tag{11-30}$$

which involves no electron transfer but simply the formation of a layer of Fe_2O_3, tending to prevent further corrosion of the metal. The equilibrium constant for this reaction is $\log K = 1.4$ (Prob. 11-7), and therefore,

$$\log \frac{(a_{H^+})^6}{(a_{Fe^{3+}})^2} = 1.4 \tag{11-31}$$

Again with $[Fe^{3+}] = 10^{-6}$ mole/liter as the criterion of corrosion,

$$-6\text{pH} + 12 = 1.4$$

and

$$\text{pH} = 1.8 \qquad (11\text{-}32)$$

This condition is represented by line 4.

The complete Pourbaix diagram for iron[1] involves the consideration of about 30 reactions and equilibria of the type illustrated above. Of course, it is somewhat more complex than the simplified diagram of Fig. 11-5, but the general form and essential concepts are the same. Diagrams are available for almost every element, and each diagram is a convenient, graphical representation of the equilibrium aspects of the various reactions involved in the corrosion of the element in question. In Sec. 11-6 we will see that these diagrams may also give information on the rate of corrosion.

11-5. Kinetic Aspects

Of the several factors that determine the rate of actual corrosion processes, the overpotential $\eta = E - E_{\text{equil}}$ is of primary importance. To understand how the departure η from the equilibrium electrode potential E_{equil} affects the kinetics of electrode reactions, consider first the condition that exists at equilibrium (Fig. 11-6*a*). Typically, the metal (M) is reactive, and its chemical free energy favors the oxidation reaction

$$M \rightarrow M^+ + e \qquad (11\text{-}33)$$

The details of this reaction involve not only ledges at the surface [analogous to vaporization (Fig. 7-1)] but also solvation of the M^+ ion, in which a cage of water molecules interacts with the electric charge on the ion. The Galvani potential φ in the boundary layer tends to force the M^+ ion back into the metal via the reduction reaction

$$M^+ + e \rightarrow M \qquad (11\text{-}34)$$

An electrical double layer, consisting of excess negative charges in the metal and excess positive ions in the solutions, is present at the metal-solution interface. The oxidation (anodic) reaction and the reduction (cathodic) reaction both occur continuously at the same rate, thus producing a condition of

[1] M. Pourbaix, "Atlas of Electrochemical Equilibria in Aqueous Solutions," Pergamon Press, New York, and Cebelcor, Brussels, 1966.

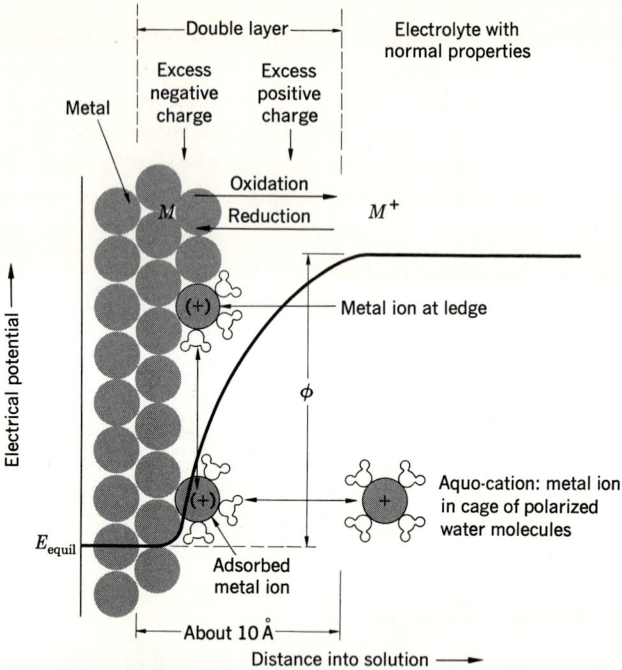

(*a*) Sketch of the dynamic equilibrium between oxidation and reduction at the electrical double layer.

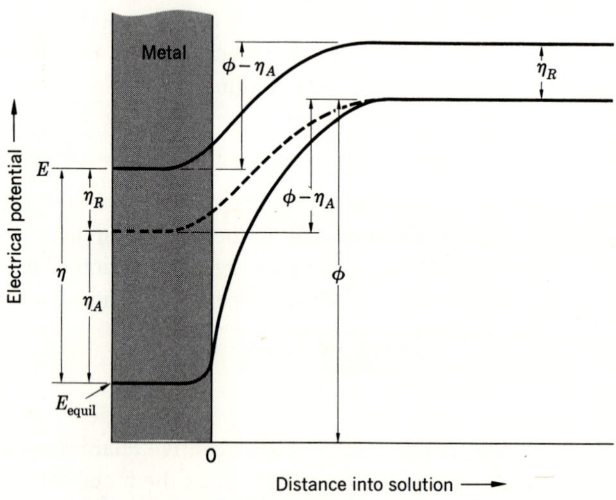

(*b*) Sketch showing the portion, η_A, of the overpotential η, that is effective in causing a net rate of corrosion.

FIGURE 11-6 Schematic illustrations emphasizing the role of the Galvani potential φ (*a*) at equilibrium and (*b*) when an overpotential η causes a net rate of corrosion at the electrode.

dynamic equilibrium. If the electrode potential is increased above E_{equil} by the amount η (Fig. 11-6*b*), the principal effect of this overpotential is to decrease the potential drop across the double layer from the equilibrium value φ to $\varphi - \eta_A$. The activation theory of reaction rates shows that η_A, called the *activation overpotential*, not only reduces the rate of the reduction reaction but also increases the rate of the oxidation reaction. The other components of the total overpotential are η_C, caused by the development of concentration gradients (Prob. 11-8), and η_R, a potential drop due to surface films and to resistance within the solution. η_C is not shown in Fig. 11-6*b*, but it can increase the measured value of E.

A modern instrument, the potentiostat, permits separate study of the anodic (oxidation) and cathodic (reduction) reactions that occur at a given electrode. Consider the simple case in which only one electrochemical reaction occurs at an electrode. If the potentiostat applies the equilibrium potential E_{equil} of this reaction to the electrode, the *net* current density is zero (Fig. 11-7). If the potential is shifted in the positive direction so that the overpotential is $\eta = +0.05$ volt, it follows from the argument of Fig. 11-6*b* that the anodic reaction is favored. The activation theory shows that

$$\eta_A = \beta \log \frac{i}{i_0} \tag{11-35}$$

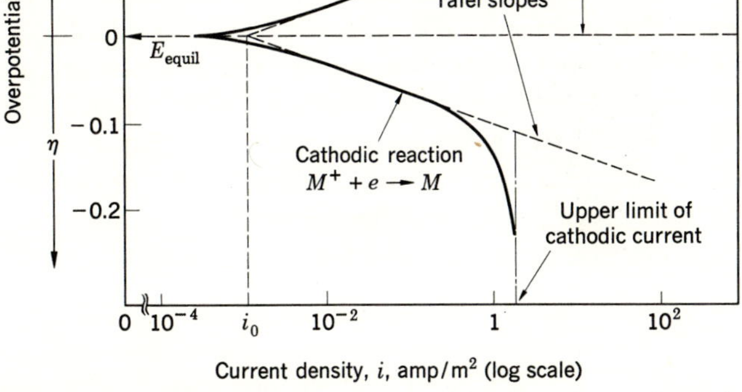

FIGURE 11-7 Schematic illustration of the dependence of reaction current on the overpotential η (conversion of units: M amp/m^2 $\times$ 10^{-4} = N amp/cm^2).

where β is called the *Tafel slope*, i is the anodic current corresponding to the activation overpotential η_A, and i_o, the *exchange current*, is the current that exists even when the net current is zero (i.e., when $\eta = 0$). Figure 11-7 shows that the curve of anodic current density versus the anodic overpotential $+\eta$ has three parts: (1) below about +0.03 volt the cathodic reaction (as well as the anodic reaction) influences the curve; (2) in the linear range above +0.03 volt the activation overpotential alone determines the experimentally measured current [Eq. (11-35)]; and (3) above the linear range, the concentration overpotential η_C becomes important and quickly sets an upper limit to the current. If the potentiostat shifts the electrode potential in the negative direction, then the cathodic reaction becomes dominant and can be investigated in the same manner as the anodic reaction.

ELECTROCHEMICAL CORROSION

As an example of a practical corrosion situation consider a steel pipe carrying drinking water. The corrosion products are iron ions and iron compounds such as hydroxides and oxides. The rate of corrosion depends on many factors, including the chemical compositions of the iron and of the water, the temperature, the rate of flow of the water, and any external potential applied to the iron. In this section we will consider how electrochemical principles can help predict the behavior of such complex systems. At present, however, empirical data are the major source of design information for practical corrosion problems. These data are tabulated in books such as the "Corrosion Handbook" and "Corrosion Guide" listed at the end of the chapter.

11-6. Anodic and Cathodic Reactions

The connection between ordinary corrosion and electrode reactions is demonstrated in Fig. 11-8 for the corrosion of zinc in an acid solution (such as about 1 molar HCl). From Table 11-3 we know that the standard potential of the $Zn|Zn^{++}$ electrode relative to the standard $H_2|H^+$ electrode is −0.763 volt. Therefore, if these two electrodes are connected (Fig. 11-8*a*), electrons will flow through the connecting wire and permit the zinc to corrode with the production of Zn^{++} ions (anodic reaction) while H^+ ions are reduced to H_2 at the hydrogen electrode (cathodic reaction). The same overall reaction

$$Zn + 2H^+ \rightarrow Zn^{++} + H_2 \qquad (11\text{-}36)$$

occurs when zinc is immersed in an acid solution (Fig. 11-8*b*). The anodic reaction occurs at certain areas of the surface of the zinc (local anodes), the cathodic reaction occurs at local cathodes, and the electrons flow in the zinc

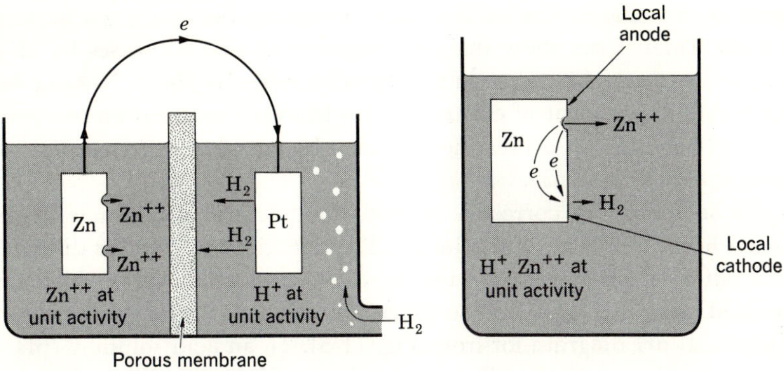

(*a*) Short-circuited cell consisting of a Zn/Zn^{++} and a H_2/H^+ electrode.

(*b*) Zinc immersed in an acid solution containing Zn^{++} ions at unit activity.

FIGURE 11-8 Illustration of the electrodes involved in the corrosion of zinc under two different experimental conditions. (*Adapted from M. G. Fontana and N. D. Greene, "Corrosion Engineering," McGraw-Hill Book Company, New York, 1967.*)

between the local anodes and cathodes. The locations of the anodes and cathodes change with time, and thus the zinc corrodes evenly over its surface.

By approximating the behavior of the local anode and cathode with that of electrodes investigated in the laboratory, one obtains a method for estimating the rate of corrosion reactions. At the local anode the reaction is $Zn \rightarrow Zn^{++} + 2e$, and Fig. 11-7 shows that a positive overpotential is then associated with this reaction. If the values $E_{equil} = -0.763$ volt and $i_o = 10^{-3}$ amp/m^2 are used for the $Zn|Zn^{++}$ electrode, approximate Tafel slopes can be sketched for this electrode in Fig. 11-9. Similarly, the cathodic reaction at the $H_2|H^+$ electrode, $2H^+ + 2e \rightarrow H_2$, can be approximated by a Tafel slope originating at $E_{equil} = 0$ volts and $i_o = 10^{-7}$ amp/m^2. [When the metallic portion of the hydrogen electrode is platinized platinum (rather than zinc), $i_o = 10$. Problem 11-9 considers the effect of this factor on the rate of corrosion.] Since the potential of the zinc must be essentially the same at cathodic and anodic regions, the anodic reaction develops a positive overpotential and the cathodic reaction a negative overpotential so that a mixed potential E_M exists on the zinc at the actual rate of corrosion given by the corrosion current i_{corr}. As shown in Prob. 11-10, $i_{corr} = 10$ amp/m^2 corresponds to a loss in weight of about 10 g/m^2-hr or to a decrease in thickness of about 2×10^{-6} m/hr.

Although internal corrosion currents are flowing in the example of Fig. 11-9, the external current is zero. If an external potential is applied to the zinc specimen in Fig. 11-8*b*, then the actual potential E can be made to differ from E_M by an overpotential $\eta = E - E_M$ analogous to that previously considered for a single electrode. If $\eta = -0.1$ volt in Fig. 11-9, readings made along the Tafel slopes show that the corrosion of zinc decreases by about a factor of ten and the liberation of hydrogen increases by about the same factor. Consequently, a current of electrons flows from the external circuit into the zinc. The technique of decreasing corrosion by use of an external potential is known as *cathodic protection.*

The nature of corrosion of iron in a neutral solution is distinctly different from that in an acid solution. The essential cause of this difference is the variation of the potential of the $H_2|H^+$ electrode with pH, which can be calculated using Eq. (11-20). These data are shown as the lower dashed line in the potential-pH diagram for iron (Fig. 11-5). In an acid solution (pH = 0, for example), iron can corrode with liberation of hydrogen gas since line 1 in Fig. 11-5 is about 0.6 volt below the $H_2|H^+$ potential. In a neutral solution (pH = 7) the potential of the $H_2|H^+$ electrode approaches that of the $Fe|Fe^{++}$ electrode. By analogy with the diagram of Fig. 11-9, the rate of reaction then

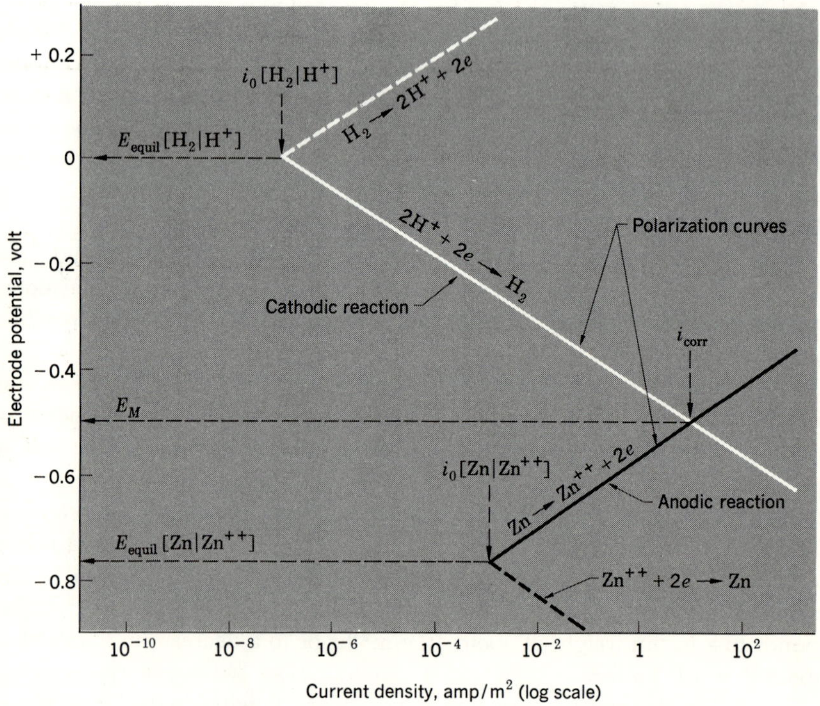

FIGURE 11-9 Schematic illustration of the mixed potential E_M developed on zinc corroding in an acid solution. (*Adapted from M. G. Fontana and N. D. Greene, "Corrosion Engineering," McGraw-Hill Book Company, New York, 1967.*)

decreases greatly; it actually becomes negligibly slow. At this stage an alternative reaction,

$$O_2 + 2H_2O + 4e \rightarrow 4OH^- \tag{11-37}$$

begins to be important as the cathodic reaction. The potential of the $O_2|OH^-$ electrode is also shown in Fig. 11-5 as a dashed line and is seen to lie well above that of the $Fe|Fe^{++}$ electrode. Although the driving force for the reaction of Eq. (11-37) is high for iron in neutral solution, the actual corrosion-current density is often limited to about 10^{-2} amp/m² because of factors such as the formation of surface films, a low value of i_o, and the action of the concentration overpotential.

Two features of the cathodic reaction involving dissolved oxygen lead to an important type of corrosion behavior termed *differential aeration* or *crevice corrosion*. Oxygen molecules have both a low solubility and a low rate of diffusion in aqueous solutions. Therefore, if a portion of solution is within a crevice (Fig. 11-10*a*), this portion soon becomes depleted in oxygen. The metal within the crevice then becomes favored for the anodic reaction (producing M^+ ions) since the metal adjacent to the body of the solution can still undergo the cathodic reaction involving oxygen. To compensate for the excess positive

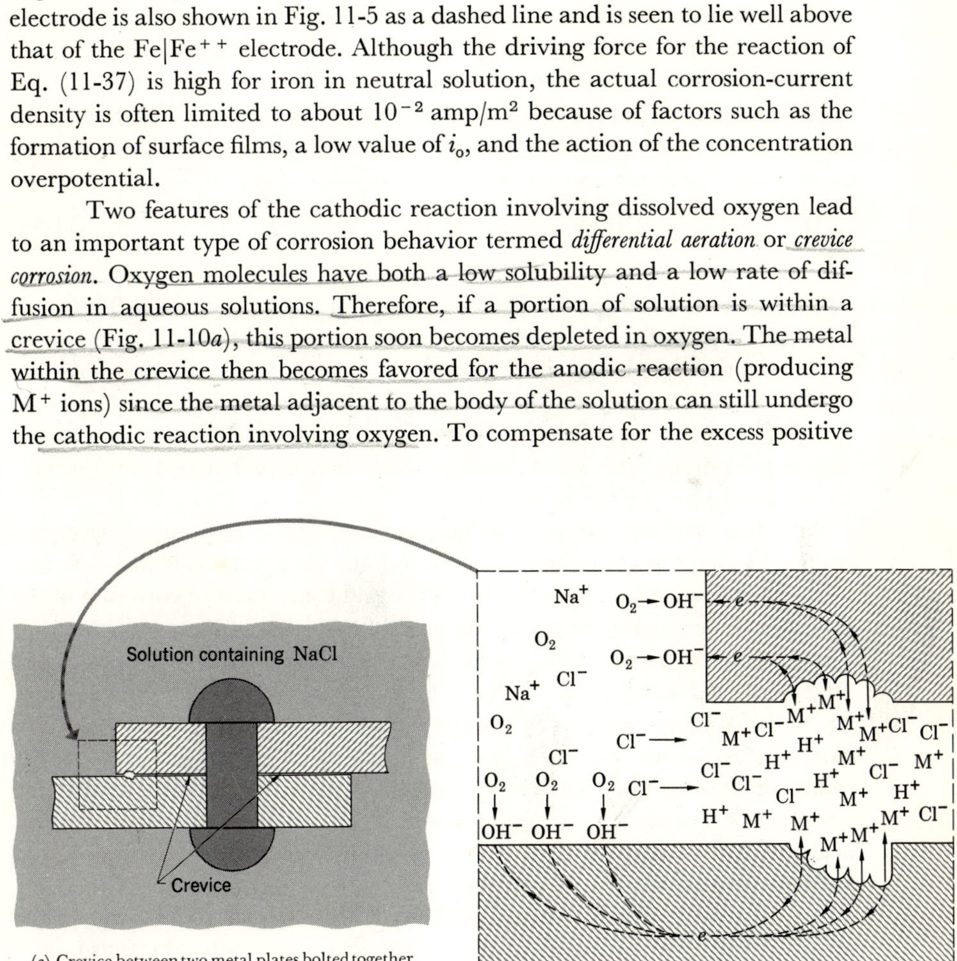

(*a*) Crevice between two metal plates bolted together.

(*b*) Enlarged view of crevice, showing the accumulation of Cl⁻ions.

FIGURE 11-10 Schematic illustration of the nature of crevice corrosion, involving depletion of the oxygen in the corrosive solution. (*Adapted from M. G. Fontana and N. D. Greene, "Corrosion Engineering," McGraw-Hill Book Company, New York, 1967.*)

charges in the crevice, the Cl^- (or similar) anions migrate into the crevice. Subsequent hydrolysis,

$$M^+ + H_2O \rightarrow MOH + H^+ \tag{11-38}$$

also increases the acidity of the solution. Then the Cl^- and H^+ ions both increase the rate of corrosion within the crevice.

Redox reactions, previously discussed in connection with Eq. (11-29), can be used either to promote corrosion [as in chemical etching (Prob. 11-11)] or to inhibit it. Consider the following redox reaction involving hydrazine, N_2H_4:

$$N_2H_4 + 4H_2O \rightarrow 2NO_2^- + 12H^+ + 10e \tag{11-39}$$

which has an electrode potential of +0.28 volt in a neutral solution. This value is considerably below the corresponding potential of the $O_2|OH^-$ electrode (+0.80 volt). Since the exchange current i_o is large, the hydrazine molecules oxidize at anodic areas on the metal and cause reduction of the oxygen at cathodic areas. In this manner hydrazine can reduce the oxygen content of a corrosive medium, such as the cooling water in a closed system, and protect the metal against serious corrosion.

When certain dissimilar metals, say copper and iron, are in electrical contact in the presence of an aqueous solution, they form a galvanic cell. The relative electrode potentials (Table 11-3) could lead to severe corrosion of the iron, but under many conditions the rate of corrosion of the iron is actually controlled by the rate of diffusion of oxygen through the solution. For example, attachment of a small piece of copper to a large iron plate immersed in water leaves the corrosion rate of the iron almost unaffected. The copper serves as a cathodic area for the reduction of oxygen, but so do many local areas on the iron. The result is quite different if the copper plate is large compared to the iron. Because the cathodic reaction now occurs on a large area, a large current of electrons can be removed from the iron and thus permit the corrosion of iron to occur very rapidly. Good corrosion design avoids galvanic cells of this type either by suitable choice of metals in contact or by electrical insulation of the metals at the joint. Examples of both deleterious and beneficial effects of galvanic cells are shown in Fig. 11-11. Since tin lies above iron in the electromotive series, the corrosion of iron is accelerated at any hole in a tin-coated steel (Fig. 11-11*a*). Zinc, on the other hand, is anodic with respect to iron, and thus the underlying steel is protected at flaws in a zinc-coated (galvanized) sheet (Fig. 11-11*b*).

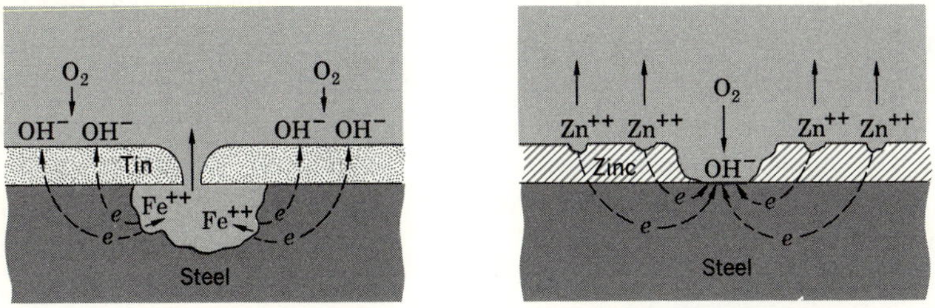

(*a*) The tin surface serves as a large cathode and accelerates corrosion of the iron.

(*b*) The zinc layer forms the anode and is sacrificially corroded to protect the iron.

FIGURE 11-11 Comparison of the effects of the galvanic cells formed at perforations in tin- and zinc-coated steels.

11-7. Surface Layers and Passivity

Common structural materials undergo corrosion, but frequently a protective layer forms during the corrosion process and limits the rate of corrosion to low values. In typical corrosion of iron (Fig. 11-12) a mixed potential of about $E_M = -0.45$ volt develops, corresponding to a position within the "corrosion" region of the Pourbaix diagram. Three methods for changing this corrosion behavior are shown in Fig. 11-12, including cathodic protection (by an overpotential of about -0.3 volt in this case), which we have considered previously. The other two methods involve the formation of protective layers. In one of them the corroding solution is made slightly alkaline. If E_M remains about the same, the thermodynamically stable compound is then Fe_3O_4, which forms [via several intermediates beginning with $Fe(OH)_2$] as a layer on the iron. Layers of this kind on various metals differ in the degree to which they protect the underlying metal from further corrosion, but in general they significantly decrease the rate compared to that characteristic of the corrosion region of the diagram.

The second method produces the type of protective layer responsible for the phenomenon of *passivity* in such technically important metals as iron, nickel, chromium, titanium, and their alloys. If the potential increases about 0.4 volt above E_M, the corroding iron moves into the "passivation" region above the dashed line in Fig. 11-12. This line is the metastable extension of the limit of thermodynamic stability of Fe_2O_3 in contact with iron and Fe^{++} ions. Two reactions are now possible at the surface of iron in contact with water: (1) the production of Fe^{++}, for which the driving force is about 0.5 volt, and (2) the production of Fe_2O_3, for which the driving force is only a fraction of a volt but which is far easier kinetically. The favorable kinetics result in the formation of a film similar to Fe_2O_3 but only about 20 Å thick. This thin film has two effects: (1) it limits its own growth by changing the distribution of the electrical potential, and (2) it greatly impedes the dissolution of iron as Fe^{++} ions. Several important corroding systems do not require a special anodic potential to become passive. When iron corrodes in strong solutions of nitric acid, E_M is sufficiently high to produce the passivating film, and therefore corrosion practically stops. Stainless irons and steels, containing more than about 11 percent chromium, develop passivity in mildly oxidizing solutions or in the atmosphere.

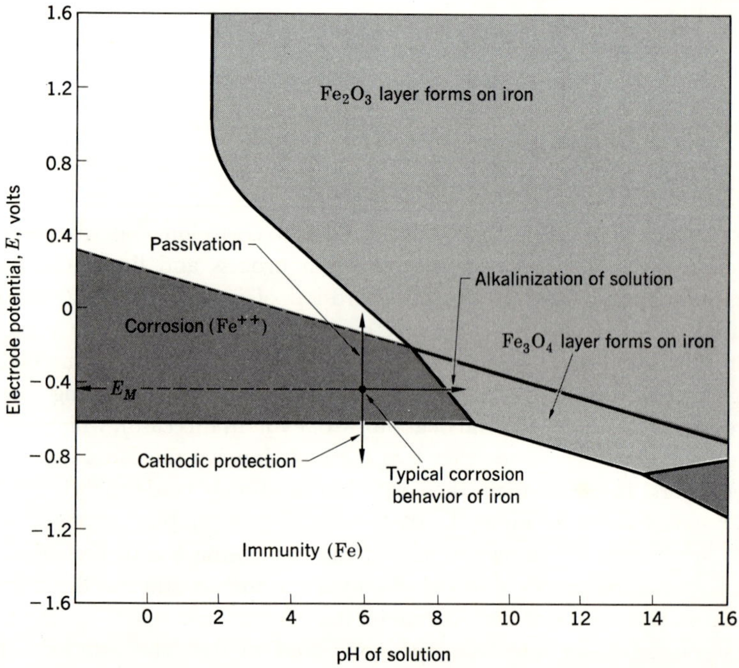

FIGURE 11-12 Schematic illustration of three methods for modifying the corrosion behavior of iron.

Passivating films can be broken down in various ways. Weak spots in the film tend to exist where the underlying metal is nonuniform because of phase boundaries, inclusions, or simply changes in contour. The presence of chloride ions accentuates effects of this kind. If the film is broken locally, then the large ratio of cathodic areas causes rapid corrosion and the formation of pits in the manner shown in Fig. 11-11*a*. Self-repair of perforated films is promoted by suitable alloying elements (chromium, nickel, and silicon in iron, for example) and by modification of the corroding environment, such as the addition of chromate ions.

11-8. The Eight Types of Corrosion

The corrosion of most metals and alloys in service falls into the eight categories described in detail by Fontana and Greene.[1] The present section summarizes the eight forms of corrosion and indicates how they are related to the principles already discussed:

1. *Uniform corrosive attack* is relatively easily controlled by better choice of material or by one of the protective measures described at the end of the chapter.
2. *Galvanic corrosion* depends both on the two metals in contact and on the corroding environment. Table 11-3 shows that the galvanic series in sea water differs considerably from the related emf series. Galvanic action can be used beneficially, as when sacrificial magnesium cathodes are buried near steel pipes and connected to them electrically to provide cathodic protection.
3. *Crevice corrosion*, an example of which is shown in Fig. 11-10, also occurs in such "crevices" as those formed under pieces of dirt on a metal surface (Prob. 11-13).
4. *Pitting* (Fig. 11-13*a*) is an especially serious type of corrosion because a pipe or vessel may develop holes from this cause even when uniform attack is slight. The addition of 2 $^w/_0$ molybdenum to stainless steel helps control pitting in this material.
5. *Intergranular corrosion* (Fig. 11-13*b*) occurs when the chemical composition at grain boundaries is significantly different from that within the grains. If a stainless steel is heated in the range 500 to 800°C (950 to 1450°F), the carbide $Cr_{23}C_6$ precipitates preferentially at grain boundaries, thus decreasing the chromium content in these regions. Such a steel is said to be *sensitized* and is susceptible to corrosion at the grain boundaries. If heating is unavoidable (as in stainless steels to be welded), sensitization can be avoided either by keeping the carbon content of the steel below 0.03 $^w/_0$ or by alloying with columbium or titanium to tie up the carbon as a more stable carbide.
6. *Selective leaching* removes one element from a solid-solution alloy. The vulnerable element may be aluminum, iron, cobalt, or chromium in various alloys, but the best known example is the dezincification of brass.
7. *Erosion corrosion* refers to the acceleration of corrosion by simultaneous abrasion by moving liquids. Impingement attack (Fig. 11-13*c*) can be especially severe.

[1] M. G. Fontana and N. D. Greene, "Corrosion Engineering," McGraw-Hill Book Company, New York, 1967.

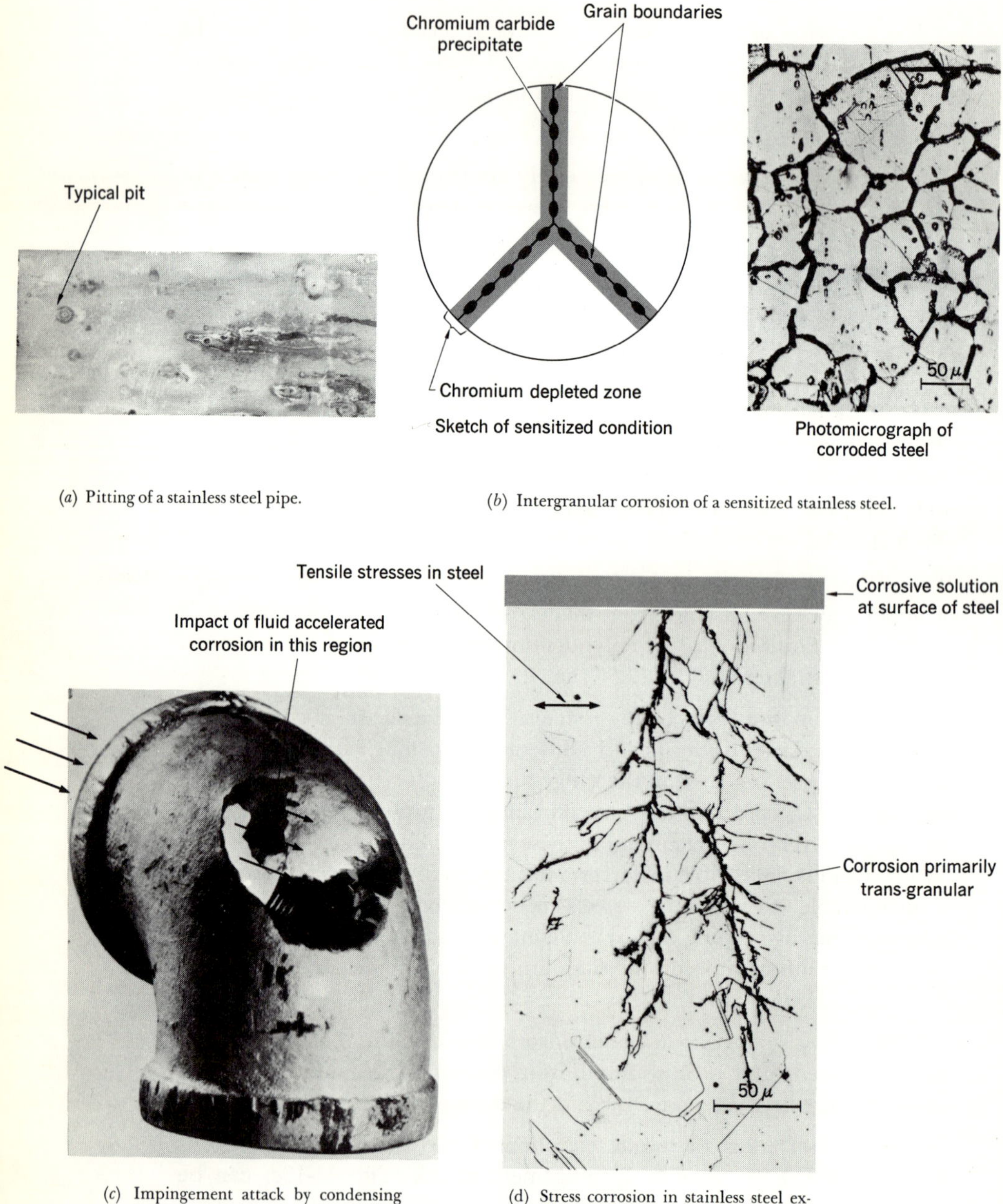

(*a*) Pitting of a stainless steel pipe.

(*b*) Intergranular corrosion of a sensitized stainless steel.

(*c*) Impingement attack by condensing steam on a steel elbow.

(d) Stress corrosion in stainless steel exposed to a boiling solution containing 42% magnesium chloride.

FIGURE 11-13 Illustrations of several forms of corrosion. (*Photographs courtesy of M. G. Fontana, Ohio State University.*)

8. *Stress corrosion* causes a metal to crack when subjected to both tensile stresses and a specific corrosive environment. In the example shown in Fig. 11-13*d*, a network of fine cracks formed within the stainless steel, otherwise little affected by the corrosive environment. Stress corrosion can drastically reduce the strength of a metal in the manner described in Sec. 10-6.

Liquid metals produce unusual types of deterioration in the materials with which they come in contact during such operations as steel making, metal casting, or heat transfer using liquid metals. The mechanisms of deterioration are not electrochemical, but instead they involve such processes as solution of the solid material, formation of chemical compounds, and diffusion of the liquid metal into the solid material. Fracture associated with a liquid-metal environment is considered in Sec. 10-6. Deterioration by mass transfer is commonly encountered when a liquid metal (sodium, for example) is the heat-transfer agent in a nuclear-power installation. The liquid has a higher solubility for the metal of the heat-exchanger tubes in the high-temperature section of the equipment; consequently, metal is removed in this region. Often a more serious effect, however, is its subsequent deposition in the low-temperature section, where it interferes with proper flow of the liquid metal.

11-9. Dissolution Processes

Processes similar to corrosion are often usefully employed in various operations on materials. Either chemical or electrochemical methods can be used to dissolve materials for such purposes as *polishing* (removing rough or distorted surface layers), *etching* (revealing structural features), or *forming* (removing relatively large amounts of material as an alternative to a machining operation). Electrochemical methods have a wider applicability than chemical methods because of their greater flexibility and ease of control. Electrochemical principles of corrosion apply also to the various dissolution processes.

If small positive overpotentials are applied to a metal (Fig. 11-14), the corrosion-current density follows a Tafel slope, as shown in Fig. 11-7. In this range of potentials the rate-limiting step in the process of dissolution is removal of atoms from terrace-ledge-kink configurations on the surface, illustrated in Fig. 7-1. Consequently, crystallographic etch figures develop on the faces of grains, and preferential attack tends to reveal grain boundaries and individual dislocation lines. When the potential reaches higher values and especially if the electrolyte has suitable properties such as high viscosity, concentration overpotential becomes the dominant factor, and the experimental curve rises sharply above the Tafel slope. The rate-controlling factor is now the removal of metal ions from the concentrated solution adjacent to the dissolving surface, and thus a smooth planar surface tends to be produced. A current density on

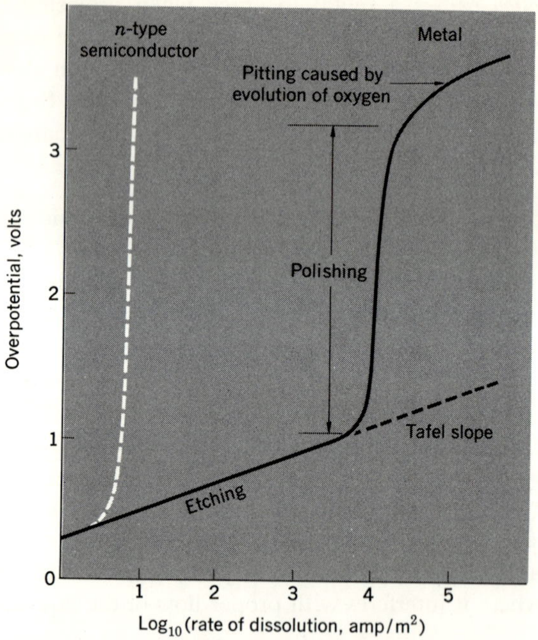

FIGURE 11-14 Schematic comparison of anodic dissolution of a metal and of an *n*-type semiconductor.

the order of 10^4 amp/m^2 gives a satisfactory rate of polishing. An excessive overpotential causes evolution of oxygen and pitting of the surface.

Semiconductors are polished in several stages of their manufacture (for example, in removing surface layers deformed during cutting), and it is important to understand the unusual behavior of *n*-type semiconductors, shown in Fig. 11-14. A current of electrons flows into a dissolving material because of the anodic reaction $M \rightarrow M^+ + e$ (Fig. 11-3). This current has a much different effect on the potential of the surface layers in an *n*-type semiconductor compared to that in a metal (or *p*-type semiconductor). The band theory (Sec. 12-2) shows that metals have vacant low-energy states that can accept the additional electrons with only a slight increase in potential. In an *n*-type semiconductor, however, no low-energy states are available for the extra electrons; the only possible sink is through reaction with the holes, $h^\bullet + e' \rightarrow 0$ (see Sec. 4-5). This reaction quickly depletes the small concentration of holes in a region near the surface, and thus diffusion of holes from within the semiconductor must occur to permit further dissolution. This process causes the potential gradient to localize in a 1-μ surface layer and encourages the production of additional holes. Nevertheless, at ordinary overpotentials the rate of dissolution is only about 10 amp/m^2. To achieve faster rates of dissolution, the hole concentration at the surface must be increased by such means as illuminating the surface (photoexcitation) or heating the semiconductor.

OXIDATION OF METALS AND ALLOYS

When materials are subjected to high temperatures (as components of gas turbines, furnaces, etc.), they may fail through loss of strength, as described in Chap. 9, or they may gradually deteriorate through reaction with the surrounding atmosphere. Reaction with oxygen is the most common cause of deterioration. With the exception of the noble metals (gold, platinum, etc.) metallic materials are prone to oxidation (Table 1-1). Most of the common ceramic materials are already oxides, and therefore they are essentially immune to further modification. In fact, during exposure to an oxidizing atmosphere the average metal develops a ceramic surface layer, such as Al_2O_3 on aluminum, which may give adequate protection to the underlying metal. Kinetic and structural factors, rather than thermodynamic ones, are decisive in determining the degree to which a surface oxide layer is protective.

11-10. Oxidation of Metals

The initial stages of oxidation can be investigated using oxygen at low pressures. If copper is heated at 500°C in oxygen at about 10^{-1} N/m² (Fig. 11-15), several minutes elapse before individual oxide nuclei appear. Before this time, however, a fraction of a monolayer of (chemisorbed) oxygen atoms is bonded chemically to the surface of the metal, often in a characteristic pattern that can be revealed by low-energy electron diffraction (Sec. 1-11). Oxide nuclei generally form epitaxially on a clean metallic surface; the Cu_2O oxide on the (111) plane of copper, for example, has (111) planes of the oxide and

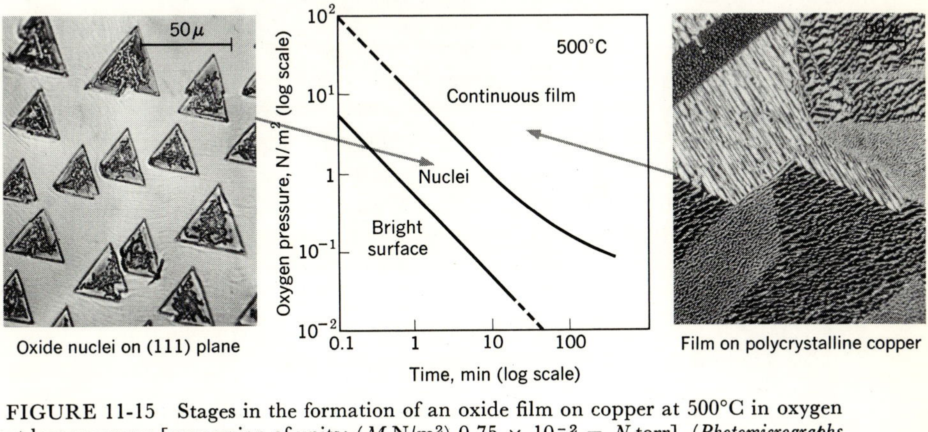

FIGURE 11-15 Stages in the formation of an oxide film on copper at 500°C in oxygen at low pressures [conversion of units: (M N/m²) 0.75×10^{-2} = N torr]. (*Photomicrographs courtesy of K. R. Lawless, University of Virginia.*)

metal parallel. The existing nuclei grow rapidly by addition of oxygen and copper atoms to their edges until finally a continuous film has been formed on the surface. On ordinary polycrystalline copper the oxide layer on each grain has a characteristic orientation and thickness. Because the atomic configuration at the surface of a metal influences the ease of nucleation and growth of its oxide, the rate of oxidation depends on crystal orientation (Fig. 11-16) both for thin films (less than about 10^3 Å in thickness) and for thick films.

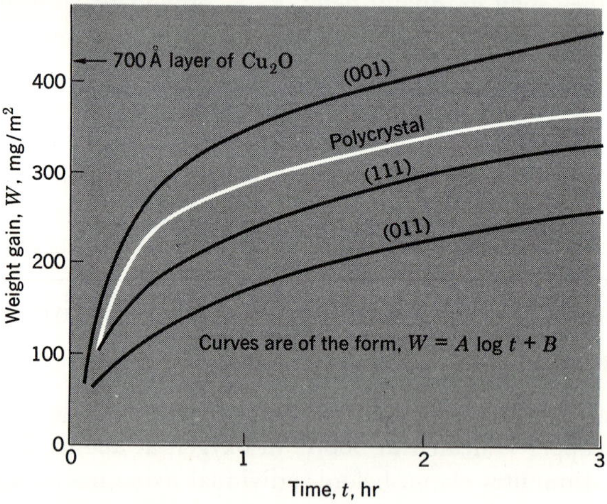

(*a*) At 200°C the films are thin and grow according to a logarithmic law.

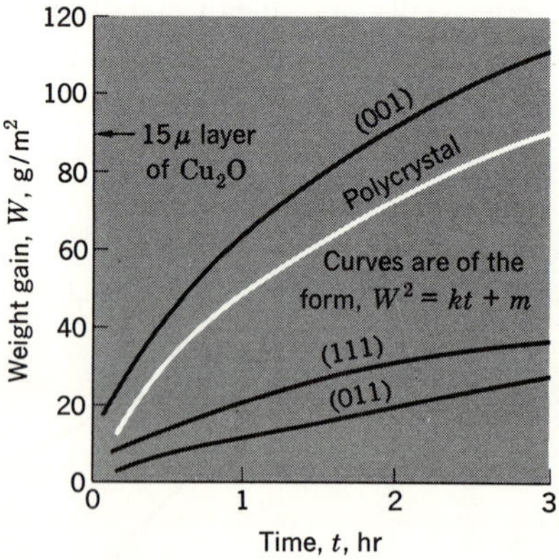

(*b*) At 600°C the films quickly become thick and grow according to the parabolic law.

FIGURE 11-16 Rate of growth of oxide films on copper at ~10^2 N/m^2 pressure of oxygen. (*Data of S. I. Ali and G. C. Wood.*)

Although many factors affect the rate at which metals oxidize, the character of the oxide layer is often decisive. If the oxide layer cracks or is porous (MgO on magnesium above 450°C, for example), then gaseous oxygen can penetrate the layer and maintain a constant rate of oxidation. The amount of oxidation then increases linearly with time, and we have the relation

$$W = Kt \tag{11-40}$$

where W is the increase in weight of the specimen. Oxides that grow according to Eq. (11-40) are called *nonprotective*. The more useful oxides are those (such as Cu_2O) that form a continuous layer over the surface of the metal and therefore tend to give greater protection the greater the thickness of the oxide layer. To see how a protective oxide can lead to the logarithmic or parabolic law of growth, illustrated in Fig. 11-16, picture the metal oxide system as an electrochemical cell (Fig. 11-17) analogous to an aqueous corrosion cell. The overall chemical reaction

$$2Cu + \tfrac{1}{2}O_2 \rightarrow Cu_2O \qquad \Delta G^\circ \approx -60 \text{ kcal/mole} \tag{11-41}$$

occurs by separate anodic and cathodic reactions at the two surfaces of the oxide layer. In accordance with Eq. (11-25), an electrical potential of approximately $E = \Delta G^\circ/2F = -1.3$ volts tends to exist between the cathodic and anodic regions. This potential has qualitatively different effects depending on whether the oxide layer is very thin (<50 Å), thin (~ 50 to 1000 Å), or thick (>1000 Å). In very thin films the potential creates such high electric fields ($1.3 \text{ volts}/10^{-9} \text{ m} = 10^9$ volts/m in a 10-Å film) that quantum-mechanical tunneling effects can aid in the growth of these films even at low temperatures. The films may virtually stop growing at a thickness corresponding to a somewhat lower electric field. At slightly higher temperatures the further growth of oxide films is strongly influenced by a concentration overpotential of the electronic defects, analogous to that described in Sec. 11-9 for the dissolution of semiconductors. A typical rate law for such thin films is the logarithmic relation shown in Fig. 11-16*a*.

The behavior of industrially important oxide layers (those that form at high temperatures and tend to become thick) depends on the characteristics of the oxide composing the film. Two types of transport must occur through an oxide layer if growth is to continue (Fig. 11-17*a*): (1) electrons and/or holes must flow as an electronic current J_{elec} between the cathodic and anodic regions to prevent excessive increase of electrical potential, and (2) cations and/or anions must diffuse as the flux J_{atomic} (or $-J_{\text{vacancies}}$) to permit cooperative building of the oxide lattice. Both J_{elec} and J_{atomic}, equal in terms of chemical equivalents, are required for growth. In a few instances, such as

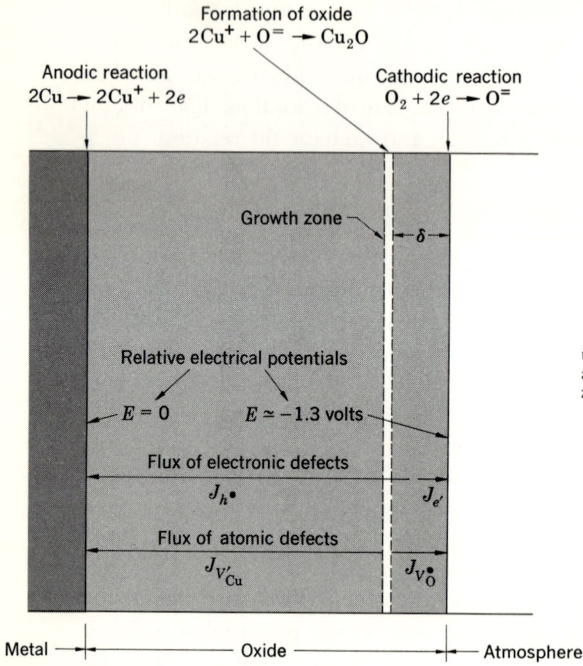

(*a*) Charge is transferred across the oxide by the fluxes of holes and electrons. Copper ions and oxygen ions are brought to the growth zone by the vacancy fluxes.

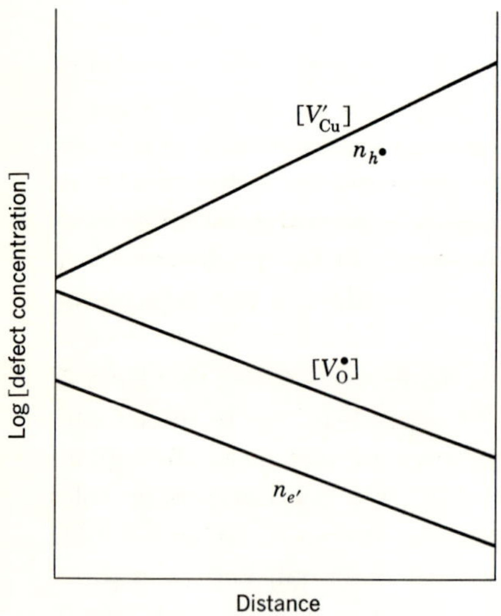

(*b*) Relative concentrations and concentration gradients of electronic and atomic defects in the oxide layer.

FIGURE 11-17 Schematic illustration of the electrochemical cell existing during oxidation of copper to Cu_2O. The distance δ is virtually zero in this case.

Al_2O_3, the oxide has low electronic conductivity, and the electronic current is the rate-limiting factor in growth. Usually, however, the oxide is a semiconductor with relatively high electronic conduction, and the ionic flux is then the slower process and determines the rate of oxidation. Cu_2O, for example, is a *p*-type semiconductor, and from the Brouwer diagram (Fig. 4-8) it follows that vacancies on copper sites, V'_{Cu}, are the principal atomic defects and are equal in concentration to hole-type electronic defects, $h^{\bullet}$. Since the oxygen concentration is highest at the oxide-oxygen interface, Fig. 4-8 shows that the concentration of copper vacancies, $[V'_{Cu}]$, is also highest at this location and decreases in going toward the metal-oxide interface. The diffusion of vacancies down this concentration gradient corresponds to the diffusion of Cu^+ ions in the opposite direction. Little diffusion of O^{--} ions occurs (primarily because of a small diffusion coefficient for this large ion), and therefore growth of Cu_2O occurs near the oxide-oxygen surface. Most of the electrons generated at the surface of the copper combine with holes, $e' + h^{\bullet} \rightarrow 0$, while the electrons needed for the reduction of oxygen are mainly supplied by the opposite of this reaction, $0 \rightarrow e' + h^{\bullet}$. Consequently, the main electronic current is a flow of holes. Problem 11-15 considers the ZnO oxide layer formed in the oxidation of zinc and shows that a flux of electrons occurs in this *n*-type semiconductor.

A theory of oxidation due to Carl Wagner is based on Ohm's law, $E = iR$; the potential E across the oxide layer drives the current of atomic and electronic defects across the total resistance R of the oxide. Wagner showed that the rate of oxidation k in the parabolic law (Fig. 11-16*b*) is given by

$$k = \frac{E\tau_e(\tau_c + \tau_a)\sigma}{F} \tag{11-42}$$

where σ is the specific electrical conductivity of the oxide and F is the Faraday constant. The transference numbers τ_e, τ_c, and τ_a specify the fraction of the total conductivity due to electrons (or holes), cations, and anions, respectively; $\tau_e + \tau_c + \tau_a = 1$. Representative values of the transference numbers might be $\tau_e = 0.90$, $\tau_c = 0.09$, and $\tau_a = 0.01$ for semiconductor oxides. τ_c (and also σ) can be changed appreciably by doping the semiconductor with a cation differing in charge from the cation of the oxide. For example, the addition of about 1 w/o aluminum to zinc decreases the rate of oxidation by a factor of 10^2 because of the incorporation of Al^{3+} ions in the layer of ZnO. By the mechanism described in Sec. 4-5, Al^{3+} ions decrease the concentration of interstitial zinc ions and therefore decrease both τ_c and σ. An ion of lower valence, such as Li^+, has the opposite effect and increases the rate of oxidation. Problem 11-16 shows that this doping effect in a *p*-type semiconductor (Cu_2O, for example) is opposite to that in ZnO, which is an *n*-type semiconductor.

11-11. Oxidation of Alloys

Alloying elements, employed to improve the strength or other properties of a metal, can influence the oxidation behavior in several ways. A moderate content of a nonoxidizable metal (10 $^w/_0$ gold in copper, for example) produces little change in the oxide layer (Cu_2O) and therefore usually has little effect on the rate of oxidation. If a metal more reactive than copper, such as aluminum, is used as the alloying element, then there is the possibility that the aluminum will be oxidized in the interior of the metallic alloy by the inward diffusion of oxygen. This process, called *internal oxidation*, occurs only under special conditions (such as aluminum contents near 0.1 $^w/_0$) and is usually undesirable. In suitable alloys, however, the internal oxide particles can be employed as a method of strengthening.

A typical beneficial effect of alloying occurs in copper alloys containing 2 to 10 $^w/_0$ aluminum. Initially a layer of Cu_2O forms at a rapid rate, but soon a protective layer of Al_2O_3 forms at the metal-oxide interface and prevents the Cu^+ ions from reaching the oxide-oxygen interface. Oxidation then virtually ceases. Chromium acts in an analogous manner to protect an iron–10 $^w/_0$ chromium alloy from oxidation, but in this case the Cr_2O_3 layer forms at the oxide-oxygen interface. Above about 15 $^w/_0$ chromium only the Cr_2O_3 layer forms. In the oxidation of some alloys a compound oxide may form and be effective in limiting the rate of oxidation. The $MgO \cdot Al_2O_3$ spinel, which is stoichiometric, forms on alloys of magnesium in aluminum and gives excellent protection. Many spinels, such as $ZnO \cdot Al_2O_3$ and $FeO \cdot Fe_2O_3$, are cation-defective and permit oxidation to continue.

PROTECTION OF MATERIALS

Resistance to deterioration is only one of the properties that are considered in choosing the best material for a given application. Cost, mechanical and physical properties, ease of fabrication, and appearance are other important considerations. Usually, therefore, structural materials must be protected against the environment in which they function. A complete program of protection should consider the following possibilities: (1) modification of the design to avoid such dangers as galvanic couples, crevice corrosion, stress corrosion, etc., (2) modification of the material by alloying or by a protective coating, (3) alteration of the environment by chemical or electrical methods, and (4) verification of the final design by tests simulating the conditions of service. Modification of a material is often of major importance, and usually specific modifications are appropriate for a given material. The industrial methods of protecting iron will serve as an example of the changes possible in a typical material.

11-12. Protection of Iron

Several alloying elements, including aluminum and silicon, increase the resistance of iron to corrosion and oxidation, but chromium is of outstanding importance. Chromium improves oxidation resistance (Fig. 11-18*a*) and also resistance to aqueous corrosion in most media (with the notable exception of chloride solutions and strong, nonoxidizing acids). The iron-chromium phase diagram (Fig. 11-18*b*) shows that a significant change in structure occurs when iron contains more than 13 $^w/_0$ chromium and lies beyond the "γ loop." The alloys are then ferritic (α phase) at all temperatures and are no longer able to undergo the $\gamma \rightarrow \alpha$ transformation that leads to martensitic hardening. Depending on the chromium and carbon contents (Fig. 11-18*c*), chromium steels may be hardenable by heat treatment, as are alloy steels (martensitic stainless), or they may be nonhardenable (ferritic stainless) but possess superior corrosion resistance because of a higher chromium content. The properties and typical applications of these and other stainless steels are listed in Table 11-4.

When nickel is added to a steel containing enough chromium (about 18 $^w/_0$) for good corrosion resistance, the corrosion resistance is further improved, but the principal effect is to change the steel from ferritic to austenitic (Fig. 11-18*d*) at nickel contents above 6.5 $^w/_0$. Once the austenite has formed at high temperatures, it persists during slow cooling to room temperature. Austenitic steels of the 18Cr–8Ni type possess an excellent combination of corrosion resistance, ductility for easy forming, and strengthening possibilities. Cold working of these steels produces an unusual degree of strengthening because the austenite tends to decompose to ferrite and produces a martensitic-type constituent. In type 301 stainless, containing about 7 $^w/_0$ nickel, the tendency toward decomposition is great, and therefore cold working produces a maximum increase in strength (Table 11-4). If the nickel content is 9 $^w/_0$ (type 302), work hardening is much reduced; 10 $^w/_0$ nickel (type 304) is employed when work-hardening effects are to be minimized. Stainless steels of the age-hardenable types (Table 11-4) can be given high strengths either with or without cold working. The heat-treating procedure concludes with an aging treatment at about 540°C (1000°F), and therefore these hardened steels can be used for prolonged periods at temperatures below this value.

A coating of some kind often provides the least expensive protection for iron and steel. Typical coatings include paints, electroplates, ceramic coatings, and chemical coatings such as that produced by reaction of steel with a dilute solution of phosphoric acid (phosphatizing). Although the corrosion principles underlying the use of a given coating are ordinarily rather simple, the technology may be extensive and involved. This subject is treated in detail in the second half of volume 2 of the "ASM Metals Handbook."

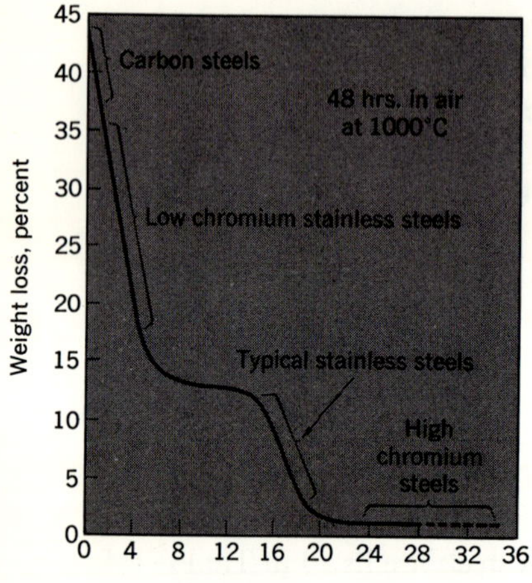

(*a*) Improvement of oxidation resistance of iron by alloying with chromium (after Tomashov).

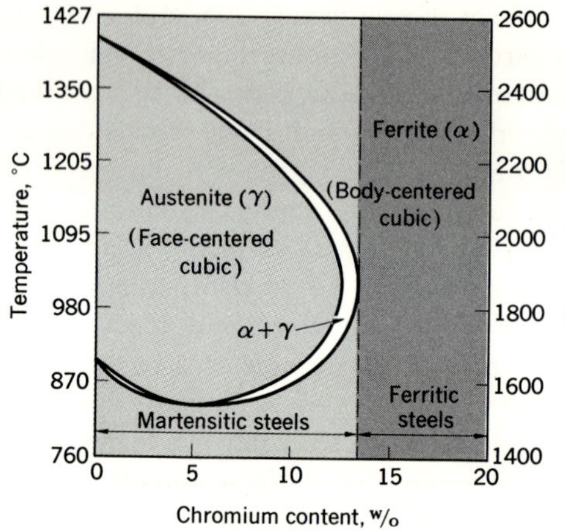

(*b*) The "loop" in the iron-chromium equilibrium diagram.

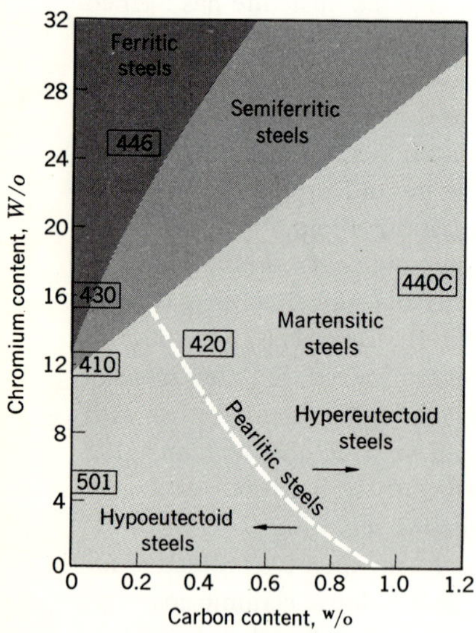

(*c*) Influence of composition on the structure of chromium stainless steels (after Tomashov).

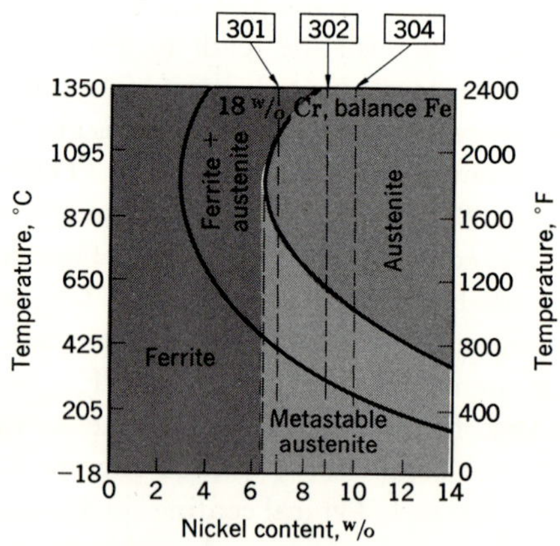

(*d*) Equilibrium diagram for alloys containing 18w% Cr plus 0 to 14% nickel. Metastability of the austenite is significent near 8% Ni.

FIGURE 11-18 Important aspects of the alloying of iron and steel for corrosion resistance. Numbers such as 501 refer to stainless steels listed in Table 11-4.

TABLE 11-4. COMPOSITION AND PROPERTIES OF TYPICAL STAINLESS STEELS

AISI type	Chemical composition, w/o C	Cr	Ni	Other	Condition	Strength, 10^7 N/m²† Tensile	Yield	Elongation, percent	Typical application
Martensitic chromium steels									
501	0.1	5	—	0.5 Mo	Annealed	48	21	28	Oil-refinery piping
410	0.1	12	—	—	Hardened	124	97	15	Turbine blades
420	0.4	13	—	—	Hardened	159	135	8	Cutlery
440C	1.1	17	—	—	Hardened	197	198	2	Cutters
Ferritic chromium steels									
430	0.1	16	—	—	Annealed	52	31	30	Auto trim
446	0.2	25	—	—	Annealed	55	35	23	Furnace parts
Austenitic chromium-nickel steels									
301	0.1	17	7	—	Annealed	81	23	68	Exterior of industrial buildings
					45% C.W.‡	155	138	7	
302	0.1	18	9	—	Annealed	65	25	61	Exterior of office buildings
					50% C.W.	122	104	6	
304	0.08	19	10	—	Annealed	59	24	55	Chemical equipment
316	0.1	17	12	2 Mo	Annealed	59	24	55	For pitting resistance
321	0.08	18	10	0.3 Ti	Annealed	60	24	55	Welded parts
Age-hardenable stainless steels									
635	0.07	17	7	0.8 Ti, 0.2 Al	Hardened	135	124	7	Aircraft and rockets
631	0.07	17	7	1.15 Al	Hardened	162	152	6	

† Conversion of units: (M N/m²) $1.450 \times 10^{-4} = N$ lb/in²; (M N/m²) $1.020 \times 10^{-7} = N$ kg/mm².
‡ Cold work.

PROBLEMS

11-1. The solubility parameter of atactic polystyrene is about 9, and that of polyethylene is about 8. In view of the difference in crystallinity of these two polymers, predict their behavior in contact with (*a*) carbon tetrachloride (solubility parameter 8.5) and (*b*) hexane (solubility parameter about 7).

11-2. Convert Eq. (11-19), the *Nernst equation*, into the form in which it is usually employed for calculations [Eq. (11-20)].

11-3. If the $Cu|Cu^{++}$ electrode (corresponding to the reaction $Cu \rightleftarrows Cu^{++} + 2e$) is compared with the standard hydrogen electrode, its standard potential is found to be $E^\circ = +0.337$ volt. (*a*) Using a sketch similar to Fig. 11-4, explain the behavior of a Cu^{++} ion if a standard $Cu|Cu^{++}$ electrode were connected directly to a standard hydrogen electrode. (*b*) Does this behavior correspond to the forward (oxidation) or backward (reduction) direction of the reaction $Cu \rightleftarrows Cu^{++} + 2e$? (*c*) Calculate the activity of Cu^{++} ions in a $Cu|Cu^{++}$ electrode that would be in equilibrium with a standard hydrogen electrode.

11-4. Using the E° value for the standard $Fe|Fe^{++}$ electrode, show that an electrode potential of -0.62 is in equilibrium with an $Fe|Fe^{++}$ electrode containing 10^{-6} mole/liter of Fe^{++} ions in solution. Recall that K in Eq. (11-20) is merely $a_{Fe^{++}}$ in this case, and that the activity can be replaced in dilute solution by the concentration in moles per liter.

11-5. The electrode reaction of Eq. (11-24) can be pictured as the sum of four simpler reactions: (1) the oxidation of Fe^{++} to Fe^{3+}, (2) the decomposition of water into OH^- and H^+, (3) the formation of $Fe(OH)_3$, and (4) the decomposition of $Fe(OH)_3$ into Fe_2O_3 and H_2O. Write these four reactions, and then sum them to obtain Eq. (11-24).

11-6. (*a*) Determine the value of E° for the reaction of Eq. (11-29), that is, when Fe^{++} and Fe^{3+} are at unit activity. (*b*) Using Eq. (11-20), write the expression for the electrode potential E. (*c*) Provided that the activities of Fe^{++} and Fe^{3+} are equal, how is E related to E°?

11-7. Using Eq. (2-21), calculate $\log K$ for the reaction of Eq. (11-30) at 25°C. [Since the calculation refers to the equilibrium condition, $\Delta G = 0$.]

11-8. Consider an $Fe|Fe^{++}$ electrode at equilibrium with a large volume of solution containing Fe^{++} ions at unit activity. Its electrode potential is then given by Table 11-3 as $E^\circ = -0.44$ volt. If an external potential increases this value to -0.34 volt, so that $\eta = 0.10$ volt, additional Fe^{++} ions will tend to form. (*a*) Explain in terms of diffusion theory why the solution near the metal surface will develop a high concentration of Fe^{++} ions. (*b*) Assume that this local concentration corresponds to $a_{Fe^{++}} = 2$, and use Eq. (11-20) to calculate the quasi-equilibrium electrode potential E. (*c*) In this case $\eta_C = E - E°$. Explain why η_C must be subtracted from the total overpotential η to obtain the activation overpotential η_A that is effective in causing a net rate of oxidation reaction. Assume that η_R is negligible.

11-9. Would the rate of corrosion of zinc be greater in the short-circuited cell of Fig. 11-8*a* or immersed in the acid solution (Fig. 11-8*b*)? The crucial factor is the difference in i_o for the $H_2|H^+$ electrode, which is 10 amp/m^2 in *a* but only 10^{-7} in *b*. By changing i_o to 10 in Fig. 11-9, estimate the rate for *a*.

11-10. (*a*) Since the Faraday constant is 96,500 coul/equivalent, show that the weight loss per second per square meter due to corrosion of zinc to Zn^{++} ions is $(i/96{,}500)(M/2)$, where i is the (anodic) corrosion current density, in amperes per square meter, and M is the molecular weight of zinc, 65.4 g. (*b*) If $i = 10$, what is the rate of corrosion of zinc expressed in grams per square meter per hour? (*c*) Obtain an expression similar to that in part *a* but for the decrease in thickness in meters per second due to corrosion. The molar volume V of zinc is 9.2 cm^3. (*d*) If $i = 10$, what is the rate of corrosion of zinc in units of meters per hour? (*e*) A common unit for rate of corrosion is *ipy*, inch per year, and a corrosion rate of more than 0.05 ipy is considered excessive for most applications. What is the corresponding value of the corrosion current density i for zinc? (*f*) The following computer program is useful for making the conversion from experimental corrosion data to a standard corrosion rate, meters per hour in this case. The initial metallic specimen is a disk of diameter D_s (DS) and thickness T. Its initial weight is W, and its weight after corrosion for TH hr has been reduced to WF. The loss in weight (WL) is $W - WF$ and is equal to (loss in volume) (density D). Therefore,

$$\text{Corrosion rate (CR)} = \frac{\text{WL}}{A \times D \times TH} \quad \text{m/hr} \tag{11-43}$$

where the surface area A is given by

$$A = 2\left(\frac{\pi D_s^2}{4}\right) + \pi D_s T = 1.571 D_s^2 + 3.142 D_s T \tag{11-44}$$

If N different materials are tested under identical conditions (giving N sets of values of DS, T, D, W, TH, and WF, all expressed in kilograms, meters, and hours), the N values of CR can be obtained.

```
C  CALCULATION OF CORROSION RATE, M/HR
    1  FORMAT (6E9.2)
    2  NUMBER = 1
C  READ DATA
    3  READ (5,1, END = 10) DS, T, D, W, TH, WF
C  CALCULATION
    4  A = 1.571 * DS * DS + 3.142 * DS * T
    5  CR = (W - WF)/(A * D * TH)
C  OUTPUT SECTION
    6  WRITE (6,7) NUMBER, CR
    7  FORMAT (1X, 'NUMBER = ', I3, 1X, 'CR = ', E9.2)
    8  NUMBER = NUMBER + 1
    9  GO TO 3
   10  STOP
       END
```

11-11. A useful etching solution for metallographic specimens of iron, copper, and other metals is an aqueous solution of $FeCl_3$. (*a*) Since the potential of the $Fe^{++}|Fe^{3+}$ redox system is +0.77 volt, explain why the Fe^{3+} ions can produce the cathodic areas on the specimen that permit the metal to go into solution by an anodic reaction. (*b*) Would you recommend using $FeCl_3$ for etching platinum? (See Table 11-3.)

11-12. Using the potential-pH diagram for iron, explain why iron corrodes more rapidly in dilute nitric acid than in either concentrated nitric acid or a neutral solution. [HINT: The mechanism of corrosion is different in each case.]

11-13. If a piece of dirt is allowed to remain on the "chrome" (usually a stainless steel) trim of an automobile (subjected to rain, dew, etc.), show by analogy with Fig. 11-10 that localized corrosion will occur.

11-14. (*a*) Explain why a *p*-type semiconductor can be electropolished at a rate comparable to that of a typical metal. (*b*) Consider a single crystal of germanium, half of which is doped *n*-type and the other half *p*-type. Explain whether the location of the *p-n* junction would be made visible after the crystal had been immersed in a solution of potassium hydroxide (which tends to dissolve germanium).

11-15. (*a*) Use the Brouwer diagram of Fig. 4-8 to sketch the analog of Fig. 11-17*b* for ZnO, which is an *n*-type semiconductor. (*b*) If inert markers are placed at the surface of a 10-μ thick ZnO layer and then the specimen is oxidized further, the markers are observed to remain 10 μ from the surface of the zinc. Explain with the aid of a sketch whether zinc ions or oxide ions are more mobile. (*c*) In addition to vacancy-type defects, interstitial zinc ions $Zn_i^{\bullet}$ are believed to be present in ZnO. Explain how this condition would help account for the result in part *b*.

11-16. NiO is a *p*-type semiconductor and behaves similarly to Cu_2O (Fig. 11-17*b*). (*a*) Show that the addition of small amounts of Cr^{3+} ions to NiO increases the concentration of V'_{Cu} vacancies and therefore increases the rate of oxidation. (*b*) What would be the effect of small additions of Li^+ ions?

REFERENCES

Biczok, Imre: "Concrete Corrosion and Concrete Protection," Chemical Publishing Company, Inc., New York, 1967.

Brydson, J. A.: "Plastics Materials," Iliffe Books, London, 1966.

Evans, U. R.: "An Introduction to Metallic Corrosion," 2d ed., Edward Arnold (Publishers) Ltd., London, 1963.

Fontana, M. G., and N. D. Greene: "Corrosion Engineering," McGraw-Hill Book Company, New York, 1967.

Godard, H. P., et al.: "The Corrosion of Light Metals," John Wiley & Sons, Inc., New York, 1967.

Hauffe, Karl: "Oxidation of Metals," Plenum Press, New York, 1965.

Kamal, M. R. (ed.): "Weatherability of Plastic Materials," Applied Polymer Symposia No. 4, Interscience Publishers, New York, 1967.

Kofstad, Per: "High-Temperature Oxidation of Metals," John Wiley & Sons, Inc., New York, 1966.

Logan, H. L.: "The Stress Corrosion of Metals," John Wiley & Sons, Inc., New York, 1967.

Rabald, E.: "Corrosion Guide," American Elsevier Publishing Company, New York, 1968.

Rosato, D. V., and R. T. Schwartz: "Environmental Effects on Polymeric Materials," vols. I and II, Interscience Publishers, New York, 1968.

Scully, J. C.: "The Fundamentals of Corrosion," Pergamon Press, Oxford, 1966.

Shreir, L. L. (ed.): "Corrosion," vols. I and II, John Wiley & Sons, Inc., New York, 1963.

Tomashov, N. D.: "Theory of Corrosion and Protection of Metals," The Macmillan Company, New York, 1966.

Uhlig, H. H.: "Corrosion and Corrosion Control," John Wiley & Sons, Inc., New York, 1963.

——— (ed.): "Corrosion Handbook," John Wiley & Sons, Inc., New York, 1948.

West, J. M.: "Electrodeposition & Corrosion Processes," D. Van Nostrand Company, Inc., Princeton, N.J., 1965.

Chapter Twelve

Electronic Structure and Physical Properties

Electronic structure is basic, in principle, for all properties of materials, but it is essential in fact for understanding their physical properties. The wide range of electrical properties exhibited by various materials, for example, is an immediate consequence of differences in their electronic-energy bands. Since the Bohr theory of the atom is part of most introductory courses in chemistry, the present discussion continues from that starting point. Essential concepts of modern theory are explained, and their importance for understanding physical properties is illustrated by specific examples.

ELECTRON THEORY OF SOLIDS

The experimental data on energy levels in isolated atoms (Fig. 12-1*a*) are explained fairly well by the Bohr theory (for a few simple atoms such as hydrogen), but more adequately by wave mechanics. The basic formulation of the latter theory includes the *uncertainty principle* (Prob. 12-1),

$$\Delta p \, \Delta x \geq h \tag{12-1}$$

The inherent uncertainties Δp and Δx in any simultaneous measurements of the momentum p and the position x of a given particle have this lower limit. Although Planck's constant h has an extremely small value, 6.63×10^{-34} J-sec, relative to ordinary objects, the uncertainty principle has basic significance for atomic events.

This fundamental restriction on experimental observations of electrons is introduced into electron theory by representing the particle as a *wave packet* (Fig. 12-1*b*), a combination of waves having various phases and wavelengths λ. The characteristic wavelength to be associated with a given particle is determined by the *de Broglie relation* (Prob. 12-2),

$$\lambda = \frac{h}{p} \tag{12-2}$$

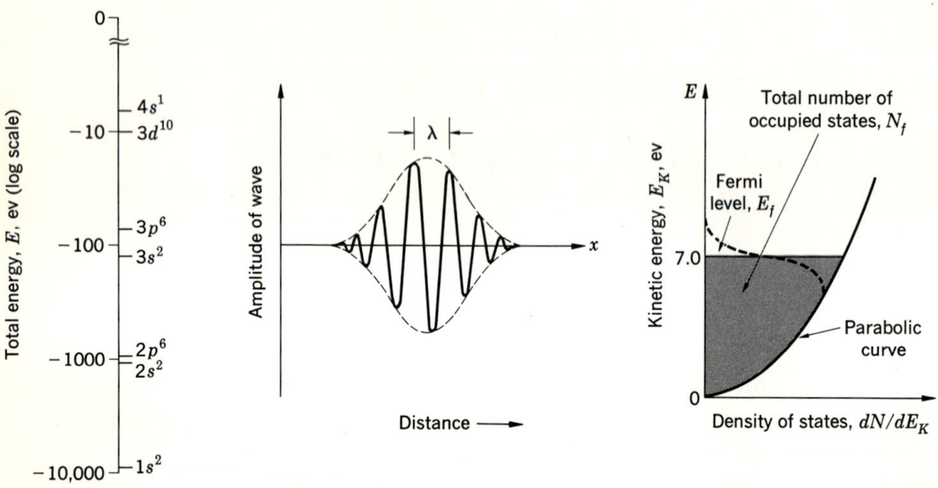

(*a*) Energy levels of the core electrons ($1s^2, 2s^2, 2p^6, 3s^2, 3p^6, 3d^{10}$) and of the valence electron ($4s^1$) in an isolated atom of copper.

(*b*) Wave packet representation of an electron moving in the *x*-direction. The *x*-coordinate of the electron is somewhere within the length of the packet.

(*c*) Filling of the energy states, E_K by the valence electrons in a 1 m^3 piece of copper at O°K. Dashed curve indicates the effect of a higher temperature.

FIGURE 12-1 Aspects of the simple theory of electrons in a typical material; copper is used as an example.

Just as the physically meaningful waves of a vibrating string in a violin are the standing waves that fit geometrically the length of the string, in a similar manner the stable states of a particle correspond to the fitting of the wavelength λ to the environment of the particle. Fitting is accomplished by solving the Schrödinger equation[1] in three dimensions, often with time as an additional variable. The purpose of this operation is to determine the energies of the valence electrons. The essential result needed for the present discussion can also be obtained by the following simplified argument leading to the main quantum number n.

We begin by considering an electron in a simple condition, circulating around the nucleus (in the case of the hydrogen atom) or moving in one

[1] The Schrödinger or wave equation is

$$\frac{-h^2}{8\pi^2 m}\left(\frac{\partial^2\Psi}{\partial x^2} + \frac{\partial^2\Psi}{\partial y^2} + \frac{\partial^2\Psi}{\partial z^2}\right) + E_P(x,y,z)\Psi = \frac{-h}{2\pi i}\frac{\partial\Psi}{\partial t} \tag{12-3}$$

where Ψ is the wave function and E_P is the potential energy. Applications of this equation are discussed in textbooks on atomic physics and advanced physical chemistry. A broad lucid treatment is given by C. W. N. Cumper, "Wave Mechanics for Chemists," Academic Press Inc., New York, 1966.

dimension along a bar or wire. The kinetic energy $E_K = mv^2/2$ can be written in terms of momentum ($p = mv$) as

$$E_K = \frac{p^2}{2m} \tag{12-4}$$

Let L represent the length into which the waves of wavelength λ are to be fit. For the hydrogen atom, $L = 2\pi r$, where r is the radius; for a bar of length l, the corresponding length is $L = 2l$. A wavelength λ can correspond to a possible energy state for an electron only if λ "fits" into L an integral number of times n, the quantum number. These significant values of wavelength are

$$\lambda_n = \frac{L}{n} \qquad n = 1, 2, 3, \ldots \tag{12-5}$$

Substitution of this value for λ in Eq. (12-2) gives

$$p_n = \frac{nh}{L} \tag{12-6}$$

The kinetic energy E_{K_n} corresponding to quantum number n is therefore, from Eq. (12-4),

$$E_{K_n} = \frac{n^2h^2}{2mL^2} \tag{12-7}$$

This basic relation gives the principal electronic energies in a bar of a solid material (Sec. 12-1), as well as in the hydrogen atom.

For the single electron in a free hydrogen atom, the kinetic energy is half as large as the potential energy E_P and opposite in sign. The total energy is (Prob. 12-3)

$$E_n = E_{P_n} + E_{K_n} = -\frac{n^2h^2}{8m\pi^2r_n^2} \tag{12-8}$$

where r_n is the radius of the "orbit" associated with quantum number n. Although the present simplified derivation considers only the principal quantum number n, quantitative calculations with wave equations show that four quantum numbers determine the state of an electron and can influence its energy under suitable conditions. The secondary quantum number l is especially important and leads to the divisions $s(l = 0)$, $p(l = 1)$, $d(l = 2)$, etc., within the energies corresponding to a given principal quantum number, for example, $3s$, $3p$, and $3d$ for $n = 3$ in a copper atom (Fig. 12-1*a*).

12-1. Free-electron Theory

A useful theory for solid materials considers only the valence electrons. This approximation is justified because the energies of the inner electrons are essentially the same whether the atom is isolated (in the gas phase) (Fig. 12-1*a*) or is densely packed with other atoms, for example, in a crystal of ordinary solid copper. A valence electron differs significantly from the core electrons in being able to transfer easily from one atom to another. The kinetic energy E_K associated with this motion is given by Eq. (12-7) for each of the three directions of translation (x, y, and z) in the crystal by the quantum numbers n_x, n_y, and n_z [Eq. (12-5)]. Because the component momenta [Eq. (12-6)] add vectorially, the translational kinetic energy of a valence electron in a cube l m on an edge is

$$E_K = \frac{(n_x^2 + n_y^2 + n_z^2)h^2}{8ml^2} \tag{12-9}$$

This simple equation permits us to relate electronic behavior to such physical properties as electrical conductivity (Sec. 12-3).

To see why a parabolic distribution of electronic energies arises (Fig. 12-1*c*), consider a 1-m^3 specimen of copper, which contains 10^{29} valence electrons shared by 10^{29} atoms. The factor $h^2/8ml^2$ in Eq. (12-9) is then about 10^{-19} ev, and thus the lowest energy states ($n_x = 1$, $n_y = 0$, $n_z = 0$, for example) have essentially zero energy (Fig. 12-1*c*). The fourth quantum number is the spin of the electron, which may be plus or minus. To accommodate the $10^{29}/2$ pairs of electrons (with opposite spin quantum numbers), the quantum numbers n_x, n_y, and n_z must each approach a value of 10^{10} (Prob. 12-4). Consequently, the electronic energies typically reach a maximum of several electron volts, for example, 7.0 ev for copper. Consideration of the increase in the number N of possible quantum states (n_x,n_y,n_z) with increase in the value of the quantity $n^2 = (n_x^2 + n_y^2 + n_z^2)$ (Prob. 12-5) shows that a parabolic relation (Fig. 12-1*c*) exists between the "density" of states, dN/dE_K, and the energy E_K. This relation is the essential result of the free-electron theory of solids and has many direct applications. It is also the starting point for the more adequate band theory.

A valence electron usually occupies the quantum state of lowest energy available to it. Accordingly, the 10^{29} valence electrons in copper tend to occupy all the low-lying energy states up to a maximum value E_f, the *Fermi level*. Various influences, such as an electric field, may disturb this simple distribution. The influence of thermal energy kT at a given temperature T is relatively small because of the *Pauli exclusion principle*, which specifies that no two electrons can occupy a particular energy state (specified by four quantum numbers). The thermal energy is a small fraction of E_f, about 1 percent at 800°K, and thus at this temperature only those electrons within about 1 percent of E_f can change their energy and move up to an unoccupied energy level. The dashed curve in Fig. 12-1*c* is an (exaggerated) indication of the change in occupied states produced at 800°K.

12-2. Brillouin Zones and Energy Bands

The free-electron theory must be modified to take into account the positively charged atomic cores (Fig. 12-2*a*) that are such a prominent part of the structure of an actual solid material. The principal interaction between the electrons and the atomic cores is electrical and can be expressed as the potential energy of the electrons (Fig. 12-2*b*). The motion of the electrons in such a varying potential can be treated by wave mechanics, and the essential result can be interpreted in a simplified manner as diffraction of the electron waves by the crystalline lattice. Recall the phenomenon of electron diffraction (Sec. 1-11). Equation (12-5) gives the correspondence between the quantum numbers of an electron and a wave of wavelength λ moving in a definite direction in the crystal lattice (Fig. 12-3*a*). When the wavelength and the direction of motion are such as to satisfy Bragg's law [Eq. (1-13)], the energy state in

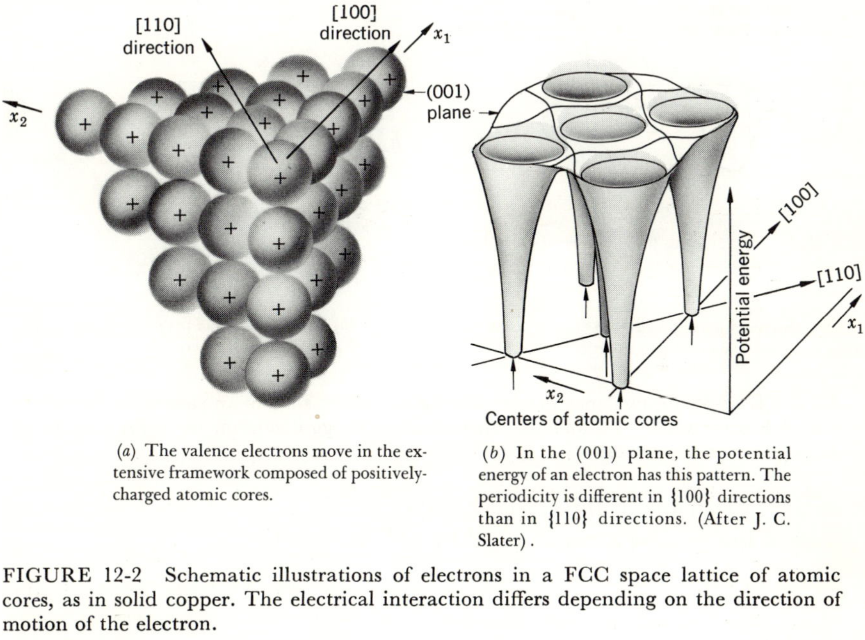

(*a*) The valence electrons move in the extensive framework composed of positively-charged atomic cores.

(*b*) In the (001) plane, the potential energy of an electron has this pattern. The periodicity is different in {100} directions than in {110} directions. (After J. C. Slater).

FIGURE 12-2 Schematic illustrations of electrons in a FCC space lattice of atomic cores, as in solid copper. The electrical interaction differs depending on the direction of motion of the electron.

question cannot exist because the electron suffers diffraction by the (*hkl*) lattice plane. This important condition for diffraction can be described more conveniently if the quantum state is described using the wave vector **k**, where $|\mathbf{k}| = 2\pi/\lambda$. From Eq. (12-2),

$$\mathbf{k} = \frac{2\pi}{h}\mathbf{p} \qquad (12\text{-}10)$$

As with the momentum, **k** is simply related to the kinetic energy. A given electron wave can be represented both in magnitude and direction by the equivalent wave vector (Fig. 12-3*b*). The component $k_\perp$ of **k** perpendicular

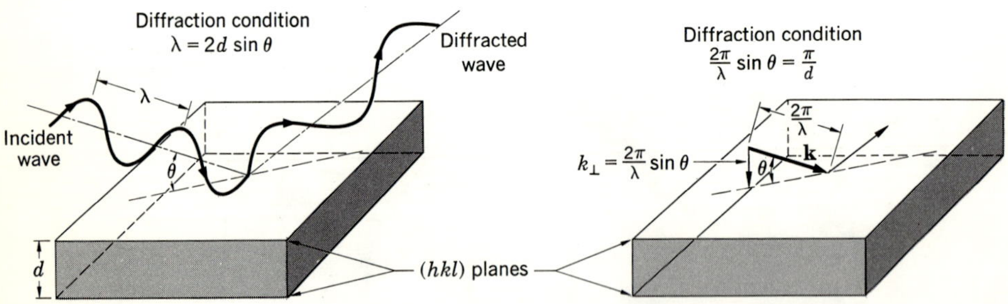

(*a*) Usual representation of an electron as a wave of wavelength λ, moving in a particular direction.

(*b*) Representation of the same electron as the wave vector, **k**, where $|\mathbf{k}| = 2\pi/\lambda$.

FIGURE 12-3 Equivalence of representing an energy state of an electron as a wave (*a*) and as a wave vector (*b*), and the corresponding Bragg condition for diffraction by the (*hkl*) planes in the crystal.

to the (*hkl*) plane in the crystal has the magnitude $(2\pi/\lambda)\sin\theta$. When both sides of Bragg's law are multiplied by π/d, the result is

$$\frac{2\pi}{\lambda}\sin\theta = \frac{\pi}{d} \qquad (12\text{-}11)$$

Therefore, diffraction of electron waves from the (*hkl*) crystal plane (Fig. 12-3*b*) occurs when the perpendicular component $k_\perp$ of the wave vector has the value $k_\perp = \pi/d$, where d is the interplanar spacing.

The relation of electron energies in solids to the critical conditions for diffraction can be shown graphically in several ways. If the energy states of the 10^{29} valence electrons in 1 m^3 of solid copper are represented in terms of their **k** values, then to a first approximation the ends of the $10^{29}/2$ **k** vectors fill a spherical volume (Fig. 12-4*a*) whose radius k_{max} is less than the critical value of π/d for diffraction from the {111} planes,† $\pi/d = \sqrt{3}\,\pi/a_0$. Accordingly, the

† The {111} planes are the first from which diffraction can occur. Recall from Problem 1-25 that no diffraction occurs in FCC lattices from {100} or {110} planes.

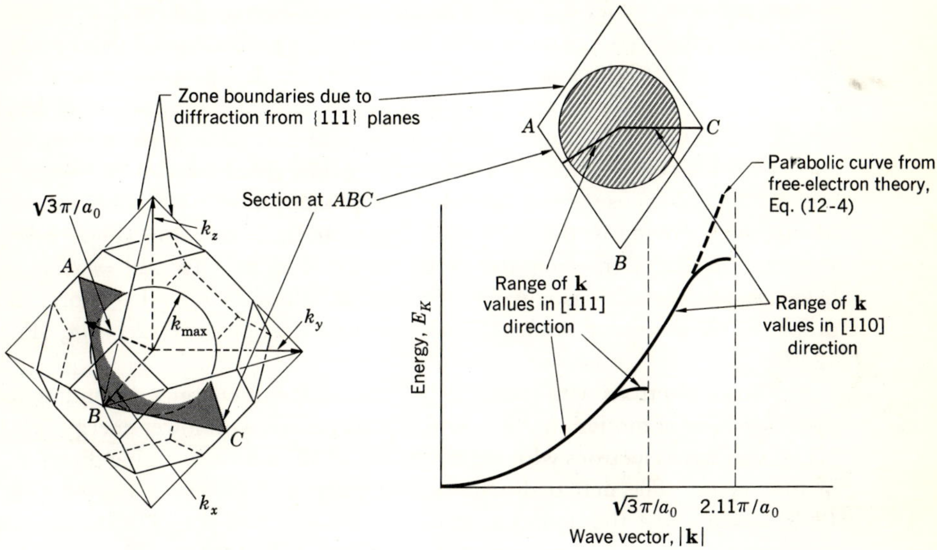

(*a*) Free-electron approximation. The Fermi surface is a sphere lying completely inside the octahedron formed by the [111] diffraction condition.

(*b*) The electron energies drop below the free-electron value as **k** approaches a zone boundary. The decrease occurs at lower **k** values in the [111] direction than in the [110] direction.

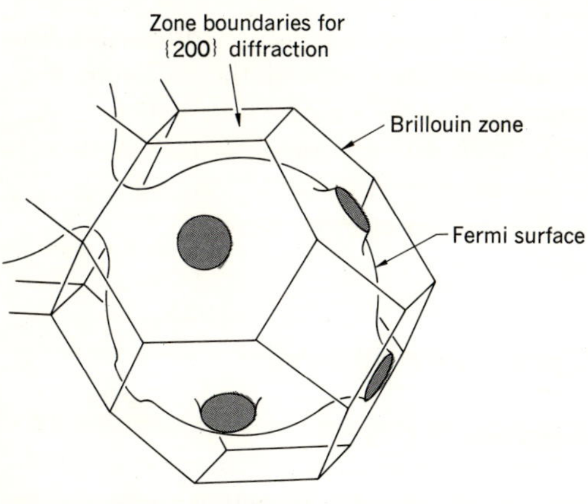

(*c*) Experimentally-determined distribution of **k** values in copper, shown within the Brillouin zone. (Data of A. B. Pippard).

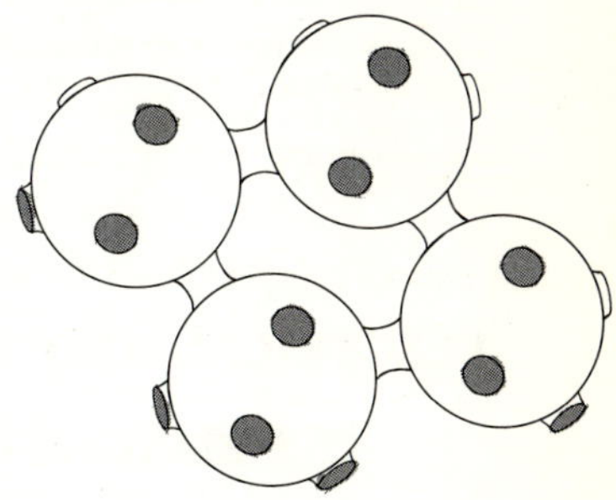

(*d*) Periodic repetition of the Brillouin zone leads to this picture of the Fermi surface as having continuity in certain portions of **k** space.

FIGURE 12-4 Representation of the energies of the valence electrons in copper by their wave vectors **k**.

Fermi surface, in this free-electron approximation, lies within the octahedron that represents values of **k** for which diffraction occurs. However, this simple picture fails to consider differences in the curve of **k** versus energy for various directions in the crystal. Figure 12-4*b* illustrates the pronounced differences for the [111] and [110] directions. In the [111] direction **k** is perpendicular to the diffracting (111) planes, and thus $|\mathbf{k}| = k_\perp = \sqrt{3}\,\pi/a_0$ is the value of **k** at which diffraction will occur. Wave-mechanical calculations show that the energy of a given electron falls below the value given by the free-electron theory (Fig. 12-1*c*) as the critical value is approached. This behavior is shown by electrons moving in the [111] direction near $|\mathbf{k}| = \sqrt{3}\,\pi/a_0$ (Fig. 12-4*b*). However, electrons moving the [110] direction must have $|\mathbf{k}| = 2.11\pi/a_0$ before $k_\perp = |\mathbf{k}| \sin 55°$ is equal to the critical value $\sqrt{3}\,\pi/a_0$. Therefore, the dropping off of energy values occurs only as $|\mathbf{k}|$ approaches $2.11\pi/a_0$ (Fig. 12-4*b*).

Since electrons with the same $|\mathbf{k}|$ have different energies depending on their direction of motion in the crystal, the directions with lower energy (near [111]) will have electrons with higher $|\mathbf{k}|$ values than the remaining directions in the crystal. The actual distribution of valence electrons in copper (Fig. 12-4*c*) reveals that the quantum states in the vicinity of the $\langle 111 \rangle$ directions are occupied all the way out to the critical value $|\mathbf{k}| = \sqrt{3}\,\pi/a_0$. Consequently, rather than being spherical, the region of occupied states has periodic protruberances. The quantum states at the surface of this figure all have the same energy, which is the Fermi energy E_f. For this reason the surface is known as the *Fermi surface.* The surfaces defined by the critical values of $k_\perp$ form a closed figure, the Brillouin zone. In the FCC lattice, diffraction from both {111} and {200} planes account for the shape of the Brillouin zone. Problem 12-6 considers the Brillouin zone in the simple-cubic lattice. Certain superpositions or extensions of the *Brillouin zone* are useful for interpreting the behavior of electrons in various materials. The periodic zone scheme (Fig. 12-4*d*) permits easy visualization of particular electron orbits.

ELECTRICAL AND THERMAL PROPERTIES

12-3. Electrical Conductivity

An electric current is a flow of charge past a reference plane in the material. Two important aspects of this phenomenon can be seen in the following expression for the current density I amp/m², based on Eq. (6-49):

$$I = \sum_{i=1}^{n} J_i z_i e \tag{12-12}$$

where e is the electronic charge, 1.602×10^{-19} coul. First, the charge may be carried by various types of particles (electrons, holes, or ions); the magnitude and sign of the charge on particles of type i are specified by the charge number z_i. Second, the total flux J_i of particles of type i may have components caused by various driving forces, such as gradients in concentration, temperature, or electrical potential V. When only the electrical potential gradient is important, the flux of particles is given by Eq. (6-48) and the current density is

$$I = -\sum c_i z_i e M_i \frac{dV}{dX} \tag{12-13}$$

Since Ohm's law can be written (Prob. 12-7)

$$I = -\sigma \frac{dV}{dX} \tag{12-14}$$

the coefficient σ connecting the driving force and the flux of charge is given by

$$\sigma = \sum c_i z_i e M_i \tag{12-15}$$

That is, the *electrical conductivity* σ of a given material may result from the motion of several different types of particles. The contribution of a given type of particle depends on (1) the number c_i of particles per unit volume, (2) the electrical charge $z_i e$ of each particle, and (3) the average mobility M_i of the particles per unit gradient of electrical potential.

Equation (12-15) explains the wide variation in electrical conductivity (or resistivity, $1/\sigma$) exhibited by different classes of materials (Fig. 12-5). Usually a given type of particle predominates, such as the many conduction electrons in metals that give this class of materials its characteristic high conductivity (low resistivity). Electrons also account for the conductivity of n-type semiconductors, but the relatively low concentration of electrons in these materials compared to metals is the major factor causing the conductivity to be much lower. In ionic crystals such as Al_2O_3 the concentration of conduction electrons is so low that a different particle, the ion, becomes comparatively significant for conduction. Electrical conduction is then so small that these materials are used as insulators.

METALLIC CONDUCTION. For those materials that are electron (or hole) conductors, the band structure has primary importance. Metals have a large conductivity because their energy bands are only partially filled. The single valence electron in a copper atom, for example, only half fills the first (s-type) energy band (Fig. 12-6a). A more diagrammatic illustration of the filling of available energy states in this band is shown in Fig. 12-1c. In metals that have

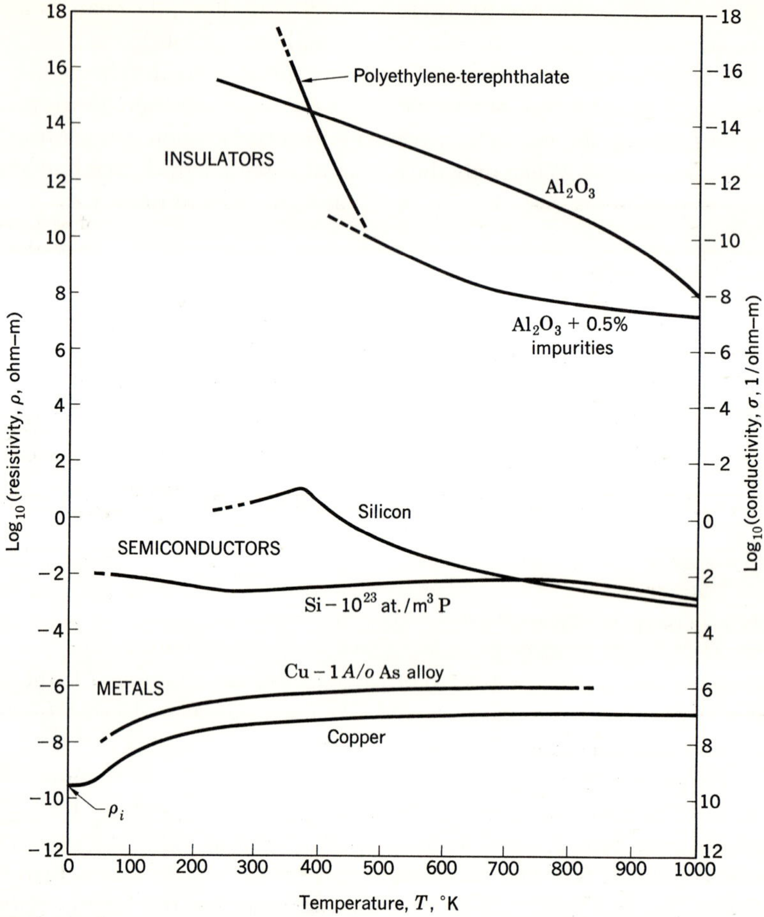

FIGURE 12-5 Data for representative materials of various types showing the wide variation in electrical resistivity. The effects of temperature and of alloying additions are also different for each type of materials.

two valence electrons, such as calcium, the variation of the *E*-vs.-**k** curve with direction (Fig. 12-4*b*) becomes significant. In the first energy band some of these curves extend to high values of **k** and of energy (Fig. 12-6*b*). Consequently, some energy states in the second band lie below the highest values in the first band. For this reason the first band is incompletely filled, and the second band contains some electrons. Since vacant energy states exist adjacent to the occupied states in metals, an electric field can accelerate the electrons (which requires changing them to a higher energy state).

The action of an electric field $E = -dV/dX$ on an electron can be quite

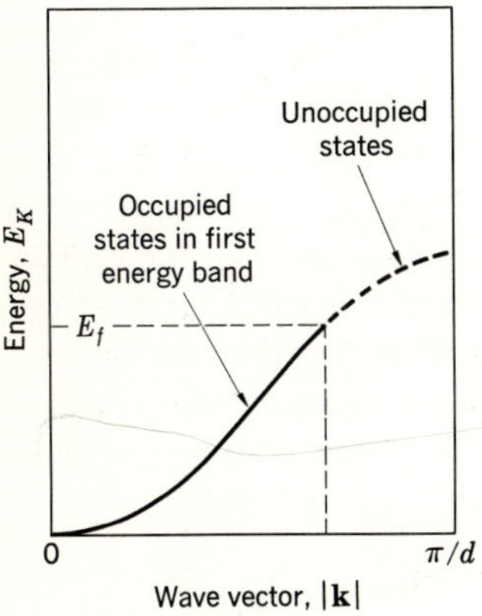

(*a*) The single valence electron in copper only half fills the first energy band.

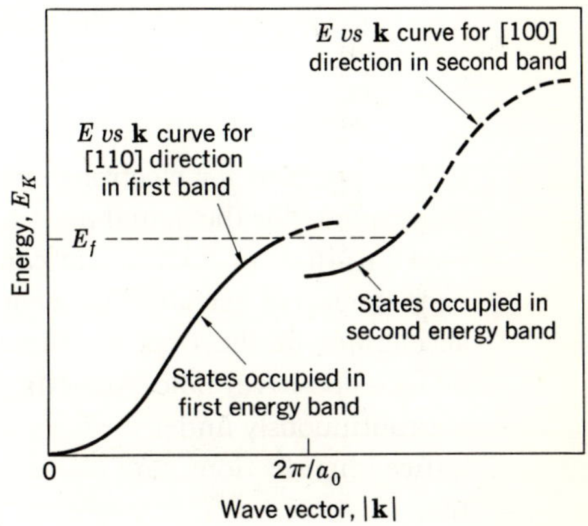

(*b*) The two valence electrons in calcium occupy two overlapping bands.

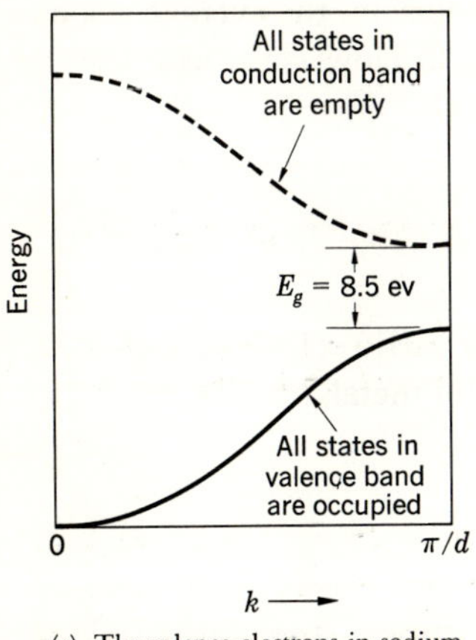

(*c*) The valence electrons in sodium chloride completely fill the valence band. The conduction band, which is empty, is shown above the other band for convenience.

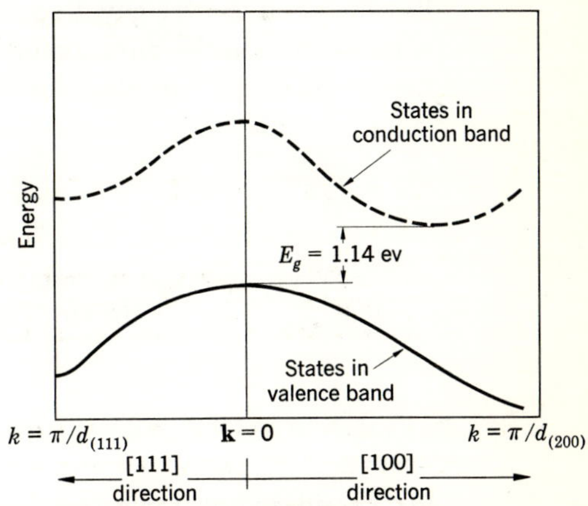

(*d*) The band structure of silicon is typical of semiconductors in having an "indirect" minimum band gap, involving different values of $\mathbf{k}$ in the two bands. (After Charles Kittel).

FIGURE 12-6 Simplified models of the band structures of various materials.

complex, but it involves the following aspects even in the simplest case. Newton's second law requires that the force $-eE$ on an electron produce a change in its momentum $\mathbf{p}$ given by

$$-eE = \frac{d\mathbf{p}}{dt} \tag{12-16}$$

The usual expression for the momentum,

$$\mathbf{p} = m^*\mathbf{v} \tag{12-17}$$

can be retained for electrons in a crystal provided that the *effective mass* m^* is substituted[1] for the actual mass m. The ratio m^*/m is near unity for the valence electrons in most metals, but can depart greatly from this value for electrons near the top of certain filled energy bands. When m^* is negative, the electron accelerates in the opposite direction and is then alternatively described as being a (positive) hole. According to Eq. (12-16), the electrons would accelerate continuously under the force $-eE$ if they did not "collide" with impurities, lattice imperfections, and phonons. If the average time between collisions is 2τ, then

$$-eE2\tau = \delta\mathbf{p} = m^* \, \delta\mathbf{v}_{\text{max}} \tag{12-18}$$

Provided that the velocity drops to zero at each collision, the average velocity of an electron in the direction of the field is $\delta\mathbf{v}_{\text{max}}/2$. However, this velocity is the product M_eE, and thus the electron mobility is

$$M_e = \frac{-e\tau}{m^*} \tag{12-19}$$

When this expression for M_e is substituted in Eq. (12-15), it gives the following equation for the electrical conductivity of metals:

$$\sigma = \frac{c_e e^2 \tau}{m^*} \tag{12-20}$$

We consider the simple case in which only electrons (and not holes) are important, and in which $m^*/m = 1$. Since each atom of a typical metal contributes its valence electron(s) to the "cloud" of conduction electrons, c_e in Eq. (12-20) is very large (about $10^{29}/\text{m}^3$) and produces the large value of σ shown by the data of Fig. 12-5. An approximate calculation (Prob. 12-8)

[1] The basis for this substitution is given by Charles Kittel, "Introduction to Solid State Physics," 3d ed., p. 277, John Wiley & Sons, Inc., New York, 1966.

shows that the average electron in copper travels only about 3×10^{-8} m between collisions at room temperature. At 4°K, however, this mean free path is much greater (about 3×10^{-3} m). Three different contributions to the resistivity of metallic materials can be distinguished. A specimen so thin that its thickness is comparable to the mean free path of the electrons has a contribution to its resistivity caused by scattering of the electrons at the surface. In specimens of normal size, however, the resistivity ρ is the sum of two terms involving only the bulk of the material; thus

$$\rho = \rho_L + \rho_i \tag{12-21}$$

ρ_L is a contribution due to thermal vibrations in the lattice; it vanishes at low temperatures (see the curve for copper in Fig. 12-5) but is the dominant factor in relatively pure metals at room temperature. ρ_i is the resistivity caused by imperfections (both defects and impurity atoms) in the lattice. The effect of a given impurity depends on its chemical nature as well as on the amount present. Arsenic strongly increases the resistivity of copper (Fig. 12-5) because these two elements are so different chemically.

MATERIALS WITH FILLED BANDS. Most electrons in nonmetallic substances behave quite differently from the conduction electrons in a metal. These substances are characterized by essentially completely full and completely empty energy bands. The electrons in a full band cannot move to states of higher energy within the band since none are available. Consequently, the electrons cannot acquire a net drift velocity and cannot exhibit metallic-type conductivity. Other mechanisms of electrical conductivity then become the only type possible, but the amount of conduction is lower by many orders of magnitude.

In ionic crystals such as sodium chloride (Fig. 12-6*c*), the empty conduction (upper) band is separated from the valence band by a large energy gap, and thus thermal energy is incapable of raising any appreciable number of electrons into the conduction band. An applied electric field exerts a force $z_i eE$ on each of the two types of ions in the crystal, but a given ion can move only by a defect mechanism, such as exchange with a vacancy of the same type. The number of cation vacancies is equal to the number of anion vacancies (because of electrical neutrality), but in sodium chloride the small cation exchanges far more easily than the large anion. Consequently, the electric current is carried predominantly by the cations. At room temperature the fraction of vacancies is only 10^{-17}, and their jump frequency is only 1/sec, and thus the electrical conductivity from this source is extremely low. Even in "pure" ionic materials, impurity ions account for the conductivity observed below roughly 1000°K. The data for Al_2O_3 (Fig. 12-5) show that an intentionally added impurity (such as an alkali oxide in this case) increases the conductivity of an ionic material by a large factor. At temperatures approaching 2000°K, however, the intrinsic ionic mobility of Al_2O_3 accounts for the observed conductivity of quite impure material. Nevertheless, even near its melting point, 2300°K, Al_2O_3 is a good electrical insulator. Ionic conduction leads to the accumulation of ions of one type at each end of the specimen where they are continuously converted to the un-ionized condition (gaseous chlorine and metallic sodium in the case of an NaCl specimen). These chemical reactions are a characteristic and potentially harmful aspect of ionic conductivity.

Semiconductors have quite unusual band structures,[1] but the essential feature is the small energy gap between the valence band and the conduction band (Fig. 12-6*d*). As explained in Chap. 4, defect electrons and holes are produced by thermal excitation and cause the increase in conductivity (decrease in resistivity) exhibited by silicon above 300 or 400°K (Fig. 12-5). Below this temperature the residual impurities in even "pure" silicon contribute an "extrinsic" effect. Two important aspects of extrinsic semiconductors are shown by the data for the (*n*-type) phosphorus-doped silicon in Fig. 12-5. As the temperature increases from near 0°K, the conductivity increases because of increasing ionization of the electrons donated by the phosphorus atoms. After ionization is complete (near room temperature), the conductivity decreases because the mobility of the electrons decreases with increasing temperature. At a sufficiently high temperature, 800°K in this case, the intrinsic behavior of the silicon begins to dominate the conduction process.

In regions of intrinsic behavior the defect electrons e' in the conduction band and the holes $h^\bullet$ left in the valence band are able to move to higher-energy states under the action of an electric field. Consequently, the total electrical conductivity is the sum

$$\sigma = (c_e z_e M_e + c_h z_h M_h)e \qquad (12\text{-}22)$$

This equation also describes the electrical conductivity in extrinsic semiconductors, although one of the two terms may be negligibly small.

Various organic polymers cover almost the entire spectrum of electrical conductivity including insulators, semiconductors, ionic conductors, electronic conductors, and perhaps even superconductors. Polymeric insulators are of greatest technical importance, however, and they are described briefly here. The band structure of the typical nonpolar polymer is similar to that of an ionic substance (Fig. 12-6*c*) in that the energy gap is too large to permit

[1] An introductory discussion of this topic is given by C. A. Wert and R. M. Thomson, "Physics of Solids," 2d ed., pp. 254ff, McGraw-Hill Book Company, New York, 1970.

appreciable thermal excitation of electrons into the conduction band. Conduction at room temperature is due to the motion of impurity ions and is extremely low in polyethylene, polystyrene, etc. Typical data for polyethylene-terephthalate are given in Fig. 12-5. Ionic mobility increases rapidly with increase in temperature, and the conductivity changes accordingly. The polymers described here obey Ohm's law fairly well, but marked deviations occur with some materials. In polypropylene, for example, additional charge carriers are produced as the electric field increases. The conductivity increases linearly with field strength in this case.

SUPERCONDUCTIVITY. The electrical resistivity of certain pure metals and compounds drops suddenly to zero when the temperature is lowered to a critical temperature T_c close to absolute zero. This phenomenon of the superconducting state, although possessing enormous technical potential, has had only limited application because of two difficulties. The first is the cost of maintaining devices at temperatures on the order of 10°K. The second is the effect of a sufficiently strong magnetic field in causing the material to revert to its normal state. A type I(a) superconductor (Fig. 12-7*a*) is in the superconducting state (or phase) in a region bounded by the critical temperature T_c (for $\mu_v H = 0$) and the critical magnetic field† $\mu_v H_c$ (for $T = 0$). In a type I(b) superconductor (Fig. 12-7*b*) the transition from the superconducting state to the normal state involves the formation at the surface of the specimen of a sheath phase having very low electrical resistivity.

Type II superconductors (Fig. 12-7*c*) are important because they usually have much higher effective values of $\mu_v H_c$. When the magnetic field reaches the critical value $\mu_v H_{c_1}(T)$ at some useful operating temperature T, domains of normal phase begin to appear within the superconducting phase. This mixture is known as the vortex phase but still has zero electrical resistance since all of the current is carried by the superconducting part of the mixture. The amount of normal phase in the mixture gradually increases as the magnetic field approaches $\mu_v H_{c_2}(T)$. At this value the entire specimen is normal except for the sheath phase, and this phase gradually disappears as $\mu_v H_{c_3}(T)$ is approached.

The magnetic behavior of superconductors has basic significance. The superconducting phase, which exists below $\mu_v H_{c_1}$ in Fig. 12-8*a*, is a perfect diamagnet (Sec. 12-6) and contains no magnetic flux. To cancel the external magnetic field $\mu_v H$, the magnetization within the superconducting phase must be of the same magnitude (except negative in sign). Electric currents flowing

† In cgs units the flux density B (in gauss) $\equiv \mu_v H$ produced by the magnetic field H (in oersteds) is numerically equal to H since the permeability of a vacuum, μ_v, is unity. In SI units $\mu_v = 4\pi \times 10^{-7}$ henry/m, and therefore it must be included in the expression for the flux density. Nevertheless, "magnetic field" continues to be employed in referring to $\mu_v H$.

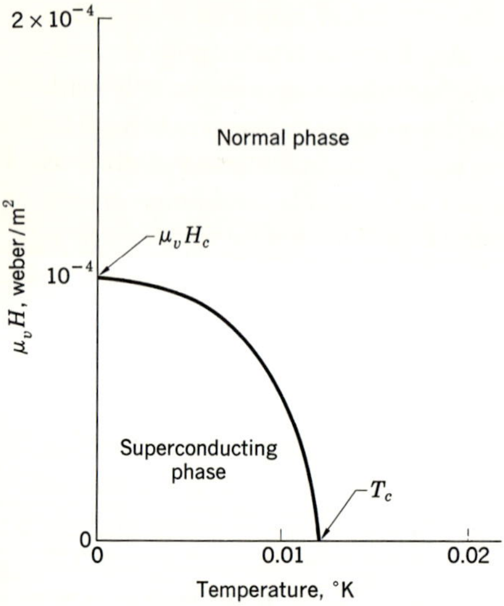

(*a*) A type I (*a*) superconductor. The properties shown are approximately those of pure tungsten.

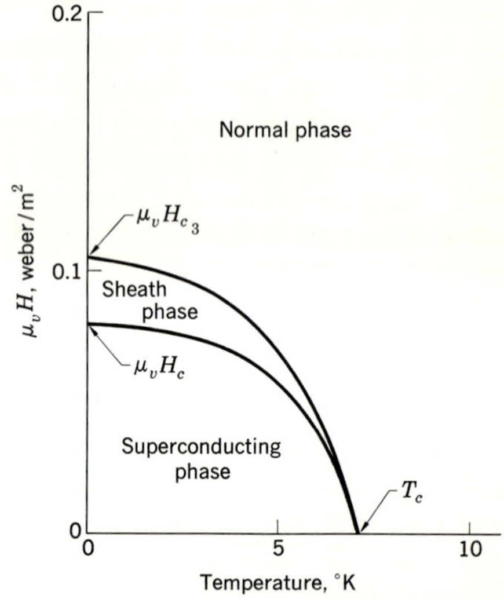

(*b*) A type I (*b*) superconductor. The properties shown are approximately those of pure lead.

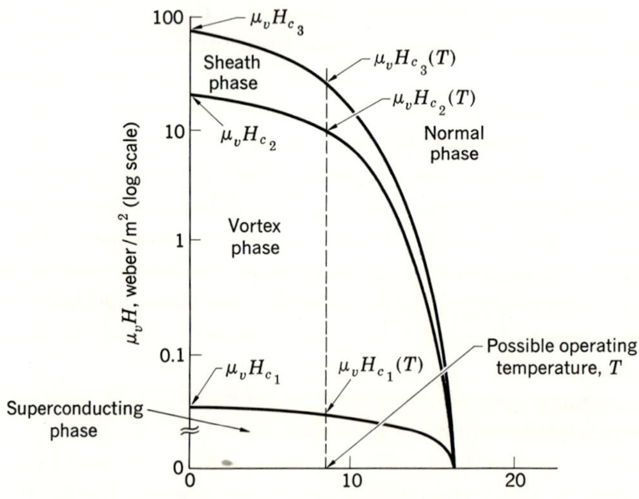

(*c*) A type II superconductor. The properties shown are approximately those of "hardened" Nb_3Sn.

FIGURE 12-7 Regions of superconducting properties of three types of superconductors as a function of temperature and magnetic field. The width of the field of the sheath phase is much smaller than shown (conversion of units: M webers/m² × 10^4 = N gauss).

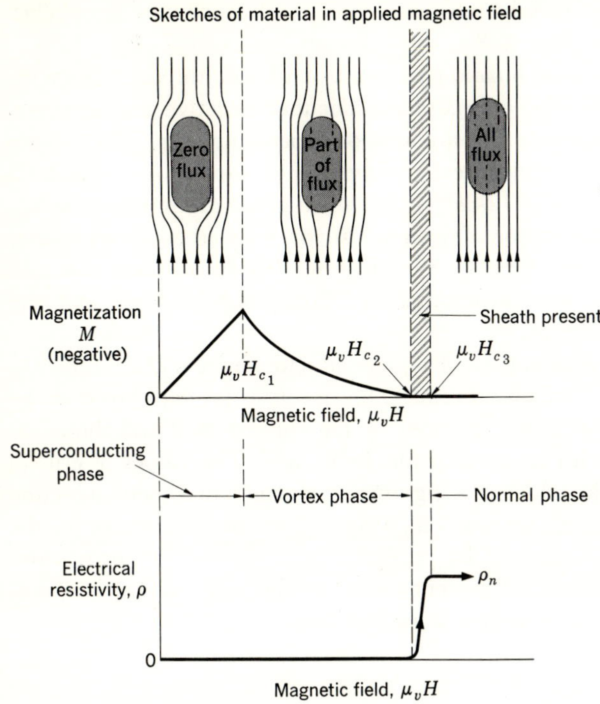

(*a*) The magnetic behavior in the various regions of superconducting behavior.

(*b*) The electrical resistivity remains essentially zero up to $\mu_v H_{c_2}$ and then rises rapidly to the normal value at $\mu_v H_{c_3}$.

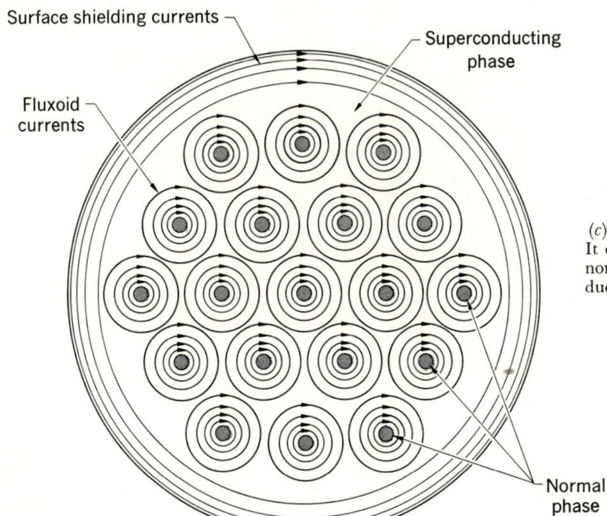

(*c*) Cross-section view of the vortex phase. It consists of an ordered array of rods of normal material embedded in superconducting phase.

FIGURE 12-8 Schematic illustration of the magnetic and electrical properties of a type II superconductor as a function of magnetic field. (*After J. K. Hulm and D. W. Deis, Westinghouse Research Laboratories.*)

in the surface of the superconducting phase exclude the magnetic flux. When the vortex phase forms (Fig. 12-8*c*), the "threads" of normal phase carry a portion of the magnetic flux; fluxoid currents around each thread exclude the flux from the superconducting matrix phase. The total negative magnetization of the specimen decreases as the amount of superconducting phase tends toward zero at $\mu_v H_{c_2}$. The electrical resistivity then rapidly increases from zero to the normal value (Fig. 12-8*b*) as the sheath phase disappears at $\mu_v H_{c_3}$. One type of potential application of superconductors is based on their capacity to store energy in magnetic flux. An electrical or mechanical input can shift the flux from one conductor to another (a "flux pump" action) and can generate large direct currents.

Many potential applications of superconductors involve their ability to carry extremely large electrical currents. This ability is limited, however, by two factors, the magnetic field generated by the current itself and the interaction of the current with an applied magnetic field. The values of critical current, J_{c_1} and J_{c_2}, at which the significant changes in superconducting behavior occur (Fig. 12-9), depend very strongly on the constitution of the material. Hardening the material by precipitation, cold working, or neutron irradiation increases the critical current (and also the $\mu_v H_c$ values) and produces a superior material for technical applications.

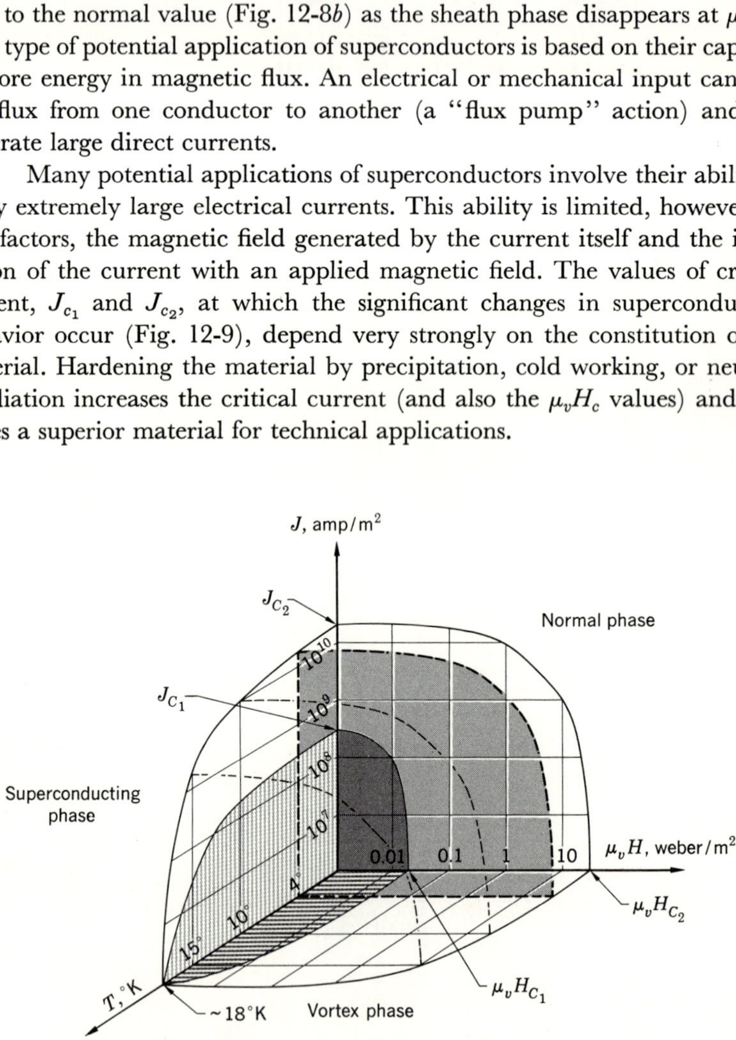

FIGURE 12-9 The effect of electric current, applied magnetic field, and temperature on the superconducting state of Nb_3Sn in a hardened condition. (*After J. K. Hulm and D. W. Deis, Westinghouse Research Laboratories.*)

12-4. Thermal Properties

Thermal energy causes effects of many kinds in materials. Phase transformations, changes in solubility, and activation of reactions are examples studied in previous chapters. We will now consider three additional thermal characteristics of solid materials:

1. Specific heat, which determines the increase in temperature produced by the absorption of a given quantity of heat.
2. Thermal conductivity, which governs the rate at which heat flows in the material.
3. Thermoelectric effects, which interconnect thermal and electrical flows and driving forces.

The phonon theory of lattice vibrations and the band theory of electrons furnish a useful basis for understanding the broad differences in thermal behavior among the various classes of solid materials.

SPECIFIC HEAT. The specific heat at constant volume, C_V, is defined by the equation

$$\delta Q = C_V \, \delta T \tag{12-23}$$

and specifies the amount of heat δQ necessary to raise the temperature of 1 g mole of material by 1°K. The heat is pictured as being transferred at constant volume so that no external work is done. The added heat causes a number of changes in the material, such as increasing the number of point defects, but usually its effect on the lattice vibrations is of major importance. If the lattice vibrations are described in terms of phonons (Sec. 4-2), the uniform behavior of all solid materials is readily understood. At temperatures above the Debye temperature (Table 4-1) all phonons have essentially the same energy [Eq. (4-7)], and the number of phonons increases linearly with temperatures [Eq. (4-6)]. Consequently, C_V has the constant value 6 cal/g mole-°K given by the well-known law of Dulong and Petit. Below the Debye temperature both the number and the average energy of the phonons decrease; consequently, the specific heat decreases gradually at first and then approaches zero rapidly as the temperature approaches 0°K. Only in metals are there enough electronic carriers to appreciably influence the specific heat. Even in this case, since only the electrons at the top of a band can be thermally excited, the electronic contribution to the specific heat is only a few percent of the total value at normal temperatures.

THERMAL CONDUCTIVITY. When a temperature gradient dT/dX exists in a material, a heat flux H cal/m^2-sec occurs in accordance with Fourier's law; thus

$$H = -K\frac{dT}{dX} \tag{12-24}$$

where K is the thermal conductivity. Since thermal energy is primarily in the form of lattice vibrations, the conduction process consists of the transfer of vibrational energy from a region of higher temperature (higher energy) to a region of lower temperature. When lattice vibrations are represented as "particles" (phonons), these particles can be expected to behave somewhat like the atoms in a gas and to roughly obey the same equation for thermal conductivity,[1] that is,

$$K = \tfrac{1}{3}Cv\lambda \tag{12-25}$$

Here C is the specific heat per unit volume, v is the velocity of the phonons, and λ is their mean free path. A typical variation of K with temperature for a nonmetallic crystal is that of germanium (Fig. 12-10). Because of the specific heat term, K is zero at 0°K. It first increases to a maximum but then decreases because interactions among the phonons cause λ to decrease continuously with increasing temperature. Glassy solids behave quite differently because their λ value is determined by structural disorder of the molecules; consequently, λ is very small (of molecular dimensions) and is insensitive to changes in temperature. In these materials K increases continuously with increase in temperature, but at room temperature K is still about a factor of 10 below the value for a crystalline nonmetallic solid.

In metals, the conduction electrons provide an alternative mechanism of thermal conduction and produce distinctly higher values of thermal conductivity, as shown by the data for copper in Fig. 12-10. The electrons acquire energy from a high-energy phonon, carry this energy to a cooler portion of the specimen, and transfer the energy to a lower-energy phonon. Equation (12-25) can also be applied to the "electron gas," and for this reason the plot of thermal conductivity versus temperature is similar to that for the phonon mechanism. Even in a good electrical conductor such as copper, phonons account for part of the thermal conductivity. In a relatively poor metallic conductor, such as a highly alloyed metal, phonons and electrons make comparable contributions to K.

THERMOELECTRIC EFFECTS. The term *cross effect* describes a phenomenon in which a driving force, say an electric field, produces some secondary effect,

[1] An elementary derivation is given by Charles Kittel, "Introduction to Solid State Physics," 3d ed., p. 186, John Wiley & Sons, Inc., New York, 1966.

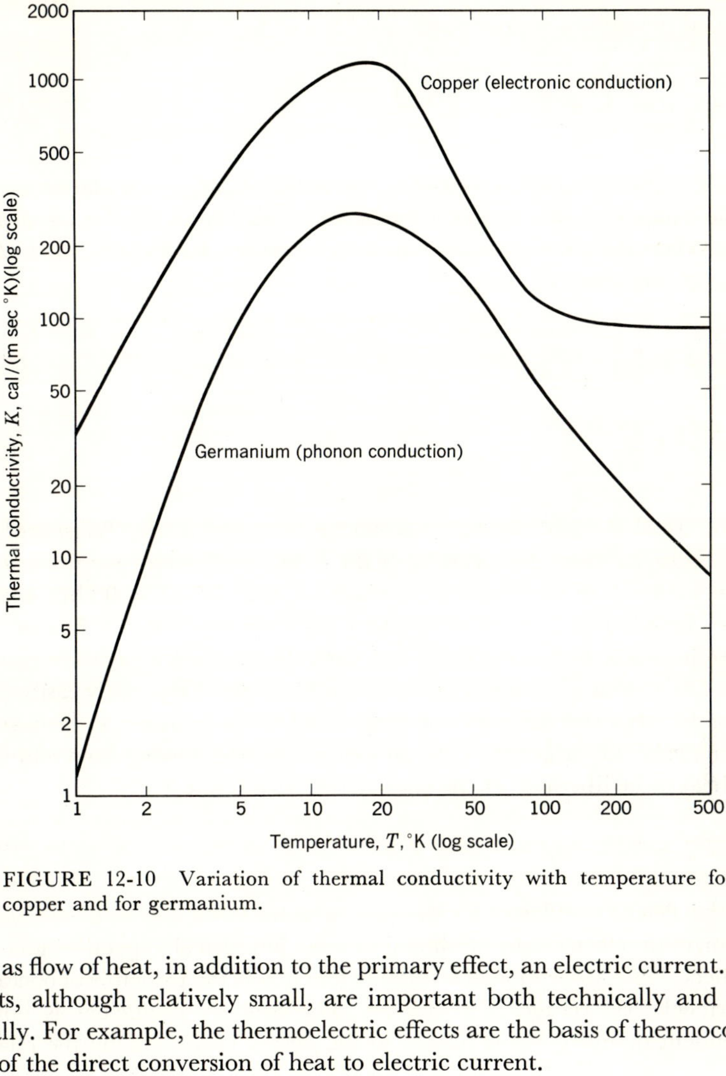

FIGURE 12-10 Variation of thermal conductivity with temperature for copper and for germanium.

such as flow of heat, in addition to the primary effect, an electric current. Cross effects, although relatively small, are important both technically and scientifically. For example, the thermoelectric effects are the basis of thermocouples and of the direct conversion of heat to electric current.

Consider first a material such as a typical metal or an *n*-type semiconductor in which electrical conduction is primarily by electrons. A flux of electrons, J_e, is produced both by an electric field $E = -dV/dX$ [Eq. (6-48)] and also by a temperature gradient dT/dX [Eq. (6-59)]. The total of these two fluxes is

$$J_e = c_e M_e E - D_e^T \frac{dT}{dX} \tag{12-26}$$

Since dT/dX and the current density $I = -J_e e$ are usually the experimentally controlled quantities, we rearrange Eq. (12-26) to the form

$$E = -(c_e M_e e)^{-1} I + D_e^T (c_e M_e)^{-1} \frac{dT}{dX} \tag{12-27}$$

From Eq. (12-15), $c_e M_e e = -\sigma = -1/\rho$, and $D_e^T (c_e M_e)^{-1}$ can be set equal to a single quantity α the Seebeck coefficient (also known as the *thermoelectric power*). When the corresponding equation is written for the heat flux H, the final set of equations is

$$E = \rho I + \alpha \frac{dT}{dX} \tag{12-28}$$

$$H = \pi I - K \frac{dT}{dX} \tag{12-29}$$

K is the usual thermal conductivity, and π is known as the *Peltier coefficient.*

To understand the meaning of the Peltier coefficient, picture a rod of a given material at constant temperature (so that $dT/dX = 0$) but with an electric current I because of an applied electrical potential. A flux of heat, $H = \pi I$, must also be flowing in the rod, even though the temperature gradient is zero. The presence of such heat fluxes can be detected especially easily at the junction between two different materials a and b. For example, if the heat flux H_a in material a is larger than H_b, an electric current flowing from a to b will cause heat to be liberated at the a-b junction at the rate

$$\Delta H = H_a - H_b = (\pi_a - \pi_b) I = \pi_{ab} I \tag{12-30}$$

π_{ab} is the Peltier coefficient for the pair of materials a and b. Changing the direction of the electric current through the a-b junction changes the sign of the heat effect. Heat absorption at a suitable junction is the principle of operation of thermoelectric refrigeration devices. The Seebeck coefficient is not an independent quantity but is related to the Peltier coefficient by the relation

$$\alpha = \frac{\pi}{T} \tag{12-31}$$

where T is the absolute temperature.

Consideration of the Peltier effect at the junction between a metal and and an n-type semiconductor (Fig. 12-11) reveals the physical nature of this phenomenon. When a voltage is applied to the device shown, the energy level

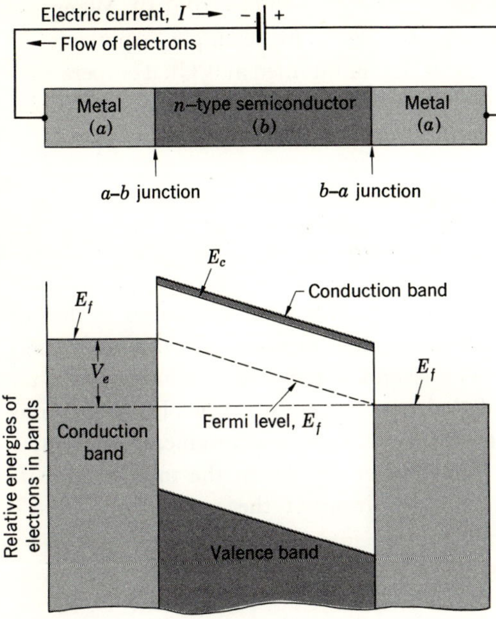

(*a*) An applied voltage causes a flow of electrons through the semiconductor and its metal contacts.

(*b*) The Fermi levels match at both junctions, so E_c lies above the Fermi level of the metal.

FIGURE 12-11 Illustration of the nature of the Peltier effect at the junctions between a metal and an *n*-type semiconductor.

of the right-hand metal contact is lowered relative to the other contact (Fig. 12-11*b*) and produces a flow of current (that is, of electrons) through the circuit, including the semiconductor. Even those electrons flowing in the metal that have the Fermi energy E_f are not able to enter the semiconductor at junction *a-b* because the only vacant states are at the higher energy E_c. Consequently, only the especially energetic electrons cross the junction, the temperature near the junction is lowered, and (the Peltier) heat flows into the junction. By convention, $E_c \equiv 0$, and the Fermi energy E_f is measured down from this reference value; therefore,

$$\pi_{ab} = E_f + 2kT \tag{12-32}$$

The additional $2kT$ of energy is required for transport of the electrons in the semiconductor. Peltier heat of the same magnitude is *liberated* at junction *b-a*.

In a metal-metal junction the Peltier effect is a second-order effect (Prob. 12-9) and is only a few percent of that possible in metal-semiconductor junctions. Nevertheless, the small Peltier and Seebeck effects are the basis of thermocouples for measuring temperature. A typical pair of thermocouple wires are chromel-alumel (both are nickel-base alloys) which can be used in continuous operation from −200 to +1200°C (−330 to 2190°F). When the welded junction of these two wires is at 1000°C, for example, a potential of 0.0413 volt exists across the cold ends (when maintained at 0°C). Temperature-potential tables for several thermocouples can be found in standard handbooks, including the "Handbook of Chemistry and Physics."

12-5. Dielectric Properties

Electrical insulators (ceramics, polymers, etc.) have two different responses to an electric field. First, as for metals, the field produces a steady flux of current (Sec. 12-3). This flux is extremely small compared to that in metals, however, and these materials can be used to electrically insulate a good conductor, such as a copper wire. The second response of these materials to an electric field is a displacement of their common centers of plus and minus electric charge q a fraction of an atomic distance d, thus producing an electric dipole $p = qd$. The sum $\sum qd$ of these atomic dipoles in the total volume V of the material is employed in the definition of the macroscopic *electric polarization*.

$$P = \sum \frac{qd}{V} \tag{12-33}$$

which is fixed in the material (in contrast to the mobile current of electrical carriers in a conductor). The term *dielectric* refers to nonconductors whose polarization is caused by an electric field. A separate class of materials, the *ferroelectrics*, have permanent atomic polarization even in the absence of an electric field.

DIELECTRIC MATERIALS. A typical use of a dielectric is in a capacitor, a device widely used in electric circuits for storing charge. The electronic mechanism of the operation of a capacitor can be understood by considering the parallel-plate type sketched in Fig. 12-12*a*. When a voltage V is applied across two metallic plates, the electrons are displaced a net amount proportional to the field E. This field is essentially zero within the body of the conductor but has the appreciable value $E = -V/l$ in the vacuum between the plates and therefore at the surface of the plates. The total charge Q produced on the surface area A of each plate by the local displacement of electrons is

$$Q_v = A\epsilon_v \frac{V}{l} = C_v V \tag{12-34}$$

ϵ_v, the dielectric constant (or *permittivity*) of a vacuum, is 8.854×10^{-12} farad/m, and C_v is the capacitance in farads of a capacitor of given geometry with a vacuum between its plates.

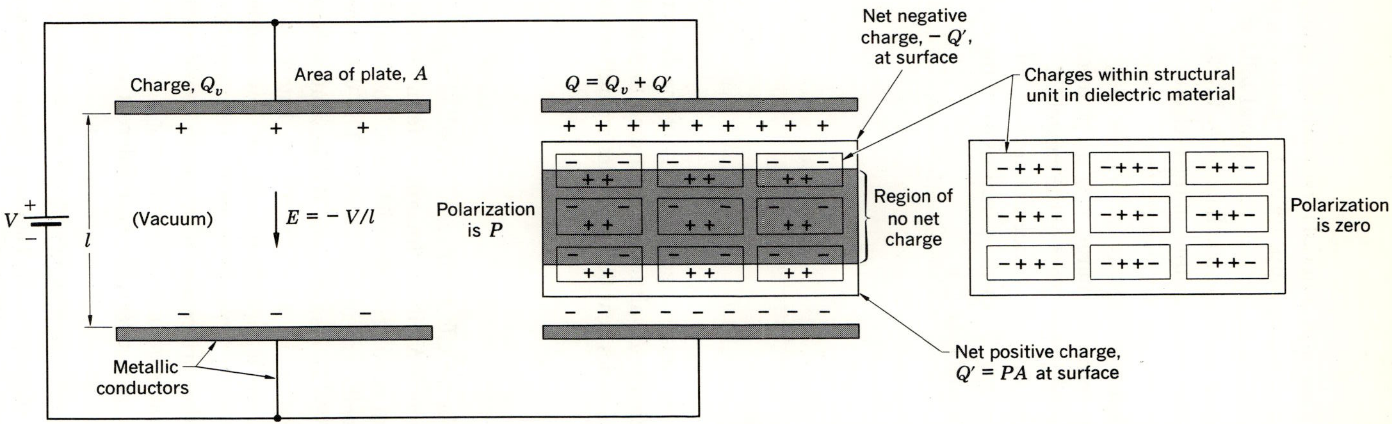

(*a*) Displacement of the electrons in the surfaces of the conducting plates produces the stored charge Q_v.

(*b*) The induced charge, Q', at the surface of the dielectric increases the stored charge to Q.

(*c*) In the absence of an electric field, the centers of the negative and positive charges coincide.

FIGURE 12-12 Schematic illustration of the effect of a dielectric material in increasing the capacity of a capacitor. The dielectric actually fills all the space between the plates.

When a sheet of dielectric material is between the plates of the capacitor (Fig. 12-12*b*), an increased charge appears on the plates for the same voltage because of polarization of the dielectric. The typical solid dielectric has zero polarization in the absence of an electric field (Fig. 12-12*c*). The applied field E causes polarization of the entire volume of the dielectric [Eq. (12-33)], which results in the production of induced charges Q'. As suggested by the units of the polarization P (coulombs per square meter), the magnitude[1] of Q' per unit area is simply P. If the induced charge $-Q'$ is pictured as neutralizing an equal positive charge in the upper plate of the capacitor, then the total charge that can be stored is $Q = Q_v + Q'$. Since the applied voltage is constant, Eq. (12-34) gives a value $\epsilon = \epsilon_v Q/Q_v$ for the permittivity of the dielectric material. Correspondingly, the capacitance C is

$$C = KC_v \tag{12-35}$$

where

$$K = \frac{\epsilon}{\epsilon_v} \tag{12-36}$$

is the dielectric coefficient (relative permittivity) of the dielectric.

The simple coefficient K, although often used to describe the behavior of dielectrics at low frequencies near 60 Hz (cycles/second), must be generalized to a complex quantity

$$K = K' - iK'' \tag{12-37}$$

to account adequately for the effects observed at radio frequencies or in optical experiments. The coefficient K'' in the imaginary term measures the electrical power loss (and resultant heating) in the dielectric. The real part K' is equivalent to the dielectric coefficient in Eq. (12-35). The variation of K' and K'' with frequency (Fig. 12-13) results from the separate operation of several mechanisms of polarization in solid dielectrics. Interfacial polarization is usually unimportant since it can occur only at low frequencies. In most materials the effect is small; it results from the relatively slow collection of extraneous charges on the surface and depends on contaminating substances. In semicrystalline polymers or in those containing a filler, however, the surfaces in question are those between the two different types of materials, and thus interfacial polarization can be appreciable at low frequencies. Similar effects occur in heterogeneous ceramic materials. As the frequency reaches

[1] A proof is given by C. A. Wert and R. M. Thomson, "Physics of Solids," 2d ed., p. 383, McGraw-Hill Book Company, New York, 1970.

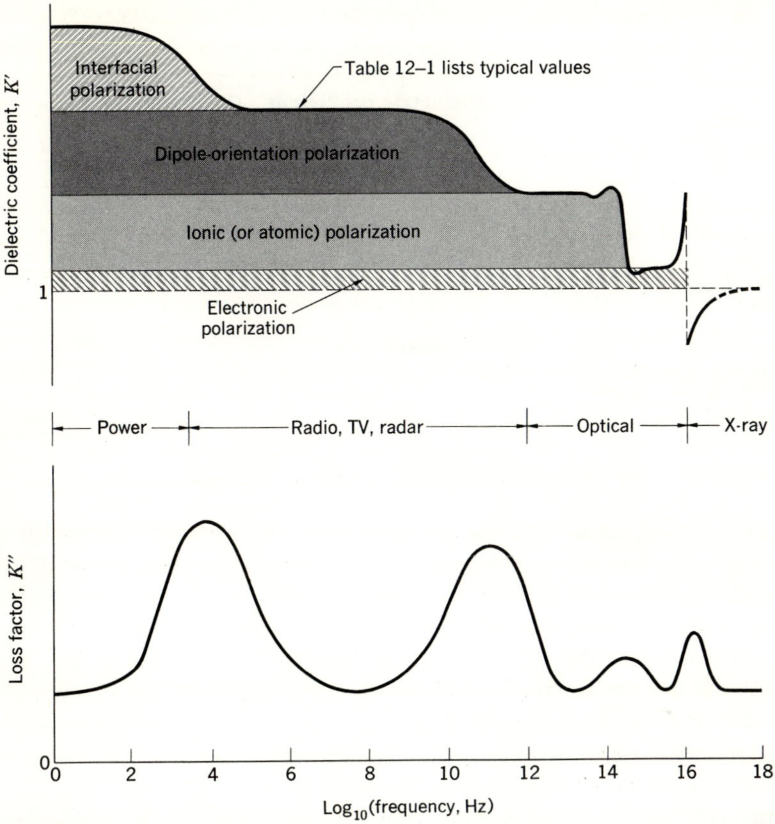

FIGURE 12-13 Schematic variation with frequency of the real part K' and the imaginary part K'' of the dielectric coefficient of a hypothetical solid dielectric having four mechanisms of polarization.

about 10^3 Hz, the rate of collection of charge begins to lag behind the variation in electric field. Consequently, the current and voltage are forced out of their usual phase relationship, and the energy loss in the dielectric increases sharply. At higher frequencies near 10^6 Hz, the interfacial mechanism is virtually inoperative, and the loss factor returns to a lower value. Representative values of the dielectric coefficient for this frequency at room temperature are listed in Table 12-1.

Dielectric materials can be divided into two classes: (1) those such as polyethylene whose basic structural unit (—CH_2—) is not a permanent dipole and (2) those such as polyvinyl-chloride that contain a permanent dipole (—CHCl—). In the absence of an electric field the dipoles are oriented at random and produce no overall polarization. An applied field tends to align the dipoles, thus making a substantial contribution to the total dielectric

TABLE 12-1. REPRESENTATIVE DIELECTRIC AND FERROELECTRIC MATERIALS

Material	Dielectric coefficient K', at 10^6 Hz and 25°C	Curie temperature, °C
	Dielectrics	
Polyethylene	2.3	
Polyvinyl-chloride	2.8	
Nylon	3.6	
Porcelain	~5	
Mica	~7	
	Ferroelectrics	
$BaTiO_3$ (ceramic)	1700	120
PZT-5A (ceramic)†	1700	365
PZT-5H (ceramic)†	3400	193

† Lead zirconate plus lead titanate (PZT) in several variations are widely applied technically.

coefficient. This effect occurs mostly in polymers. An increase in temperature has two effects on dipole orientation. Thermal energy tends to oppose alignment and therefore tends to lower the dielectric coefficient. Molecular mobility increases with temperature, and therefore the dipole-orientation mechanism can operate at higher frequencies as the temperature increases. Near 10^{10} Hz, however, this mechanism gradually becomes inoperative, with an accompanying peak in the loss factor (Fig. 12-13).

In most ceramic materials the arrangement of positive and negative ions produces no overall polarization. However, an applied electric field shifts the centers of charge and can produce large ionic polarization. This mechanism can operate up to frequencies of 10^{13} Hz. Another mechanism, electronic polarization, operates in all materials but makes only a relatively small contribution. It consists of the shifting of the electrons relative to the nucleus in each atom of the material. This mechanism ceases to operate above about 10^{16} Hz. Since the ionic and electronic mechanisms operate in the range of optical frequencies, they strongly influence the optical properties of materials (Sec. 12-8).

FERROELECTRIC MATERIALS. Dielectric coefficients 10^3 times larger than in ordinary dielectrics occur in materials known as *ferroelectrics* [by analogy with ferromagnetism (Sec. 12-6)]. Barium titanate, $BaTiO_3$, a representative material of this class, can serve as an example of the interesting mechanisms that account for this unusual behavior. Above 120°C, $BaTiO_3$ has a cubic crystal structure (Fig. 12-14*a*) and no polarization since the centers of its positive and negative charges coincide. At lower temperatures appreciable electrical energy is gained by the production of electrical dipoles; therefore,

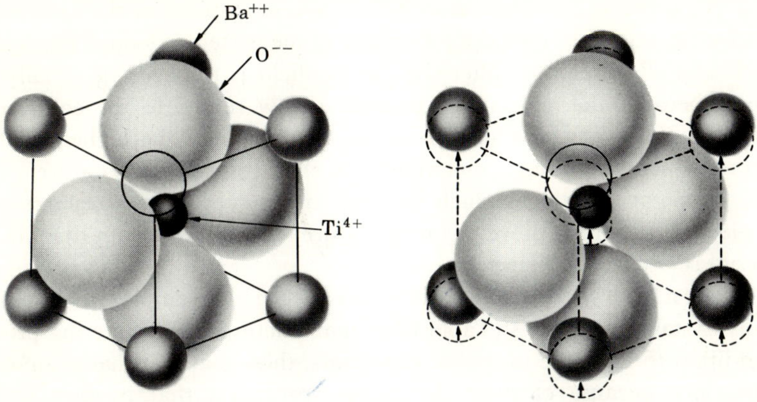

(*a*) Unpolarized $BaTiO_3$ has a cubic structure and the cations touch each of the surrounding oxygen ions.

(*b*) Polarization develops when the cations move upwards relative to the oxygen ions.

FIGURE 12-14 Illustration of the changes in atomic arrangement that accompany the development of the ferroelectric state in barium titanate.

considerable mechanical energy can be expended to distort the main framework, composed of oxygen ions (Fig. 12-14*b*). The sketch shows how the titanium ions move upward and increase the spacing of the oxygen ions in the vertical direction; the crystal structure changes from cubic to tetragonal as a result. The barium ions behave similarly. This transformation occurs in an entire region of the crystal, called a *domain*, and a high degree of spontaneous polarization $P_s = 0.26$ coul/m^2 develops within this domain. No applied electric field is involved in this polarization mechanism.

The spontaneous polarization of ferroelectrics explains their unusual response to an applied electric field compared to dielectrics. A typical dielectric material simply increases or decreases its polarization reversibly as the electric field increases or decreases, but even at high fields its maximum polarization is only about 10^{-4} of the P_s value of barium titanate. At equilibrium, a sample of barium titanate has zero net polarization because of a domain structure comparable to that of a ferromagnetic material (Fig. 12-18*b*). A domain with polarization in one direction has neighboring domains polarized in other directions, usually at 180° or at 90° to the first domain. The result over an array of many domains is mutual cancellation of the individual P_s values. When such a material is subjected to an electric field, the process of initial polarization consists of the increase in volume of favorably oriented domains at the expense of the remaining domains. This increase can result from the movement of the domain wall between two differently oriented domains, or a thin layer of favorably oriented structure may nucleate within an existing domain and then grow as the electric field increases.

An increase in temperature affects several aspects of the behavior of ferroelectric materials, including the shape of the hysteresis loop. The remnant polarization P_r remaining after the electric field is removed, gradually disappears because thermal energy permits a slow return to the equilibrium condition. This relaxation process occurs more rapidly the higher the temperature. Each ferroelectric material has a characteristic temperature T_c (its Curie temperature) above which thermal energy prevents the occurrence of ferroelectric domains. Above this temperature, 120°C for barium titanate, the material is essentially dielectric.

For most technical applications, ferroelectrics are employed in the form of polycrystalline ceramics (Table 12-1) manufactured by a sintering process. In addition to their use in special capacitors, these materials have important applications because of their *piezoelectric* properties; that is, they deform mechanically when subjected to an electric field, and conversely, they change their polarization when undergoing a mechanical strain. Devices that use piezoelectric materials to convert electrical impulses to mechanical impulses (or vice versa) are examples of transducers.

MAGNETIC PROPERTIES

The effect of a magnetic field on materials has a rough parallelism to the effect of an electric field. In particular, a few materials (ferromagnetic and ferrimagnetic) exhibit large magnetic effects of technical importance; most materials have a negligibly small magnetic response. Although electric charge exists as individual, separable positive and negative units (such as the electron), no corresponding separate magnetic units exist. Magnetic flux has its origin in an electric current; no experimental justification exists for the term "magnetic poles."

12-6. Types of Magnetism

Magnetic effects in materials originate in the tiny electric currents associated either with electrons in atomic orbits or with the spin of electrons. In *diamagnetic* materials (Fig. 12-15*a*), the numerous currents within an atom (or ion) cancel, thus leaving zero magnetic moment for the atom as a whole in the absence of an external magnetic field. In *paramagnetic* materials, the individual atoms possess net magnetic moments (Fig. 12-15*b*), but their random orientations result in zero net magnetization for a group of atoms. Most materials are either diamagnetic or paramagnetic and can be considered "nonmagnetic" for most purposes. The cause of large magnetization in a few materials is the spontaneous alignment of adjacent atomic magnetic moments

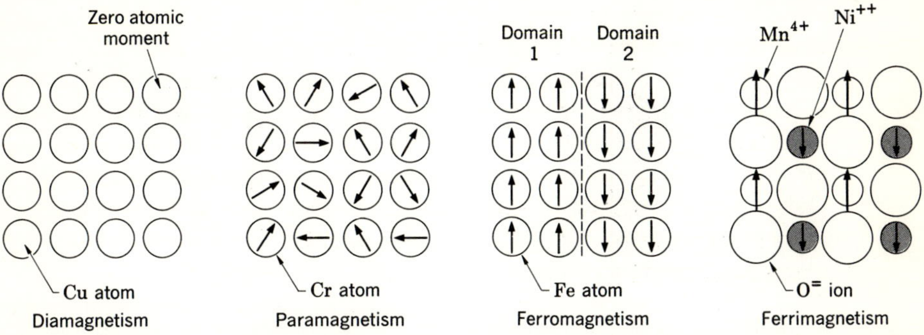

(*a*) Each atom has zero magnetic moment; copper is an example.

(*b*) Magnetic moments of atoms are random and cancel; chromium is an example.

(*c*) Magnetic moments are aligned within a given domain; iron is an example.

(*d*) Magnetic moments of each type are aligned; $NiMnO_3$ is an example.

FIGURE 12-15 Schematic illustrations of the atomic origins of the four types of magnetic behavior. No external magnetic field is present.

within a domain. *Ferromagnetism* (Fig. 12-15*c*) corresponds to parallel alignment, and *ferrimagnetism* (Fig. 12-15*d*) corresponds to antiparallel alignment in a domain. If cancellation occurs among neighboring domains, as indicated in Fig. 12-15*c*, the overall magnetization can be zero. We now consider the behavior of these four types of materials in an external magnetic field.

DIAMAGNETISM AND PARAMAGNETISM. Just as the essential dielectric effect can be demonstrated with a capacitor, so the essential magnetic effect can best be visualized within a magnetic coil (a coiled wire carrying an electric current). If the current flows in a helical coil (Fig. 12-16*a*), then the relation between the electrical current and the magnetic field (within the coil) is especially simple. The intensity of the magnetic field H is defined as the product of the current in amperes times the number of turns per meter in the coil. Typical numerical values are shown in Fig. 12-16*a*. In the SI system of units, H is pictured as producing a flux density $B = \mu H$ in any given medium, where μ is the permeability of the medium. Even if the medium is a vacuum, B differs from H by the factor $\mu_v = 4\pi \times 10^{-7}$.

If a diamagnetic or a paramagnetic material is placed in a magnetic field (Fig. 12-16*b*), the flux density within the material is almost identical with that in a vacuum. The permeability μ of paramagnetic materials is larger than μ_v by roughly 1 part in 10^5; for diamagnetic materials μ is smaller than μ_v by roughly 1 part in 10^6. The magnetic behavior of a typical material in each of these two classes (Fig. 12-16*c*) is conveniently[1] shown as a plot of B versus $\mu_v H$. As $\mu_v H$ increases, the curve for each material essentially coincides with

[1] In the cgs system of units this plot is labeled B (in gauss) versus H (in Oersteds) and is identical with the one shown, except that both ordinate and abscissa are larger by a factor of 10^4.

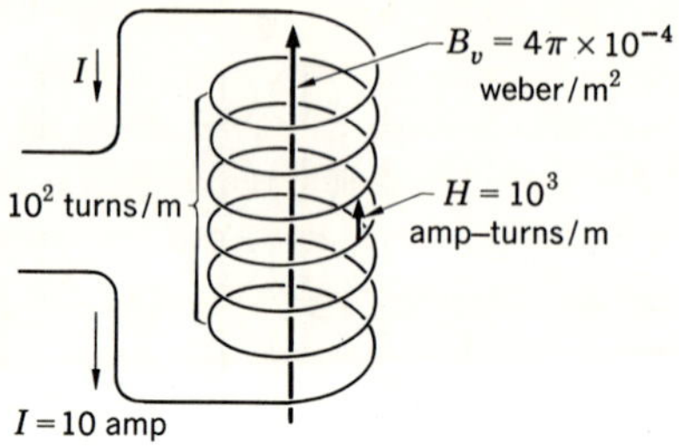

(*a*) An electric current, *I*, is accompanied by a magnetic field, *H*. The corresponding magnetic flux in vacuum is $B_v = \mu_v H$; $\mu_v = 4 \times 10^{-7}$ henry/m. Typical values are shown.

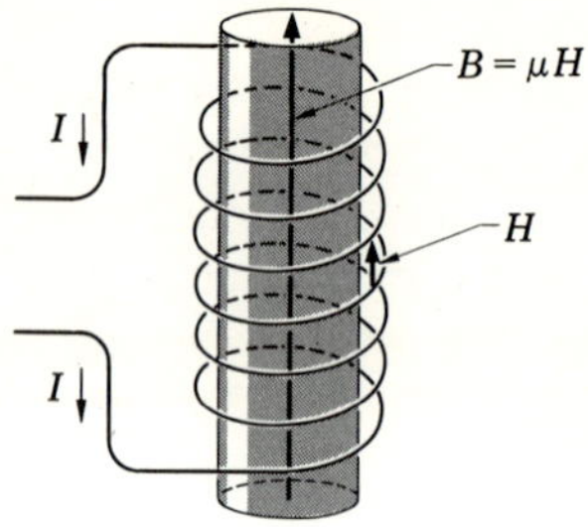

(*b*) The magnetic flux inside a solid material is $B = \mu H$, where μ is its permeability. $\mu \simeq \mu_v$ except for ferromagnetic materials.

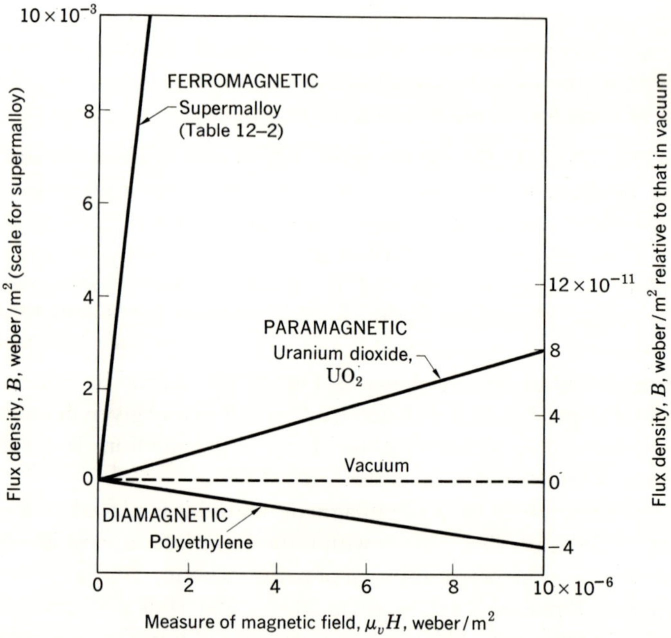

(*c*) Initial portion of $B - \mu_v H$ curves for three materials that show the three different types of magnetic behavior.

FIGURE 12-16 Schematic illustration of the physical significance of magnetic field strength H, flux density in vacuum B_v, and flux density B in the three types of magnetic materials (conversion of units: M webers/m² × 10^4 = N gauss).

that for a vacuum; the small positive or negative deviation has been magnified by a greatly enlarged scale in the figure. In diamagnetic materials the balance in atomic magnetic effects leading to zero atomic magnetic moment (Fig. 12-15*a*) is perturbed by an applied field and results in a very small opposing field. Thus, the magnetic flux within a piece of copper, for example, is slightly less than that in a vacuum. The random atomic moments in a paramagnetic material (Fig. 12-15*b*) become partially oriented in the direction of an applied magnetic field; consequently, a material such as chromium has a slightly greater magnetic flux than the corresponding vacuum.

FERROMAGNETISM. The ferromagnetic materials are iron, nickel, cobalt, six rare-earth elements, and numerous alloys. Their response to a magnetic field is about 10^9 times greater than that of ordinary (diamagnetic or paramagnetic) substances, as shown by the data in Fig. 12-16*c* for the initial magnetization of permalloy, a technically important ferromagnetic alloy (Table 12-2). The slope μ_r of this plot of B versus $\mu_v H$ is the relative permeability,[1] corresponding to the equation

$$B = \mu_r(\mu_v H) \tag{12-38}$$

Since $B = \mu H$,

$$\mu_r = \frac{\mu}{\mu_v} \tag{12-39}$$

The relative permeabilities of several ferromagnetic materials are listed in Table 12-2; μ_r for diamagnetic and paramagnetic substances is virtually unity.

The action of an applied field in developing strong magnetism in a ferromagnetic material is analogous to the behavior in ferroelectrics. Each domain (Fig. 12-15*c*) has a large magnetic flux density, on the order of 1 weber/m². The overall magnetization, however, may be essentially zero because of mutual cancellation among the multitude of domains. The effect of an applied magnetic field on an unmagnetized specimen is described by an initial curve and, if the field reverses periodically, by a hysteresis loop. Figure 12-17 shows a typical behavior for each of the two kinds of ferromagnetic materials, *soft* and *hard*. An applied magnetic field causes the domains to change in size and orientation, as described below, thus producing a net magnetization. The essential requirement for soft magnetic materials (Table 12-2) is that a high saturation magnetization B_s be produced by a small

[1] μ_r is a convenient quantity in SI units because it is numerically the same as the corresponding quantity in the cgs system (the permeability, μ).

TABLE 12-2. PROPERTIES OF REPRESENTATIVE SOFT MAGNETIC MATERIALS

Material	Composition	Relative permeability, μ_r		Saturation magnetization B_s, weber/m²	Residual magnetization B_r, weber/m²	Coercive force $(\mu_v H)_c$, weber/m²
		Initial	Maximum			
Metallic alloys (ferromagnetic)						
Pure iron	~0.1 w/o impurities	500	20,000	2.15	1.0	5×10^{-5}
Silicon steel	4.25 w/o Si	1,500	9,000	1.95	0.7	4×10^{-5}
Grain-oriented silicon steel	3.25 w/o Si	2,000	40,000	2.0	—	1×10^{-5}
Supermalloy	79 w/o Ni 16 w/o Fe 5 w/o Mo	10^4	10^5	0.8	0.4	10^{-6}
Ceramic ferrites (ferrimagnetic)						
Ferroxcube, type A6	(Mn, Zn) Fe_2O_4	1,500	—	0.4	—	4×10^{-5}
Ferroxcube, type B2	(Ni, Zn) Fe_2O_4	300	—	0.3	—	

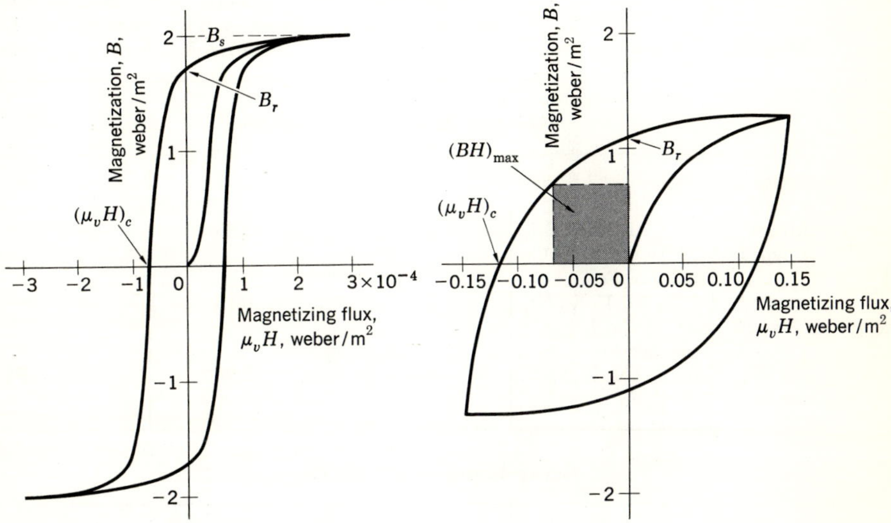

(*a*) Soft magnetic materials require only a small magnetizing field and achieve a high B_s.

(*b*) The essential property of hard magnetic materials is resistance to demagnetization, measured by $(BH)_s$.

FIGURE 12-17 Representative curves of initial magnetization and hysteresis loops for the two types of ferromagnetic materials.

applied field (Fig. 12-17*a*). The correspondingly small field $(\mu_v H)_c$ required to cause demagnetization, insures a low energy loss per cycle, that is, a small area of the hysteresis loop. This property is useful in applications, such as transformers, involving continual change in the direction of magnetization. In contrast to a soft magnetic material, which operates in the presence of a magnetic field, a hard (or permanent) magnet (Table 12-3) is given its magnetization during manufacture. Consequently, such a magnet should retain in service the residual flux density B_r remaining after the strong magnetizing field has been removed (Fig. 12-17*b*). The coercive force $(\mu_v H)_c$ is one indication of the resistance to demagnetization; a more precise measure is $(BH)_{max}$, which will be described more fully in Fig. 12-19. A large hysteresis loop is characteristic of hard magnetic materials.

Several aspects of their domain structure underlie the behavior of ferromagnetic materials. Magnetization occurs in a preferred crystallographic direction, for example, the $\langle 100 \rangle$ directions in iron (Fig. 12-18). In contrast to the analogous directionality in ferroelectrics, however, the direction of magnetization can rotate into other directions in the crystal, provided that the necessary *anisotropy energy* is supplied. Initially (at $B = 0$) the domains in a

TABLE 12-3. PROPERTIES OF REPRESENTATIVE HARD MAGNETIC MATERIALS

Material	Composition	Residual magnetization B_r, weber/m²	Coercive force $(\mu_v H)_c$, weber/m²	Maximum energy product $(BH)_{max}$, J/m³
Metallic alloys (ferromagnetic)				
Tungsten steel	5 w/o W, 0.7 w/o C	1.0	0.007	2.6×10^3
Alnico, type 5B	8 w/o Al, 14 w/o Ni, 24 w/o Co, 3 w/o Cu balance Fe	1.3	0.065	4.4×10^4
Alnico, type 9	7 w/o Al, 15 w/o Ni, 35 w/o Co, 4 w/o Cu balance Fe	1.1	0.15	8×10^4
Cunife, type 1	60 w/o Cu, 20 w/o Ni, 20 w/o Fe	0.6	0.06	1.5×10^4
Ceramic materials (ferrimagnetic)				
Ceramic, type 5	$BaO \cdot 6Fe_2O_3$	0.4	0.24	2.8×10^4
Ceramic, type 7	$SrO \cdot 6Fe_2O_3$	0.34	0.33	2.3×10^4

crystal of iron are all in the direction of easy (preferred) magnetization (Fig. 12-18*b*). When a magnetic field is first applied in some direction, all of the domains retain their ⟨100⟩ orientations, but those most nearly in the direction of the field increase in size. The low-energy process of domain-boundary migration permits one domain to grow at the expense of its neighbors. If the field is along a ⟨100⟩ direction, boundary migration can lead to an alignment of all domains in the direction of the field. The saturation value of magnetization B_{max} is then achieved with a low magnetizing field (Fig. 12-18*a*). However, a field along a ⟨111⟩ direction can use the boundary-migration process only to reverse the direction of those domains (about half of the total) pointed in the opposite direction. Beyond this initial rapid rise in B, any further increase requires that the direction of magnetization within a domain rotate from an easy ⟨100⟩ direction toward a hard (high-energy) ⟨111⟩ direction. A much higher magnetic field is required to reach B_{max} in this case.

The domain structure is strongly influenced by the energy associated with domain boundaries. A very high energy is associated with a bounding surface of a domain if this surface is perpendicular to the domain's direction of magnetization. The photomicrograph of Fig. 12-18*b* illustrates the special

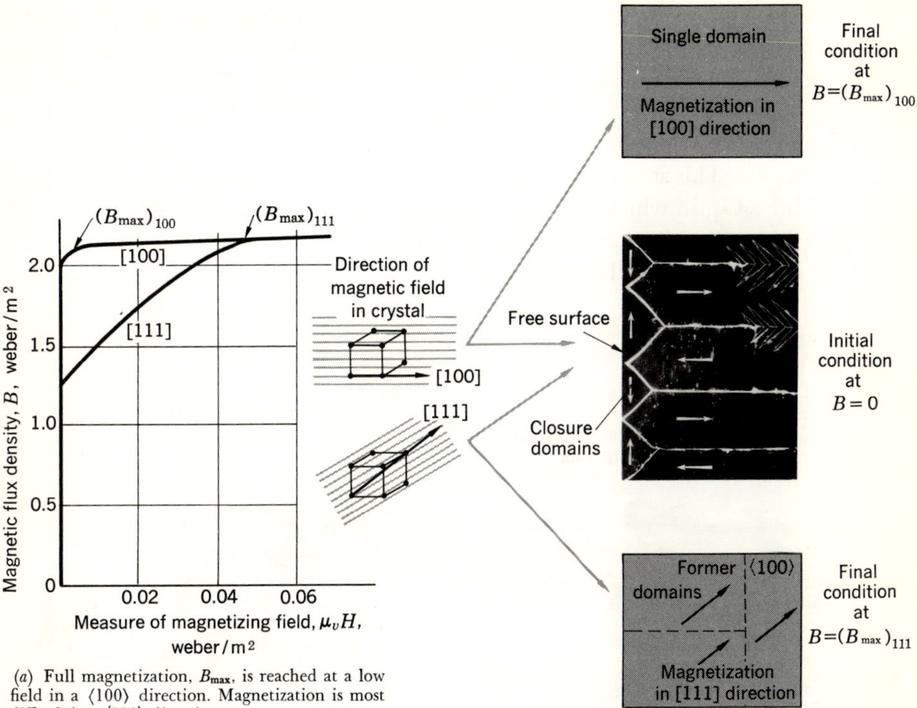

(*a*) Full magnetization, B_{max}, is reached at a low field in a ⟨100⟩ direction. Magnetization is most difficult in a ⟨111⟩ direction.

(*b*) The initial domains in ⟨100⟩ directions are revealed by a special microscopic technique. Analogous sketches show the final domain structures.

FIGURE 12-18 Effect of magnetocrystalline anisotropy on the magnetization of a single crystal of iron. (*Photomicrograph courtesy E. A. Nesbitt, Bell Telephone Laboratories.*)

"closure" domains that eliminate such configurations and maintain the magnetization parallel to free surfaces. The boundary between two domains called a *Bloch wall*, is a region about 10^3 Å wide in which the direction of magnetization gradually changes from that of one domain to that of the other (Prob. 12-11). The energy of the boundary is great enough to allow very fine particles of ferromagnetic materials (about 300 Å in diameter) to exist as single domains. In films only 10^3 Å thick, the direction of magnetization in a domain boundary must rotate in the plane of the film to avoid the existence of large areas of free surface perpendicular to the direction of magnetization. Boundaries with this type of rotation are termed *Néel walls*.

FERRIMAGNETISM. This type of magnetism results from antiparallel coupling of atomic magnetic moments, but it is comparable in magnitude to ferromagnetism, which involves parallel coupling. The most important type of soft ferrimagnetic materials are the ferrites (Table 12-2), which are ceramic materials of the *spinel* group. A spinel discussed in previous chapters, $MgAl_2O_4$, is sketched in Fig. 12-19*a*. The face-centered cubic framework of oxygen atoms contains tetrahedral and octahedral interstitial sites, a fraction of which are occupied by the Mg^{++} and Al^{3+} ions. In response to valence requirements, the large Mg^{++} ion occupies the tetrahedral site rather than the larger octahedral site. This structure is called a *normal* spinel. Consider now a spinel such as $MnFe_2O_4$, in which the two cations possess net atomic magnetic moments. The Mn^{++} ion is enough larger than the Mg^{++} ion (of a normal spinel) to require an octahedral site (Fig. 12-19*b*). Consequently, half of the Fe^{3+} ions

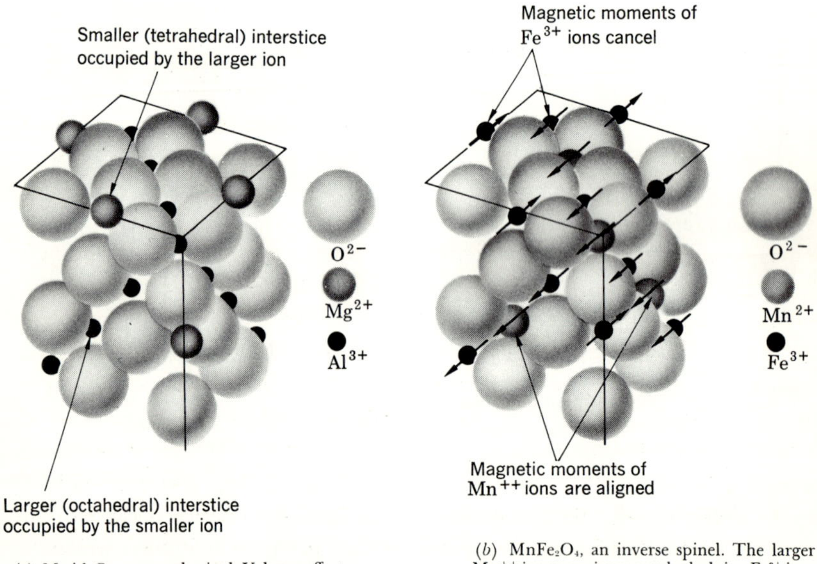

(*a*) $MgAl_2O_4$, a normal spinel. Valency effects cause the Mg^{++} ion to occupy a tetrahedral site.

(*b*) $MnFe_2O_4$, an inverse spinel. The larger Mn^{++} ion occupies an octahedral site. Fe^{3+} ions occupy both types of site.

FIGURE 12-19 Schematic illustrations of the crystal structures of a normal spinel (*a*) and an inverse spinel (*b*).

are "forced out" of octahedral sites and occupy tetrahedral sites. This structure is called an *inverse* spinel. Ions in the octahedral sites have their magnetic moments antiparallel to those in tetrahedral sites. Consequently, the moments of the Fe^{3+} ions cancel, but the moments of the Mn^{++} ions are aligned and give a large magnetization within a domain. Substitution of Mn^{++} by other magnetic ions such as Ni^{++} or partially by nonmagnetic Zn^{++} (Table 12-2) produces a range of ferrites having a variety of magnetic properties.

Typical of the hard ferrimagnetic ceramics (Table 12-3) is $BaO \cdot 6Fe_2O_3$, known as Ceramic 5 (or Ferroxdure). Its complex crystal structure (magnetoplumbite) leads to a partially antiparallel arrangement of the 12 Fe^{3+} ions that results in a net magnetization equal to that of four ions. The domain structure is similar to that for ferromagnetism, but the critical size for the existence of a single domain is much larger, about 1 μ. Consequently, when a magnet is prepared by sintering a mixture of finely ground powder, it consists of single-domain grains and has the unusual properties discussed in Sec. 12-7. A magnetic field applied during the pressing of such a sintered ceramic magnet aligns the powder particles and increases the residual magnetization in the direction of interest. In contrast to metallic ferromagnetic materials, ferrimagnetic ceramics are electrical insulators and therefore have negligible energy loss due to parasitic currents (eddy-current losses).

12-7. Magnetic Materials

Commercial magnetic materials are ferro- (or ferri-)magnetic and are capable of a high flux density, about 1 weber/m^2. Other requirements for the two classes of magnetic materials, soft and hard magnets, are now briefly reviewed.

MAGNETICALLY SOFT MATERIALS. These materials must permit reversal of the direction of magnetization with little loss of energy per cycle. The crucial factor is the area of the hysteresis loop (Fig. 12-17*a*) which is determined primarily by the coercive force $(\mu_v H)_c$. This value is below 10^{-4} weber/m^2 for typical soft magnetic materials (Table 12-2), corresponding to an energy loss of less than 10^{-2} J/cycle-kg of material. Commercial magnetic materials are polycrystalline, but special manufacturing procedures can align each grain in a direction of easy magnetization. When thin-film magnetic strips are prepared by vapor deposition, the presence of a suitable magnetic field causes the preferred crystallographic direction to lie in the long direction of the strip. For ordinary metallic sheet, suitable control of cold working and recrystallization during mechanical working leads to the desired preferred orientation (Sec. 8-1). In grain-oriented silicon-iron, for example, the cube texture (100)[100] provides a direction of easy magnetization in both the rolling and transverse direction in grain-oriented sheet.

Other aspects of magnetically soft materials are important in certain applications. The efficiency of a transformer, for example, depends on the total loss of energy, which includes eddy-current loss as well as hysteresis loss. The eddy-current loss for a given design can be reduced by increasing the electrical resistivity; silicon accomplishes this purpose in silicon-iron. Elastic strains increase the coercive force $(\mu_v H)_c$ and therefore adversely effect soft magnetic properties. These strains act through the *magnetostriction effect*, which is analogous to but much smaller than the piezoelectric effect in ferroelectrics. This effect is the principle of operation for magnetostriction generators, which convert alternating electrical currents into mechanical vibrations. Thin-film magnets have the advantage of requiring a very short time, about 10^{-9} sec, to reverse the direction of magnetization. They are an excellent switching device in high-speed equipment such as computers.

MAGNETICALLY HARD MATERIALS. In principle, a magnetized hard magnet has a permanent source of energy for operation in radio speakers, small motors, magnetic latches, and many other devices. This energy is added in a manufacturing operation involving strong magnetizing fields which leaves the magnet with a (permanent) residual flux density B_r (Fig. 12-17*b*). The amount

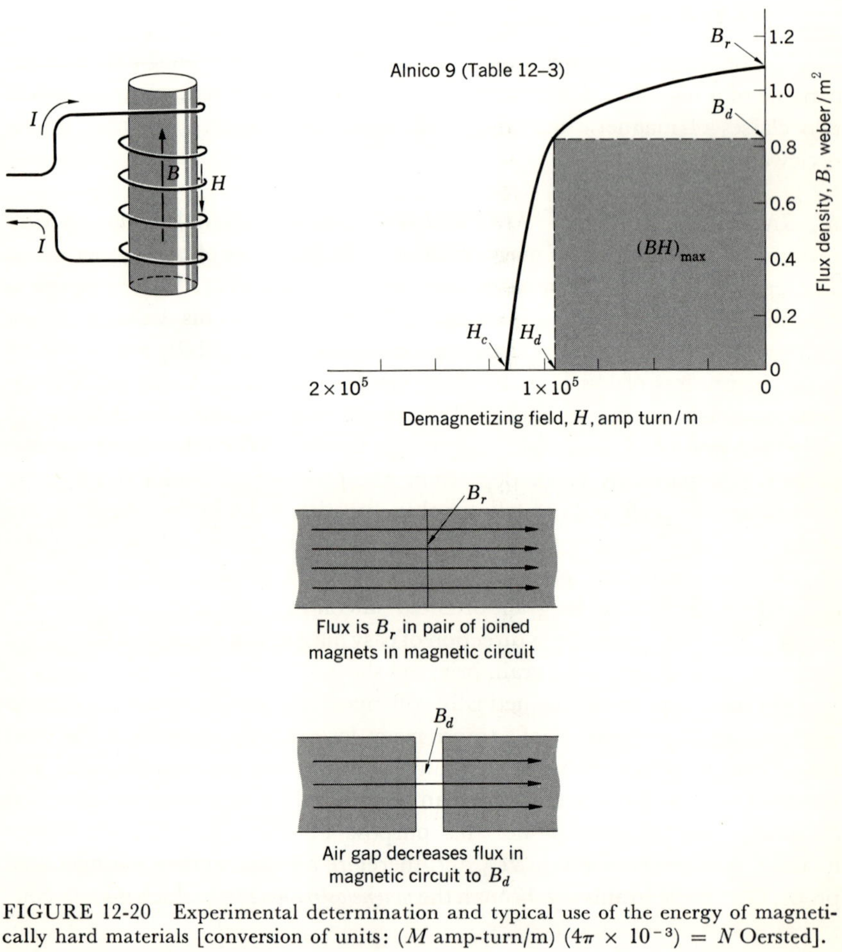

FIGURE 12-20 Experimental determination and typical use of the energy of magnetically hard materials [conversion of units: (M amp-turn/m) ($4\pi \times 10^{-3}$) = N Oersted].

of magnetic energy available can be determined by subjecting the fully magnetized material to a demagnetizing field H (Fig. 12-20*a*). The energy of the demagnetizing field is the product BH J/m^3 and passes through a maximum at H_d as H increases from 0 to H_c (Fig. 12-20*b*). The value of B corresponding to the maximum energy product is denoted B_d. As an example of the application of $(BH)_{\max}$ in design, consider the use of a pair of magnets to produce a usable magnetic field in an air gap (Fig. 12-20*c*). The maximum flux B_r exists when the air gap is negligibly small, that is, when the two magnets are in contact. As the magnets are separated to make an air gap, the flux decreases roughly in proportion to the width of the gap; the air gap acts as a large reluctance (resistance) in the magnetic circuit. If the gap is set so that $B = B_d$, then the magnetic energy available is being used most effectively in producing a space containing a magnetic flux.

Both the Alnico and Ceramic permanent magnets listed in Table 12-3 owe their high coercive force to the presence of isolated domains in their structure. In Alnico 9, special heat treatment in a magnetic field causes tiny aligned rods of ferromagnetic material to precipitate within a nonferromagnetic matrix. Each rod is so small that it exists as a single domain; only a large energy can reverse its direction of magnetization. Growth and rotation of domains in magnetic materials is inhibited by the presence of inclusions, such as precipitate particles or a second phase. Cold working also tends to increase the coercive force. In general, structural changes that increase mechanical hardness also increase the magnetic hardness (permanence) of a material. Various factors tend to destroy magnetism in a permanent magnet; high temperatures and mechanical blows are especially deleterious.

OPTICAL PROPERTIES

The human eye is extremely sensitive to the manner in which materials behave when subjected to a certain range of electromagnetic waves, the visible spectrum. Although any treatment of optical properties must give appropriate emphasis to this part of the subject, optical phenomena cover a much broader range of wavelengths and are independent of the characteristics of our organs of vision. Appropriately, then, we begin by considering the reasons a material is "transparent" to an electromagnetic wave of a given wavelength and also the general nature of the refractive index. These topics then furnish a basis for discussing such optical materials as glass (and plastic) for lenses.

12-8. Electromagnetic Waves in Materials

The major part of the electromagnetic spectrum has wavelengths much longer than the size of an atom ($\sim$4 Å). Such waves are not diffracted by the crystalline structure, but they tend to propagate in a straight path through a homogeneous solid phase. We begin with the behavior of insulators and make use of the two following basic relations[1] between the dielectric constants [Eq. (12-37)] and the optical constants n and k:

$$2nk = K'' \qquad (12\text{-}40)$$

$$n^2 + k^2 = K' \qquad (12\text{-}41)$$

The index of refraction, n, is the real part of the coefficient of refraction, n^*; thus

$$n^* = n - ik \qquad (12\text{-}42)$$

and k (index of absorption) is the imaginary part. As with K'', k is the major indicator of loss of energy as an electromagnetic wave passes through a material. At frequencies in the radio band an insulator has low absorption (Fig. 12-21) and thus its coefficient of refraction [Eq. (12-41)] depends essentially on the index of refraction, n. Near 10^{12} Hz, however, the ionic component of the polarization is disappearing, and both n and k [and K' and K'' (Fig. 12-13)] show important variations with frequency. The previously transparent insulator absorbs the infrared radiation and becomes opaque. Its refractive behavior must be described by n^* (that is, by n and k), not by n alone.

As the frequency increases toward and through the visible spectrum, the absorption again drops to a low value, and the refractive index n dominates the optical behavior. An insulating crystal, such as MgO, is transparent to these electromagnetic waves, and its index of refraction is almost 2; that is, the velocity of propagation is reduced by almost a factor of 2 in the crystal compared to the velocity in vacuum. The properties in the visible spectrum are the basis of the usual applications of optical materials (Sec. 12-9). The next absorption maximum is in the ultraviolet (Fig. 12-21), but the existence of multiple peaks shows that the cause is more complex than simply a polarization phenomenon. An electron from the valence band is elevated to a special energy state below the conduction band, roughly comparable to a donor level in an n-type semiconductor (Fig. 4-6). The electron remains bound electrically to the "hole" left in the valence band; this electron-hole pair, known as an *exciton*, does not contribute to electrical conductivity. At still higher frequencies in the x-ray range absorption effects are caused by electron transitions from an inner shell up to the conduction band. The corresponding influence on the index of refraction is too small to appear in Fig. 12-21.

The absorption in metals is very high at low frequencies (Fig. 12-21) because the conduction electrons are able to absorb small photons of energy

[1] A relatively elementary derivation of these relations is given in G. Joos, "Theoretical Physics," 2d ed., pp. 331–332, Hafner Publishing Company, Inc., New York, 1950.

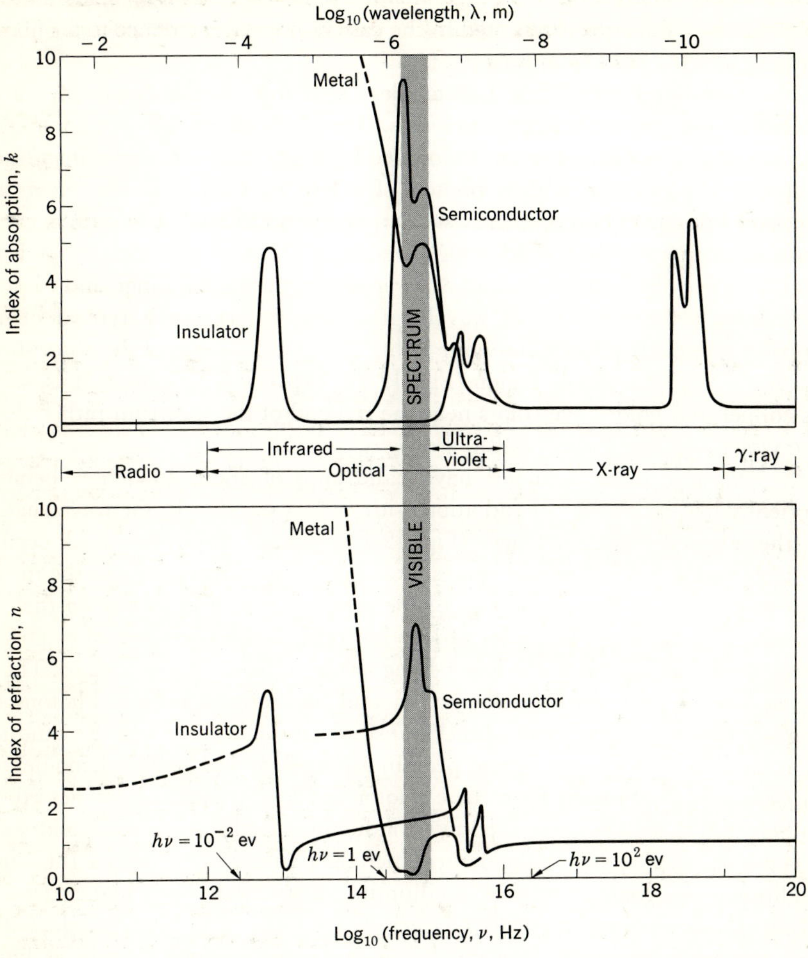

FIGURE 12-21 Schematic variation with frequency of the index of absorption and index of refraction of insulators, metals, and semiconductors.

$h\nu$ from an electromagnetic wave. An electron then moves from its position close to the Fermi surface to a higher, unoccupied energy state in the same band. In such an *intraband transition* the electron must change its momentum by an amount larger than that carried by the photon, and thus a phonon must take part in the process of energy exchange. The importance of high absorption of metals even in the visible range lies in the following parallel relation between reflectivity R and the index of absorption, k:

$$R = \frac{(n - 1)^2 + k^2}{(n + 1)^2 + k^2} \tag{12-43}$$

Since n may be about 10^{-1}, a typical value of $k = 4$ gives $R = 0.98$; that is, 98 percent of the light is reflected in the case of normal incidence to a smooth surface.

Intraband transitions become less probable as the frequency (and therefore the photon energy) increases. The decrease in absorption in the vicinity of the visible spectrum is modified by a new factor, *interband transitions.* Typically, interaction with a photon of a few electron volts can move an electron from an occupied energy state in one band to an empty state of comparable momentum in a band of higher energy. This type of transition tends to maintain the absorption at a higher level within the visible range and to provide good reflectivity for all wavelengths. For this reason a typical metal appears white or "silvery" in white light. The general trend of the absorption curve through the visible region (Fig. 12-21) explains why some metals, such as copper and gold, have colors near the red end of the spectrum rather than the violet end (see Prob. 12-12).

Because semiconductors have band gaps of about 1 ev, they behave optically like insulators for radiation with $h\nu < 1$ ev, that is, for wavelengths in the infrared region or longer (Fig. 12-21). A large absorption peak occurs just beyond $\lambda = 10^{-6}$ m and causes a change from transparency in the infrared to opacity in the visible range; the reflectivity is also similar to that of metals. The cause of the major absorption peak is the transition of electrons from the valence band to the lower-lying levels in the conduction band. These transitions require phonon interaction and are termed *indirect.* Photons of slightly higher energy permit direct transitions to higher levels in the conduction band and produce a second peak in absorption. The useful phenomenon of *photoconductivity* results from these photon-induced transitions. As described in Sec. 4-5, thermal energy produces conduction electrons in a semiconductor (even if the specimen is in the dark). Since the photons in ordinary light cause additional electronic transitions, an illuminated specimen has a higher density of current carriers and therefore a larger conductivity. Cadmium sulfide, CdS, is a typical commercial photoconductor employed in such devices as light meters. *Luminescence* refers to the reemission of energy in or near the visible spectrum, regardless of the initial form of the energy. If reemission occurs instantly, the phenomenon is called *fluorescence*; delayed emission leads to the more useful form of luminescence termed *phosphorescence.* The operation of a television screen depends on electron stimulation of the phosphor zinc sulfide, ZnS, suitably doped with copper.

12-9. Optical Materials

Many properties are desirable in materials for optical devices, but most important is the ability to transmit light. The percent of an initial light beam remaining after passing through 10 mm of a material is defined as the percent transmittance. Since transmittance is (inversely) related to the index of absorption, k, it also varies significantly with the wavelength of the light (Fig. 12-22). Glass is the principal optical material for use in the visible range, and hundreds of compositions and "qualities" are available. Resistance to deterioration, suitability for grinding of lenses, and stability of optical properties are some of the major advantages of glass. A number of plastics have the proper index of refraction and high transmittance for optical applications, but their principal disadvantage is instability of various kinds. They scratch easily, suffer changes in their optical properties during use, and also are more difficult to machine. Inexpensive plastic lenses can, however, be made by molding (rather than grinding). The nonshattering quality of plastics and their light weight make them useful in safety glasses and shields.

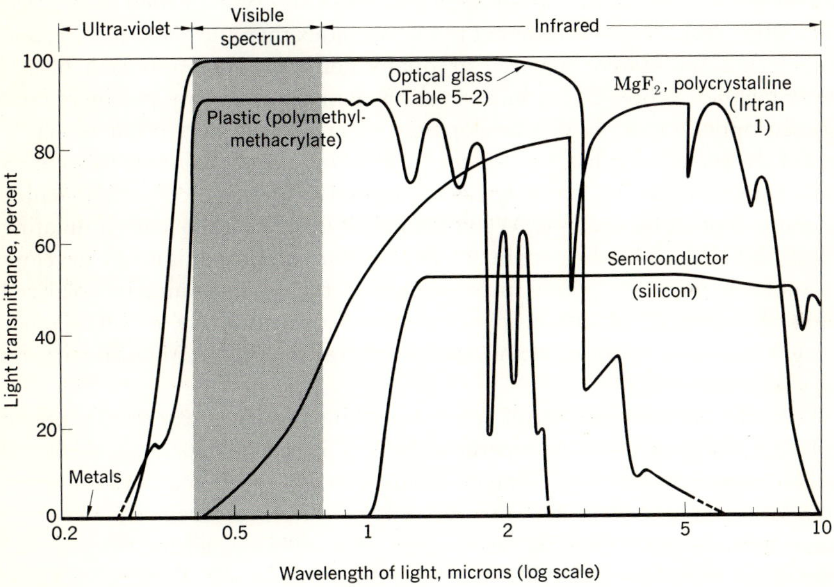

FIGURE 12-22 A comparison of light transmittance for three types of commercial optical materials; values for a semiconductor and for metals are shown for comparison. Data obtained from specimens 2 mm thick (polymethyl-methacrylate, 3.2 mm). Transmittance refers to a standard thickness of 10 mm, and the main plateau is relatively insensitive to the actual thickness of the specimen; the absorption minima, however, increase strongly with increase in thickness.

Glass must be replaced by other materials for instruments employing ultraviolet light because the transmittance of glass falls off seriously near 0.4 μ. Single crystals of quartz, SiO_2, and of calcium fluoride, CaF_2, permit operation to an appreciably lower wavelength, but they are expensive and difficult to fabricate. Ordinary polycrystalline ceramics are unsuitable for optical purposes because the light beam is scattered at internal interfaces, across which a change in refractive index occurs. This change is especially great at the surface of a pore, and the usual ceramic has sufficient porosity to make it opaque. Even if porosity is completely eliminated, such polycrystalline ceramics as Al_2O_3 are only translucent, rather than transparent, because their crystalline anisotropy is accompanied by anistropy of the refractive index. Consequently, a ray of light, in passing from a crystal of one orientation into a crystal of a different orientation, is partially scattered by the small difference in refractive index. MgF_2 has the C4(TiO_2) structure, but it has a low index of refraction and is almost optically isotropic. Consequently, its polycrystalline form has unimpaired transmittance, provided that porosity is almost absent. This material, having trade names such as *Irtran-1* and *IR-50*, is useful both in the ultraviolet and infrared ranges (Fig. 12-22).

Glassy polymers, which are optically isotropic and often transparent, are subject to change in both of these properties. If crystallization occurs, the crystalline regions have a different refractive index than the glassy matrix, and therefore a transparent material gradually changes to a translucent one as crystallization proceeds. An important phenomenon in glassy polymers (and also in optical glass) is the development of anisotropy as a result of elastic strain. In polymers it leads to the useful technique of *photoelasticity* in which a transparent plastic (such as polystyrene) is used to make a model of a machine element, a gear, for example. When the model is subjected to loads simulating those experienced by the gear in actual operation, optical anisotropy develops in each region of the plastic in proportion to the elastic strain. In polarized light the amount of anisotropy is observed quantitatively as interference fringes, and thus the location and amount of strain in the actual metal part can be deduced.

The aesthetic appeal of colored materials leads to extensive artificial coloring, not to mention the superficial finishes furnished by paints and colored oxide (anodized) coatings. Some materials such as copper (red) and cadmium sulfide (yellow) possess an intrinsic color because of their characteristic absorption spectra. Other materials are given a uniform color by two principal means—a soluble impurity or a finely dispersed dye (stain). A crystal of Al_2O_3 is colorless, but if a trace of Ti^{3+} ion becomes incorporated in the crystal structure, the blue color of the gem stone sapphire appears. The Ti^{3+} ions permit an electron transition that causes appreciable absorption in the red portion of the visible spectrum, leaving blue as the predominant intensity. Transition-metal ions (including the rare earths) are generally effective in producing color by this solution mechanism. A stain simply imparts its own intrinsic color to the material in which it is dispersed in small amount. Particles of stain as small as 5 μ in size are able to give brilliant coloration and also to impart opacity efficiently to a transparent matrix material. A suitable stain must be compatible with the matrix and must be stable under the conditions of temperature and environment to which it is exposed during manufacture and service.

PROBLEMS

12-1. The Heisenberg uncertainty principle [Eq. (12-1)] can be demonstrated by a hypothetical diffraction experiment on a single particle (Fig. 12-23). (*a*) At the instant the particle is passing through the slit of width s, what is the uncertainty Δx in its position? (*b*) The slit disturbs the wave and produces diffraction; the angle α of the first-

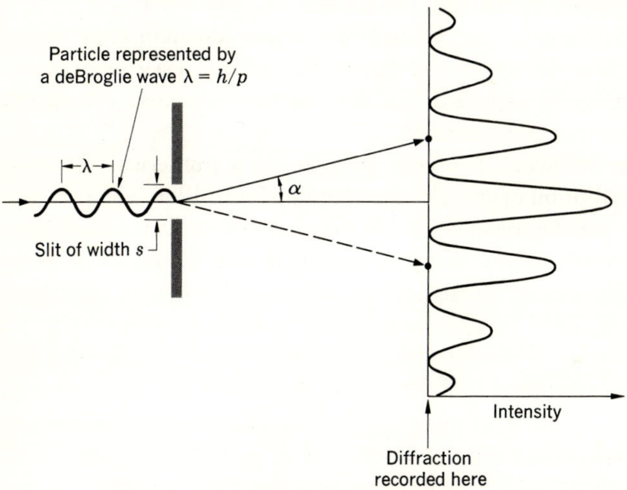

FIGURE 12-23

order diffraction peak is roughly λ/s. Show that the corresponding *transverse* momentum acquired by the particle is about $p\lambda/s$, where p is the original forward momentum. (*c*) Show that the values in parts *a* and *b* are consistent with Eq. (12-1).

12-2. *Wave-particle duality* applies both to light *waves* and to electrons (*particles*). To explain such phenomena as the photoelectric effect, a light wave of wavelength λ must be considered a "packet" (photon) having momentum $p = mc$, where c is the velocity of light. The relation between λ and p is $p = h/\lambda$. Analogously, the fact that electrons can be diffracted (just as a light wave) shows that the same relation [Eq. (12-2)] attributes a wave character to objects ordinarily considered to be particles. (*a*) Calculate the momentum of electrons accelerated by 10,000 volts. (*b*) Such an electron beam is found experimentally to be diffracted by the (111) planes of nickel ($a_0 = 3.52$ Å) at $\theta = 1.72°$. Use Bragg's law to determine the corresponding λ. (*c*) Show that the results of parts *a* and *b* satisfy Eq. (12-2).

12-3. Compare Eq. (12-8) with the expression for E_n given by the usual derivation of the Bohr theory. Show that $r_n = n^2 r_0$, where r_0 is the Bohr radius, 0.529 Å.

12-4. The counting of the number N of energy states as a function of $n^2 = (n_x^2 + n_y^2 + n_z^2)$, and therefore of energy [Eq. (12-9)], can be reduced to the determination of the volume of an octant of a sphere of radius n. If the values $n_x = 1, 2, 3, \ldots$ are plotted along the x axis and n_y and n_z are plotted similarly, all possible values of (n_x,n_y,n_z) are represented by points in the positive octant. (*a*) Show that the total number of energy states (n_x,n_y,n_z) having energies less than some particular value E_K is given by the volume of the octant of a sphere of radius n. (*b*) Obtain the expression for the "density" of energy states, dN/dE_K, and show that it depends on $(E_K)^{1/2}$.

12-5. The following computer program is a numerical illustration of the distribution of energies discussed in Prob. 12-4. The present example is also for the free-electron theory, but computer methods are applicable for more general distributions of electron energies. (*a*) Show that Eq. (12-9) can be written

$$E_K = \frac{h^2 k^2}{8m\pi^2} \tag{12-44}$$

where $k^2 = (k_x^2 + k_y^2 + k_z^2)$ and k_x is the x component of the wave vector **k**. (*b*) The computer program first calculates an energy value EK (I,J,K) for a set of values $I \equiv k_x$, $J \equiv k_y$, and $K \equiv k_z$. Relative values of energy are sufficient for the present illustration, and thus the constants in Eq. (12-44) need not be considered. The computer program next determines in which one of 10 energy groups the given value of EK falls. The 10 ranges of energy are 0 to 40, 41 to 80, etc., up to 361 to 400. EK values for I, J, K each ranging from 1 to 20 (a total of 8,000 values) are computed and grouped. Plot the number of energy values versus the average value of EK for each of the 10 ranges of energy. Compare with the curve of Fig. 12-1*c*.

```
C   CALCULATION OF DISTRIBUTION OF ELECTRON ENERGIES
      IMPLICIT INTEGER (A-Z)
      DIMENSION GROUP (10)
  1   DO 2 I = 1,10
  2   GROUP (I) = 0
```

```
C  CALCULATION OF A VALUE OF EK
    3  DO 18 I = 1,20
       ISQ = I ** 2
    4  DO 18 J = 1,20
    5  JSQ = J ** 2
    6  DO 18 K = 1,20
    7  EK = ISQ + JSQ + (K ** 2)
    8  RIGHT = 40
    9  LEFT = 0
C  ASSIGNMENT OF EK TO ONE OF TEN ENERGY GROUPS
   10  DO 17 L = 1,10
   11  IF (EK.GT.LEFT.AND.EK.LE.RIGHT) GO TO 13
   12  GO TO 15
   13  GROUP(L) = GROUP(L) + 1
   14  GO TO 18
   15  RIGHT = RIGHT + 40
   16  LEFT = LEFT + 40
   17  CONTINUE
   18  CONTINUE
C  OUTPUT SECTION
   19  RIGHT = 40
   20  LEFT = 0
   21  DO 26 I = 1,10
   22  WRITE (6,23) LEFT, RIGHT, GROUP(I)
   23  FORMAT (1X, 'NUMBER OF VALUES GREATER THAN', I4,
       1X, 'BUT LESS THAN OR EQUAL TO', I4, 1X, '=', I5)
   24  RIGHT = RIGHT + 40
   25  LEFT = LEFT + 40
   26  CONTINUE
   27  STOP
       END
```

12-6. Sketch the Brillouin zone (diffraction from {100} planes) for the simple-cubic lattice, and determine the critical value $k_\perp$ that characterizes this zone.

12-7. The usual form of Ohm's law,

$$i = \frac{V}{R} \tag{12-45}$$

describes the current i (amperes) flowing in a conductor of cross section A and length l. A voltage drop V occurs in the direction of current flow; that is, dV/dx is negative. The resistance R is related to the resistivity (and conductivity) by the equations

$$R = \frac{\rho l}{A} = \frac{l}{\sigma A} \tag{12-46}$$

(*a*) Show that Eq. (12-14) is equivalent to Eq. (12-45). (*b*) The "conventional current carriers" in Ohm's law are positive particles; for example, they are pictured as flowing from a positive terminal, +50 volts, to a negative terminal, −50 volts. Explain why the current density I in Eqs. (12-13) or (6-49) is also expressed in terms of positive particles, even though the actual current carriers are usually electrons (which are negatively charged).

12-8. (*a*) Using the value of electrical conductivity for copper at 300°K, calculate the corresponding value of τ. Assume that $m^* = m$. (*b*) From the Fermi energy of copper, 7.0 ev, calculate the velocity of the electrons at the Fermi surface. (*c*) From the results of parts *a* and *b*, calculate the mean free path of the electrons between successive collisions.

12-9. At the junction of two metals, the Fermi levels must coincide, just as in the metal-semiconductor junction shown in Fig. 12-11*b*. Using sketches similar to those in Fig. 12-11, explain why the mechanism that operates at a metal-semiconductor junction cannot be used to explain the Peltier effect in a metal-metal junction.

12-10. Cobalt, one of the ferromagnetic metals, has a hexagonal crystal structure near room temperature. The direction of easy magnetization is [0001], and the directions of hard magnetization lie in the basal plane. Explain why curves of B versus $\mu_v H$ for an easy and a hard direction do *not* coincide initially, unlike the corresponding curves for iron (Fig. 12-18*a*).

12-11. (*a*) In a bulk ferromagnetic material the Bloch wall is only about 10^{-3} of a typical dimension in a domain. Explain why the direction of magnetization rotates about an axis *perpendicular* to the Bloch wall in changing from the direction of one domain to that of the neighboring domain. [HINT: Consider the amount of high-energy bounding surface existing in the wall.] (*b*) In a thin-film ferromagnetic material the wall is almost as large as one dimension of a domain, the thickness of the film. Explain why the direction of magnetization rotates about an axis lying in the wall (perpendicular to the plane of the film). This is a Néel wall.

12-12. For copper, k decreases from 5 to 2 and n increases from 0.1 to 0.9 as the wavelength of light changes from the red end of the spectrum to the violet end. Use Eq. (12-43) to explain why copper appears reddish when seen in white light.

REFERENCES

Alfrey, T., and E. F. Gurnee: "Organic Polymers," Prentice-Hall, Inc., Englewood Cliffs, N.J., 1967.

Blakemore, J. S.: "Solid State Physics," W. B. Saunders Company, Philadelphia, 1969.

Brandup, J., and E. H. Immergut (eds.): "Polymer Handbook," Interscience Publishers, New York, 1966.

Cohen, M. H. (ed.): "Superconductivity in Science and Technology," University of Chicago Press, Chicago, 1968.

Craik, D. J., and R. S. Tebble: "Ferromagnetism and Ferromagnetic Domains," North Holland Publishing Company, Amsterdam, 1965.

Kingslake, R. (ed.): "Applied Optics and Optical Engineering," 5 vols., Academic Press Inc., New York, 1965.

Kittel, Charles: "Introduction to Solid State Physics," 3d ed., John Wiley & Sons, Inc., New York, 1966.

Meaden, G. T.: "Electrical Resistance of Metals," Plenum Press, New York, 1965.

Miller, M. L.: "The Structure of Polymers," Reinhold Publishing Corporation, New York, 1966.

Morrish, A. H.: "The Physical Principles of Magnetism," John Wiley & Sons, Inc., New York, 1965.

Wert, C. A., and R. M. Thomson: "Physics of Solids," 2d ed., McGraw-Hill Book Company, New York, 1970.

Woodington, M.: The Selection of Materials for Electrical and Magnetic Properties, chap. 9 in Sharp, H. J. (ed.), "Engineering Materials," American Elsevier Publishing Company, New York, 1966.

Index